Heavy Oil Recovery and Upgrading

Heavy Oil Recovery and Upgrading

James G. Speight
Laramie, WY, United States

ELSEVIER

Gulf Professional Publishing
An imprint of Elsevier

Gulf Professional Publishing is an imprint of Elsevier
50 Hampshire Street, 5th Floor, Cambridge, MA 02139, United States
The Boulevard, Langford Lane, Kidlington, Oxford, OX5 1GB, United Kingdom

Library of Congress Cataloging-in-Publication Data
A catalog record for this book is available from the Library of Congress

British Library Cataloguing-in-Publication Data
A catalogue record for this book is available from the British Library

ISBN: 978-0-12-813025-4

Publisher: Brian Romer
Senior Acquisitions Editor: Katie Hammon
Editorial Project Manager: Aleksandra Packowska
Production Project Manager: Anitha Sivaraj
Designer: Victoria Pearson

Typeset by SPi Global, India

Contents

About the Author

James G. Speight, BSc (chemistry), PhD (chemistry), CChem. He also holds a DSc (geological sciences) and a PhD (petroleum engineering).

Dr Speight is the author of more than 80 books in petroleum science, petroleum engineering, and environmental sciences. Formerly the CEO of the Western Research Institute (now an independent consultant), he has served as adjunct professor in the Department of Chemical and Fuels Engineering at the University of Utah and in the Department of Chemistry and Chemical and Petroleum Engineering at the University of Wyoming. In addition, he has also been a visiting professor in the Department of Chemical Engineering at the following universities: the University of Missouri-Columbia, the Technical University of Denmark, and the University of Trinidad and Tobago.

Dr Speight was awarded The Diploma of Honor, National Petroleum Engineering Society, for Outstanding Contributions to the Petroleum Industry, 1993. He was elected to the Russian Academy of Sciences in 1996 and awarded the Gold Medal of Honor that same year for outstanding contributions to the field of petroleum sciences. He has also received the Scientists without Borders Medal of Honor of the Russian Academy of Sciences. In 2001, the Academy also awarded Speight the Einstein Medal for outstanding contributions and service in the field of geological sciences. In 2007, he was also appointed as the Methanex Distinguished Professor, University of Trinidad and Tobago in Recognition of Excellence in Research.

Preface

In spite of the reserves of crude oil in tight formations, conventional crude oil is a declining share of the energy reserves, and interest is turning to other resources such as heavy oil, extra heavy oil, and tar sand bitumen. As presented in this book, heavy oil differs from conventional crude oil by its higher viscosity (resistance to flow) at reservoir temperature and the need to apply enhanced oil recovery methods. Extra heavy oil and tar sand bitumen are neither conventional crude oil nor heavy oil (as defined by the United States government) and are immobile in the deposit that requires different methods for recovery. Extra heavy oil has similar properties to tar sand bitumen but, because of a relatively high reservoir temperature, has some fluidity in the reservoir. For convenience in this book, these feedstocks are collectively referred to as heavy feedstocks.

Heavy oil and extra heavy oil have been used as refinery feedstocks for considerable time, usually blending with more conventional feedstocks, but have commanded lower prices because of its lower quality relative to conventional oil. Tar sand bitumen has been the subject of much development in Alberta, Canada. These heavy feedstocks have also been available at lower prices because of its lower quality relative to conventional oil. But both of these heavy feedstocks are sources of liquid fuels and other products that are used in large amount, and both of these feedstocks are used to supplement to supplies of conventional crude oil to refineries.

There is an immense resource base; heavy oil, extra heavy oil, and tar sand bitumen are costlier to produce and transport. The extra production, transportation, and upgrading costs explain why the development and production of extra heavy oil and bitumen are still limited. However, their abundance, geographic distribution, quality, and costs will shape their role in the future oil supply. Nevertheless, in preparation for the near future and before biofuels can be processed to meet the impending fuel shortage, refineries must and indeed are eager to adapt to changing circumstances and are amenable to trying new technologies that are radically different in character.

Currently, refineries are also looking to exploit heavy (more viscous) feedstocks such as heavy oil, extra heavy oil, and tar sand bitumen provided they have the refinery technology capable of handling such feedstocks. Transforming the higher-boiling constituents of these feedstock components into liquid fuels is becoming a necessity. It is no longer a simple issue of mixing the heavy feedstock with conventional petroleum to make up a blended refinery feedstock. Incompatibility issues arise that can, if not anticipated, close down a refinery or, at best, a major section of the refinery. Therefore, handling such feedstocks requires technological change, including more effective and innovative use of hydrogen within the refinery.

Difficult-to-refine feedstocks, such as heavy oil, extra heavy oil, and tar sand bitumen, are characterized by low API gravity (high density) and high viscosity, high initial boiling point, high-carbon residue, high nitrogen content, high sulfur content, and high metal content. In addition to these properties, the heavy feedstocks also have an increased molecular weight and reduced hydrogen content with a relatively low content of volatile saturated and aromatic constituents and a relatively high content of asphaltene and resin constituents that is accompanied by a high heteroatom (nitrogen, oxygen, sulfur, and metals) content. Thus, such feedstocks are not typically subject to distillation unless contained in the refinery feedstock as a blend with other crude oils.

It is not surprising that there has been a growing interest and research in the potential to expand enhanced oil recovery methods to heavy oil reservoirs. In fact, during the past five decades, a variety of enhanced oil recovery (EOR) methods have been developed and applied to mature and mostly depleted oil reservoirs. These methods improve the efficiency of oil recovery compared with primary (pressure depletion) and secondary (waterflooding) oil methods. Overall, enhanced oil recovery development has expanded successfully into heavy oil recovery, and some projects offer additional benefits such as sites for disposing (sequestering) carbon dioxide at modest costs or even full-cycle profit.

Finally, it is essential to realize that in the current context of recovery operations, there are several parameters that can influence properties and recovery. Readers will find the issue extremely useful and informative as a means of defining and understanding these parameters, the most prominent of which are (i) the properties of the feedstock, (ii) the properties of the reservoir, and (iii) the production method. The concept of site specificity cannot be overemphasized and must be given serious attention. These issues are presented in Part I of the book.

The limitations of processing these heavy feedstocks depend to a large extent on the tendency for coke formation and the deposition of metals and coke on the catalyst die to the higher molecular weight (low volatility) and heteroatom content. However, the essential step required of refineries is the upgrading of heavy feedstocks. In fact, the increasing supply of heavy crude oils is a matter of serious concern for the petroleum industry. In order to satisfy the changing pattern of product demand, significant investments in refining conversion processes are necessary and will continue to be necessary in order to profitably utilize these heavy feedstocks. The most efficient and economical solution to this problem will depend to a large extent on individual country and company situations.

But there are challenges when refining heavy feedstocks such as the deposition of solids (phase separation) that is a direct consequence of high asphaltene and any inorganic solids. One of the most notorious effects of asphaltene constituents is the pronounced tendency to form aggregates in the liquid (oil) medium and also under unfavorable solvent conditions leading to separation form the liquid medium. Inorganic fine solids are generally associated with asphaltene constituents, and as a result, the separated asphaltene constituents often contain high concentration of inorganic fine solids. The separation of organic and/or inorganic solids during processing poses severe problems leading to coking in the reactor and in the refinery lines and catalyst deactivation. This issue is addressed by inclusion of a chapter on instability and incompatibility.

Other challenges in heavy feedstock processing can be traced to the high content of heteroatoms (sulfur, nitrogen, and oxygen) and heavy metals (particularly nickel and vanadium). Although the concentration of these elements may be quite small, their impact is significant. For example, the presence of heteroatoms may also be responsible for objectionable characteristics in finished products causing environmental concerns, so the levels of heteroatoms in finished products have to be reduced following more and more stringent environmental regulations. Also, the deposition of trace heavy metals (vanadium and nickel) and chemisorption of nitrogen-containing compounds on the catalysts are the main reasons for catalyst passivation and/or poisoning in catalytic operations and thus necessitate frequent regeneration of the catalyst or its replacement. These issues are presented in Part II of the book.

The reader might also be surprised at the number of older references that are included. The purpose of this is to remind the reader that there is much valuable work cited in the older literature. As a work that is still of value and, even though in some cases, there has been similar work performed with advanced equipment, the older work has stood the test of time. This is particularly true of some of the older concepts of the chemical and physical structure of petroleum. Many of the ideas are still pertinent and should not be forgotten in terms of the valuable contributions they have made to petroleum science and technology. However, many of the older references included in previous versions of this book have been deleted—unavailability of the source for the general scientific researcher and the current lack of substantiated sources (other than the files collected by the author) have been the root cause of such omissions.

It is the purpose of this book to describe heavy oil, extra heavy oil, and tar sand bitumen in terms of composition and recovery followed by descriptions of the current and future methods of refining. This book also presents viable options to the antiquated definitions of the heavy feedstocks (heavy oil, extra heavy oil, and tar sand bitumen) and an introduction to the various aspects of heavy feedstock refining in order for the reader to place each feedstocks in the correct context of properties, behavior, and refining needs. In addition, the book also includes descriptions of the environmental impact of recovering and refining heavy feedstocks as well as the future of the industry and the effects of refining such feedstocks on the environment.

The book is designed to be suitable for undergraduate students, graduate students, technicians, professionals, and managers who are working with heavy oil and tar sand bitumen. Each chapter includes a list of references that will guide the reader to more detailed information. In addition, a detailed glossary is so included to assist the reader with any unknown or difficult terminology.

James G. Speight
Laramie, WY, United States

Part I

Recovery

Chapter 1

Heavy Oil, Extra Heavy Oil, and Tar Sand Bitumen

1. Introduction

In any text related to the properties and behavior (recovery or refining) of a natural resource (i.e., conventional crude oil, heavy oil, extra heavy oil, and tar sand bitumen), it is necessary to understand the resource first through the name or terminology or definition. *Terminology* is the means by which various subjects are named so that reference can be made in conversations and in writings and so that the meaning is passed on. *Definitions* are the means by which scientists and engineers communicate the nature of a material to each other and to the world, through either the spoken or the written word. Thus, the definition of a material can be extremely important and have a profound influence on how the technical community and the public perceive that material. Thus, part of the text attempts to alleviate much of the confusion that exists, but it must be remembered that the terminology of crude oil is unfortunately still open to personal choice and historical use of the various names.

Recovery and refining technologies were developed for conventional crude oil, and to some extent, heavy oil does not always relate to the issues of producing and refining extra heavy oil and tar sand bitumen—often collective referred to (along with crude oil residua) as viscous feedstocks in many refineries (Speight, 2013a, 2014a, 2017). While to some observers, viscosity is the key to recovery and refining, there are other aspects of behavior that influence recovery and refining. Every new development eventually requires some form of enhanced oil recovery (EOR), which generally means steam, solvents, or a combination of both or mining. Without knowledge in these areas, the recovery factor might be as little as 1% (by volume, 1% v/v) of the original oil in place and no more than 10% v/v of the total resource.

In the context of this book, heavy oil typically has relatively low proportions of volatile compounds with low molecular weights and quite high proportions of high-molecular-weight compounds of lower volatility. The high-molecular-weight

Heavy Oil Recovery and Upgrading. https://doi.org/10.1016/B978-0-12-813025-4.00001-5

fraction of a heavy oil is composed of complex assortment of different molecular and chemical types—a complex mixture of compounds and not necessarily just paraffin derivatives or asphaltene constituents—with high melting points and high pour points that greatly contribute to the poor fluid properties of the heavy oil, thereby contributing to low mobility (compared with conventional crude oil). The same is true for extra heavy oil and tar sand bitumen (Speight, 2013b,c, 2014a).

More generally, heavy oil typically has low levels (if any at all) of paraffin derivatives (straight-chain alkanes) with moderate-to-high levels of asphaltene constituents. The asphaltene constituents are not necessarily the primary cause for the high specific gravity (low American Petroleum Institute (API) gravity) of the oil, nor are they always the prime cause for production problems. It is essential to consider the content of the resin constituents and the aromatic constituents—briefly, the asphaltene constituents are those constituents of oil that are insoluble in *n*-heptane or *n*-pentane, while the resin constituents are those constituents of oil that are soluble in *n*-heptane or *n*-pentane but are adsorbed from these hydrocarbon solutions onto an adsorbent such as clay or alumina. In the separation procedure (Fig. 1.1), the hydrocarbon (*n*-heptane or *n*-pentane) must be specified since each liquid hydrocarbon gives different yield of the asphaltene fraction (Fig. 1.1). Both the resin constituents and the asphaltene constituents produce coke in thermal processes (Speight, 2013a,b,

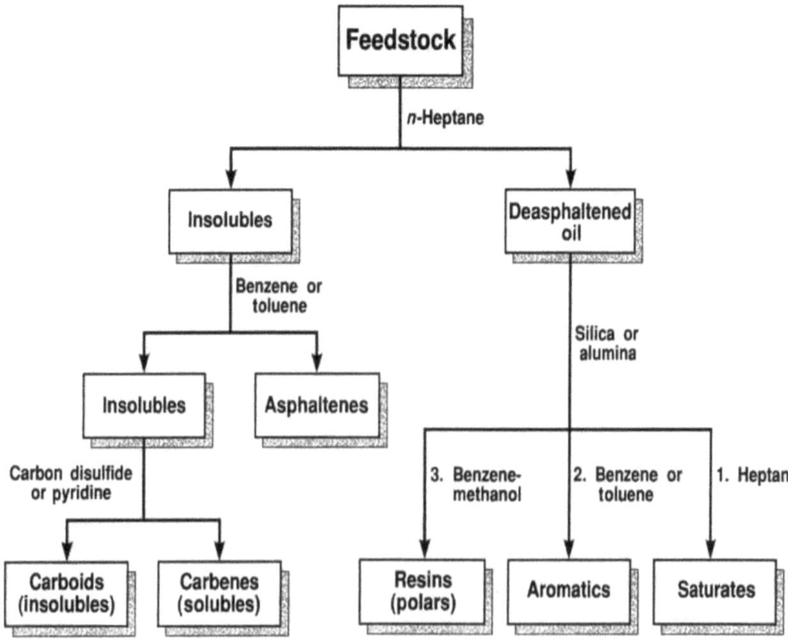

FIG. 1.1 Schematic of the separation of oil into various bulk fractions.

2014a, 2017) and, therefore, are capable of interfering with the production process by means of sediment formation in the reservoir and in the production well. It is only when the asphaltene constituents separate from the oil as separate phase that they deposit in the formation or in the production train.

2. History

Crude oil, in various forms, is not a recent discovery (Abraham, 1945; Forbes, 1958a,b, 1959, 1964; Speight, 1978, 2007; Totten, 2007). More than 4000 years ago, bitumen from natural seepages was employed in the construction of the walls and towers of Babylon. Ancient writing tablets indicate the medicinal and lighting uses of crude oil in various societies. In terms of recovery, the earliest known wells were drilled in China in 347 BC to depths of 800 ft (240 m) and were drilled using bits attached to bamboo poles. The oil was burned to evaporate brine and produce salt. By the 10th century, extensive bamboo pipelines connected oil wells with salt springs.

The use of crude oil in the Middle East was established by the 8th century, when the streets of the newly constructed Baghdad were paved with the nonvolatile residue derived from accessible crude oil and seepages (particularly Hit) in the region. In the 9th century, crude oil was distilled at Baku, Azerbaijan, to produce naphtha that formed the basis of the incendiary *Greek fire* (Cobb and Goldwhite, 1995). These Baku experiences were reported by the geographer Masudi in the 10th century and by Marco Polo in the 13th century, who described the output of those wells as hundreds of shiploads.

The earliest mention of crude oil in the Americas occurs in Sir Walter Raleigh's documentation of the Trinidad Asphalt Lake (also called the Trinidad Pitch Lake) in 1595. In 1632, the journal of a Franciscan, Joseph de la Roche d'Allion, who described his visit to the oil springs of New York was published in Sagard's *Histoire du Canada*. A Russian traveler, Peter Kalm, in his work on America published in 1748 showed on a map the oil springs of Pennsylvania. In 1854, Benjamin Silliman, a science professor at Yale University in New Haven, Connecticut, followed the work by Arabic alchemists and fractionated crude oil by distillation. These discoveries rapidly spread around the world, and Meerzoeff built the first Russian refinery in the then-mature oil fields at Baku in 1861, at which time about 90% of the world's oil was produced at Baku.

The first commercial oil well drilled in North America was in Oil Springs, Ontario, Canada, in 1858 by James Miller Williams. The crude oil industry in the United States began with Edwin Drake in 1859 who drilled a 69 ft (21 m) oil well at a place aptly named Oil Creek near Titusville, Pennsylvania, for the Seneca Oil Company. The well originally yielded 25 barrels per day, and by the end of the year, output was at the rate of 15 barrels per day. The industry grew through the 1800s, driven by the demand for kerosene and for oil lamps. Crude oil refining became even more popular, perhaps essential, in the early part of the 20th century with the introduction of the internal combustion engine, which provided a demand that has largely sustained the industry during the past 100 years. Early finds like those in Pennsylvania and Ontario were quickly outpaced by demand leading to oil booms in Texas, Oklahoma, and California.

By 1910, significant oil fields had been discovered in Canada, the Dutch East Indies (1885, in Sumatra), Iran, (1908, in Masjed Soleiman), Venezuela, and Mexico, which were being developed at an industrial level. Even until the mid-1950s, coal was still the world's foremost fuel, but oil quickly took over. The 1973 energy crisis and the 1979 energy crisis brought to light the concern that oil is a limited resource that will diminish, at least as an economically viable energy source. At the time, the most common and popular predictions were spectacularly dire.

The value of crude oil as a portable, dense energy source powering the vast majority of vehicles and as the base of many industrial chemicals makes it one of the most important commodities available to the world economies. Access to it was a major factor in several military conflicts including World War II and the more recent wars in the Persian Gulf of the 20th and early 21st centuries. Approximately 80% of the readily accessible reserves in the world are located in Middle Eastern countries with the majority of the reserves located in Saudi Arabia. However, when the reserves of heavy oil and tar sand bitumen are taken into account, the balance shifts. Venezuela and Canada have substantial reserves of heavy oil, extra heavy oil, and tar sand bitumen, which are sufficient to shift the balance of oil reserves from the Middle East to North America and South America. But first, it is necessary through an examination of the properties and behavior to understand the nature of heavy oil, extra heavy oil, and tar sand bitumen as compared with conventional crude oil (Table 1.1).

In spite of the apparent plentiful supply of oil in tight formation (such as the Bakken formation in the northern states of the United States), the oil industry is planning for the future since some of the most prolific basins have begun to experience reduced production rates and are reaching or already into maturity. At the same time, the demand for oil continues to grow every year, because of increased demands by the rapidly growing economies of China and India. This declining availability of conventional oil combined with this rise in demand for oil and oil-based products has put more pressure on the search for alternate energy sources (Speight, 2008, 2011a,b,c). But these sources are not yet ready for full commercialization to replace oil-based fuels and products and may not be so for the next several decades (Speight and Islam, 2016) during which time the production of oil-based fuels and products will have to fulfill the demand.

In order to satisfy this demand, it will be necessary to develop the reservoirs (of heavy oil) and deposits (of extra heavy oil and tar sand bitumen) that are located in the Western hemisphere. These resources are more difficult and costly to extract, so they have barely been touched in the past. However, though these resources, the world could soon have access to oil sources almost equivalent to those of the Middle East. In fact, with the variability and uncertainty of crude oil supply due to a variety of geopolitical issues (Speight, 2011b),

TABLE 1.1 Simplified Differentiation Between Conventional Crude Oil, Tight Oil, Heavy Oil, Extra Heavy Oil, and Tar Sand Bitumen

Conventional crude oil
Mobile in the reservoir
API gravity: > 25°
High-permeability reservoir
Primary recovery
Secondary recovery
Tight oil
Similar properties to the properties of conventional crude oil
API gravity: > 25°
Immobile in the reservoir
Low-permeability reservoir
Horizontal drilling into reservoir
Fracturing (typically multifracturing) to release fluids/gases
Medium crude oil
Similar properties to the properties of conventional crude oil
API gravity: 20–25°
High-permeability reservoir
Primary recovery
Secondary recovery
Heavy crude oil
More viscous than conventional crude oil
API gravity: 10–20°
Mobile in the reservoir
High-permeability reservoir
Secondary recovery
Tertiary recovery (enhanced oil recovery (EOR); e.g., steam stimulation)
Extra heavy oil
Fluid and/or mobile in the reservoir
Similar properties to the properties of tar sand bitumen
API gravity: < 10°

Continued

TABLE 1.1 Simplified Differentiation Between Conventional Crude Oil, Tight Oil, Heavy Oil, Extra Heavy Oil, and Tar Sand Bitumen—cont'd

High-permeability reservoir
Secondary recovery
Tertiary recovery (enhanced oil recovery (EOR), such as steam stimulation)
Tar sand bitumen
Immobile (solid to near solid) in the deposit
API gravity: < 10°
High-permeability reservoir
Mining (often preceded by explosive fracturing)
Steam-assisted gravity drainage (SAGD)
Solvent methods (VAPEX)
Extreme heating methods
Innovative methods[a]

[a]Innovative methods exclude tertiary recovery methods and methods, such as SAGD and VAPEX, but do include variants or hybrids thereof.

investments in the more challenging reservoirs tend to be on a variable acceleration-deceleration slope. Nevertheless, the importance of heavy oil, extra heavy oil, and tar sand bitumen will continue to emerge as the demand for crude oil products remains high. As this occurs, it is worth moving ahead with heavy oil, extra heavy oil, and tar sand bitumen resources on the basis of obtaining a measure (as yet undefined and country-dependent) of oil independence.

3. Origin

The origins of extra heavy oil and tar sand bitumen are the same as the origins of conventional oil, and a brief discussion of the mean by which oil is formed is warranted here as a point of reference for heavy oil, extra heavy oil, and tar sand bitumen properties as illustrated by the general description of these feedstocks by the use of API gravity data (Fig. 1.2) and behavior during production and refining (Yui and Chung, 2001; Speight, 2013a,b,c, 2014a, 2017).

There are two theories on the origin of crude oil and the associated heavy oil, extra heavy oil, and tar sand bitumen: (i) the abiogenic theory and (ii) the biogenic theory, and both theories have been intensely debated since the 1860s, shortly after the discovery of widespread occurrence of crude oil. It is not the intent of this section to sway the reader in his or her views of the origin

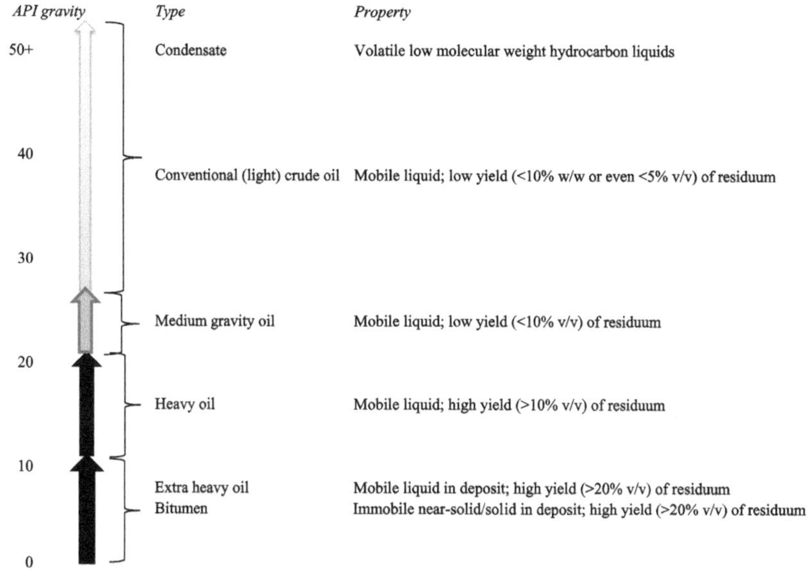

FIG. 1.2 General description of various feedstocks by API gravity.

of crude oil and natural gas. The intent is to place before the reader both points of view from which the reader can do further research and decide.

Generally, heavy oil and extra heavy oil and tar sand bitumen were oil that migrated from deep source rocks or deep reservoirs to the near surface, where the oil was biologically degraded and weathered by water. It is postulated that bacteria feeding on the migrated conventional oil removed hydrogen and produced the denser, more viscous heavy oil. Lower-boiling hydrocarbon derivatives may also have evaporated from the shallow formations.

3.1 Abiogenic Origin

From the chemical point of view, the inorganic theories are interesting because of their historical importance, but these theories have not received much attention. Geologic and chemical methods have demonstrated the optical activity of crude oil constituents, the presence of thermally labile organic compounds, and the almost exclusive occurrence of oil in sedimentary rocks.

There have been several attempts at formulating theories that describe the detail of the origin of crude oil, of which the early postulates started with inorganic substances as source material. For example, in 1866, Berthelot considered acetylene the basic material, and crude oil constituents were produced from the acetylene.

$$CaCO_3 + M^+ \text{ (alkali metal)} \rightarrow CaC_2 \text{ (calcium carbide)}$$

$$CaC_2 + H_2O \rightarrow HC{\equiv}CH \text{ (acetylene)} \rightarrow \text{crude oil}$$

The idea of abiogenic origin of crude oil proposes that large amounts of carbon exist naturally, some in the form of hydrocarbons. Hydrocarbon derivatives are less dense than aqueous pore fluids and migrate upward through deep fracture networks. Thermophilic, rock-dwelling microbial life-forms are in part responsible for the biomarkers found in crude oil. However, their role in the formation, alteration, or contamination of the various hydrocarbon deposits is not yet understood. Thermodynamic calculations and experimental studies confirm that *n*-alkane derivatives (common components of crude oil) do not spontaneously evolve from methane at pressures typically found in sedimentary basins, and so, the theory of an abiogenic origin of hydrocarbons suggests deep generation (below 120 mi) of the oil.

3.2 Biogenic Origin

It is now generally accepted, but not conclusively proved, that the formation of crude oil predominantly arises from the decay of organic matter in the earth. It is therefore from this scientific aspect that crude oil formation is referenced in this text. Nevertheless, alternative theories should not be dismissed until it can be conclusively established that crude oil formation is due to one particular aspect of geochemistry.

It has been proposed that the formation of crude oil constituents occurs through the progressive chemical change of materials provided by microscopic aquatic organisms that were incorporated over eons in marine or near-marine sedimentary rocks. In fact, the details of crude oil genesis (diagenesis, catagenesis, and metagenesis) have long been a topic of interest. However, the details of this transformation and the mechanism by which crude oil is expelled from the source sediment and accumulates in the reservoir rock are still uncertain.

Transformation of this sedimentary material to crude oil probably began soon after deposition, with bacteria playing a role in the initial stages and clay particles serving as catalysts. Heat within the strata may have provided energy for the reaction, temperatures increasing more or less directly with depth. Some evidence indicates that most crude oil has formed at temperatures not exceeding about 100–120°C (210–250°F), with the generation of crude oil hydrocarbons beginning as low as 65°C (150°F). Thus, it is possible to form heavy oil, extra heavy oil, and tar sand bitumen by several processes. First, the oil may be expelled from its source rock as immature oil.

There is general agreement that immature oils account for a small portion of the heavy oil, extra heavy oil, and tar sand bitumen (Larter et al., 2006). Some observers are of the opinion that heavy oil, extra heavy oil, and tar sand bitumen are thought to be expelled from source rocks as light crude oil or medium crude oil and subsequently migrated to a trap. If oxidation occurs, several processes can convert the oil to heavy oil, extra heavy oil, and tar sand bitumen. These processes include (i) water washing, (ii) bacterial degradation, and (iii) evaporation of the lower-boiling constituents. It is also postulated that biodegradation can occur at depth in subsurface reservoirs (Head et al., 2003; Larter et al., 2003, 2006). This concept allows for biodegradation to occur in any reservoir that has not been heated to temperatures in excess of more than 80°C (176°F). In fact, biodegradation depends on local temperature and pressure parameters factors rather than basin-wide temperature and pressure parameters.

In addition, like conventional or light oil, the composition of heavy oil, extra heavy oil, and tar sand bitumen is greatly influenced not only by the nature of the precursors that eventually form the heavy oil, extra heavy oil, and tar sand bitumen but also by the relative amounts of these precursors that occur in the source material that, in turn, are dependent upon the mix of the local flora and fauna. Hence, it is not surprising that heavy oil, extra heavy oil, and tar sand bitumen and light conventional crude oil can vary in composition with the location and age of the reservoir or the deposit. The lower mobility of heavy oil, extra heavy oil, and tar sand bitumen also makes it extremely likely that two wells in the same reservoir or deposit will produce heavy oil, extra heavy oil, and tar sand bitumen with different characteristics.

With this in mind, the following sections present descriptions of the various members of the crude oil family (i.e., crude oil, opportunity crude oil, high-acid crude oil, foamy oil, and heavy oil) as well as extra heavy oil and tar sand bitumen. But before proceeding any further, a discussion of the definitions and terminology applied to the various oils is warranted.

4. Definitions and Terminology

Because of the need for a thorough understanding of crude oil and the associated technologies, it is essential that the definitions and the terminology of crude oil science and technology be given prime consideration. While there is standard terminology that is recommended for crude oil and crude oil products (ASTM D4175, 2017), there is little in the way of standard terminology for heavy oil, extra heavy oil, and tar sand bitumen (Speight, 2013a,b,c, 2014a). At best, the terminology is ill-defined and subject to changes from one governing body company to another. Particularly troublesome and more confusing are those terms that are applied to the more viscous materials, for example, the use of the terms *bitumen* and *asphalt*. An example of an irrelevant term is *black oil* that, besides the color of the oil, offers nothing in the way of explanation of the properties of the oil material and certainly adds nothing to any scientific and/or engineering understanding of the oil. Thus, this term (i.e., *black oil*) is not used in this text.

4.1 Crude Oil

Crude oil is a naturally occurring, unrefined liquid (which also may occur in gaseous and/or solid form) composed of hydrocarbon derivatives and other organic materials containing so-called heteroatoms nitrogen, oxygen, and sulfur and metals such as iron, copper, nickel, and vanadium. Typically, crude oil can be refined to produce usable products such as gasoline, diesel fuel, fuel oils, lubricating oil, wax, and various forms of petrochemicals (Niu and Hu, 1999; Parkash, 2003; Gary et al., 2007; Speight, 2014a, 2017; Hsu and Robinson, 2017). However, crude oil is a nonrenewable resource that cannot be replaced naturally at the rate that it is consumed and is, therefore, a limited resource but a current lifetime on the order of 50 years (Speight, 2011a; Speight and Islam, 2016).

Crude oil is referred to generically as a *fossil energy resource* and is further classified as a *hydrocarbon resource*. The crude oil family includes crude oil itself, opportunity crude oil, high-acid oil, family oil, and heavy oil. Each of these oils bears a resemblance to conventional crude oil but differs in terms of physical properties and the method of recovery. However, a note of caution must be added here insofar as in order to classify cured oil into the various categories or types, the use of a single parameter such as viscosity is not sufficiently accurate to define the nature and properties of crude oil and crude oil-related materials. The general classification of crude oil into conventional crude oil and heavy oil involves not only an inspection of several properties but also some acknowledgment of the method of recovery.

Crude oil (petroleum, conventional crude oil, and conventional petroleum) is found in the microscopic pores of sedimentary rocks such as sandstone and limestone (Niu and Hu, 1999; Parkash, 2003; Gary et al., 2007; Speight, 2014a, 2017; Hsu and Robinson, 2017). Not all of the pores in a rock contain crude oil, and some pores will be filled with water or brine that is saturated with minerals. However, discovered oil fields (heavy oil fields, extra heavy oil fields, and tar sand bitumen deposits—the term oil is used in the current content as a generic term to include heavy oil, extra heavy oil, and tar sand bitumen, and it is not intended to be a means for the definition of these resources) are not always developed for oil production since (i) the formation that contains the oil may be too deep for efficient economic exploitation of the resource, (ii) the available volume of the oil may be insufficient for economic exploitation of the resource, or (iii) the oil field may be so remote that transport costs would be high to balance the economics of development.

The *definition* of crude oil has been varied, unsystematic, diverse, and often archaic. In fact, there has been a tendency to define crude oil and heavy oil on the basis of a single property. While this may be suitable for a general understanding, it is by no means accurate and does not reflect the true nature of crude oil or heavy oil or the characterization of the material. Unfortunately, this form of *identification* or differentiation is a product of many years of growth and its long established use, however general or inadequate it may be, is altered with difficulty, and a new term, however precise, and is adopted only slowly. In fact, fluid typing, composition, gravity, viscosity, asphaltene content, and mineral content are important in planning a suitable production method. This requires good laboratory data on representative fluid samples (Speight, 2015) as input parameters to the reservoir simulator. Laboratory data obtained at representative temperature and pressure are needed, possibly also in the presence of solvents or during combustion.

Crude oil is a naturally occurring mixture of hydrocarbons, generally in a liquid state, which may also include compounds of sulfur, nitrogen, oxygen, metals, and other elements (ASTM D4175, 2017). Thus, the common use of the term crude oil includes all gaseous, liquid, and solid hydrocarbon derivatives. In the reservoir, the proportions of crude oil constituents that occur in the gaseous phase, the liquid phase, and the solid phase depend on subsurface temperature and pressure. However, under the temperature and pressure conditions prevalent on the surface (i.e., at the wellhead), the lower-boiling hydrocarbon derivatives (methane, ethane, propane, and butane) emerge from the crude oil as gases, while pentane and the higher-boiling (higher-molecular-weight) hydrocarbon derivatives remain in the liquid phase. Higher-molecular-weight hydrocarbon derivatives can occur in the solid phase (i.e., wax derivatives) and remain as solids dissolved in the liquid phase.

Typically, an oil well produces predominantly crude oil, with some of the low-boiling hydrocarbons dissolved in the liquid phase, but at the surface where the pressure is lower than the pressure in the reservoir, much of the gaseous constituents (methane, ethane, propane, and butane) come out of solution and are recovered as *associated gas* (also called *solution gas*). However, because the underground temperature and pressure are higher than at the surface, the gas may contain higher-boiling hydrocarbon derivatives up to and including decane derivatives (C_5 to C_{10} derivatives), which condenses out of the gas as natural gas condensate (Speight, 2014a, 2018). Natural gas condensate (*condensate*) resembles naphtha (a blend stock for gasoline manufacture) in properties and appearance (Speight, 2007, 2018). However, the proportion of low-boiling hydrocarbon derivatives (boiling point $< 200°C$ or $< 390°F$) in the crude oil mixture varies considerably among different crude oil fields and ranges ranging from as much as 75% v/v in the conventional lighter oils to less than 40% v/v in the heavier (more viscous) crude oil and especially in extra heavy oil and tar sand bitumen.

The hydrocarbon derivatives in crude oil are predominantly mostly alkane derivatives, cycloalkane derivatives, and various aromatic hydrocarbon derivatives, while the other organic compounds contain nitrogen, oxygen, sulfur, and trace amounts of metal derivatives such as iron, copper, nickel, and vanadium. The exact molecular composition of crude oil varies widely from formation to formation, but the proportion of chemical elements varies over fairly narrow limits (Speight, 2014a), thus,

Carbon: 83%–85% w/w
Hydrogen: 10%–14% w/w
Nitrogen: 0.1%–2% w/w
Oxygen: 0.05%–1.5% w/w

Sulfur: 0.05%–6% w/w
Metals: 100–5000 ppm w/w

Thus, the term *crude oil* covers a wide assortment of materials consisting of mixtures of hydrocarbons and other compounds containing variable amounts of sulfur, nitrogen, and oxygen, which may vary widely in specific gravity, API gravity, and the amount of residuum (Table 1.2; Speight, 2012). Metal-containing constituents, notably those compounds that contain vanadium and nickel, usually occur in the more viscous crude oils in amounts up to several thousand parts per million and can have serious consequences during the processing of these feedstocks (Speight, 1984, 2014a). Because crude oil is a mixture of widely varying constituents and proportions, its physical properties also vary widely, and the color varies from near colorless to black.

The fuels derived from crude oil contribute approximately one-third to one-half of the total world energy supply and are used not only for transportation fuels (i.e., gasoline, diesel fuel, and aviation fuel) but also to heat buildings. Crude oil products have a wide variety of uses that vary from gaseous and liquid fuels to near-solid machinery lubricants. In addition, the residua of many refinery processes (Fig. 1.3) can be used to produce residua—once-maligned by-products—from which asphalt can be manufactured and is now a premium

TABLE 1.2 Typical Variations in the Properties of Crude Oil

Petroleum	Specific Gravity	API Gravity	Residuum > 1000°F % v/v
US Domestic			
California	0.858	33.4	23.0
Oklahoma	0.816	41.9	20.0
Pennsylvania	0.800	45.4	2.0
Texas	0.827	39.6	15.0
Texas	0.864	32.3	27.9
Foreign			
Bahrain	0.861	32.8	26.4
Iran	0.836	37.8	20.8
Iraq	0.844	36.2	23.8
Kuwait	0.860	33.0	31.9
Saudi Arabia	0.840	37.0	27.5
Venezuela	0.950	17.4	33.6

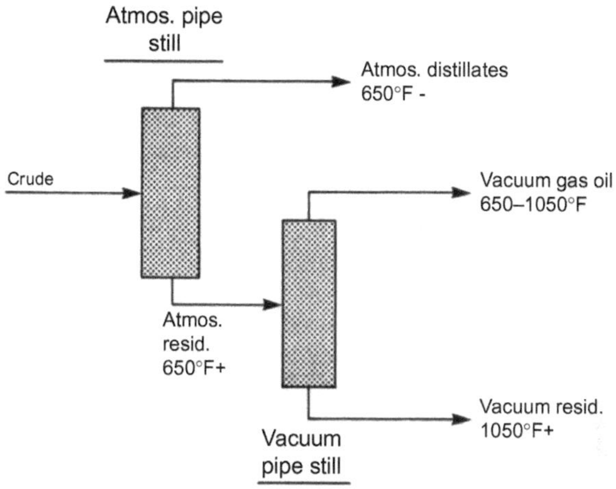

FIG. 1.3 Crude oil residua are different from residua insofar as residua are manufactured products—heavy oil, extra heavy oil, and tar sand bitumen are naturally occurring products.

value product for highway surfaces, roofing materials, and miscellaneous waterproofing uses (Parkash, 2003; Gary et al., 2007; Speight, 2014a, 2017; Hsu and Robinson, 2017).

The molecular boundaries of crude oil cover a wide range of boiling points and carbon numbers of hydrocarbon compounds and other compounds containing nitrogen, oxygen, and sulfur, as well as metal-containing (porphyrin) constituents. However, the actual boundaries of such a *crude oil map* can only be arbitrarily defined in terms of boiling point and carbon number. In fact, crude oil is so diverse that materials from different sources exhibit different boundary limits, and for this reason alone, it is not surprising that crude oil has been difficult to *map* in a precise manner.

Since there is a wide variation in the properties of crude oil, the proportions in which the different constituents occur vary with the origin and the relative amounts of the source materials that form the initial *protopetroleum* and the maturation conditions. Thus, some crude oils have higher proportions of the lower-boiling components, and others (such as heavy oil, extra heavy oil, and tar sand bitumen) have higher proportions of higher-boiling components (a high-boiling gas-oil fraction, resin constituents, and asphaltene constituents).

Crude oil occurs underground, at various pressures depending on the depth. Because of the pressure, it contains considerable natural gas in solution. Crude oil underground is much more fluid than it is on the surface and is generally mobile under reservoir conditions because the elevated temperatures (the *geothermal gradient*) in subterranean formations decrease the viscosity. Although the geothermal gradient varies from place to place, it is generally on the order of 25–30°C/km (15°F/1000 ft or 120°C/1000 ft, that is, 0.015°C per foot of depth or 0.012°C per foot of depth).

Crude oil is derived from aquatic plants and animals that lived and died hundreds of millions of years ago. Their remains mixed with mud and sand in layered deposits that, over the millennia, were geologically transformed into sedimentary rock. Gradually, the organic matter decomposed and eventually formed crude oil (or a related precursor), which migrated from the original source beds to more porous and permeable rocks, such as *sandstone* and *siltstone*, where it finally became entrapped. Such entrapped accumulations of crude oil are called *reservoirs*. A series of reservoirs within a common rock structure or a series of reservoirs in separate but neighboring formations are commonly referred to as an *oil field*. A group of fields is often found in a single geologic environment known as a *sedimentary basin* or *province*.

The major components of crude oil are *hydrocarbon derivatives*, compounds of hydrogen and carbon that display great variation in their molecular structure. The simplest hydrocarbons are a large group of chain-shaped molecules known as the *paraffins*. This broad series extends from methane, which forms natural gas, to liquids that are refined into gasoline, to crystalline waxes. A series of ring-shaped hydrocarbons, known as the *naphthenes*, range from volatile liquids such as *naphtha* to high-molecular-weight substances isolated as the *asphaltene* fraction. Another group of ring-shaped hydrocarbons is known as the *aromatics*; the chief compound in this series is benzene, a popular raw material for making petrochemicals. *Nonhydrocarbon constituents* of crude oil include organic derivatives of nitrogen, oxygen, sulfur, and the metals nickel and vanadium. Most of these impurities are removed during refining.

Geologic techniques can determine only the existence of rock formations that are favorable for oil deposits, not whether oil is actually there. Drilling is the only sure way to ascertain the presence of oil. With modern rotary equipment, wells can be drilled to depths of more than 30,000 ft (9000 m). Once oil is found, it may be recovered (brought to the surface) by the pressure created by natural gas or water within the reservoir. Crude oil can also be brought to the surface by injecting water or steam into the reservoir to raise the pressure artificially or by injecting such substances as carbon dioxide, polymers, and solvents to reduce crude oil viscosity. Thermal recovery methods are frequently used to enhance the production of heavy crude oils, whose extraction is impeded by viscous resistance to flow at reservoir temperatures (Fig. 1.4).

Crude oil is typically recovered from the reservoir by the application of primary and secondary recovery techniques. Although covered elsewhere (Chapter 2), mention of primary, secondary, and tertiary recovery is warranted here in terms of the general description of definitions of these techniques.

Primary recovery refers to the process in which the crude oil in the reservoir trap is forced to the surface by the natural pressure contained in the trap. This pressure may result from several forces: (i) When the reservoir is penetrated, the pressure release allows the water layer to expand and push the oil upward and replaces it in the rock pores—this is the most effective technique and is known as a water drive system; (ii) if the drill penetrates into a layer of oil that has a gas cap above it, the release of pressure allows the gas layer to expand rapidly

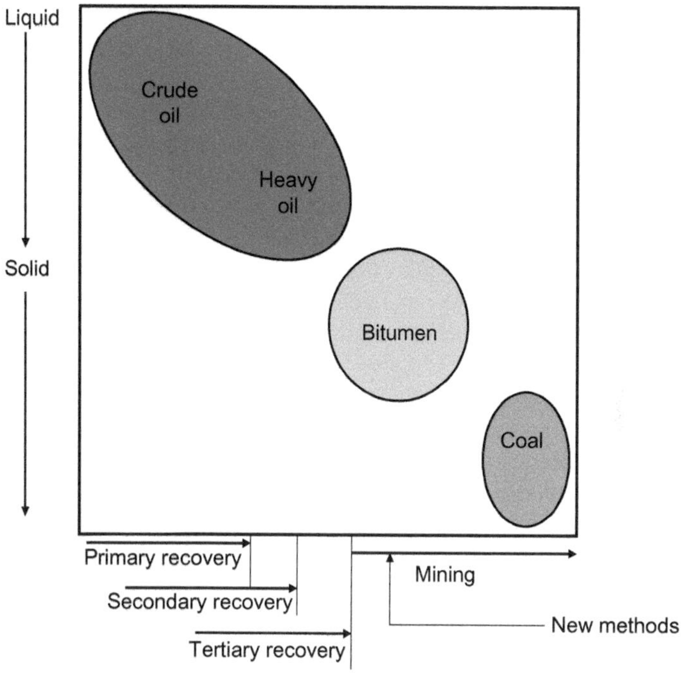

FIG. 1.4 Schematic representation of the properties and recovery methods for crude oil, heavy oil, and tar sand bitumen. Coal, typically recovered by mining methods, is included as an example.

causing a downward pressure on the oil forcing it to move up through the well (gas-cap drive); and (iii) gas dissolved in the oil may be released as bubbles when the trap is pierced; as the oil moves up, the gas in the oil expands, and the growing bubbles push the oil to the surface (solution-gas drive). In most reservoir traps, these pressures are sufficient to initially force the crude oil to the surface.

At some point in time, this pressure will fall. Crude oil production decreases because (i) there is less force driving the oil toward the well; and/or (ii) the gas that moves into the emptied pore spaces reduces the permeability of the rock, making it more difficult for oil to flow through; and/or (iii) the fall in pressure and the loss of dissolved gas increase the surface tension and viscosity of the oil. Thus, primary recovery techniques usually account for less than 30% of the total volume of crude oil recovered.

Secondary recovery involves trying to maintain reservoir pressure. One technique is to inject natural gas into the reservoir above the oil, forcing the oil downward, and then inject water below the oil so forcing it upward. Sometimes, the gas that is used is that that has just been released during *primary recovery*. The disadvantage of using the released gas is that this gas is a marketable product in its own right. However, this is a good method to use if transporting the gas would be costly, and in any case, the reinjected gas can always be collected again if necessary. Alternative secondary techniques involve injecting carbon dioxide or nitrogen into the oil. This makes the oil more fluid, and the gas pushes the oil upward.

Tertiary recovery is the most expensive approach and involves injecting steam, detergents, solvents, bacteria, or bacterial nutrient solutions into the remaining oil. When high-pressure steam is injected, it heats the oil, decreasing its density and viscosity and increasing its flow rate. Sometimes, some of the oil in the reservoir rock is deliberately set on fire. This is used to increase the flow rate of the oil ahead of the combustion front. Detergents that can be injected reduce the viscosity of the oil and act as surfactants reducing the ability of the oil to stick to the rock surface and thus making it easier for it to be flushed up to the surface.

Another *tertiary recovery* technique involves injecting bacteria into the oil field. Some bacteria produce polysaccharides that reduce the permeability of the water-filled pores of the reservoir rock, and this effectively forces injected water into the oil-filled pores, pushing the oil out. Other bacteria produce carbon dioxide that helps to increase pressure within the rock pores, forcing out the oil. Other bacteria produce surfactants and/or chemicals that reduce the viscosity of the oil.

After recovery, crude oil is transported to refineries by pipelines, which can often carry more than 500,000 barrels per day, or by ocean-going tankers. The basic refinery process is distillation, which separates the crude oil into fractions of differing volatility. After the distillation, other physical methods are employed to separate the mixtures, including absorption, adsorption, solvent extraction, and crystallization. After physical separation into such constituents as light and heavy naphtha, kerosene, and light and heavy gas oils, selected crude oil fractions may be subjected to conversion processes, such as thermal cracking (i.e., coking) and catalytic cracking. In the most general terms, cracking breaks the large molecules of heavier gas oils into the smaller molecules that form the lighter, more valuable naphtha fractions. *Reforming* changes the structure of straight-chain paraffin molecules into branched-chain isoparaffin derivatives and ring-shaped aromatics. The process is widely used to raise the octane number of gasoline obtained by the distillation of paraffinic crude oils.

Before passing on to heavy oil, there are three types of conventional crude oil that need to be addressed: (i) opportunity crude oil, (ii) high-acid crude oil, and (iii) foamy oil.

4.1.1 Opportunity Crude Oil

There is also the need for a refinery to be configured to accommodate *opportunity crude oils* and/or *high-acid crude oils* that, for many purposes, are often included with heavy feedstocks (Speight, 2014a,b; Yeung, 2014). *Opportunity crude oils* are either new crude oils with unknown or poorly understood properties relating to processing issues or are existing crude oils with well-known properties and processing concerns (Ohmes, 2014). Opportunity crude oils are often, but not always, heavy crude oils but in either case are more difficult to process due to high levels of solids (and other contaminants) produced with the oil, high levels of acidity, and high viscosity. These crude oils may also be incompatible with other oils in the refinery feedstock blend and cause excessive equipment fouling when processed either in a blend or separately (Speight, 2015).

In addition to taking preventative measure for the refinery to process these feedstocks without serious deleterious effects on the equipment, refiners need to develop programs for detailed and immediate feedstock evaluation so that they can understand the qualities of a crude oil very quickly and it can be valued appropriately, and the management of the crude processing can be planned meticulously (Babich and Moulijn, 2003; Speight, 2014a). For example, the compatibility of opportunity crudes with other opportunity crudes and with conventional crude oil and heavy oil is a very important property to consider when making decisions regarding which crude to purchase. Blending crudes that are incompatible can lead to extensive fouling and processing difficulties due to unstable asphaltene constituents (Speight, 2014a, 2015). These problems can quickly reduce the benefits of purchasing the opportunity crude in the first place. For example, extensive fouling in the crude preheat train may occur resulting in decreased energy efficiency, increased emissions of carbon dioxide, and increased frequency at which heat exchangers need to be cleaned. In a worst-case scenario, crude throughput may be reduced leading to significant financial losses.

Opportunity crude oils, while offering initial pricing advantages, may have composition problems that can cause severe problems at the refinery, harming infrastructure, yield, and profitability. Before refining, there is the need for comprehensive evaluations of opportunity crudes, giving the potential buyer and seller the needed data to make informed decisions regarding fair pricing and the suitability of a particular opportunity crude oil for a refinery. This will assist the refiner to manage the ever-changing crude oil quality input to a refinery—including quality and quantity requirements and situations, crude oil variations, contractual specifications, and risks associated with such opportunity crudes.

4.1.2 High Acid Crude Oil

High-acid crude oils are crude oils that contain considerable proportions of naphthenic acids that, as commonly used in the crude oil industry, refer collectively to all of the organic acids present in the crude oil (Shalaby, 2005; Speight, 2014b). In many instances, the high-acid crude oils are actually the heavier crude oils (Speight, 2014a,b), and it is interesting to note that the acidic fraction has the greatest impact on decreasing the interfacial tension with the potential for facilitating recovery (Khulbe et al., 1996). The total acid matrix is therefore complex, and it is unlikely that a simple titration, such as the traditional methods for measurement of the total acid number, can give meaningful results to use in predictions of problems. An alternative way of defining the relative organic acid fraction of crude oils is therefore a real need in the oil industry, both upstream and downstream.

By the original definition, a naphthenic acid is a monobasic carboxyl group attached to a saturated cycloaliphatic structure. However, it has been a convention accepted in the oil industry that all organic acids in crude oil are called naphthenic acids. Naphthenic acids in crude oils are now known to be mixtures of low- to high-molecular-weight acids, and the naphthenic acid fraction also contains other acidic species. Naphthenic acids, which are not *user-friendly* in terms of refining (Kane and Cayard, 2002; Ghoshal and Sainik, 2013), can be either (or both) water-soluble to oil-soluble depending on their molecular weight, process temperatures, salinity of waters, and fluid pressures. In the water phase, naphthenic acids can cause stable reverse emulsions (oil droplets in a continuous water phase). In the oil phase with residual water, these acids have the potential to react with a host of minerals, which are capable of neutralizing the acids. The main reaction product found in practice is the calcium naphthenate soap (the calcium salt of naphthenic acids).

Normally, the end result of the formation of low-molecular-weight acidic species is treated in the overheads in refineries. A combined approach to front-end treating at crude inlet to heaters and preheat exchangers should be considered. It is commonly assumed that acidity in crude oils is related to carboxylic acid species, that is, components containing a —COOH functional group. While it is clear that carboxylic acid functionality is an important feature (60% of the ions have two or more oxygen atoms), a major portion (40%) of the acid types are not carboxylic acids. In fact, naphthenic acids are a mixture of different compounds that may be polycyclic and may have unsaturated bonds, aromatic rings, and hydroxyl groups. Even the carboxylic acids are more diverse than expected, with ~85% containing more heteroatoms than the two oxygen atoms needed to account for the carboxylic acid groups. Examining the distribution of component types in the acid fraction reveals that there is a broad distribution of species.

High-acid crude oils cause corrosion in the refinery—corrosion is predominant at temperatures in excess of 180°C (355°F) (Kane and Cayard, 2002; Ghoshal and Sainik, 2013; Speight, 2014c) and occurs particularly in the atmospheric distillation unit (the first point of entry of the high-acid crude oil) and also in the vacuum distillation units. In addition, overhead corrosion is caused by the mineral salts, magnesium, calcium, and sodium chloride that are hydrolyzed to produce volatile hydrochloric acid, causing a highly corrosive condition in the overhead exchangers. Therefore, these salts present a significant contamination in opportunity crude oils. Other contaminants in opportunity crude oils that are shown to accelerate the hydrolysis reactions are inorganic clays and organic acids.

Corrosion by naphthenic acids typically has a localized pattern, particularly at areas of high velocity and, in some cases, where condensation of concentrated acid vapors can occur in crude distillation units. The attack also is described as lacking corrosion products. Damage is in the form of unexpected high corrosion rates on alloys that would normally be expected to resist sulfidic corrosion (particularly steels with more than 9% Cr). In some cases, even very highly alloyed materials (i.e., 12% Cr, type 316 stainless steel (SS), type 317 SS, and in severe cases even 6% Mo stainless steel) have been found to exhibit sensitivity to corrosion under these conditions.

The corrosion reaction processes involve the formation of iron naphthenates:

$$Fe + 2RCOOH = Fe(RCOO)_2 + H_2 \ (4)$$

$$Fe(RCOO)_2 + H_2S = FeS + 2RCOOH$$

The iron naphthenates are soluble in oil, and the surface is relatively film free. In the presence of hydrogen sulfide, a sulfide film is formed, which can offer some protection depending on the acid concentration. If the sulfur-containing compounds are reduced to its hydrogen sulfide, the formation of a potentially protective layer of iron sulfide occurs on the unit walls, and corrosion is reduced (Kane and Cayard, 2002). When the reduction product is water, coming from the reduction of sulfoxides, the naphthenic acid corrosion is enhanced.

Thermal decarboxylation can occur during the distillation process (during which the temperature of the crude oil in the distillation column can be as high as 400°C, 750°F):

$$RCO_2H \rightarrow R - H + CO_2$$

However, not all acidic species in crude oil are derivatives of carboxylic acids (—COOH), and some of the acidic species are resistant to high temperatures. For example, acidic species appear in the vacuum residue after having been subjected to the inlet temperatures of an atmospheric distillation tower and a vacuum distillation tower (Speight and Francisco, 1990). In addition, for the acid species that are volatile, naphthenic acids are most active at their boiling point, and the most severe corrosion generally occurs on condensation from the vapor phase back to the liquid phase.

In addition to taking preventative measure for the refinery to process these feedstocks without serious deleterious effects on the equipment, refiners will need to develop programs for detailed and immediate feedstock evaluation so that they can understand the qualities of a crude oil very quickly and it can be valued in terms of the application of the necessary refinery operations.

4.1.3 Foamy Oil

Foamy oil is oil-continuous foam that contains dispersed gas bubbles produced at the wellhead from heavy oil reservoirs under solution-gas drive. The nature of the gas dispersions in oil distinguishes foamy oil behavior from conventional heavy oil. The gas that comes out of solution in the reservoir does not coalesce into large gas bubbles nor into a continuous flowing gas phase. Instead, it remains as small bubbles entrained in the crude oil, keeping the effective oil viscosity low while providing expansive energy that helps drive the oil toward the producing well. Foamy oil accounts for unusually high production in heavy oil reservoirs under solution-gas drive (Sun et al., 2013).

Foamy oil behavior is a unique phenomenon associated with production of heavy oils. It is believed that this mechanism contributes significantly to the abnormally high production rate of heavy oils observed in the Orinoco Belt. During production of heavy oil from solution-gas-drive reservoirs, the oil is pushed into the production wells by energy supplied by the dissolved gas. As fluid is withdrawn from the production wells, the pressure in the reservoir declines and the gas that was dissolved in the oil at high pressure starts to come out of solution (hence, foamy oil). As pressure declines further with continued removal of fluids from the production wells, more gas is released from solution and the gas already released expands in volume. The expanding gas, which at this point is in the form of isolated bubbles, pushes the oil out of the pores and provides energy for the flow of oil into the production well. This process is very efficient until the isolated gas bubbles link up and the gas itself starts flowing into the production well. Once the gas flow starts, the oil has to compete with the gas for available flow energy. Thus, in some heavy oil reservoirs, due to the properties of the oil and the sand and also due to the production methods, the released gas forms foam with the oil and remains subdivided in the form of dispersed bubbles much longer.

Thus, foamy oil is formed in solution-gas-drive reservoirs when gas is released from solution with a decline in reservoir pressure. It has been noted that the oil at the wellhead of these heavy oil reservoirs resembles the form of foam, hence the term *foamy oil*. The gas initially exists in the form of small bubbles within individual pores in the rock. As time passes and pressure continues to decline, the bubbles grow to fill the pores. With further declines in pressure, the bubbles created in different locations become large enough to coalesce into a continuous gas phase. Once the gas phase becomes continuous (i.e., when gas saturation exceeds the critical level—the minimum saturation at which a continuous gas phase exists in porous media), traditional two-phase (oil and gas) flow with classical relative permeability occurs. As a result, the production gas-oil ratio (GOR) increases rapidly after the critical gas saturation has been exceeded.

However, it has been observed that many heavy oil reservoirs in Alberta and Saskatchewan exhibit foamy oil behavior that is accompanied by sand production, leading to anomalously high oil recovery and lower gas-oil ratio (Chugh et al., 2000). These observations suggest that the foamy oil flow might be physically linked to sand production. It is apparent that some additional factors, which remain to be discovered, are involved in making the foamy solution gas possible at field rates of decline. One possible mechanism is the synergistic influence of sand influx into the production wells. Allowing 1%–3% w/w sand to enter the wellbore with the fluids can result in the propagation of a front of sharp pressure gradients away from the wellbore. These sharp pressure gradients occur at the advancing edge of solution-gas drive. It is still unknown how far from the wellbore the dilated zone can propagate.

However, the actual structure of foamy oil flow and its mathematical description are still not well understood. Much of the earlier discussion of such flows was based on the concept of microbubbles (i.e., bubbles that are much smaller than the average pore-throat size and are thus free to move with the oil during flow) (Sheng et al., 1999). Dispersion of this type can be produced only by the nucleation of a very large number of bubbles (explosive nucleation) and by the availability of a mechanism that prevents these bubbles from growing into larger bubbles with decline in pressure. Another hypothesis for the structure of foamy oil flow is that much larger bubbles migrating with the oil, with the dispersion created by the breakup of bubbles during migration. The major difference between conventional solution-gas drive and foamy solution-gas drive is that the pressure gradient in the latter is strong enough to mobilize gas clusters once they have grown to a certain size (Maini, 1999).

Reservoirs that exhibit foamy oil behavior are typically characterized by the appearance of an oil-continuous foam at the wellhead. When oil is produced as this nonequilibrium mixture, reservoirs can perform with higher than expected rates of production: up to 30 times that predicted by Darcy's law, and lower than expected production gas-oil ratios (Poon and Kisman, 1992). Moreover, foamy oil flow is often accompanied by sand production along with the oil and gas—the presence of sand at the wellhead leads to sand dilation and the presence of high-porosity, high-permeability zones (wormholes) in the reservoir (Maini, 1999, 2001). It is generally believed that in the field, the high rates and recoveries observed are the combination of the foamy oil mechanism and the presence of these wormholes.

Thus, when, for example, heavy oil contains associated gas, foaming may occur at any point in the reservoir, wellbore, flow lines, or production equipment at the surface. Foaming occurs when gas reaches the bubble point and comes out of suspension as the pressure and temperature of the fluids change. Gas separates from lighter crude oil much faster and more predictably than it does from heavy oil. The ability of the oil to foam depends on its viscosity, so higher viscosity oil is more likely to foam, and the foam it makes will be longer lasting.

Foamy oil behavior is a unique phenomenon associated with the production of heavy oil. It is believed that this mechanism contributes significantly to the abnormally high production rate of heavy oils observed in the Orinoco Belt (Sun et al., 2013). However, while foaming can be a problem, it can also assist in the recovery operations since any foam that forms in the reservoir increases pressure; it can serve as a temporary gas drive, pushing oil toward the wellbore.

This requires an understanding of the fluid properties in order to know where and when the foaming is likely to occur and to ensure that foaming does not occur in the oil-water separator.

In addition, detailed reservoir characterization or deposit characterization are also the foundation for future decisions, such as selecting the right time and follow-up production method to improve overall recovery. That improved ability to predict how a reservoir will behave benefit the overall project.

4.2 Heavy Oil

The name *heavy oil* can often be misleading as it has also been used in reference to (i) fuel oil that contains residuum leftover from distillation, that is, heavy fuel oil or residual fuel oil, (ii) coal tar creosote, or (iii) viscous crude oil. Thus, for the purposes of this text, the term is used to mean *viscous crude oil*. Extra heavy oil has been included in many of these categories. On the other hand, tar sand bitumen (often called simply *bitumen*) is often confused with manufactured asphalt (confusingly referred to as bitumen in many countries). To add further to this confusion, in some countries, tar sand bitumen is referred to as natural as asphalt!

Heavy oil is a viscous type of crude oil that contains higher level of sulfur than conventional crude oil and which occurs in similar locations to crude oil (IEA, 2005; Ancheyta and Speight, 2007). The nature of heavy oil is a problem for recovery operations and for refining—the viscosity of the oil may be too high, thereby leading to difficulties in recovery and/or difficulties in refining the oil (Speight, 2016, 2017). However, tar sand bitumen, in terms of properties and behavior, are a world apart from conventional crude oil (Table 1.3). Success with this material and with extra heavy oil depends as much on understanding the fluid (or nonfluid) properties of the material and the behavior of the fluids in the deposit in which they occur as it does on knowing the geology of the deposit (Speight, 2013a,b,c, 2014a). The reason is that the chemical and physical differences between heavy oil, extra heavy oil, and tar sand bitumen oil ultimately affect the viscosity and other relevant properties that, in turn, influence the individual aspects of recovery and refining operations.

There are large resources of *heavy oil* in Canada, Venezuela, Russia, the United States, and many other countries (Niu and Hu, 1999; Meyer et al., 2007; Attanasi and Meyer, 2010). The resources in North America alone provide a small percentage of current oil production (~2%); existing commercial technologies could allow for significantly increased production. Under some economic conditions, heavy oil can be profitably produced but at a smaller profit margin than for conventional oil, due to higher production costs and upgrading costs in conjunction with the lower market price for heavier crude oils.

Heavy oil is a *type* of crude oil that is different from conventional crude oil insofar as they are much more difficult to recover from the subsurface reservoir.

TABLE 1.3 Comparison of the Properties of Conventional Crude Oil With Athabasca Bitumen[a]

Property	Athabasca Bitumen	Conventional Crude Oil
Specific gravity	1.03	0.85–0.90
Viscosity, cp		
38°C/100°F	750,000	< 200
100°C/212°F	11,300	
Pour point, °F	> 50	ca −20
Elemental analysis (% w/w)		
Carbon	83.0	86.0
Hydrogen	10.6	13.5
Nitrogen	0.5	0.2
Oxygen	0.9	< 0.5
Sulfur	4.9	< 2.0
Ash	0.8	0.0
Nickel (ppm)	250	< 10.0
Vanadium (ppm)	100	< 10.0
Fractional composition (% w/w)		
Asphaltenes (pentane)	17.0	< 10.0
Resins	34.0	< 20.0
Aromatics	34.0	> 30.0
Saturates	15.0	> 30.0
Carbon residue (% w/w)		
Conradson	14.0	< 10.0

[a]*Extra heavy oil has a similar analysis to tar sand bitumen.*

These materials have a much higher viscosity (and lower API gravity) than conventional crude oil, and the recovery of these crude oil types usually requires thermal stimulation of the reservoir. When crude oil occurs in a reservoir that allows the crude material to be recovered by pumping operations as a free-flowing dark-to-light-colored liquid, it is often referred to as *conventional crude oil.*

Very simply, heavy oil is a type of crude oil that is very viscous and does not flow easily. The common characteristic properties (relative to conventional crude oil) are high specific gravity, low hydrogen-to-carbon ratios, high carbon residues, and high contents of asphaltenes, heavy metal, sulfur, and nitrogen. Specialized recovery and refining processes are required to produce more useful fractions, such as naphtha, kerosene, and gas oil.

Heavy oil is an oil resource that is characterized by high viscosity of the oil (i.e., a higher resistance to flow than conventional crude oil) and high density compared with conventional oil. Most heavy oil reservoirs originated as conventional oil that formed in deep formations but migrated to the surface region where they were degraded by bacteria and by weathering and where the lightest hydrocarbons escaped. Heavy oil is deficient in hydrogen and has high carbon, sulfur, and heavy metal content. Hence, heavy oil requires additional processing (upgrading) to become a suitable refinery feedstock for a normal refinery.

Heavy oil accounts for more than double the resources of conventional oil in the world, and heavy oil offers the potential to satisfy current and future oil demand. Not surprisingly, heavy oil has become an important theme in the crude oil industry with an increasing number of operators getting involved or expanding their plans in this market around the world. However, heavy oil is more difficult to recover from the subsurface reservoir than conventional or light oil. A very general definition of heavy oils has been and remains based on the API gravity or viscosity, and the definition is quite arbitrary although there have been attempts to rationalize the definition based upon viscosity, API gravity, and density.

For example, heavy oil is considered (arbitrarily) to be those crude oils that had gravity somewhat less than 20° API with the heavy oil falling into the API gravity range 10–15°. For example, Cold Lake heavy crude oil has an API gravity equal to 12°, and extra heavy oil, such as tar sand bitumen, usually has an API gravity in the range 5–10° (Athabasca bitumen = 8° API). Residua would vary depending upon the temperature at which distillation was terminated, but usually vacuum residua are in the range 2–8° API (Speight, 2000, 2017).

Heavy oil has a much higher viscosity (and lower API gravity) than conventional crude oil, and the recovery of heavy oil usually requires thermal stimulation of the reservoir. The generic term *heavy oil* is often applied to a crude oil that has less than 20° API and usually, but not always, sulfur content higher than 2% by weight (Speight, 2000). Furthermore, in contrast to conventional crude oils, heavy oils are darker in color and may even be black.

The term *heavy oil* has also been arbitrarily (incorrectly) used to describe both the heavy oils that require thermal stimulation of recovery from the reservoir and the bitumen in bituminous sand (tar sand) formations from which the heavy bituminous material is recovered by a mining operation. *Extra heavy oil* is a nondescript term (related to viscosity) of little scientific meaning that is usually applied to tar sand bitumen, which is generally incapable of free flow under reservoir conditions.

The methods outlined in this book for the recovery of heavy oil, extra heavy oil, and tar sand bitumen (Chapters 2–4) focus on oil with an API gravity of less

TABLE 1.4 API Gravity and Sulfur Content of Selected Heavy Oils

	API	Sulfur (wt.%)
Bachaquero	13.0	2.6
Boscan	10.1	5.5
Cold Lake	13.2	4.1
Huntington Beach	19.4	2.0
Kern River	13.3	1.1
Lagunillas	17.0	2.2
Lloydminster	16.0	2.6
Lost Hills	18.4	1.0
Merey	18.0	2.3
Midway Sunset	12.6	1.6
Monterey	12.2	2.3
Morichal	11.7	2.7
Mount Poso	16.0	0.7
Pilon	13.8	1.9
San Ardo	12.2	2.3
Temblador	19.0	0.8
Tia Juana	12.1	2.7
Wilmington	17.1	1.7
Zuata Sweet	15.7	2.7

For reference, Athabasca tar sand bitumen has API = 8° and sulfur content = 4.8% by weight.

than 20° (Table 1.4). However, it must be recognized that recovery depends not only on the characteristics of the oil but also on the characteristics of the reservoir—including the temperature of the reservoir and the pour point of the oil (see also Chapter 2). These heavy oils fall into a range of high viscosity (Fig. 1.5), and the viscosity is subject to temperature effects (Fig. 1.6), which is the reason for the application of thermal methods to heavy oil recovery.

4.3 Extra Heavy Oil

An example is the Zaca-Sisquoc extra heavy oil (sometimes referred to as the Zaca-Sisquoc bitumen) that has an API gravity on the order of 4.0–6.0°. The reservoir has an average depth of 3500 ft, average thickness of 1700 ft, average

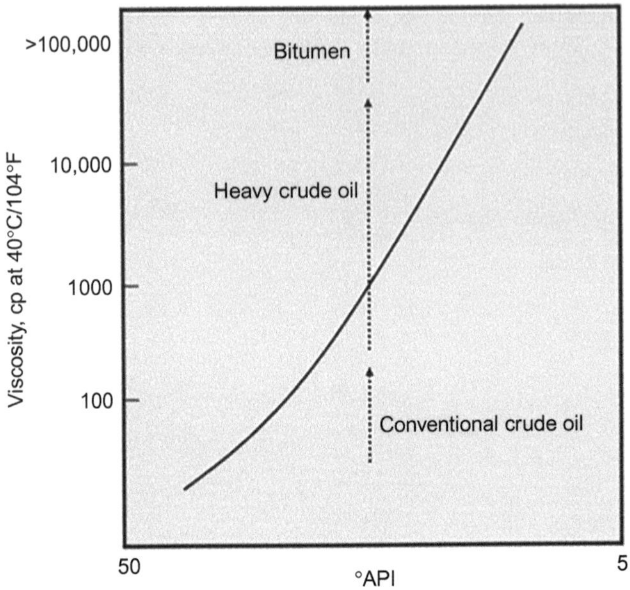

FIG. 1.5 General relationship of viscosity to API gravity.

temperature of 51–71°C (125–160°F), and sulfur in the range 6.8%–8% (Isaacs, 1992; USGS, 2006; Villarroel and Hernández, 2013). The deposit temperature is certainly equal to or above the pour point of the oil (Isaacs, 1992). This renders the oil capable of being pumped as a liquid from the deposit because of the high deposit temperature (which is higher than the pour point of the oil). The same rationale is applied to the extra heavy oil found in the Orinoco deposits.

For example, the field, which is located on the southern flank of the Eastern Venezuela Basin, produces extra heavy oil from porous unconsolidated sandstones of the lower Miocene Oficina Formation. The properties of the Zuata extra heavy oil (Table 1.5) are similar to the properties of Athabasca bitumen (but which lacks mobility in the Athabasca deposit), but the Zuata heavy oil is mobile in the deposit but is generally incapable of free flow under surface conditions. The Orinoco Oil Belt area has a reservoir section depth range on the order of 800–4000 ft with temperatures on the order of 38–60°C (100–140°F) and initial reservoir pressures of 450–1200 psi. The in situ oil viscosity ranges from 1000 to 5000 cp, and the API gravity of the oil ranges from 6.5 to 10.5°. The net sand thickness values of 20–300 ft are observed with the porosity of 28%–34% and permeability of 1–20 or more darcys (Villarroel and Hernández, 2013).

The Orinoco Belt essentially contains extra heavy crude: crude oil that flows only with difficulty under ambient (surface, i.e., above ground at normal temperature and under atmospheric pressure) conditions but is liquid at reservoir conditions. Examples are oils from the Morichal (8.5° API), Jobo (8–9°

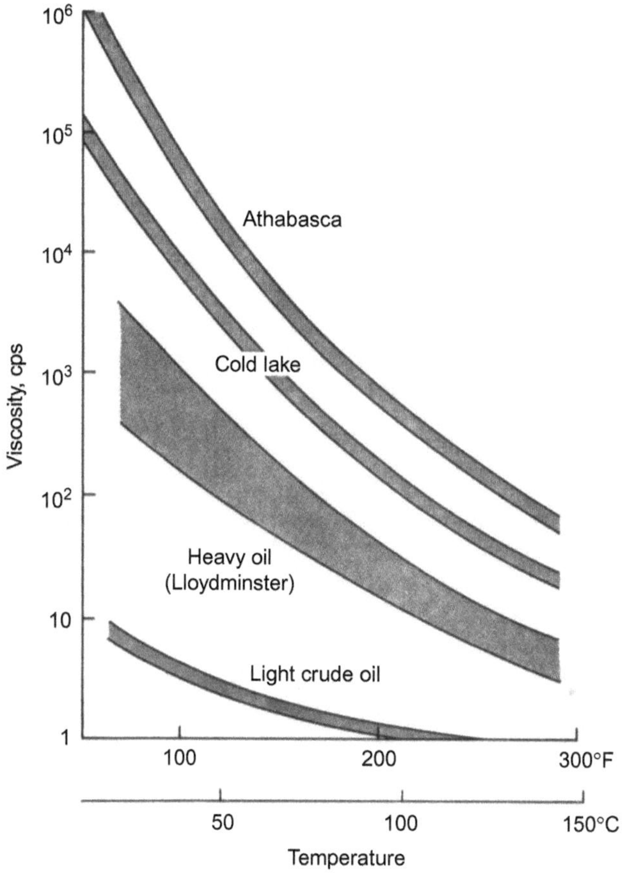

FIG. 1.6 General relationship of viscosity to temperature.

TABLE 1.5 Comparison of Selected Properties of Athabasca Tar Sand Bitumen (Alberta, Canada) and Zuata Extra Heavy Oil (Orinoco, Venezuela)

		Athabasca Bitumen	Zuata Extra Heavy Oil
Whole oil	API gravity	8	8
	Sulfur, % w/w	4.8	4.2
650F+	% v/v	85	86
	Sulfur, % w/w	5.4	4.6
	Ni + V, ppm	420	600
	CCR, % w/w[a]	14	15

[a]*Conradson carbon residue.*
1: Canada.
2: Venezuela.

API), and Pilon (13° API) fields. Compared with the Alberta tar sand deposits, reservoirs in the Orinoco Oil Belt have higher reservoir temperatures, greater reservoir permeability, and higher gas-to-oil ratios, which give the oil lower viscosity and greater mobility (Dusseault, 2001). These extra heavy oils are produced by conventional methods including alternated steam soaking and were considered as part of the traditional oil reserves of Venezuela. The Morichal and Jobo oils were diluted with gas oil at the wellhead to reduce their high viscosity and facilitate their transport by pipeline and reach their sale specification as the commercial blend known as the Morichal Segregation with 12.5° API that was used to obtain a high-quality asphalt. Pilon and Temblador were also blended to make the Pilon Segregation with 13.5° API for general refining feedstock. In addition, by mixing the extra heavy oil with water to solve a transport problem, it was discovered that this mixture could actually be used as a fuel in power stations. For this objective, the optimal mix was 70% extra heavy crude with 30% water and 1% surfactant in order to stabilize the emulsion (*Orimulsion*). However, burning extra heavy crude directly without refining it at least to some degree generates gases and ashes, which are very polluting.

The general difference is that extra heavy oil, which may have properties similar to tar sand bitumen in the laboratory but unlike immobile tar sand bitumen in the deposit, has some degree of mobility in the reservoir or deposit (Delbianco and Montanari, 2009; Speight, 2014a). Extra heavy oils can flow at reservoir temperature and can be produced economically, without additional viscosity-reduction techniques, through variants of conventional processes such as long horizontal wells or multilaterals. This is the case, for instance, in the Orinoco basin (Venezuela) or in offshore reservoirs of the coast of Brazil, but once outside of the influence of the high reservoir temperature, these oils are too viscous at surface to be transported through conventional pipelines and require heated pipelines for transportation. Alternatively, the oil must be partially upgraded or fully upgraded or diluted with a light hydrocarbon (such as aromatic naphtha) to create a mix that is suitable for transportation (Speight, 2014a).

The Orinoco Belt (Faja) in Venezuela contains one of the largest resources in the world of heavy oil and extra heavy oil. Projects in the area focus on increasing the recovery of heavy oil and extra heavy oil by the use of nonthermal and thermal methods. Steam-based thermal recovery processes are more efficient in low-pressure reservoirs or deposits. However, because of the depth below the surface, the initial pressures of the reservoirs in the Faja are relatively high, ranging from 600 to 1500 psi with the oil viscosity on the order of 2000 cp and higher. Thus, it is often necessary to decrease the pressure of the reservoir (deposit) with primary production techniques in order to facilitate the economical implementation of steam-injection-based methods.

Thus, extra heavy oil is a material that occurs in the solid or near-solid state and generally has mobility under reservoir conditions. While this type of oil resembles tar sand bitumen and does not flow easily, extra heavy oil is generally recognized as having mobility in the reservoir compared with tar sand bitumen, which is typically incapable of mobility (free flow) under reservoir conditions. It is likely that the mobility of extra heavy oil is due to a high reservoir temperature (that is higher than the pour point of the extra heavy oil) or due to other factors that are variable and subject to localized conditions in the reservoir.

4.4 Tar Sand Bitumen

The expression *tar sand* is commonly used in the crude oil industry to describe sandstone reservoirs that are impregnated with a heavy, viscous black crude oil that cannot be retrieved through a well by conventional production techniques (FE-76-4, above). However, the term *tar sand* is actually a misnomer; more correctly, the name *tar* is usually applied to the heavy product remaining after the destructive distillation of coal or other organic matter (Speight, 2013d). Current recovery operations of bitumen in tar sand formations are predominantly focused on a mining technique.

The term *bitumen* (also, on occasion, referred to as *native asphalt* and *extra heavy oil*) includes a wide variety of reddish brown to black materials of semisolid, viscous to brittle character that can exist in nature with no mineral impurity or with mineral matter contents that exceed 50% by weight and are often structurally dissimilar (Kam'Yanov et al., 1995; Kettler, 1995; Ratov, 1996; Parnell et al., 1996; Niu and Hu, 1999; Meyer et al., 2007; Speight, 2014a). Bitumen is found in deposits where the permeability is low and passage of fluids through the deposit can only be achieved by prior application of fracturing techniques. Tar sand bitumen is a high-boiling material with little, if any, material boiling below 350°C (660°F) and the boiling range approximately the same as the boiling range of an atmospheric residuum. *Tar sands* have been defined in the United States (FE-76-4) as

> *...the several rock types that contain an extremely viscous hydrocarbon that is not recoverable in its natural state by conventional oil well production methods including currently used enhanced recovery techniques. The hydrocarbon-bearing rocks are variously known as bitumen-rocks oil, impregnated rocks, oil sands, and rock asphalt.*

The term *natural state* cannot be defined out of context and in the context of Federal Energy Administration (FEA) Ruling 1976-4, and the term is defined in terms of the composition of the heavy oil or bitumen—extra heavy oil adds a further dimension to this definition as it can be ascribed to the properties of the oil in the reservoir vis-à-vis the properties of the oil under ambient conditions. The final determinant of whether a reservoir is a tar sand deposit is the character of the viscous phase (bitumen) and the method that is required for recovery. From this definition and by inference crude oil and heavy oil are recoverable by well production methods and currently used enhanced recovery techniques.

For convenience, it is assumed that before depletion of the reservoir energy, conventional crude oil is produced by primary and secondary techniques, while heavy oil requires tertiary enhanced oil recovery (EOR) techniques and the recovery of tar sand bitumen requires more advanced methods (Speight, 2014a). While this is an oversimplification, it may be used as a general guide for the recovery of the different materials.

There has been the suggestion that tar sand bitumen differs from gave oil by using an arbitrary ill-conceived limit of 10,000 cp as the upper limit for heavy oil and the lower limit of tar sand bitumen. Such a system based on one physical property (viscosity) is fraught with errors! For example, this requires that a tar sand bitumen could have a viscosity on the order of 10,050 cp and an oil with a viscosity of 9950 cp is heavy oil. Both numbers fall within the limits of experimental difference of the method used to determine viscosity. The limits are the usual laboratory experimental difference be (±3%) or more likely the limits of the accuracy of the method (±5% to ±10%); there is the question of accuracy when tax credits for recovery of heavy oil, extra heavy oil, and tar sand bitumen are awarded. In fact, the inaccuracies (i.e., the limits of *experimental difference*) of the method of measuring viscosity also increase the potential for misclassification using this (or any) single property for classification purposes.

The recovery of the bitumen depends to a large degree on the composition and construction of the sands. Generally, the bitumen found in tar sand deposits is an extremely viscous material that is *immobile under reservoir conditions* and cannot be recovered through a well by the application of secondary or enhanced recovery techniques.

It is incorrect to refer to native bituminous materials as *tar* or *pitch*. Although the word tar is descriptive of the black, heavy bituminous material, it is best to avoid its use with respect to natural materials and to restrict the meaning to the volatile or near-volatile products produced in the destructive distillation of such organic substances as coal (Speight, 2013d). In the simplest sense, pitch is the distillation residue of the various types of tar. Thus, alternative names, such as *bituminous sand* or *oil sand*, are gradually finding usage, with the former name (bituminous sands) more technically correct. The term *oil sand* is also used in the same way as the term *tar sand*, and these terms are used interchangeably throughout this text.

Bituminous rock and *bituminous sand* are those formations in which the bituminous material is found as a filling in veins and fissures in fractured rocks or impregnating relatively shallow sand, sandstone, and limestone strata. These terms are, in fact, one correctly geologic description of *tar sand*. The deposits contain as much as 20% bituminous material, and if the organic material in the rock matrix is bitumen, it is usual (although chemically incorrect) to refer to the deposit as *rock asphalt* to distinguish it from bitumen that is relatively mineral free. A standard test (ASTM D4, 2017) is available for determining the bitumen content of various mixtures with inorganic materials, although the use of word *bitumen* as applied in this test might be questioned and it might be more appropriate to use the term *organic residues* to include *tar* and *pitch*. If the material is of the asphaltite type or asphaltoid type, the corresponding terms should be used: rock asphaltite or rock asphaltoid.

Bituminous rocks generally have a coarse, porous structure, with the bituminous material in the voids. A much more common situation is that the organic material is present as an inherent part of the rock composition insofar as it is a diagenetic residue of the organic material detritus that was deposited with the sediment. The organic components of such rocks are usually refractory and are only slightly affected by most organic solvents. A special class of bituminous rocks that has achieved some importance is the so-called oil shale. These are argillaceous, laminated sediments of generally high organic content that can be thermally decomposed to yield appreciable amounts of oil, commonly referred to as *shale oil*. Oil shale does not yield shale oil without the application of high temperatures and the ensuing thermal decomposition that is necessary to decompose the organic material (*kerogen*) in the shale.

Sapropel is an unconsolidated sedimentary deposit rich in bituminous substances. It is distinguished from peat in being rich in fatty and waxy substances and poor in cellulosic material. When consolidated into rock, sapropel becomes oil shale, bituminous shale, or boghead coal. The principal components are certain types of algae that are rich in fats and waxes. Minor constituents are mineral grains and decomposed fragments of spores, fungi, and bacteria. The organic materials accumulate in water under reducing conditions.

Generally, bitumen is solid or near solid at room temperature and is solid or near solid at reservoir temperature. In other words, tar sand bitumen is immobile in the reservoir and requires conversion or extreme stimulation for recovery. Moreover, by this definition (FE-76-4), tar sand bitumen is not crude oil, and it is set apart from conventional crude oil and heavy crude oil insofar as it cannot be recovered from a deposit by the use of conventional (including enhanced) oil recovery techniques as set forth in the June 1979 Federal Energy Regulations. To emphasize this point, bitumen has been recovered commercially by mining, and the hot water process is currently upgraded (converted to synthetic crude oil) by a combination of a thermal or hydrothermal process followed by product hydrotreating to produce a low-sulfur hydrocarbon product known as *synthetic crude oil*.

Since the most significant property of tar sand bitumen is its *immobility* under the conditions of temperatures and pressure in the deposit, the interrelated properties of API gravity (ASTM D287, 2017) and viscosity (ASTM D445, 2017) may present an *indication* (but only an indication) of the mobility of oil or immobility of bitumen, but these properties only offer subjective descriptions of the oil in the reservoir. The most pertinent and objective representation of this oil or bitumen mobility is the *pour point* (ASTM D97, 2017), which can be compared directly with the reservoir/deposit temperature (Table 1.6; Speight, 2014a).

TABLE 1.6 Simplified Use of Pour Point and Reservoir/Deposit Temperature to Differentiate Between Heavy Oil, Extra Heavy Oil, and Tar Sand Bitumen

Oil	Location	Temperature	Effect on Oil
Heavy oil	Reservoir or deposit	Higher than oil pour point	Fluid and/or mobile
			Mobile
	Surface/ ambient	Higher than oil pour point	Fluid and/or mobile
			Mobile
Extra heavy oil	Reservoir or deposit	Higher than oil pour point	Fluid and/or mobile
			Mobile
	Surface/ ambient	Lower than oil pour point	Solid to near solid
			Fluidity much reduced
			Immobile
Tar sand bitumen	Reservoir or deposit	Lower than oil pour point	Solid to near solid
			Not fluid
			Immobile
	Surface/ ambient	Lower than oil pour point	Solid to near solid
			Not fluid
			Immobile

The *pour point* is the lowest temperature at which oil will move, pour, or flow when it is chilled without disturbance under definite conditions (ASTM D97, 2017). In fact, the pour point of an oil when used in conjunction with the reservoir temperature gives a better indication of the condition of the oil in the reservoir than the viscosity. Thus, the pour point and reservoir temperature present a more accurate assessment of the condition of the oil in the reservoir, being an indicator of the mobility of the oil in the reservoir. Indeed, when used in conjunction with reservoir temperature, the pour point gives an indication of the liquidity of the heavy oil, extra heavy oil, or bitumen and, therefore, the ability of the heavy oil and extra heavy oil to flow under reservoir conditions (Table 1.6). In summary, the pour point is an important consideration because, for efficient production, additional energy must be supplied to the reservoir by a thermal process to increase the reservoir temperature beyond the pour point.

For example, Athabasca bitumen with a pour point in the range 50–100°C (122–212°F) and a deposit temperature of 4–10°C (39–50°F) is a solid or near solid in the deposit and will exhibit little or no mobility under deposit conditions. Pour points of 35–60°C (95–140°F) have been recorded for the bitumen in Utah with formation temperatures on the order of 10°C (50°F). This indicates that the bitumen is solid within the deposit and therefore immobile. The injection of steam to raise and maintain the reservoir temperature above the pour point of the bitumen and to enhance bitumen mobility is difficult, in some cases almost impossible. Conversely, when the reservoir temperature exceeds the pour point, the oil is fluid in the reservoir and therefore mobile. The injection of steam to raise and maintain the reservoir temperature above the pour point of the bitumen and to enhance bitumen mobility is possible, and oil recovery can be achieved.

A method that uses the pour point of the oil and the reservoir temperature (Table 1.6) adds a specific qualification to the term extremely viscous as it occurs in the definition of tar sand. In fact, when used in conjunction with the recovery method, pour point offers more general applicability to the conditions of the oil in the reservoir or the bitumen in the deposit, and the comparison of the two temperatures (pour point and reservoir temperatures) shows much promise and warrants further consideration as a means of defining the fluidity and mobility of the oil.

5. Resources and Reserves

Over the past several decades, the assessment of crude oil, heavy oil, extra heavy oil, and tar sand bitumen resources and reserves has gone through several iterations, partly because of the evolution of the definitions of resources and reserves that are key factors in the continued development of resources (SPE, 2001). It is essential for future planning that governments and industry have a clear assessment of the quantities of crude oil, heavy oil, extra heavy oil, and tar sand bitumen that are available for production and quantities that are anticipated to become available within a practical time frame through additional field development, technological advances, or exploration. To achieve such an assessment, it is imperative that the industry adopt a consistent nomenclature for assessing the current and future quantities of crude oil expected to be recovered from naturally occurring underground accumulations. Such quantities are defined as reserves, and their assessment is of considerable importance to governments, international agencies, economists, bankers, and the international energy industry.

A conscious effort has been made to keep the recommended terminology as close to current common usage as possible in order to minimize the impact of previously reported quantities and changes required to bring about wide acceptance. Due to the many forms of occurrence of crude oil, heavy oil, extra heavy oil, and tar sand bitumen, the wide range of characteristics of these materials, the uncertainty associated with the geologic environment, and the constant evolution of evaluation technologies, a precise classification and definition system is not always practical, and the complexity required for a precise system would detract from its understanding by those involved in the development of crude oil, heavy crude oil, and tar sand bitumen resources.

In fact, the terminology used in classifying crude oil, heavy oil, extra heavy oil, and tar sand bitumen is still in evolutionary stages with new terminologies being added to the crude oil lexicon on a regular basis, sometimes with justification but in many cases without any form of scientific justification. Thus, it is not surprising that the various categories of reserves have been the subject of much study and discussion for many years. In fact, attempts to standardize reserve terminology began in the mid-1930s when the American Petroleum Institute considered classification for crude oil and definitions of various reserve categories. Since then, the evolution of technology has yielded more precise engineering methods to determine reserves and has intensified the need for an improved nomenclature to achieve consistency among professionals working with reserve terminology.

Oil reserves are the estimated quantities of conventional crude oil and/or heavy crude oil that are claimed to be recoverable under existing economic and operating conditions. Many oil-producing nations do not reveal their reservoir engineering field data and instead provide unsubstantiated claims for their oil reserves. In most cases, *oil* refers to conventional oil, but depending on the source, tar sand bitumen has not always been included but now is recognized as a resource (BP, 2017). However, the exact definitions and data vary from country to country, which means that national statistics are not always comparable and may even be manipulated for political reasons (Speight, 2011b).

The total amount of conventional oil or heavy oil in a reservoir is known as oil in place of which estimates vary from year to year because of the variations in production-use and the discovery and production of new reserves, including crude oil from tight formations such as shale formations. However, no matter what the size of the resource, because of reservoir characteristics and limitations in production technology, only a fraction of this oil can be brought to the surface, and it is only this producible fraction that is considered to be *reserves*. The ratio of reserves to oil in place for a given field is often referred to as the *recovery factor*.

The recovery factor of a field may change over time based on operating history and in response to changes in technology and economics. The recovery factor may also rise over time for conventional oil if additional investment is made in enhanced oil recovery techniques such as gas injection or waterflooding. For heavy oil and extra heavy oil, the evolution of older methods and the use of newer and innovative methods may increase the recovery factor.

Because the geology of the subsurface cannot be examined directly, indirect techniques must be used to estimate the size and recoverability of the resource.

While new technologies have increased the accuracy of these techniques, significant uncertainties still remain. In general, most early estimates of the reserves of an oil field are conservative and tend to grow with time; this phenomenon is referred to as *reserve growth* (Morehouse, 1997). In addition, the estimates of the crude oil in a reservoir can only be approximate with year-to-year variations because of variations in production rate and also because of the variations of the character of the reservoir and the means by which the crude oil can be brought to the surface.

5.1 Resources

The term *resource* as used in this text is intended to encompass all quantities of crude oil that are naturally occurring on or within the crust of the Earth, discovered and undiscovered (recoverable and unrecoverable), plus those quantities have already been produced. Further, the term includes all types of crude oil (conventional crude oil, opportunity crude oil, high-acid crude oil, foamy oil, and heavy oil unless otherwise specified separately).

Thus, crude oil and heavy oil (as is extra heavy oil and tar sand bitumen) are a *resource*: in particular, a *fossil fuel resource*. A *resource* is the entire commodity that exists in the sediments and strata, whereas the *reserves* represent that fraction of a commodity that can be recovered economically. Reserves may be for a well, for a reservoir, for a field, for a nation, or for the world. Different classifications of reserves are related to their degree of certainty that, unfortunately, may include much speculation and many political statements from the government of the country that owns the resource. It is, however, no matter how useful (or useless) the terminology is, in order to give some consideration to the terminology of the various subcategories of resource.

Unconventional resources include extra heavy oil and tar sand bitumen, and unlike conventional resources, in which the crude oil is recovered through wellbores and typically requires minimal processing prior to sale, unconventional resources require specialized extraction technology to produce. For example, steam and/or solvents are used to mobilize bitumen for in situ recovery, and moreover, the extracted material may require significant processing prior to sale (e.g., bitumen upgraders). The total amount of unconventional resources in the world considerably exceeds the amount of conventional oil reserves but is much more difficult and expensive to develop.

5.2 Reserves

The definitions that are used to describe crude oil and heavy oil *reserves* are often misunderstood because they are not adequately defined at the time of use (Speight, 2014a). Therefore, as a means of alleviating this problem, it is pertinent at this point to consider the definitions used to describe the amount of crude oil that remains in subterranean reservoirs. However, the use of the term *reserves* as being descriptive of the resource is subject to much speculation. In fact, it is subject to word variations! Nevertheless, the various terminology is presented here as an aid to the reader if she/he encounters such terminology. For example, reserves are classed as *proved*, *unproved*, *probable*, *possible*, and *undiscovered*. By the same rationale, tar sand bitumen, although not considered to be a member of the crude oil family by virtue of the methods used for recovery, is also a fossil fuel resource, and the reserves can also be classed as *proved*, *unproved*, *probable*, *possible*, and *undiscovered*.

Proved reserves are those reserves of crude oil that are actually found by drilling operations and are recoverable by means of current technology. They have a high degree of accuracy and are frequently updated as the recovery operation proceeds. They may be updated by means of reservoir characteristics, such as production data, pressure transient analysis, and reservoir modeling.

Probable reserves are those reserves of crude oil that are nearly certain about which a slight doubt exists. *Possible reserves* are those reserves of crude oil with an even greater degree of uncertainty about recovery about which there is some information. An additional term *potential reserves* is also used on occasion; these reserves are based upon geologic information about the types of sediments where such resources are likely to occur, and they are considered to represent an educated guess. Then, there are so-called undiscovered reserves, which are little more than figments of the imagination! The terms *undiscovered reserves* or *undiscovered resources* should be used with caution, especially when applied as a means of estimating reserves of crude oil reserves. The data are very speculative and are regarded by many energy scientists as having little value other than unbridled optimism.

The term *inferred reserves* is also commonly used in addition to or in place of *potential reserves*. Inferred reserves are regarded as of a higher degree of accuracy than potential reserves, and the term is applied to those reserves that are estimated using an improved understanding of reservoir frameworks. The term also usually includes those reserves that can be recovered by further development of recovery technologies. Furthermore, in any definition, the quantities of, for example, the crude oil estimated to be initially in place are (i) total crude oil initially in place, (ii) discovered crude oil initially in place, and (iii) undiscovered crude oil initially in place. In any case and whatever the definition, reserves are a subset of resources and are those quantities of crude oil that are discovered (i.e., in known accumulations), recoverable, commercial, and remaining.

The *total crude oil initially in place* is that quantity of crude oil (or heavy oil) that is estimated to exist originally in naturally occurring accumulations. The *total crude oil initially in place* is, therefore, that quantity of crude oil that is estimated, on a given date, to be contained in known accumulations, plus those quantities already produced therefrom, plus those estimated quantities in accumulations yet to be discovered. The *total crude oil initially in place* may be subdivided into *discovered crude oil initially in place* and *undiscovered crude oil initially in place*, with *discovered crude oil initially in place* being limited to known accumulations.

It is recognized that the quantity of *crude oil initially in place* may constitute *potentially recoverable resources* since the estimation of the proportion that may be recoverable can be subject to significant uncertainty and will change with variations in commercial circumstances, technological developments, and data availability. A portion of those quantities classified as *unrecoverable* may become recoverable resources in the future as commercial circumstances change, technological developments occur, or additional data are acquired.

Discovered crude oil initially in place is that quantity of crude oil that is estimated, on a given date, to be contained in known accumulations, plus those quantities already produced therefrom. *Discovered crude oil initially in place* may be subdivided into *commercial* and *subcommercial* categories, with the estimated potentially recoverable portion being classified as *reserves* and *contingent resources*, respectively (as defined below).

Reserves are those quantities of crude oil that are anticipated to be commercially recovered from known accumulations from a given date forward (BP, 2017). Estimated recoverable quantities from known accumulations that do not fulfill the requirement of commerciality should be classified as *contingent resources* (as defined below). The definition of commerciality for an accumulation will vary according to local conditions and circumstances and is left to the discretion of the country or company concerned. However, reserves must still be categorized according to specific criteria, and therefore, proved reserves will be limited to those quantities that are commercial under current economic conditions, while probable and possible reserves may be based on future economic conditions. In general, quantities should not be classified as reserves unless there is an expectation that the accumulation will be developed and placed on production within a reasonable timeframe.

Contingent resources are those quantities of crude oil that are estimated, on a given date, to be potentially recoverable from known accumulations, but that are not currently considered as commercially recoverable. Some ambiguity may exist between the definitions of contingent resources and unproved reserves. This reflects variations in current industry practice, but if the degree of commitment is such that the accumulation is not expected to be developed and placed on production within a reasonable timeframe, the estimated recoverable volumes for the accumulation will be classified as contingent resources. *Contingent resources* may include, for example, accumulations for which there is currently no viable market, or where commercial recovery is dependent on the development of new technology, or where the evaluation of the accumulation is still at an early stage.

Undiscovered crude oil initially in place is that quantity of crude oil that is estimated, on a given date, to be contained in accumulations yet to be discovered. The estimated potentially recoverable portion of *undiscovered crude oil initially in place* is classified as *prospective resources*, which are those quantities of crude oil that are estimated, on a given date, to be potentially recoverable from undiscovered accumulations. In addition, the *estimated ultimate recovery* (EUR) is the quantity of crude oil that is estimated, on a given date, to be potentially recoverable from an accumulation, plus those quantities already produced therefrom. *Estimated ultimate recovery* is not a resource category, but it is a term that may be applied to an individual accumulation of any status/maturity (discovered or undiscovered).

The differences between the data obtained from the various estimates of the subcategories of the resource can be considerable, but it must be remembered that any data about the reserves of crude oil (and, for that matter, about any other fuel or mineral resource) will always be open to questions about the degree of certainty. Thus, in reality and in spite of the use of self-righteous word manipulation, *proved reserves* may be a very small part of the total hypothetical and/or speculative amounts of a resource.

At some time in the future, certain resources may become reserves for righteous or unrighteous reasons (Speight, 2011b). Such a reclassification can arise as a result of improvements in recovery techniques that may either make the resource accessible or bring about a lowering of the recovery costs and render winning of the resource an economical proposition. In addition, other uses may also be found for a commodity, and the increased demand may result in an increase in price. Alternatively, a large deposit may become exhausted and unable to produce any more of the resource, thus forcing production to focus on a resource that is lower grade but has a higher recovery cost.

5.3 Reserve Estimation

As with the term *resource*, the subcategories are many and varied with each definition being subject to the country or company that owns the reserves. Thus, crude oil quantities classified as *reserves*, *contingent resources*, or *prospective resources* should not be aggregated with each other without due consideration of the significant differences in the criteria associated with their classification. In particular, there may be a significant risk that accumulations containing *contingent resources* or *prospective resources* will not achieve commercial production.

Furthermore, the estimation of reserves is more than just a periodic, statutory calculating and reporting of company assets (although those are important functions); it is an essential element of investment planning and resource management for every prudent operator. Estimation of reserves begins with identifying a drillable prospect, and it continues while the prospect is developed and placed on production and thereafter as warranted by well and/or reservoir performance, new geologic data, competitor (offset) operations, unitization, contract renegotiation, improved technology, and/or changing economic conditions, which contribute to the range of uncertainty (Morehouse, 1997; Kovarik, 2003).

The *range of uncertainty* reflects a reasonable range of estimated potentially recoverable volumes for an individual accumulation. Any estimation of resource quantities for an accumulation is subject to both technical and commercial uncertainties and should, in general, be quoted as a range. In the case of reserves and where appropriate, this range of uncertainty can be reflected in estimates for *proved reserves* (1P), *proved plus probable reserves* (2P), and *proved plus probable plus possible reserves* (3P) scenarios. For other resource categories, the terms *low estimate*, *best estimate*, and *high estimate* are recommended.

The term *best estimate* is used as a generic expression for the estimate considered to be the closest to the quantity that will actually be recovered from the accumulation between the date of the estimate and the time of abandonment. If probabilistic methods are used, this term would generally be a measure of central tendency of the uncertainty distribution. The terms *low estimate* and *high estimate* should provide a reasonable assessment of the range of uncertainty in the *best estimate*.

For undiscovered accumulations (*prospective resources*), the range will, in general, be substantially greater than the ranges for discovered accumulations. In all cases, however, the actual range will be dependent on the amount and quality of data (both technical and commercial) that is available for that accumulation. As more data become available for a specific accumulation (e.g., additional wells and reservoir performance data), the range of uncertainty in the *estimated ultimate recovery* for that accumulation should be reduced.

The *low estimate*, *best estimate*, and *high estimate* of potentially recoverable volumes should reflect some comparability with the reserve categories of *proved reserves*, *proved plus probable reserves*, and *proved plus probable plus possible reserves*, respectively. While there may be a significant risk that subcommercial or undiscovered accumulations will not achieve commercial production, it is useful to consider the range of potentially recoverable volumes independently of such a risk.

However, there are uncertainties that exist in estimating the remaining ultimately recoverable resources of crude oil (Speight and Islam, 2016). The encompassing terms *conventional crude oil* and *unconventional crude oil* are subject to separation into the numerous categories of oil that exist within each. In fact, the estimation of reserves is inherently uncertain as, regardless of the calculation method used, data will be limited; it will not be possible to specifically determine all required factors, and various assumptions will be necessary. Reserve estimates are therefore frequently given to three levels of confidence, namely, proved reserves (1P); proved and probable reserves (2P); and proved, probable, and possible reserves (3P). These levels of confidence reflect the uncertainties that are an aspect of any normal reservoir evaluation. Nevertheless, there is a large degree of variability in published figures for a number of nontechnical reasons, and many potential problems exist with the use of reserve data. The first problem occurs because the definitions of 1P, 2P, and 3P reserves are found to vary widely from one country to another and from one company to another.

In addition, different organizations use different categories as part of the estimates: one group may include conventional crude oil, condensate, natural gas liquids, extra heavy oil, and tar sand bitumen as parts of the estimate, while another group may report conventional crude oil, condensate, and natural gas liquids in the estimate. While many data sources exist, few use a consistent basis, and so, attempting to extract purely conventional oil reserve estimates from the reserve figures available is difficult.

On the other hand, another uncertainty arises when the reported reserves include any *political reserves* (Speight, 2011a,b,c). Political crude oil reserves are the crude oil reserves declared by a country or by a company that correspond not to the reserves it possesses but to those that it would like the rest of the world to believe it possesses. There is particular and much reported concern with the reserves declared by the member states of OPEC as a major increase was witnessed in the reserves of many OPEC countries between 1985 and 1990 despite no new discoveries being reported (Yergin, 1991; Morehouse, 1997; Speight, 2011b; McGlade, 2012).

Thus, present and future assessments of crude oil resources and projections of crude oil production should hence acknowledge the various uncertainty issues and acknowledge the effects that the uncertainties have on results and present ranges in any estimates produced or provided (McGlade, 2012). As a result of the various uncertainties, the international crude oil industry has increased the emphasis on the need to understand and develop more reliable estimates of crude oil reserves and to quantify the uncertainty associated with the classifications of reserves. Furthermore, the regulatory authorities of many countries, particularly of the United States, have found it necessary to accommodate the increasing internationalization of the industry and to manage their increasing involvement and influence in the industry.

6. Conclusions

In summary, supplies of conventional crude oil are depleted, and (tight oil reserves notwithstanding) heavy oil, extra heavy oil, and tar sand bitumen resources are becoming more important as contributors to liquid fuel production. On the other hand, extra heavy oil and tar sand bitumen are unconventional resources that are characterized by high viscosity (i.e., they have a strong resistance to flow) and high density compared with conventional oil. Most extra heavy oil and tar sand bitumen deposits are very shallow. Also, extra heavy oil and tar sand bitumen are deficient in hydrogen and have high carbon, sulfur, and heavy metal content. Hence, they require additional processing (upgrading) to become a suitable feedstock for a typical refinery that was built around the concept of conventional crude oil as the refinery feedstock.

Finally, it is essential to realize that (in the current context) of conventional crude oil and heavy oil, there are several parameters that can influence properties and recovery. These properties are usually site-specific to the particular reservoir in which the crude oil or heavy oil is located (Table 1.7).

TABLE 1.7 Representation of Site-Specific Parameters

Oil	Composition	
	Change in composition	Wax deposition
		Asphaltene deposition
	Properties	Pour point-reservoir temperature
		Viscosity
	Change in properties	Wax deposition
		Asphaltene deposition
Reservoir	Porosity	
	Change in porosity	
	Permeability	
	Change in permeability	Wax deposition
		Asphaltene deposition
Production method	Oil properties	Wax deposition
		Asphaltene deposition
	Reservoir properties	Sandstone
		Carbonate
		Mudstone (clay minerals)

References

Abraham, H., 1945. Asphalts and Allied Substances. Van Nostrand Scientific Publishers, New York.

Ancheyta, J., Speight, J.G., 2007. Hydroprocessing of Heavy Oils and Residua. CRC-Taylor and Francis Group, Boca Raton, FL.

ASTM D287, 2017. Standard test method for API gravity of crude petroleum and petroleum products (hydrometer method). In: Annual Book of Standards. ASTM International, West Conshohocken, PA.

ASTM D4, 2017. Standard test method for bitumen content. In: Annual Book of Standards. ASTM International, West Conshohocken, PA.

ASTM D4175, 2017. Standard terminology relating to petroleum, petroleum products, and lubricants. In: Annual Book of Standards. ASTM International, West Conshohocken, PA.

ASTM D445, 2017. Standard test method for kinematic viscosity of transparent and opaque liquids (and calculation of dynamic viscosity). In: Annual Book of Standards. ASTM International, West Conshohocken, PA.

ASTM D97, 2017. Standard test method for pour point of petroleum products. In: Annual Book of Standards. ASTM International, West Conshohocken, PA.

Attanasi, E.D., Meyer, R.F., 2010. Natural bitumen and extra-heavy oil. In: Trinnaman, J., Clarke, A. (Eds.), 2010 Survey of Energy Resources. World Energy Council, London, pp. 123–150.

Babich, I.V., Moulijn, J.A., 2003. Science and technology of novel processes for deep desulfurization of oil refinery streams: a review. Fuel 82, 607–631.

BP, 2017. Statistical Review of World Energy. British Petroleum Company, London.

Chugh, S., Baker, R., Telesford, A., Zhang, E., 2000. Mainstream options for heavy oil: part I—cold production. J. Can. Pet. Technol. 39 (4), 31–39.

Cobb, C., Goldwhite, H., 1995. Creations of Fire: Chemistry's Lively History From Alchemy to the Atomic Age. Plenum Press, New York.

Delbianco, A., Montanari, R., 2009. Encyclopedia of Hydrocarbons, Volume III/New Developments: Energy, Transport, Sustainability. Eni S.p.A, Rome, Italy.

Dusseault, M.B., 2001. Comparing Venezuelan and Canadian heavy oil and tar sands. In: Processing, Paper No. 2001-061, Petroleum Society of the CIM, International Conference, Alberta, Canada, June 12–14. Petroleum Society, Canadian Institute of Mining and Metallurgy, Westmount. http://www.southportland.org/files/3713/9387/3165/Heavy_Oil_and_Tar_Sands.pdf (web archive link, 12 March 2018). (Accessed 12 March 2018) https://open.alberta.ca/dataset/3c5689e3-8602-44d3-a7f3-ac60c52ad0ad/resource/941dabcc-3927-4450-9e64-cb50f9dcfaa0/download/00_main_page.pdf (Accessed 12 March 2018).

Forbes, R.J., 1958a. A History of Technology. Oxford University Press, Oxford.

Forbes, R.J., 1958b. Studies in Early Petroleum Chemistry. E. J. Brill, Leiden.

Forbes, R.J., 1959. More Studies in Early Petroleum Chemistry. E. J. Brill, Leiden.

Forbes, R.J., 1964. Studies in Ancient Technology. E. J. Brill, Leiden.

Gary, J.G., Handwerk, G.E., Kaiser, M.J., 2007. Petroleum Refining: Technology and Economics, fifth ed. CRC Press, Taylor & Francis Group, Boca Raton, FL.

Ghoshal, S., Sainik, V., 2013. Monitor and minimize corrosion in high-TAN crude processing. Hydrocarb. Process. 92 (3), 35–38.

Head, I.M., Jones, D.M., Larter, S.R., 2003. Biological activity in the deep subsurface and the origin of heavy oil. Nature 426 (20), 344–352.

Hsu, C.S., Robinson, P.R. (Eds.), 2017. Handbook of Petroleum Technology. Springer International Publishing AG, Cham.

IEA, 2005. Resources to Reserves: Oil & Gas Technologies for the Energy Markets of the Future. International Energy Agency, Paris.

Isaacs, C.M., 1992. Preliminary petroleum geology background and well data for oil samples in the Cooperative Monterey organic geochemistry study, Santa Maria and Santa Barbara-Ventura Basins, California. Open-File Report No. USGS 92-539-F, United States Geological Survey, Reston, VA.

Kam'Yanov, V.F., Braun, A.Y., Gorbunova, L.V., Shabotkin, L.G., 1995. Natural bitumens of mortuk. Pet. Chem. 35 (5), 377–389 (Also: Neftekhimiya, 35(5), 397–409).

Kane, R.D., Cayard, M.S., 2002. A Comprehensive Study on Naphthenic Acid Corrosion. Corrosion 2002. NACE International, Houston, TX.

Kettler, R.M., 1995. Incipient bitumen generation in miocene siliceous sedimentary rocks from the Japan sea. Org. Geochem. 23 (7), 699–708.

Khulbe, K.C., Mann, R.S., MacPhee, J.A., 1996. Separation of acidic fraction from the Cold Lake bitumen asphaltenes and its relationship to enhanced oil recovery. Fuel Process. Technol. 46, 63–69.

Kovarik, W., 2003. The Oil Reserve Fallacy: Proven Reserves Are Not a Measure of Future Supply. http://www.runet.edu/~wkovarik/oil/3unconventional.html (Accessed 31 March 2018).

Larter, S., Whilhelms, A., Head, I., Koopmans, M., Aplin, A., Di Primio, R., Zwach, C., Erdmann, M., Telnaes, N., 2003. The controls on the composition of biodegraded oils in the deep subsurface—part 1: biodegradation rates in petroleum reservoirs. Org. Geochem. 34 (3), 601–613.

Larter, S., Huang, H., Adams, J., Bennett, B., Jokanola, O., Oldenburg, T., Jones, M., Head, I., Riediger, C., Fowler, M., 2006. The controls in the composition of biodegraded oils in the deep subsurface: part II. Geological controls on subsurface biodegradation fluxes and constraints on reservoir-fluid property prediction. Am. Assoc. Pet. Geol. Bull. 90 (6), 921–938.

Maini, B.B., 1999. Foamy oil flow in primary production of heavy oil under solution gas drive. In: Paper No. SPE 56541. Proceedings Annual Technical Conference and Exhibition, Houston, Texas USA, October 3–6. Society of Petroleum Engineers, Richardson, TX.

Maini, B.B., 2001. Foamy oil flow. In: Paper No. SPE 68885. SPE J. Pet. Tech., Distinguished Authors Series. Society of Petroleum Engineers, Richardson, TX, pp. 54–64.

McGlade, C.E., 2012. A review of the uncertainties in estimates of global oil resources. Energy 47 (1), 262–270.

Meyer, R.F., Attanasi, E.D., Freeman, P.A., 2007. Heavy oil and natural bitumen resources in geological basins of the world. USGS Open File Report No. 2007-1084, United States Geological Survey, Reston, VA.

Morehouse, D.F., 1997. The Intricate Puzzle of Oil and Gas Reserves Growth. Natural Gas Monthly. US Energy Information Administration, Washington, DC.

Niu, J., Hu, J., 1999. Formation and distribution of heavy oil and Tar Sands in China. Mar. Pet. Geol. 16, 85–95.

Ohmes, R., 2014. Characterizing and Tracking Contaminants in Opportunity Crudes. Digital Refining. http://www.digitalrefining.com/article/1000893,Characterising_and_tracking_contaminants_in_opportunity_crudes_.html#.VJhFjV4AA (Accessed 1 November 2014).

Parkash, S., 2003. Refining Processes Handbook. Gulf Professional Publishing, Elsevier, Amsterdam.

Parnell, J., Monson, B., Geng, A., 1996. Maturity and petrography of bitumens in the carboniferous of Ireland. Int. J. Coal Geol. 29, 23–38.

Poon, D., Kisman, K., 1992. Non-newtonian effects on the primary production of heavy oil reservoirs. J. Can. Pet. Technol. 31 (7), 1–6.

Ratov, A.N., 1996. Physicochemical nature of structure formation in high-viscosity crude oils and natural bitumens and their rheological differences. Pet. Chem. 36, 191–206. (also: Neftekhimiya, 1996, 36(3), 195208).

Shalaby, H.M., 2005. Refining of Kuwait's heavy crude oil: materials challenges. In: Proceedings Workshop on Corrosion and Protection of Metals, Arab School for Science and Technology, December 3–7, Kuwait.

Sheng, J.J., Maini, B.B., Hayes, R.E., Tortike, W.S., 1999. Critical review of foamy oil flow. Transp. Porous Media 35, 157–187.

SPE, 2001. Guidelines for the Evaluation of Petroleum Reserves and Resources: A Supplement to the SPE/WPC Petroleum Reserves Definitions and the SPE/WPC/AAPG Petroleum Resources Definitions. Society of Petroleum Engineers, Richardson, TX.

Speight, J.G., 1978. Personal Observations at Archeological Digs at the Cities of Babylon, Calah, Nineveh, and Ur. University Library, Mosul, Iraq.

Speight, J.G., 1984. Characterization of heavy crude oils and petroleum residues. In: Kaliaguine, S., Mahay, A. (Eds.), Characterization of Heavy Crude Oils and Petroleum Residues. Elsevier, Amsterdam, p. 515.

Speight, J.G., 2000. The Desulfurization of Heavy Oils and Residua, second ed. Marcel Dekker, New York.

Speight, J.G., 2007. Natural Gas: A Basic Handbook. GPC Books, Gulf Publishing Company, Houston, TX.

Speight, J.G., 2008. Synthetic Fuels Handbook: Properties, Processes, and Performance. McGraw-Hill, New York.

Speight, J.G., 2011a. The Refinery of the Future. Gulf Professional Publishing, Elsevier, Oxford.

Speight, J.G., 2011b. An Introduction to Petroleum Technology, Economics, and Politics. Scrivener Publishing, Salem, MA.

Speight, J.G. (Ed.), 2011c. The Biofuels Handbook. Royal Society of Chemistry, London.

Speight, J.G., 2012. Crude Oil Assay Database. Knovel, Elsevier, New York. Online version available from:http://www.knovel.com/web/portal/browse/display?_EXT_KNOVEL_DISPLAY_bookid=5485&VerticalID=0 (Accessed 4 June 2018).

Speight, J.G., 2013a. Heavy Oil Production Processes. Gulf Professional Publishing, Elsevier, Oxford.

Speight, J.G., 2013b. Heavy and Extra Heavy Oil Upgrading Technologies. Gulf Professional Publishing, Elsevier, Oxford.

Speight, J.G., 2013c. Oil Sand Production Processes. Gulf Professional Publishing, Elsevier, Oxford.

Speight, J.G., 2013d. The Chemistry and Technology of Coal, third ed. CRC Press, Taylor & Francis Group, Boca Raton, FL.

Speight, J.G., 2014a. The Chemistry and Technology of Petroleum, fifth ed. CRC Press, Taylor & Francis Group, Boca Raton, FL.

Speight, J.G., 2014b. High Acid Crudes. Gulf Professional Publishing, Elsevier, Oxford.

Speight, J.G., 2014c. Oil and Gas Corrosion Prevention. Gulf Professional Publishing, Elsevier, Oxford.

Speight, J.G., 2015. Handbook of Petroleum Product Analysis, second ed. John Wiley & Sons, Hoboken, NJ.

Speight, J.G., 2016. Introduction to Enhanced Recovery Methods for Heavy Oil and Tar Sands, second ed. Gulf Professional Publishing, Elsevier, Oxford.

Speight, J.G., 2017. Handbook of Petroleum Refining. CRC Press, Taylor & Francis Group, Boca Raton, FL.

Speight, J.G., 2018. Handbook of Natural Gas Analysis. John Wiley & Sons, Hoboken, NJ.

Speight, J.G., Francisco, M.A., 1990. Studies in petroleum composition IV: changes in the nature of chemical constituents during crude oil distillation. Rev. Inst. Fr. Pétrol. 45, 733.

Speight, J.G., Islam, M.R., 2016. Peak Energy: Myth or Reality. Scrivener Publishing, Beverly, MA.

Sun, X., Zhang, Y., Li, X., Cui, G., Gu, J., 2013. A case study on foamy oil characteristics of the Orinoco Belt, Venezuela. Adv. Pet. Explor. Dev. 5 (1), 37–41.

Totten, G.E., 2007. A Timeline of Highlights From the Histories of ASTM D2 Committee and the Petroleum Industry. http://www.astm.org/COMMIT/D02/to1899_index.html (Accessed 30 October 2014).

USGS, 2006. Natural Bitumen Resources of the United States. Fact Sheet 2006-3133. United States Geological Survey, Reston, VA.

Villarroel, T., Hernández, R., 2013. Technological developments for enhancing extra heavy oil productivity in fields of the Faja Petrolifera del Orinoco (FPO), Venezuela. In: Proceedings, AAPG Annual Convention and Exhibition, Pittsburgh, PA, May 19–22. American Association of Petroleum Geologists, Tulsa, OK.

Yergin, D., 1991. The Prize: The Epic Quest for Oil, Money, and Power. Simon & Schuster, New York.

Yeung, T.W., 2014. Evaluating Opportunity Crude Processing. Digital Refining. http://www.digitalrefining.com/article/1000644 (Accessed 25 October 2014).
Yui, S., Chung, K.H., 2001. Processing oil sands bitumen. Oil Gas J. 23, 46–52.

Further Reading

Speight, J.G., 2005. Upgrading and refining of natural bitumen and heavy oil. In: Coal, Oil Shale, Natural Bitumen, Heavy Oil and Peat, Encyclopedia of Life Support Systems (EOLSS), Developed Under the Auspices of the UNESCO. EOLSS Publishers, Oxford. http://www.eolss.net.

Chapter 2

Nonthermal Methods of Recovery

1. Introduction

Crude oil is found in the microscopic pores of sedimentary rocks such as sandstone and limestone. Not all of the pores in a rock will contain crude oil—some will be filled with water or brine that is saturated with minerals. Seismic surveys are used to try to predict where oil fields may be, but the only way of making certain is by drilling. In the search for new recoverable oil and gas fields, crude oil companies use their vast knowledge of how and when crude oil was formed, their knowledge of geologic structures that may have entrapped crude oil, and of course their knowledge of how to recover the resources.

The term *recovery*, as applied in the crude oil industry, is the production of oil from a reservoir. There are several methods by which this can be achieved that range from recovery due to reservoir energy (i.e., the oil flows from the wellhole without assistance) to enhanced recovery methods in which considerable energy must be added to the reservoir to produce the oil. However, the effect of the method on the oil and on the reservoir must be considered before application.

In terms of recovery operations, there are three phases of recovering reserves (Table 2.1 and Fig. 2.1). Production rates from reservoirs depend on a number of factors, such as reservoir geometry (primarily formation thickness and reservoir continuity), reservoir pressure, reservoir depth, rock type and permeability, fluid saturations and properties, extent of fracturing, number of wells and their locations, and the ratio of the permeability of the formation to the viscosity of the oil (Taber and Martin, 1983; DOE, 1996; Jayasekera, and Goodyear, 1999). Operators can increase production over that that would naturally occur by such methods as fracturing the reservoir to open new channels for flow, injecting gas and water to increase the reservoir pressure, or lowering oil viscosity with heat or chemicals. These supplementary techniques are expensive, and the extent to which they are used depends on such external factors as the operator's economic condition, sales prospects, and perceptions of future prices.

Heavy Oil Recovery and Upgrading. https://doi.org/10.1016/B978-0-12-813025-4.00002-7

TABLE 2.1 Methods for Oil Recovery

Recovery Method	Comments
Primary recovery	Occurs due to natural energy from expansion of gas and water within the producing formation, pushing fluids into the wellbore and lifting them to the surface
Secondary recovery	Occurs by application of artificial energy to lift fluids to surface
	Accomplished by injecting gas down a hole to lift fluids to the surface, installation of a subsurface pump, or injecting gas or water into the formation itself
Tertiary recovery	Occurs when means of increasing fluid mobility in oil reservoirs within the reservoir are introduced in addition to secondary techniques
	Accomplished by introducing additional heat into the formation to lower the viscosity (thin oil) and improve its ability to flow to the wellbore
	Heat may be introduced by either injecting steam in a steam flood or injecting oxygen to enable the ignition and combustion of oil within the reservoir in a fireflood

The extraordinary geologic variability of different reservoirs means that production profiles differ from field to field. Heavy oil, extra heavy oil, and tar sand bitumen reservoirs can be developed to significant levels of production and maintained for a period of time by supplementing natural drive force, while gas reservoirs normally decline more rapidly. On this basis, an oil reservoir with the seemingly large reserve of a million barrels might produce only 200–400 bbl per day during the best production years.

The anatomy of a reservoir is complex and is site-specific, microscopically and macroscopically (Table 2.2), which, along with the properties of the oil, dictates the selection of the recovery process (Fig. 2.1; Chapter 1). Because of the various types of accumulations and the existence of wide ranges of both rock and fluid properties, reservoirs respond differently and must be treated individually. These factors need recognition when planning a recovery project (Table 2.3) because of the adverse effects that can arise not only because of the properties of the oil and the reservoir but also because of the nature of the process (Table 2.4).

Crude oil development and production can include up to three distinct phases: primary, secondary, and tertiary (or enhanced) recovery (Table 2.3). During primary recovery, the natural pressure of the reservoir or gravity drives

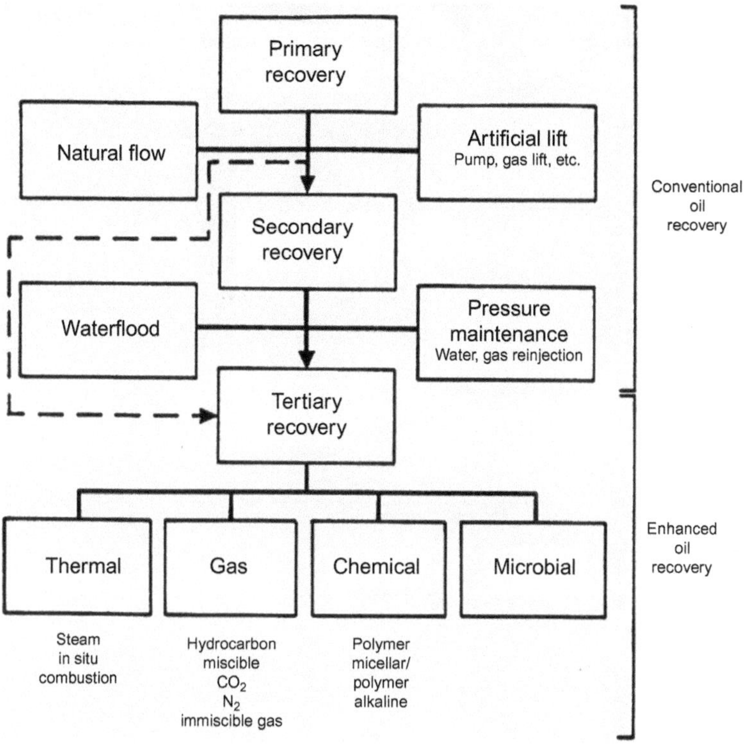

FIG. 2.1 Methods for oil recovery.

oil into the wellbore, combined with artificial lift techniques (such as pumps) that bring the oil to the surface. But only about 10% of a reservoir's original oil in place is typically produced during primary recovery. Secondary recovery techniques extend the field's productive life generally by injecting water or gas to displace oil and drive it to a production wellbore, resulting in the recovery of 20%–40% of the original oil in place.

Generally, the first stage in the extraction of crude oil is to drill a well into the underground reservoir. Often, many wells (multilateral wells) will be drilled into the same reservoir, to ensure that the extraction rate will be economically viable. Also, some wells (secondary wells) may be used to pump water, steam, acids, or various gas mixtures into the reservoir to raise or maintain the reservoir pressure and so maintain an economic extraction rate. If the underground pressure in the oil reservoir is sufficient, the oil will be forced to the surface under this pressure (primary recovery).

Over the lifetime of the well, the pressure will fall, and at some point, there will be insufficient underground pressure to force the oil to the surface. Secondary oil recovery uses various techniques to aid in recovering oil from depleted or low-pressure reservoirs. Sometimes, pumps, such as balanced beam pumps

TABLE 2.2 Representation of Site-Specific Parameters

Oil	**Composition**	
	Change in composition	Wax deposition
		Asphaltene deposition
	Properties	Pour point-reservoir temperature
		Viscosity
	Change in properties	Wax deposition
		Asphaltene deposition
Reservoir	Porosity	
	Change in porosity	
	Permeability	
	Change in permeability	Wax deposition
		Asphaltene deposition
Production method	Oil properties	
	Reservoir properties	Sandstone
		Carbonate
		Mudstone (clay minerals)

TABLE 2.3 Methods for Recovery of Oil

Primary recovery	
	Natural flow
	Artificial lift
Secondary recovery	
	Waterflood
	Gas flood
Tertiary recovery[a]	
	Nonthermal
	Thermal

[a]*Enhanced oil recovery (EOR).*

TABLE 2.4 Recovery Process Effects and the Potential Adverse Effects Leading to Sludge and Sediment Formation

Effect	Comments
Carbon dioxide injection	Lowers pH; can change oil composition leading to phase separation of sludge or sediment and blocking of channels
Miscible flooding	Hydrocarbon-rich gases lower the solubility parameter and solvent power of the oil and cause separation of asphaltenes
Organic chemicals	Can lower the solubility parameter of the oil and cause separation of asphaltene constituents and blocking of channels
Acidizing	Interaction of crude oil constituents upsetting molecular balance and deposition of sediment; blocking of channels
Pressure decrease	Can change composition of oil medium leading to phase separation of asphaltene material; blocking of channels
Temperature decrease	Can change composition of oil medium leading to phase separation of asphaltene material; blocking of channels

(also called horsehead pumps) and electric submersible pumps, are used to bring the oil to the surface. Other secondary recovery techniques increase the reservoir's pressure by water injection, natural gas reinjection, and gas lift, which injects air, carbon dioxide, or another suitable gas into the reservoir.

Tertiary oil recovery (enhanced oil recovery, EOR) methods take oil recovery one step further (Chakma et al., 1991; Alvarado and Eduardo Manrique, 2010; Mandal, 2015) and rely on methods that cause the viscosity of the oil to increase, compared with the natural or induced energy methods of primary recovery and secondary recovery. Tertiary recovery is started before secondary oil recovery techniques are no longer enough to sustain production. For example, thermally enhanced oil recovery methods are recovery methods in which the oil is heated making it easier to extract; usually, steam is used for heating the oil. In addition, the term *improved oil recovery* (IOR) (Barillas et al., 2008) is also often used and includes enhanced oil recovery methods and drilling and well technologies, advanced reservoir monitoring techniques, and the application of different enhancements of primary and secondary recovery processes.

Conventional primary and secondary recovery processes are ultimately expected to produce about one-third of the original oil discovered, although recoveries from individual reservoirs can range from less than 5% to as high as 80% of the original oil in place. This broad range of recovery efficiency is a result of variations in the properties of the specific rock and fluids involved from reservoir to reservoir and the kind and level of energy that drives the oil to producing wells, where it is captured.

Conventional oil production methods may be unsuccessful because the management of the reservoir was poor or because reservoir heterogeneity has prevented the recovery of crude oil in an economical manner. Reservoir heterogeneity, such as fractures and faults, can cause reservoirs to drain inefficiently by conventional methods. Also, highly cemented or shale zones can produce barriers to the flow of fluids in reservoirs and lead to high residual oil saturation. Reservoirs containing crude oils with low API gravity often cannot be produced efficiently without the application of *enhanced oil recovery* methods because of the high viscosity of the crude oil. In some cases, the reservoir pressure was depleted prematurely by poor reservoir management practices to create reservoirs with low energy and high oil saturation.

As might be expected, the type of exploration technique employed depends upon the nature of the site. In other words and as for many environmental operations, the recovery techniques applied to a specific site are dictated by the nature of the site and are, in fact, *site-specific*. For example, in areas where little is known about the subsurface, preliminary reconnaissance techniques are necessary to identify potential reservoir systems that warrant further investigation. Techniques for reconnaissance that have been employed to make inferences about the subsurface structure include *satellite and high-altitude imagery* and *magnetic and gravity surveys*.

Once an area has been selected for further investigation, more detailed methods (such as the *seismic reflection* method) are used. Drilling is the final stage of the exploratory program and is in fact the only method by which a crude oil reservoir can be conclusively identified. However, in keeping with the concept of site specificity, drilling may be the only option in some areas for commencement of the exploration program. The risk involved in the drilling operation depends upon previous knowledge of the site subsurface. Thus, there is the need to relate the character of the exploratory wells at a given site to the characteristics of the reservoir.

Thus, once the well is completed, the flow of oil into the well is commenced. For limestone reservoir rock, acid is pumped down the well and out the perforations. The acid dissolves channels in the limestone that lead oil into the well. For sandstone reservoir rock, a specially blended fluid containing *proppants* (sand, walnut shells, and aluminum pellets) is pumped down the well and out the perforations. The pressure from this fluid makes small fractures in the sandstone that allow oil to flow into the well, while the proppants hold these fractures open. Once the oil is flowing, the oil rig is removed from the site, and production equipment is set up to extract the oil from the well.

A well is always carefully controlled in its flush stage of production to prevent the potentially dangerous and wasteful *gusher*. This is actually a dangerous condition and is (hopefully) prevented by the blowout preventer and the pressure of the drilling mud. In most wells, acidizing or fracturing the well starts the oil flow.

As already noted, crude oil accumulates over geologic time in porous underground rock formations called reservoirs that are at varying depths in the Earth's crust, and in many cases elaborate, expensive equipment is required to get it from there. The oil is usually found trapped in a layer of porous sandstone, which lies just beneath a dome-shaped or folded layer of some nonporous rock such as limestone. In other formations, the oil is trapped at a fault or break in the layers of the crust.

Generally, crude oil reservoirs sometimes exist with an overlying *gas cap*, in communication with aquifers, or both. The oil resides together with water and free gas in very small holes (pore spaces) and fractures. The size, shape, and degree of interconnection of the pores vary considerably from place to place in an individual reservoir. Below the oil layer, the sandstone is usually saturated with salt water. The oil is released from this formation by drilling a well and puncturing the limestone layer on either side of the limestone dome or fold. The oil in such formations is usually under such great pressure that it flows naturally and, sometimes with great force, from the well. However, in some cases, this pressure later diminishes so that the oil must be pumped from the well. Natural gas or water is sometimes pumped into the well to replace the oil that is withdrawn. This is called repressurizing the oil well.

Conventional crude oil is a brownish green to black liquids of specific gravity in a range from about 0.810 to 0.985 and having a boiling range from about 20°C (68°F) to above 350°C (660°F), above which active decomposition ensues when distillation is attempted. The oils contain from 0% to 35% or more of gasoline and varying proportions of kerosene hydrocarbons and higher-boiling constituents up to the viscous and nonvolatile compounds present in lubricant oil and in asphalt. The composition of the crude oil obtained from the well is variable and depends not only on the original composition of the oil in situ but also on the manner of production and the stage reached in the life of the well or reservoir.

For a newly opened formation and under ideal conditions, the proportions of gas may be so high that the oil is, in fact, a solution of liquid in gas that leaves the reservoir rock so efficiently that a core sample will not show any obvious oil content. A general rough indication of this situation is a high ratio of gas to oil produced. This ratio may be zero for fields in which the rock pressure has been dissipated. The oil must be pumped out to as much as 50,000 ft^3 or more of gas per barrel of oil in so-called condensate reservoirs, in which a very light crude oil (0.80 specific gravity or lower) exists as vapor at high pressure and elevated temperature.

New methods to drill for oil are continually being sought, including directional or horizontal drilling techniques, to reach oil under ecologically sensitive areas, and using lasers to drill oil wells. *Directional drilling* is also used to reach formations and targets not directly below the penetration point or drilling from shore to locations under water (Fig. 2.2). A controlled deviation may also be used from a selected depth in an existing hole to attain economy in drilling costs.

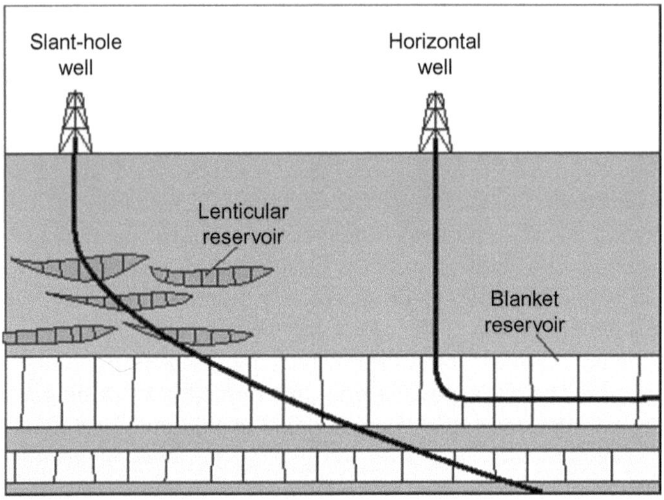

FIG. 2.2 Directional drilling.

Various types of tools are used in directional drilling along with instruments to help orient their position and measure the degree and direction of deviation; two such tools are the *whipstock* and the *knuckle joint*. The whipstock is a gradually tapered wedge with a chisel-shaped base that prevents rotation after it has been forced into the bottom of an open hole. As the bit moves down, it is deflected by the taper about 5° from the alignment of the existing hole.

Approximately one-third of the world's crude oil is produced from offshore fields, usually from steel drilling platforms set on the ocean floor. In shallow, calm waters, these may be little more than a wellhead and workspace. The larger ocean drilling rigs, however, include not only the well equipment but also processing equipment and extensive crew quarters. Recent developments in ocean drilling include the use of floating tension-leg platforms that are tied to the seafloor by giant cables and drill ships, which can hold a steady position above a seafloor well using constant, computer-controlled adjustments. Subsea satellite platforms, where all of the necessary equipment is located on the ocean bed at the well site, have been used for small fields located in producing areas. In Arctic areas, islands are built from dredged gravel and sand to provide platforms capable of resisting drifting ice fields.

Drilling does not end when production commences and continues after a field enters production. Extension wells must be drilled to define the boundaries of the crude oil pool. In-field wells are necessary to increase recovery rates, and service wells are used to reopen wells that have become clogged. Additionally, wells are often drilled at the same location but to different depths, to test other geologic structures for the presence of crude oil.

Once the final depth has been reached, the well is completed to allow oil to flow into the casing in a controlled manner. First, a *perforating gun* is lowered into the well to the production depth. The gun has explosive charges to create holes in the casing through which oil can flow. After the casing has been perforated, a small-diameter pipe (*tubing*) is run into the hole as a conduit for oil and gas to flow up the well, and a *packer* is run down the outside of the tubing. When the packer is set at the production level, it is expanded to form a seal around the outside of the tubing. Finally, a multivalve structure—the *Christmas tree* (Fig. 2.3)—is installed at the top of the tubing and cemented to the top of the casing and allows control of the flow of oil from the well.

Tight formations are occasionally encountered, and it becomes necessary to encourage flow. Several methods are used, one of which involves setting off small explosions to fracture the rock. If the formation is mainly limestone, hydrochloric acid is sent down the hole to dissolve channels in the rock. The acid is inhibited to protect the steel casing. In sandstone, the preferred method is hydraulic fracturing that is used to create additional passageways in the oil reservoir that can facilitate the flow of oil to a producing well. Tight reservoirs and deposits, those whose oil-containing rocks have restricted pore volume and connectivity that impede the flow of oil through the reservoir or deposit, are commonly fractured by injecting a fluid containing sand or other proppant under sufficient pressure to create fractures in the rock through which the oil can more easily flow (Speight, 2016). Care is also taken to contain the fracturing within the reservoir or deposit to avoid intersecting adjoining aquifers that would introduce excess water into the oil-producing zone.

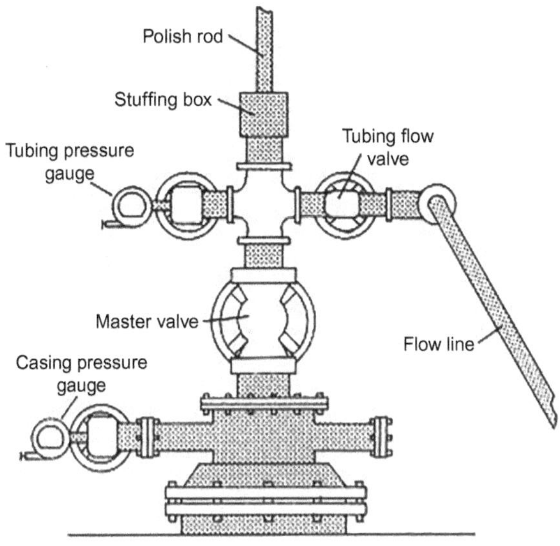

FIG. 2.3 The Christmas tree.

Another form of fracturing, *tailored pulse fracturing*, is employed to control the extent and direction of the produced fractures. Precise quantities of solid propellants are ignited in the wellbore to create a controlled pressure pulse that creates fractures in a more predictable pattern. In addition, *foam fracturing* that uses foam under high pressure in gas reservoirs has the advantage over high-pressure water injection because it does not create as much damage to the formation. *Carbon dioxide-sand fracturing* increases production by eliminating much of the inhibiting effects of pumped fluids such as plugging by solids, water retention, and chemical interactions. Finally, the drilling job is complete when the drill bit penetrates the reservoir and the reservoir is evaluated to see whether the well represents the discovery of a *prospect* or whether it is a dry hole.

Evaluation is usually initiated by examining the cuttings from the wellbore for evidence of hydrocarbon derivatives while the drill bit passes through a reservoir trap. The evaluation of these cuttings helps pinpoint the possible producing intervals in the wellbore. At this time, a wire line, lowered into the hole, and an electric log are run to help define possible producing intervals, the presence of hydrocarbon derivatives, and detailed information about the different formations throughout the wellbore. Further tests can also be run on individual formations within the wellbore such as pressure tests, formation fluid recovery, and sidewall core analysis. On the other hand, if the hole is dry, it is plugged and abandoned.

It is the differential pressure between the reservoir and the open hole that moves the hydrocarbon derivatives out of the reservoir, into the well, and up to the surface. The pressure may be the result of a number of forces. For example, water located below the oil layer may be pressing upward; when this occurs, it is referred to as a water drive system (see image). If the gas cap located above the oil is causing a downward pressure, it is referred to as a gas-cap drive system.

The vast heavy oil reserves (including extra heavy oil reserves and tar sand bitumen reserves) available in various parts of the world are becoming increasingly important as a secure future energy source (Safinya, 2008; Shah et al., 2010). In fact, the heavy oil resource is measured in trillions to quadrillion of barrels (1 trillion barrels = 1×10^{12} bbl and 1 quadrillion = 1×10^{15}). Whether the potential and promise of these viscous feedstocks are realized depends on the evolution of recovery technologies that are appropriate for the wide range of reservoir and oil-phase conditions. Such technologies also need to be comparatively benign from an environmental aspect (Table 2.5).

Heavy oil production has been increasing in recent years and is expected to increase in the future because of expected supply shortfall in conventional oil and an abundance of relatively large and known heavy oil reservoirs. Heavy oil is commercially produced by primary recovery, water injection, and thermal enhanced oil recovery methods. A majority of the recent primary production is in Venezuela's Orinoco Belt, and long horizontal wells have made these projects possible.

TABLE 2.5 Examples of Recovery Processes for Heavy Oil, Extra Heavy Oil, and Tar Sand Bitumen

Method	Description
VAPEX	Uses solvent in place of steam in SAGD-type wells
Hybrid	Solvent plus steam in SAGD, CSS, and steam flood wells
In situ combustion	Uses vertical and horizontal wells
	Injected air (oxygen) to support combustion
	Use part of oil as fuel for combustion
Downhole heating	Resistance induction heating
	Radio-frequency heating
Gasification	Used for steam generation and hydrogen production

However, heavy oil in a reservoir is often very viscous (unless the reservoir temperature is high) and does not flow easily; therefore, traditional methods of oil recovery, such as primary and secondary methods, are not usually applicable to heavy oil, or if they are applicable, it is often decided to bypass such methods and proceed to enhanced methods of recovery. The route to use is, of course, reservoir-specific, and what is applied to a heavy oil reservoir cannot necessarily be applied to another heavy oil reservoir.

The amount of oil that is recoverable is determined by a number of factors including the permeability of the rocks, the strength of natural drives (the gas present and pressure from adjacent water or gravity), and the viscosity of the oil. When the reservoir rocks are "tight" such as shale, oil generally cannot flow through, but when they are permeable such as in sandstone, oil flows freely. The flow of oil is often helped by natural pressures surrounding the reservoir rocks including natural gas that may be dissolved in the oil, natural gas present above the oil, water below the oil, and the strength of gravity. Heavy oils tend to span a large range of viscosity from manageable liquids to poorly mobile materials.

Thus, the quest to produce heavy oil is a global issue. However, in many cases, the resource does not resemble the conditions that are thought to be typical for heavy oil: viscous oil held in relatively permeable, shallow sands and the fields of interest have evolved to include fractured carbonates; offshore settings; and deeper, more geologically heterogeneous heavy oil resources. Such new settings introduce new challenges, which include highly permeable pathways and limits on reservoir access, in addition to those of large oil-phase viscosity and low reservoir energy.

Waterflooding is more usually used as a secondary recovery process but has also been successfully employed as a primary technique in some heavy oil fields. In the process, water is injected in a well and pushes the oil toward the producing wells where submersible pumps help raise the oil to the surface. Eventually, the waterfront will reach the production well, and increasingly larger quantities of water will be produced, making the process less economical until the producer well becomes nonproductive. The use of horizontal injector and producer wells allows in-line drive with improved efficiency of the process by providing a more uniform pressure.

Unfortunately, waterflooding tends to be less effective in reservoirs with an absence of underlying water as the injected water has a propensity to slump and channel along the base of the reservoir before coning into the producer. Recovery factors of up to 20%–40% have been achieved in conventional crude oil reservoirs, even up to 60%, but this decreases significantly with increasing oil viscosity due to channeling of the waterflood front, resulting in poor sweep efficiency for all but the relatively low-viscosity heavy oils. As a result, the waterflooding process for recovery of heavy oil is often summarily dismissed from consideration because of adverse mobility ratios.

However, waterflooding methods have been used with heavy oil to enhance formation pressures and help displace the heavy crude oil (Charles and Startzman, 1996; NPC, 2007). Heavy oil viscosity decreases rapidly with increasing temperature; therefore, the presence of internal heat is an asset for production. In fact, reservoir temperature (the mobility of the heavy oil is increased at high reservoir temperature) and the temperature of the water may be deciding factors when considering heavy oil recovery by waterflooding. In cold and/or offshore environments, waterflooding and perhaps polymer-augmented waterflooding may present the most attractive recovery option following primary recovery (Taylor and Nasr-El-Din, 1998).

This chapter describes the various types of recovery processes in common use but with a focus on the nonthermal processes that can be used to recover heavy oil, extra heavy oil, and tar sand bitumen. Thermal recovery methods include cyclic steam injection, steam flooding, and in situ combustion (Chapter 3). The steam processes are the most advanced of all enhanced oil recovery methods in terms of field experience and thus have the least uncertainty in estimating performance, provided that a good reservoir description is available. Steam processes are most often applied in reservoirs containing viscous oils and tars, usually in place rather than following secondary or primary methods. Commercial application of steam processes has been underway since the early 1960s. In situ combustion has been field-tested under a wide variety of reservoir conditions, but few projects have proved economical and advanced to commercial scale.

2. Primary Recovery Methods

Primary recovery of heavy oil, extra heavy oil, and tar sand bitumen from reservoirs and deposits is not the usual approach to recovery. In most reservoirs, initial pressure is sufficient to push the oil to the surface of the production well with only minimal help from a downhole pump. But, with declining well pressures, it becomes more difficult to get the hydrocarbon to the surface. Sometimes, artificial oil lift is needed.

If the underground pressure in the oil reservoir is sufficient, then this pressure will force the oil to the surface. Gaseous fuels, natural gas, or water are usually present, which also supply needed underground pressure. In this situation, it is sufficient to place a complex arrangement of valves (the Christmas tree) on the wellhead to connect the well to a pipeline network for storage and processing. Usually, about 20% of the oil in a reservoir can be extracted using primary recovery methods.

Thus, primary recovery techniques rely entirely on natural forces within the reservoir, and although primary recovery is widely applicable to the recovery of conventional crude oil, it is less applicable to heavy oil. At high reservoir temperature that keep the heavy oil sufficiently fluid, primary recovery may undoubtedly be applicable. For example, production mechanisms of the heavy oil solution-gas-drive process are not completely elucidated, and consequently, it is difficult to optimize primary recovery.

Crude oil reservoirs usually start with a formation pressure high enough to force crude oil into the well and sometimes to the surface through the tubing (Lake and Walsh, 2004). However, since production is invariably accompanied by a decline in reservoir pressure, *primary recovery* through natural drive soon comes to an end. In addition, many oil reservoirs enter production with a formation pressure high enough to push the oil into the well but not up to the surface through the tubing. In these cases, some means of artificial lift must be installed.

The most common method used for artificial life uses a pump at the bottom of the production tubing that is operated by a motor and a *walking beam* (an arm that rises and falls like a seesaw) on the surface. A string of solid metal *sucker rods* connects the walking beam to the piston of the pump. Another method (gas lift) uses gas bubbles to lower the density of the oil, allowing the reservoir pressure to push it to the surface. The gas is usually injected down the annulus between the casing and production tubing and through a special valve at the bottom of the tubing. In a third type of artificial lift, the produced oil is forced down the well at high pressure to operate a pump at the bottom of the well.

With the artificial lift methods described above, oil may be produced as long as there is enough nearby reservoir pressure to create flow into the wellbore. Inevitably, however, a point is reached at which commercial quantities no longer flow into the well. In most cases, less than one-third of the oil originally present can be produced by naturally occurring reservoir pressure alone, and in some cases (e.g., where the oil is quite viscous and at shallow depths), primary production is not economically possible at all.

Crude oil moves out of the reservoir into the well by one or more of three processes. These processes are *dissolved gas drive*, *gas-cap drive*, and *water drive*. Early recognition of the type of drive involved is essential to the efficient development of an oil field.

In *dissolved gas drive* (*solution gas drive*) (Fig. 2.4), the propulsive force is the gas in solution in the oil, which tends to come out of solution because of the pressure release at the point of penetration of a well. Dissolved gas drive is the least efficient type of natural drive as it is difficult to control the gas-oil ratio; the bottom-hole pressure drops rapidly, and the total eventual recovery of crude oil from the reservoir may be less than 20%.

If gas overlies the oil beneath the top of the trap, it is compressed and can be utilized to drive the oil into wells situated at the bottom of the oil-bearing zone (*gas-cap drive*) (Fig. 2.5). By producing oil only from below the gas cap, it is possible to maintain a high gas-oil ratio in the reservoir until almost the very end of the life of the pool. If, however, the oil deposit is not systematically developed so that bypassing of the gas occurs, an undue proportion of oil is left behind. The usual recovery of crude oil from a reservoir in a gas-cap field is 40%–50%.

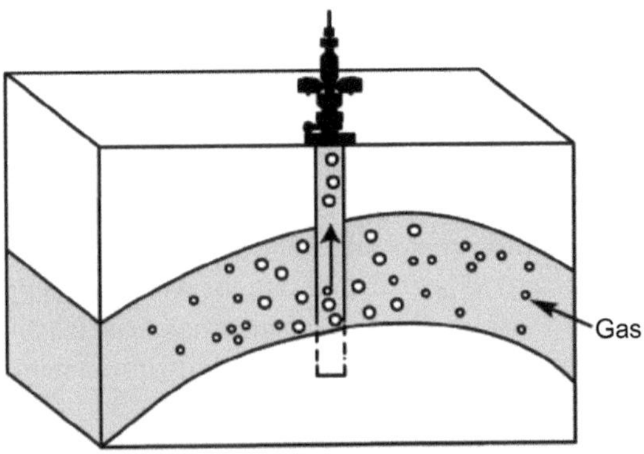

FIG. 2.4 Solution gas drive.http://intranet.southdowns.ac.uk/sdc/sciencemaths/environmental/013%20Crudeoil.pdf

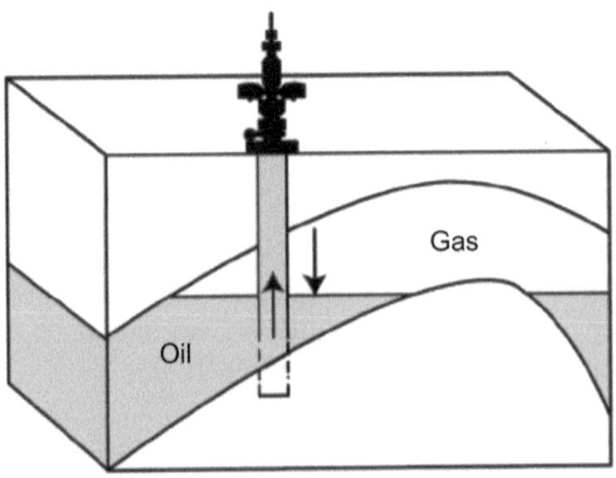

FIG. 2.5 Gas-cap drive.http://intranet.southdowns.ac.uk/sdc/sciencemaths/environmental/013%20Crudeoil.pdf

Usually, the gas in a gas cap (*associated natural gas*) contains methane and other hydrocarbon derivatives that may be separated out by compressing the gas. A well-known example is *natural gasoline* that was formerly referred to as *casinghead gasoline* or *natural gasoline*. However at high pressures, such as those existing in the deeper fields, the density of the gas increases, and the density of the oil decreases until they form a single phase in the reservoir. These are the so-called retrograde condensate pools because a decrease (instead of an increase) in pressure brings about condensation of the liquid hydrocarbon derivatives. When this reservoir fluid is brought to the surface and the condensate is removed, a large volume of residual gas remains. The modern practice is to cycle this gas by compressing it and inject it back into the reservoir, thus maintaining adequate pressure within the gas cap, and condensation in the reservoir is prevented. Such condensation prevents recovery of the oil, for the low percentage of liquid saturation in the reservoir precludes effective flow.

The most efficient propulsive force in driving oil into a well is the natural *water drive* (Fig. 2.6) in which the pressure of the water forces the lighter recoverable oil out of the reservoir into the producing wells. In anticlinal accumulations, the structurally lowest wells around the flanks of the dome are the first to come into water. Then, the oil-water contact plane moves upward until only the wells at the top of the anticline are still producing oil; eventually, these also must be abandoned as the water displaces the oil.

In a water drive field, it is essential that the removal rate be adjusted so that the water moves up evenly as space is made available for it by the removal of the hydrocarbon derivatives. An appreciable decline in bottom-hole pressure is necessary to provide the pressure gradient required to cause water influx. The pressure differential needed depends on the reservoir permeability; the greater the permeability, the less the difference in pressure necessary. The recovery of crude oil from the reservoir in properly operated water drive pools may run

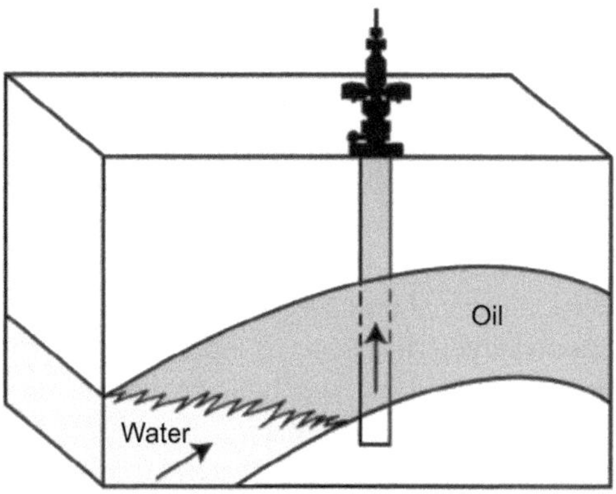

FIG. 2.6 Water drive.http://intranet.southdowns.ac.uk/sdc/sciencemaths/environmental/013%20Crudeoil.pdf

as high as 80%. The force behind the water drive may be the hydrostatic pressure, the expansion of the reservoir water, or a combination of both. Water drive is also used in certain submarine fields.

Gravity drive is an important factor when oil columns of several thousands of feet exist, as they do in some North American fields. Furthermore, the last bit of recoverable oil is produced in many pools by gravity drainage of the reservoir. Another source of energy during the early stages of withdrawal from a reservoir containing undersaturated oil is the expansion of that oil as the pressure reduction brings the oil to the bubble point (the pressure and temperature at which the gas starts to come out of solution).

For primary recovery operations, no pumping equipment is required. If the reservoir energy is not sufficient to force the oil to the surface, then the well must be pumped. In either case, nothing is added to the reservoir to increase or maintain the reservoir energy or to sweep the oil toward the well. The rate of production from a flowing well tends to decline as the natural reservoir energy is expended. When a flowing well is no longer producing at an efficient rate, a pump is installed.

The recovery efficiency for primary production is generally low when liquid expansion and solution gas evolution are the driving mechanisms. Much higher recoveries are associated with reservoirs with water and gas-cap drives and with reservoirs in which gravity effectively promotes drainage of the oil from the rock pores. The overall recovery efficiency is related to how the reservoir is delineated by production wells. Thus for maximum recovery by primary recovery, it is often preferable to sink several wells into a reservoir, thereby bringing about recovery by a combination of the methods outlined here.

When production begins to drop off, it may be time for the well to receive a workover (a major repairing and cleaning out of all pipes). Producing wells are like anything else; they require periodic maintenance. Corrosion can roughen pipe walls or cause failure, allowing product to leak onto the surface. Pieces of rock from the side of the well may break off and fall into the well clogging it. Natural gas pipes tend to accumulate paraffin (hydrate compounds that build up inside the pipe causing restrictions). Maintenance can result in everything from cleaning fluids being injected into the pipes to wire brushes being inserted to brush the pipes clean. Residues are flushed from the system before it is reconnected.

However, workover is not restricted only to the hardware; it may also be applied to the downhole portion of the rock formation. Often, the formation through which the hydrocarbon derivatives are flowing becomes clogged, which diminishes the volume of product reaching the well. Two processes used to improve formation characteristics are (i) acidizing and (ii) fracturing. Acidizing involves injecting an acid into a soluble formation, such as a carbonate, where it dissolves rock. This process enlarges the existing voids and increases permeability. Fracturing (*hydraulic fracturing* and *fracking*) involves injecting a fluid into the formation under significant pressure that makes existing small fractures larger and creates new fractures.

Thus, in conventional oil production, the concept of applying more than one recovery technology, one after the other, to a reservoir is well established. When primary production declines and becomes less economic, producers investigate the opportunity to waterflood the reservoir as a secondary recovery technology. Finally, tertiary recovery methods are usually applied when the waterflood method yields diminishing returns of oil. Heavy oil, extra heavy oil, and tar sand bitumen have a shorter history, and generally, reservoirs have only been subject to only one recovery technology. In the case of tar sand bitumen, primary and secondary recovery technologies, as defined for conventional oil, are not applicable because bitumen is immobile at deposit conditions. Therefore, tar sand development generally starts with a thermal recovery technology that would be considered a tertiary method or enhanced recovery method for conventional oil. However, as the development of heavy oil, extra heavy oil, and tar sand bitumen matures, the concept of applying more than one recovery technology in a specific order is likely to also be applied.

Given the option, operators will try to produce as much heavy oil as possible by primary recovery methods (*cold production*), but this is dependent upon the fluidity of the oil that, in turn, is dependent upon the reservoir temperature. The term *cold production* refers to the use of operating techniques and specialized pumping equipment to aggressively produce heavy oil reservoirs without applying heat. This encourages the associated production of large quantities of the unconsolidated uncemented reservoir sand that in turn results in significantly higher oil production.

A cold production oil reservoir is a *solution gas drive reservoir* if the major reservoir energy for primary depletion is supplied by the release of gas from the oil and the expansion of the in-place fluids as reservoir pressure declines. The fraction of original oil in place that can be recovered by solution gas drive decreases with increasing oil viscosity. For heavy oil reservoirs, the expected recovery factor by solution gas drive is typically about 5%. A number of heavy oil reservoirs under solution gas drive, however, have obtained anomalous primary performance results: low-production gas-oil ratios, high oil production rates, and recovery of unexpectedly large amounts of oil. This unusual behavior has been attributed to (i) the expansion of gas bubbles giving the oil a foamy character as the bubbles are trapped by the oil and recovery is then enhanced by solution—ultimate oil recovery with primary techniques can be as high as 20% for some heavy foamy oil reservoirs—and (ii) internal erosion in unconsolidated sand reservoirs, which can create a network of high-permeability channels (*wormholes*) that can enhance drainage by a factor of 10 or more (Chen, 2006).

The Faja del Orinoco belt in Venezuela is the world's largest heavy oil accumulation in excess of 1 trillion barrels (1×10^{12} bbl). There are a few factors and technical advances that allow this heavy oil to be produced: (i) The viscosity is low enough with the existing solution gas that the heavy oil can flow at reservoir temperatures; (ii) horizontal wells up to 5000 ft allow the heavy oil to be produced at economic rates while maintaining sufficiently low drawdown pressures to prevent extensive sand production—more complex well geometries have been drilled with several horizontal branches (multilateral wells); (iii) the horizontal legs are placed precisely in the target sands using logging-while-drilling (LWD) and measurement-while-drilling (MWD) equipment, enabling more cost-effective placement of the wells; (iv) sand production from the unconsolidated formation is minimized using slotted liners and other sand control methods; and (v) a low drawdown pressure in a long multilateral can also reduce the need for significant sand control.

The main disadvantage of the cold production method is the low recovery factor, typically 6%–15% v/v, for primary production (Curtis et al., 2002). Fields are not being developed with future, secondary processes in mind. For example, wells, cement, and completions are not designed for high temperatures encountered in steam injection and other thermal recovery processes. Horizontal and fishbone wells should be drilled in the optimum location with regard to permeability, porosity, oil composition, distances above water or below gas, and the length of the laterals. Characterizing the formation and hydrocarbon derivatives in real time while drilling is essential for well placement. Drilling, measurement-while-drilling, and logging-while-drilling technologies are key enablers for this. In horizontal wells and multilateral wells, being able to monitor, understand, control, and ensure the flow from different sections of the well will improve production and reduce unwanted water and/or natural gas production.

In addition, depending on the character of the reservoir *and* the properties of the oil in the reservoir, primary production with artificial lift, for example, injection of a light oil or diluent, to reduce viscosity, may find use. Many fields produce most efficiently with horizontal production wells. In some cases, exploiting foamy oil behavior and/or encouraging the production of sand along with the oil turn out to be the preferred production strategy. Choosing the optimal cold production strategy requires an understanding of rock and fluid properties (Chen, 2006). Primary production in some heavy oil reservoirs may be larger than that estimated by conventional calculations. The main driving force behind primary recovery is pressure depletion through solution gas drive.

In the case of the Faja oil, (i) the viscosity is low enough with the existing solution gas that the heavy oil can flow at reservoir temperatures; (ii) horizontal wells up to 1500m long allow the heavy oil to be produced at economic rates while maintaining sufficiently low drawdown pressures to prevent extensive sand production; (iii) the horizontal legs are placed precisely in the target sands using logging-while-drilling and measurement-while-drilling equipment, enabling more cost-effective placement of the wells; and (iv) in some locations, sand production from the unconsolidated formation is minimized using slotted liners and other sand control methods.

The main issue for cold production is the low recovery factor for primary production. Fields are not being developed with future, secondary processes in mind. For example, wells, cement, and completions are not designed for high temperatures encountered in steam injection and other thermal recovery processes. Horizontal and fishbone wells should be drilled in the optimum location with regard to permeability, porosity, oil composition, distances above water or below gas, and the length of the laterals.

In the Orinoco field, natural gas production is an issue because of interference with progressive cavity pumps' ability to lift the heavy oil.

However, primary production in some heavy oil reservoirs is larger than that estimated by conventional calculations (Kovscek, 2002). Conventionally, the main driving force behind primary recovery is pressure depletion through solution gas drive. Solution gas drive is the mechanism, whereby the lowering of reservoir pressure through production in an undersaturated reservoir reaches the bubble point where gas starts to evolve from solution. The evolved gas does not begin to flow until the critical gas saturation has been reached. Once the critical gas saturation point is reached, there is an increase in rate of pressure drop due to the production of the gas phase. In such a case, the oil at the wellhead of these heavy oil reservoirs resembles the form of foam, hence the term *foamy oil.*

3. Secondary Recovery Methods

Over the lifetime of the well, the pressure will fall, and at some point, there will be insufficient underground pressure to force the oil to the surface. If economical, and it often is, the remaining oil in the well is extracted using secondary oil recovery methods. It is at this point that secondary recovery methods must be applied.

Even in conventional oil reservoirs, a large amount of the oil can be left behind after primary production since the natural reservoir pressure has dwindled to the point where it can't force the oil to the surface. Thus, as fluid withdrawal continues from the reservoir, the pressure within the reservoir gradually decreases, and the amount of gas in solution decreases. As a result, the flow rate of fluid into the wellbore decreases, and less gas is liberated. The fluid may not reach the surface, so a pump (artificial lift) must be installed in the wellbore to continue producing the crude oil. If this is allowed to continue, the flow rate of the crude oil becomes so small, and the cost of lifting the oil to the surface becomes so great that the well costs more to operate than the revenues that can be gained from selling the crude oil (after discounting the price for operating costs, taxes, insurance, and return on capital). The well's economic limit has then been reached, and it is abandoned. At this stage, there is insufficient underground pressure to force the oil to the surface. If economical, as often is, the remaining oil in the well is extracted using secondary oil recovery methods. However, in order to maintain well production, the flow of oil should not be allowed to decrease to the minimum. Operators usually apply secondary recovery at a point before reservoir energy is depleted.

When a large part of the crude oil in a reservoir cannot be recovered by primary means, a method for supplying extra energy must be found. Most often, secondary recovery is accomplished by injecting gas or water into the reservoir to replace produced fluids and thus maintain or increase the reservoir pressure. When gas alone is injected, it is usually put into the top of the reservoir, where crude oil gases normally collect to form a gas cap. Gas injection can be a very effective recovery method in reservoirs where the oil is able to flow freely to the bottom by gravity. When this gravity segregation does not occur, however, other means must be sought.

The most common follow-up or *secondary recovery* operations usually involve the application of pumping operations or of injection of materials into a well to encourage movement and recovery of the remaining crude oil. The pump, generally known as the *horsehead pump* (*pump jack*, *nodding donkey*, or *sucker rod pump*) (Fig. 2.7), provides mechanical lift to the fluids in the reservoir. The pump is powered by an electric motor, but some older, less producing wells use propane as an alternative source of power rather than a large power grid. Some of these wells can even use the natural gas from the casing as fuel, and the well can be completely self-sufficient. The engine of the pump runs a set of pulleys to the transmission that in turn drives a pair of cranks, generally with counter weights on them to assist the motor in lifting the heavy string of sucker rods. Electrically powered centrifugal pumps and submersible pumps (both pump and motor are in the well at the bottom of the tubing) have proved their production capabilities in numerous applications.

There are also *secondary oil recovery* operations that involve the injection of water or gas into the reservoir. When water is used, the process is called a

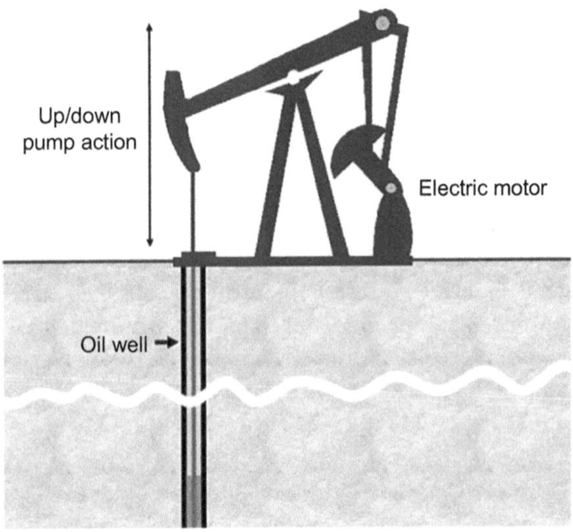

FIG. 2.7 A horsehead pump.

waterflood and with gas a *gas flood*. Separate wells are usually used for injection and production. The injected fluids maintain reservoir pressure or repressure the reservoir after primary depletion and displace a portion of the remaining crude oil to production wells. In fact, the first method recommended for improving the recovery of oil was probably the reinjection of natural gas, and there are indications that gas injection was utilized for this purpose before 1900 (Craft and Hawkins, 1959; Frick, 1962). These early practices were implemented to increase the immediate productivity and are therefore classified as pressure maintenance projects. Recent gas injection techniques have been devised to increase the ultimate recovery, thus qualifying as secondary recovery projects.

The wells to be used for injecting water are usually located in a pattern that will best push oil toward the production wells. Water injection often increases oil recovery to twice that expected from primary means alone. Some oil reservoirs (e.g., the East Texas field) are connected to large, active water reservoirs, or aquifers, in the same formation. In such cases, it is necessary only to reinject water into the aquifer in order to help maintain reservoir pressure.

In secondary recovery, the injected fluid must dislodge the oil and propel it toward the production wells. Reservoir energy must also be increased to displace the oil. Using techniques such as gas and water injection, there is no change in the state of oil. Similarly, there is no change in the state of the oil during miscible fluid displacement technologies. The analogy that might be used is that of a swimmer (in water) in which there is no change to the natural state of the human body.

Thus, the success of secondary recovery processes depends on the mechanism by which the injected fluid displaces the oil (displacement efficiency) and on the volume of the reservoir that the injected fluid enters (conformance or sweep efficiency). In most proposed secondary projects, water does both these things more effectively than gas. It must be decided if the use of gas offers any economic advantages because of availability and relative ease of injection. In reservoirs with high permeability and high vertical span, the injection of gas may result in high recovery factors as a result of gravity segregation, as described in a later section. However, if the reservoir lacks either adequate vertical permeability or the possibility for gravity segregation, a frontal drive similar to that used for water injection can be used (dispersed gas injection). Thus, dispersed gas injection is anticipated to be more effective in reservoirs that are relatively thin and have little dip. Injection into the top of the formation (or into the gas cap) is more successful in reservoirs with higher vertical permeability (200 md or more) and enough vertical relief to allow the gas cap to displace the oil downward.

Vaporization is another recovery mechanism used to inject gas into oil reservoirs. A portion of the oil affected by the *dry* injection gas is vaporized into the oil and transported to the production wells in the vapor phase. In some instances, this mechanism has been responsible for a substantial amount of the secondary oil produced.

During the withdrawal of fluids from a well, it is usual practice to maintain pressures in the reservoir at or near the original levels by pumping either gas or water into the reservoir as the hydrocarbon derivatives are withdrawn. This practice has the advantage of retarding the decline in the production of individual wells and considerably increasing the ultimate yield. It also may bring about the conservation of gas that otherwise would be wasted and the disposal of brines that otherwise might pollute surface and near-surface potable waters.

In older fields, it was not the usual practice to maintain the reservoir pressure, and it is now necessary to obtain crude oil from these fields by means of secondary recovery projects. Thus, several methods have been developed to obtain oil from reservoirs where previous economic policies dictated that ordinary production systems were no longer viable. Considerable experimentation has been carried on the use of different types of input gas. Examples are *wet casinghead gas*; *enriched gas*; *liquefied crude oil gas* (*LPG*), such as butane and propane; high-pressure gas; and even nitrogen. High-pressure gas not only pushes oil through the reservoir but also may produce a hydrocarbon exchange so that the concentration of liquid crude oil gases in the oil is increased.

Whenever it is feasible, the injection design should take advantage of gravity, that is to say, dipping or inclined reservoirs, underlying aquifers, or gas caps. A variety of injection-production patterns have been employed in waterflooding operations. In pattern flooding, the injectors are distributed among the producers in some repeating fashion. Pattern selection is important because it can affect the area swept by the injected fluid. Well locations frequently do not coincide with the ideal well locations associated with the well patterns used. Generally, the choice of pattern (Fig. 2.8) for waterflooding must be consistent with the existing wells. The objective is to select the proper pattern that will provide the injection fluid with the maximum possible contact with the crude oil to minimize bypassing by the water.

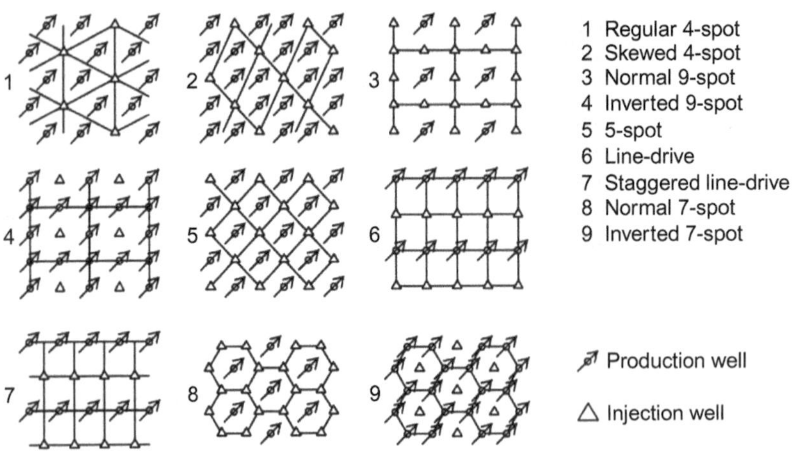

FIG. 2.8 Well patterns.

In a *four-spot pattern*, the distance between all like wells is constant. Any three injection wells form an equilateral triangle with a production well at the center. The four-spot may be used when the injectivity is high or the heterogeneity is minimal. In a *five-spot pattern*, the distance between all like wells is constant. Four injection wells form a square with a production well at the center. If existing wells were drilled on square patterns, five-spot patterns (as well as nine-spot patterns) are most commonly used since they allow easy conversion to a five-spot waterflood. If the injectivity is low or the heterogeneity is large, the five-spot pattern may not be the best choice, and a nine-spot may be more efficient.

In the *seven-spot pattern*, the injection wells are located at the corner of a hexagon with a production well at its center. If the reservoir characteristics yield lower than preferred injection rates, either a seven-spot (or a nine-spot) pattern should be considered because there are more injection wells per pattern than producing wells.

In the *nine-spot pattern*, the arrangement is similar to that of the five-spot but with an extra injection well drilled at the middle of each side of the square. The pattern essentially contains eight injectors surrounding one producer. If existing wells were drilled on square patterns, nine-spot patterns (as well as five-spot patterns) are most commonly used. If the reservoir characteristics yield lower injection rates than those desired, one should consider using either a nine-spot pattern (or a seven-spot pattern) where there are more injection wells per pattern than producing wells.

In the *inverted seven-spot pattern*, the arrangement is similar to the normal seven-spot pattern except where the position of the producer well was in the normal seven-spot pattern, there is now an injector well. Likewise, where the injector wells were in the normal seven-spot pattern, there are now producer wells. The inverted seven-spot pattern may be used when the injectivity is high or the heterogeneity is minimal.

In the *inverted nine-spot pattern*, the arrangement of the wells is similar to the normal nine-spot pattern except the position of the producer well in the normal nine-spot pattern is occupied by an injector well. Likewise, where the positions of the injector wells were in the normal nine-spot, there are now producer wells. If the reservoir is fairly homogenous and the mobility ratio is unfavorable, the inverted nine-spot pattern may be promising.

In the *direct line-drive pattern*, the lines of injection and production are directly opposite to each other. If the injectivity is low or the heterogeneity is large, direct line drive is a good option. Anisotropic permeability, permeability trends, or oriented fracture systems favor line-drive patterns.

In the *staggered line-drive pattern*, the wells are in lines as in the direct line, but the injectors and producers are no longer directly opposed but laterally displaced by a specified distance that is dependent upon the distance between wells of the same type and the distance between the lines of injector wells and producer wells. The staggered line-drive pattern is also effective for reservoirs; there is anisotropic permeability or where permeability trends or oriented fracture systems.

Reservoir uniformities also dictate the choice of pattern, and mobility ratio has an important influence on pattern selection. If the ratio is unfavorable, the injectivity of an injector will exceed the productivity of a producer, and water injection will supersede oil production. Hence, to balance the production with the water injection, more producers than injectors are required. On the other hand, if the mobility ratio is favorable, the injectivity is impaired, and the pattern should have more injectors than producers. Consideration should be given to the desired oil production level after waterflooding, which will influence the ratio of injection wells to producing wells (Table 2.6) as an aid to increasing the areal sweep efficiency.

In the ideal case, the displacement of oil by water takes place in the form of a homogenous front, and depending on the mobility ratio of oil and water, the front starts to generate fingers resulting in bypass oil. After the waterfront breaks through at the producer, high water cut and low oil production characterize the production behavior of this producer. The velocity of the frontal movement depends on the amount of water injected in the wells surrounding the producer and the amount of total fluid that is produced from the well in the center of the water injectors.

Furthermore, the recovery of oil through water displacement can be considered as follows: (1) displacement, that is, the fraction of oil displaced from each pore by the physical contact between the displacing water and oil, and (2) conformance, that is, the fraction of the total volume (area × thickness) that is contacted by the injected water.

Displacement efficiency is the fraction of the oil displaced from each pore, which is contacted by the encroaching water. Areal sweep efficiency is the fraction of oil displaced from the area physically contacted by the encroaching water. The vertical sweep efficiency is the fraction of oil displaced from the

TABLE 2.6 The Ratio of Injection Wells to Producing Wells for Various Well Patterns

Pattern	Ratio[a]
Four-spot	2
Five-spot	1
Seven-spot	0.5
Inverted seven-spot	2
Nine-spot	0.33
Inverted nine-spot	3

[a]*(Injection wells)/(producing wells).*

vertical cross-sectional coverage by the encroaching water. Hence, the recovery efficiency is the product of displacement efficiency and sweep efficiency (areal efficiency × vertical efficiency).

Waterflooding remains a predominantly secondary recovery process (*waterflood*). Probably, the principal reason for this is that reservoir formation water is ordinarily not available in volume during the early years of an oil field and pressure maintenance water from outside the field may be too expensive. When a young field produces considerable water, it may be injected back into the reservoir primarily for the purpose of nuisance abatement, but reservoir pressure maintenance is a valuable by-product.

Nevertheless, some passages in the formation are larger than others, and the water tends to flow freely through these, bypassing smaller passages where the oil remains. A partial solution to this problem is possible by *miscible fluid flooding*. Liquid butane and propane are pumped into the ground under considerable pressure, dissolving the oil and carrying it out of the smaller passages; additional pressure is obtained by using natural gas.

When dealing with heavy oil reservoirs, 90% or more of the original oil in place can be left in the reservoir after attempts at primary or cold production (Curtis et al., 2002).

Secondary oil recovery uses various techniques to aid in recovering oil from depleted or low-pressure reservoirs. Pumps, such as horsehead pumps and electric submersible pumps (ESPs), are used to bring the oil to the surface. Other secondary recovery techniques increase the reservoir's pressure by water injection (waterflooding), natural gas injection (gas flooding), and gas lift, which injects air, carbon dioxide, or some other gases into the reservoir.

3.1 Waterflooding

Waterflooding is a form of oil recovery wherein the energy required to move the oil from the reservoir rock into a producing well is supplied from the surface by means of water injection and the induced pressure from the presence of additional water. Water injection is used to prevent low pressure in the reservoir—the water replaces the oil that has been taken, keeping the production rate and the pressure the same over the long term. Waterfloods are essentially artificial water drives and, at one time, were considered to be a form of enhanced recovery.

In a completely developed oil field, the wells may be drilled anywhere from 200 to 2000 ft (60–600 m) from one another, depending on the nature of the reservoir. If water is pumped into alternate wells in such a field, the pressure in the reservoir as a whole can be maintained or even increased. In this way, the rate of production of the crude oil also can be increased; in addition, the water physically displaces the oil, thus increasing the recovery efficiency. In some reservoirs with a high degree of uniformity and little clay content, waterflooding may increase the recovery efficiency to as much as 60% or more of the original oil in place. Waterflooding was first introduced in the Pennsylvania oil fields, more or less accidentally, in the late nineteenth century, and it has since spread throughout the world.

Waterflood has been conducted successfully in a few high-viscosity reservoirs in the past, and several projects are currently ongoing and planned around the world. Incremental recovery of approximately 2%–20% of the original oil in place has been reported (Kumar, 2006). However, it should be noted that the ratio of oil recovered by waterflood oil to oil recovered by primary methods varies greatly with depth for heavy oil and for conventional oil. Shallow reservoirs tend to have both low pressures and small amounts of dissolved gas. This low available drive energy often translates into low primary recovery, sometimes less than 5% v/v of the oil in place.

Produced water is often used as an injection fluid. This reduces the potential of causing formation damage due to incompatible fluids, although the risk of scaling or corrosion in injection flow lines or tubing remains. Also, the produced water, being contaminated with hydrocarbon derivatives and solids, must be disposed of in some manner, and disposal to sea or river will require a certain level of cleanup of the water stream first. However, the processing required to render produced water fit for reinjection may be equally costly. As the volumes of water being produced are never sufficient to replace all the production volumes (oil and gas, in addition to water), additional "make-up" water must be provided. Mixing waters from different sources exacerbates the risk of scaling.

Seawater is obviously the most convenient source for offshore production facilities, and it may be pumped inshore for use in land fields. Where possible, the water intake is placed at sufficient depth to reduce the concentration of algae; however, filtering, deoxygenation, and biociding are generally required. *Aquifer water* from water-bearing formations other than the oil reservoir but in the same structure has the advantage of purity where available. *River water* will always require filtering and biociding before injection.

The filters must clean the water and remove any impurities, such as shells and algae. Typical filtration is to 2 μm. The filters are so fine so as not to block the pores of the reservoir. Sand filters are the easiest to use, because there is an automatic system with Delta P that cleans the filter with a backwash when the sand filter is dirty. The sand filter has different beds with various sizes of sand granules. The seawater traverses the first, finest, layer of sand down to the coarsest, and to clean the filter, the process is inverted. After the water is filtered, it continues on to fill the deoxygenation tower. Oxygen must be removed from the water because it promotes corrosion and growth of certain bacteria. Bacterial growth in the reservoir can produce toxic hydrogen sulfide, a source of serious production problems, and block the pores in the rock.

A *deoxygenation tower* brings the injection water into contact with a dry gas stream (gas is always readily available in the oil field). The filtered water drops into the deoxygenation tower, splashing onto a series of trays, causing dissolved oxygen to be lost to the gas stream. An alternative method, also used as a backup to deoxygenation towers, is to add an oxygen scavenging agent such as sodium bisulfite ($NaHSO_3$).

3.2 Gas Injection

Gas injection (also called reinjection or *gas repressurization*) is the reinjection of natural gas into an underground reservoir, typically one already containing both natural gas and crude oil, in order to increase the pressure within the reservoir and thus induce the flow of crude oil or else sequester gas that cannot be exported. This should not to be confused with gas lift, where gas is injected into the annulus of the well rather than the reservoir. After the crude has been pumped out, the natural gas is once again recovered. Since many of the wells found around the world contain heavy crude, this process increases their production.

The basic difference between light conventional crude oil and heavy crude oil is its viscosity and pumpability—the lighter the crude, the easier it is to pump. Recovery of hydrocarbon derivatives in a well is generally limited to 50% (heavy crudes) and 75%–80% (light crudes). Recycling of natural gas or other inert gases causes the pressure to rise in the well, thus causing more gas molecules to dissolve in the oil lowering its viscosity and thereby increasing the well's output. Air is not suitable for repressuring wells because it tends to cause deterioration of the oil; thus, carbon dioxide or natural gas is used to repressure the well. The term "gas reinjection" is also sometimes referred to as repressuring—the term being used only to imply that the pressure inside the well is being increased to aid recovery.

Injection or reinjection of carbon dioxide also takes place in order to reduce the emission of CO_2 into the atmosphere, a form of carbon sequestration. This has been mooted as a major weapon in the future fight against climate change, allowing mass storage of CO_2 over a geologic timescale.

Inert gas injection (IGI) is a technology for conventional oils in reservoirs where good vertical permeability exists or where it can be created through propped hydraulic fracturing. It is generally viewed as a *top-down* process with nitrogen or methane injection through vertical wells at the top of the reservoirs, creating a gas-oil interface that is slowly displaced toward long horizontal production wells. Thus, the process (sometimes referred to as *gravity drainage* or *gravity enhancement*) involves the use of the gravity drainage potential of a dipping or thick hydrocarbon zone (Alagorni et al., 2015). Nitrogen, which has a lower density than the reservoir fluids, when injected into the crest or allowed to migrate to the crest, will enhance the down-dip displacement and production of the reservoir fluids or of a gravity-stable miscible slug. One of the most common gravity drainage processes is the double displacement process (DDP) that is accomplished by injecting gas up-dip and producing oil down-dip. By using gravity drainage, pistonlike displacement is obtained; therefore, gas fingering is avoided. In addition, the following results are obtained: (i) horizontal gas-oil contact; (ii) gravity domination of the gas flow; (iii) optimized time between gas injection and oil production as fast as possible; (iv) the greater the dip angle, the higher the injection and production rates without gas fingering; and (v) the greater the dip, the more effective the gravity drainage. As with all gravity drainage processes, it is essential to balance the injection and production volumes precisely so that the system does not become pressure driven but remains in the gravity-dominated flow regime.

In one aspect of inert gas injection, methane (CH_4) enriched with ethane (C_2H_6), propane (C_3H_8), or butane (C_4H_{10}) (10%–20% v/v) and followed by lean gas and/or water is injected from water-injection well into the reservoir. When the injected gas contacts virgin reservoir oil, the methane, ethane, and propane are quenched from the injected gas and absorbed into the oil. The injected hydrocarbon solvent is usually displaced with cheaper chase leaner or inert gas such as methane or nitrogen.

In the process, gas that is generally injected single or intermittently with water (water-alternating-gas, WAG) has become widely practiced (Alagorni et al., 2015). According to miscibility between gas injected and oil displaced, gas injection can be classified into two major types: miscible gas injection and immiscible gas injection. In *miscible gas injection*, the gas is injected at or above minimum miscibility pressure that causes the gas to be miscible in the oil. In contrast, the *immiscible gas injection process* involves flooding by the gas that is conducted below the minimum miscibility pressure. This low-pressure injection of gas is used to maintain reservoir pressure to prevent production cutoff and thereby increase the rate of production. In miscible flooding, the incremental oil recovery is obtained by one of the three mechanisms: oil displacement by solvent through the generation of miscibility (i.e., zero interfacial tension between the oil and the solvent—hence infinite capillary number), oil swelling, and reduction in oil viscosity.

At reservoir conditions, the most usual problem that occurs with the hydrocarbon miscible flood is the gravity override because of its lighter density than the oil and water. So that in any miscible flood, the minimum miscibility pressure plays the most major role to overcome this problem. As a remedial factor, the solvent is to be injected at or above the minimum miscibility pressure of the reservoir fluid. Once it becomes miscible, then it improves the sweep efficiency and fallouts in optimum recovery (Alagorni et al., 2015).

High recovery ratios are achieved because, in the absence of elevated pressure gradients, a thin oil film is maintained between the gas and water phases. The film is maintained because the sum of the oil-water and gas-oil surface tensions is always less than the water-gas surface tension. Thus, the thin-film configuration is thermodynamically stable and allows the oil to drain to values far lower than the *residual oil saturation* value.

The interfaces in inert gas injection are gravity-stabilized because of the difference in phase densities, so that at slow drainage rates, the interfaces remain approximately horizontal without viscous fingering. In one configuration, the horizontal wells are produced under a back pressure equal to the pressure in the underlying water phase, so no water coning can occur. During production, if the water cut increases, the production rate is reduced so that the interface becomes stable. Alternatively, if gas is injected too quickly, gas coning can develop, and if this is observed, the gas injection rate must be reduced to sustain stability. The process is continued until the oil zone is "pinched down" to the horizontal well, achieving the high recovery ratios possible with gravity drainage methods. These principles are fundamental to all gravity-dominated processes, and failure to adhere to them will drive the system into conditions of instability (such as coning and fingering).

3.3 Cold Production

The term cold production refers to the use of operating techniques and specialized pumping equipment to aggressively produce heavy oil reservoirs without applying heat (Chugh et al., 2000). This encourages the associated production of large quantities of the unconsolidated uncemented reservoir sand that in turn results in significantly higher oil production. In contrast, conventional practices of primary heavy oil production discourage sand production and result in minimized initial unit operating costs. This practice, however, may prevent many wells from achieving their maximum oil production rate and reserve potential.

The basis of cold production is that the oil production and recovery improve when sand production occurs naturally. Field production data indicate that heavy oil flows more efficiently when sand is produced from unconsolidated reservoirs. In the Elk Point and Lindberg reservoirs, sand production from well averages occurs regularly (Loughead, 1992; McCaffrey and Bowman, 1991). Sand production is thought to be a function of (i) the absence of clays and cementation materials, (ii) the viscosity of the oil, (iii) the producing water cut and gas/oil ratio, and (iv) the rate of pressure drawdown (Chugh et al., 2000).

The presence of clay stabilizes the sand grains and reduces sand movement. Higher-viscosity oil increases the frictional drag between the oil and the sand grains that promotes sand movement. High water or gas production inhibits sand production because gas/water is produced instead of an oil/sand mixture. Increasing the drawdown rate also promotes sand movement because of the increase in the velocity of the fluid into the wellbore and hence increased frictional drag on the sand grains. It has been reported that gross near-wellbore failure of the formation due to sand production results in excellent productivity.

The produced sand creates a modified wellbore geometry that could have several configurations including piping tubes (wormholes), dilated zones, sheared zones, or possibly cavities. Porosity in the dilated zones may increase leading to large increases in reservoir permeability. In addition, the flow of sand with the oil has the potential to reduce the frictional drag forces on the oil and result in increased productivity in the porous region. Furthermore, fine migration, which occurs during oil production, can block pore throats and reduce the number of flow paths available for the oil. Producing sand helps to eliminate many of these bottlenecks, and the dilation of the sand also creates larger pore throats that are more difficult to block.

Cold heavy oil production with sand (CHOPS) is now widely used as a production approach in unconsolidated sandstones. The process modifies the more usual cold production technique where the completion zone of the well is designed to retain the crushed rock and sand that is carried to the well by the oil flow. This makes it easier to process the oil at the surface. Instead, the CHOPS process allows the sand to erode out of the well, thus extending highly permeable flow paths, or wormholes, through the reservoir to the well and thereby increasing production rates to 10–30 times that achieved in traditional cold production. The technique takes advantage of the fact that heavy oil reservoirs are predominantly in high porosity (>28%), unconsolidated sandstones that possess no tensile strength from grain-to-grain cementing and has been employed extensively. Gas evolving from the depressurization of the oil helps disturb and set the sand particles in motion, and initially, a high percentage of sand is produced, as much as 40% sand by volume. As sand and oil production continues, these wormholes continue to grow larger and extend further through the reservoir.

Typically, over 12% of oil in place can be recovered that is an improvement on traditional cold production. A major issue with CHOPS, however, is the separation of sand that is both mechanically and economically intensive due to the specialist fluid handling equipment that is necessary. Moreover, if the expanding wormholes reach a water source, water will flow through these pathways, and the production well will be filled of water, forcing its abandonment. A number of methods have been proposed to shut off these watered-out wormholes by means of clay, polymer gels, and gel foams with moderate success.

The process results in the development of high-permeability channels (*wormholes*) in the adjacent low cohesive strength sands, facilitating the flow of oil foam that is caused by solution gas drive. The process is used for thin subsurface tar sand (typically 3–20 ft thick) in Canada, provided the tar sand is unconsolidated and provided the heavy oil contains sufficient solution gas to power the production process. Foamy oil production may occur without sand production in other areas, such as in the Faja belt, Venezuela.

Thus, instead of blocking sand ingress by screens or gravel packs, sand is encouraged to enter the wellbore by aggressive perforation and swabbing strategies. Vertical or slightly inclined wells (vertical to 45°) are operated with rotary progressive cavity pumps (rather than reciprocating pumps), and old fields are converting to higher-capacity progressive cavity pumps, giving production boosts to old wells. Productivity increases over conventional production, and a CHOPS process can produce as much as 12% to perhaps as much as 25% of the original oil in place that can be recovered, rather than the 0%–5% typical of primary production without sand in such cases. Finally, because massive sand production creates a large disturbed zone, the reservoir may be positively affected for later implementation of thermal processes.

The CHOPS process increases productivity for the following reasons: (i) If the sand can move or is unconsolidated, the basic permeability to fluids is enhanced. (ii) As more sand is produced, a growing zone of greater permeability is generated, similar to a large-radius well that gives better production. (iii) Gas coming out of solution in heavy oil does not generate a continuous gas phase; rather, bubbles flow with the fluid and do not coalesce, but expand down-gradient, generating an "internal" gas drive, referred to as *foamy flow*. This also helps to locally destabilize the sand, sustaining the process. (iv) Continuous sand production means that asphaltene or fines plugging of the near-wellbore environment potentially do not occur, so there is no possibility of an effect to impair productivity. (v) As sand is removed, the overburden weight acts to shear and destabilize the sand, helping to drive sand and oil toward the wellbore.

Typically, a well placed on CHOPS production will initially produce a high percentage of sand, greater than 20% by volume of liquids. However, this generally drops after some weeks or months. The huge volumes of sand are disposed of by slurry fracture injection or salt cavern placement or by sand placement in a landfill in an environmentally acceptable manner.

CHOPS is used for thin subsurface tar sand (typically 1–7 m thick) in Canada, provided the tar sand is unconsolidated and provided the heavy oil contains sufficient solution gas to power the production process. To have any natural gas in solution, the tar sand must be at least a few hundred meters deep. For example, there are a large number of CHOPS wells located near Lloydminster, Alberta. CHOPS wells (by definition) require sand production. Foamy oil production may occur without sand production in other areas, such as in the Faja belt, Venezuela. Alternatively, oil may be produced with sand but without solution gas in still other areas.

It is believed that CHOPS production occurs with the formation of "wormholes," tunnels that may extend some distance into the formation. There are no current methods for predicting the distribution, location, length, or diameter of wormholes, and there are very limited means of measuring them once formed. Surface seismic may give an indication of their distribution and density. Hence, there is considerable uncertainty about the behavior of CHOPS wells.

CHOPS wells are vertical or slightly deviated wells; they are cased and perforated, and a downhole pump is deployed to create an aggressive pressure differential between formation and wellbore pressures. This causes natural gas to break out of solution from the heavy oil, resulting in the production of foamy oil. Gas bubbles evolving at the wormhole-sand interface destabilize sand grains, and the expanding gas helps move the mixture through the wormholes. Gravity drive on the unconsolidated sands also provides energy for production. At the start-up of production, up to 10% sand by volume is produced along with oil, water, and gas. Sand production eventually falls to under 2% during the well lifetime.

The surface footprint for CHOPS wells is small, only requiring space for the wellhead, a storage tank, and a small doghouse. Any produced gas is used on-site to power equipment or to heat the storage tank. Because a large volume of sand is produced, pipelines cannot be used for transportation. Instead, trucks are required to move oil, water, and sand for processing or disposal. During spring breakup, the CHOPS wells in Alberta must be shut in since trucks cannot navigate the roads.

The technical challenges for CHOPS wells include a better understanding of their behavior and more predictive performance models. Progressive cavity pumps have increased production rates, but increased reliability and longer maintenance-free periods would improve economics. A method for water shut-off would bring some unproductive wells back to life.

A major breakthrough would be a secondary recovery method to tap the remaining ~90% of the original oil in place. Two possible secondary recovery methods are in situ combustion and solvent flood. In addition, a primary production method that has a higher recovery factor would also have a significant impact. A high recovery factor and oil production without sand might replace trucking with pipelines, thus reducing carbon dioxide emissions and manpower costs and allowing year-round production.

3.4 Pressure Pulse Technology

Pressure pulse technology (PPT) is a technology that can be used to enhance the recovery rate of nonaqueous phase liquid (NAPL) and to reduce solids clogging in wells, permeable reactive barriers, and fractured media (CRA, 2003). Pressure pulse flow enhancement technology is based on the discovery that large amplitude pressure pulses that are dominated by low-frequency wave energy generate enhanced flow rates in porous media. For example, in preliminary experiments in heavy oil reservoirs in Alberta, pressure pulse technology has reduced the rate of depletion, increased the oil recovery ratio, and prolonged the life of wells.

The technology uses steady, nonseismic pulse vibrations (e.g., 15 pulses per minute) that generate a low velocity wave effect to encourage flow of oils and small solid particles. It is effective in geologic formations exhibiting elastic properties, such as unconsolidated sediments and sedimentary rocks. It must be applied in a downhole manner in order to be effective. It has been used by the oil industry to improve oil recovery from otherwise exhausted reserves for many years. Also, it has been found that very large amplitude pressure pulses applied for 5–30 h to a blocked producing well can reestablish economic production in a CHOPS well for many months, even years. Pulsing has been applied in injector wells for improving the efficiency of waterflood patterns and has shown indications of increased oil production and decreased water cut. Additional potential applications include improving the effectiveness of matrix acidizing and diversion.

The mechanism by which pressure pulse technology works is to generate a porosity dilation wave (a fluid displacement wave similar to a tidal wave); this generates pore-scale dilation and contraction so that oil and water flow into and out of pores, leading to periodic fluid accelerations in the pore throats. As the porosity dilation wave moves through the porous medium at a velocity of about 50–100 ft per second (40–80 m per second), the small expansion and contraction of the pores with the passage of each packet of wave energy help unblock pore throats, increase the velocity of liquid flow, overcome part of the effects of capillary blockage, and reduce some of the negative effects of instability due to viscous fingering, coning, and permeability streak channeling.

3.5 Solvent Processes

Solvent-based methods have been developed to move heavy oil. A diluent such as naphtha or light oil may be injected near the pump to reduce the viscosity of the heavy oil and allow it to be more easily pumped. Alternatively, diluent may be added at the surface to facilitate pipeline transport.

Vapor-assisted petroleum extraction (VAPEX) is a nonthermal solvent-based relatively cold (40°C), low-pressure process in which two parallel horizontal wells are drilled with about a 15 ft vertical separation (Yazdani and Maini, 2008). The VAPEX process is similar to the steam-assisted gravity drainage (SAGD) process. However, in the VAPEX process, the viscosity of the oil is reduced through mass transfer phenomena by use of a low-boiling hydrocarbon liquid instead of heat transfer phenomena in the SAGD process. Low-boiling hydrocarbon derivatives, from propane to heptane, are utilized as solvents in the process (Butler and Mokrys, 1991, 1995a,b; Das and Butler, 1998; Butler and Jiang, 2000). The use of noncondensable gases, such as methane, nitrogen, or carbon dioxide (CO_2), in combination with pure solvents to solve both economic and operational problems has also been recommended (Butler et al., 1995). According to this recommendation, there are three advantages to applying a noncondensable gas: (i) The process would become more economically feasible by minimizing the solvent requirements; (ii) solvent diffusion rate would be promoted, which could lead to oil upgrading during the process; and (iii) the higher density difference between the solvent mixture and the heavy oil would improve gravity drainage during the VAPEX process.

The process can be applied in paired horizontal wells, single horizontal wells, or a combination of vertical and horizontal wells. The physics of the VAPEX process are essentially the same as for the SAGD process (Chapters 3 and 4), and the configuration of wells is generally similar. The key benefits are claimed to be (i) significantly lower energy costs, (ii) the potential for in situ upgrading, and (iii) application to thin reservoirs, with bottom water or reactive mineralogy.

In addition to concern over the cost of the solvent, there are the usual concerns about the interaction of the solvent with the reservoir minerals (clay is known to adsorb organic solvents very strongly) and the integrity of the reservoir formations and associated strata (a minor fault can cause loss of the solvent and environmental havoc). There is also the concern over the deposition of asphaltic material and its effect on reservoir permeability. While the function of the solvents might be to extract soluble components of heavy oil, initial contact between the solvent and the oil at a low solvent-to-oil ratio will cause solubilization of the asphaltic constituents (Mitchell and Speight, 1973) with later deposition of these constituents as the solvent-to-oil ratio increases in the later stages of the process.

Because of the slow diffusion of gases and liquids into viscous oils, this approach, used alone, perhaps will be suited only for less viscous oils although preliminary tests indicate that there are micromechanisms that act so that the VAPEX dilution process is not diffusion rate limited and the process may be suitable for the highly viscous heavy oil and tar sand bitumen (Yang and Gu, 2005a,b).

In addition to the VAPEX concept, another concept advocates the use of liquefied petroleum gas as the flooding medium (Alagorni et al., 2015; Ortega et al., 2017). Miscible liquefied petroleum gases such as ethane, propane, or butane have first-contact miscibility, which means they will be miscible from the first contact with oil. This process uses a slug of propane or other liquefied petroleum gas (2%–5% pore volume) followed by natural gas, inert gas, and/or water. Thus, the solvent will bank oil and water ahead and fully displace all contacted oil. However, the actual amount of the liquefied petroleum gas must be carefully monitored to avoid precipitation of the asphaltene constituents.

Hybrid approaches that involve the simultaneous use of several technologies are evolving and will see greater applications in the future. Some of the evolving options that will be tried at the field scale in the next decade include the following:

1. A mixture of steam and miscible and noncondensable hydrocarbon derivatives is being field-tested as a hybrid SAGD-VAPEX approach, with apparent good success and reduction of steam-oil ratios.
2. Single horizontal laterally offset wells can be operated as moderate pressure cyclic steam stimulation wells in combination with SAGD pairs to widen the steam chamber and reduce steam-oil ratios by about 20%.
3. Simultaneous CHOPS and SAGD with CHOPS are used in offset wells until steam breakthrough occurs. Then, the CHOPS wells are converted to slow gas and hot-water (or steam) injection wells to control the process. The high-permeability zones generated by CHOPS should accelerate the SAGD recovery process.
4. Incorporating pressure pulse technology along with CHOPS has already been field-tested with economic success, and pressure pulse technology has potential applications in other hybrid approaches.
5. Pressure pulse technology may aid in partially stabilizing waterflood through reducing the viscous fingering and coning intensity.

In addition to hybrid approaches, the new production technologies, along with older, pressure-driven technologies, will be used in successive phases to extract more oil from reservoirs, even from reservoirs that have been abandoned after primary exploitation. Old reservoirs can be redeveloped with horizontal wells, even linking up the wells to bypassed oil because of the physics of oil film spreading between water and gas phases. These staged approaches hold the promise of significantly increasing recoverable reserves worldwide, not just in heavy oil cases.

4. Enhanced Oil Recovery Methods

Traditional primary and secondary recovery methods typically recover only one-third of the original oil in place. It is at some point before secondary recovery ceases to remain feasible that tertiary recovery methods (also called enhanced recovery methods) must be applied if further oil is to be recovered (Lake, 1989). For tax purposes, the Internal Revenue Service of the United States has listed the projects that qualify as enhanced oil recovery projects (CFR, 2018; Table 2.7).

Several processes leading to the formation of heavy oil and bitumen have been proposed (Shah et al., 2010). Most heavy oil, extra heavy oil, and tar sand bitumen are thought to be expelled from source rocks as light or medium oil and later converted to higher-boiling components by bacterial degradation in subsurface reservoirs. These oils are asphaltic, dense, and viscous oils having an API gravity typically between 10° and 20° API and a viscosity greater than 100 cp. Extra heavy oil and tar sand bitumen share similar attributes to heavy oil but are yet more dense and viscous with API gravity less than 10° API and viscosity several tens of thousands cp. The elevated viscosity and density of these nonconventional crude oils demand more energy-intensive operations not only for their production and upgrading but also for transportation, and consequently, they are more costly to extract.

To exploit heavy oil and bitumen reservoirs, the crude oil industry has in the past relied on cold production methods, predominantly surface mining. The enhanced oil recovery techniques can be broadly classified into either thermal or nonthermal methods (Table 2.8), and their suitability to a particular project is usually dependent on the reservoir location and oil. More recently, a variety of novel enhanced oil recovery technologies have been developed. Success has varied with recovery levels ranging from as low as 10% (waterflooding), to 70% with steam-assisted gravity drainage (SAGD), potentially to greater than 80% with the *toe-heel-air-injection process* (THAI process) (Chapters 3 and 4).

Enhanced oil recovery (*tertiary oil recovery*) is the incremental ultimate oil that can be recovered from a crude oil reservoir in addition to the oil that can be obtained by primary and secondary recovery methods (Table 2.9; Lake, 1989; Arnarnath, 1999; Al Adasani and Bai, 2011; Kamari and Mohammadi, 2014; CFR, 2018). Enhanced oil recovery is often synonymous with *tertiary recovery*

TABLE 2.7 Examples of Methods Used for Enhanced Oil Recovery (Tertiary Oil Recovery)

Method	Description
Thermal recovery methods	
Steam drive injection	Continuous injection of steam into one set of wells (injection wells) or other injection source to effect oil displacement toward and production from a second set of wells (production wells)
Cyclic steam injection	Alternating injection of steam and production of oil with condensed steam from the same well or wells
In situ combustion	Combustion of part of the oil in the reservoir sustained by injection of air, oxygen-enriched air, oxygen, or supplemental fuel supplied from the surface to displace unburned oil toward producing wells; may include the concurrent, alternating, or subsequent injection of water
Gas flood recovery methods	
Miscible fluid displacement	Injection of gas (e.g., natural gas, enriched natural gas, a liquefied crude oil slug driven by natural gas, carbon dioxide, nitrogen, or flue gas) or alcohol into the reservoir at pressure levels such that the gas or alcohol and reservoir oil are miscible
Carbon dioxide-augmented waterflooding	Injection of carbonated water, or water and carbon dioxide, to increase waterflood efficiency
Immiscible carbon dioxide displacement	Injection of carbon dioxide into an oil reservoir to effect oil displacement under conditions in which miscibility with reservoir oil is not obtained; may include the concurrent, alternating, or subsequent injection of water
Immiscible nonhydrocarbon gas displacement	Injection of nonhydrocarbon gas (e.g., nitrogen) into an oil reservoir, under conditions in which miscibility with reservoir oil is not obtained, to obtain a chemical or physical reaction (other than pressure) between the oil and the injected gas or between the oil and other reservoir fluids; may include concurrent, alternating, or subsequent injection of water
Chemical flood recovery methods	
Surfactant flooding	A multiple slug process involving the addition of surface-active chemicals to water to reduce the capillary forces that trap the oil in the pores of the rock
Microemulsion flooding	Also known as surfactant-polymer flooding

TABLE 2.7 Examples of Methods Used for Enhanced Oil Recovery (Tertiary Oil Recovery)—cont'd

Method	Description
	Involves injection of a surfactant system (e.g., a surfactant, hydrocarbon, cosurfactant, electrolyte, and water) to enhance the displacement of oil toward producing wells
Caustic flooding	Injection of water that has been made chemically basic by the addition of alkali metal hydroxides, silicates, or other chemicals
Polymer-augmented waterflooding	Injection of polymeric additives with water to improve the areal and vertical sweep efficiency of the reservoir by increasing the viscosity and decreasing the mobility of the water injected

and to some extent *improved oil recovery* (IOR) and *advanced oil recovery* (AOR) although these terms also apply to primary and secondary methods. Since that time, enhanced oil recovery methods have focused on recovering the remaining oil from a reservoir that had been depleted of energy during the application of primary and secondary recovery methods. Much of the easy-to-produce oil has already recovered from many oil fields, and producers have attempted several tertiary or enhanced oil recovery techniques that offer prospects for ultimately producing 30%–60% or more of the reservoir's original oil in place.

To understand the phenomenon of enhanced oil recovery, it is helpful to understand the condition in the reservoir after other recovery operations have been exhausted. The oil remaining after conventional recovery operations is retained in the pore space of reservoir rock at a lower concentration than originally existed. In portions of the reservoir that have been contacted or swept by the injection fluid, the residual oil remains as droplets (or ganglia) trapped in either individual pores or clusters of pores. It may also remain as films partly coating the pore walls. Entrapment of this residual oil is predominantly due to capillary and surface forces and to pore geometry.

In the pores of those volumes of reservoir rock that were not well swept by displacing fluids, the oil continues to exist at higher concentrations and may exist as a continuous phase. This macroscopic bypassing of the oil occurs because of reservoir heterogeneity, the placement of wells, and the effects of viscous, gravity, and capillary forces, which act simultaneously in the reservoir. The resultant effect depends upon conditions at individual locations. The higher the mobility of the displacing fluid relative to that of the oil (i.e., the higher the *mobility ratio*), the greater the propensity for the displacing fluid to bypass oil.

TABLE 2.8 General Summary of Enhanced Oil Recovery Methods

Method	Type	Agent	Detail
Nonthermal			
	Chemical flooding		
		Alkaline	
		Emulsion	
		Micellar	
		Polymer	
		Surfactant	
	Miscible displacement		
		Alcohol	
		CO_2 miscible	
		Enriched gas drive	
		N_2 miscible	
		Vaporizing gas drive	
	Immiscible displacement		
		Flue gas	
		Inert gas	
		CO_2 immiscible	
	MEOR		
Thermal			
	Steam		
		Conduction heating	
		Steam flood	
		CSS	
		SAGD	
			VAPEX
			VAPEX/ steam
	Hot water		
	In situ combustion		

TABLE 2.8 General Summary of Enhanced Oil Recovery Methods—cont'd

Method	Type	Agent	Detail
		Forward	
			Dry
			Wet
			THAI
			CAPRI
		Reverse	
	Electric heating		

TABLE 2.9 General Approximation of Oil Recovery by the Various Methods

Type of Oil	Method	Recovery[a]
Conventional crude oil		
	Primary	25%
	Secondary	30%
	EOR target	45%
Heavy oil		
	Primary	5%
	Secondary	5%
	EOR target	90%
Tar sand bitumen		
	Primary	0%
	Secondary	0%
	EOR target	100%

[a]*General approximation.*

Because of fluid density differences, gravity forces cause vertical segregation of the fluids in the reservoir so that water tends to underrun and gas to override the oil-containing rock. These mechanisms can be controlled or utilized only to a limited extent in primary and secondary recovery operations.

For example, during waterflood, the capillary forces that cause the displacement of oil by water can also result in the trapping of residual oil. Particularly important is the faster movement of water through the smaller pore because (1) its smaller diameter increases the capillary force and (2) the oil volume displaced by water is far less. Water moving more rapidly through the small pore will reach a common outlet before all the oil is displaced from the upper large pore. A net capillary force then is exerted on the downstream end of the large pore, which can be likened to closing a back door. Further displacement of oil ceases, trapping oil between the two interfaces. Thus, most of the emphasis in developing chemically enhanced methods has been toward recovering such residual oil as might remain after waterflood.

Certain reservoir types, such as those with very viscous crude oils and some low-permeability carbonate (limestone, dolomite, or chert) reservoirs, respond poorly to conventional secondary recovery techniques. The viscosity (or the API gravity) of crude oil is an important factor that must be considered when heavy oil, extra heavy oil, and tar sand bitumen are recovered from a reservoir. In these reservoirs, it is desirable to initiate *enhanced oil recovery* operations as early as possible. This may mean considerably abbreviating conventional secondary recovery operations or bypassing them altogether. Thermal floods using steam and controlled in situ combustion methods are also used. Thermal methods of recovery reduce the viscosity of the crude oil by heat so that it flows more easily into the production well. Thus, several methods used for enhanced oil recovery are variations of secondary methods with a goal of improving the sweeping action of the invading fluid.

Gas flooding has been the most widely used recovery methods of light, condensate, and volatile oil reservoirs. Although nitrogen (N2) injection has been proposed to increase oil recoveries under miscible conditions favoring the vaporization of light fractions of light oils and condensates, similar to nitrogen injection, hydrocarbon gas injection projects in onshore sandstone reservoirs have made a relatively marginal contribution in terms of total oil recovered. On the other hand, carbon dioxide flooding has been the most widely used enhanced oil recovery method for medium and light oil production in sandstone reservoirs during last decades in both sandstone and carbonate reservoirs (Alvarado and Eduardo Manrique, 2010).

Enhanced recovery begins when secondary oil recovery isn't enough to continue adequate production, but only when the oil can still be extracted profitably (Shah et al., 2010) by methods that, in the United States are defined by the Treasury Department for tax purposes (CFR, 2018). This depends on the cost of the extraction method and the current price of crude oil. When prices are high, previously unprofitable wells are brought back into production, and when they are low, production is curtailed. Tertiary recovery allows another 5%–15% of the reservoir's oil to be recovered. Tertiary oil recovery methods reduce the viscosity of the oil to increase oil production (Guo et al., 2016).

Thermally enhanced oil recovery methods (TEOR) are tertiary recovery techniques that heat the oil and make it easier to extract (Chapter 3). Steam injection is the most common form of thermally enhanced oil recovery and is often done with a cogeneration plant. In this type of cogeneration plant, a gas turbine is used to generate electricity, and the waste heat is used to produce steam, which is then injected into the reservoir. In situ combustion is another form of thermally enhanced oil recovery, but instead of steam, some of the oil is burned to heat the surrounding oil. Occasionally, detergents are also used to decrease oil viscosity as a tertiary oil recovery method. Another method to reduce viscosity is carbon dioxide flooding. However, there is a renewed interest in chemical enhanced oil recovery (Fig. 2.1) because of diminished reserves and advances in surfactant and polymer technology. Greater understanding of the chemical reactions involved has led to good results in the field (Krumrine and Falcone Jr., 1987). Combinations of chemicals may be applied as premixed slugs or in sequence (Mandal, 2015).

The choice of the method and the expected recovery depends on many considerations, economic and technological. Some methods are commercially successful, while others remain largely of academic interest. Only a few recovery methods have been commercially successful, such as steam-injection-based processes in heavy oils (if the reservoir offers favorable conditions for such applications) and miscible carbon dioxide for light oil reservoirs (Thomas, 2008). Methods for improving oil recovery, in particular those concerned with lowering the interstitial oil saturation, have received a great deal of attention both in the laboratory and in the field. From the vast amount of literature on the subject, one gets the impression that it is relatively simple to increase oil recovery beyond secondary (assuming that the reservoir lends itself to primary and secondary recovery), but this is not the case (Thomas, 2008).

The typical flooding techniques (such as waterflooding) of oil recovery are not applicable to the recovery of heavy oil, extra heavy oil, and tar sand bitumen because the viscosity difference between the oil and water is too large, which results in a serious issue known as viscous fingering (Thomas and Farouq Ali, 1999). In the case of waterflooding, the less viscous water tends to bypass once it breaks through the viscous heavy oils because of the high interfacial tension and immiscibility between these two phases. This fingering instability thus results in poor sweep and displacement efficiencies. In order to prevent this phenomenon, different chemicals are incorporated to modify the properties of injected water.

The injection of chemicals (*chemical flooding*) involves the use of injected chemicals to decrease interfacial tension and improve sweep efficiency. The three main groups are (i) alkaline flooding, (ii) surfactant flooding, and (iii) polymer flooding although they can be combined to obtain the best characteristics of each. Reservoir characteristics place a particular restriction as carbonates and clays absorb the chemicals. More recent developments include the use of emulsions, foams, and microbes, the latter producing surfactants in situ as a result of their normal biological processes.

Moreover, chemical injection is mostly used for enhanced recovery of heavy oil but is not very suitable for immobile bitumen in tar sand formations. In addition, instead of reducing the oil viscosity, which is the main mechanism of the thermal methods, lowering the interfacial tension, mobility control, and wettability alteration are the alternative solutions to improve the mobility of viscous oils in all chemical-based recovery technologies.

For rocks exhibiting a strong wettability preference for a specific phase, experiments have shown that flow characteristics depend on whether the wetting or nonwetting phase saturation is increasing. The flow dependence on the saturation direction is contributed largely to nonwetting phase entrapment and leads to the development of flow properties for imbibition and drainage flows. One of the flow characteristics that plays a crucial role in flow prediction is the relative permeability. Relative permeability relations should reflect the saturation direction effects for imbibition and drainage flows. As a result, relative permeability should be a function of saturation and also a function of saturation history that describes the way that saturation is approached (Furati, 1998).

In particular, when immiscible water and oil are in contact, an interface will form between the two phases together with the inevitable interfacial tension. The high interfacial is responsible for the resistance of the flow of residual oils, and thus, reduction of the interfacial tension is a key factor in the mobility improvement of heavy oils.

4.1 Alkaline Flooding

In alkaline flooding (*caustic flooding*), an aqueous solution of an alkaline chemical, most usually a sodium hydroxide, sodium carbonate, or sodium orthosilicate, is injected in a slug form. The alkali reacts with acidic components of the crude oil and generates the surfactant in situ; therefore, a sufficiently high organic acid content is necessary (Cooke et al., 1974; Chiwetelu et al., 1994; Khulbe et al., 1996; Mandal, 2015). The alkaline chemical reacts with certain types of oils, forming surfactants inside the reservoir. However, the process is not suitable for use in carbonate formations due to abundance of calcium that can react with the alkali to form precipitates that can damage the formation.

Eventually, the surfactants reduce the interfacial tension between oil and water and trigger an increase in oil production. However, alkaline flooding is not recommended for carbonate reservoirs because of the abundance of calcium: the mixture between the alkaline chemical and the calcium ions can produce hydroxide precipitation that may damage the formation.

Alkaline flooding is one of several chemical-based enhanced oil recovery (CEOR) methods that are an advancement on conventional secondary waterflooding operations. Chemical-based enhanced oil recovery techniques produce fewer amounts of greenhouse gases than thermal-based enhanced oil recovery approaches. A modification to the process is the addition of surfactant and polymer to the alkali, giving rise to an alkaline-surfactant polymer (ASP) enhanced oil recovery method, essentially a form of micellar-polymer flooding.

The alkaline agents aid in the displacement of crude oil by raising the pH of the injected flood water. The alkali reacts with acidic components in crude oil forming an in situ surfactant at the oil/brine interface. This mixture then mobilizes the crude oil and removes it from the pore spaces in the reservoir. Alkaline flooding has been used only in reservoirs containing specific types of high-acid crude oils. Although surfactant flooding is expensive, it has been used in a few large-scale projects and has been demonstrated to have excellent potential for improving the recovery of low-to-moderate viscosity oils. Chemically enhanced oil recovery is commercially available under limited conditions, determined by reservoir characteristics, including depth, salinity, and pH. The high cost of chemicals and reservoir characterization studies need to be reduced, to allow expanded use of chemical enhanced oil recovery methods before full commercialization can take place.

The addition of silicates is an enhancement to alkaline flooding insofar as the silicates play two major functions: (i) as a buffer, maintaining a constant high pH level to produce a minimum interfacial tension, and (ii) by improving surfactant efficiency through the removal of hardness ions from reservoir brines, thus reducing adsorption of surfactants on rock surfaces.

Alkaline-surfactant-polymer flooding takes advantage of the individual alkali, surfactant, and polymer methods. The major mechanisms are interfacial tension reduction and improved reservoir sweep. Field results are encouraging—recovery factors on the order of 25%–30% v/v of the oil in place.

4.2 Surfactant Flooding

Chemical-based enhanced oil recovery methods include polymer flooding, surfactant flooding, and alkaline flooding processes; the most widely applicable of which is surfactant flooding (Mandal, 2015). Polymer flooding is commercially available but too expensive for most applications.

Surfactant flooding refers to the recovery method in which a small amount of surfactant is added to the aqueous injection fluid to displace the oil. The surfactants are a blend of surface-active substances that are usually amphiphilic compounds (e.g., crude oil sulfonates). The rationale behind the use of surfactant is to recover the residual oil trapped by capillary pressure, which is the major factor controlling the oil-water distribution in porous reservoir rocks (Fletcher et al., 2015).

With surfactant flooding, surface-active agents (such as crude oil sulfonate derivatives) are mixed with other chemicals such as alcohol and salt in water and injected into the formation to mobilize the crude oil. These surface-active materials are effective in lowering interfacial tension between oil and water. The major difficulty in the past has been due to excessive surfactant loss to the porous medium, and the presence of clay minerals can be particularly problematic, and therefore, good understanding of the reservoir conditions is essential.

The blend of the surface-active materials has the potential to reduce interfacial tension between oil and water because of its amphiphilic character, boosting the dispersion of oil in water, making oil droplets more readily deformed and easing the flow through pores, lowering the threshold pressure and residual oil saturation in the oil-bearing formation upon injection, thereby increasing the oil production rate. In some cases, cosurfactants such as alcohol and water-soluble polymers are injected to improve the interactions between surfactant and sandstone rocks. However, the porous medium geology in addition to the reservoir pore wettability greatly influences the performance of surfactant-based flooding for enhanced oil recovery. Besides, the slug size and surfactant concentration are limited by cost. Additional difficulties facing surfactant flooding are (i) excessive surfactant loss to the porous medium of the reservoir incurring more cost, (ii) treatment and disposal of emulsions, and (iii) the adsorption and reactions of surfactant with rock.

4.3 Carbon Dioxide Flooding

Carbon dioxide (CO_2) flooding is a process whereby carbon dioxide is injected into an oil reservoir in order to increase output when extracting oil (Mandal, 2015). Ideally, carbon dioxide flooding is particularly effective in reservoirs deeper than 2,000 ft, where carbon dioxide will be in a supercritical state, with the oil gravity on the order of greater than 22–25°. Carbon dioxide flooding is not effected by the lithology of the reservoir area but simply by the reservoir characteristics. First tried in 1972 in Scurry County, Texas, carbon dioxide injection has been used successfully throughout the Permian Basin of West Texas and Eastern New Mexico and is now being pursued to a limited extent in Kansas, Mississippi, Wyoming, Oklahoma, Colorado, Utah, Montana, Alaska, and Pennsylvania.

The use of carbon dioxide in the enhanced production of oil has been gaining prominence in recent years, partly due to the possibility of simultaneous sequestration of carbon dioxide. Carbon dioxide is a particularly attractive displacing agent because it has a relatively low minimum miscibility pressure in a wide range of crude oils. Strictly speaking, carbon dioxide is not truly miscible with the oil; instead, the gas extracts light-to-intermediate components (C_5–C_{30}) from the reservoir oil and develops miscibility after multiple contacts. This leads to lowering of both the surface tension and viscosity of the oil to improve displacement.

Because of the very low viscosity of carbon dioxide, mobility control is poor, and the gas tends to finger and break through to the producer well, leaving large areas of the reservoir unswept. To counter this, alternating volumes of gas and water (or brine) can be injected as slugs (water-alternating-gas process,

WAG process). This approach tends to reduce the viscous instabilities. The main problems associated with carbon dioxide miscible flooding are the high gas requirements and the cost of repressurizing recycled gas. Asphaltene precipitation may also be a problem in some cases.

Carbon dioxide flooding is commonly used to recover oil from reservoirs in which the initial pressure has been depleted through primary production and possibly waterflooding (Orr Jr et al., 1982). Water is injected into the reservoir until pressure is restored to a desired level, and then, carbon dioxide is introduced into the reservoir through these same injection wells. As the carbon dioxide is forced into the reservoir, a zone of miscible carbon dioxide and low-boiling hydrocarbon derivatives forms a front that is soluble with the oil, making it easier to move toward production wells.

If a well has been produced before and has been designated suitable for carbon dioxide flooding, the first action is to restore the pressure within the reservoir to one suitable for production. This is done by injecting water (with the production well shutoff) that will restore pressure within the reservoir to a suitable pressure for carbon dioxide flooding. Once the reservoir is at this pressure, the next step is to inject the carbon into the same injection wells used to restore pressure. The carbon dioxide gas is forced into the reservoir and is required to come into contact with the oil. This creates a miscible zone that can be moved easier to the production well. The carbon dioxide injection may be alternated with more water injection, and the water acts to sweep the oil toward the production zone.

The initial carbon dioxide slug is typically followed by alternate water and carbon dioxide injection—the water serving to improve sweep efficiency and to minimize the amount of carbon dioxide required for the flood. Production is from an oil bank that forms ahead of the miscible front. As reservoir fluids are produced through production wells, the carbon dioxide reverts to a gaseous state and provides a "gas lift" similar to that of original reservoir natural gas pressure.

Until recently, most of the carbon dioxide used for enhanced oil recovery has come from naturally occurring reservoirs. But new technologies are being developed to produce carbon dioxide from industrial applications such as natural gas processing, fertilizer, ethanol, and hydrogen plants in locations where naturally occurring reservoirs are not available. One demonstration at the Dakota Gasification Company's plant in Beulah, North Dakota, is producing carbon dioxide and delivering it by a new 204 mi pipeline to the Weyburn oil field in Saskatchewan, Canada. EnCana, the field's operator, is injecting the carbon dioxide to extend the field's productive life, hoping to add another 25 years and as much as 130 million barrels of oil that might otherwise have been abandoned.

Additional work has examined potential improvements in carbon dioxide-enhanced oil recovery technologies beyond the state of the art that can further increase this potential. This work evaluating the potential of improvements in oil recovery efficiency for carbon dioxide-enhanced oil recovery illustrates that the wide-scale implementation of next-generation carbon dioxide-enhanced oil recovery technology advances has the potential to increase domestic oil recovery efficiency from about one-third to over 60%.

The presence of an oil-bearing transition zone beneath the traditionally defined base (oil-water contact) of an oil reservoir is well established. What is now clear is that under certain geologic and hydrodynamic conditions, an additional residual oil zone (ROZ) exists below this transition zone, and this resource could add another 100 billion barrels of oil resource in place in the United States, and an estimated 20 billion barrels could be recoverable with state-of-the-art carbon dioxide-enhanced oil recovery technologies.

Large volumes of technically recoverable domestic oil resources remain undeveloped and are yet to be discovered in the United States, and this potential associated with carbon dioxide-enhanced oil recovery represents just a portion, albeit large, of this potential. Undeveloped domestic oil resources still in the ground (in-place) total 1124 billion barrels. Of this large in-place resource, 430 billon barrels is estimated to be technically recoverable. This resource includes undiscovered oil, "stranded" light oil amenable to carbon dioxide-enhanced oil recovery technologies, unconventional oil (deep heavy oil and deep tar sand formations), and new crude oil concepts (residual oil in reservoir transition zones).

Cyclic carbon dioxide stimulation, also known as the "huff-and-puff" method, is a single-well operation, which is developing as a method of rapidly producing oil (Mandal, 2015). Similar to the cyclic steam process, carbon dioxide is injected into an oil reservoir after which the well is shut in for a time (providing for a *soak period*) and the well is then opened, allowing the oil and fluids to be produced. The dissolving of the carbon dioxide in the oil reduces the oil's viscosity and causes it to swell, allowing the oil to flow more easily toward the well. The process can also be used in heavy oil reservoirs by high-pressure injection of carbon dioxide to facilitate miscibility between the oil and carbon dioxide and in cases where thermal methods are not feasible.

Miscible carbon dioxide-enhanced oil recovery is a multiple-contact process, involving the injected carbon dioxide and the oil in the reservoir (Chukwudeme and Hamouda, 2009). During this multiple-contact process, carbon dioxide will vaporize the lighter oil fractions into the injected carbon dioxide phase, and carbon dioxide will condense into the reservoir's oil phase. This leads to two reservoir fluids that become miscible (mixing in all parts), with favorable properties of low viscosity, a mobile fluid, and low interfacial tension.

The primary objective of miscible carbon dioxide-enhanced oil recovery is to remobilize and dramatically reduce the after waterflooding residual oil saturation in the reservoir's pore space.

When insufficient reservoir pressure is available or the oil composition in the reservoir is less favorable (i.e., heavy oil), the injected carbon dioxide will not become miscible with the reservoir's oil. As such, another oil displacement mechanism, immiscible carbon dioxide flooding, occurs. The main mechanisms involved in immiscible carbon dioxide flooding are (i) oil-phase swelling, as the oil becomes carbon dioxide saturated with carbon dioxide; (ii) viscosity reduction of the swollen oil and carbon dioxide mixture; (iii) extraction of lighter hydrocarbon into the carbon dioxide phase; and (iv) fluid drive plus pressure.

This combination of mechanisms enables a portion of the reservoir's remaining oil to be mobilized and produced. In general, immiscible carbon dioxide-enhanced oil recovery is less efficient than miscible carbon dioxide-enhanced oil recovery in recovering the oil remaining in the reservoir.

4.4 Nitrogen Flooding

Nitrogen flooding (nitrogen injection) can be used to recover light oil that is capable of absorbing added gas under reservoir conditions, which are low in methane, and at least 5000ft deep to withstand the high injection pressure necessary for the oil to mix with the nitrogen without fracturing the producing formation (Mandal, 2015). This process is similar to the carbon dioxide process in principle although miscibility of nitrogen is poorer than carbon dioxide. At high pressure, the lower-boiling components (C_2–C_6) in the oil are vaporized to generate miscibility. However, the necessity of high pressure limits the process to reservoirs at a depth of at least 4500ft.

When nitrogen is injected into a reservoir, it forms a miscible front by vaporizing lighter oil components. As the front moves away from the injection wells, its leading edge goes into solution or becomes miscible, with the reservoir oil. Continued injection moves the bank of displaced oil toward production wells. Water slugs are injected alternately with the nitrogen to increase the sweep efficiency and oil recovery. Nitrogen can be manufactured on-site at relatively low cost by extraction from air by cryogenic separation, and being totally inert, it is noncorrosive.

In *nitrogen-carbon dioxide flooding*, because of its lower cost, the nitrogen can be used in a carbon dioxide flood to displace the CO_2 slug and its oil bank.

4.5 Polymer Flooding

Polymer flooding is an enhanced oil recovery method that uses polymer solutions to increase oil recovery by decreasing the water/oil mobility ratio by increasing the viscosity of the displacing water (Mandal, 2015). Water-soluble polymer solutions such as polysaccharide derivatives and polyacrylamide derivatives are injected into the oil-bearing formation to improve the water-oil mobility ratio, increase viscosity of water, reduce the permeability of the aqueous phase, and improve displacement sweep efficiency in order to enhance oil recovery. The method is sometimes referred to as polymer-augmented waterflooding and is applied as a slug process and driven using dilute brine.

The method produces additional oil by improving the displacement efficiency and increasing the volume of reservoir that is contacted by increasing the viscosity of the water. This reduces the probability of the flood breaking through to the production well while also producing oil at a higher rate. Polymer flooding has considerably higher oil recovery potentials than waterflooding—typically on the order of 6%–12% higher, which gives a 40%–50% recovery factor of initial oil in place (IOIP) after application.

Polymer flooding is used under certain reservoir conditions that lower the efficiency of a regular waterflood, such as fractures or high-permeability regions that channel or redirect the flow of injected water, or heavy oil that is resistant to flow. Adding a water-soluble polymer to the waterflood allows the water to move through more of the reservoir rock, resulting in a larger percentage of oil recovery. Polymer gel is also used to shut off high-permeability zones. In the process, the volumetric sweep is improved, and the oil is more effectively produced. Often, injectivity will be one of the critical factors. The polymer solution should therefore be a non-Newtonian and shear thinning fluid, that is, the viscosity decreases with increasing shear rate.

Reservoir permeability needs to be higher than for gas displacement techniques although can be lower than that necessary for thermal methods. Recovery levels, however, are only a little higher than that achieved with just water, and so, thorough understanding of the reservoir is necessary to make the process profitable. Loss of polymer to the porous medium, particularly in reservoirs with high clay content, is particularly problematic as can be polymer degradation.

In the process, a water-soluble polymer is added into the floodwater. This increases the viscosity of water. There are three potential ways in which a polymer flooding makes the oil recovery process more efficient: (i) through the effects of polymers on fractional flow, (ii) by decreasing the water/oil mobility ratio, and (iii) by diverting injected water from zones that have been swept. The most important preconditions for polymer flooding are reservoir temperature and the chemical properties of reservoir water. At high temperature or with high salinity in reservoir water, polymer cannot be kept stabile and polymer concentration will lose most of its viscosity.

Polymer flooding has been found to be suitable for reservoirs with high mobile oil saturation and moderate reservoir heterogeneity. Polymer flooding of heavy oil reservoirs is gaining some feasibility and attention with the introduction of horizontal producer well and current rise in oil prices. However, the technique is limited by polymer degradation in the reservoir and loss of polymer to porous matrix of the reservoir, and for heavy oil, a higher concentration of polymer solution is required to reduce the mobility ratio.

4.6 Micellar Flooding

Micellar flooding refers to a fluid injection process in which a stable solution of oil, water, and one or more surfactants along with electrolytes of salts is injected into the formation and is displaced by a mobility buffer solution. The main components of this method are a microemulsion slug (also known as a micellar slug) and a polymer slug. These two slugs are driven using brine. Microemulsions are surfactant-stabilized, oil-water dispersions with small drop size distributions on the order of 10^{-4}–10^{-6} mm that are miscible with reservoir oil and water. The two chemical slugs are designed to promote very low interfacial to increase the mobility of the oil. Recovery factors have ranged between 35% and 50% v/v of the oil in place. However, the downside is the requirement of small well spacing and the considerable delay in response. Moreover, the technique is unsuitable for reservoirs with high salinity, temperature, and clay content.

The *micellar-polymer flooding* method uses the injection of a micellar slug into a reservoir (Mandal, 2015). Since micellar solution makes fluids miscible in the reservoir, almost 100% of oil can be displaced especially in the presence of alkaline (sodium carbonate). However, due to reservoir rock nonuniformity in the field, the amount of oil recovered is reduced. The main objective of micellar injection is to reduce interfacial tension to enhance oil recovery. Micellar solutions are mixtures of surfactants, cosurfactants, electrolytes, hydrocarbon, and water. Surfactants are substances known as surface-active agents, such as soap. Cosurfactants are used for stability such as alcohols. Electrolytes are salts used to control viscosity and interfacial tension such as sodium chloride (NaCl) or ammonium sulfate [$(NH_4)_2SO_4$] (Alagorni et al., 2015).

This method is one of the most efficient enhanced oil recovery methods but is also one of the most costly to implement. The slug is a solution usually containing a mixture of a surfactant, cosurfactant, alcohol, brine, and oil that acts to release oil from the pores of the reservoir rock much as a dishwashing detergent releases grease from dishes so that it can be flushed away by flowing water. As the micellar solution moves through the oil-bearing formation in the reservoir, it releases much of the oil trapped in the rock. To further enhance production, polymer-thickened water for mobility control (as described in the polymer flooding process) is injected behind the micellar slug. Here again, a buffer of fresh water is normally injected following the polymer and ahead of the drive water to prevent contamination of the chemical solutions. This method has one of the highest recovery efficiencies of the current enhanced oil recovery methods, but it is also one of the most costly to implement.

4.7 Hydrocarbon Miscible Flooding

A number of gas drive methods make use of light hydrocarbon derivatives, the simplest of which is analogous to carbon dioxide and nitrogen flooding except that methane or liquefied petroleum gas (e.g., propane) is used. Enriched (or condensing) gas drive consists of injecting slugs of methane enriched with higher-molecular-weight hydrocarbon derivatives (C_2–C_6) dispersed by larger methane slugs. The enriching components are transferred from the gas to the oil, increasing miscibility but also decreasing the viscosity of the oil. The inverse of this process (vaporization gas drive) consists of injecting dry methane at high pressure to vaporize the C_2–C_6 components. In this case, the expanding vapor increases the mobility of the reservoir. In order to contain the high pressure of vaporizing gas drive, the process is limited to reserves at a depth of at least 5000 ft.

4.8 Microbial Enhanced Oil Recovery

Microbial enhanced oil recovery (MEOR) is one of the tertiary enhanced oil recovery methods that utilizes microorganisms and their bioproducts to increase oil production from an otherwise sluggish reservoir or deposit (Lazar et al., 2007; Al-Sulaimani et al., 2011; Sun et al., 2011). While microbial enhanced oil recovery is not specially a refinery process, it can play a role in refining insofar a change made to the heavy oil prior to and during the recovery process that can have a beneficial effect on the refinability of the oil.

The process relies on microbes to ferment hydrocarbon derivatives and produce a by-product that is useful in the recovery of oil (Monticello and Haney III, 1996). Nutrients such as sugars, nitrates, or phosphates are regularly injected to stimulate the growth of the microbes, which are indigenous to some reservoirs, and aid their performance. The microbes then generate surfactants and carbon dioxide that help to displace the oil in a similar way to other displacement methods. Since growth occurs at exponential rates, the process quickly generates considerable surfactant in a cost-effective manner.

Microbial enhanced oil recovery processes involve the use of reservoir microorganisms or specially selected natural bacterial to produce specific metabolic events that lead to enhanced oil recovery (Tables 2.10 and 2.11). However, some bacteria produce polysaccharides that reduce the permeability of the water-filled pores of the reservoir rock, and this effectively forces injected water into the oil-filled pores, pushing the oil out. Also, some species produce carbon dioxide gas that helps to increase pressure within the rock pores, forcing out the oil, and other species produce surfactants or other substances that reduce the viscosity of the oil. The processes that facilitate oil production are complex and may involve multiple biochemical processes.

Microbial biomass or biopolymers may plug high-permeability zones and lead to a redirection of the waterflood, produce surfactants that lead to increased mobilization of residual oil, increase gas pressure by the production of carbon dioxide, or reduce the oil viscosity due to digestion of large molecules. Referred to as bacterial plugging, this fluid-flow modification is achieved by an increase in microbial cell mass within the reservoir. Production in biomass can be achieved by stimulating either indigenous microbial populations or injected microorganisms with growth nutrients. The injected nutrient and microbes preferentially go into the high-permeability zones of the reservoir. As cell growth occurs, the biomass selectively plugs these zones to a greater extent than the moderate- or low-permeability zones. This results in an increase in sweep

TABLE 2.10 Types of Microbial Processes for Oil Recovery

Process	Production Problem	Types of Activity or Product Needed
Wellbore cleanup (improve oil drainage into wellbore)	Paraffin deposits Scale deposits	Emulsifiers, biosurfactants Solvents, acids, hydrocarbon degradation
Well stimulation (stimulate release of oil entrapped by capillaries and brine)	Formation damage Pore damage High water production	Gas, acids, solvents Biosurfactants Biomass and polymer production
Enhanced waterflood (reduce permeability variation and block water channels)	Poor displacement efficiency Poor sweep efficiency Scoring	Biosurfactants, solvents, polymers Biomass and polymer production Nitrate reduction

efficiency by forcing the injected water to pass through previously bypassed zones of the reservoir (Bryant et al., 1998).

Overall, the use of microbes in oil recovery operations has the advantage that microbes do not consume large amounts of energy, but with increasing subsurface depth, temperature appears to be the principal factor limiting microbial life, besides availability of suitable nutrients. They are also susceptible to salinity that limits their use. Chemically, a slow anaerobic degradation process takes place that leads to the sequential removal of different classes of compounds from the oil. Removal of n-alkanes generates methane, which could be artificially accelerated to provide a relatively clean gaseous fuel from oil fields where recovery by alternative methods would be difficult. This method does not release significant amounts of carbon dioxide that would be generated through mining and upgrading, although the methane gas generated for use as a fuel is itself a greenhouse gas.

Microbial enhanced oil recovery has become an important, fast developing tertiary recovery method that uses microorganisms or their metabolites to enhance the recovery of heavy oil (Banat, 1995; Xu et al., 2009; Gao and Zekri, 2011; Sivasankar and Kumar, 2017). Microbial enhanced oil recovery is different from the more conventional enhanced oil recovery methods, such

TABLE 2.11 Microbial Products and Their Contribution to Enhanced Oil Recovery

Microbial	Effect
Acids	Modification of reservoir rock
	Improvement of porosity and permeability
	Reaction with calcareous carbon dioxide production
Biomass	Selective or nonselective plugging
	Emulsification through adherence to hydrocarbons
	Modification of solid surfaces, for example, wetting
	Degradation and alteration of oil
	Reduction of oil viscosity and oil pour point
	Desulfurization of oil
Gases (CO_2, CH_4, H_2)	Reservoir repressurization
	Oil swelling
	Viscosity reduction
	Increase of permeability due to solubilization of carbonate rocks by CO_2
Solvents	Oil dissolution
Surface-active agent	Lowering interfacial tension
Polymers	Mobility control
	Selective or nonselective plugging

as carbon dioxide injection, steam injection, chemical surfactant, and polymer flooding, in that it involves injecting live microorganisms and nutrients into the reservoir so that bacteria and their metabolic products mobilize the heavy oil. The process is considered to be a more environmentally friendly method since it does not involve any toxic chemicals and it is easy to carry out in fields since it does not need any modifications of existing water-injection amenities (Lazar et al., 2007; Brown, 2010). Microbial enhanced oil recovery takes place by different mechanisms, such as reduction of oil-water interfacial tension and alteration of wettability by surfactant production, selective plugging by microorganisms and their metabolites, oil viscosity reduction by gas production or degradation or biotransformation of long-chain saturated hydrocarbon derivatives, and production of acids that improves absolute permeability by dissolving minerals in the rock (Nielsen et al., 2010). The microbial metabolic products include biosurfactants, biopolymers, acids, solvents, gases, and enzymes. The bacteria used for microbial enhanced oil recovery are typically hydrocarbon-utilizing, nonpathogenic, and naturally occurring in many reservoirs (Almeida et al., 2004).

In microbial enhanced oil recovery processes, microbial technology is exploited in oil reservoirs to improve recovery (Clark et al., 1981; Stosur, 1991; Banat, 1995; Bryant et al., 1998). From a microbiologist's perspective, microbial enhanced oil recovery processes are somewhat akin to in situ bioremediation processes. Injected nutrients, together with indigenous or added microbes, promote in situ microbial growth and/or generation of products that mobilize additional oil and move it to producing wells through reservoir repressurization, interfacial tension/oil viscosity reduction, and selective plugging of the most permeable zones (Bryant et al., 1989; Bryant and Lindsey, 1996). Alternatively, the oil-mobilizing microbial products may be produced by fermentation and injected into the reservoir.

Organisms that participate in oil recovery produce a variety of fermentation products, for example, carbon dioxide, methane, hydrogen, biosurfactants, and polysaccharides from crude oil, pure hydrocarbon derivatives, and a variety of nonhydrocarbon substrates. Organic acids produced through fermentation readily dissolve carbonates and can greatly enhance permeability in limestone reservoirs, and attempts have been made to promote anaerobic production.

Specific microorganisms produce biosurfactants that reduce oil/water interfacial tension (IFT) and cause formation of stable oil-in-water emulsions. In addition, these surfactants increase the relative permeability of the reservoir to oil by changing the wettability of the reservoir core to a more water-wet condition. Microorganisms also produce acids, primarily low-molecular-weight fatty acids, which cause rock dissolution and decrease interfacial tension between the oil and reservoir brine. In situ acid production by microorganisms can be used to treat various production problems, such as formation damage, low oil relative permeability, trapped oil due to capillary forces, and paraffin and scaling problems (Bryant et al., 1998).

The microbial enhanced oil recovery process may modify the immediate reservoir environment in a number of ways that could also damage the production hardware or the formation itself. Certain sulfate reducers can produce hydrogen sulfide, which can corrode pipeline and other components of the recovery equipment. Thus, despite numerous microbial enhanced oil recovery tests, considerable uncertainty remains regarding process performance. Ensuring success requires an ability to manipulate environmental conditions to promote growth and/or product formation by the participating microorganisms. Exerting such control over the microbial system in the subsurface is itself a serious challenge. In addition, conditions vary from reservoir to reservoir, which calls for reservoir-specific customization of the microbial enhanced oil recovery process, and this alone has the potential to undermine microbial process economic viability.

Studies have shown that several microbially produced biosurfactants compare favorably with chemically synthesized surfactants.

This technology requires consideration of the physicochemical properties of the reservoir in terms of salinity, pH, temperature, pressure, and nutrient availability (Khire and Khan, 1994a,b). Only bacteria are considered promising candidates for microbial enhanced oil recovery. Molds, yeasts, algae, and protozoa are not suitable due to their size or inability to grow under the conditions present in reservoirs. Many crude oil reservoirs have high concentrations of sodium chloride (Jenneman, 1989) and require the use of bacteria that can tolerate these conditions (Shennan and Levi, 1987). Bacteria producing biosurfactants and polymers can grow at sodium concentrations up to 8% and selectively plug sandstone to create a biowall to recover additional oil (Raiders et al., 1989).

Microbial enhanced oil recovery differs from chemical enhanced oil recovery in the method by which the enhancing products are introduced into the reservoir. Thus, in oil recovery by the *cyclic microbial method*, a solution of nutrients and microorganisms is introduced into the reservoir during injection. The injection well is the shut for an incubation period allowing the microorganisms to produce carbon dioxide gas and surfactants that assist in mobilization of the oil. The well is then opened, and oil and oil products resulting from the treatment are produced. The process is repeated as often as oil can be produced from the well. Oil recovery by *microbial flooding* also involved the use of microorganisms, but in this case, the reservoir is usually conditioned by a water flush after which a solution of microorganisms and nutrients is injected into the formation. As this solution is pushed through the reservoir by water drive, gases and surfactants are formed; the oil is mobilized and pumped through the well. However, even though microbes produce the necessary chemical reactions in situ whereas surface injected chemicals may tend to follow areas of higher permeability, resulting in decreased sweep efficiency, there is need for caution and astute observation of the effects of the microorganisms on the reservoir chemistry.

The mechanism by which microbial enhanced oil recovery processes work can be quite complex and may involve multiple biochemical processes. In selective plugging approaches, microbial cell mass or biopolymers plug high-permeability zones and lead to a redirection of the waterflood. In other processes, biosurfactants are produced in situ that leads to increased mobilization of residual oil. In still other processes, microbial production of carbon dioxide and organic solvents reduces the oil viscosity as the primary mechanism for enhanced oil recovery. In particular, for both recovery and refining operations, the asphaltene constituents have drawn considerable attention due to problems caused by their detrimental effects in the extraction, transportation, and processing of residua because of their viscous and flocculating nature and their relative resistance to biodegradation following spill of crude oil that contains the asphaltene constituents (Speight and Arjoon, 2012; Speight, 2014).

The asphaltene constituents have extremely complex and variable molecular structures containing sulfur, nitrogen, and metal elements, such as iron, nickel, and vanadium, with a high molecular weight (Speight, 1994, 2014). The asphaltene constituents are considered to be the products of complex heteroatomic aromatic macrocyclic structures. Breaking the asphaltene constituents into smaller molecules and cutting an internal aliphatic linkage (sulfides, esters, and ethers) of an asphaltene molecule can lead to a reduction in viscosity.

A complex set of multiple biochemical reactions between selected microorganisms and heavy crude oils under controlled conditions have been reported by Premuzic et al. (1999) that led to a significant lowering (24%–40%) of the nitrogen. The reactions are both biocatalyst and crude oil dependent and, in terms of chemical mechanisms, appeared to involve the asphaltene and the associated polar fractions. Asphaltene constituents from a crude oil rich in heavy metals (Castilla crude oil) have been fractionated, and the biocatalytic modifications of these fractionated asphaltene constituents by three different hemoproteins chloroperoxidase (CPO), cytochrome C peroxidase (CCP), and lignin peroxidase (LPO) have been evaluated in both aqueous buffer and organic solvents. However, only the chloroperoxidase-mediated reactions were effective in eliminating the Soret peak in both aqueous and organic solvent systems, and the chloroperoxidase has been reported to be able to alter components in the heavy fractions of crude oil and remove 53% and 27% of total heavy metals (Ni and V, respectively) from petroporphyrin-rich fractions and asphaltene constituents (Mogollón et al., 1998; Premuzic and Lin, 1999; Premuzic et al., 1999).

The conditions for microbial metabolism are supported via injection of nutrients. In some processes, this involves injecting a fermentable carbohydrate into the reservoir. Some reservoirs also require inorganic nutrients as substrates for cellular growth or for serving as alternative electron acceptors in place of oxygen or carbohydrates.

The stimulation of oil production by in situ bacterial fermentation is thought to proceed by one or a combination of the following mechanisms: (i) improvement of the relative mobility of oil to water by biosurfactants and biopolymers; (ii) partial repressurization of the reservoir by methane and carbon dioxide; (iii) reduction of oil viscosity through the dissolution of organic solvents in the oil phase; (iv) increase in reservoir permeability and widening of the fissures and channels through the etching of carbonaceous rocks in limestone reservoirs by organic acids produced by anaerobic bacteria; (v) cleaning the wellbore region through the acids and gas from in situ fermentation in which the gas pushes oil from dead space and dislodge debris that plugs the pores and the average pore size is increased and, as a result, the capillary pressure near the wellbore is made more favorable for the flow of oil; and (vi) selective plugging of highly permeable zones by injecting specific bacteria to improve aerial sweep efficiency.

The target for enhanced oil recovery processes is the quantity of unrecoverable oil in known reservoirs and bitumen and known deposits. One of the major attributes of microbial enhanced oil recovery technologies is its low cost, but there must be the recognition that microbial enhanced oil recovery is a single process. Furthermore, reports on the deleterious activities of microorganisms in the oil field contribute to the skepticism of employing technologies using microorganisms. It is also clear that scientific knowledge of the fundamentals of microbiology must be coupled with an understanding of the geologic and engineering aspects of oil production in order to develop microbial enhanced oil recovery technology.

Finally, recent developments in *upgrading* of heavy oil, extra heavy oil, and tar sand bitumen (Speight, 2011, 2014, 2017) indicate that the near future could see a reduction of the differential cost of upgrading heavy oil. These processes are based on a better understanding of the issues of asphaltene solubility effects at high temperatures, incorporation of a catalyst that is chemically precipitated internally during the upgrading, and improving hydrogen addition or carbon rejection.

5. Oil Mining

Oil mining is the term applied to the surface or subsurface excavation of crude oil-bearing formations for subsequent removal of the heavy oil or bitumen by washing, flotation, or retorting treatments. Oil mining also includes recovery of heavy oil by drainage from reservoir beds to mine shafts or other openings driven into the rock or by drainage from the reservoir rock into mine openings driven outside the tar sand but connected with it by boreholes or mine wells.

Oil mining is not new. Mining of crude oil and tar sand bitumen has occurred in the Sinai Peninsula, the Euphrates Valley, and in Persia prior to 5000 BC. In addition, subsurface oil mining was used in the Pechelbronn oil field in Alsace, France, as early as 1735. This early mining involved the sinking of shafts to the reservoir rock, only 100–200 ft (30–60 m) below the surface, and the excavation of the tar sand in short drifts driven from the shafts. The tar sand was brought to the surface and washed with boiling water to release the bitumen. The drifts were extended as far as natural ventilation permitted. When these limits were reached, the pillars were removed and the openings filled with waste. This type of mining continued at Pechelbronn until 1866, when it was found that oil could be recovered from deeper and more prolific sands by letting it drain in place through mine openings with no removal of sand to the surface for treatment. Nevertheless, mining for crude oil is a new challenge facing the crude oil industry.

Oil mining methods should be applied in reservoirs that have significant residual oil saturation and have reservoir or fluid properties that make production by conventional methods inefficient or impossible. The high well density in improved oil mining usually compensates for the inefficient production caused by reservoir heterogeneity. However, close well spacing can also magnify the deleterious effects of reservoir heterogeneity. If a high-permeability streak exists with a lateral extent that is less than the interwell spacing of conventional wells but is comparable with that of improved oil mining, the channeling is more unfavorable for the improved oil mining method.

Engineering a successful oil mining project must address a number of items because there must be sufficient recoverable resources, the project must be conducted safely, and the project should be engineered to maximize recovery within economic limits. The use of a reliable screening technique is necessary to locate viable candidates. Once the candidate is defined, this should be followed by an exhaustive literature search covering the local geology, drilling, production, completion, and secondary and tertiary recovery operations.

The reservoir properties, which can affect the efficiency of heavy oil, extra have oil, or tar sand bitumen production by mining technology, can be grouped into three classes (Table 2.12). There are also important rock mechanical parameters of the formation in which a tunnel is to be mined and from where all oil mining operations will be conducted. These properties are mostly related to the mining aspects of the operations and not all are of equal importance in their influence on the mining technology. Their relative importance also depends on the individual reservoir.

Many of the candidate reservoirs for application of improved oil mining are those with high oil saturation resulting from the adverse effects of reservoir heterogeneity. Faulting, fracturing, and barriers to fluid flow are features that cause production of shallow reservoirs by conventional methods to be inefficient. Production of heterogeneous reservoirs by underground oil production methods requires consideration of the manner in which fractures alter the flow of fluids.

TABLE 2.12 Reservoir or Deposit That Can Affect the Efficiency of Heavy Oil, Extra Heavy Oil, or Tar Sand Bitumen Production by Mining Technology

Property	Description
Primary properties	Properties that influence fluid flow and fluid storage properties and include rock and fluid properties
	Include porosity, permeability, wettability, crude oil viscosity, and pour point
Secondary properties	Properties that significantly influence the primary properties
	Include pore size distribution, clay type, and clay content
Tertiary properties	Properties that mainly influence oil production operation
	Include fracture breakdown pressure, hardness, and thermal properties as well as the mining operations such as subsidence potential and fault distribution in the formation

In a highly fractured formation with low matrix permeability, the fluid conductivity of the fracture system may be many times that of the matrix rock. In a highly fractured reservoir with low matrix permeability and reasonably high porosity, the fracture system provides the highest permeability to the flow of oil, but the matrix rock contains the greater volume of the oil in place. The rate of the flow of oil from the matrix rock into the fracture system, the extent and continuity of the fracture system, and the degree to which the production wells effectively intersect the fracture system determine the production rate. Special consideration must be given to these factors in predicting production rates in fractured reservoirs. Under favorable circumstances, higher production rates may be achieved in fractured reservoirs by improving mining methods than in less heterogeneous reservoirs. Other reservoirs that are good candidates for oil mining are those that are shallow, have high oil saturation, have a nearby formation that is competent enough to support the mine, and cannot be efficiently produced by conventional methods.

The first step in surface mining is the removal of muskeg and overburden. Muskeg is a water-soaked area of decaying plant material that is 1–3 m thick and lies on top of the overburden material. Before the muskeg can be removed, it must be drained of its water content. The process can take up to 3 years to complete. Once the muskeg has been drained and removed, the overburden must also be removed. Overburden is a layer of clay, sand, and silt that lies directly above the tar sand deposit. Overburden is used to build dams and dikes around the mine and will eventually be used for land reclamation projects. When all of the overburden is removed, the tar sand is exposed and is ready to be mined.

There are two methods of mining currently in use in the tar sand of the Athabasca region of Alberta (Canada). Suncor Energy uses the truck and shovel method of mining, whereas Syncrude uses the truck and shovel method of mining, draglines, and bucket-wheel reclaimers. These enormous draglines and bucket-wheels are being phased out and soon will be completely replaced with large trucks and shovels. The shovel scoops up the tar sand and dumps it into a heavy hauler truck. The heavy hauler truck takes the tar sand to a conveyor belt that transports the tar sand from the mine to the extraction plant. Presently, there are extensive conveyor belt systems that transport the mined tar sand from the recovery site to the extraction plant. With the development of new technologies these conveyors are being phased out and replaced with hydrotransport technology. Hydrotransport is a combination of ore transport and preliminary extraction. After the bituminous sands have been recovered using the truck and shovel method, it is mixed with water and caustic soda to form a slurry and is pumped along a pipeline to the extraction plant. The extraction process thus begins with the mixing of the water and agitation needed to initiate bitumen separation from the sand and clay.

Mine spoils need to be disposed of in a manner that assures physical stabilization. This means appropriate slope stability for the pile against not only gravity but also earthquake forces. Since return of the spoils to the mine excavations is seldom economical, the spoil pile must be designed as a permanent structure whose outline blends into the landscape. Straight, even lines in the pile must be avoided.

Even though estimates of the recoverable oil from the Athabasca deposits are only of the order of 27 billion barrels (27×10^9 bbl) of synthetic crude oil (representing < 10% of the total in-place material), this is, for the Canadian scenario, approximately six times the estimated volume of recoverable conventional crude oil. In addition, the comparative infancy of the development of the alternative options almost ensured the adoption of the mining option for the first two (and even later) commercial ventures.

Underground mining options have also been proposed but for the moment have not been developed because of the fear of collapse of the formation onto any operation/equipment. This particular option should not, however, be rejected out of hand because a novel aspect or the requirements of the developer (which remove the accompanying dangers) may make such an option acceptable. Currently, bitumen is recovered commercially from tar and deposits by a mining technique. This produces tar sand that is sent to the processing plant for separation of the bitumen from the sand prior to upgrading.

The oil mining method of recovery has received considerable attention since it was chosen as the technique of preference for the only two commercial bitumen recovery plants in operation in North America. In situ processes have been tested many times in the United States, Canada, and other parts of the world and are ready for commercialization. There are also conceptual schemes that are a combination of both mining (aboveground recovery) and in situ (nonmining recovery) methods.

The bitumen occurring in tar sand deposits poses a major recovery problem. The material is notoriously immobile at formation temperatures and must therefore require some stimulation (usually by thermal means) in order to ensure recovery. Alternately, proposals have been noted that advocate bitumen recovery by solvent flooding or by the use of emulsifiers. There is no doubt that with time, one or more of these functions may come to fruition, but for the present, the two commercial operations rely on the mining technique. After the tar sand has been recovered using the truck and shovel method, it is mixed with water and caustic soda to form a slurry and is pumped along a pipeline to the extraction plant.

The *hot-water process* (Clark, 1944) is, to date, the only successful commercial process to be applied to bitumen recovery from mined tar sand in North America (Clark, 1944). Many process options have been tested with varying degrees of success, and one of these options may even supersede the hot-water process. The process utilizes the linear and the nonlinear variation of bitumen density and water density, respectively, with temperature so that the bitumen that is more dense than water at room temperature becomes lighter than water at 80°C (180°F). Surface-active materials in the tar sand also contribute to the process. The essentials of the hot-water process involve a conditioning, separation, and scavenging.

Thus, in the hot-water extraction process, the tar sand feed is introduced into a *conditioning* drum. In this step, the tar sand is heated and mixed with water, and agglomeration of the oil particles begins. The conditioning is carried out in a slowly rotating drum that contains a steam-sparging system for temperature control and mixing devices to assist in lump size reduction and a size ejector at the outlet end. The tar sand lumps are reduced in size by ablation and mixing action. The conditioned *pulp* has the following characteristics: (i) solids 60%–85% and (ii) pH 7.5–8.5.

In the conditioning step, also referred to as mixing or pulping, tar sand feed is heated and mixed with water to form a pulp of 60% by weight to 85% by weight solids at 80–90°C (17–196°F). First, the lumps of tar sand as-mined are reduced in size by ablation, that is, successive layers of lump are warmed and sloughed off revealing cooler layers. The conditioned pulp is screened through a double-layer vibrating screen. Water is then added to the screened material (to achieve more beneficial pumping conditions), and the pulp enters the separation cell through a central feed well and distributor. The bulk of the sand settles in the cell and is removed from the bottom as tailing, but the majority of the bitumen floats to the surface and is removed as froth. A middling stream (mostly of water with suspended fines and some bitumen) is withdrawn from approximately midway up the side of the cell wall. Part of the middlings is recycled to dilute the conditioning drum effluent for pumping. Clays do not settle readily and generally accumulate in the middling layer.

The combined froth from the separation cell and scavenging operation contains an average of about 10% by weight mineral material and up to 40% by weight water. The dewatering and demineralizing are accomplished in two stages of centrifuging; in the first stage, the coarser mineral material is removed, but much of the water remains. The feed then passes through a filter to remove any additional large-size mineral matter that would plug up the nozzles of the second stage centrifuges.

Bituminous froth from the hot-water process may be mixed with a hydrocarbon diluent, for example, coker naphtha, and centrifuged. The Suncor process employs a two-stage centrifuging operation, and each stage consists of multiple centrifuges of conventional design installed in parallel. The bitumen product contains 1% by weight to 2% by weight mineral (dry bitumen basis) and 5% by weight to 15% by weight water (wet diluted basis). Syncrude also utilizes a centrifuge system with naphtha diluent.

Environmental regulations in Canada or the United States will not allow the discharge of tailing streams into (i) the river, (ii) onto the surface, or (iii) onto any area where groundwater domains or the river may be contaminated. The tailing stream is essentially high in clays and contains some bitumen, hence the current need for tailing ponds, where some settling of the clay occurs. In addition, an approach to acceptable reclamation of the tailing ponds will have to be accommodated at the time of site abandonment.

The bitumen isolated by the hot-water separation process is then sent for upgrading to a synthetic crude oil.

References

Al Adasani, A., Bai, B., 2011. Analysis of EOR projects and update screening criteria. J. Pet. Sci. Eng. 79, 10–24.

Alagorni, A.H., Bin Yaacob, Z., Nour, A.H., 2015. An overview of oil production stages: enhanced oil recovery techniques and nitrogen injection. Int. J. Environ. Sci. Dev. 6 (9), 693–701.

Almeida, P., Moreira, R., Almeida, R., Guimaraes, A., Carvalho, A., Quintella, C., Esperidia, M., Taft, C., 2004. Selection and application of microorganisms to improve oil recovery. Eng. Life Sci. 4, 319–325.

Al-Sulaimani, H., Joshi, S., Al-Wahaibi, Y., Al-Bahry, S., Elshafie, A., Bemani, A.A., 2011. Microbial biotechnology for enhancing oil recovery: current developments and future prospects. Biotechnol. Bioinformatics Bioeng. 1 (2), 147–158.

Alvarado, V., Eduardo Manrique, M., 2010. Enhanced oil recovery: an update review. Energies 3, 1529–1575.

Arnarnath, A., 1999. Enhanced oil recovery scoping study. Report No. TR-113836Electric Power Research Institute, Palo Alto, CA.

Banat, I.M., 1995. Biosurfactant production and possible uses in microbial enhanced oil recovery and oil pollution remediation. Bioresour. Technol. 51, 1–12.

Barillas, J.L.M., Dutra Jr., T.V., Mata, W., 2008. Improved oil recovery process for heavy oil: a review. Braz. J. Pet. Gas 2 (1), 45–54.

Brown, L.R., 2010. Microbial enhanced oil recovery (MEOR). Curr. Opin. Microbiol. 13, 316–320.

Bryant, R.S., Lindsey, R.P., 1996. World-wide applications of microbial technology for improving oil recovery. In: Proceedings of the SPE Symposium on Improved Oil Recoverypp. 27–134.

Bryant, R.S., Donaldson, E.C., Yen, T.F., Chilingarian, G.V., 1989. Microbial enhanced oil recovery. In: Donaldson, E.C., Chilingarian, G.V., Yen, T.F. (Eds.), Enhanced Oil Recovery II: Processes and Operations. Elsevier, Amsterdam, pp. 423–450.

Bryant, R.S., Bailey, S.A., Stepp, A.K., Evans, D.B., Parli, J.A., Kolhatkar, A.R., 1998. Biotechnology for Heavy Oil Recovery. Paper No. 1998.110Society of Petroleum Engineers, Richardson, TX.

Butler, R.M., Jiang, Q., 2000. Improved recovery of heavy oil by VAPEX with widely spaced horizontal injectors and producers. J. Can. Pet. Technol. 39, 48–56.

Butler, R.M., Mokrys, I.J., 1991. J. Can. Pet. Technol. 30 (1), 97–106.

Butler, R.M., Mokrys, I.J., 1995a. Process and Apparatus for the Recovery of Hydrocarbons From a Hydrocarbon Deposit. United States Patent 5,407,009. April 18.

Butler, R.M., Mokrys, I.J., 1995b. Process and Apparatus for the Recovery of Hydrocarbons From a Hydrocarbon Deposit. United States Patent 5,607,016. March 4.

Butler, R.M., Mokrys, I.J., Das, S.K., 1995. The solvent requirements for VAPEX recovery. In: Paper No. SPE 30293, Proceedings. SPE International Heavy Oil Symposium, Calgary, Alberta, Canada, June 19–21. Society of Petroleum Engineers, Richardson, TX.

CFR, 2018. Internal Revenue Service, Department of the Treasury. United States Code of Federal Regulations. 26 CFR 1.43-2-Qualified Enhanced Oil Recovery Projects. Government of the United States, Washington, DC. https://www.law.cornell.edu/cfr/text/26/1.43-2(accessed 24 January 2018).

Chakma, A., Islam, M.R., Berruti, F. (Eds.), 1991. Enhanced oil recovery. AIChE Symposium Series No. 280. In: 87, American Institute of Chemical Engineers, New York, p. 147.

Charles, D.D., Startzman, R.A., 1996. Horizontal well stream function model for improved waterflood performance and evaluation. J. Pet. Sci. Eng. 16, 109–130.

Chen, X., 2006. Heavy oils, part 1. SIAM News. 39(3).

Chiwetelu, C.L., Neale, G.H., Hornof, V., George, A.E., 1994. Recovery of a Saskatchewan heavy oil using alkaline solutions. J. Can. Pet. Technol. 33 (4), 37–42.

Chugh, S., Baker, R., Telesford, A., Zhang, E., 2000. Mainstream options for heavy oil: part I–cold production. J. Can. Pet. Technol. 39 (4), 31–39.

Chukwudeme, E.A., Hamouda, A.A., 2009. Enhanced Oil Recovery (EOR) by miscible CO2 and water flooding of asphaltenic and non-asphaltenic oils. Energies 2, 714–737.

Clark, K.A., 1944. Trans. Can. Inst. Min. Met. 47, 257.

Clark, J.B., Munnecke, D.M., Jenneman, G.E., 1981. In situ microbial enhancement of oil production. Dev. Ind. Microbiol. 15, 695–701.

Cooke, C.E.J., Williams, R.E., Kolodzie, P.A., 1974. Oil recovery by alkaline flooding. J. Pet. Technol. 26 (12), 1365–1374.

CRA, 2003. Pressure Pulse Technology. vol. 3. Innovative Technology Group, Conestoga-Rovers and Associates, Niagara Falls, NY. No. 1 (January)http://www.craworld.com/en/newsevents/resources/inn_2003_jan.pdf.

Craft, B.C., Hawkins, M.F., 1959. Applied Crude Oil Reservoir Engineering. Prentice-Hall, Englewood Cliffs, NJ.

Curtis, C., Kopper, R., Decoster, E., Guzmán-Garcia, A., Huggins, C., Knauer, L., Minner, M., Kupsch, N., Linares, L.M., Rough, H., Waite, M., 2002. Heavy-oil reservoirs. Oilfield Rev. 14 (3), 42–46.

Das, S.K., Butler, R.M., 1998. Mechanism of the vapor extraction process for heavy oil and bitumen. J. Pet. Sci. Eng. 21, 43–59.

DOE, 1996. Maintaining Oil Production from Marginal Fields: A Review of the Department of Energy's Reservoir Class Program. Panel on the Review of the Oil Recovery Demonstration Program of the Department of Energy, Committee on Earth Resources, Board on Earth Sciences and Resources, Commission on Geosciences, Environment, and Resources, National Research CouncilNational Academy Press, Washington, DC.

Fletcher, P.D., Savory, L.D., Woods, F., Clarke, A., Howe, A.M., 2015. Model study of enhanced oil recovery by flooding with aqueous surfactant solution and comparison with theory. Langmuir 31 (10), 3076–3085.

Frick, T.C., 1962. Crude Oil Production Handbook. vol. II. McGraw-Hill, New York.

Furati, K.M., 1998. History effects on oil recovery efficiency. J. Pet. Sci. Eng. 19, 295–308.

Gao, C.H., Zekri, A., 2011. Applications of microbial-enhanced oil recovery technology in the past decade. Energy Sources, Part A 33 (10), 972–989.

Guo, K., Li, H., Yu, Z., 2016. In-situ heavy and extra-heavy oil recovery: a review. Fuel 185, 886–902.

Jayasekera, A.J., Goodyear, S.G., 1999. The development of heavy oil, extra heavy oil, and tar sand bitumen fields in the UK continental shelf: past, present, and future. SPE 54623.

Jenneman, G.E., 1989. The potential for in-situ microbial applications. Dev. Pet. Sci. 22, 37–74.

Kamari, A., Mohammadi, A.H., 2014. Screening of enhanced oil recovery methods. In: Ambrosio, J. (Ed.), Handbook on Oil Production Research.pp. 285–295(Chapter 10).

Khire, J.M., Khan, M.I., 1994a. Microbially enhanced oil recovery (MEOR). Part 1. Importance and mechanisms of microbial enhanced oil recovery. Enzym. Microb. Technol. 16, 170–172.

Khire, J.M., Khan, M.I., 1994b. Microbially enhanced oil recovery (MEOR). Part 2. Microbes and the subsurface environment for microbial enhanced oil recovery. Enzym. Microb. Technol. 16, 258–259.

Khulbe, K.C., Mass, R.S., MacPhee, J.A., 1996. Separation of acidic fraction from the cold lake bitumen asphaltenes and its relationship to enhanced oil recovery. Fuel Process. Technol. 46, 63–69.

Kovscek, A.R., 2002. Heavy and thermal oil recovery production mechanisms. In: Quarterly Technical Progress Report, Reporting Period: April 1 through June 30, 2002, DOE Contract Number: DE-FC26-00BC15311, July.

Krumrine, P.H., Falcone Jr., J.S., 1987. Beyond alkaline flooding: design of complete chemical systems. In: SPE International Symposium in Oil Field Chemistry 16280.

Kumar, M., 2006. Heavy Oil Recovery—Recent Developments and Challenges. Los Angeles Monthly Petroleum Technology Forum, L.A. Basin Section, SPEhttp://www.laspe.org/petrotech/petrooct10906.html.

Lake, L.W., 1989. Enhanced Oil Recovery. Prentice-Hall, Englewood Cliffs, NJ.

Lake, L.W., Walsh, M.P., 2004. Primary Hydrocarbon Recovery. Elsevier, Amsterdam.

Lazar, I., Petrisor, I., Yen, T.F., 2007. Microbial enhanced oil recovery (MEOR). Pet. Sci. Technol. 25, 1353–1366.

Loughead, D.J., 1992. Lloydminster heavy oil production—why so unusual? In: Ninth Annual Heavy Oil and Oil Sands Technology Symposium, Calgary, Alberta, Canada, March 11.

Mandal, A., 2015. Chemical flood enhanced oil recovery: a review. Int. J. Oil, Gas Coal Technol 9 (3), 241–264.

McCaffrey, W.J., Bowman, R.D., 1991. Recent successes in primary bitumen production. In: HOOS Technical Symposium Challenges and Innovation, Annual Technical Meeting, Petroleum Society of the Canadian Institute of Mining.

Mitchell, D.L., Speight, J.G., 1973. The solubility of asphaltenes in hydrocarbon solvents. Fuel 52, 149.

Mogollón, L., Rodríguez, R., Larrota, W., Ortiz, C., Torres, R., 1998. Biocatalytic removal of nickel and vanadium from petroporphyrins and asphaltenes. Appl. Biochem. Biotechnol. 70/72, 765–777.

Monticello, D.J., Haney III, W.M., 1996. Biocatalytic Process for Reduction of Petroleum Viscosity. United States Patent 5,529,930. June 25.

Nielsen, S.M., Shapiro, A.A., Michelsen, M.L., Stenby, E.H., 2010. 1D simulations for microbial enhanced oil recovery with metabolite partitioning. Transp. Porous Media 85, 785–802.

NPC, 2007. Heavy oil working document. In: Topic Paper #22, NPC Global Oil & Gas Study, National Petroleum Council, Washington, DC, July 18.

Orr Jr., F.M., Heller, J.P., Taber, J.J., 1982. Carbon dioxide flooding for enhanced oil recovery: promise and problems. J. Am. Oil Chem. Soc. 59(10).

Ortega, A., Hernández, A., Puello, J., Marin-Batista, J., 2017. Effect of liquefied petroleum gas (LPG) on heavy oil recovery process. Chem. Eng. Trans. 57, 1297–1302.

Premuzic, E.T., Lin, M.S., 1999. Induced biochemical conversion of heavy crude oils. J. Pet. Sci. Eng. 22 (1–3), 171–180.

Premuzic, E.T., Lin, M.S., Bohenek, M., Zhou, W.M., 1999. Bioconversion reactions in asphaltenes and heavy crude oils. Energy Fuel 13 (2), 297–304.

Raiders, R.A., Knapp, R.M., McInerney, M.J., 1989. Microbial selective plugging and enhanced oil recovery. J. Ind. Microbiol. 4, 215–230.

Safinya, K., 2008. Heavy oil recovery—the road ahead. Alberta Oil 4 (1), 14–19.

Shah, A., Fishwick, R., Wood, J., Leeke, G., Rigby, S., Greaves, M., 2010. A review of novel techniques for heavy oil and bitumen extraction and upgrading. Energy Environ. Sci. 3, 700–714.

Shennan, J.L., Levi, J.D., 1987. In situ microbial enhanced oil recovery. In: Kosaric, N., Cairns, W.L., Gray, N.C.C. (Eds.), Biosurfactants and Biotechnology. Marcel Dekker, New York, pp. 163–180.

Sivasankar, P., Kumar, G.S., 2017. Influence of pH on the dynamics of microbial enhanced oil recovery processes using biosurfactant producing pseudomonas putida: mathematical modeling and numerical simulation. Bioresour. Technol. 224, 498–508.

Speight, J.G., 1994. Chemical and physical studies of petroleum asphaltene constituents. In: Yen, T.-F., Chilingarian, G.V. (Eds.), Asphaltene constituents and asphalts. I. Developments in petroleum science. In: vol. 40. Elsevier, Amsterdam (Chapter 2).

Speight, J.G., 2011. The Refinery of the Future. Gulf Professional Publishing, Elsevier, Oxford.
Speight, J.G., 2014. The Chemistry and Technology of Petroleum, fifth ed. CRC Press, Taylor and Francis Group, Boca Raton, FL.
Speight, J.G., 2016. Handbook of Hydraulic Fracturing. John Wiley & Sons, Hoboken, NJ.
Speight, J.G., 2017. Handbook of Petroleum Refining. CRC Press, Taylor & Francis Group, Boca Raton, FL.
Speight, J.G., Arjoon, K.K., 2012. Bioremediation of Petroleum and Petroleum Products. Scrivener Publishing, Beverly, MA.
Stosur, G.J., 1991. Unconventional EOR concepts. Crit. Rep. Appl. Chem. 33, 341–373.
Sun, S., Zhang, Z., Luo, Y., Zhong, W., Xiao, M., Yi, W., Yub, L., Fu, P., 2011. Exopolysaccharide production by a genetically engineered *Enterobacter cloacae* strain for microbial enhanced oil recovery. Bioresour. Technol. 102, 6153–6158.
Taber, J.J., Martin, F.D., 1983. Technical screening guides for the enhanced recovery of oil. In: SPE 12069 Proceedings, 58th SPE Annual Technical Conference and Exhibition, San Francisco, CA, October 5–8. Society of Petroleum Engineers, Richardson, TX.
Taylor, K.C., Nasr-El-Din, H.A., 1998. Water-soluble hydrophobically associating polymers for improved oil recovery: a literature review. J. Pet. Sci. Eng. 19, 265–280.
Thomas, S., 2008. Enhanced oil recovery: an overview. Oil Gas Sci. Technol. 63 (1), 9–19.
Thomas, S., Farouq Ali, S.M., 1999. Status and assessment of chemical oil recovery methods. Energy Sources 21 (1–2), 177–189.
Xu, T., Chen, C., Liu, C., Zhang, S., Wu, Y., Zhang, P., 2009. A novel way to enhance the oil recovery ratio by Streptococcus sp. BT-003. J. Basic Microbiol. 49, 477–481.
Yang, C., Gu, Y., 2005a. A novel experimental technique for studying solvent mass transfer and oil swelling effect in a vapor extraction (VAPEX) process. In: Paper No. 2005-099, Proceedings 56th Annual Technical Meeting, The Canadian International Crude oil Conference, Calgary, June 7–9.
Yang, C., Gu, Y., 2005b. Effects of solvent-heavy oil, extra heavy oil, and tar sand bitumen interfacial tension on gravity drainage in the VAPEX process. In: Paper No. SPE 97906, Society of Crude Oil Engineers International Thermal Operations and Heavy Oil, Extra Heavy Oil, and Tar Sand Bitumen Symposium, Calgary, Alberta, Canada, November 1–3.
Yazdani, A., Maini, B.B., 2008. Modeling the VAPEX process in a very large physical model. Energy Fuel 22, 535–544.

Further Reading

Meyer, R.F., Attanasi, E.D., 2003. Fact Sheet 70-03, Heavy Oil and Natural Bitumen—Strategic Petroleum Resources. United States Geological Survey.http://pubs.usgs.gov/fs/fs070-03/.
Pourabdollah, K., Zarringhalam Moghaddam, A., Kharrat, R., Mokhtari, B., 2011. Improvement of heavy oil recovery in the VAPEX process using montmorillonite nanoclays. Oil Gas Sci. Technol. 66, 1005–1016.

Chapter 3

Thermal Methods of Recovery

1. Introduction

Cold production of heavy oil, extra heavy oil, and tar sand bitumen typically achieves oil recovery factors of <20% v/v of the oil in place. In such situations, reducing the viscosity of the heavy oil (the extra heavy oil or the tar sand bitumen) is the most efficient and immediate means to produce the oil or bitumen left in the reservoir after cold production. In such cases, thermally enhanced oil recovery methods offer the potential for secondary higher recovery since the methods are designed to supply heat to the reservoir, which induces temperature rise in the oil-bearing matrix, thereby reducing the viscosity of the oil in situ by several orders of magnitude while vaporizing the lower-boiling components of the oil and thus mobilizing the oil so that it (the oil) can be more effectively driven to producing wells.

The thermal methods include (i) hot-fluid injection and (ii) in situ combustion in which the heat generated is in situ by burning small fraction of the oil. The objective of thermally enhanced oil recovery processes is to reduce the viscosity of the heavy oil to improve the mobility of the oil from the reservoir to the producer well, as such increase the well productivity index.

In addition to adding heat, these processes provide permanent and significant changes to the oil and the driving force (pressure) to move oil to producing wells (Henderson and Weber, 1965). The most common thermal methods are steam flooding and steam cycling that have experienced significant development and use in the past five decades and have grown to be the major contributors to oil production by enhanced oil recovery methods. However, the process efficiency (the reservoir sweep efficiency) of the steam flooding process is limited by gravity segregation and each process requires consideration of the site parameters (Table 3.1).

Generally, the thermal recovery processes used today fall into two classes: (i) processes in which a hot fluid such as steam is injected into the reservoir and (ii) processes in which heat is generated within the reservoir itself (Fig. 3.1). The former are usually the steam-based processes, whereas the latter are processes such as in situ combustion or fire flooding (Kök and Ocalan, 1995). Thermal recovery processes also can be classified as thermal drives or

Heavy Oil Recovery and Upgrading. https://doi.org/10.1016/B978-0-12-813025-4.00003-9

TABLE 3.1 Representation of Site Specific Parameters

Oil	Composition	
	Change in composition	Wax deposition
		Asphaltene deposition
	Properties	Pour point-reservoir temperature
		Viscosity
	Change in properties	Wax deposition
		Asphaltene deposition
Reservoir	Porosity	
	Change in porosity	
	Permeability	
	Change in permeability	Wax deposition
		Asphaltene deposition
Production method	Oil properties	Wax deposition
		Asphaltene deposition
	Reservoir properties	Sandstone
		Carbonate
		Mudstone (clay minerals)

stimulation treatments. A thermal drive not only reduces the viscosity but also provides a force to increase the flow rates of the oil to the production well. In thermal stimulation treatments, only the reservoir near production wells is heated.

Heavy oil, extra heavy oil, and tar sand bitumen are characterized by high viscosity [i.e., resistance to flow measured in centipoises (cp)] and high densities compared with conventional oil (IEA, 2005). Although there are many definitions for heavy oil, extra heavy oil, and tar sand bitumen, the most appropriate definition is drawn from US Federal Energy Administration (FE-76-4), which relates to tar sand:

> *...the several rock types that contain an extremely viscous hydrocarbon which is not recoverable in its natural state by conventional oil well production methods including currently used enhanced recovery techniques. The hydrocarbon-bearing rocks are variously known as bitumen-rocks oil, impregnated rocks, oil sands, and rock asphalt.*

By inference, crude oil and heavy oil, extra heavy oil, and tar sand bitumen are recoverable by well production methods and currently used enhanced recovery

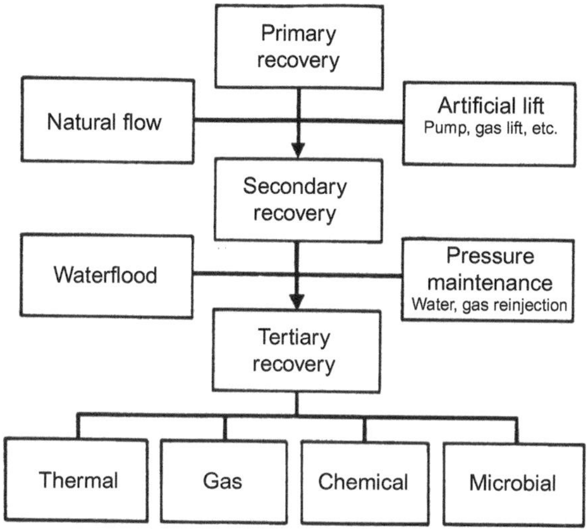

FIG. 3.1 Methods for oil recovery.

techniques. For convenience, it is assumed that before the depletion of the reservoir energy, conventional crude oil is produced by primary and secondary techniques, whereas heavy oil, extra heavy oil, and tar sand bitumen require tertiary (enhanced) oil recovery techniques. While this is an oversimplification, it may be used as a general guide. The term *natural state* cannot be defined out of context; in the context of Federal Energy Administration (FEA) Ruling 1976-4, the term is defined in regard to the composition of the heavy oil, the extra heavy oil, or the tar sand bitumen. The final determinant of whether or not a reservoir is a tar sand deposit is the character of the viscous phase (bitumen) and the method that is required for recovery. Using this definition removes reliance upon a single property such as viscosity or API gravity (subject to experimental difference in the analytic method and the interpretation of the user of the data), which varies from country to country. While a single property might serve to generally define heavy oil, extra heavy oil, and tar sand bitumen, it is by no means accurate and should not be taken as a conclusive definition.

The majority of subsurface heavy oil, extra heavy oil, and tar sand bitumen will not flow toward a wellbore in sufficient quantity to be economically viable. Thermal recovery involves heating up the reservoir, thereby lowering the viscosity of the heavy oil, extra heavy oil, and tar sand bitumen and enabling the oil to flow to the wellbore. The API gravity of extra heavy oil and tar sand bitumen is typically <20°, depending upon the reservoir, and viscosity is very high at reservoir temperature. The viscosity of heavy oil, extra heavy oil, and tar sand bitumen viscosity decreases rapidly with increasing temperatures; therefore, external heat may be required for production. High-temperature steam is commonly used to deliver heat to the formation. The steam/oil ratio (SOR) or fuel oil ratio is an important measure of the energy required to produce heavy oil; a steam/oil ratio of three means that three barrels of water (converted to steam) are needed to produce one barrel of oil.

Thermal methods are the main means of recovering heavy oil, extra heavy oil, and tar sand bitumen, typically the heavy oil, extra heavy oil, and tar sand bitumen with and API gravity $<20°$, which typically are not amenable to recovery by primary production or waterflooding so initial oil saturation is typically high at the start of a thermal recovery project. The principle of thermal recovery is simple: (i) to increase the temperature dramatically and reduce the viscosity of the oil and (ii) to improve the mobility ratio. The two primary methods of heating reservoir oil are injection of fluid heated at the surface or production of heat directly within the reservoir by burning some of the oil in place. The recovery mechanism of the steam flooding process is complex—in addition to viscosity reduction, another important criterion (not always recognized and fully appreciated) is steam distillation of the lower-boiling constituents of the heavy oil, extra heavy oil, and tar sand bitumen (Wu and Brown, 1975; Shu, 1984).

In fact, thermal processes have been the only practical means of improving heavy oil, extra heavy oil, and tar sand bitumen recovery performance of reservoirs containing viscous crudes. The use of heat in wellbores to decrease the crude viscosity and increase crude production rates has been accepted as viable for many decades. Hot water, hot gas, steam (hot-fluid injection processes), and various kinds of combustion drives have gained varying degrees of acceptance. Because of its success, relative simplicity, and low cost, the use of heat to improve oil recovery has gained wide acceptance by the oil industry.

Thus, thermal methods are another way to alter the viscosity of oil and to subject the oil to thermal decomposition by (i) visbreaking, (ii) thermal cracking, or (iii) pyrolysis. The pyrolysis process causes destruction of the oil and produces products such as lower-molecular-weight less viscous liquids, higher-molecular-weight more viscous liquids, and varying amounts of solid carbonaceous residue that is often arbitrarily referred to as coke. Visbreaking is a mild form of pyrolysis that decreases viscosity without significantly changing other crude oil properties of interest. Some molecular alteration of the oil constituents occurs in all thermal recovery processes, but a significant amount of conversion by pyrolysis occurs only at high temperatures (Parkash, 2003; Gary et al., 2007; Speight, 2014, 2017; Hsu and Robinson, 2017).

Although the development of thermal recovery processes was prompted initially by difficulties in displacing and producing heavy crudes, the drive processes technically can be applied also to low-viscosity crudes. Steam- and combustion-drive processes have recovered low-viscosity crudes in several fields. Even though the amount of oil recovered may be higher, these processes seldom can compete commercially with cold-water flooding. Hot-water flooding and thermal stimulation methods are somewhat ineffective for low-viscosity crudes.

By definition, stimulation processes decrease resistance to flow and, thus, allow the driving forces present in the reservoir to increase crude production. In thermal stimulation, the reduction in flow resistance is achieved by heating the wellbore and the reservoir near it. One mechanism that is often preferred is thermal stimulation, which leads to a reduction in the viscosity of the oil. A second mechanism is wellbore cleanup, in which the following might occur: organic solids near a wellbore may be melted or dissolved, clays may be stabilized, the absolute permeability may be increased by the high temperatures, or fines that could be inhibiting flow in gravel packs may be flushed away. Thermal stimulation currently is an effective treatment for viscous oil reservoirs with poor lateral continuity. Because the effects are confined to the neighborhood of the wellbore, thermal stimulation improves oil production rates rather quickly.

Thus, thermal recovery (thermal stimulation) methods are based on adding heat to the oil, mainly to decrease the viscosity (CFR, 2018). In this way, the mobility ratio between oil and the displacing fluids becomes more favorable. For example, heavy oil, extra heavy oil, and tar sand bitumen are usually highly viscous (hence the use of the adjective *heavy*), with a viscosity ranging from ~100 cp to several million centipoises at the reservoir conditions. In addition, oil viscosity is also a function of temperature and API gravity (Speight, 2014). Thus, for heavy crude oil samples with API gravity ranging from 4° to 21° API (1.04–0.928 kg/m^3),

$$\log \log (\mu\sigma + \alpha) = A - B \log (T + 460)$$

In this equation, $\mu\sigma$ is oil viscosity in centipoise, T is temperature in Fahrenheit, A and B are constants, and α is an empirical factor used to achieve a straight-line correlation at low viscosity. This equation is usually used to correlate kinematic viscosity in centistokes, in which case an α of 0.6 to 0.8 is suggested (dynamic viscosity in centipoise equals kinematic viscosity in cSt times density in grams per milliliter). An alternative equation for correlating viscosity data is

$$\mu = ae^{b/T^*}$$

where a and b are constants and T^* is the absolute temperature.

However, it is an oversimplification to state that thermal recovery processes are successful because they tend to reduce the viscosity of the crude oil; there are several important factors related to oil chemistry and oil rock interactions that play a role in heavy oil, extra heavy oil, and tar sand bitumen recovery (Prats, 1986; Lake and Walsh, 2004). For example, the role of steam-distillation and other physical effects in steam-based processes has long been ignored or not understood. It is only when there is a thorough knowledge of such phenomena that heavy oil, extra heavy oil, and tar sand bitumen recovery will be fully understood.

In thermal processes, fluid is injected continuously into a number of injection wells to displace oil and obtain production from other wells. The pressure required to maintain the fluid injection also increases the driving forces in the reservoir, increasing the flow of crude oil. Driving forces present in the reservoir, such as gravity, solution gas, and natural water drive, effect the improved recovery rates once the flow resistance is reduced. Stimulation treatments also can be combined with thermal drives, in which case the driving forces are both natural and imposed.

However, the successful recovery technique that is applied to one heavy oil, extra heavy oil, and tar sand bitumen reservoir is not necessarily the technique that will guarantee success for another reservoir. General applicability of the techniques is not guaranteed. Caution is advised when applying the knowledge gained from one resource to the issues of another resource. Although the principles may at first sight appear to be the same, the technology must be adaptable.

In fact, each production method must be tailored for the particular resource and its fluid properties. A method that works in one situation may fail utterly in a different one. It is essential that the properties of the resource (in this case, of the reservoir and the oil) be fully understood before selecting a production scheme. Essential properties include the geologic setting; the depth, areal extent, and thickness of the resource; oil composition, density, viscosity, and gas content; the presence of bottom water or top gas zones; petrophysical and geomechanical properties such as porosity, permeability, and rock strength; the presence of shale layers; vertical and horizontal permeability; and the variation of these properties across the reservoir.

In fact, geomechanics (measuring and understanding the formation and overburden mechanical properties under drilling and production conditions) is an important aspect of prerecovery operations that is a necessity for optimum process performance. Borehole stability is always an issue when drilling horizontal wells in unconsolidated formation (e.g., for SAGD process). In addition, fracturing the formation may be desirable or undesirable depending on the process. For example, fracturing the overburden and allowing steam to escape would adversely affect the SAGD process, the cyclic steam stimulation process, and the steam-flood process.

Thermal processes have been the only practical means of improving the oil recovery from heavy crude oil reservoirs, and the use of heat in wellbores to increase production rates for heavy crude oils has been accepted as viable for several decades. For example, cyclic steam injection has been commercially successful since it was first applied in the early 1960s. Processes using hot water, hot gas, steam, and various kinds of combustion drives have gained varying degrees of acceptance. Even though the application of thermal recovery processes was initiated because of the difficulties encountered in recovering heavy crude oil (specifically, the difficulties in encouraging flow to a production well), the drive processes can also be applied to the recovery of the residual oil in energy-depleted reservoirs that hold considerable amounts of conventional (low-viscosity) crude oil.

Thermal decomposition (visbreaking or thermal cracking) can also be used to decrease the viscosity of heavy crude oil. The intramolecular balance of the crude oil is changed with the resulting deposition of a carbonaceous residue that separates from the oil, leaving a lower-viscosity (easier-flowing) product phase. Visbreaking is a surface process, and its chemistry offers opportunities for thermal recovery of heavy oil, extra heavy oil, and tar sand bitumen. The partial upgrade of oil in situ with the deposition of the carbonaceous residue, which contains most of the nitrogen and metals originally in the oil, and the production of partially upgraded oil at the surface are to be preferred. However, the disadvantage is that the deposition of a carbonaceous solid has the potential to result in the blockage of the pores and flow channels within the reservoir. Such interference with the recovery process is to be avoided.

Processes combining injection and in situ generation of heat have been tested[1] but currently are not practiced to any great extent. Thermal recovery processes also can be classified as thermal drives or stimulation treatments. In thermal drives, fluid is injected continuously into a number of injection wells to displace oil and obtain production from other wells. The pressure required to maintain the fluid injection also increases the driving forces in the reservoir, increasing the flow of crude. Thus, a thermal drive not only reduces the flow resistance but also provides a force that increases flow rates. In thermal stimulation treatments, only the reservoir near production wells is heated. Stimulation treatments also can be combined with thermal drives, in which case the driving forces are both natural and imposed. In thermal stimulation treatments, the reduction in flow resistance also may result from the removal of organic or other solids from openings in the casing, the liner, the screen, and even from the pores of the reservoir rock.

Finally, important technical considerations for any thermal process are those that affect (i) the ability to generate heat within the formation, (ii) the ability to inject heat into an oil-containing formation reservoir at an efficient rate, (iii) the ability to displace the oil, and (iv) the ability to recover the oil, all in a controlled manner. Attempts to translate these considerations into a specific property or a group of properties that are applicable to a thermal process are not always convenient and may even be difficult.

For example, in a steam-drive process, the API gravity (specific gravity or density) of the oil may not play a significant role in the technical considerations of the process, but it might affect the ease (or difficulty) with which the oil can plug the formation and, thus, interfere the ability to maintain adequate communication between wells during the process, especially in the formation that contain not only heavy oil but also extra heavy oil and tar sand bitumen. With regard to the loss in injectivity or productivity, gravity override (or bypassing) of steam reduces the tendency of the formation to plug but results in the loss of efficiency of the recovery process. However, additional process considerations such as hydraulic fracturing, control of the injection temperature, and cyclic steam injection have been used successfully in thermal operations to avoid or minimize formation plugging through the use of multiphase flow patterns (Charles and Startzman, 1996; Jabbour, et al., 1996). Thus, it seems impractical to place a limit on the range of API gravity of crudes to be considered for steam-drive processes when alteration of the formation structure changes in the process parameters and flow patterns might be more effective.

It is the purpose of this chapter to present the various types of thermal recovery processes in common use. Also, it must be recognized that it is generally the case that some degree of crude oil alteration occurs in all thermal recovery processes, but a significant amount of conversion by thermal changes occurs only at the high temperatures encountered during combustion processes.

2. Thermal Stimulation

In terms of heavy oil, extra heavy oil, and tar sand bitumen recovery, it is more traditional to apply thermal oil recovery techniques of which steam injection has been the most popular. Thermal recovery processes may be used for either stimulation or flooding, and the choice is usually dictated by the properties of the oil and the reservoir. In reservoirs that are small or that have relatively poor continuity, it may not be feasible to drill enough wells to ensure adequate communication to implement a flooding operation. It may be necessary to consider a stimulation treatment to increase both recovery rates and ultimate economic recovery. Cyclic steam injection has proved to be a successful stimulation technique for heavy oil, extra heavy oil, and tar sand bitumen as long as there is the means to produce the crude oil once its viscosity is reduced. Combustion stimulation has been found to be successful in burning solid organic particles, in stabilizing clay minerals, and in increasing absolute permeability near the treated well. These effects are particularly attractive in low-permeability reservoirs.

Thus, heavy oil, extra heavy oil, and tar sand bitumen recovery efforts include thermal methods (steam floods, cyclic steam stimulation, and SAGD) and nonthermal methods (cold flow with sand production, cyclic solvent process, and VAPEX). Significant improvements to the effectiveness of these methods can be achieved by developing a basic understanding of the complex displacement mechanisms and by developing new techniques for in situ characterization of fluid and reservoir characteristics or by the use of additives such as clay minerals to adsorb asphaltene constituents. Development of optimal strategies for recovering these reserves requires the development of state-of-the-art reservoir flow simulator(s) that incorporate techniques for coupling geomechanics and fluid flow besides accurately representing thermal, phase equilibria and mass transfer effects. Improved recovery efficiency can also be achieved by various combinations of thermal and nonthermal processes. In addition, recent advances in drilling and production from unconsolidated sands allow the development of heavy oil, extra heavy oil, and tar sand bitumen recovery strategies that were not possible three decades ago.

Thermally enhanced oil recovery processes add heat to the reservoir to reduce oil viscosity and vaporizing lower-boiling components, making the oil more mobile (Shah et al., 2010). Thermal methods can be used for most oils but are highly suitable for the hard-to-extract, high-viscosity heavy oil, extra heavy oil, and tar sand bitumen. The use of electric heaters or hot water in a process analogous to water flooding have been used, but the two major groups use either steam or the in situ combustion of a proportion of the reservoir to heat the oil.

By definition, stimulation processes decrease resistance to flow and, thus, allow the driving forces present in the reservoir to increase crude production. In thermal stimulation, the reduction in flow resistance is achieved by heating the wellbore and the reservoir near it. One mechanism that is always in force in thermal stimulation is a reduction in the viscosity of the crude and of the water; reducing the viscosities tends to reduce the flow resistance. A second mechanism is wellbore cleanup, in which the following might occur: organic solids near a wellbore may be melted or dissolved. Clay minerals may be stabilized, the absolute permeability may be increased by the high temperatures, or fines that could be inhibiting flow in gravel packs may be flushed away. Wellbore cleanup usually has a rather minor effect after the first stimulation cycle.

Thermal stimulation currently is the only effective treatment for viscous oil reservoirs with poor lateral continuity. Because the effects are confined to the neighborhood of the wellbore, thermal stimulation improves oil production rates rather quickly. Three possible processes that can be effective for the recovery of viscous oil are (i) wellbore heating, (ii) downhole heating, and (iii) hot-fluid injection.

2.1 Wellbore Heating

In wellbore heating, the wellbore is heated at the producing interval. Generally, production and heating are performed concurrently, but sometimes, they are alternated. The reservoir is heated by thermal conduction countercurrent to the transfer of heat out of the reservoir by the produced fluids. The wellbore normally is heated either by using a gas-fired downhole burner or a downhole electric heater hanging on an electric power cable or by circulating fluids heated at the surface. Heating by passing electric currents through a reservoir also has been proposed.

Wellbore heating largely has been supplanted by cyclic steam injection, which is generally much more effective. As is the case with many recovery processes, the mechanisms involved in the production of oil during cyclic steam injection are diverse and complex, but there is little doubt that the reduction in the viscosity of the crude in the heated zone near the well greatly affects the production process and adds to the process efficiency.

2.2 Downhole Heating

Electric heating consists of providing electric currents to generate heat and increase the temperature near the wellbore. Initially, most downhole heating devices were designed for combustion or downhole steam generation

(Jamaluddin et al., 1998; Cerutti et al., 2013). Currently, electric downhole heating systems are finding common use to lower the viscosity of heavy oil, extra heavy oil, and tar sand bitumen in the reservoirs to increase the flow rates of crude and improve recovery. Generally, the systems are very efficient, reliable, and easily controlled solution to allow the production of heavy, high-viscosity oils. Various methods of electric heating have been used for more than 25 years to control not only heavy viscous oil but also wax and hydrates. Currently, two kinds of downhole heaters in use by the oil industry are as follows: (i) inductive heaters and (ii) resistive heaters. In the latter case, the heat transfer is by conduction, and it requires an extended time period to heat the reservoir. The key parameters to be understood in downhole heating process are the variation in the viscosity of the heavy oil, extra heavy oil, and tar sand bitumen with temperature and the rate of heat provided by the heater associated with the generated temperature gradient in the volume around the well (Rodriguez et al., 2008).

Because of the production decline of conventional light crude oil, projects must focus on increasing the recovery of heavy oil, extra heavy oil, and tar sand bitumen by the use of thermal and nonthermal methods. Steam-based thermal recovery processes are more efficient in low-pressure reservoirs; however, because of their depth, the initial pressures of the reservoirs in the Faja are relatively high, ranging from 600 to 1500 psi with viscosities typically >2000 cp. For these reasons, it is important to decrease the pressure of the reservoirs with primary production techniques to facilitate the economical implementation of steam-injection-based methods. In such cases, the initial production of heavy and viscous oils can be accelerated by the use of downhole heaters that, by providing energy to the vicinity of the well, decrease oil viscosity and increase the oil production rate. A consequential advantage of using downhole heaters as a prelude to a steam injection process is that they accelerate early production and reservoir-pressure depletion (Rodriguez et al., 2008).

The Orinoco belt (Faja) in Venezuela contains one of the largest resources in the world of heavy oil, extra heavy oil, and tar sand bitumen. Thus, as an example of heavy oil, extra heavy oil, and tar sand bitumen recovery, the method has been applied to the recovery of Venezuelan crude oil (API gravity ranging from 8° to 9°) and a viscosity of 18,000 cp at the reservoir temperature of 47°C (117°F). The heating varied over the horizontal length from the toe to the heel with more heat (higher watt density) delivered to the toe to equalize the wellbore pressure drop in the long horizontal producing zone. The temperature increased to 76°C (168°F) at the toe. In another example, a California crude oil (API gravity ranging from 8.5° to 13°), heating the oil to temperatures above 93°C (200°F), substantially reduced the viscosity, resulting in increased production (Parman and Ojeda, 2013).

2.3 Hot Fluid Injection

As the name implies, hot-fluid injection processes involve the injection of preheated fluids into a relatively cold reservoir. Generally, the injected fluids are heated at the surface, although wellbore heaters have been used on occasion. Fluids range from water (both liquid and vapor) and air to others such as natural gas, carbon dioxide, exhaust gases, and even solvents. The choice is controlled by cost, expected effect on crude production response, and availability of fluids. The fluids range from water (both liquid and vapor) and air to others such as natural gas, carbon dioxide, exhaust gases, and even solvents. It should be noted here that if water vapor is injected into the reservoir, the process might be more correctly designated as a steam-based process.

In many cases, the choice of injection fluid is governed by the nature of the reservoir. Hot gas or even steam (but not hot water) may be considered for injection into the crest of a reservoir with a small, low-pressure gas cap—making sure to maintain gravitational stability by not exceeding the critical rate. Injection of hot water near zones swept by an encroaching natural water drive may prevent oil resaturation of the zones already swept. But injection of steam under similar circumstances should be considered carefully, as the cooling by any encroaching water could have a significantly adverse effect on the rate of growth of the steam zone. Also, steam or fresh water should not be injected, without due consideration of the possible consequences, into reservoirs containing significant amounts of swelling clay minerals.

For an oil of high viscosity, such as heavy oil, extra heavy oil, and tar sand bitumen, the drop in oil saturation at the displacement front is much smaller, and a substantial amount of mobile oil may be bypassed by the leading edge of the displacement front. Put simply, because of the high viscosity of the oil, the water has a lesser resistance to flow than the oil, and thus, some of the potentially mobile oil is not recovered. Thus, for the same volume of water injected, the oil displacement front would be closer to a producer in a heavy oil reservoir, but for the more viscous oils such as extra heavy oil and tar sand bitumen, water will break through at the producer earlier at the same water injection rate and injected volume. The more viscous oil will continue to be produced after water breakthrough, but the average oil cut (the fraction of oil in the produced stream) will be lower. With the high-viscosity oil, such as extra heavy oil and tar sand bitumen, many pore volumes of injected water are required to reduce the remaining oil to a near nonrecoverable amount. In the case of highly viscous oil, not only the displacement efficiency is poor, but also the volumetric sweep efficiency is decreased. This decrease in volumetric sweep efficiency is the result of two different mechanisms.

In the first mechanism, during the displacement of the viscous oil by water, the macroscopic displacement front is never planar but varies according to the velocity of the flow lines. Moreover, since the flow resistance of a flow channel is least where the displacement front is most advanced, the flow velocity in the channels are the highest. The result is that the displacement front arrives at the producer soonest along the flow line of highest velocity and that breakthrough occurs earlier (for a given injection rate) for the more viscous oil. The increasing oil viscosity (which is proportional to mobility ratio) is reflected in the sweep efficiency at the time of water breakthrough. Layers of different permeability and gravity tongues are examples of other conditions leading to bypassing of the oil by the water.

In the second mechanism, reducing the volumetric sweep efficiency is associated with small-scale viscous instability, which in the displacement front will develop in the most homogeneous porous media. The viscous instability arises from perturbations—caused, for example, by small differences in the shape of grains or pores or even by the presence of clay lenses in a nonclay formation—in an otherwise uniform flow pattern. The distance between the viscous fingers that are most likely to form decreases with increasing viscosity of the oil, and the fingering is likely to occur at relatively low flow rate. In addition, some of the oil between adjacent fingers may be bypassed and trapped more easily when, because of high viscosity, the fingers are close together and, thus, the displacement efficiency is lower the higher the viscosity of the viscosity. Thus, the presence of heterogeneity in a reservoir not only ensures the presence of microscopic viscous fingering but also results in bypassing and trapping of oil—the effect is more pronounced the more viscous the oil.

The effective mobility ratio associated with hot-fluid injection is very unfavorable for noncondensable gases, less so for hot water, and the least unfavorable (or even favorable) for condensable gases such as steam. The effectiveness of steam displacement processes is related to the fact that the ratio of the pressure gradients upstream and downstream of the condensation front, evaluated at the condensation front, is affected greatly by the condensation mechanism itself. This effect, which will be discussed in more detail in Chapter 5, does not occur with hot-water and hot-gas injection. Since thermal drives involve moving hot fluids from one well to another, they require reservoir continuity over distances somewhat greater than the well spacing.

All processes in which hot fluid is injected through the wellhead suffer from heat losses from the injection wellbore to the formations overlying the reservoir (the overburden). Such heat losses can be a significant fraction of the injected heat when the wells are deep or poorly insulated and the injection rates are low. Under such conditions, the temperature of an injected noncondensable fluid entering the formation may be significantly lower than at the wellhead. When the injected fluid is condensable, as in the case of steam, the heat losses cause some of the vapors to be condensed. However, the temperature remains approximately constant as long as there is vapor present, unless there is a large friction pressure drop or a large gravity head—two effects that generally tend to offset each other and are negligible in most current operations. Surface lines generally are insulated, and injection well can be completed in such a way as to reduce heat losses. Because of the thermal stresses on the casing, it is usual to use new wells, cemented to surface, for injection. Existing wells in good condition often are used for production in drive projects.

For example, in the hot-water process, the water is filtered, treated to control corrosion and scale, heated, and if necessary treated to minimize the swelling of clay minerals in the reservoir. The primary role of the hot water is to reduce the viscosity of the oil to improving the displacing efficiency over that obtainable from a conventional waterflood. The design and operation of hot-water drives have many elements in common with conventional waterflooding (Chapter 2). Thus, the *hot-water drive* (*hot-water flood*) is a displacement process in which oil is displaced immiscibly by both hot and cold water (Dietz, 1967). In this process, the water is filtered, treated to control corrosion and scale, heated, and if necessary treated to minimize the swelling of clays in the reservoir. The primary role of the heated water is to reduce the oil viscosity and, thereby, improve the displacing efficiency over that obtainable from a conventional waterflood.

In the simplest form, a hot-water drive involves the flow of only two phases: water and oil. On the other hand, steam-based processes and combustion-based processes typically involve a third phase: gas. Thus, the elements of hot-water flooding are relatively easy to describe, it being basically a displacement process in which oil is displaced immiscibly by both hot and cold water. Except for the effect of temperature and the fact that hot-water floods are applied to relatively viscous crudes, these processes have many elements in common with conventional waterfloods. However, because of the ever-present water in all crude oil reservoirs, displacement by hot water must occur to some extent in all thermal recovery processes. In fact, hot water is known to contribute to the displacement of oil in the zones downstream of both steam-drive processes and combustion processes.

In the process, the leading edge of the injected hot water loses heat so rapidly that it quickly reaches the initial reservoir temperature. Thus, at the leading edge of the displacement front, the oil mobility is that of the unheated oil. On the other hand, the viscosity of the injected hot water is lower than in conventional waterfloods. Thus, the mobility ratio of the oil ahead of the displacement front and the injected water near the injection well is more unfavorable in hot-water floods than in conventional waterfloods. This should result in somewhat earlier water breakthrough in hot-water floods. And there is some evidence for this. On the other hand, the mobility ratio of the fluids in the heated zones is more favorable in the hot-water floods than in conventional waterfloods. This results in better displacement efficiency from the heated zone and would improve the ultimate recovery even where residual oil saturation does not decrease with increasing temperature.

Except for temperature effects and the fact that they generally are applied to relatively viscous crude oils, hot-water drive processes have many elements in common with conventional waterfloods. The process is a displacement process in which oil is displaced immiscibly by both hot- and cold-water steam and combustion processes and, on the other hand, always involves a third phase: gas. Except for temperature effects and the fact that they generally are applied to relatively viscous crudes, hot-water floods have many elements in common with conventional waterflood.

Because of the pervasive presence of water in all crude oil reservoirs, displacement by hot water must occur to some extent in all thermal recovery processes. It is known to contribute to the displacement of oil in the zones downstream of both steam-drive projects and combustion-based projects. The complex mineralogy of clay minerals (Grim, 1968) must be considered wherever clay occurs in a reservoir.

In a hot-water flood, the leading edge of the injected hot water loses heat so rapidly that it quickly reaches the initial reservoir temperature, and the oil mobility is that of the unheated oil at the leading edge of the displacement front. On the other hand, the viscosity of the injected hot water is lower than in conventional waterfloods. Thus, the mobility ratio of the oil ahead of the displacement front and the injected water near the injection well is less favorable in hot-water floods than in conventional waterfloods (which should result in somewhat earlier water breakthrough than hot-water floods). On the other hand, the mobility ratio of the fluids in the heated zones is more favorable in the hot-water floods than in conventional waterfloods. This results in better displacement efficiency from the heated zone and would improve the ultimate recovery even where residual oil saturation does not decrease with increasing temperature.

Thermal expansion of oil also contributes to the improved displacement efficiency of thermal projects. Gravity separation usually is accentuated by the increased density difference between water and oil as temperatures increase. In addition, most of the reservoir heat is in the zones from which most of the oil already has been displaced. Indeed, all thermal drives are characterized by the presence of large amounts of heat in oil-depleted portions of the reservoir; this has prompted a number of modifications aimed at scavenging or recycling the heat to improve the efficiency of the process (Lake and Walsh, 2004).

The amount of oil displaced in a hot-water drive is invariably larger than the amount produced. The oil that is displaced but not produced is stored in unswept portions of the reservoir. In the case of viscous crude oils, especially, the mobility ratio between the advancing oil and any gas or water in the reservoir is very favorable. This means that crude oil tends to fill regions of the reservoir initially filled with mobile gas and water before it is produced. Where an oil bank forms, consideration of these effects allows an estimation of the recovery history from the estimates of the oil displacement history.

Although there are no simple methods for predicting oil recovery from hot-water floods, an approach is suggested that (i) is based on conventional waterflood technology, (ii) appears to have some of the elements necessary to describe hot-water floods, and (iii) considers only the effects of permeability variations and mobility ratio. It is by no means a proved method and is offered merely as an example of how it may be possible to adapt the existing technology from a related field when there is no other method for making the desired estimates of oil recovery.

In addition to cost, the choice of the fluid is controlled by the expected effect of the fluid on the production response of the crude oil and on and availability of fluids. For example, seawater may be used for injection into an undersea reservoir where the cost of water delivery to the platform would be prohibited. On the other hand, choice of injection fluid is governed by the nature of the reservoir, with each reservoir being considered a specific entity subject to the character of the oil and the mineralogy of the reservoir rock (Table 3.1). For example, steam or hot water should not be injected without first considering the possible consequences of its effects on reservoirs containing significant amounts of swelling clay minerals.

The effective mobility ratio associated with hot-fluid injection is very unfavorable for noncondensable gases, less so for hot water, and the least unfavorable (or even favorable) for condensable gases such as steam. The effectiveness of steam displacement processes is related to the fact that the ratio of the pressure gradients upstream and downstream of the condensation front, evaluated at the condensation front, is affected greatly by the condensation mechanism itself. This effect does not occur with hot-water injection or with hot-gas injection. Since thermal drives involve moving hot fluids from one well to another, they require reservoir continuity over distances somewhat greater than the well spacing.

Injection of hot noncondensable gases has been attempted also. The low viscosity of noncondensing gases results in poor mobility ratios and, consequently, poor displacement efficiencies. Stripping of light ends also may be important in hot-gas drives. Here, condensation of the stripped components results from cooling of the gas stream (which is analogous to meteorological precipitation) rather than, as in the case of steam, from the condensation of the carrier vapor phase. Because of the low density of gases, high volumetric gas injection rates (referred to standard conditions) are required to provide heat injection rates comparable with those obtainable with water or steam. At the injection rates typically encountered in field operations, wellbore heat losses cause a significant reduction in the temperature of the hot gas.

In fact, all processes in which hot medium (gas or fluid) fluid is injected through the wellhead into a formation result in heat loss from the injection wellbore, to the formation, to the formations overlying the reservoir (the overburden). The heat loss can be a significant fraction of the injected heat when at low injection rates and when the wells are deep or are poorly insulated. Under such conditions, the temperature of an injected noncondensable fluid entering the formation will, more than likely, be significantly lower than the temperature of an injected fluid at the wellhead.

When the injected fluid is condensable, as in the case of steam, the heat loss can cause some of the vapors to condense. On the other hand, the temperature of the fluid may remain approximately constant as long as there is vapor present. Typically, the surface lines are insulated, and the injection wells can be completed in such a manner that heat loss will be reduced. In addition, it is usual to use new wells, cemented to surface, for injection because of the thermal stresses placed on the well casing and existing wells that have remained in good condition may be used as production wells in steam-drive projects.

3. Steam-Based Processes

Steam can also be injected into one or more wells with production coming from other wells (*steam drive*). This technique is very effective in heavy oil, extra heavy oil, and tar sand bitumen formations but has found little success during application to tar sand deposits because of the difficulty in connecting injection and production wells. However, once the flow path has been heated, the steam pressure is cycled, alternately moving steam up into the oil zone, then allowing oil to drain down into the heated flow channel to be swept to the production wells.

In *steam stimulation* processes, heat and drive energy are supplied in the form of steam injected through wells into the heavy oil, extra heavy oil, and tar sand bitumen reservoir. In most instances, the injection pressure must exceed the formation fracture pressure in order to force the steam into the reservoir and into contact with the oil. When sufficient heating has been achieved, the injection wells are closed for a soak period of variable length and then allowed to produce, first applying the pressure created by the injection and then using pumps as the wells cool and production declines.

If the viscous oil (the word oil is used in the general sense) in a formation can be made mobile by admixture of either a hydrocarbon diluent or an emulsifying fluid, a relatively low-temperature secondary recovery process is possible (*emulsion steam drive*). If the formation is impermeable, communication problems exist between injection and production wells. However, it is possible to apply a solution or dilution process along a narrow fracture plane between injection and production wells.

The steam-based processes are the most advanced of all enhanced oil recovery methods in terms of field experience and thus have the least uncertainty in estimating performance (EPRI, 1999; Curtis et al., 2002; Advanced Resources International, 2005; Shah et al., 2010). However, as with all enhanced oil recovery processes, a good reservoir description is a necessity (Lake et al., 1992). Steam processes are most often applied in reservoirs containing viscous oils and tars, usually in place of rather than following secondary or primary methods. Commercial application of steam processes has been underway since the early 1960s. In contrast, although in situ combustion has been field tested under a wide variety of reservoir conditions, few projects have proved economical and advanced to commercial scale.

Steam can also be injected into one or more wells with production coming from other wells (*steam drive*). This technique is very effective in heavy oil, extra heavy oil, and tar sand bitumen formations but has found little success during application to heavy oil, extra heavy oil, and tar sand bitumen reservoirs because of the difficulty in connecting injection and production wells. However, once the flow path has been heated, the steam pressure is cycled, alternately moving steam up into the oil zone, and the oil is allowed to drain down into the heated flow channel to be swept to the production wells.

The concept behind the steam-based processes is generally viscosity reduction so that the heavy oil, extra heavy oil, and tar sand bitumen can flow to the production well. Other effects such as emulsion drive, thermal expansion, solution gas drive, and steam distillation are also believed to occur. The extent of each is unknown and oil-dependent. In the steam-stripping processes, steam removes relatively low-boiling components from the heavy oil, extra heavy oil, and tar sand bitumen. In fact, steam stripping removes a larger fraction of the crude oil than would be suggested by the boiling-point distribution because of *steam distillation* (Willman et al., 1961; Volek and Pryor, 1972; Shu, 1984; Northrop and Venkatesan, 1993). There is no chemical change to the constituents of the oil, although there may be favorable compositional changes to the oil insofar as lower-boiling fractions are recovered and higher-boiling materials remain in the reservoir.

As the steam condenses, most of the stripped or steam-distilled components in the steam also condense and form a solvent front that will assist in displacing heavy oil, extra heavy oil, and tar sand bitumen (as a solvent/oil mix) toward the production well. These effects help to improve the displacement efficiency. Whether or not steam distillation occurs and the role it plays in oil recovery depend on the character of the heavy oil, extra heavy oil, and tar sand bitumen as well as the downhole conditions. Steam distillation can indeed augment the process when the steam-distilled material moves with the steam front and acts as a solvent for oil ahead of the steam front (Willman et al., 1961; Volek and Pryor, 1972; Shu, 1984; Prats, 1986; Northrop, and Venkatesan, 1993).

Typically, steam is produced using once-through boilers that also produce varying amounts of hot water. The hot water usually is injected into the reservoir along with the steam, although the amount of hot water to be injected with the steam may be reduced by separating the vapor and liquid phases upstream of the wellhead. Any hot water that is separated is usually recycled through the boiler feed system (as long as the water is sufficiently clean and will not cause corrosion or fouling of the lines) or used to preheat the feed water to the boiler. Heating the water can promote the deposition of dissolved solids (depending upon the chemical nature of the solids), and thus, it is especially important in steam-injection operations to ensure the use of clean water. Furthermore, because of the condensation resulting from heat losses, the water content of the steam entering a formation is generally higher than that at the wellhead. Injected or condensed fresh (hot) water may create problems in reservoirs that contain clay minerals that are subject to swelling by intake (absorption) of water.

In addition to possessing the displacement mechanisms and thermal phenomena present in hot-water floods (which lead to improvements in displacement efficiency resulting from a decrease in viscosity, a reduction in capillary forces, and thermal expansion of the crude), steam drives also (i) cause boiling and steam stripping of the light components in the crude, (ii) promote the formation of a low-viscosity oil bank near the condensation front, (iii) yield unusually low residual oil saturations, and (iv) improve the effective mobility ratio of the displacing process.

When dealing with heavy oil, extra heavy oil, and tar sand bitumen, it can be reasonably assumed that the steam-distilled material is composed of aromatics and naphthenic constituents that are excellent solvents for oil (Mitchell and Speight, 1973). On the other hand, if the steam-distilled constituents have a predominantly paraffinic character (which is not considered likely for heavy oil, extra heavy oil, and tar sand bitumen but very likely for residual oil in an energy-depleted conventional oil reservoir), this will cause deposition of asphaltic material ahead of the steam front. The ever-present disadvantage is that any deposited material will block the reservoir flow channels, thereby restricting oil flow to the production well. The distinct advantage is that if flow to the recovery well is not restricted, there is the potential for recovery of partially upgraded heavy oil, extra heavy oil, and tar sand bitumen.

Another effect is that the oil bypassed by the advancing condensation front is so volatile that it shrinks significantly as steam continues to pass by, resulting in low recovery of the heavy oil, extra heavy oil, and tar sand bitumen. These effects would be more pronounced with the residual oil from conventional crude oil recovery, which contains relatively high levels of volatile constituents. Overall, the condensation steam-stripping and steam-distillation effects result in more favorable recovery than is expected on the basis of viscosity reduction alone.

If the viscous bitumen in a tar sand formation can be made mobile by admixture of either a hydrocarbon diluent or an emulsifying fluid, a relatively low-temperature secondary recovery process is possible (*emulsion steam drive*). If the formation is impermeable, communication problems exist between injection and production wells. However, it is possible to apply a solution or dilution process along a narrow fracture plane between injection and production wells.

All of the processes in which hot fluid is injected through the wellhead suffer from heat losses from the injection wellbore to the overburden formations. These heat losses can be a significant fraction of the injected heat when the wells are deep or poorly insulated and the injection rates are low. Under such conditions, the temperature of an injected noncondensable fluid entering the formation may be significantly lower than at the wellhead. When the injected fluid is condensable, as in the case of steam, the heat losses cause condensation of the vapor. If the steam is unable to maintain the heat and enters the reservoir as hot water, the oil in the reservoir actually comes into contact with hot water rather than steam. Whether or not the reservoir is injected with steam is subject to the depth of the reservoir (i.e., the distance the surface-generated steam has to travel to enter the reservoir). Cooling of the steam by any encroaching water could have a significantly adverse effect not only on the rate of growth of the steam zone but also on the initiation of the steam zone. However, there are several schools of thought that contend that the majority of steam-based processes are, in fact, hot-water-based processes. To overcome the steam-or-hot-water issue, surface lines generally are insulated, and injection well is completed in a manner that reduces heat losses.

3.1 Steam Drive

The *steam-drive* process involves the injection of steam through an injection well into a reservoir and the production of the mobilized heavy oil, extra heavy oil, and tar sand bitumen and steam condensate from a production well. Steam drive is usually a logical follow-up to cyclic steam injection. Steam drive requires sufficient effective permeability to allow the injection of the steam at rates sufficient to raise the reservoir temperature to move the heavy oil, extra heavy oil, and tar sand bitumen to the production well.

The process occurs in two steps: (i) steam stimulation of production wells, that is, direct steam stimulation, and (ii) steam drive by steam injection to increase production from other wells (indirect steam stimulation). When there is some natural reservoir energy, steam stimulation normally precedes steam drive. In steam stimulation, heat is applied to the reservoir by the injection of high-quality steam into the production well. When natural reservoir drive energy is depleted and productivity declines, most cyclic steam-injection projects are converted to steam drives. In some projects, producing wells are periodically steam-stimulated to maintain high production rates. Normally, stream-drive projects are developed on relatively close well spacing to achieve thermal communication between adjacent injection and production wells. To date, steam methods have been applied almost exclusively in relatively thick reservoirs containing viscous crude oil.

Thus, the *steam-drive process* (*steam-flood process*) is a process that requires continuous injection of steam into the oil layer to supply thermal energy and so to reduce the viscosity of the heavy oil, extra heavy oil, and tar sand bitumen to enhance the recovery of viscous crude oils. A suitable pattern of injection wells typically surrounding a production well is selected, with the pattern largely dependent on the geology of the formation to be flooded. Steam is continuously injected into the selected injection wells, forming a steam zone, which both lowers the viscosity of the oil and also provides the pressure to drive the heavy oil, extra heavy oil, and tar sand bitumen toward the producer. Typical recovery factors are on the order of 50%–60% v/v of the oil in place.

A steam-drive process differs significantly from a hot-water drive—the difference in performance being solely due to the presence and effects of the condensing vapor. The presence of the gas phase causes light components in the crude to be distilled and carried along as hydrocarbon components in the gas phase. Where the steam condenses, the condensable hydrocarbon components do likewise, thus reducing the viscosity of the crude at the condensation front. Moreover, the condensing steam makes the displacement process more efficient and improves the sweep efficiency. Thus, the net effect is that recovery from steam drives is significantly higher than from hot-water drives. In addition, a steam-drive process can also promote steam distillation of some of the components of the oil, thereby having the potential to create a solvent front that will dissolve part or all of the oil as it passed along the flow channel(s) (Willman et al., 1961; Volek and Pryor, 1972; Shu, 1984; Northrop and Venkatesan, 1993).

In the *steam-stripping process*, steam removes relatively low-boiling constituents from the oil. The process recovers a larger fraction of the crude than would be suggested by its boiling-point distribution (e.g., the boiling-point distribution of a particular crude might indicate that 20% of the crude boils at 400°F, but when the vapor pressure is maintained at a low level by the flowing vapor, more than 20% of the crude would be removed). As the steam condenses, most of the stripped components in the steam also condense, and these light ends condensing at the steam condensation front help to improve the displacement efficiency. The mixing of the condensing light ends with crude reduces the viscosity of the oil contacted by a subsequent advance of the condensation front.

Eventually, the oil bypassed by the advancing condensation front is so light and distillable that it shrinks significantly as steam continues to pass by. Such shrinkage of the bypassed oil can result in very low ultimate values for the residual oil in steam drives. These effects would be more pronounced with crudes containing a high percentage of light fractions but would occur to some extent with any crude. In addition, the condensation phenomenon itself, quite apart from any distillation effects, results in more favorable effective mobility ratios than would have been expected on the basis of viscosity ratios alone.

Two expected problems inherent in the steam-drive process are steam override and reservoir plugging. Any in situ thermal process tends to override (migrate to the top of the effected interval) because of differential density of the hot and cold fluids. These problems can be partially mitigated by rapid injection of steam at the bottom or below the target interval through a high-permeability water zone or fracture. Each of these options will raise the temperature of the entire reservoir by conduction and, to a lesser degree, by convection, and the effectiveness of the following injection of steam into the target interval will be enhanced.

Other variations on this theme include the use of steam and the means of reducing interfacial tension by the use of various solvents. The *solvent extraction* approach has had some success when applied to bitumen recovery from mined tar sand, but when applied to unmined material, losses of solvent and bitumen are always a major obstacle. This approach should not be rejected out of hand since a novel concept may arise, which guarantees minimal (acceptable) losses of bitumen and solvent.

3.2 Cyclic Steam Injection

Cyclic steam injection is the alternating injection of steam and production of oil with condensed steam from the same well or wells. The process is also known as the *steam soak* and *huff and puff*, and the method of heavy oil recovery has been used to produce heavy oil from Cold Lake, Alberta, Canada. The recovery mechanism is similar to other steam flooding techniques that include heating oil-bearing matrix, steam distillation, viscosity reduction, gas drive, and mobilization of the heavy oil toward the well.

Thus, steam generated at surface is injected in a well, and the same well is subsequently put back on production. A cyclic steam injection process includes three stages. In the first stage is injection, during which a measured amount of steam is introduced into the reservoir. The second stage (*the soak period*) requires that the well be shut in for a period of time (usually several days) to allow uniform heat distribution to reduce the viscosity of the oil (alternatively, to raise the reservoir temperature above the pour point of the oil). Finally, during the third stage, the now-mobile oil is produced through the same well. The cycle is repeated until the flow of oil diminishes to a point of no returns. One mechanism that aids the production of the oil is the flashing of hot water (originally condensed from steam injected under high pressure) back to steam as pressure is lowered when a well is put back on production. This process is also known by a number of names: steam soak, steam stimulation, huff and puff, and cyclic steam injection.

In the process, a single well for both injection and production and steam is injected into a well for several days or weeks. Then, the well is shut in for several days to a month or more, the soak period. After the soak period, the well is produced for up to 6 months, after which the process is repeated. The steam heat the rock and fluids surrounding the wellbore and also provides some drive pressure—by the time production resumes, the steam has condensed, and oil and water are produced. Other fluids can be used instead of steam, but none have been found to be as effective. Hot water introduces a larger volume of water per unit of heat injected, and its use results in higher water saturations. This can affect the producing oil cut adversely.

The advantage of the huff-and-puff process is the relatively short time required for steam-injection time so that the well is producing most of the time. The disadvantage is that only the reservoir near the wellbore is stimulated. As a result, this method therefore often has to be followed by continuous steam injection to drive oil toward a separate production well. Reservoir pressure, dictated by depth, imposes a limit on steam flooding: higher pressures require more fuel to generate higher temperature on the surface to produce saturated steam needed for efficient steam flood. The higher temperature also leads to greater heat losses.

Cyclic steam injection is popular because the production response is obtained earlier and the amount of recovered oil per amount of steam injected is often higher than in thermal drives. Also, relatively small steam boilers can be used, and they can be moved from well to well. Wells can be subject to steam soaking several times, the main requirements are natural driving forces, such as solution gas or gravity drainage, and sufficient oil near the wells. The process employs a pattern of vertical wells. Typically, pressurized steam is pumped down a well for weeks and sometimes months, thoroughly heating the reservoir near the well. The process is then halted, usually for several weeks; this allows the heavy oil, extra heavy oil, and tar sand bitumen to become separated from the reservoir sand. Then, the heavy oil, extra heavy oil, and tar sand bitumen is artificially lifted from the same well. This technology has been successfully used for many years but is fast being replaced in Canada by the SAGD method. It is still considered applicable in heavy oil, extra heavy oil, and tar sand bitumen regions where SAGD is not suitable.

In practice, the heat injected warms the heavy oil, extra heavy oil, and tar sand bitumen and lowers its viscosity. A heated zone is created through which the warmed heavy oil, extra heavy oil, and tar sand bitumen can flow back into the well. This is a well-developed process; the major limitation is that <30% (usually <20%) of the initial oil in place can be recovered. The oil production rate increases quickly to a high rate, staying at that level for a short time before gradually declining over several months. The cycle is then repeated when the oil rate becomes uneconomic. Steam-to-oil ratios are initially in the region of 2:1, but this increases as cycles are repeated and is typically in the range of 3:1 to 4:1 across the lifetime of the process. For this reason, the process is often used for the first cycle of production only before switching the reservoir to steam flooding. Cyclic steam stimulation works best when there are thick pay zones (> 10 m) with high-porosity sands (>30%). Shale layers that reduce vertical permeability are not a problem for vertical wells that penetrate thick pay zones. However, good horizontal permeability (>1 darcy) is important for production.

In addition to heating the reservoir, the high-pressure steam creates fractures in the formation, thereby improving fluid flow, but for this reason, an overburden cover typically on the order of 100 ft is required to contain the pressure. The process is also sensitive to the geology near the well, which should aid heat distribution and the capture of the mobilized oil. The process is particularly attractive because it has quick payout; however, recovery factors are low compared with other thermal techniques, typically on the order of 10%–40% v/v of the oil in place as are recovery rates due to the noncontinuous nature of the process.

The cyclic steam injection process reduces the flow resistance near the wellbore, but it also (i) enhances the depletion mechanism by causing gas dissolved in the crude to become less soluble as the temperature increases and (ii) increases the amount of water retained in the reservoir through a combination of temperature-induced wettability changes, including the separation of asphaltene constituents as a separate phase (Table 3.2), and hysteresis in the relative permeability curves in a manner analogous to a countercurrent imbibition displacement process. Recognition of some of the mechanisms that occur naturally during cyclic steam injection has led to modifications to enhance them—for example, the addition of a noncondensable gas to the injected steam.

Cyclic steam stimulation is applied to viscous oil reservoirs to boost recovery during the primary production phase. Cyclic steam stimulation is a

TABLE 3.2 Effects of the Separation of Asphaltene Constituents

Cause	Effect
Asphaltene separation	Reservoir plugging
	Wettability alteration
	Fouling production facilities
	Pipeline fouling
	Destruction of refinery catalysts

single-well process and consists of three stages. In the initial stage, steam injection is continued for about a month. The well is then shut in for a few days for heat distribution, denoted by soak. Following that, the well is put on production. The rate of oil production increases quickly to a high rate, stays at that level for a short time, and declines over several months. Cycles are repeated when the oil rate becomes uneconomic. The process assists natural reservoir energy by thinning the oil, so it will more easily move through the formation to the injection/production wells. The minimum depth for applying cyclic steam stimulation is on the order of 1000 ft; depth varies depending upon the type and structure of the overlying formations.

Thus, to recover heavy oil, extra heavy oil, and tar sand bitumen, a predetermined amount of high-pressure, high-temperature steam (estimated to be at ~350°C, 650°F) is injected into wells that have been drilled or converted for injection purposes into the reservoir or deposit. The pressure of the steam fractures the oil sand, while the heat of the steam melts the heavy oil, extra heavy oil, and tar sand bitumen. The wells are then shut in to allow the steam to heat (soak period) the producing formation around the well. As the steam soaks into the deposit, the heated material flows to a producing well and is pumped to the surface. After a sufficient time has elapsed to allow adequate heating, the injection wells are returned to production until the heat is dissipated with the produced fluids. This cycle of soak and produce (or huff and puff) may be repeated until the response becomes marginal because of declining natural reservoir pressure and increased water production. At this time, a continuous steam flood is usually initiated to continue the heating and thinning of the oil and to replace declining reservoir pressure so that production may continue. When the steam flooding is started, some of the original injection wells will be converted for use as production wells, along with the others drilled or designated for that purpose. This process can be repeated several times in a formation, and it can take between 120 days and 2 years to complete a steam stimulation cycle.

Typical recovery factors for cyclic steam stimulation are 20%–35% with steam-to-oil ratios (SOR) of 3 to 5.22. Steam floods may follow cyclic steam stimulation. While cyclic steam stimulation produces the heavy oil, extra heavy oil, and tar sand bitumen around a single wellbore, steam flooding recovers the heavy oil, extra heavy oil, and tar sand bitumen between wells. For example, a five-spot pattern with four producing wells surrounding a central steam-injection well is a common configuration. The well spacing can be less than two acres for a field in steam flood. The steam heats the oil to lower its viscosity and provides pressure to drive the heavy oil, extra heavy oil, and tar sand bitumen toward the producing wells. In most steam-flood operations, all of the wells are steam-stimulated at the beginning of the flood. In a sense, cyclic steam stimulation is always the beginning phase of a steam flood. In some cases, even the steam-flood injection wells are put on production for one or two cyclic steam stimulation cycles to help increase initial project production and pay out the high steam-flood capital and operating costs.

Cyclic steam stimulation and steam floods are used in California, Western Canada, Indonesia, Oman, and China. California's Kern River production rose from <20,000 bbl/day in the late 1950s before cyclic steam stimulation to over 120,000 bbl/day by 1980 after the introduction of cyclic steam stimulation. The Duri field in Indonesia is the world's largest steam flood and produces 230,000 bbl/day with an estimated ultimate recovery factor of 70% in some locations.

Cyclic steam injection is used extensively in viscous oil reservoirs and tar sand deposits and, in some cases, to improve injectivity prior to steam flooding or in situ combustion operations. The technique has also been applied to the California tar sand deposits (Bott, 1967) and in some heavy oil, extra heavy oil, and tar sand bitumen reservoirs north of the Orinoco deposits (Ballard et al., 1976). The steam flooding technique has been applied, with some degree of success, to the Utah tar sands (Watts et al., 1982) and has been proposed for the San Miguel (Texas) tar sands (Hertzberg et al., 1983).

Technical challenges for cyclic steam stimulation and steam flooding are primarily related to reducing the cost of steam, which is generated in most locations using natural gas. The economics may be improved by also generating and selling electricity and by using waste heat for cogeneration. Alternative fuels (coal, heavy ends, and coke) are discussed separately below; they could also reduce the cost of steam generation. Monitoring and controlling the steam front could also reduce costs by redirecting steam to zones where the heavy oil, extra heavy oil, and tar sand bitumen has not been produced. Steam could be shut off from zones that have been successfully swept and directed toward unswept regions.

Gravity override is a natural occurrence in every steam flood. The steam breaks through to the producers, at which time the process turns into a gravity drainage process. The steam chest at the top of the formation expands downward, and the heated heavy oil, extra heavy oil, and tar sand bitumen drain by gravity to the producing wells. Although the geometry configuration is different to that of the SAGD process, the basic effects are the same.

The measurement of the produced fluids (oil, water, and natural gas) at the surface for each well can be used to optimize production by adjusting artificial lift rates and steam-injection rates. Downhole fluid-flow measurements could be used to identify which zones are producing oil, water, or gas in a producing well. Monitoring may involve drilling observation wells where permanent sensors may be deployed or where logging can be periodically performed. Downhole temperature and pressure sensors may use fiber-optic or wire line technology. Water and steam saturation outside the casing of an observation well can be measured with nuclear spectroscopy logs. Time-lapse, cross well electric imaging can be used to identify bypassed heavy oil, extra heavy oil, and tar sand bitumen zones between closely spaced (500m) observation wells. Cross well seismic and surface seismic measurements might be used to locate steam fronts. High-resolution imaging of the formation and the fluid saturations before completing the wells and during production, however, is an open technical challenge. Technologies must be reliable and have long operating periods between service periods. High-temperature- (up to 300°C, 570°F) and corrosion-resistant equipment including pumps (artificial lift), cements, completions, liners, packers, valves, electronics, and sensors are needed. Thermal expansion of the formation can also cause the casing to fail.

Most cyclic steam stimulation and steam-flood wells have been vertical wells. More recently, vertical wells with multilateral branches and horizontal wells are being tried. The advantage is a reduced footprint while tapping large subsurface regions. Optimal control and configuration of these wells for cyclic steam stimulation and steam-flood recovery processes are still being developed.

Cyclic steam injection also is used as a precursor to steam-drive technology. In reservoirs containing heavy crude oil, the resistance to flow between the wells may be sufficiently high that steam-injection rates are severely limited, making steam drive technically inefficient. Cyclic steam injection reduces the flow resistance near wells, where the resistance is most pronounced; this alone improves the injection rate attainable during steam drives by reducing the resistance to flow between wells. Repeated cyclic steam injection reduces the flow resistance still farther from the wells and may lead to connecting the heated zones of adjacent wells and further improving the operability of the steam drive.

A drawback to cyclic steam injection is that the ultimate recovery may be low relative to the total oil in place in the reservoir. Ultimate recoveries from steam drives are generally much larger than those from cyclic steam injection. Thus, cyclic steam injection followed by a steam drive is an attractive combination; crude production is accelerated quickly, and the ultimate recovery is quite high.

Recently, cyclic steam stimulation has been applied to wells with multilateral horizontal legs with some degree of success (Perez and Escobar, 2016). This technology, (HASD technology) was designed to be applied in a set of horizontal and parallel wells located at the same level in the reservoir sand.

The wells are cyclically switching between injection and production phases. The oil recovery mechanism in place is a combination of horizontal steam flooding and cyclic steam stimulation of each of the horizontal wells. In this technology, a steam chamber is created, while each well is injecting, and it is laterally driven by the pressure differentials created between the injectors and the adjacent producers. Therefore, the method combines the cyclic steam stimulation and steam flood, thereby generating a process that, at efficient operating conditions, has the potential for higher oil production than the typical processes of individual cyclic steam stimulation followed by steam drive.

3.3 Steam Assisted Gravity Drainage

The development of horizontal drilling technology led to the development of the steam-assisted gravity drainage process (SAGD process), which is another steam-injection technique but with the potential (and reality) of higher recovery of heavy oil, extra heavy oil, and tar sand bitumen compared with other steam processes. The gravity-assisted mechanism allows the expansion of a stable steam chamber and the short distance oil displacement (Butler and Stephens, 1980; Butler, 1985).

Steam-assisted gravity drainage (SAGD) was initially developed to recover bitumen from the Canadian oil sands and other tar sand reservoirs (Butler, 1985; Rose and Deo, 1995). The key element of SAGD is that the two wells need to be parallel and horizontal. It is only in the last 10–15 years that directional drilling technology has been able to achieve these two characteristics with any degree of certainty.

In the simplest or typical form of the process, the technology requires the drilling of two parallel horizontal wells through the oil-bearing formation. Into the upper well, steam is injected creating a high-temperature steam chamber. The increased heat loosens the thick crude oil causing it to flow downward in the reservoir to the second horizontal well. This second well is located parallel to and below the steam-injection well. This heated, thinner oil is then pumped to the surface via the second horizontal or production well. Water is injected into the bitumen-drained area to maintain the stability of the deposit.

The process is sensitive to operational and geometric reservoir parameters with high vertical permeability being particularly crucial to the success of the process. Rock permeability, reservoir heterogeneity, oil reservoir thickness, and operational conditions such as well separation and length and steam rate are also important. In addition, the process is not applicable to thin reservoirs that do not allow sufficient separation of the two wells and is not truly suitable for reservoirs of <20 ft in thickness.

Several variations of this process have been developed—one variation uses a single horizontal well, with steam injection through a central pipe and production along the annulus, while another variation involves steam injection through existing vertical wells and production through an underlying horizontal well.

The key benefits of the SAGD process are an improved steam-oil ratio and high ultimate recovery (on the order of 60%–70%). The outstanding technical issues relate to low initial oil rate; artificial lifting of bitumen to the surface; horizontal well operation; and the extrapolation of the process to reservoirs having low permeability, low pressure, or bottom water.

Even though the injection and production wells can be very close (between 5 and 7 m), the mechanism causes the steam-saturated zone (known as the steam chamber) to rise to the top of the reservoir, expand gradually sideways, and eventually allow drainage from a very large area. The method is claimed to significantly improve heavy oil, extra heavy oil, and tar sand bitumen recovery by between 50% and 60% of the original oil in place (OOIP) and is therefore more efficient than most other thermal recovery methods.

The key benefits of the SAGD process are an improved steam-oil ratio and high ultimate recovery (on the order of 60%–70%). Steam, perhaps along with a mixture of hydrocarbon derivatives that dissolve into the oil and help reduce its viscosity, is injected into the upper well. The heat reduces the oil viscosity to values as low as 1–10 cp (depending on temperature and initial conditions) and develops a "steam chamber" that grows vertically and laterally. The steam and gases rise because of their low density, and the oil and condensed water are removed through the lower well. The gases produced during SAGD tend to be methane, with some carbon dioxide and traces of hydrogen sulfide.

To a small degree, the noncondensable gases tend to remain high in the structure, filling the void space and even acting as a partial insulating blanket that helps to reduce vertical heat losses as the chamber grows laterally. At the pore scale and at larger scales as well, flow is through countercurrent, gravity-driven flow, and a thin and continuous oil film is sustained, giving recoveries estimated to be on the order of 70%–80% v/v of the oil in place in suitable reservoirs.

Operating the production and injection wells at approximately the same pressure as the reservoir eliminates viscous fingering and coning processes and also suppresses water influx or oil loss through permeable streaks. This keeps the steam chamber interface relatively sharp and reduces heat losses considerably. Injection pressures are much lower than the fracture gradient, which means that the chances of breaking into a thief zone (an instability problem that plagues all high-pressure steam injection processes, such as cyclic steam soak) are essentially zero.

The SAGD process, like all gravity-driven processes, is extremely stable because the process zone grows only by gravity segregation, and there are no pressure-driven instabilities such as channeling, coning, and fracturing. It is vital in the SAGD process to maintain a volume balance, replacing each unit volume withdrawn with a unit volume injected to maintain the processes in the gravity-dominated domain. If bottom-water influx develops, this indicates that the pressure in the water is larger than the pressure in the steam chamber, and steps must be taken to balance the pressures. Because it is not possible to reduce the pressure in the water zone, the pressure in the steam chamber and production well region must be increased. This can be achieved by increasing the operating pressure of the steam chamber through the injection rate of steam or through reduction of the production rate from the lower well.

Clearly, a low pressure gradient between the bottom water and the production well must be sustained. If pressure starts to build up in the steam chamber zone, then the loss of hot water can take place as well. In such cases, the steam chamber pressure must be reduced and perhaps the production rate must be increased slightly to balance the pressures. In all these cases, the system tends to return to a stable configuration because of the density differences between the phases.

The SAGD process seems to be relatively insensitive to shale streaks and similar horizontal barriers, even up to 3–6 ft thick, that otherwise would restrict vertical flow rates. This occurs because as the rock is heated, differential thermal expansion causes the shale to be placed under a tensile stress, and vertical fractures are created, which serve as conduits for steam (up) and liquids (down). As high temperatures hit the shale, the kinetic energy in the water increases, and adsorbed water on clay particles is liberated. Thus, instead of thermal expansion, dehydration (the loss of water) occurs, and this leads to volumetric shrinkage of the shale barriers. As the shale shrinks, the lateral stress (fracture gradient) drops until the pore pressure exceeds the lateral stress, which causes vertical fractures to open. The combined processes of gravity segregation and shale thermal fracturing make the SAGD process highly efficient with recovery on the order of 60%–70% v/v of the viscous oil in place, which are achievable even in cases where there are many thin shale streaks.

Heat losses and deceleration of lateral growth mean that there is an economic limit to the lateral growth of the steam chamber. This limit is thought to be a chamber width of four times (4 ×) the vertical zone thickness. For thinner zones, horizontal well pairs would therefore have to be placed close together, increasing costs and providing lower total resources per well pair. Consequently, the zone thickness limit (net pay thickness) must be defined for all reservoirs.

The cost of heat is a major economic constraint on all thermal processes. Currently, steam is generated with natural gas, and when the cost of natural gas rises, operating costs rise considerably. The SAGD process is approximately twice as thermally efficient as the cyclic steam stimulation process, and steam-oil ratios are on the order of two. If high recovery ratios are possible, the process has the potential to replace pressure-driven thermal process in all cases where the reservoir is reasonably thick and favorable to the process. Also, because of the lower pressures associated with the SAGD process in comparison with high-pressure processes such as cyclic steam soak and steam drive, greater wellbore stability reduces substantially the number of sheared wells that are common in cyclic steam soak projects.

However, the SAGD process is not suitable for all reservoirs, and the process performs best in clean, continuous sand formations with continuous vertical permeability.

Recently, one of the new improved techniques in SAGD is FAST-SAGD, which combines conventional SAGD wells with cyclic steam stimulation wells (Shin and Polikar, 2004, 2005). These offset wells are drilled at the same depth as that of SAGD production well. After short period of time of operation of the SAGD, which is started to operate similarly to the conventional SAGD, steam is injected through cyclic steam stimulation wells with high injection rate under high operation pressure to make heat easily transfer to reservoir.

However, in the FAST-SAGD operation, steam tends to bypass other wells during the injection period due to the operation of the cyclic steam stimulation wells. Hence, live steam will be produced at producer, which has an adverse effect on the thermal efficiency of the process, thereby causing the production rate to decrease significantly. In order to offset this condition, there are several constrains that need to be considered: (i) The location of the offset wells should be calculated carefully so that the combination of the steam chambers at the expected time to get the effective thermal process, and (ii) the cyclic steam stimulation wells should be planned with consideration of reservoir parameters—permeability, oil saturation, formation heat conduction, and heterogeneity (such as the presence of a shale barrier or a clay formation) in order to achieve high efficiency.

As a result of recent innovations, a collection of hybrid variants is available for thermal recovery processes (Table 3.3). Recognizing that there is no one process that will satisfy the requirements for heavy oil, extra heavy oil, and tar sand bitumen recovery, a hybrid process (as the name suggests) is based on a recovery concept (that has been tested and authenticated) and consists of an added option to assure increased recovery. For example, the hybrid SAGD process (HSAGD process) uses a similar well configuration to the fracture-assisted steam technology coupled with the SAGD process (FAST-SAGD method). However, the wells are operated very differently in terms of the operating conditions. In FAST-SAGD process, the SAGD wells are operated first, and cyclic steam stimulation wells (offset wells) start later and require higher injection pressure and injection rate. Therefore, steam is easy bypass to other wells but the HSAGD process can improve this phenomenon. In the HSAGD process, all cyclic steam stimulation wells are placed in a staggered pattern, and the wells are operated at the same pressure and placed in operation earlier than SAGD wells.

The *expanding-solvent SAGD process* (ES-SAGD process) is a combination of solvent and steam injection that take advantage of the benefits from the heat provided by steam and the miscibility offered by the solvent (Nasr and Ayodele, 2005). The basic concept of ES-SAGD is to inject a light hydrocarbon additive at low concentration together with steam, in a process whereby the dominant force is gravity. The additive is selected in such way that can evaporate and

TABLE 3.3 Hybrid Variants Proposed for Thermal Recovery Processes

Steam-assisted gravity drainage (SAGD)
FAST-SAGD
Expanding-solvent SAGD (ES-SAGD)
Horizontal alternate steam drive (HASD)
Steam and gas push (SAGP)
Steam-assisted gas-oil gravity drainage (SAGOGD)
In situ combustion
Toe-to-heel air injection (THAI)
Basal combustion
Top-down combustion
Solvent-based injection
Vapor extraction (VAPEX)
Liquid addition to steam for enhancing recovery (LASER)
Steam-alternating solvent process (SAS)
Carbon dioxide-based VAPEX process

condense at water conditions. In this way, solvent can condense with steam close to the steam chamber interface. The added hydrocarbon is injected in the vapor form. The condensed solvent dilutes in oil and, aided by the heat, reduces the oil viscosity in the reservoir, thereby enabling flow to the production well. A further way to reduce the high demand of steam is to replace a proportion with a noncondensable gas such as natural gas or nitrogen [often referred to as steam and gas push (SAGP)] or to add a surfactant to the process (Adibhatla and Mohanty, 2008).

Another concept related to the VAPEX process but used in a different manner can be used in conjunction with a steam-based process, such as SAGD. In fact, VAPEX (Das and Butler, 1995) is the nonthermal counterpart of SAGD, and it works on the same principles as SAGD. Instead of steam, a solvent gas, or a mixture of solvents, such as ethane, propane, and butane, is injected along with a carrier gas such as nitrogen or carbon dioxide. Solvent selection is based upon the reservoir pressure and temperature. The solvent gas is injected at its dew point. The carrier gas is intended to raise the dew point of the solvent vapor so that it remains in the vapor phase at the reservoir pressure. A vapor chamber is formed, and it propagates laterally. The main mechanism is viscosity reduction. The process relies on molecular diffusion and mechanical dispersion for the transfer of solvent to the bitumen for viscosity reduction. Dispersion and diffusion are inherently slow and, therefore, are much less efficient than heat for viscosity reduction.

4. In Situ Combustion Processes

The other significant thermal recovery process, in situ combustion (sometimes referred to as *fire flooding*), attempts to recover oil by igniting a portion of the in-place crude by injecting air or oxygen or by chemical or electric means (Moore et al., 1995; Shah et al., 2010). In situ combustion is an oil recovery process in which energy is generated in the reservoir by reactions between oxygen and either original or modified fractions of the crude oil.

The process is initiated by heating the oil-bearing sand around an injection well with a gas burner, with an electric heater, or by the injection of a hot fluid such as steam. An oxygen-containing gas is injected to ignite the oil and to propagate the combustion zone outward from the injector. The elevated-temperature zone displaces reservoir fluids, which are advanced toward the production wells by mechanisms such as vaporization and distillation, steam displacement, hot- and cold-water displacement, and gas flooding. In the process, the characteristic regions are as follows:

1. *Burned zone:* the region that has been swept by the combustion zone. The temperature in the burned zone increases in the direction of the combustion front, and a significant proportion of the generated energy either remains in this region or is lost in the surrounding strata. Under efficient high-temperature burning conditions, this area is essentially devoid of fuel.
2. *Combustion zone:* the main energy generation region, in which oxygen reacts with residual hydrocarbon derivatives to produce carbon oxides and water. Temperature levels in this relatively narrow region are largely determined by the amount of fuel consumed per unit volume of reservoir rock.
3. *Cracking and vaporization region:* assuming that the oxygen has been removed in the combustion zone, hydrocarbon derivatives contacted by the leading edge of the high-temperature region undergo thermal cracking and vaporization. The mobilized light ends are transported downstream and are mixed with the native crude. The heavy residue, nominally defined as coke, is deposited on the core matrix and is the main fuel source for the combustion process.
4. *Steam zone:* connate water and water of combustion move ahead of the high-temperature region. The temperature in the steam zone is dictated by the operating pressure and the concentration of combustion gases.
5. *Condensation front:* the leading edge of the steam bank is the primary area of oil mobilization. Only the residual oil remaining behind the condensation front and steam bank undergoes vaporization and thermal cracking.
6. *Native reservoir:* cold native reservoir altered mainly by the presence of a gas saturation.

The generated coke supports continuous combustion with air and the heat generated by the combustion process increases the temperature of the oil-bearing formation, which causes viscosity reduction by several orders of magnitude and mobilizes the flow of the unburned oil fraction (Castanier and Brigham, 2003). The conventional in situ combustion technique for heavy oil recovery uses vertical injection well and vertical production well arrangement. Therefore, the mobilized oil in a conventional in situ combustion accumulates in the colder region of the oil reservoir before moving to the producer well.

An inefficiency associated with dry in situ combustion is that a significant proportion of the generated energy remains behind the combustion front and cannot be used to increase the temperature in the native portion of the reservoir. If water is injected either simultaneously or alternately with the oxygen-containing gas, some of this energy will be scavenged, since the effective heat capacity of water is several orders of magnitude larger than that of air. The injected water-air ratio determines whether water is transported through the oxidation zone in a vapor or liquid state. At low water-air ratios, water passes through the combustion front as superheated steam, and the process is defined as normal wet combustion. A distinct vaporization front, with a velocity that increases with the water-air ration, is formed in the swept region. Partially quenched combustion occurs when the velocity of this front increases to the extent that liquid water enters but vaporizes before passing all the way through the oxidation region. At elevated water-air ratios, liquid water passes through the oxidation zone, and the process is then defined as superwet or quenched combustion (Moore et al., 1995).

The process has always appeared to be attractive—especially for the recovery of bitumen from tar sand deposits and the potential for in situ upgrading of heavy oil, extra heavy oil, and tar sand bitumen (Chapter 6). In addition to providing the heat to mobilize the oil, in situ combustion of heavy oil, extra heavy oil, and tar sand bitumen can provide some in situ upgrading through the use of minerals or additives (Dabbous and Fulton, 1972; Fassihi and Brigham, 1984, Fassihi et al., 1984; He, 2004; He et al., 2005; Shallcross et al., 1991; Strycker et al., 1999; Castanier and Kovscek, 2005). During in situ combustion of heavy oil, extra heavy oil, and tar sand bitumen, temperatures of up to 700°C (1290°F) can be observed at the combustion front, which is sufficient to promote some upgrading (depending upon the residence time of the heavy oil, extra heavy oil, and tar sand bitumen in the hot zone).

In situ combustion is normally applied to reservoirs containing low-gravity oil but has been tested over perhaps the widest spectrum of conditions of any enhanced oil recovery process. In the process, heat is generated within the reservoir by injecting air and burning part of the formation oil. This reduces the oil viscosity and partially vaporizes the oil in place, and the oil is driven out of the reservoir by a combination of steam, hot water, and gas drive. Forward combustion involves movement of the hot front in the same direction as the injected air; reverse combustion involves movement of the hot front opposite to the direction of the injected air. The performance of in situ combustion is predominantly determined by the four following factors: (i) the quantity of oil that initially resides in the rock to be burned, (ii) the quantity of air required to burn the portion of the oil that fuels the process, (iii) the distance to which vigorous combustion can be sustained against heat losses, and (iv) the mobility of the air or combustion product gases.

The relatively small portion of the viscous oil that remains after, say, the application of steam-based processes or the viscous oil that cannot be moved by steam-based processes becomes the target resource for the in situ combustion process. Production is obtained from wells offsetting the injection locations. In some applications, the efficiency of the total in situ combustion operation can be improved by alternating water and air injection. The injected water tends to improve the utilization of heat by transferring heat from the rock behind the combustion zone to the rock immediately ahead of the combustion zone. Furthermore, the appeal of an in situ combustion process is the potential for partial upgrading of the oil in the reservoir, providing the undesirable constituents of the oil remain in the reservoir (Chapter 6). In addition, the process has the rapid kinetics of a thermal process, and there is no need to generate energy at the surface.

In the process, oxygen is injected into a reservoir, the crude in the reservoir is ignited, and part of that crude is burned in the formation to generate heat. Air injection is by far the most common way to introduce oxygen into a reservoir. Since the injected air normally is cool (except as compression would warm it), the surface lines need to be designed only as would be required to conform with prudent practice. Injection wells generally require special design considerations only where they are specifically vulnerable. Production wells, on the other hand, can be expected to be assaulted to varying degrees by corrosion, erosion, and high temperature from the producing interval to the wellhead, the most severe conditions being at the producing interval. The standard well equipment, therefore, must be modified accordingly. The degree of modification will depend on the crude and water in the reservoir, the friability of the sand, the steel used in the tubing and casing, the method of well completion, the amount of heat and free oxygen, and the type of in situ combustion process or control measures used.

Energy is generated in the formation by igniting bitumen in the formation and sustaining it in a state of combustion or partial combustion. The high temperatures generated decrease the viscosity of the oil and make it more mobile. Some cracking of the bitumen also occurs and an upgraded product rather than bitumen itself is the fluid recovered from the production wells. In some applications, the efficiency of the total in situ combustion operation can be improved by alternating water and air injection. The injected water tends to improve the utilization of heat by transferring heat from the rock behind the combustion zone to the rock immediately ahead of the combustion zone.

Injection of air alone is known as dry underground combustion, in situ combustion, or fire flooding. Generally, unless otherwise specified, underground or in situ combustion means dry forward combustion. In the dry forward combustion process, the combustion front moves out from the injection well as air injection is continued. The combustion front moves in the same direction as the air. Reverse combustion occurs when the combustion front moves in a direction opposite that of the injected air. Reverse combustion is achieved by igniting the crude near a production well while temporarily injecting air into it. Upon resumption of the normal air-injection program, the combustion front will move toward the injection wells. Wet and partially quenched combustion, also known by the acronym COFCAW (combination of forward combustion and waterflooding) (Dietz Weijdema, 1968a; Parrish and Craig, 1969), uses water injection during the combustion process to recuperate the heat from the burned zone and adjacent strata. In this process, the ratio of injected water to air is used to control the rate of advance of the combustion front, the size of the steam zone, and the temperature distribution.

Both forward and reverse combustion methods have been used with some degree of success when applied to tar sand deposits. The forward combustion process has been applied to the Orinoco deposits and in the Kentucky sands. The reverse combustion process has been applied to the Orinoco deposit (Burger, 1978) and the Athabasca. In tests such as these, it is essential to control the airflow and to mitigate the potential for spontaneous ignition (Burger, 1978). A modified combustion approach has been applied to the Athabasca deposit. The technique involved a heat-up phase and a production (or blowdown phase) followed by a displacement phase using a fireflood-waterflood (COFCAW) process.

The main parameters required in the design of an in situ combustion project are, in addition to operating costs, (i) the fuel concentration per unit reservoir volume burned, (ii) the composition of the fuel, (iii) the amount of air required to burn the fuel, (iv) the volume of reservoir swept by the combustion zone, (v) the required air-injection rates and pressures, and (vi) the oil production rate (Alexander et al., 1962). Injection of air alone is known as dry underground combustion, in situ combustion, or fire flooding (Kuhn and Koch, 1953). However, there are several variants of the in situ combustion process.

The combustion zone is burned out as the combustion front advances and any water formed or injected will turn to steam in this zone due to residual heat. This steam flows on into the unburned area of the formation, helping to heat it. The high temperature just ahead of the combustion zone causes lower-boiling constituents of the oil or products to vaporize, leaving a heavy residual coke or carbon deposit as fuel for the advancing combustion front. In the vaporizing that contains combustion products, vaporized low-boiling hydrocarbon derivatives, and steam, cooling (owing to the distance of the zone from the combustion front) causes low-boiling hydrocarbon derivatives to condense and steam to revert back to hot water. This action displaces oil and, with combustion gases, aid in driving the oil to production wells. The oil bank (an accumulation of displaced oil) cools as it moves toward production wells, and temperatures will drop to that near initial reservoir temperature. When the oil bank reaches the production wells, the oil, water, and gases will be brought to the surface and separated—the oil to be sold and the water and gases sometimes reinjected. The process is terminated by stopping air injection when predesignated areas are burned out or the burning front reaches production wells.

Ideally, the fire propagates uniformly from the air-injection well to the producing well, moving oil and combustion gases ahead of the front, and any coke remaining behind the moved oil provides the fuel. Temperatures in the thin combustion zone may reach several hundred degrees centigrade, sufficient to crack the heavy oil, extra heavy oil, and tar sand bitumen into lower-boiling products. The oil is subjected to a combination of miscible displacement by the condensed lower-boiling hydrocarbon derivatives, hot-water drive, vaporization, and steam and gas drive. As the temperature in the volume element exceeds about 345°C (650°F), the oil will more than likely undergo thermal cracking to form a volatile fraction and a low-volatility, coke-like residue. The volatile products are carried in the gas stream, while the coke-like residue is burned as fuel in the combustion zone. The heat generated at the combustion zone is transported ahead of the front by conduction through the formation matrix and by convection of the vapors and liquids (Wu and Fulton, 1971).

The thickness of the combustion zone is variable, and the temperature is usually on the order of 345–650°C (650–1200°F) but may be difficult to control (this is the issue with many combustion-related recovery processes). As the combustion front moves forward, a zone of clean sand is left behind where only air flows. As a result of distillation and thermal cracking, the quality of the produced oil is improved. For example, in South Belridge (California), the produced oil gravity was as high as 18° API, compared with 12.9° API for the original oil.

Heavy oil, extra heavy oil, and tar sand bitumen upgrading is of major economic importance. Numerous field observations have shown upgrading of 2–6° API for heavy oil, extra heavy oil, and tar sand bitumen undergoing combustion (Ramey et al., 1992). For the most past, in situ combustion processes (also known as fireflood processes) are not new; work on various aspects of the processes has continued since at least 1923 (Howard, 1923; Wolcott, 1923; Kuhn and Koch, 1953: Grant and Szasz, 1954). In fact, in situ combustion has been applied in over a hundred fields (Farouq Ali, 1972; Chu 1977, 1982; Brigham et al., 1980). The South Belridge (California) project, which began commercial operations in 1964, is of special significance since it was a commercial success, and analyses of the field data yielded a number of useful, new concepts and correlations. Certain ideas carried over from waterflooding were found to be inappropriate to the in situ combustion process (Gates and Ramey, 1958, 1980; Ramey et al., 1992).

Furthermore, in the South Belridge (California) field, it became evident that reservoir lithology is an important parameter in fuel deposition (Gates and Ramey, 1958), and the amount of fuel deposited increases with the addition of clay to the sample of oil and sand (Bousaid and Ramey, 1968). This agrees chemically with the known catalyst activity of minerals in refining processes and the ability of clay to adsorb polar constituents such as the carbenes and carboids that are the thermal precursors to coke (Parkash, 2003; Gary et al., 2007; Speight, 2014, 2017; Hsu and Robinson, 2017).

In terms of the chemistry of the process, at temperatures in excess of ~345°C (650°F), reactions between oxygen and the organic fuel result in the production of carbon dioxide, carbon monoxide, and water as the principal reaction products. However, if the temperature is less than ~345°C (650°F), water and oxygenated organic compounds (such as phenol derivatives) are the principal reaction products along with carboxylic acids, aldehydes, ketones, and alcohols, which may be the result of low-temperature oxidation (Burger and Sahuquet, 1972). Moreover, in many cases, it is often assumed that the minerals and the water in the reservoir are inert. However, the presence of pyrite (FeS_2) increases air requirements, and naturally occurring carbonates (MCO_3, M is a divalent metal) and sulfates (MSO_4, M is a divalent metal) decompose at high temperatures and react with acid products formed by low-temperature oxidation (Poettmann et al., 1967).

The fire front can be difficult to control, and it may propagate in a haphazard manner, resulting in premature breakthrough to a producing well. There is a danger of a ruptured well with hot gases escaping to the surface. The produced fluid may contain an oil-water emulsion that is difficult to break, and contrary to expectations, it may also contain heavy-metal compounds that are difficult to remove in the refinery.

The wellbore near the pay zone or for that matter any part of the injection well that might come in contact with free oxygen and fuel (crude oil) should be designed for high thermal stresses. Crude oil is likely to enter the wellbore by gravity drainage where the air enters the formation preferentially over a short segment of a large open interval that has adequate vertical permeability. This crude inflow may be increased as the reservoir temperature near the wellbore increases as a result of the heat generated either by the ignition system used in the wellbore or by the combustion process itself (including reverse combustion following spontaneous ignition a short distance into the reservoir). When designing injection wells, precautions should be taken against any likelihood of combustion in the wellbore.

During in situ combustion or fire flooding, energy is generated in the formation by igniting heavy oil, extra heavy oil, and tar sand bitumen in the formation and sustaining it in a state of combustion or partial combustion. The high temperatures generated decrease the viscosity of the oil and make it more mobile. Some cracking of the heavy oil, extra heavy oil, and tar sand bitumen occurs, and an upgraded product rather than heavy oil, extra heavy oil, and tar sand bitumen itself is the fluid recovered from the production wells.

In many field projects, the high gas mobility has limited recovery through its adverse effect on the sweep efficiency of the burning front. Because of the density contrast between air and reservoir liquids, the burning front tends to override the reservoir liquids. To date, combustion has been most effective for the recovery of viscous oils in moderately thick reservoirs in which reservoir dip and continuity provide effective gravity drainage or operational factors permit close well spacing.

The use of combustion to stimulate oil production is regarded as attractive for deep reservoirs. In contrast to steam injection, it usually involves no loss of heat. The duration of the combustion may be <30 days or as much as 90 days, depending on the requirements. In addition, backflow of the oil through the hot zone must be prevented, or coking will occur. However, the combustion process must be controlled because of the varied pathways by which the combustion reaction can proceed (Fig. 3.2).

The process has (i) the potential to be thermally more efficient than steam-based processes; (ii) has no depth restriction, other than advised caution in shallow-depth reservoirs/deposits; and is (iii) well suited to relatively thin (<25 ft) reservoir/deposit sands. However, the process is complicated insofar as it is difficult to predict or control, and operational problems from the high temperature include cement failures, sanding/erosion, corrosion at both injection and production well because of oxygen and moisture, and high gas production rate.

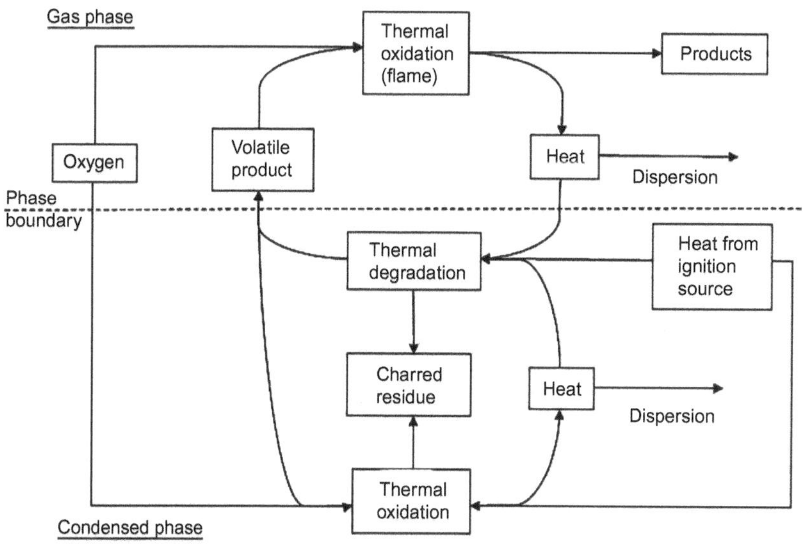

FIG. 3.2 The potential reaction pathways by which the combustion process can proceed.

4.1 Forward Combustion

The most common form of in situ combustion is *dry forward combustion*, which involves movement of the hot front in the same direction as the injected air, while reverse combustion involves movement of the hot front opposite to the direction of the injected air. In the process, the combustion front moves in the same direction as the air flow, but *reverse combustion* occurs when the combustion front moves in a direction opposite to the flow of the injected air. Reverse combustion is achieved by igniting the crude near a production well while temporarily injecting air into it. Upon resumption of the normal air-injection program, the combustion front will move toward the injection wells.

Temperature levels in dry forward combustion are affected by the amount of fuel burned per unit bulk volume of reservoir rock. The temperature levels in turn affect the displacement, distillation, stripping, cracking, and boiling of the crude and formation of "solid" fuel downstream of the combustion front. Temperatures in the range of 650–1500°F have been observed frequently both in the laboratory and in the field. At high temperatures, the combustion zone is very thin. At moderate temperatures, the combustion reaction proceeds slowly enough to allow significant leakage of free oxygen in the direction of flow, thus increasing the thickness of the reaction zones. At lower temperatures, a smoldering reaction with the bypassed air may occur over distances of several feet. This usually happens when air is injected into previously unheated crude-containing reservoirs; the ensuing smoldering reaction generates heat and ultimately causes spontaneous ignition. Air bypassing may also occur in any part of the reservoir if the local air flux is very great, even when fuel is present, and the temperature is high. A characteristic of the dry forward combustion process is that the temperature of the burned zone remains quite high because the heat capacity of the air injected is too low to transfer a significant amount of heat. For this reason, water sometimes is used during or after the combustion process to help transfer the heat from the burned zone and to use it efficiently downstream, where the oil is.

In this process, air is injected into a heavy oil, extra heavy oil, and tar sand bitumen reservoir; the crude is ignited in situ; and the resulting combustion front moves away from the injection well. The heat generated at the combustion front propagates through the reservoir, reduces the oil viscosity, and thereby increases the oil production rate and recovery. The propagation of a combustion front in a reservoir is the most rapid method of thermal recovery. The combustion front can move more rapidly than heat can be moved by conduction and convection in a reservoir, and the convective heat wave velocity for the case of air injection is about one quarter that of the combustion front (Martin et al., 1958; Ramey, 1971). The hydrocarbon products released from the zone of combustion move into a relatively cold portion of the formation. Thus, there is a definite upper limit of the viscosity of the liquids that can be recovered by a forward combustion process. On the other hand, since the air passes through the hot formation before reaching the combustion zone, burning is complete; the formation is left completely cleaned of hydrocarbon derivatives.

The fuel actually burned in forward in situ combustion is not the crude oil in the reservoir but is primarily the carbon-rich residue resulting from thermal cracking and distillation of the residual crude near the combustion front. Naturally occurring coal, if present in the rock, also can contribute to the fuel available for combustion. The amount of fuel present per unit bulk volume of reservoir is an extremely important parameter in combustion operations, for it generally determines the air required to burn a unit bulk volume of reservoir. In wet combustion operations, however, the addition of sufficient water may reduce the temperature to the point that the fuel will not burn completely. Where the combustion of all the fuel is unnecessary, incomplete combustion may be an advantage since it reduces the amount of air required to burn through a unit bulk volume of reservoir. The process also can be carried out under partially quenched conditions so that a fraction of the crude is burned or oxidized at low temperatures. The reverse combustion process also consumes a fraction of the crude oil in place.

However, low-temperature oxidation that occurs when the oxygen can bypass the flame front can cause compositional changes in the oil. For example, low-temperature oxidation can significantly increase the asphaltene content of the oxidized oil. The low-temperature oxidation does not significantly affect the volatility or carbon number distributions of either the light or heavy oils (Babu and Cormack, 1984; Fassihi et al., 1990). This is consistent with the higher-molecular-weight polar and/or the aromatic species in the oil reacting readily with the oxygen to convert nonasphaltene material to asphaltene material (Moschopedis and Speight, 1975; Speight, 2014).

The fuel burned in a formation also is known as the fuel content, the fuel consumption, or the fuel availability. For combustion at high temperatures, the terms are interchangeable. Not all companies have the equipment required for determining the air requirements and the fuel burned. Some service companies and consulting firms have those facilities. The field user of the combustion process generally is interested in (i) the amount of air required to burn a unit bulk volume of reservoir rock and (ii) the amount of crude available for displacement from the burned zone. Although it is assumed that any unburned crude is displaced, careful analyses of the burned matrix usually show some organic residue. In such cases, the estimate of fuel available—the organic residue plus the fuel burned—cannot, in principle, be used interchangeably with the estimate of fuel burned. However, this organic residue is generally negligible where combustion occurs at high temperatures. Organic residues also are found after conventional core extraction.

Generally, forward combustion is referred to as *dry forward combustion,* and the effects of any reservoir water are, for unknown reasons, ignored. Temperature levels in dry forward combustion, which affect the displacement, distillation, stripping, cracking, and formation of solid fuel downstream of the combustion front, are affected by the amount of fuel burned per unit volume of reservoir rock. At high temperatures (~815°C, 1500°F), the combustion zone is very thin, whereas at lower temperatures (~345°C, 650°F), a smoldering reaction with the bypassed air may occur over distances of several feet and generate heat and ultimately cause spontaneous ignition. At intermediate temperatures (~600°C, 1100°F), the combustion reaction proceeds slowly enough to allow significant leakage of free oxygen in the direction of flow, thus increasing the thickness of the reaction zones.

A characteristic of the dry forward combustion process is that the temperature of the burned zone remains quite high because the heat capacity of the air injected is too low to transfer a significant amount of heat. For this reason, water sometimes is used during or after the combustion process to help transfer the heat from the burned zone to downstream areas.

Forward combustion is particularly applicable to reservoirs containing heavy oil, extra heavy oil, and tar sand bitumen and/or with a high effective permeability. Even though lower effective reservoir permeability is required for air injection compared with steam injection, the reservoir ahead of the combustion front is subject to plugging as the vaporized fluids cool and condense. Consequently, a relatively high permeability (400–1000 md) and relatively low bitumen saturation (45%–65% of pore volume) are most favorable for this process. The combustion process yields a partially upgraded product because the temperature gradient ahead of the combustion front mobilizes the lower-boiling hydrocarbon components that move toward the cooler portion of the reservoir and mix with unheated bitumen. This mixture is eventually produced through a production well. The higher-boiling components (e.g., coke) are left on the sand grains and are consumed as fuel for the combustion. Under certain operating conditions, a significant cost saving is attained by injecting oxygen or oxygen-enriched air rather than atmospheric air because of reduced compression costs and a lower produced gas/oil ratio.

Another form of in situ combustion is the *wet combustion* method, in which air and water are injected concurrently or alternately. The purpose of injecting water is to recuperate and transport heat from the burned zone to the colder regions downstream of the combustion front. This method may be considered for thin reservoirs, where heat loss to adjacent formations is significant (Dietz and Weijdema, 1963; Parrish and Craig, 1969; Dietz, 1970; Beckers and Harmsen, 1970; Burger and Sahuquet, 1973).

Wet combustion is the general name given to the process in which water passes through the combustion front along with the air (or other reaction gas). The process always has been applied to forward combustion. The water entering the combustion zone may be in either the liquid or the vapor phase or both. Ideally, the water is injected along with the air but is injected intermittently with the air when the flow resistance to two-phase flow near the injection well is too high to achieve the desired injection rates. In addition to reducing the amount of air required to move a combustion front through a reservoir unit volume, the wet combustion process also generates a significant steam zone ahead of the combustion front. The steam zone usually leaves only residual oil in its swept path, so that it is generally uneconomical to continue a wet combustion project once the response to the steam zone becomes unattractive and the process can be terminated.

Wet combustion should be considered an alternative to dry combustion in all cases, since it could reduce the air requirements and accelerate the production response. However, the process should not be used in formations where flow resistance is marginally acceptable for dry combustion, because the addition of water will increase the flow resistance further. Neither should the process be used where the injected water would interact adversely with formation clays or other minerals to reduce the reservoir injectivity. The effectiveness of the wet combustion process decreases where gravity override is expected to be important, especially in thick, massive intervals having good vertical continuity and high permeability.

The addition of water during the combustion process has interesting consequences, such as the following: (i) Heat is transferred more effectively than with air alone and (ii) the steam zone ahead of the combustion front is larger, and thus, the reservoir is swept more efficiently than with air alone. The improved displacement from the steam zone results in lower fuel availability and consumption in the combustion zone, so that a greater volume of the reservoir is burned for a given volume of air injected. Water must be injected in the wet combustion process. In low-permeability reservoirs, it may be difficult to inject both air and water simultaneously at the desired rates in which case the air and water may be injected alternately with control of the duration of the air and water injection periods so that the desired average water/air ratio is achieved.

The water/air ratio also is controlled to obtain desired improvements in combustion-front velocities or temperature levels. At low "water/air ratios," all the water that reaches the combustion front already has been converted to steam. If the water/air ratios are kept sufficiently high, most of the water reaching the combustion front still will be in the liquid phase. This reduces the maximum temperature level, in some cases to the temperature level corresponding to the partial pressure of steam in the steam/gas mixture that is adequate for thermal drive processes.

A benefit of the addition of water during the combustion process is that heat is transferred more effectively than with air alone. In addition, the steam zone ahead of the combustion front is larger, and the reservoir is swept more efficiently than with air alone. The improved displacement from the steam zone results in lower fuel availability and consumption in the combustion zone, so a greater volume of the reservoir is burned for a given volume of air injected. Another benefit and more pertinent to heavy oil, extra heavy oil, and tar sand bitumen technology is the production of a partially upgraded product. The temperature gradient ahead of the combustion front either causes the lower-molecular-weight (more volatile) constituents to distill and move toward the cooler portion of the reservoir and mix with unheated oil or some of the higher-molecular-weight (less volatile) constituents crack (thermally decompose), and the volatile products also move toward the cooler portion of the reservoir and mix with unheated oil. The cracking process also produces a carbonaceous residue that deposits on the reservoir rock and is consumed as fuel during the combustion process.

The operative term for any in situ combustion project is *process control* or *flame front control*, which related to the amount of air and the amount and type of fuel in the subterranean formation(s). The fuel burned in a reservoir also is known as the *fuel content*, the *fuel consumption*, or the *fuel availability*. In terms of the fuel itself, the fuel that is burned during the in forward in situ combustion process is not the crude oil in the reservoir, but it is, for the most part, the carbonaceous residue that was produced by thermal cracking and distillation (near the combustion front) of the residual crude oil—in the current context, the fuel will be part of the heavy oil, extra heavy oil, or tar sand bitumen. The amount of fuel present per unit bulk volume of reservoir is an extremely important parameter in combustion operations and is a determinant in the amount of air required to burn a unit bulk volume of reservoir. In wet combustion operations, however, the addition of sufficient water may reduce the temperature to the point that the fuel will not burn completely. Where the combustion of all the fuel is unnecessary, incomplete combustion may be an advantage since it reduces the amount of air required to burn through a unit bulk volume of reservoir. The process also can be carried out under partially quenched conditions so that a mid-boiling fraction of the crude oil is burned or oxidized at low temperatures. On the other hand, the reverse combustion process also consumes a mid-boiling fraction of the crude oil.

Air consumption is affected not only by reactions with organic fuel but also by reactions between oxygen and some minerals (such as pyrite—iron sulfide [FeS_2]) in the reservoir. However, when minerals are involved in the process and react, special attention must be paid to the amount of fuel burned.

In addition to the fuel available and/or consumed, the parameters of interest are (i) the amount of air required to burn a unit bulk volume of reservoir rock and (ii) the amount of crude available for displacement from the burned zone. The general assumption is that any unburned crude oil is displaced; this is not typically the case, and there is usually a (partially combusted or thermally generated) carbonaceous residue, similar to coke or semicoke. In such cases, the estimate of fuel available—the carbonaceous residue plus the fuel burned—cannot, in principle, be assumed to be the fuel burned. However, where combustion occurs at high temperatures, the carbonaceous residue may be negligible. In fact, there is a reasonable correlation between fuel burned and air consumed when the combustion occurs at high temperatures. At lower temperatures, some fraction of the consumed oxygen generally is reacted with the crude without generating carbon oxide gases or water.

4.2 Reverse Combustion

Another variation of the in situ combustion process is the *reverse combustion* method. In this technique, the combustion zone is initiated at a production well. In this process, the combustion front moves opposite to the direction of air flow. Combustion is initiated at the production well, and the combustion front moves against the air flow. In a frontal displacement model, the oil that is displaced must pass through the burning combustion zone and through the hot burned zone. The extent of the high-temperature region in the burned zone depends (among other things) on the rate of heat losses to the adjacent formations. As the crude is displaced through the combustion front, it is cracked; the light ends vaporize, and the heavy ends contribute a residue. As the vapors approach cooler sections of the burned zone, some condensation occurs, and liquid oil and water may exist near the outlet. The region upstream of the combustion zone is heated by heat conduction, which leads to low-temperature oxidation reactions and the generation of considerable heat at significant rates.

Thus, the reverse combustion front travels countercurrent to the air toward the injection well where air is injected. The oil flows toward the production well, through the combustion zone. Since no oil bank is formed, the total flow resistance decreases with time, and thus, this method is particularly suitable for reservoirs containing very viscous crude oils. One disadvantage of this method is the likelihood of spontaneous ignition. Spontaneous ignition would result in oxygen being consumed near the injector, and the process would change to forward combustion (Dietz and Weijdema, 1968b). Another disadvantage of reverse combustion is the inherent instability of the process, which results in narrow combustion channels being formed and therefore an inefficient burn (Gunn and Krantz, 1980; Johnson et al., 1980). Historically, reverse combustion has been difficult to maintain because the oxygen is depleted not far from the injection well (Dietz and Weijdema, 1968b). Furthermore, sustained air injection into an unheated reservoir generally leads to spontaneous ignition near the injection well (Elkins et al., 1974; Burger, 1976; Tadema and Weijdema, 1970).

In reverse combustion the combustion front is initiated at the production well and moves backward against the air flow. As the combustion front reaches the heavy oil, extra heavy oil, and tar sand bitumen, a significant amount of cracking occurs, and a relatively large amount of solid fuel (compared with the amount formed during forward combustion) is deposited on the reservoir rock. However, recovery of the partially upgraded product is lower than in forward combustion because some is burned in the process, and a high degree of *equivalent oil saturation* occurs in the burned zone (including unburned solid fuel). On the other hand, the API gravity of the recovered product is increased significantly by extensive cracking because the products flow through the hot burned zone and are subject to secondary, tertiary, or higher level cracking reactions.

The addition of water or steam to an in situ combustion process can result in a significant increase in the overall efficiency of the process. Modes of application include injection of alternate slugs of air (oxygen) and water or coinjection of air (oxygen) and steam. Again, the combination of air (oxygen) injection and steam or water injection increases injectivity costs that may be justified by increased bitumen recovery. Two major benefits may be derived. Heat transfer in the reservoir is improved because the steam and condensate have greater heat-carrying capacity than combustion gases and gaseous hydrocarbon derivatives. Sweep efficiency may also be improved because of the more favorable mobility ratio of steam-oil compared with gas-oil. However, a major difficulty with the process is that of maintaining an active front since sustained air injection into an unheated reservoir generally leads to spontaneous ignition near the injection well. As a result, the oxygen then is depleted not far from the injection well; reverse combustion cannot be maintained.

The reverse combustion process, in contrast to the forward combustion process, is particularly applicable to reservoirs with lower effective permeability. The process is more effective because the lower permeability would cause the reservoir to be plugged by the mobilized fluids ahead of a forward combustion front. In the reverse combustion process, the vaporized and mobilized fluids move through the heated portion of the reservoir behind the combustion front. The reverse combustion partially cracks the oil, consumes a portion of the oil as fuel, and deposits residual coke on the sand grains, leaving 40%–60% of the oil as recoverable oil. This coke deposition serves as a cementing material, reducing movement and production of sand.

Process efficiency is affected by reservoir heterogeneity that reduces horizontal sweep. The underburden and overburden must provide effective seals to avoid the loss of injected air and produced oil. Process efficiency is enhanced by the presence of some interstitial water saturation. The water is vaporized by the combustion and enhances the heat transfer by convection. The combustion processes are subject to override because of differences in the densities of injected and reservoir fluids. Production wells should be monitored for and equipped to cool excessively high temperatures (>1095°C, >2000°F) that may damage downhole production tools and tubulars.

Applying a preheating phase before the recovery phase may significantly enhance the steam or combustion extraction processes. Preheating can be particularly beneficial if the saturation of the more viscous oil (API gravity = 12° or lower) is sufficiently great to lower the effective permeability to the point that production is precluded by reservoir plugging. Preheating increases the mobility of the oil by raising its temperature and lowering its viscosity, and the outcome is a lower required pressure to inject steam or air to recover the oil.

Using combustion to stimulate oil production is regarded as attractive for deep reservoirs. In contrast to steam injection, it usually involves no loss of heat. The duration of the combustion may be <30 days or ~90 days, depending upon the requirements. In addition, backflow of the oil through the hot zone must be prevented, or coking occurs.

When it is determined that the reservoir should be preheated, there are several methods by which this can be accomplished. Conducting a reverse combustion phase in a zone of relatively high effective permeability and low oil saturation is one method. Steam or hot gases may be rapidly injected into a high-permeability zone in the lower portion of the reservoir. In the *fracture-assisted steam technology* (FAST) process, steam is injected rapidly into an induced horizontal fracture near the bottom of the reservoir to preheat the reservoir. This process has been applied successfully in three pilot projects in Southwest Texas. Shell has accomplished the same preheating goal by injecting steam into a high-permeability bottom-water zone in the Peace River (Alberta) field. Electric heating of the reservoir by radio-frequency waves may also be an effective method.

A variation of the combustion process involves the use of a heat-up phase, a blowdown (production) phase, and then a displacement phase using a fire-water flood (a *combination of forward combustion and waterflood* (COFCAW)). In the process, water is injected simultaneously or alternately with air into the formation and is actually wet forward combustion, which was developed to use the great amount of heat that would otherwise be lost in the formation. The injected water recovers the heat from behind the burning front and transfers it to the oil bank ahead. Because of this additional energy, the oil displacement is more efficient and requires less air. In spite of these advantages, a wet combustion process cannot avoid liquid-blocking problems, and the use of wet combustion is limited by the oil viscosity.

In any field in which primary recovery operations are followed by secondary or enhanced recovery operations, there is a change in product quality. Product oils recovered by the thermal stimulation of heavy oil, extra heavy oil, and tar sand bitumen reservoirs show some improvement in properties over those of the heavy oil, extra heavy oil, and tar sand bitumen in place. Although this improvement in properties may not appear to be too drastic, nevertheless it usually is sufficient to have major advantages for refinery operators. Any incremental increase in the units of the hydrogen/carbon ratio can save amounts of costly hydrogen during upgrading. The same principles are also operative for reductions in the nitrogen, sulfur, and metal content. This removal of nitrogen, sulfur, and metals from the products also improves catalyst life and activity when the product oil is refined. In short, in situ recovery processes may have the added benefit of *leaving* some of the more obnoxious constituents (from the processing objective) in the ground.

A combustion displacement process may be more attractive than a steam-drive process, assuming that a downhole steam generator is also not available to the reservoir. Conditions that might preclude a steam drive and be in favor of a combustion process include (i) high sustained injection pressures (above 1500 psi), (ii) excessive heat losses from the injection well in reservoirs more than 4000 ft deep, (iii) a lack of a supply of fresh water or treatment costs that make the use of steam prohibitively expensive, (iv) serious clay swelling problems, and (v) thin or low-porosity sands where heat management using wet combustion processes tends to make them more efficient than steam drives (Wilson and Root, 1966).

Wet combustion would be considered instead of dry combustion where there is ample available water and where water/air injectivity is favorable. Wet combustion would not be used where there is little likelihood that the water would move through the burned zone to recuperate heat effectively, as in gravity-dominated operations (Koch, 1956; Gates and Sklar, 1971).

Applying a preheating phase before the bitumen recovery phase may significantly enhance the steam or combustion extraction processes. Preheating can be particularly beneficial if the saturation of highly viscous bitumen is sufficiently great as to lower the effective permeability to the point of production being precluded by reservoir plugging. Preheating partially mobilizes the bitumen by raising its temperature and lowering its viscosity. The result is a lower required pressure to inject steam or air and move the bitumen.

Preheating may be accomplished by several methods.

4.3 The THAI Process

In situ combustion may make a comeback through a relatively recent concept or at least a derivative thereof. Toe-to-heel air injection (THAI) is based on the geometry of horizontal wells that may solve the problems that have plagued conventional in situ combustion. The well geometry is based on a short flow path so that any instability issues that are associated with conventional combustion are reduced or, as best as can be determined, eliminated.

One of the prerequisites for in situ combustion is fuel (i.e., coke from the combusted oil) availability in the reservoir, which is sufficient to support the propagation of the combustion front. The fuel accessibility is a function of thermal cracking of high-molecular-weight hydrocarbon molecules. This complex process is primarily a function of (i) oil composition, (ii) cracking temperature, (iii) reservoir rock properties, (iv) reservoir pressure, (v) fluid flow, and (vi) temperature. The main distinguishing feature between THAI process and a conventional in situ combustion process is that the THAI process uses a horizontal producer well(s) instead of vertical producer well as in conventional in situ combustion process. The propagation of the combustion front from toe to heel due to continuous injection of air creates different zones, namely, (i) the burned zone, (ii) the combustion front, (iii) the coke production zone, (iv) the mobile oil zone, (v) and the cold oil zone.

In the process, a vertical air-injection well is combined with a horizontal production well and offers a way to control the forefront and produce a product that is partially upgraded and ready for the refinery. In the process, a combustion front is created where part of the oil in the reservoir is burned, generating heat that reduces the viscosity of the oil, allowing it to flow by gravity to the horizontal production well. The combustion front sweeps the oil from the toe to the heel of the horizontal producing well, recovering an estimated 80% of the original oil in place while partially upgrading the crude oil in situ.

The temperature ranges from 500°C to 600°C (930–1110°F) in the combustion zone, to aid thermal cracking (or pyrolysis) of high-molecular-weight hydrocarbon derivatives. A mobile oil zone is created ahead of the combustion front where most of the thermal upgrading reactions occur. The mobilized and thermally upgraded oil in mobile oil zone does not pass through the cold region as in a conventional in situ combustion process but is drawn in a downward direction into the horizontal production well. This eliminates the problem of gas overriding and bypass, as the horizontal well seals as a result of coke formation during the propagation of the combustion front. The recovery factor of the THAI process ranges from 70% to 80% v/v of oil in place.

The THAI process has the potential to operate in reservoirs that are lower in pressure, of a lower quality, thinner, and deeper than required for the SAGD process. In addition, the THAI process integrates existing proved technologies and provides the opportunity to create a step change in the development of heavy oil, extra heavy oil, and tar sand bitumen resources globally. The horizontal well arrangement provides a unique gravity and pressure drawdown geometry. Another key feature is that oil recovery occurs via a short displacement mechanism, which requires oil to move downwards (with the help of gravity) typically just 15–30 ft, as opposed to the lateral movement of several hundred feet in the usual combustion processes.

The THAI process begins with preheating both wellbores using steam to initiate oil mobility and clear pore space between the injector and the toe of the producing well. After ignition (autoignition with some oils), the energy to sustain in situ combustion comes from the burning of the coke that is continuously laid down within the reservoir. Product sulfur is reduced, as are heavy metals, which are left as an inert residue on the reservoir rock. No water or gas fuel is required during production, and the produced water can be treated to usable industrial quality.

A further benefit of the THAI process is that it performs in situ upgrading through thermal cracking of the heavy oil, extra heavy oil, and tar sand bitumen. Laboratory tests achieved upgrading by up to 10° API. On this basis, a 10° API oil could be expected to yield an 18°–20° API oil at the surface. This is a very desirable feature of any recovery process since every increase of 1° API can mean refinery savings of several dollars per barrel.

The reservoir matrix is an important aspect of the THAI process; some clay content is required to help thermal cracking of the oil ahead of the combustion front. The process can break down intermittent shale breaks that could otherwise obstruct permeability. The THAI process can also cope with a bottom-water zone by effectively steaming it out. Tests also indicate that the THAI process can handle a gas cap; gravity and pressure draw down all the gases ahead of the combustion front.

Ahead of the combustion front (typically at a temperature of ~600°C, 1112°F) is the coking zone, in front of which is a 10–15 ft-wide mobile oil zone through which drainage takes place into the horizontal well. These zones move through the reservoir at about 1–3 ft/day, depending on the air-injection rate. The temperature drops to between 200°C and 350°C at the front of the mobile oil zone, with a corresponding reduction in the rate of drainage. Ahead of the mobile oil zone is the cold immobile virgin oil layer, through which there is no communication for gas. This characteristic of the process geometry means that the only way out is down into the open section of the horizontal well. The horizontal well trajectory is thus a built-in self-controlling guidance system for fluid flow. This makes the process much more controllable than conventional in situ recovery systems, in which fluids are less controlled and can move and penetrate anywhere in the reservoir.

The THAI technology offers many potential advantages over the SAGD technology, including higher resource recovery of the original oil in place, lower production and capital costs, minimal usage of natural gas and fresh water, a partially upgraded crude oil product, reduced diluent requirements for transportation, and significantly lower greenhouse gas emissions. The THAI process also has potential to operate in reservoirs that are lower in pressure, containing more shale, lower in quality, thinner, and deeper than SAGD. This type of technology could be utilized in deep heavy oil, extra heavy oil, and tar sand bitumen resources both onshore and offshore.

In summary, the THAI process has many potential technical benefits including the following: (i) higher source recovery that is estimated at 70%–80% of the oil in place and is potentially feasible for use over a broader range of reservoirs, including reservoirs having low pressure, thin reservoirs, and previously steamed reservoirs; (ii) well geometry that enforces a short flow path so that the instabilities associated with the longer flow path in conventional combustion methods are reduced or even eliminated; and (iii) a lower environmental impact insofar as there is negligible freshwater use, less greenhouse gas emissions, a smaller surface footprint, and easier reclamation.

As an extension of the THAI process, the CAPRI process involves a layer of refinery-type catalyst along the outside of the horizontal producer well; it is therefore the catalytic variant of the THAI process. It uses an annular sheath of solid catalyst surrounding the horizontal producer well in the bottom of the oil layer. The thermally cracked oil produced drains into the horizontal producer well, first passing through the layer of catalyst where the high pressure and temperature in the reservoir enable thermal cracking and hydroconversion reactions to take place so that only light, converted oil is produced at the surface.

5. Comparison of Recovery Methods

Thermal recovery processes are used for either stimulation or flooding with the choice typically being governed by the properties of the formation (Shah et al., 2010). In formations that are small or that have relatively poor continuity, it may not be feasible to drill enough wells to implement a flooding operation. In fact, if drilling additional wells to ensure adequate communication over close spacing is prohibitively costly, there may be no choice but to consider a stimulation treatment to increase both recovery rates and ultimate economic recovery. If the viscosity of the oil is high, cyclic steam injection has proved to be a successful stimulation technique as long as natural driving forces are available to produce the crude once its viscosity is reduced. Combustion stimulation has been found to be successful in burning solid organic particles, in stabilizing clay minerals, and in increasing absolute permeability near the treated well. These effects are particularly attractive in low-permeability reservoirs.

Cyclic steam injection also is used as a precursor to steam drives. In reservoirs containing very viscous crudes, the flow resistance between wells may be so great that steam-injection rates are severely limited, making steam drives both technically inefficient and uneconomic. Cyclic steam injection reduces the flow resistance near wells, where the resistance is most pronounced; this alone improves the injection rate attainable during steam drives by reducing the resistance to flow between wells. Repeated cyclic steam injection reduces the flow resistance still farther from the wells and may lead to connecting the heated zones of adjacent wells and further improving the operability of steam drives.

In addition to its sometimes being a necessary precursor to a steam drive, cyclic steam injection also is preferred for economic reasons, and because it is a stimulation treatment, oil production is accelerated. Furthermore, since steam injection into any well lasts only a few weeks and the resulting production lasts several months, portable steam generators have been developed. A single portable steam generator can be moved from well to well to apply the cyclic steam injection process to several wells at a reasonable capital cost. A disadvantage to cyclic steam injection, however, is that the ultimate recovery may be low relative to the total oil in place in the reservoir. The ultimate oil recovery from a steam-drive process is typically much larger than the ultimate oil recovery from a cyclic steam injection process. Thus, cyclic steam injection followed by a steam drive is an attractive combination in that crude production is accelerated quickly and the ultimate recovery is quite high.

The effective mobility ratio of a steam-drive process is more favorable than that of a combustion process. Also, for combustion processes, there is a rather high reported incidence of well failure associated with high temperatures, corrosion, and erosion. These factors indicate that when the economics are the same (laying aside considerations of risk), a steam injection process is to be preferred to a combustion drive process. However, each process has its limitations, and sometimes, the conditions may point to only one thermal process, which may well be a combustion process.

In the *modified* in situ *extraction* process, a combination of in situ recovery and mining techniques is used to access the deposit reservoir when the bitumen (or even heavy oil and extra heavy oil) proves to be too difficult to move to the production well. The process hails back to the heavy oil mining concept in which the heavy oil drains to a collection chamber. This type of process, with the inclusion of the thermal stage to enable the bitumen to flow to the collection area of which the SAGD process, is amenable to the recovery of bitumen from tar sand deposits (Fig. 3.3). A portion of the reservoir rock must be removed to enable application of the in situ extraction technology. The most common method is to enter the reservoir through a large-diameter vertical shaft, excavate horizontal drifts from the bottom of the shaft, and drill injection and production wells horizontally from the drifts. Thermal extraction processes are then applied through the wells. When the horizontal wells are drilled at or near the base of the tar sand reservoir, the injected heat rises from the injection wells through the reservoir, and drainage of produced fluids to the production wells is assisted by gravity.

As an addendum to this section, mention must be made to the proposed recovery processes that involve microwave heating and the potential for in situ upgrading. Microwave technology has been identified as having potential application to tar sand recovery (Bosisio et al., 1977; Mutyala et al., 2010; Robinson et al., 2010). The ability of microwaves to heat the deposit selectively and volumetrically leads to the possibility for the extraction of a product in high yield and that the oil can be upgraded in the process (Bosisio et al., 1977). The potential for the success of microwave heating for heavy oil or bitumen

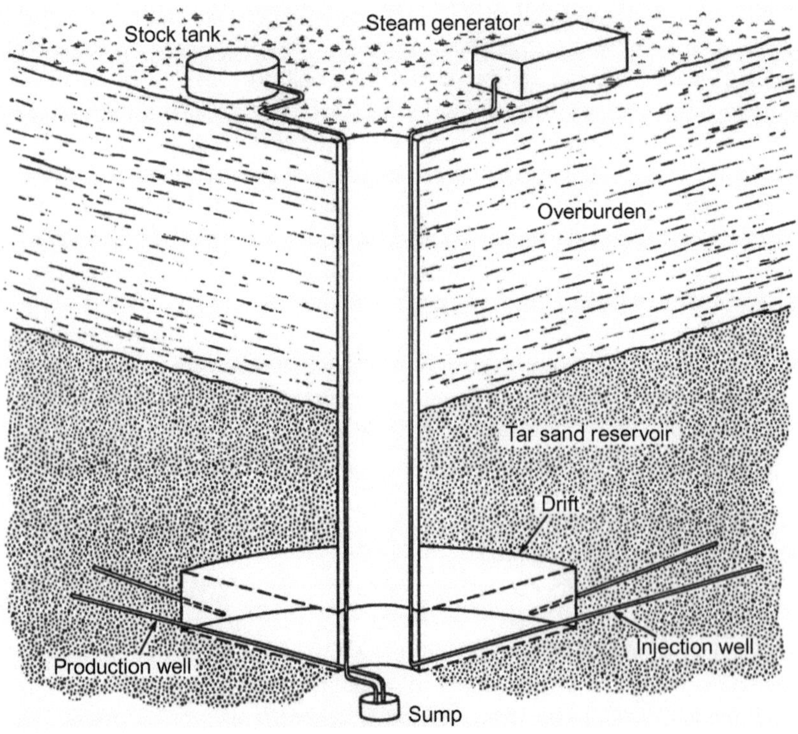

FIG. 3.3 Schematic of the modified in situ recovery process.

extraction leads to the hypothesis that microwave heating can remove the hydrocarbonaceous products (bitumen or a bituminous product) from tar sand deposits by selectively heating the water phases that are present interstitially, and within the hydrophilic clay minerals (Robinson et al., 2010). Preliminary analysis has shown the product quality to be favorable, but a rigorous assessment of the concept is still necessary. While the preliminary analysis has shown that the oil quality can be favorable, a rigorous assessment is needed to understand the requirements for further upgrading of the product. Furthermore, it is possible that the clay in the sand also played a role as a catalyst.

It must be emphasized that there is no single enhanced oil recovery technique that is the cure-all for oil recovery from heavy oil, extra heavy oil, and tars sand formations. Most reservoirs and deposits are complex, and the oil reservoir system must be considered as a whole rather than as individual, but equally complex, entities. Selection of a method or combination of methods for heavy oil, extra heavy oil, or tar sand bitumen recovery is best achieved by a detailed study of each reservoir or deposit, and the reservoir or the deposit should be examined carefully (Table 2.7). Reservoir/deposit evaluation should proceed in carefully planned stages with the objective of reducing any uncertainty that might arise during the application of the enhanced oil recovery process.

Finally, by all definitions, the quality of the product from a heavy oil, extra heavy oil, and bitumen from tar sand formations is poor as a refinery feedstock. As in any field in which primary recovery operations are followed by secondary or enhanced recovery operations and there is a change in product quality, such is also the case for tar sand recovery operations. Thus, product oils recovered by the thermal stimulation of tar sand deposits show some improvement in properties over those of the bitumen in place.

Although this improvement in properties may not appear to be too drastic, nevertheless it usually is sufficient to have major advantages for refinery operators. Any incremental increase in the units of hydrogen/carbon ratio can save amounts of costly hydrogen during upgrading. The same principles are also operative for reductions in the nitrogen, sulfur, and oxygen contents. This latter occurrence also improves catalyst life and activity and reduces the metal content. In short, in situ recovery processes (although less efficient in terms of bitumen recovery relative to mining operations) may have the added benefit of *leaving* some of the more obnoxious constituents (from the processing objective) in the ground.

References

Adibhatla, B., Mohanty, K.K., 2008. Oil recovery from fractured carbonates by surfactant-aided gravity drainage: laboratory experiments and mechanistic simulations. SPE Reserv. Eval. Eng. 11, 119–130.

Alexander, J.D., Martin, W.L., Dew, J.N., 1962. Factors affecting fuel availability and composition during in-situ combustion. J. Pet. Technol. 4, 1154–1164.

Advanced Resources International, 2005. Basin Oriented Strategies for CO_2 Enhanced Oil Recovery: Onshore California Oil Basins. US Department of Energy, Washington, DC.

Babu, D.R., Cormack, D.E., 1984. Effect of oxidation on the viscosity of Athabasca bitumen. Can. J. Chem. Eng. 62, 562–564.

Ballard, J.R., Lanfranchi, E.E., Vanags, P.A., 1976. Proceedings of the Twenty-Seventh Annual Meeting Petroleum Society, Canadian Institute of Mining, Calgary, Alberta, Canada, June.

Beckers, H.L., Harmsen, G.J., 1970. The effect of water injection on sustained combustion in a porous medium. Soc. Pet. Eng. J. 10, 145–163.

Bosisio, R., Cambon, J., Chavarie, C., Klvana, D., 1977. Experimental results on the heating of Athabasca tar sand samples with microwave power. J. Microwave Power 12, 301–307.

Bott, R.C., 1967. Cyclic steam project in a virgin tar reservoir. J. Pet. Technol. 19, 585–591.

Bousaid, I.S., Ramey Jr., H.J., 1968. Oxidation of crude oil in porous media. Soc. Petrol. Eng. J. 8, 137–148.

Brigham, W.E., Satman, A., Solitaire, M.Y., 1980. Recovery correlations for in-situ combustion field projects and application to combustion pilots. J. Pet. Technol. 2132–2138.

Burger, J.G., Sahuquet, B.C., 1972. Chemical aspects of in-situ combustion: heat of combustion and kinetics. Soc. Pet. Eng. J. 12, 410–422.

Burger, J.G., Sahuquet, B.C., 1973. Laboratory research on wet combustion. J. Pet. Technol. 25, 1137–1146.

Burger, J.G., 1976. Spontaneous ignition in oil reservoirs. Soc. Pet. Eng. J. 16, 73–81.

Burger, J.G., 1978. In situ combustion. Developments in petroleum science. In: Chilingarian, G.V., Yen, T.F. (Eds.), Bitumens, Asphalts and Tar Sands. In: vol. 7. Elsevier, New York, p. 191.

Butler, R.M., Stephens, D.J., 1980. The gravity-drainage of steam-heated heavy-oil to parallel horizontal wells. In: Proceedings of the 31st Annual Technical Meeting Petroleum Society of CIM, Calgary, Alberta, Canada May 25–28. Canadian Institute of Mining, Metallurgy and Petroleum, Westmount.

Butler, R.M., 1985. A new approach to the modeling of steam-assisted gravity drainage. J. Can. Pet. Technol. 24, 42–50.

Castanier, L.M., Brigham, W.E., 2003. Upgrading of crude oil via in situ combustion. J. Pet. Sci. Eng. 39, 125–136.

Castanier, L.M., Kovscek, A.R., 2005. Heavy-oil upgrading in situ via solvent injection and combustion: a "new" method. In: Proceedings of the European Association of Geoscientists and Engineers 67th Conference and Exhibition, Madrid, Spain, June 13–16.

Cerutti, A., Bandinelli, M., Bientinesi, M., Petarca, L., De Simoni, M., Manotti, M., Maddinelli, G., 2013. A new technique for heavy oil recovery based on electromagnetic heating: system design and numerical modelling. Chem. Eng. Trans. 32, 1255–1260.

CFR, 2018. Internal Revenue Service, Department of the Treasury. United States Code of Federal Regulations, 26 CFR 1.43-2—Qualified Enhanced Oil Recovery ProjectGovernment of the United States, Washington, DC.https://www.law.cornell.edu/cfr/text/26/1.43-2. (Accessed 24 January 2018).

Charles, D.D., Startzman, R.A., 1996. Horizontal well stream function model for improved waterflood performance and evaluation. J. Pet. Sci. Eng. 16, 109–130.

Chu, C., 1977. A study of fireflood field projects. J. Pet. Technol. 29, 171–179.

Chu, C., 1982. State-of-the-art review of fireflood field projects. J. Pet. Technol. 34, 19–36.

Curtis, C., Kopper, R., Decoster, E., Guzmán-Garcia, A., Huggins, C., Knauer, L., Minner, M., Kupsch, N., Linares, L.M., Rough, H., Waite, M., 2002. Heavy oil reservoirs. Oilfield Rev. 30–51.

Dabbous, M.K., Fulton, P.F., 1972. Low temperature oxidation kinetics and effects on the in-situ combustion process. In: Paper No. SPE 4143, Society of Petroleum Engineers-American Institute of Mechanical Engineers 47th Annual Fall Meeting, San Antonio, TA, October 8–11.

Das, S.K., Butler, R.M., 1995. Extraction of heavy oil and bitumen using solvents at reservoir pressure. In: Paper No. 95-118, Proceedings of the 6th Petroleum Conference of the South Saskatchewan Section of the Petroleum Society of CIM, Regina, October 16–18, 1995. Canadian Institute of Mining, Westmount.

Dietz, D.N., Weijdema, J., 1963. Wet and partially quenched combustion. J. Pet. Technol. 20, 411–415.

Dietz, D.N., 1967. Hot water drive. In: Proceedings of the Seventh World Petroleum Congress, Mexico City, Mexicovol. 3. pp. 451–457.

Dietz, D.N., Weijdema, J., 1968a. Wet and partially quenched combustion. J. Pet. Technol. 20, 411–413.

Dietz, D.N., Weijdema, J., 1968b. Reverse combustion seldom feasible. Prod. Mon. 32 (5), 10.

Dietz, D.N., 1970. Wet underground combustion, state of the art. J. Pet. Technol. 22, 605–617.

Elkins, L.F., Skov, A.M., Martin, P.J., Lutton, D.R., 1974. Experimental fireflood–Carlyle field, Kansas. In: Paper No. SPE 5014, Society of Petroleum Engineers Annual Meeting, Houston, TX, October 6–9.

EPRI, 1999. Enhanced oil recovery scoping study. Report TR-113836Electric Power Research Institute, Palo Alto, CA.

Farouq Ali, S.M., 1972. A current appraisal of in-situ combustion field tests. J. Pet. Technol. 24, 477–486.

Fassihi, M.R., Brigham, W.E., Ramey, H.H., 1984. Reaction kinetics of in-situ combustion: part 1—observations. Soc. Pet. Eng. J. (August), 399–416.

Fassihi, M.R., Brigham, W., 1984. Reaction kinetics of in-situ combustion. Soc. Pet. Eng. J. 24 (September), 399–416.

Fassihi, M.R., Meyers, K.O., Basile, 1990. Low-temperature oxidation of viscous crude oils. SPE Reserv. Eng. 5 (4), 609–616SPE Paper No. 15648, Society of Petroleum Engineers, Richardson, TX.

Gary, J.G., Handwerk, G.E., Kaiser, M.J., 2007. Petroleum Refining: Technology and Economics, fifth ed. CRC Press, Taylor & Francis Group, Boca Raton, FL.

Gates, G.F., Ramey Jr., H.J., 1958. Field results of south Belridge thermal recovery experiment. Trans. Am. Inst. Mech. Eng. 213, 236–244.

Gates, G.F., Ramey Jr., H.J., 1980. Method of engineering in-situ combustion oil-recovery projects. J. Pet. Technol. 32, 285–294.

Gates, C.F., Sklar, I., 1971. Combustion as a primary recovery process–midway sunset field. J. Pet. Technol. 23, 981–986.

Grant, B.F., Szasz, S.E., 1954. Development of an underground heat wave for oil recovery. J. Pet. Technol. 6, 22–23.

Grim, R.E., 1968. Clay Mineralogy. McGraw-Hill, New York.

Gunn, R.D., Krantz, W.B., 1980. Reverse combustion instabilities in tar sands and coal. Soc. Pet. Eng. J. 20, 267–277.

He, B., 2004. The effect of metallic salt additives on in-situ combustion performance. M.S. Report-Stanford University, Stanford, CA.http://ekofisk.stanford.edu/pereports/web/default.htm.

He, B., Chen, Q., Castanier, L.M., Kovscek, A.R., 2005. Improved in-situ combustion performance with metallic salt additives. In: Paper No. SPE 93901, Proceedings of the Society of Petroleum Engineers Western Regional Meeting, Irvine, CA, March 30–April 1.

Henderson, J.H., Weber, L., 1965. Physical upgrading of heavy crude oils by the application of heat. J. Can. Pet. Technol. 4, 206–212.

Hertzberg, R., Hojabri, F., Ellefson, L., 1983. Preprint No. 35e. In: Summer National Meeting of the American Institute of Chemical Engineers, Denver, Colorado, August 28–31.

Howard, F.A., 1923. Method of Operating Oil Wells. US Patent No. 1,473,348.

Hsu, C.S., Robinson, P.R. (Eds.), 2017. Handbook of Petroleum Technology. Springer International Publishing AG, Cham.

IEA, 2005. Resources to Reserves—Oil and Gas Technologies for the Energy Markets of the Future. International Energy Agency, Paris.https://www.iea.org/Textbase/publications/free_new_Desc.asp?PUBS_ID=1568. (Accessed 31 March 2015).

Jabbour, C., Quintard, M., Bertin, H., Robin, M., 1996. Oil recovery by steam injection: three-phase flow patterns. J. Pet. Sci. Eng. 16, 109–130.

Jamaluddin, A.K.M., Mehta, S.A., Moore, R.G., 1998. Downhole heating device to remediate near-wellbore formation damage related to clay swelling and fluid blocking. In: Paper No. 98-73, Proceedings of the Annual Technical Meeting, Petroleum Society of the Canadian Institute of Mining, Calgary, Alberta, Canada, June 8–10. Canadian Institute of Mining, Westmount.

Johnson, L.A., Fahy, L.J., Romanowski, L.J., Barbour, R.V., Thomas, K.P., 1980. An echoing in-situ combustion oil recovery project in a utah tar sand. J. Pet. Technol. 32, 295–305.

Koch, R.L., 1956. Practical use of combustion drive at west Newport field. Pet. Eng. 37, 72.

Kök, M.V., Ocalan, R., 1995. Modeling of in situ combustion for Turkish heavy crude oil. Fuel 74, 1057–1060.

Kuhn, C.S., Koch, R.L., 1953. In-situ combustion—newest method of increasing oil recovery. Oil Gas J. 52 (14), 92.

Lake, L.W., Schmidt, R.L., Venuto, P.B., 1992. A niche for enhanced oil recovery in the 1990s. Oilfield Rev. 4, 55–61.

Lake, L.W., Walsh, M.P., 2004. Primary Hydrocarbon Recovery. Elsevier, Amsterdam.

Martin, W.L., Alexander, J.D., Dew, J.N., 1958. Process variables of in-situ combustion. Trans. AIME 213, 28–35.

Mitchell, D.L., Speight, J.G., 1973. The solubility of asphaltenes in hydrocarbon solvents. Fuel 52, 149.

Moore, R.G., Laureshen, C.J., Belgrave, J.D.M., Ursenbach, M.G., Mehta, S.A., 1995. In situ combustion in Canadian heavy oil reservoirs. Fuel 74 (8), 1169–1175.

Moschopedis, S.E., Speight, J.G., 1975. Oxidation of a bitumen. Fuel 54, 210–212.

Mutyala, S., Fairbridge, C., Paré, J.R.J., Bélanger, J.M.R., Ng, S., Hawkins, R., 2010. Microwave applications to oil sands and petroleum: a review. Fuel Process. Technol. 91, 127–135.

Nasr, T.N., Ayodele, O.R., 2005. Thermal techniques for the recovery of heavy oil bitumen. In: Paper No. SPE 97488, Proceedings of the SPE International Improved Oil Conference in Asia Pacific, Kuala Lumpur, Malaysia, December 5–6. Society of Petroleum Engineers, Richardson, TX.

Northrop, P.S., Venkatesan, V.N., 1993. Analytical steam distillation model for thermal enhanced oil recovery processes. Ind. Eng. Chem. Res. 32 (9), 2039–2046.

Parkash, S., 2003. Refining Processes Handbook. Gulf Professional Publishing, Elsevier, Amsterdam.

Parman, D., Ojeda, S., 2013. Use of electric downhole heaters to improve production and recovery of heavy, viscous oil in California and Venezuela. In: Paper No. SPE 167347, Proceedings of the SPE Kuwait Oil and Gas Show and Conference, Kuwait City, Kuwait, October 8–10. Society of Petroleum Engineers, Richardson, TX.

Parrish, D.R., Craig Jr., F.F., 1969. Laboratory study of a combination of forward combustion and waterflooding—the COFCAW process. J. Pet. Technol. 21, 753–761.

Perez, S., Escobar, E., 2016. Insight of HASD technology in an extra heavy oil field in comparison to traditional thermal EOR processes. In: Paper No. SPE 180170-MS, Proceedings of the 78th

EAGE Conference and Exhibition Held in Vienna, Austria, May 30–June 2. Society of Petroleum Engineers, Richardson, TX.

Poettmann, F.H., Schilson, R.E., Surkalo, H., 1967. Philosophy and technology of in-situ combustion in light oil reservoirs. In: Proceedings of the 7th World Petroleum Congressvol. 3. p. 487.

Prats, M., 1986. Thermal Recovery. Society of Petroleum Engineers, Richardson, TX.

Ramey Jr., H.J., 1971. In-situ combustion. In: Proceedings of the World Petroleum Congress, Moscowpp. 253–262.

Ramey Jr., H.J., Stamp, V.V., Pebdani, F.N., 1992. Case history of south Belridge, California, in-situ combustion oil recovery. In: Paper No. SPE 24200, Proceedings of the 9th Society of Petroleum Engineers/US Department of Energy Enhanced Oil Recovery Symposium, Tulsa, Oklahoma, April 21–24.

Robinson, J., Binner, E., Saeid, A., Al-Harahsheh, M., Kingman, S., 2010. Microwave processing of oil sands and contribution of clay minerals. Fuel 93, 153–161.

Rodriguez, R., Bashbush, J.L., Rincón, A., 2008. Feasibility of using electrical downhole heaters in Faja heavy-oil reservoirs. In: Paper No. SPE 117682, Proceedings of the SPE International Thermal Operations and Heavy Oil Symposium, Calgary, Alberta, Canada. Society of Petroleum Engineers, Richardson, TX.

Rose, P.E., Deo, M.D., 1995. Steam-assisted gravity drainage in oil sand reservoirs using a combination of vertical and horizontal wells. Fuel 34, 1180–1184.

Shah, A., Fishwick, R., Wood, J., Leeke, G., Rigby, S., Greaves, M., 2010. A review of novel techniques for heavy oil and bitumen extraction and upgrading. Energy Environ. Sci. 3, 700–714.

Shallcross, D.C., Rios, C.F.D.L., Castanier, L.M., 1991. Modifying in-situ combustion performance by the use of water-soluble additives. SPE Reservoir Eng. 6, 287–294.

Shin, H., Polikar, M., 2004. Review of reservoir parameter to optimize SAGD and FAST-SAGD operating conditions. In: Paper No. 2004-211, Proceedings of the Canadian International Petroleum Conference (CIPC), Petroleum Society, Canadian Institute of Mining, Metallurgy & Petroleum, Calgary, Alberta, Canada, June.

Shin, H., Polikar, M., 2005. Optimizing the SAGD process in three major Canadian oil-sand areas. In: Paper No. SPE 95754, Proceedings SPE Annual Technical Conference and Exhibition, Dallas, TX, October.

Shu, W.R., 1984. A viscosity correlation for mixtures of heavy oil, bitumen, and petroleum fractions. Soc. Pet. Eng. J. 24, 277–292.

Speight, J.G., 2014. The Chemistry and Technology of Petroleum, fifth ed. CRC Press, Taylor and Francis Group, Boca Raton, FL.

Speight, J.G., 2017. Handbook of Petroleum Refining. CRC Press, Taylor and Francis Group, Boca Raton, FL.

Strycker, A., Sarathi, P., Wang, S., 1999. Evaluation of in situ combustion for Schrader bluff. Topical ReportNational Petroleum Technology Office, United States Department of Energy, Washington, DC.http://www.osti.gov/bridge/.

Tadema, H.J., Weijdema, J., 1970. Spontaneous ignition in oil sands. Oil Gas J. 68, 77–80.

Volek, C.W., Pryor, J.A., 1972. Steam distillation drive—Brea field, California. J. Pet. Technol. 24, 899–906.

Watts, K.C., Hutchinson, H.L., Johnson, L.A., Barbour, R.V., Thomas, K.P., 1982. Proceedings of the 54th Annual Fall Meeting, Society of Petroleum Engineers, American Institute of Mechanical Engineers, New Orleans, September 26–29.

Willman, B.T., Valleroy, V.V., Runberg, G.W., Cornelius, A.J., Powers, L.W., 1961. Laboratory studies of oil recovery by steam injection. J. Pet. Technol. 13, 681–690.

Wilson, L.A., Root, P.J., 1966. Cost comparison of reservoir heating using steam or air. J. Pet. Technol. 18, 233–239.
Wolcott, E.R., 1923. Method of Increasing the Yield of Oil Wells. US Patent No. 1,457,479.
Wu, C.H., Brown, A., 1975. Steam distillation in porous media. In: Paper No. SPE 5569, Proceedings of the Fall Meeting of the Society of Petroleum Engineers of AIME, Dallas, TX, September 28–October 1. Society of Petroleum Engineers, Richardson, TX.
Wu, C.H., Fulton, P.F., 1971. Experimental simulation of the zones preceding the combustion front of an in-situ combustion process. Soc. Pet. Eng. J. 11, 38–46.

Further Reading

Craig Jr., F.F., 1971. The Reservoir Engineering Aspects of Waterflooding. Society of Petroleum Engineers, Dallas, TX.

Chapter 4

Recovery of Tar Sand Bitumen

1. Introduction

By definition, tar sand bitumen (also referred to as *oil sand bitumen* in Canada) is not crude oil (Chapter 1) and, because of this, is deserving of a separate chapter relating to the recovery of bitumen. Even though extra heavy crude oil is related to bitumen insofar as under ambient (surface) conditions, flow characteristics are absent, and it is only the relatively high reservoir temperature by means of a steam-based recovery process that enables the oil to flow and recovery (Chapters 1 and 3).

There are extensive well-defined resources of heavy oil, extra heavy oil, and bitumen in Canada, Venezuela, the United States, and many other countries. The resources in Canada and the United States are readily accessible to oil companies, and the political and economic environments are stable. Moreover, existing commercial technologies could allow for significantly increased production (Table 4.1). In situ production of heavy oil, extra heavy oil, and tar sand bitumen with sand and thermal production using injected steam provides the remainder of Canada's production. In particular, the application of the steam-assisted gravity drainage (SAGD) method (or hybrid methods of SAGD) is receiving increased attention. In Venezuela, because of the high reservoir temperatures that give the oil some mobility of the deposit, cold production with horizontal and multilateral wells predominates and has proved to be a successful method for the recovery of the extra heavy oil.

For example, Athabasca bitumen with a pour point in the range 50–100°C (122–212°F) and a deposit temperature of 4–10°C (39–50°F) is a solid or near solid in the deposit and will exhibit little or no mobility under deposit conditions. Pour points of 35–60°C (95–140°F) have been recorded for the bitumen in Utah with formation temperatures on the order of 10°C (50°F). This indicates that the bitumen is solid within the deposit and therefore immobile (Chapter 1). The injection of steam to raise and maintain the reservoir temperature above the pour point of the bitumen and to enhance bitumen mobility is difficult, in some cases almost impossible. Conversely, when the reservoir temperature exceeds the pour point (as it does in many of the Venezuelan extra heavy oil deposits), the oil is fluid in the reservoir and, therefore, has some mobility. The injection of steam to raise

Heavy Oil Recovery and Upgrading. https://doi.org/10.1016/B978-0-12-813025-4.00004-0

TABLE 4.1 Production Methods for Heavy Oil, Extra Heavy Oil, and Tar Sand Bitumen

Method	Current Usage
Cold production	Used in Venezuela, some use for use in North Sea
Cold heavy oil production with sand (CHOPS)	Used in western Canada to exploit thin layers
Cyclic steam stimulation (CSS)	Used in the United States, Canada, Indonesia, and many other countries
Downhole heating with electricity	Resistance, induction, radio frequency (RF)
Hybrid	Solvent plus steam in SAGD, CSS, and steam flood wells
In situ combustion with vertical and horizontal wells	Uses heavy oil in reservoir and injected air
Open-pit mining	Used in Canada for shallow tar sands
Steam flood	Used in the United States, Canada, Indonesia, and many other countries
Steam-assisted gravity drainage (SAGD)	Used in Canada
VAPEX	Uses solvent rather than steam in SAGD-type wells

and maintain the reservoir temperature above the pour point of the bitumen and to enhance oil mobility is possible, and oil recovery can be achieved.

Where the oil viscosity is sufficiently low to flow at reservoir conditions, such as the oil in the Orinoco region of Venezuela, the extra heavy oil can be produced using (in the Venezuelan case) horizontal and multilateral wells that are drilled in order to contact as much of the reservoir as possible, and diluents, such as naphtha, are injected to decrease fluid viscosity further. In order to lift the oil to the surface, electric submersible pumps and progressing cavity pumps are employed. The key advantage of the cold production method is the considerably lower capital expenditure relative to thermally assisted techniques; however, it is severely limited by the suitability of reservoirs (i.e., the temperature of the reservoir vis-à-vis the pour point of the oil). A considerable quantity of unrecovered oil is left in place, and reservoirs may be viable for further exploitation by more efficient methods.

In the case of the Athabasca deposits, the tar sand is a combination of clay; sand; water; and bitumen, a high-molecular-weight black viscous oil. Tar sand can be mined and processed to extract the bitumen, which is then refined into oil

(Speight, 2013a,b,c, 2014, 2017). The bitumen in tar sand deposits cannot be pumped from the ground in its natural state, but instead, tar sand deposits are mined, usually using by a strip mining technique or an open-pit mining technique, or the bitumen oil is extracted by underground heating with additional upgrading.

The Athabasca deposits are largely unconsolidated sand deposits with a porosity ranging up to 45% and have good intrinsic permeability. However, the deposits in the United States in Utah range from predominantly low-porosity, low-permeability consolidated sand to, in a few instances, unconsolidated sand deposits. In addition, the bitumen properties are not conducive to fluid flow under normal reservoir conditions in either the Canadian or United States deposits. Nevertheless, where the general nature of the deposits prohibits the application of a mining technique (as in many of the United States deposits), a nonmining technique may be the only feasible bitumen recovery option.

One additional aspect of the character of Athabasca tar sand deposits is that the sand grains are not uniform in character. Grain-to-grain contact is variable, and such a phenomenon influences attempts to repack mined sand, as may be the case in studies involving bitumen removal from the sand in laboratory-type in situ studies. This phenomenon also plays a major role in the expansion of the sand during processing where the sand to be returned to the mine site might occupy 120%–150% of the volume of the original as-mined material.

The tar sand mass can be considered a four-phase system composed of solid phase (siltstone and clay), liquid phase (from fresh to more saline water), gaseous phase (natural gases), and viscous phase (black and dense bitumen, about 8° API). The bitumen content of the tar sand of the United States varies from zero to as much as 22% by weight. There are, however, noted relationships between the bitumen, water, fines, and mineral contents for the Canadian tar sand deposits. Similar relationships may also exist for the tar sand deposits in the United States, but an overall difficulty of bitumen recovery and reduced interest in the deposits have prevented the further development.

In a typical sandstone deposit, the sand grains are in grain-to-grain contact, but tar sand is thought to have no grain contact due to the surrounding of individual grains by fines with a water envelope and/or a bitumen film. The remaining void space might be filled with water, bitumen, and gas in various proportions. The sand material in the formation is represented by quartz and clays (99% by weight), where fine content is ~30% by weight; the clay content and clay size are important factors that affect the bitumen content.

Recovery of bitumen from the Athabasca tar sand deposits is hindered primarily by the high viscosity of the oil in the deposit (as well as the relatively high pour point and the low deposit temperature). For example, the intrinsic viscosity of the bitumen in the Athabasca oil sand deposit has been estimated to be more than 1×10^6 cp. The application of heat is the simplest and most efficient way of lowering bitumen viscosity. The methods of heating the reservoir oil include well-known fluid injection methods—such as cyclic steam stimulation, steam flooding, and fireflooding—and the newer techniques of heating the reservoir with electromagnetic energy. Steam-injection and in situ combustion techniques are now applied commercially to heavy oil deposits, but they are still technically difficult and usually uneconomical in very viscous oil sand deposits. All fluid injection methods in oil sand deposits are faced with (i) very low initial injectivity, (ii) difficulty in establishing communication paths between wells, (iii) poor control of injected fluid movement, (iv) reservoir inhomogeneity, (v) steam override, and (vi) very poor mobility ratio leading to a poor sweep efficiency (Hiebert et al., 1986).

The shallow depth of Canadian reservoirs is another limitation for the steam drive methods like SAGD that are limited to deep reservoirs.

One of the known methods of in situ heat generation that overcomes these difficulties is the electric heating method in which there is not any need for injection at the initial time, so the problem of low injectivity, communication paths, and poor mobility that is common in other heating methods doesn't occur. Also, the depth and thickness are not the limitation factors for electric heating. This method can also be used as a preheating for steam drive process to mobilize the oil for steam injection.

Since the Athabasca tar sand went to full commercialization in 1967, mining has been the predominant process by which bitumen has been recovered from the Athabasca tar sand (which involves excavation of the bitumen-rich sand using open-pit mining methods), although other processes have been investigated with some degree of success (Table 4.2). This is the most efficient method of extraction when there are large deposits of bitumen with little overburden. In situ methods involve processing the tar sand deposit so that the bitumen is removed while the sand remains in place. These methods are used for tar sand deposits that are too deep to support surface mining operations to an economical degree. Eighty percent of the resource in Northern Alberta lies deep below the surface.

Mining heavy oil is not a new concept (*oil mining*), but the manner in which mining is applied to the recovery of tar sand bitumen is new. This is the term applied to the surface or subsurface excavation of crude oil-bearing formations for subsequent removal of the heavy oil, extra heavy oil, and tar sand bitumen by washing, flotation, or retorting treatments. For example, mining of crude oil and tar sand bitumen occurred in the Sinai Peninsula, in the Euphrates valley, and in Persia prior to 5000 BC. In addition, subsurface oil mining was used in the Pechelbronn oil field in Alsace, France, as early as 1735. This early mining involved the sinking of shafts to the reservoir rock, only 100–200 ft (30–60 m) below the surface, and the excavation of the tar sand in short drifts driven from the shafts. The tar sand was brought to the surface (by hoist) and washed with boiling water to release the bitumen. The drifts were extended as far as natural ventilation permitted. When these limits were reached, the pillars were removed, and the openings were filled with waste. This type of mining continued at Pechelbronn until 1866, when it was found that oil could be recovered from deeper and more prolific tar sand deposits by letting the sand drain in place through mine

TABLE 4.2 Examples of Recovery Processes (Speight, 1990, 2008, 2013a,b, 2014)

Nonthermal Methods[a]	Thermal Methods[a]	Mining Methods[a]
Carbon dioxide	Combustion	Surface
Emulsification	Forward	Plus hot-water separation of the bitumen from the sand
Inert gas injection	Reverse	Subsurface
MEOR (bacterial)	Electromagnetic heating	Modified in situ process
Miscible displacement	Hot water	
Solvent	SAGD	
	Plus hybrids	
	Steam	
	Various options	

[a]*Listed alphabetically rather than by efficiency or preference (see Chapters 2 and 3).*

openings with no removal of sand to the surface for treatment. In the modern sense, oil mining also includes the recovery of heavy oil, extra heavy oil, and tar sand bitumen by drainage from reservoir beds to mine shafts or other openings driven into the rock or by drainage from the reservoir rock into mine openings driven outside the tar sand but connected with it by boreholes or mine wells.

Even though estimates of the recoverable oil from the Athabasca deposits are only of the order of 27 billion barrels (27×10^9 bbl) of synthetic crude oil (representing <10% of the total in-place material), this is, for the Canadian scenario, approximately six times the estimated volume of recoverable conventional crude oil. In addition, the comparative infancy of the development of the alternative options almost ensured the adoption of the mining option for the first two (and even later) commercial ventures.

The mineralogy of tar sand deposits is also worthy of note as it does affect the potential for the recovery of the bitumen. Usually, more than 99% by weight of the tar sand mineral is composed of quartz sand and clays. In the remaining 1%, more than 30 minerals have been identified, mostly calciferous or iron-based. Particle size ranges from large grains (99.9% is finer than 1000 μm) down to 44 μm (325 mesh), the smallest size that can be determined by dry screening. The size between 44 and 2 μm is referred to as silt; sizes below 2 μm (equivalent spherical diameter) are clay.

Finally, in a highly fractured formation with low matrix permeability, the fluid conductivity of the fracture system may be many times that of the matrix rock. In a highly fractured reservoir with low matrix permeability and reasonably high porosity, the fracture system provides the highest permeability to the flow of fluids, but the matrix rock contains the greater volume of the oil in place. The rate of the flow of fluids from the matrix rock to the fracture system, the extent and continuity of the fracture system, and the degree to which the production wells effectively intersect the fracture system determine the production rate. Special consideration must be given to these factors in predicting production rates in fractured reservoirs. Under favorable circumstances, higher production rates may be achieved in fractured reservoirs by improving mining methods than in less heterogeneous reservoirs. Other reservoirs that are good candidates for oil mining are those that are shallow, have high oil saturation, have a nearby formation that is competent enough to support the mine, and cannot be efficiently produced by conventional methods.

2. Nonmining Methods

Production methods that are applied to tar sand can be classified as (i) surface mining or (ii) production through a well. Primary subsurface production methods include cold production (horizontal and multilateral wells, waterflood, and cold heavy oil production with sand) and thermal production (cyclic steam stimulation, steam flood, and steam-assisted gravity drainage). However, ~20% w/w of the economically recoverable tar sand bitumen reserves are close enough to the surface to make mining feasible. The first step in surface mining is the removal of overburden and the muskeg, a water-soaked area of decaying plant material that is 3–10ft thick and lies on top of the overburden. First, the muskeg must be drained of its water content before it is removed. Overburden, which is used to build dams and dikes around the mine, is a layer of clay, sand, and silt that lies directly above the tar sand deposit. After all of the overburden is removed, the tar sand is exposed and can be mined.

The gravity of tar sand bitumen is usually less than 10° API depending upon the deposit, and viscosity is very high. Whereas conventional crude oils may have a viscosity of several poise (at 40°C, 105°F), the tar sand bitumen has a viscosity of the order of 50,000–1,000,000 cp or more at formation temperatures (~0–10°C, 32–50°F depending upon the season). This offers a formidable (but not insurmountable) obstacle to bitumen recovery.

The successful recovery technique that is applied to one deposit/resource is not necessarily the technique that will guarantee success for another deposit. There are sufficient differences between the tar sand deposits of the United States and Canada that general applicability is not guaranteed. Hence, caution is advised when applying the knowledge gained from one resource to the issues of another resource. Although the principles may at first sight appear to be the same, the technology must be adaptable.

In principle, the *nonmining recovery of bitumen from tar sand deposits* is an enhanced recovery technique and requires the injection of a fluid into the formation through an injection well. This leads to the in situ displacement of the bitumen from the recovery and bitumen production at the surface through an egress (production well). There are, however, several serious constraints that are particularly important and relate to bulk properties of the tar sand and the bitumen. In fact, both must be considered in toto in the context of bitumen recovery by nonmining techniques. For example, such processes need a relatively thick layer of overburden to contain the driver substance within the formation between injection and production wells.

One of the major deficiencies in applying mining techniques to bitumen recovery from tar sand deposits is (next to the immediate capital costs) the associated environmental problems. Moreover, in most of the known deposits, the vast majority of the bitumen lies in formations in which the overburden/pay zone ratio is too high. Therefore, it is not surprising that over the last two decades, a considerable number of pilot plants have been applied to the recovery of bitumen by nonmining techniques from tar sand deposits where the local terrain and character of the tar sand may not always favor a mining option.

In principle, the nonmining recovery of bitumen from tar sand deposits requires the injection of a fluid into the formation through an injection well, the in situ displacement of the bitumen from the reservoir, and bitumen production at the surface through an egress (production well). There are, of course, variants around this theme, but the underlying principle remains the same.

There are, however, several serious constraints that are particularly important and relate to bulk properties of the tar sand and the bitumen. In fact, both must be considered in the context of bitumen recovery by nonmining techniques. For example, the Canadian deposits are unconsolidated sand deposits with a porosity ranging up to about 45%.

In principle, the nonmining recovery of bitumen from tar sand deposits is an enhanced oil recovery technique and requires the injection of a fluid into the formation through an injection well. This leads to the in situ displacement of the bitumen from the reservoir and bitumen production at the surface through an egress (production) well.

Another general constraint to bitumen recovery by nonmining methods is the relatively low injectivity of tar sand formations. It is usually necessary to inject displacement/recovery fluids at a pressure such that fracturing (parting) is achieved. Such a technique, therefore, changes the reservoir profile and introduces a series of channels through which fluids can flow from the injection well to the production well. On the other hand, the technique may be disadvantageous insofar as the fracture occurs along the path of least resistance giving undesirable (i.e., inefficient) flow characteristics within the reservoir between the injection and production wells that leave a part of the reservoir relatively untouched by the displacement or recovery fluids.

3. Mining Methods

Surface mining is the mining method that is currently being used by Suncor Energy and Syncrude Canada Ltd to recover tar sand from the ground. Surface mining can be used in minable tar sand areas that lie under 75 m (250 ft) or less of overburden material. Only 7% of the Athabasca tar sand deposit can be mined using the surface mining technique, as the other 93% of the deposit has more than 75 m of overburden. This other 93% will have to be mined using different mining techniques.

Oil mining methods should be applied in reservoirs that have significant residual oil saturation and have reservoir or fluid properties that make production by conventional methods inefficient or impossible. The high well density in improved oil mining usually compensates for the inefficient production caused by reservoir heterogeneity. However, close well spacing can also magnify the deleterious effects of reservoir heterogeneity. If a high-permeability streak exists with a lateral extent that is less than the interwell spacing of conventional wells but is comparable with that of improved oil mining, the channeling is more unfavorable for the improved oil mining method.

The equipment employed at a *tar sand mine* is a combination of mining equipment and an on-site transportation system that may (currently) be either conveyor belts or large trucks. The shovel scoops up the tar sand and dumps it into a heavy hauler truck. The heavy (400 ton) hauler truck takes the tar sand to a conveyor belt that transports the tar sand from the mine to the extraction plant. Presently, there are extensive conveyor belt systems that transport the mined tar sand from the recovery site to the extraction plant. With the development of new technologies, the conveyors have been replaced by hydrotransport technology, which is a combination of tar sand transport and preliminary extraction of the bitumen.

The mining equipment must be durable and strong enough to withstand extreme climate and the abrasive nature of the tar sand. Mining never stops; the trucks and other equipment work day and night, every day of the year. Planning is an essential and continuous part of the process. After the tar sand has been recovered using the truck and shovel method, it is mixed with water and caustic soda to form a slurry and is pumped along a pipeline to the extraction plant.

3.1 Tar Sand Mining

The bitumen occurring in tar sand deposits poses a major recovery problem. The material is notoriously immobile at formation temperatures and must therefore require some stimulation (usually by thermal means) in order to ensure recovery. Alternately, proposals have been noted that advocate bitumen recovery by solvent flooding or by the use of emulsifiers. There is no doubt that with time, one or more of these functions may come to fruition, but for the present, the two commercial operations rely on the mining technique.

There are two methods of mining that have been used to recover tar sand from the Athabasca deposits. Suncor Energy uses the truck and shovel method of mining, whereas Syncrude uses the truck and shovel method of mining and draglines and bucket-wheel reclaimers. These enormous draglines and bucket wheels are being phased out and soon will be completely replaced with large trucks and shovels. The shovel scoops up the tar sand and transfers it to a heavy hauler truck. The heavy hauler truck takes the tar sand to a conveyor belt that transports the tar sand from the mine to the extraction plant.

To transfer the tar sand to the bitumen recovery plant, an extensive conveyor belt system is used to transport the mined tar sand from the recovery site to the extraction plant. With the development of new technologies, these conveyors are being phased out and replaced with hydrotransport technology. Hydrotransport is a combination of ore transport and preliminary extraction. After the sand has been recovered using the truck and shovel method, it is mixed with water and caustic soda to form a slurry and is pumped along a pipeline to the extraction plant. The extraction process thus begins with the mixing of the water, and agitation is needed to initiate bitumen separation from the sand and clay.

The advantages of the hydrotransport include (i) the breakdown of large lumps of tar sand in the ore and some separation of bitumen from the sand as the slurry moves through the pipeline; (ii) much more flexibility than conveyor belt systems, because pipelines can follow circuitous routing and be placed on uneven terrain; and (iii) low-energy extraction—because of separation during hydrotransport, extraction plant operating temperatures can be reduced to 50°C (122°F) or less. The reduced energy requirements will result in lower emissions (McColl et al., 2008).

Suncor uses hydrotransport to bring ore across the Athabasca River from the Steepbank Mine. Syncrude uses hydrotransport to bring ore to the Mildred Lake upgrader from the North Mine. At more remote mines, primary extraction occurs at the mine site. Bitumen froth is then transported to a central site by pipeline for secondary extraction and upgrading. Syncrude has remote primary extraction at the Aurora Mine, 35 km north of the Mildred Lake upgrader. Suncor has remote primary extraction at its Millennium Mine on the east side of the Athabasca River. Some of the major challenges faced by hydrotransport operations include the effects of fine solids (clay minerals), temperature, bitumen content (tar sand grade), and average sand grain size on the preconditioning process and on pipeline friction losses.

Tar sand properties that are of general interest are bulk density, porosity, and permeability. Porosity is, by definition, the ratio of the aggregate volume of the interstices between the particles to the total volume and is expressed as a percentage. High-grade tar sand usually has clay mineral porosity in the range from 30% to 35% that is somewhat higher than the porosity (5%–25%) of most reservoir sandstone. The higher porosity of the tar sand has been attributed to the relative lack of mineral cement (chemically precipitated material that binds adjacent particles together and gives strength to the sand, which in most sandstone occupies a considerable amount of what was void space in the original sediment).

Permeability is a measure of the ability of a sediment or rock to transmit fluids. It is, to a major extent, controlled by the size and shapes of the pores and the channels (throats) between the pores; the smaller the channel, the more difficult it is to transmit the reservoir fluid (water and bitumen). Fine-grained sediments invariably have a lower permeability than coarse-grained sediments, even if the porosity is equivalent. It is not surprising that the permeability of the bitumen-free sand from the Alberta deposits is quite high. On the other hand, the bitumen in the deposits, immobile at formation temperatures (~4°C, 40°F) and pressures, actually precludes any significant movement of fluids through the tar sand deposits under unaltered formation conditions.

For the Canadian tar sand deposits, bitumen contents from 8% to 14% by weight may be considered as normal (or average). Bitumen contents above or below this range have been ascribed to factors that influence impregnation of the sand with the bitumen (or the bitumen precursor). There are also instances where bitumen contents in excess of 12% by weight have been ascribed to gravity settling during the formative stages of the bitumen. Bitumen immobility then prevents further migration of the bitumen itself or its constituents.

Surface mining of tar sand bitumen has been operated extensively from opencast mines in Canada since 1967, although heavy oil, extra heavy oil, and tar sand bitumen have also been recovered by subsurface mining in Russia. The Athabasca (Alberta, Canada) tar sand deposit is composed of sand, silt, clay, water, and ~10%–12% w/w bitumen, and ~12% v/v of the total reserves are located in sufficient quantity at shallow depths to make the process economical. However, mining is strictly limited by the depth of the reservoir, and it is estimated that ~80% of the Alberta tar sand deposits and nearly all of the Venezuelan sand deposits are too far below the surface to allow opencast mining. In addition, the process is associated with severe environmental problems: (i) Wetlands need to be drained, (ii) rivers need to be diverted, and (iii) all trees and vegetation need to be stripped from the surface.

A further concern regarding surface mining is the vast areas of land that need to be disturbed in order to extract the bitumen. Because of the large scale and time frame of the projects related to the recovery of the bitumen, reclamation of the land does not happen quickly. However, on the positive side, much of the disturbed areas will be reclaimed, although the wetland environments that they replaced may be impossible to fully restore.

Engineering a successful oil mining project must address a number of items because there must be sufficient recoverable resources, the project must be conducted safely, and the project should be engineered to maximize recovery within economic limits. The use of a reliable screening technique is necessary to locate viable candidates. Once the candidate is defined, this should be followed by an exhaustive literature search covering the local geology, drilling, production, completion, and secondary and tertiary recovery operations.

The properties of the reservoir or deposit, which can affect the efficiency of heavy oil, extra heavy oil, and tar sand bitumen production by mining technology, can be grouped into three classes: (i) primary properties, (ii) secondary properties, and (iii) tertiary properties (Table 4.3). There are also important rock mechanical parameters of the formation in which a tunnel is to be mined and from where all oil mining operations will be conducted. These properties are mostly related to the mining aspects of the operations, and not all are of equal importance in their influence on the mining technology. Their relative importance also depends on the mineralogy and character of the individual reservoir.

Many of the candidate reservoirs for the application of improved oil mining are those with high oil saturation resulting from the adverse effects of reservoir heterogeneity. Faulting, fracturing, and barriers to fluid flow are features that cause the production of shallow reservoirs by conventional methods to be inefficient. Production of heterogeneous reservoirs by underground oil production methods requires consideration of the manner in which fractures alter the flow of fluids.

However, overburden depth alone is not a true indicator of whether an area is capable of sustaining an economically viable surface mining operation. The tar sand thickness, grade, clay content, and the extent of reject zones are also important parameters to be considered in the economic evaluation of a potential tar sand mining project. The thickness of overburden, tar sand deposit (often referred to as the *ore*), and center reject can be combined to give an overburden-ore ratio (thickness of overburden plus reject zones divided by the ore thickness), which can be used as an economic indicator of the cost of

TABLE 4.3 Properties of the Athabasca Tar Sand Deposits

Designated Property	Comment
Primary	Properties that have an influence on the fluid flow and fluid storage properties and include rock and fluid properties, such as porosity, permeability, wettability, crude oil viscosity, and pour point
Secondary	Properties that significantly influence the primary properties, including pore size distribution, clay type, and content
Tertiary	Properties that mainly influence oil production operation (fracture breakdown pressure, hardness, and thermal properties) and the mining operations (e.g., temperature, subsidence potential, and fault distribution)

delivering a unit of ore to the extraction plant. The bitumen content of the tar sand (referred to as the *tar sand grade*) and the content of clay minerals give an indication of the amount of bitumen that can be recovered from the unit of the ore, which is an indication of the value of that tar sand stratum. It is not unusual to use only the bitumen content to define the expected processability of tar sand, even though tar sand mineralogy is an important aspect of the project.

The rate of the flow of oil from the matrix rock into the fracture system, the extent and continuity of the fracture system, and the degree to which the production wells effectively intersect the fracture system determine the production rate.

The oil mining method of recovery has received considerable attention since it was chosen as the technique of preference for the only two commercial bitumen recovery plants in operation in North America. In situ processes have been tested many times in the United States, Canada, and other parts of the world and are ready for commercialization. There are also conceptual schemes that are a combination of both mining (aboveground recovery) and in situ (nonmining recovery) methods.

Since the 1920s, open-pit mining has been central to the development of tar sand deposits, although less than 10% of the Athabasca tar sand deposit can be mined using the surface mining technique. Surface mining is the mining method that is currently being used by Suncor Energy (formerly Great Canadian Oil Sands) and Syncrude Canada Ltd to recover tar sand from the ground. Great Canadian Oil Sands was originally owned by the Sun Oil Company of Marcus Hook, Pennsylvania.

By way of information, the Sun Oil Company (owners of Suncor Energy) has become Sunoco LP, a master limited partnership (controlled by the Energy Transfer Partners) registered and organized in Delaware and headquartered in Dallas (Texas). The partnership is a wholesale distributor of motor fuels and distributes fuel to gas stations under the Sunoco brand stations, almost all of which are owned and operated by third parties. The partnership was formerly known as Sun Company Inc. (1886–1920 and 1976–98) and Sun Oil Company (1920–76). The company was formerly engaged in oil refining and the chemical industry and retail sales.

On the other hand, the Syncrude project is a joint venture undertaking among Imperial Oil Resources Ltd., Nexen Oil Sands Partnership, Sinopec Oil Sands Partnership, and Suncor Energy Inc. (with the Suncor interest held by Canadian Oil Sands Partnership No. 1 and Suncor Energy Ventures Partnership, both wholly owned affiliates of Suncor Energy Inc.) as the project owners and Syncrude as the project operator. The Mildred Lake facility is located 25 mi north of Fort McMurray (Alberta).

Originally, tar sand was mined using draglines to excavate the face of the formation. Bucket-wheel excavators and long conveyor belts moved the raw bitumen to on-site processing facilities. This method has been replaced by shovel and truck mining, which gives greater flexibility. Surface mining can be used in minable tar sand areas that lie under 250 ft or less of overburden material. The first large-scale commercial operation used by Suncor Energy introduced bucket-wheel excavators from the coal mining industry when they opened in 1967. Syncrude Canada Ltd. opened in 1978 and introduced draglines with 360 ft booms. Substantial reductions in costs have been achieved through continual process improvement but more dramatically through two major innovations in the 1990s. First, there was a move toward replacing the draglines and bucket-wheel reclaimers with more flexible, robust, and energy-efficient trucks and power shovels. Second, hydrotransport systems were introduced to replace the conveyor belts used to transport tar sand to the processing plant. For hydrotransport, the tar sand ore is mixed with heated water (and chemicals in some cases) at the ore preparation plant to create an aqueous slurry of the tar sand that is pumped via pipeline to the extraction plant. Hydrotransport preconditions the ore for the extraction of crude bitumen and improves energy efficiency and environmental performance compared with conveyor systems.

The first step in surface mining is the removal of muskeg and overburden. Muskeg is a water-soaked area of decaying plant material that is 1–3 m thick and lies on top of the overburden material. Before the muskeg can be removed, it must be drained of its water content. The process can take up to 3 years to complete. Once the muskeg has been drained and removed, the overburden must also be removed. Overburden is a layer of clay, sand, and silt that lies directly above the tar sand deposit. Overburden is used to build dams and dikes around the mine and will eventually be used for land reclamation projects. When all of the overburden is removed, the tar sand is exposed and is ready to be mined.

Mine spoils need to be disposed of in a manner that assures physical stabilization. This means appropriate slope stability for the pile against not only gravity but also earthquake forces. Since the return of the spoils to the mine excavations is seldom economical, the spoil pile must be designed as a permanent structure whose outline blends into the landscape. Straight, even lines in the pile must be avoided.

For tar sand reserves with a depth of less than 160 ft of overburden, the overburden is stripped to expose the ore body, and tar sand will still have to be mined and processed in a water-based extraction process. In general, no revolutionary changes in mining tar sand deposits are expected to emerge, although incremental advances are expected in several fields concerning tar sand mining. Improved materials and equipment that are more durable and better suited to the tar sand industry combined with better monitoring systems for mechanical equipment to reduce production interruptions are expected. Some reductions in bitumen loss through primary separation and reductions in the energy intensity of the extraction process may be achieved. Finally, continued improvement in the performance of existing upgrading technologies including increased energy efficiency, catalyst development, and reductions in hydrogen use is likely to take place.

3.2 Improved Mining

Underground mining options have also been proposed but for the moment have not been developed because of the fear of collapse of the formation onto any operation/equipment. This particular option should not, however, be rejected out of hand because a novel aspect or the requirements of the developer (which remove the accompanying dangers) may make such an option acceptable. Currently, bitumen is recovered commercially from tar sand deposits by a mining technique. This produces tar sand that is sent to the processing plant for the separation of the bitumen from the sand prior to upgrading. The underground mining options often fall under the option known as *improved mining*.

In *improved mining*, directional (horizontal or slant) wells are drilled into the reservoir from a mine in an underlying formation to drain oil by pressured depletion and gravity drainage. In the process of gravity drainage extraction of liquid crude oil, the wells are completed so that only the forces acting within the reservoir are used. The forces acting on the reservoir are left intact, perhaps maintained or increased. A large number of closely spaced wells can be drilled into a reservoir from an underlying tunnel more economically than the same number of wells from the surface. In addition, only one pumping system is required in underground drainage, whereas at the surface, each well must have a pumping system. The objective of using a large number of wells is to produce each well slowly so that the gas/oil and water/oil interfaces move toward each other efficiently. By maintaining the reservoir pressures because of forces acting on the reservoir, it is then assured that the oil production is provided by the internal forces due to gravity (the buoyancy effect) and capillary effects.

The recovery efficiency is improved by applying an enhanced oil recovery method. The production drain holes should be surveyed while drilling, and the drilling should be accomplished using mud-operated drills with a bent sub for control of direction. The drain holes must be controlled so that a network of uniformly spaced holes conforming to the data output from the computer modeling can be drilled in the production zone. The drill string should be equipped with check valves to minimize the backflow of mud during the installation of additional drill pipe. The return mud-line should be equipped with a blow line to safely vent to the surface any formation gas encountered. All drilling should be accomplished working through a blowout preventer or diverter. Drill cuttings should be contained in a closed system and not allowed to encumber the mine atmosphere.

Large vertical shafts sunk from the surface are generally the means through which underground openings can be excavated. These shafts are one means of access to offer an outlet for the removal of excavated rock, provide sufficient opening for equipment, provide ventilation, and allow the removal of oil and gas products during later production. These requirements plus geologic conditions and oil reservoir dimensions determine the shaft size. It is expected that an access shaft will range from 8 to 20 ft in diameter.

Completing wells from a level of drifts beneath the reservoir is the most economical application of gravity drainage. In this case, only one pumping system is required rather installing a pump in each well as is necessary in wells drilled from the surface. When wells are drilled from beneath the producing formation in an oil-water system, the developer has two options for completion. The casing may be set totally through the formation and the region opposite the oil saturation perforated to permit production. If desired, the easing may pass only through the water-saturated zone, and then, a jet slotting process can be used to wash out or drill slots in the oil-saturated zone to increase the productivity of the individual well. Other shaft configurations include those for mining in weak rock and shafts for pumping drainage.

Excavation of shafts and tunnels during the mining process results in waste rock that must be stored or disposed of at the ground surface. They cannot be dumped in a heap near the mine shaft as an eyesore for future generations and a source of air and water pollution for untold years. Air, water, and aesthetics must be conserved not only for the present but also against reasonable future contingencies, either natural or man-made. The cautions necessary for underground mining options include the fear of collapse of the unconsolidated formation onto any operation/equipment.

4. The Hot Water Process

In terms of bitumen separation and recovery, the hot-water process is, to date, the only successful commercial process to be applied to bitumen recovery from mined tar sand in North America (Table 4.2). Many process options have been tested with varying degrees of success, and one of these options may even supersede the hot-water process (Clark, 1944; Fear and Innes, 1967; Speight, 2013a,b, 2014).

Tar sand, as mined commercially in Canada, contains an average of 10%–12% bitumen, 83%–85% mineral matter, and 4%–6% water. A film of water coats most of the mineral matter, and this property permits extraction by the *hot-water process* (Carrigy, 1963a,b).

The process utilizes the linear and the nonlinear variation of bitumen density and water density, respectively, with temperature so that the bitumen that is heavier than water at room temperature becomes lighter than water at 80°C (180°F). Surface-active materials in the tar sand also contribute to the process. The essentials of the hot-water process involve a conditioning, separation, and scavenging.

The extraction of bitumen from tar sand mines involves the liberation and separation of bitumen from the associated sand in a form that is suitable for further processing in order to produce a marketable product. Among several processes for bitumen extraction, the hot-water extraction process is a well-developed commercial recovery technique (Speight, 2013b, 2014). In the process, mined tar sand is mixed with hot water to create slurry suitable for extraction. Caustic soda is added to adjust the slurry pH to a desired level in order to enhance the efficiency of the separation of the bitumen. Recent development has shown the feasibility of operating at lower temperatures and without caustic addition in the slurring processes. The extract typically comprises an inorganic predominant phase (known as tailing stream) that is made up of coarse solids, some fine solids, and water. A typical composition of bitumen froth stream is about 60% w/w bitumen, 30% w/w water, and 10% w/w mineral matter (solids). The water and mineral matter in the froth are considered as contaminants.

The process to reject the water and mineral matter contaminants is known as froth treatment process. Due to the high viscosity of bitumen, the first step in the treatment is usually the introduction of a solvent. There are two major commercial approaches to reject the froth contaminants, namely, naphtha solvent-based froth treatment and paraffinic solvent-based froth treatment. Solvent addition (bitumen dilution) increases the density differential between bitumen, water, and mineral matter. Rejection of the contaminants can be achieved and carried out by a number of methods, such as centrifugation or gravity separation. The separation scheme generally results in a product effluent stream of diluted bitumen and a reject or tailing stream, commonly referred to as the froth treatment tailing (which is composed of mineral matter, water, residual solvent, and some residual bitumen). In the paraffinic froth treatment process, the solvent dilution reduces the precipitation of asphaltenes from the bitumen as an additional contaminant that results in an improvement in the efficiency of the contaminant rejection process.

Thus, in the hot-water extraction process, the tar sand feed is introduced into a *conditioning* drum in which the tar sand is heated and mixed with water, and the agglomeration of the oil particles begins. The conditioning is carried out in a slowly rotating drum that contains a steam-sparging system for temperature control and mixing devices to assist in lump size reduction and a size ejector at the outlet end. The tar sand lumps are reduced in size by ablation and mixing.

In the conditioning step, also referred to as mixing or pulping, tar sand feed is heated and mixed with water to form a pulp of 60%–85% by weight solids at 80–90°C (175–196°F) and a pH on the order of 7.5–8.5. First, the lumps of tar sand as-mined are reduced in size by ablation, that is, successive layers of lump are warmed and sloughed off revealing cooler layers. The conditioned pulp is screened through a double-layer vibrating screen. Water is then added to the screened material (to achieve more beneficial pumping conditions), and the pulp enters the separation cell through a central feed well and distributor. The bulk of the sand settles in the cell and is removed from the bottom as tailing, but the majority of the bitumen floats to the surface and is removed as froth. A middling stream (mostly of water with suspended fines and some bitumen) is withdrawn from approximately midway up the side of the cell wall. Part of the middlings is recycled to dilute the conditioning drum effluent for pumping. Clays do not settle readily and generally accumulate in the middling layer. High concentrations of clays increase the viscosity and can prevent normal operation in the separation cell. Thus, it is necessary to withdraw a drag stream to act as a purge; this is usually done at high clay concentrations but may not be as essential with a low-clay tar sand charge.

Under certain operating conditions, it may be necessary to withdraw a middling stream to the scavenger cells (air flotation cells to recover bitumen from the drag stream). The froth from the scavenger unit(s) usually has a high mineral and water content that can be removed by gravity settling in froth settlers after which the froth is combined with the froth from the main separation cell from centrifuge plant for dewatering and demineralizing. Before the centrifuging operation, the froth is deaerated, and naphtha is added to lower the viscosity for a more efficient water and mineral removal operation.

The separation cell acts like two settlers, one on top of the other. In the lower settler, the sand settles down, whereas in the upper settler, the bitumen floats. The bulk of the sand in the feed is removed from the bottom of the separation cell as tailings. A large portion of the feed bitumen floats to the surface of the separation cell and is removed as froth. A middling stream consists mostly of water with some suspended fine minerals and bitumen particles. A portion of the middlings may be returned for mixing with the conditioning drum effluent in order to dilute the separation-cell feed for pumping. The remainder of the middlings is called the drag stream, which is withdrawn from the separation cell to be rejected after processing in the scavenger cells.

Tar sand feed contains a certain portion of fine minerals that, if allowed to build up in concentration in the middlings, increases the viscosity and eventually disrupts settling in the separation cell. The drag stream is required as a purge in order to control the fine concentration in the middlings. The amounts of water that can enter with the feed and leave with the separation-cell tailings and froth are relatively fixed. Thus, the size of the drag stream determines the makeup water requirement for the separation cell. The separation cell is an open vessel with straight sides and a cone bottom. Mechanical rakes on the bottom move the sand toward the center for discharge. Wiper arms rotating on the surface push the froth to the outside of the separation cell where it overflows into launders for collection.

The combined froth from the separation cell and scavenging operation contains an average of about 10% by weight mineral material and up to 40% by weight water. The dewatering and demineralizing are accomplished in two stages of centrifuging; in the first stage, the coarser mineral material is removed, but much of the water remains. The feed then passes through a filter to remove any additional large-size mineral matter that would plug up the nozzles of the second-stage centrifuges.

The third step in the hot-water process is scavenging. Depending on the drag stream size and composition, enough bitumen may leave the process in the drag stream to make another recovery step economical. Froth flotation with air is usually employed. The scavenger froth is combined with the separation-cell froth to be further treated and upgraded to synthetic crude oil. Tailings from the scavenger cell join the separation-cell tailing stream and go to waste. Conventional froth flotation cells are suitable for this step.

Froth from the hot-water process may be mixed with a hydrocarbon diluent, for example, coker naphtha and centrifuged. The Suncor process employs a two-stage centrifuging operation, and each stage consists of multiple centrifuges of conventional design installed in parallel. The bitumen product contains 1% by weight to 2% by weight mineral (dry bitumen basis) and 5% by weight to 15% by weight water (wet diluted basis). Syncrude also utilizes a centrifuge system with naphtha diluent.

The first commercial operations (GCOS/Suncor and Syncrude) used the Clark hot-water extraction process. In this hot-water process, the tar sand is mixed with hot water (at 70–80°C, 158–176°F) and caustic (sodium hydroxide) in large rotating tumblers to begin the separation of the bitumen from the sand. The slurry from the tumblers is fed into large primary separation vessels (PSVs) where the bitumen is separated from the sand by gravity and floats to the surface of the primary separation vessels as a bituminous froth with the sand settling to the bottom primary separation vessels. The froth is subjected to further processing (froth treatment) for water and solid removal.

The froth treatment step is necessary in order to minimize the amount of water and solids going to the upgrader, so at this point, naphtha (produced from the bitumen in the upgrading unit) is added as a diluent, and the mixture enters a high-speed centrifuge to complete the cleaning/separation. The diluted bitumen is moved to the upgrading unit, while the sand and other materials that settle during the separation process are removed as a slutty (tailing slurry) for disposal in large tailing ponds.

One of the major problems that comes from the hot-water process is the disposal and control of the tailings that consist of ~49%–50% w/w sand, 1% w/w bitumen, and 50% w/w water. Furthermore, each ton of tar sand in place typically has a volume of ~16 ft^3, which will generate ~22 ft^3 of tailings giving a volume gain on the order of 40%. If the mine produces tar sand on the order of 200,000 tons of tar sand per day, the volume expansion represents a considerable solid disposal problem. The average particle size of the sand is about 200 μm, and it is a suitable material for dike building. Accordingly, Suncor used this material to build the sand dike, but for fine sand, the sand must be well compacted.

Environmental regulations in Canada or the United States will not allow the discharge of tailing streams into (i) the river, (ii) onto the surface, or (iii) onto any area where groundwater domains or the river may be contaminated. The tailing streams are essentially high in clays and contain some bitumen, hence the current need for tailing ponds, where some settling of the clay occurs. In addition, an approach to acceptable reclamation of the tailing ponds will have to be accommodated at the time of site abandonment.

The structure of the dike may be stabilized on the upstream side by beaching, which gives a shallow slope but consumes sand during the season when it is impossible to build the dike. In remote areas such as the Fort McMurray (Alberta) site, the dike can only be built in above-freezing weather because (i) frozen water in the pores of the dike will create an unstable layer and (ii) the vapor emanating from the water creates a fog, which can create a work hazard. The slope of the tailing dike is ~2.5:1 depending on the amount of fines in the material. It may be possible to build with 2:1 slopes with coarser material, but steeper slopes must be stabilized quickly. After discharge from the hot-water separation system and separation of the sand, sludge, and water, the sand is used to build dikes, and the runoff that contains the silt, clay, and water collects in the pond. Silt and some clay settle out to form sludge, and some of the water is recycled to the plant.

In a technique that is relatively new to the tar sand industry, the bituminous froth is sent to countercurrent decantation unit (McColl et al., 2008). In this unit, a solvent is added that separates the remaining solids, water, and heavy asphaltene constituents in a three-stage, dual-circuit process. The process yields clean, diluted bitumen, which is low in contaminants and with a viscosity that enables the bitumen to be transported by pipeline. Moreover, considerable effort is underway to reduce the energy required for bitumen extraction. At the Aurora Mine, opened in 2000, Syncrude installed a low-energy extraction process that operates at ~35°C (95°F) and is designed to consume about one-third of the energy of the traditional 80°C (176°F) process. Success in this area would result in substantial cost reductions and have considerable environmental benefits.

The US tar sand deposits have received considerably less attention than the Canadian deposits. Nevertheless, approaches to recover the bitumen from US tar sand deposits have been made. An attempt has been made to develop the hot-water process for the Utah tar sand deposits. The process differs significantly from that used for the Athabasca tar sand deposits due to the oil-wet Utah sand contrasting to the water-wet Athabasca sand. This necessitates disengagement by hot-water digestion in a high-shear force field under appropriate conditions of pulp density and alkalinity. The dispersed bitumen droplets can also be recovered by aeration and froth flotation.

In summary, the hot-water separation process involves extremely complicated surface chemistry with interfaces among various combinations of solids (including both silica sand and aluminosilicate clays), water, bitumen, and air. The control of pH seems to be critical with the preferred range being 8.0–8.5, which is achievable by the use of any of the monovalent bases. Polyvalent cations must be excluded because they tend to flocculate the clays and thus raise the viscosity of the middlings in the separation cell.

The main challenges for the surface mining process and the accompanying hot-water process are minimization of the environmental impact, land reclamation, and forest restoration. For every unit of synthetic crude produced, 6 units of sand and 1.5 units of fine tailings must be transported. Stockpiled overburden, sand, and tailing ponds can occupy a significant area of the mine and may have to be moved to access tar sand deposits beneath them. Land reclamation has only started on a small scale. Bitumen contains ~5% w/w sulfur, which is removed in the upgrading process. Large quantities of water have to be captured during spring flood and stored in reservoirs for use during the low-water months of the year. Large quantities of coke produced during the upgrading process must also be stored and eventually disposed.

5. Other Processes

It is conceivable that the problems related to bitumen mining and bitumen recovery for the tar sand may be alleviated somewhat by the development of process options that require considerably less water in the sand/bitumen separation step. Such an option would allow a more gradual removal of the tailing ponds.

5.1 The Cold Water Process

The cold-water bitumen separation processes have been developed to the point of small-scale continuous pilot plants (Miller and Misra, 1982). The proposed *cold-water process* for bitumen separation from mined tar sand has also been recommended (Misra et al., 1981; Miller and Misra, 1982). The process uses a combination of cold water and solvent, and the first step usually involves disintegration of the tar sand charge that is mixed with water, diluent, and reagents. The diluent may be a crude oil distillate fraction such as kerosene and is added in approximately a 1:1 weight ratio to the bitumen in the feed. The pH is maintained at 9–9.5 by the addition of wetting agents and ~0.77 kg of soda ash per ton of tar sand. The effluent is mixed with more water, and in a raked classifier, the sand is settled from the bulk of the remaining mixture. The water and oil overflow the classifier and are passed to thickeners where the oil is concentrated. Clay in the tar sand feed has a distinct effect on the process; it forms emulsions that are hard to break and are wasted with the underflow from the thickeners.

5.2 Solvent Extraction

An anhydrous *solvent extraction process* for bitumen recovery has been attempted and usually involves the use of a low-boiling hydrocarbon. The process generally involves up to four steps. In the mixer step, fresh tar sand is mixed with recycle solvent that contains some bitumen and small amounts of water and mineral. Solvent-to-bitumen weight ratio is adjusted to ~0.5. The drain step consists of a three-stage countercurrent wash. Settling and draining time is ~30 min for each stage. After each extraction step, a bed of sand is formed, and the extract is drained through the bed until the interstitial pore volume of the bed is emptied. From time to time, the bed is plugged with fine mineral or emulsion. In these cases, the drainage rate is essentially zero, and the particular extraction stage is ineffective. The last two steps of the process are devoted to solvent recovery. Stripping of the solvent from the bitumen is straightforward. The solvent recovery from the solids holds the key to the economic success of an anhydrous process.

Although solvent extraction processes have been attempted and demonstrated for the Athabasca, Utah, and Kentucky tar sand deposits, solvent losses influence the economics of such processes, and they have not yet been reduced to commercial practice.

5.3 The Sand Reduction Process

The *sand-reduction process* is a cold-water process without solvent. In the first step, the tar sand feedstock is mixed with water at ~20°C (68°F) in a screw conveyor in a ratio of 0.75:3 tons per ton of tar sand (the lower range is preferred). The mixed pulp from the screw conveyor is discharged into a rotary-drum screen, which is submerged in a water-filled settling vessel. The bitumen forms agglomerates that are retained by an 840 μm (20 mesh) screen. These agglomerates settle and are withdrawn as oil product. The sand readily passes through the 840 μm (20 mesh) screen and is withdrawn as waste stream. The process is called sand reduction because its objective is the removal of sand from the tar sand to provide a feed suitable for a fluid coking process; ca. 80% of sand is removed. Nominal composition of the oil product is 58% by weight bitumen, 27% by weight mineral matter, and 15% by weight water.

5.4 The Spherical Agglomeration Process

The *spherical agglomeration process* resembles the sand-reduction process. Water is added to tar sand, and the mixture is ball milled. The bitumen forms dense agglomerates of 75%–87% by weight bitumen, 12%–25% by weight sand, and 1%–5% by weight water.

5.5 The Oleophilic Sieve Process

An *oleophilic sieve process* (Kruyer, 1983, 2012) offers the potential for reducing tailing pond size because of a reduction in the water requirements. The process is based on the concept that when a mixture of an oil phase and an aqueous phase is passed through a sieve made from oleophilic materials, the aqueous phase and any hydrophilic solids pass through the sieve, but the oil adheres to the sieve surface on contact. The sieve is in the form of a moving conveyor, the oil is captured in a recovery zone, and recovery efficiency is high. Removing the dispersed bitumen and oleophilic minerals (i) improves dewatering of the tailings and yields a higher-quality bitumen for upgrading, (ii) reduces the release of methane and other gases to the air from the ponds, (iii) and frees tailing water faster, resulting in more effective fluid-tailing compaction for permanent storage or for site remediation (Kruyer, 2012).

The sieve uses an endless cable made from steel or polymer rope looped multiple times in wraps around a separation zone, comprising a revolving apertured drum filled with medium-density balls, and around a bitumen-removal zone, comprising one or more revolving squeeze rollers. An endless cable is wrapped multiple times around two or more roller supports. Unlike a mesh belt, the oleophilic sieve does not have cross members and, as a result, has an improved flexibility and eliminates the need for expensive belt tracking and automatic control of belt alignment. Spaces between sequential cable wraps form the apertures of the sieve, and these can be tailored for width to optimize sieving of the feedstock of various compositions and temperatures. A cable guide keeps the wraps from falling off the revolving supports. The wraps walk across the revolving roller surfaces until reaching a roller's edge, and then, the last wrap is guided and redirected to again become the first wrap.

The process does not need caustic and uses ambient temperature in a rotating separation cage with long oleophilic rods to partition oleophilic phase from hydrophilic phase and transfer the oleophilic phase to an oleophilic sieve while discarding the hydrophilic phase. The loaded sieve travels through a hot zone to release the captured oleophilic phase as a flowing liquid before the empty sieve revolves back to the cage to capture more oleophilic phase. Heating only the product instead of the feed, coupled with only one-tenth the required process residence time, makes this a very energy-efficient technology. The process opens the way for mine-face tar sand extraction to eliminate a large amount of material transport from mine to central processing and of sand back to the mine for eventual reclamation.

5.6 The Direct Heating Process

Another aboveground method of separating bitumen from mined tar sand involves *direct heating of the tar sand* without previous separation of the bitumen such as in the case of the use of a specialized retort (Taciuk et al., 1994). Thus, the bitumen is not recovered intact or unchanged but is recovered as an upgraded overhead product. In the process, the sand is crushed and introduced into a vessel, where it is contacted either with hot (spent) sand or with hot product gases that furnish part of the heat required for cracking and volatilization. The volatile products are passed out of the vessel and are separated into gases and (condensed) liquids. The coke that is formed as a result of the thermal decomposition of the bitumen remains on the sand, which is then transferred to a vessel for coke removal by burning in air. The hot flue gases can be used either to heat incoming tar sand or as refinery fuel. As expected, processes of this type yield an upgraded product but require various arrangements of pneumatic and mechanical equipment for solid movement around the refinery.

Electromagnetic heating techniques, which heat the reservoir by means of the ohm losses of electric current flowing in the connate water and the oil of the reservoir, can overcome many of these difficulties. Because no fluid is injected, low initial injectivity is not a problem. Heating occurs both near the well and deep into the problem. Heating occurs both near the well and deep into the formation. In addition, with proper attention to the reservoir lithology and electromagnetic properties, the reservoir may be heated relatively uniformly. Production may occur during or immediately after electromagnetic heating if the formation pressure is sufficient, or electromagnetic heating may be used to preheat the formation for a steam flood or fireflood.

5.7 Hybrid Technologies

One approach toward recovering a significant portion of heavy oil, extra heavy oil, and tar sand bitumen from a reservoir or deposit is to use a combination of crude oil recovery and mining technologies. Through the use of mining technology, access is developed below the crude oil reservoir within or beneath a permeability barrier. Underground development consists of providing room for subsurface drilling and crude oil production operations as well as a life-supporting atmosphere and safe working conditions. Wells are drilled at inclinations from the horizontal to the vertical upward into the reservoir. Wells are produced as a result of a combination of pressure depletion and gravity drainage.

References

Carrigy, M.A., 1963a. Bulletin No. 14. Alberta Research Council, Edmonton.

Carrigy, M.A., 1963b. The Oil Sands of Alberta. Information Series No. 45, Alberta Research Council, Edmonton.

Clark, K.A., 1944. Hot-water separation of Alberta bituminous sand. Trans. Can. Inst. Min. Met. 47, 257–274.

Fear, J.V.D., Innes, E.D., 1967. Canada's first commercial tar sand development. In: Proceedings of the 7th World Petroleum Congress. vol. 3. Elsevier, Amsterdam, pp. 633–650.

Hiebert, A.D., Vermeulen, F.E., Chute, F.S., Capjack, C.E., 1986. Numerical simulation results for the electrical heating of athabasca oil-sand formations. SPE Reserv. Eng. 1 (1), 76–84Paper No. SPE 13013-PA. Society of Petroleum Engineers, Richardson, Texas.

Kruyer, J., 1983. Oleophilic separation of tar sands. Oil-water mixtures and minerals. In: Preprint No. 3d, Summer National Meeting, August 28-31. American Institute of Chemical Engineers, Denver, CO.

Kruyer, J., 2012. Processing oil sands fluid tailings by oleophilic sieve technology. Paper No. 1112-0033J, Petrol. Technol.

McColl, D., Mei, M., Millington, D., Kumar, C., 2008. Green bitumen: the role of nuclear, gasification, and CCS in Alberta's oil sands: part I—introduction and overview. Study No. 119, Canadian Energy Research Institute, Calgary.

Miller, J.C., Misra, M., 1982. Hot water process development for Utah tar sands. Fuel Process. Technol. 6, 27–59.

Misra, M., Aguilar, R., Miller, J.D., 1981. Surface chemistry features in the hot water processing of Utah Tar sand. Sep. Sci. Technol. 16 (10), 1523–1544.

Speight, J.G., 1990. Part II: tar sands. In: Speight, J.G. (Ed.), Fuel Science and Technology Handbook. Marcel Dekker, New York (Chapters 12–16).

Speight, J.G., 2008. Synthetic Fuels Handbook: Properties, Processes, and Performance. McGraw-Hill, New York (Chapter 4).

Speight, J.G., 2013a. Heavy Oil Production Processes. Gulf Professional Publishing, Elsevier, Oxford.

Speight, J.G., 2013b. Oil Sand Production Processes. Gulf Professional Publishing, Elsevier, Oxford.

Speight, J.G., 2013c. Heavy and Extra-Heavy Oil Upgrading Technologies. Gulf Professional Publishing, Elsevier, Oxford.

Speight, J.G., 2014. The Chemistry and Technology of Petroleum, fifth ed. CRC Press, Taylor & Francis Group, Boca Raton, FL.

Speight, J.G., 2017. Handbook of Petroleum Refining. CRC Press, Taylor & Francis Group, Boca Raton, FL.

Taciuk, W., Caple, R., Goodwin, S., Taciuk, G., 1994. Dry Thermal Processor. United States Patent 5,366,596, November 22.

Further Reading

Kruyer, J., 1982. Oleophilic separation of tar sands. Oil-water mixtures and minerals. In: Proceedings Second International Conference on Heavy Crude and Tar Sands, Caracas, Venezuela, February 7-17.

Zaki, N.N., Poindexter, M.K., Kilpatrick, P.K., 2002. Factors contributing to petroleum foaming. 2. Synthetic crude oil systems. Energy Fuel 16, 711–717.

Chapter 5

Instability and Incompatibility

1. Introduction

The study of the chemical reactivity of crude oil and the heavy feedstocks (heavy oil, extra heavy oil, and tar sand bitumen as well as residua) would not be complete unless attention is given to the phenomena of instability and incompatibility. Both of these phenomena are a result of chemical reactions that form degradation products or other undesirable changes in the original chemical properties of crude oil and crude oil products. Furthermore, it is the various analytic methods (also discussed here) that provide the data that point to the reason for problems in the refinery or for the failure of products to meet specifications and to perform as desired.

Deposition of asphaltene constituents (so-named by virtue of the separation process, Fig. 5.1) is the consequence of instability of the crude oil or the crude oil product (e.g., fuel oil). The asphaltene constituents are stabilized by resin constituents (Koots and Speight, 1975; Speight, 1992, 1994, 2014) and maintained in the crude oil due to this stabilization. Asphaltene dispersants are substitutes for the natural resin constituents and are able to keep the asphaltene constituents dispersed to prevent flocculation/aggregation and phase separation (Speight, 1994; Browarzik et al., 1999). Dispersants will also clean up sludge in the fuel system, and they have the ability to adhere to surface of materials that are insoluble in the oil and convert them into stable colloidal suspensions.

At the other end of the molecular weight scale and refining, the heteroatoms (particularly nitrogen, sulfur, and trace metals) (Chapter 1) are present in crude oil and might be also expected to be present in liquid fuels and other products from crude oil. And, indeed, this is often the case, although there may have been some skeletal changes induced by the refining processes. Oxygen is much more difficult to define in crude oil and liquid fuels. However, it must be stressed that instability/incompatibility is not directly related to the total nitrogen, oxygen, or sulfur content. The formation of color/sludge/sediment is a result of several factors. Perhaps, the main factor is the location and nature of the heteroatom that, in turn, determines reactivity (Por, 1992; Mushrush and Speight, 1995; Speight, 2014).

Heavy Oil Recovery and Upgrading. https://doi.org/10.1016/B978-0-12-813025-4.00005-2

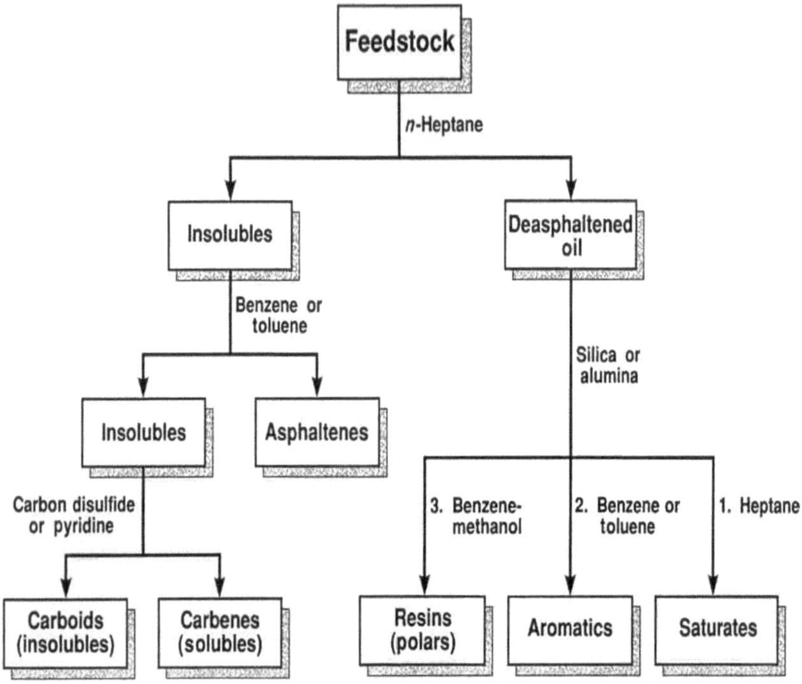

FIG. 5.1 General fractionation scheme and nomenclature of crude oil fractions—carbenes and carboids are thermally generated product fractions.

In general, the chemical reaction sequence for instability and/or incompatibility can be envisaged as being dependent upon the most reactive of the various heteroatomic species that are present in fuels (Pedley et al., 1987). The worst-case scenario would consist of a high-olefin fuel with both high indole concentration and a catalytic trace of sulfonic acid species. This reaction matrix would lead to rapid degradation. However, just as there is no *one* specific distillate product, there is also no *one* mechanism of degradation. In fact, the mechanism and the functional groups involved will give a general but not specific mode of incompatibility (Hiley and Pedley, 1986; Mushrush et al., 1990, 1991).

The focus of incompatibility studies has usually been on the whole crude oil and specifically on the asphaltene fraction. The asphaltene constituents are the highest molecular weight and most polar fractions found in crude oil. The characteristics of the asphaltene constituents and the amount in crude oil depend to a greater or lesser extent on the crude oil source. During crude oil refining, the majority of the asphaltene constituents will end up in the residual fuel products since the lower-boiling fractions (such as naphtha and kerosene) are removed from the oil through cracking and visbreaking. The problems of asphaltene constituents in a feedstock have increased due to the need to extract even the most viscous oils and the trend to extract large amounts of lower-boiling fractions out of crude oil by visbreaking and cracking processes. For example, when various feedstocks are blended at the refinery, incompatibility can be explained by the onset of acid-base catalyzed condensation reactions of the various organonitrogen compounds in the individual blending stocks. These are usually very rapid reactions with practically no observed induction time period (Hardy and Wechter, 1990).

Crude oil and the heavy feedstocks (heavy oil, extra heavy oil, and tar sand bitumen) are delicately balanced molecular systems insofar as the different fractions are compatible as long as there are no significant disturbances or changes made to the system (Koots and Speight, 1975; Speight, 1992, 1994, 2014). Any changes to the system that can occur that will upset the balance of the crude oil system are (i) alteration of the natural abundances of the different fractions, as might occur when gases are dissolved in the crude oil under reservoir pressure and/or when these dissolved gases are released at the time when a reservoir is first penetrated; (ii) chemical alteration of the constituents as might occur during recovery processes, especially changes that might be brought on by thermal recovery processes; and (iii) alteration of the distribution of the polar functional groups as might occur during oxidation or the elimination of polar functions during recovery operations, as might occur when exposure of crude oil to air occurs.

When such disturbances occur, it is the higher-molecular-weight constituents that are most seriously affected. This can lead to incompatibility and instability of the oil, which is variously referred to as precipitation, sludge formation, sediment formation, and deposition of asphaltene constituents or asphaltene-type material, depending on the circumstances. When incompatibility and instability occur, it is the higher-molecular-weight polar constituents (either the natural asphaltene constituents or, in some instances, the natural resin constituents or, in the case of a thermal process, either the reacted asphaltene constituents or the reacted resin constituents) that form a separate phase as a result of the loss of dispersability in the oil medium that has become more aliphatic in character (Fig. 5.2). In the reservoir, the incompatibility of the asphaltene constituents (and even, in some cases, the incompatibility of the resin constituents depending upon the process) can cause blockages of the pores and channels through which the oil must move during recovery operations.

Thus, the dispersability of the higher-molecular-weight constituents becomes an issue that needs attention when consideration is given to the mode of recovery or the mode of refining (Speight, 2014, 2017). And one of the ways by which this issue can be understood is to be aware of the chemical and physical character of the polar constituents of crude oil. By such means, the issue of dispersability, fluid-fluid interactions, oil-rock interactions, and the attending issue of incompatibility can be understood and perhaps predicted.

Crude oil is rarely used in its raw form but must instead be processed into its various products, generally as a means of forming products with hydrogen content different from that of the original feedstock. Similarly and because of the

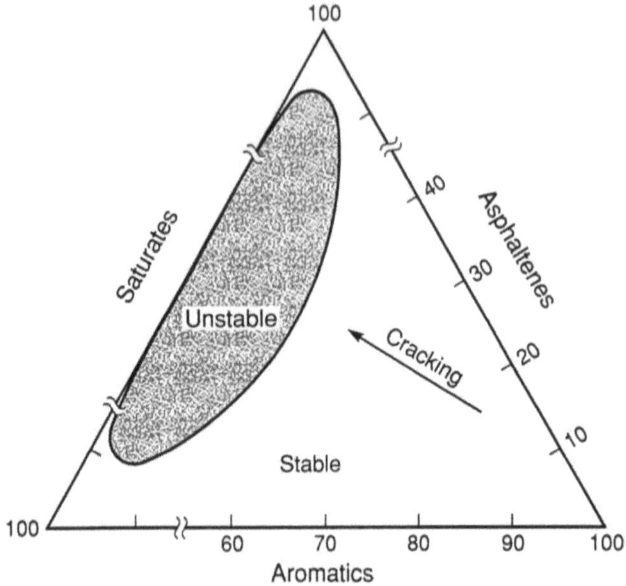

FIG. 5.2 Incompatibility occurs when the oil medium has become more aliphatic in character.

additional complexity, heavy oil, extra heavy oil, and tar sand bitumen also require processing into the various products, no matter what the extent of the upgrading that has occurred as part of the recovery process (Chapter 6; Speight, 2013a,b).

In the natural state crude oil, heavy oil, extra heavy oil, and tar sand bitumen are not always homogeneous materials, and the physical characteristics of each oil differ depending on where the material was produced and the formation or the deposit from which the oil was recovered. This is due to the fact that any of these feedstocks from different geographic locations will naturally have unique properties. In its natural, unrefined state, crude oil, heavy oil, extra heavy oil, and tar sand bitumen range in density and consistency from very thin, lightweight, and volatile fluidity to an extremely thick, semisolid (Speight, 2014). There is also a considerable gradation in the color of the oil that ranges from a light-colored golden yellow (conventional crude oil) to black (tar sand bitumen) (Speight, 2012, 2014).

Thus, crude oil, heavy oil, extra heavy oil, and tar sand bitumen are not (within the individual categories) uniform materials, and the chemical and physical (fractional) composition of each of these refinery feedstocks can vary not only with the location and age of the reservoir or deposit but also with the depth of the individual well within the reservoir or deposit. On a molecular basis, the three feedstocks are complex mixtures containing (depending upon the feedstock) hydrocarbons with varying amounts of hydrocarbonaceous constituents that contain sulfur, oxygen, and nitrogen and constituents containing metallic constituents, particularly those containing vanadium nickel, iron, and copper. The hydrocarbon content may be as high as 97% w/w, for example, in a light (low viscosity) crude oil or <50% w/w in heavy crude oil and bitumen (Speight, 2014).

Heavy oil, extra heavy oil, and tar sand bitumen contain many thousands of different compounds that vary in molecular weight up to several hundred or, in some case, several thousand (Speight et al., 1985; Speight, 1994, 2014). This broad range in molecular weights results in a boiling points that range up to temperatures on the order of 1100°C (2000°F). The constituents of heavy oil, extra heavy oil, and tar sand bitumen can undergo a diversity of thermal and catalytic reactions, thereby influencing the chemistry of the recovery processes and the chemistry of the refining processes (Speight, 2013a,b,c, 2014, 2017).

Understanding refining chemistry not only allows an explanation of the means by which these products can be formed from crude oil but also offers a chance of predictability. This is very necessary when the different types of heavy oil, extra heavy oil, and tar sand bitumen accepted by refineries are considered. And the major processes by which the various products are produced from heavy oil, extra heavy oil, and tar sand bitumen constituents involve thermal decomposition either during a thermal recovery process or during thermal and catalytic refining processes. The changes that occur during thermal recovery and refining processes give rise to the potential (often the reality) of instability and incompatibility of heavy oil, extra heavy oil, and tar sand bitumen not only with each other but also with their respective thermal products.

Thus, the study of the properties of heavy oil, extra heavy oil, and tar sand bitumen would not be complete without some attention to the phenomena of instability and incompatibility. Furthermore, the occurrence of instability and/or incompatibility during recovery operations and/or during refinery operations can have serious consequences for the producer or the refiner. Both result in the formation of degradation products or other undesirable changes in the original properties of heavy oil, extra heavy oil, and tar sand bitumen products. It is only by understanding the properties of the heavy oil, extra heavy oil, and tar sand bitumen through application of analytic methods that provide the data that point to the reason for problems during production and during refining (Parkash, 2003; Gary et al., 2007; Speight, 2014, 2017; Hsu and Robinson, 2017).

In addition to the thermal (noncatalytic products) of the decomposition of heavy oil, extra heavy oil, and tar sand bitumen, the presence of hydrogen (even in the presence of high proportions of asphaltene constituents) changes the nature of the products (especially the decreasing coke yield) by preventing the buildup of precursors that are incompatible in the liquid medium and form coke (Magaril and Akensova, 1967, 1968, 1970a,b, 1972; Magaril and Ramazaeva, 1969; Magaril et al., 1970, 1971; Schucker and Keweshan, 1980; Speight and Moschopedis, 1979; Speight, 1990, 1998, 2014; Wiehe, 1992, 1993, 1994, 1995).

Nitrogen species—of which there is sufficient in the asphaltene and resin constituents to initiate the reaction—contribute to the pattern of the thermolysis insofar as the carbon-carbon bonds adjacent to ring nitrogen undergo thermolysis quite readily (Fitzer et al., 1971; Speight, 1998, 2014). Thus, the initial reactions of asphaltene decomposition involve thermolysis of aromatic-alkyl bonds that are enhanced by the presence of heterocyclic nitrogen. Thus, the molecular species within the asphaltene fraction, which contain nitrogen and other heteroatoms (and have lower volatility than the pure hydrocarbons), are the prime movers in the production of coke. Such species, containing various polynuclear aromatic systems, can be denuded of the attendant hydrocarbon moieties and are undoubtedly insoluble in the surrounding hydrocarbon medium. The next step is gradual carbonization of these heteroatom-rich entities to form coke.

In contrast to the reaction of heavy oil, extra heavy oil, and tar sand bitumen under visbreaking conditions (Speight, 2015a, 2017) in which the general principle is the production of products for use as fuel oil, the presence of hydrogen can reduce the amount of coke formed during a thermal process. Nevertheless, the decomposition of asphaltene constituents is, again (or always), an issue and frequent leads to instability and/or incompatibility of the thermal products. Even when this occurs during recovery operations (Waxman et al., 1980), changes in the structure and molecular association of the asphaltene constituents and the resin constituents can lead to phase separation of the reacted asphaltene constituents and deposition of these molecular species on the reservoir rock. While appearing to be an advantage insofar, the asphaltene constituents—the most difficult to refine constituents of the oil—remain in the ground; blockage of flow channels by the deposited material can result in poor recovery of the oil.

Furthermore, the deposition of solids (due to instability or incompatibility of the products) is still possible when asphaltene constituents interact with catalysts (especially acidic support catalysts) in the subsurface during recovery operations or in the refinery during upgrading operations (Speight, 2013c), through the functional groups, for example, the basic nitrogen species just as they interact with adsorbents. And there is a possibility for interaction of the asphaltene with the catalyst through the agency of a single functional group in which the remainder of the asphaltene molecule remains in the liquid phase. There is also a less desirable option in which the asphaltene reacts with the catalyst at several points of contact, causing immediate incompatibility on the catalyst surface.

In the refinery, the reaction of asphaltene constituents (and resin constituents) with the catalyst is to be deplored, unless the catalyst is a guard catalytic and placed in a location to protect the more expensive, say, hydrogenation catalyst. On the other hand, during recovery operations, the reaction of asphaltene constituents (and resin constituents) with the catalyst such as an in situ *clay* mineral or a specially designed adsorbent catalyst is to be encourage so that the main coke former (the asphaltene constituents and the resin constituents) remains in the formation with the result being a much more refinable feedstock.

Thus, it is not surprising that instability and incompatibility during oil recovery and oil refining have become a major issue over the past four decades (Speight, 2014, 2015a, 2017). Differences in oil composition and in the nature of the products from refinery operations have all served to emphasize the issue of instability and incompatibility. The best means of approaching the issue of instability and incompatibility is, in the context of this book, from an understanding of the nature of heavy oil, extra heavy oil, and tar sand bitumen. In fact, it is essential to understand not only the nature of these feedstocks but also the chemical and physical nature of the oils. Understanding these phenomena is crucial to understanding the means by which instability and incompatibility can arise.

Feedstock complexity and the means by which the feedstock is evaluated (Speight, 2001, 2015a; ASTM, 2017) have made the industry unique among industries. But product complexity has also brought to the fore issues such as *instability* and *incompatibility*. Product complexity becomes even more disadvantageous when various fractions from different types of crude oil are blended or are allowed to remain under conditions of storage (prior to use), and a distinct phase separates from the bulk product. The adverse implications of this for refining the fractions to salable products increase (Batts and Fathoni, 1991; Mushrush and Speight, 1995).

Therefore, it is appropriate here to define some of the terms that are used in the stability-incompatibility field so that their use later in the text will be more apparent and will also alleviate some potential for misunderstanding. The general scientific areas of *instability* and *incompatibility* are complex and have been considered to be a chemical unknown because not all of the reactions that contribute to instability and incompatibility have been defined in detail. Nevertheless, recent studies over the past three decades have made valuable contributions to our understanding of instability and incompatibility in fuels. But, for the most part, gaps remain in our knowledge of the chemistry and physics of instability and incompatibility.

Briefly, the term *instability* is often used in reference to the formation of color, sediment, or gum in the liquid over a period of time and is usually due to chemical reactions, such as oxidation, and is chemical rather than physical. This term may be used to contrast the formation of a precipitate in the near term (almost immediately). On the other hand, the term *incompatibility* refers to the formation of a *precipitate* (or *sediment*) or *separate phase* when two liquids are mixed.

The phenomenon of *instability* is often referred to as *incompatibility* and more commonly known as *sludge formation*, *sediment formation*, or *deposit formation*. In heavy oil, extra heavy oil, and tar sand bitumen and its products, *instability* often manifests itself in various ways (Table 5.1; Stavinoha and Henry, 1981; Hardy and Wechter, 1990; Power and Mathys, 1992).

TABLE 5.1 Examples of Properties Related to Instability in Crude Oil and Crude Oil Products

Property	Comments
Asphaltene constituents	Influence oil-rock interactions
	Separates from oil when gases are dissolved
	Thermal alteration can cause phase separation
Heteroatom constituents	Provide polarity to oil
	Preferential reaction with oxygen
	Preferential thermal alteration
Aromatic constituents	May be incompatible with paraffinic medium
	Phase separation of paraffinic constituents
Nonasphaltene constituents	Thermal alteration causes changes in polarity
	Phase separation of polar species

Hence, there are different ways of defining each of these terms, but the terms are often used interchangeably.

Gum formation (ASTM D525) alludes to the formation of soluble organic material, whereas *sediment* is the insoluble organic material. *Storage stability* (or *storage instability*) (ASTM D381 and ASTM D4625) is a term used to describe the ability of the liquid to remain in storage over extended periods of time without appreciable deterioration as measured by gum formation and/or the formation sediment. *Thermal stability* is also defined as the ability of the liquid to withstand relatively high temperatures for short periods of time without the formation of sediment (i.e., carbonaceous deposits and/or coke). *Thermal-oxidative stability* is the ability of the liquid to withstand relatively high temperatures for short periods of time in the presence of oxidation and without the formation of sediment or deterioration of properties (ASTM D3241), and there is standard equipment for various oxidation tests (ASTM D4871). *Stability* is also as the ability of the liquid to withstand long periods at temperatures up to 100°C (212°F) without degradation. Determination of the *reaction threshold temperature* for various liquid and solid materials might be beneficial (ASTM D2883).

Existent gum is the name given to the nonvolatile residue present in the oil as received for test (ASTM D381). In this test, the sample is evaporated from a beaker maintained at a temperature of 160–166°C (320–331°F) with the aid of a similarly heated jet of air. This material is distinguished from the *potential gum* that is obtained by aging the sample at an elevated temperature.

Thus, *potential gum* is determined by the *accelerated gum test* (ASTM D873) that is used as a safeguard of storage stability and can be used to predict the potential for gum formation during prolonged storage. In this test, the oil is heated for 16h with oxygen under pressure in a bomb at 100°C (212°F). After this time, both the gum content and the solids precipitate are measured. A similar test, using an accelerated oxidation procedure, is also in use for determining the oxidative stability of oil (ASTM D942, ASTM D2274, ASTM D2272, and ASTM D2274).

Dry sludge is defined as the material separated from heavy oil, extra heavy oil, and tar sand bitumen and heavy oil, extra heavy oil, and tar sand bitumen products by filtration, which is insoluble in heptane. *Existent dry sludge* is the dry sludge in the original sample as received and is distinguished from the accelerated dry sludge obtained after aging the sample by chemical addition or heat. The *existent dry sludge* is distinguished from the *potential dry sludge* that is obtained by aging the sample at an elevated temperature. The *existent dry sludge* is the material separated from the bulk of a crude oil or crude oil product by filtration, which is insoluble in heptane. The test is used as an indicator of process operability and as a measure of potential downstream fouling.

An analogous test, the *thin-film oven test* (*TFOT*) (ASTM D1754) and an aging test (IP 390) are used to indicate the rate of change of various physical properties such as penetration (ASTM D5), viscosity (ASTM D2170), and ductility (ASTM D113) after a film of asphalt or bitumen has been heated in an oven for 5h at 163°C (325°F) on a rotating plate. A similar test is available for the stability of engine oil by thin-film oxygen uptake (*TFOUT*) (ASTM D4742).

This test establishes the effects of heat and air based on changes incurred in the above physical properties measured before and after the oven test. The allowed rate of changes in the relevant bitumen properties after the exposure of the tested sample to the oven test is specified in the relevant specifications (ASTM D3381).

Attractive as they may be, any tests that involve *accelerated oxidation of* the sample must be used with caution and consideration of the chemistry. Depending on the constituents of the sample, it is quite possible that the higher temperature and extreme conditions (oxygen under pressure) may not be truly representative of the deterioration of the sample under storage conditions. The higher temperature and the oxygen under pressure might change the chemistry of the system and produce products that would not be produced under ambient storage conditions. An assessment of the composition of the heavy oil, extra heavy oil, and tar sand bitumen prior to storage and application of the test will assist in this determination.

In general, *instability* and *incompatibility* can be related to the heteroatom-containing compounds (i.e., nitrogen-, oxygen-, and sulfur-containing compounds) that are present in the heavy oil, extra heavy oil, and tar sand bitumen. The degree of unsaturation of the heavy oil, extra heavy oil, and tar sand bitumen or the respective products caused by thermal recovery processes or by refining processes (i.e., the level of olefinic species) also plays a role in determining instability/incompatibility. And recent investigations have also implicated catalytic levels of various oxidized intermediates and acids as especially deleterious.

The term *incompatibility* can have many meanings. The most obvious example of incompatibility (*nonmiscible*) is the inability of heavy oil, extra heavy oil, and tar sand bitumen or their respective products and water to mix. In the present context, incompatibility usually refers to the presence of various polar functions (i.e., heteroatom function groups containing nitrogen, oxygen, or sulfur and even various combinations of the heteroatoms) in the crude oil.

Instability reactions are usually defined in terms of the formation of filterable and nonfilterable sludge (sediments, deposits, and gums), an increased peroxide level, and the formation of color bodies. Color bodies in and of themselves do not predict instability. However, the reactions that initiate color-body formation can be closely linked to heteroatom-containing (i.e., nitrogen-, oxygen-, and sulfur-containing) functional group chemistry. For example, blending crude oils or heavy feedstocks that are incompatible can lead to extensive fouling and processing difficulties due to unstable asphaltene constituents (Speight, 2014, 2015a). For example, extensive fouling in the crude preheat train may occur resulting in decreased energy efficiency, increased emissions of carbon dioxide, and increased frequency at which heat exchangers need to be cleaned. In a worst-case scenario, crude throughput may be reduced leading to significant financial losses.

Thus, heavy oil, extra heavy oil, and tar sand bitumen constituents are *incompatible* when sludge, semisolid, or solid particles (for convenience here, these are termed *secondary products* to distinguish them from the actual heavy oil, extra heavy oil, and tar sand bitumen product) are formed during and after blending. This phenomenon usually occurs prior to use.

If the secondary products are marginally soluble in the blended heavy oil, extra heavy oil, and tar sand bitumen product, use might detract from solubility of the secondary products, and they will appear as sludge or sediment that can be separated by filtration or by extraction (ASTM D4310). When the secondary products are truly insoluble, they separate and settle out as a semisolid or solid phase floating in the heavy oil, extra heavy oil, and tar sand bitumen or are deposited on the walls and floors of containers. In addition, secondary products usually increase the viscosity of the heavy oil, extra heavy oil, and tar sand bitumen product. Standing at low temperatures will also cause a viscosity change in heavy oil, extra heavy oil, and tar sand bitumen (ASTM D2532).

In the current context, the meaning of the term *incompatibility* is found when it is applied to heavy oil, extra heavy oil, and tar sand bitumen recovery and heavy oil, extra heavy oil, and tar sand bitumen refining. The application occurs when a product is formed in the formation or well pipe (recovery) or in a reactor (refining) and the product is incompatible with (immiscible with or insoluble in) the original heavy oil, extra heavy oil, and tar sand bitumen or its products. Such an example is the formation and deposition of other solids during recovery or the formation of coke precursors and even of coke during many thermal and catalytic operations. Coke formation is considered to be an initial *phase separation* of an insoluble, solid, coke precursor prior to coke formation proper. In the case of crude oils, *sediments* and *deposits* are closely related to *sludge*, at least as far as compositions are concerned. The major difference appears to be in the character of the material.

There is also the suggestion (often, but not always, real) in that the sediments and deposits originate from the inorganic constituents of heavy oil, extra heavy oil, and tar sand bitumen. They may be formed from the inherent components of the crude oil (i.e., the metalloporphyrin constituents) or from the ingestion of contaminants by the crude oil during the initial processing operations. For example, crude oil is known to *pick up* iron and other metal contaminants from contact with pipelines and pumps.

Sediments can also be formed from organic materials, but the usual inference is that these materials are usually formed from inorganic materials. The inorganic materials can be salt, sand, rust, and other contaminants that are insoluble in the crude oil and that settle to the bottom of the storage vessel. For example, *gum* typically forms by way of a hydroperoxide intermediate that induces polymerization of olefins. The intermediates are usually soluble in the liquid medium. However, gum that has undergone extensive oxidation reactions tends to be higher in molecular weight and much less soluble. In fact, the high-molecular-weight sediments that form in heavy oil, extra heavy oil, and tar sand bitumen are usually the direct result of *autoxidation* reactions. Active oxygen species involved include both molecular oxygen and hydroperoxides. These reactions proceed by a free-radical mechanism, and the solids produced tend to have increased incorporation of heteroatom and are thus also more polar and increasingly less soluble in the heavy oil, extra heavy oil, and tar sand bitumen.

The most significant and undesirable instability change in heavy oil, extra heavy oil, and tar sand bitumen is the formation of solids, termed *filterable sediment*. Filterable sediments can plug nozzles, filters, and coat heat exchanger surfaces and otherwise degrade engine performance. These solids are the result of free-radical autoxidation reactions. Although slight thermal degradation occurs in nonoxidizing atmospheres, the presence of oxygen or active oxygen species, such as hydroperoxides, will greatly accelerate oxidative degradation and significantly lower the temperature at which undesirable products are formed. Solid deposits that form as the result of short-term high-temperature reactions share many similar chemical characteristics with filterable sediment that form in storage.

The soluble sludge/sediment precursors that form during processing or use may have a molecular mass in the several hundred range. For this soluble precursor to reach a molecular weight sufficient to precipitate (or to phase-separate), one of two additional reactions must occur. Either the molecular weight must increase drastically as a result of condensation reactions leading to the higher-molecular-weight species, or the polarity of the precursor must increase (without necessarily increasing the molecular weight) by incorporation of additional oxygen, sulfur, or nitrogen functional groups. Additionally, the polarity may increase because of the removal of nonpolar hydrocarbon moieties from the polar core, as occurs during cracking reactions. In all three cases, insoluble material will form and separate from the liquid medium.

Additives are chemical compounds intended to improve some specific properties of heavy oil, extra heavy oil, and tar sand bitumen or their respective products. Different additives, even when added for identical purposes, may be incompatible with each other, for example, react and form new compounds. Consequently, a blend of heavy oil, extra heavy oil, and tar sand bitumen with other liquids, containing different additives, may form a system in which the additives react with each other and so deprive the blend of any beneficial effect.

The chemistry and physics of incompatibility can, to some extent, be elucidated (Power and Mathys, 1992; Mushrush and Speight, 1995), but many unknowns remain. In addition to the chemical aspects, there are also aspects such as the attractive force differences:

1. Specific interactions between like/unlike molecules (e.g., hydrogen bonding and electron donor-acceptor phenomena).
2. Field interactions such as dispersion forces and dipole-dipole interactions.
3. Any effects imposed on the system by the size and shape of the interacting molecular species.

Such interactions are not always easy to define, and thus, the measurement of instability and incompatibility has involved visual observations, solubility tests, hot filtration sediment (HFS), and gum formation. However, such methods are often considered to be *after-the-fact methods* insofar as they did not offer much in the way of predictability. In refinery processes (Parkash, 2003; Gary et al., 2007; Speight, 2014, 2017; Hsu and Robinson, 2017), predictability is not just a luxury, it is a necessity. The same principle must be applied to the measurement of instability and incompatibility. Therefore, methods are continually being sought to aid in achieving this goal.

In addition to the gravimetric methods, there have also been many attempts to use crude oil and/or product characteristics and their relation to the sludge and deposit formation tendencies. In some cases, a modicum of predictability is the outcome, but in many cases, the data appear as *preferred ranges* and are subject to personal interpretation. Therefore, caution is advised.

Furthermore, although the terminology applied to oil instability and incompatibility has evolved over the decades, confusion still exists. Another part of the problem (not to be underestimated) arises because of variations in terminology. The nonspecific application of various names is confusing and needs to be resolved. Part of the goal of this chapter is to attempt to resolve the issue of nomenclature. The terminology may seem confusing unless some general rules are accepted. To alleviate some of the confusion, the terms instability and incompatibility will be used with caution throughout the text, and the meanings will be made clear at the time of use. Particularly troublesome and more confusing are those terms that are applied to the more viscous materials; for example, in the context of this book, the use of the terms *bitumen* (a naturally occurring material found in tar sand deposits) (Chapter 1) and *asphalt* (a manufactured material used for road construction and waterproofing) is preferred.

It is the purpose of this chapter to present a brief overview of the types of constituents that are found in crude oil, heavy oil, and bitumen and also to present brief descriptions of the chemistry and physics of thermal decomposition. It is also the purpose of this chapter to document some of the more prominent methods used for determining instability and incompatibility. No preference will be shown, and none will be given to any individual methods. It is the choice of the individual experimentalist to choose the method on the basis of the properties of the heavy oil, extra heavy oil, and tar sand bitumen; the immediate needs; and the projected utilization of the data (Speight, 2015a). It is often the use (*misuse* is a better word) of the data that often detracts from an otherwise sound method. Also, it is appropriate at this point to define terms that are used in these important aspects of oil recovery and refining.

2. General Terminology

The general scientific areas of the instability and incompatibility of heavy oil, extra heavy oil, and tar sand bitumen are complex and have been considered to be nothing better than a black art because not all of the reactions that contribute to instability and incompatibility have been defined. Nevertheless, studies over the past three decades have made valuable contributions to our understanding of instability and incompatibility in heavy oil, extra heavy oil, and tar sand bitumen. For the most part, however, gaps remain in our knowledge of the chemistry and physics of instability and incompatibility, especially when applied to heavy oil, extra heavy oil, and tar sand bitumen.

Briefly, the term *instability* is often used in reference to the formation of color, sediment, or gum in the liquid over a period of time. This term may be used in contrast to the formation of a precipitate in the near, almost immediate term. However, the terms are often used interchangeably. On the other hand, the term *incompatibility* refers to the formation of a precipitate (or sediment) or phase separation when two liquids are mixed. The term *gum formation* alludes to the formation of soluble organic (usually considered to be polymeric) material, whereas sediment is considered to be insoluble organic (usually polymeric) material. By way of often archaic terminology, sludge that settles as a bottom layer in storage facilities is often semisolid, whereas a sediment and a deposit are usually solid materials. These are mere semantics but are often very real differences to the operators of storage facilities.

Storage stability (or *storage instability*) is a term used to describe the ability of a liquid oil to remain in storage over extended periods of time without appreciable deterioration as measured by gum formation and the depositions of insoluble material. *Thermal stability* is the ability of the liquid oil to withstand relatively high temperatures for short periods of time without the formation of carbonaceous deposits (coke). There is also the tendency to define stability as the ability of the liquid to withstand long periods at temperatures up to 100°C (212°F) without degradation.

In general, instability and incompatibility can be related to the existence of heteroatoms (i.e., containing nitrogen, oxygen, and sulfur). The degree of unsaturation of the oil (i.e., the level of olefinic species) also plays a role in determining instability and incompatibility. In addition, the phenomenon of instability (often variously referred to sludge formation and sediment formation or deposit formation) in crude oil and other products often manifests itself in various ways. Hence, there are different ways of defining each of these terms. For example, instability may be observed during crude oil storage, refining, product storage, and product utilization. The composition of the sludge/sediment often provides some insights into the mode of formation. In the areas of heavy oil science, extra heavy oil science, and tar sand bitumen science, the reactions leading to instability are usually defined in terms of the formation of filterable and nonfilterable sludge (sediments, deposits, and gums).

Both stability and incompatibility result in the formation and separation of degradation products or in other undesirable changes in the original properties of heavy oil, extra heavy oil, and tar sand bitumen. In the case of instability, there is a low resistance of the product to environmental influences during storage or to its susceptibility to oxidative or other degradation processes. In the case of incompatibility, degradation products form, or changes occur due to an interaction of some chemical groups present in the components of the final blend.

The components of heavy oil, extra heavy oil, and tar sand bitumen can be defined as being incompatible when sludge, semisolid, or solid particles are formed when such oils are blended. The products so formed can be either soluble or insoluble in the final blend. In the former case, they usually increase the viscosity of the product; in the latter case, they settle out as semisolid or solid matter floating in the heavy oil, extra heavy oil, and tar sand bitumen or being deposited on the walls and floors of containers.

Incompatibility phenomena, although not always recognized as such, can be associated with heavy oil, extra heavy oil, and tar sand bitumen, and the phenomenon of incompatibility is invariably associated with the chemical composition of the components. In most cases, a certain component in one of the oils reacts with another component in the heavy oil, extra heavy oil, and tar sand bitumen with which it is blended. This chemical reaction results in the formation of a new product that, when soluble, affects the properties of the blend and, when insoluble, settles out in the form of semisolid or solid matter.

In general terms, studies of the composition of the incompatible materials often involve determination of the distribution of the organic components by selective fractionation, analogous to the deasphalting procedure (Speight, 2014, 2015a): (i) heptane-soluble material, often called maltenes; (ii) heptane-insoluble material and benzene (or toluene)-soluble material, often referred to as the asphaltene fraction; (iii) or toluene-insoluble material, referred to as carbenes plus carboids when the fraction is a thermal product; and (iv) pyridine-soluble material (carbenes) and pyridine-insoluble material (carboids).

Whichever solvent separation scheme is employed, there should be ample description of the procedure that the work can be repeated not only in the same laboratory by a different researcher but also in different laboratories by different researchers (Speight, 2015a). Thus, for any particular feedstock and solvent separation scheme, the work should repeatable and reproducible within the limits of experimental error.

The sludge formed by the instability of heavy oil, extra heavy oil, and tar sand bitumen often contains inorganic and organic components and sometimes both intermingled in a heterogeneous matrix. The inorganic components are represented often by aqueous solutions of salts, sand, corrosion products, bacterial slime, and other contaminants. The organic components in crude oil sludge are usually polar constituents or condensed high-molecular-weight constituents (such as asphaltene material) and a variety of other organic (degradation) products. The degradation products formed during the long-term storage of heavy oil, extra heavy oil, and tar sand bitumen are often the result of air oxidation, photooxidation, and condensation. Any high-molecular-weight polar substances that stay solid at ambient temperatures are so formed and, consequently, form sludge.

To summarize, part of the problem of incompatibility arises from the manner in which the constituents of heavy oil, extra heavy oil, and tar sand bitumen can vary during production and refining or by reaction with oxygen during storage. The remainder of the problem may be due to the character of the individual constituents. Many of the constituents may be incompatible/immiscible with each other, thereby causing phase separation and, hence, incompatibility.

3. Instability and Incompatibility

Incompatibility during refining can occur in a variety of processes, either by intent (such as in the deasphalting process) or inadvertently when the separation is detrimental to the process. Thus, separation of solids occurs whenever the solvent characteristics of the liquid phase are no longer adequate to maintain polar and/or high-molecular-weight material in solution. Examples of such occurrences are (i) asphaltene separation that occurs when the paraffinic nature of the liquid medium increases, (ii) wax separation that occurs when there is a drop in temperature or the aromaticity of the liquid medium increases, (iii) sludge/sediment formation in a reactor that occurs when the solvent characteristics of the liquid medium change so that asphaltic or wax materials separate, (iv) coke formation that occurs at high temperatures and commences when the solvent power of the liquid phase is not sufficient to maintain the coke precursors in solution, and (v) sludge/sediment formation in fuel products that occurs because of the interplay of several chemical and physical factors (Mushrush and Speight, 1995).

When fractions of crude oil are heated to temperatures in excess of 350°C (>660°F), the thermal or free-radical reactions start to give way to cracking of the constituents of the oil at significant (measurable) rates. Thermal conversion does not require the addition of a catalyst, and therefore, this approach is the oldest technology available for conversion of crude oil to salable products. The severity of thermal processing determines the conversion and the product characteristics.

The thermal treatment of heavy oil, extra heavy oil, and tar sand bitumen ranges from mild, for reduction of viscosity, to ultrapyrolysis, for complete conversion to olefins and lower-boiling constituents (often referred to as *light ends*) (Parkash, 2003; Gary et al., 2007; Speight, 2014, 2017; Hsu and Robinson, 2017). The higher the temperature, the shorter the time to achieve a given conversion. The severity of the process conditions is defined as the combination of reaction time and temperature to achieve a given conversion. If side reactions are absent, very long times at low temperature are often claimed to be equivalent to very short times at high temperature. Nothing could be further from reality!

Thermal reactions, however, can give rise to a variety of different reactions, so that selectivity for a given product changes with temperature and pressure. Mild thermal processes and high-severity thermal processes are frequently used for processing higher-molecular-weight feedstocks such as heavy oil, extra heavy oil, and tar sand bitumen. More extreme conditions are only used commercially for cracking ethane (CH_3CH_3), propane ($CH_3CH_2CH_3$), butane [$CH_3CH_2CH_2CH_3$ or $(CH_3)_2CHCH3$, or $(CH3)_3CCH_3$], and low-boiling distillate feedstocks (such as naphtha, kerosene, and gas oil) to produce ethylene ($CH_2{=}CH_2$) and higher-molecular-weight olefins.

The main limitation on thermal processes is that the products can be unstable and may even be incompatible with the original feedstock and/or other products. Thermal cracking at low pressure produces olefins, which give a highly unstable product that is prone to oxidation leading to polymerization reactions to form tars and gums. The high-molecular-weight fraction can form solids or sediments, which also feed composition, which also determines the maximum conversion allowable.

The instability/incompatibility of heavy oil, extra heavy oil, and tar sand bitumen is manifested in the formation of sludge and/or sediment. Sludge (or sediment) formation takes one of the following forms: (i) material dissolved in the liquid, (ii) precipitated material, and (iii) material emulsified in the liquid. Under favorable conditions, sludge or sediment will dissolve in the heavy oil, extra heavy oil, and tar sand bitumen with the potential of increasing the viscosity. Sludge or sediment, which is not soluble in the heavy oil, extra heavy oil, and tar sand bitumen may either settle at the bottom of the storage tanks or remain in the crude oil as an emulsion. In most of the cases, the smaller part of the sludge/sediment will settle satisfactorily; the larger part will stay in the heavy oil, extra heavy oil, and tar sand bitumen as emulsions. In any case, there is a need of breaking the emulsion, whether it is a water-in-oil emulsion or whether it is the sludge itself, which has to be separated into the oily phase and the aqueous phase. The oily phase can be then processed with the crude oil, and the aqueous phase can be drained out of the system.

Phase separation can be accomplished by either the use of suitable surface-active agents allowing for sufficient settling time or the use of a high-voltage electric field for breaking such emulsions after admixing water at a rate of about 5% and at a temperature of about 100°C (212°F).

Emulsion breaking, whether the emulsion is due to crude oil-sludge emulsions and crude oil-water emulsions, or breaking the sludge themselves into organic (oily) and inorganic components are of a major importance from operational and commercial aspects. In the case of heavy oil, extra heavy oil, and tar sand bitumen, phase separation difficulties often arise. Also, some heavy oil, extra heavy oil, and tar sand bitumen emulsions may be stabilized by naturally occurring substances in the oil. Many of these polar particles accumulate at the oil-water interface, with the polar groups directed toward the water and the hydrocarbon groups toward the oil. A stable interfacial skin may be so formed—particles of clay or similar impurities present in the heavy oil, extra heavy oil, and tar sand bitumen may be embedded in this skin and make the emulsion very difficult to break.

Chemical and electric methods for sludge removal and for water removal, often combined with chemical additives, have to be used for breaking such emulsions. Each emulsion has its own structure and characteristics: water-in-oil emulsions where the oil is the major component or oil-in-water emulsions where the water is the major component. The chemical and physical nature of the components of the emulsion plays a major role in their susceptibility to the various surface-active agents used for breaking them.

Therefore, appropriate emulsion-breaking agents have to be chosen very carefully, usually with the help of previous laboratory evaluations. Water- or oil-soluble demulsifiers, the latter being often nonionic surface-active alkylene oxide adducts, are used for this purpose. But, as had been said in the foregoing, the most suitable demulsifier has to be chosen for each case from a large number of such substances in the market, by a prior laboratory evaluation.

4. Factors Influencing Instability-Incompatibility

The initial stages of the thermal decomposition of asphaltenes have been least well understood of the thermal decomposition process. It is known that the overall thermal decomposition involves scission of carbon-carbon bonds, aromatization of naphthenic ring systems, formation of lower-molecular-weight material, and formation of coke. The presence of hydrogen during the thermal decomposition of asphaltenes can alter the product slate. However, the chemistry remains complex, and the representation of the chemistry of asphaltene thermolysis is speculative and has been done simply by the use of nonchemical formulas.

If no side reactions occur, very residence long times at low temperature should be equivalent to very short times at high temperature. In fact, the nature of the thermal processes varies considerably.

One of the postulates of coke formation involves the production of coke by a sequence of polymerization and condensation steps from the lower boiling (lower molecular weight) to the heaviest (higher molecular weight) fractions. While this concept is acceptable for pure aromatic hydrocarbons, the formation of coke by thermal cracking the constituents of the asphaltene constituents of crude oil requires a modification of this concept. Indeed, present evidence (Speight, 1994, 2014; Wiehe, 1992, 1993) supports the concept that the reaction pathway of the asphaltene constituents to coke is much more complex and involves at least several reaction stages. Various findings offer some indication of the proposed chemistry of coke or sediment formation during crude oil processing. However, in each case, the aromatization of naphthene rings and the potential or reality of cyclodehydrogenation reactions are ignored. In addition, the thermal reactions of the resin constituents are not addressed.

The asphaltene constituents rely upon the resin constituents for stability within the oil medium. Removal or thermal alteration of the resin constituents will also cause the asphaltene constituents to separate as a distinct phase. Such processes would also lead to increased aromaticity of the polynuclear aromatic-naphthenic (often referenced as redistribution of hydrogen within constituents) core and further enhance the insolubility (phase separation) of the dealkylated polynuclear product.

The available data suggest that the thermal cracking of the high-molecular-weight constituents of crude oil (leading to the formation of volatile products and coke) is a complex process involving both chemical reactions and thermodynamic behavior. Reactions that contribute to this process are (i) cracking of side chains from aromatic groups, (ii) dehydrogenation of naphthenes to form aromatics, (iii) condensation of aliphatic structures to form aromatics, (iv) condensation of aromatics to form higher fused-ring aromatics, and (v) dimerization or oligomerization reactions. Loss of side chains always accompanies thermal cracking, and dehydrogenation and condensation reactions are favored by hydrogen deficient conditions.

The importance of the use of solvents to mitigate coke formation during the thermal decomposition of crude oil has been recognized for many years, but the effects have been correctly ascribed to hydrogen donor reactions that have an inhibiting effect on the formation of the molecular species that are prone to phase separation. The separation of the phases depends on the solvent characteristics of the liquid. Addition of aromatic solvents suppresses phase separation, whereas paraffins enhance separation. Microscopic examination of coke particles often shows evidence for mesophase, spherical domains that exhibit the anisotropic optical characteristics of liquid crystals.

The phase-separation phenomenon that is the prelude to coke formation can also be explained by use of the solubility parameter (δ) for crude oil fractions and for the solvents (Speight, 1992, 1994, 2014 and references cited therein). Although little is known about the solubility parameter of crude oil fractions, there has been a noteworthy attempt to define the solubility parameter ranges for different fossil fuel liquids.

When catalytic processes are employed, complex molecules (such as those that may be found in the original resin fraction and in the original asphaltene fraction or those formed during the process), are not sufficiently mobile (or are too strongly adsorbed by the catalyst) to be saturated by the hydrogenation components. Hence, these molecular species continue to condense and eventually degrade to coke. These deposits deactivate the catalyst sites and eventually interfere with the process.

As these concepts show, characterization data can be used as an integral part of understanding the thermal chemistry of asphaltenes as a start in predicting product yield and product distribution; predictions of product quality will follow.

A number of experimental methods are available for estimation of the factors that influence instability/incompatibility. These factors have been explored, and attempts are made to estimate the character of the heavy oil, extra heavy oil, and tar sand bitumen or their respective products with varied results.

4.1 Acidity

The acidity of heavy oil, extra heavy oil, and tar sand bitumen or heavy oil, extra heavy oil, and tar sand bitumen products is usually measured in terms of the *acid number* that is defined as the number of milliequivalents per gram of alkali required to neutralize the acidity of the heavy oil, extra heavy oil, and tar sand bitumen sample (ASTM D664, ASTM D974, and ASTM D3242).

Acidity due to the presence of inorganic constituents is not expected to be present in crude oils, but organic acidity might be found. Acidic character is composed of contributions from strong organic acids and other organic acids. Values above 0.15 mg potassium hydroxide per gram are considered to be significantly high. Crude oils of higher acidities may exhibit a tendency of instability.

The acid-imparting agents in crude oils are naphthenic acids and hydrosulfides (thiols, mercaptans, and R-SH). These are sometimes present in the crude oil originally in small and varying concentrations. Normally, the total acidity of crude oils is in the range of 0.1–0.5 mg potassium hydroxide per gram, although higher values are not exceptional.

Free hydrogen sulfide is often present in crude oils, a concentration of up to 10 ppm being acceptable in spite of its toxic nature. However, higher hydrogen sulfide concentrations are sometimes present, 20 ppm posing serious safety hazards. Additional amounts of hydrogen sulfide can form during the crude oil processing, when hydrogen reacts with some organic sulfur compounds converting them to hydrogen sulfide. In this case, it is referred to as potential hydrogen sulfide, contrary to free hydrogen sulfide.

Acidity can also form by bacterial action insofar as some species of aerobic bacteria can produce organic acids from organic nutrients. On the other hand, anaerobic sulfate-reducing bacteria can generate hydrogen sulfide, which, in turn, can be converted to sulfuric acid (by bacterial action).

4.2 Density/Specific Gravity

In the earlier years of the heavy oil, extra heavy oil, and tar sand bitumen industry, density and specific gravity (with the API gravity) were the principal specifications for feedstocks and refinery products (Parkash, 2003; Gary et al., 2007; Speight, 2014, 2017; Hsu and Robinson, 2017). They were used to give an estimate of the most desirable product, that is, kerosene, in crude oil. At the present time, a series of standard tests exist for determining density and specific gravity (Speight, 2001, 2015a).

There is no direct relation between the density and specific gravity of crude oils to their sludge forming tendencies, but crude oil having a higher density (thus, a lower API gravity) is generally more susceptible to sludge formation, presumably because of the higher content of the polar/asphaltic constituents.

4.3 Elemental Analysis

The ultimate analysis (elemental composition) of heavy oil, extra heavy oil, and tar sand bitumen and its products is not reported to the same extent as for coal (Speight, 2013d). Nevertheless, there are standard test methods (ASTM, 2017) for the ultimate analysis of heavy oil, extra heavy oil, and tar sand bitumen and heavy oil, extra heavy oil, and tar sand bitumen products, but many such methods may need some form of modification since they may have been designed for other materials.

Of the data that are available, the proportions of the elements in heavy oil, extra heavy oil, and tar sand bitumen vary only slightly over narrow limits: carbon 83.0–87.0% w/w, hydrogen 10.0–14.0% w/w, nitrogen 0.10–2.0% w/w, oxygen 0.05–1.5% w/w, and sulfur 0.05–6.0% w/w (Speight, 2013a,b, 2014).

And yet, there is a wide variation in physical properties from the low-viscosity more mobile crude oils at one extreme to the extra heavy crude oil at the other extreme (Chapter 1). In terms of the instability and incompatibility of heavy oil, extra heavy oil, and tar sand bitumen and heavy oil, extra heavy oil, and tar sand bitumen products, the heteroatom content appears to represent the greatest influence. In fact, it is not only the sulfur and nitrogen content of heavy oil, extra heavy oil, and tar sand bitumen that are important parameters in respect of the processing methods that have to be used in order to produce liquid fuels that meet, for example, sulfur specifications, but also the type of sulfur and nitrogen species in the oil. There could well be a relation between nitrogen and sulfur content and crude oil (or product) stability; higher nitrogen and sulfur crude oils are suspect of higher sludge forming tendencies.

4.4 Metals Content

The majority of crude oils contain metallic constituents that are often determined as combustion ash (ASTM D482). This is particularly so for the heavier feedstocks. These constituents, of which nickel and vanadium are the principal metals, are very influential in regard to feedstock behavior in processing operations.

The metal (inorganic) constituents of heavy oil, extra heavy oil, and tar sand bitumen or a liquid fuel product arise from either the inorganic constituents present in the crude oil originally or those picked up by the heavy oil, extra heavy oil, and tar sand bitumen during storage. The former are mostly metallic substances like vanadium, nickel, sodium, iron, and silica; the latter may be contaminants such as sand, dust, and corrosion products.

Incompatibility leading to deposition of the metals (in any form) onto the catalyst leads to catalyst deactivation whether it is by physical blockage of the pores or destruction of reactive sites. In the present context, the metals must first be removed if erroneously high-carbon residue data are to be avoided. Alternatively, they can be estimated as ash by complete burning of the coke after carbon residue determination.

Metal content above 200 ppm is considered to be significant, but the variations are very large. The higher the ash content, the higher is the tendency of the crude oil to form sludge or sediment.

4.5 Pour Point

The *pour point* (Speight, 2015a) is used to define the cold properties of crude oils and heavy oil, extra heavy oil, and tar sand bitumen and heavy oil, extra heavy oil, and tar sand bitumen products, that is, the minimal temperature at which they still retain their fluidity (ASTM D97). Therefore, pour point also indicates the characteristics of crude oils: the higher the pour point, the more paraffinic is the oil and vice versa. Higher pour point crude oils are waxy, and therefore, they tend to form waxlike materials that enhance sludge formation.

To determine the *pour point* (ASTM D97, ASTM D5327, ASTM D5853, ASTM D5949, ASTM D5950, ASTM D5985, IP 15, IP 219, and IP 441), the sample is contained in a glass test tube fitted with a thermometer and immersed in one of three baths containing coolants. The sample is dehydrated and filtered at a temperature 25°C (45°F) higher than above the anticipated cloud point. It is then placed in a test tube and cooled progressively in coolants held at −1 to +2°C (30–35°F), −18 to −20°C (−4–0°F), and −32 to −35°C (−26 to −31°F). The sample is inspected for cloudiness at temperature intervals of 1°C (2°F). If conditions or oil properties are such that reduced temperatures are required to determine the pour point, alternate tests are available that accommodate the various types of samples.

4.6 Viscosity

The viscosity of a feedstock varies with the origin and type of the crude oil and also with the amounts of the various chemical constituents, particularly the polar functions where intermolecular interactions can occur. For example, there is a gradation of viscosity between conventional crude oil, heavy oil, and bitumen (Speight, 2001, 2015a).

Viscosity is a measure of fluidity properties and consistencies at given temperatures. Heavier crude oil, that is, crude oil having lower API gravity, typically has higher viscosity. Increases of viscosity during storage indicate either an evaporation of volatile components or formation of degradation products dissolving in the crude oil.

4.7 Volatility

Heavy oil, extra heavy oil, and tar sand bitumen can be subdivided by distillation into a variety of fractions of different *boiling ranges* or *cut points* (Parkash, 2003; Gary et al., 2007; Speight, 2014, 2017; Hsu and Robinson, 2017). In fact, distillation was, and still is, the method for feedstock evaluation for various refinery options. Indeed, volatility is one of the major tests for heavy oil, extra heavy oil, and tar sand bitumen products, and it is inevitable that the majority of all products will, at some stage of their history, be tested for volatility characteristics (Parkash, 2003; Gary et al., 2007; Speight, 2014, 2017; Hsu and Robinson, 2017).

As an early part of characterization studies, a correlation was observed between the quality of heavy oil, extra heavy oil, and tar sand bitumen fractions and their heteroatom content (Speight, 2014). Thus, it is not surprising that test to determine the volatility of heavy oil, extra heavy oil, and tar sand bitumen and heavy oil, extra heavy oil, and tar sand bitumen products was among the first to be defined. Heavy oil, extra heavy oil, and tar sand bitumen, yielding higher amounts of residua, tend to form more sludge during storage compared with low-viscosity crude oil.

4.8 Water Content, Salt Content, and Bottom Sediment/Water (BS&W)

Water content (ASTM D4006, ASTM D4007, ASTM D4377, and ASTM D4928), salt content (ASTM D3230), and bottom sediment/water (ASTM D96, ASTM D1796, and ASTM D4007) indicate the concentrations of aqueous contaminants, present in the crude either originally or picked up by the crude during handling and storage. Water and salt content of crude oils produced at the field can be very high, forming sometimes its major part. The salty water is usually separated at the field, usually by settling and draining; surface-active agents and electric methods for the destruction of emulsions (emulsion breakers and desalters) are sometimes employed. The water and salt contents of crude oil supplied to the buyers is function of the production field. Water content below 0.5%, salt content up to 20 lb per 1000 barrels, and bottom sediment and water up to 0.5% are considered to be satisfactory.

Although the centrifuge methods are still employed (ASTM D96, ASTM D1796, ASTM D2709, and ASTM D4007), many laboratories prefer the Dean and Stark adaptor (ASTM D95) (Speight, 2015a). The apparatus consists of a round-bottom flask with capacity of 50 mL connected to a Liebig condenser by a receiving tube with capacity of 25 mL, graduated in 0.1 mL. A weighed amount, corresponding to ~100 mL of oil, is placed in the flask with 25 nil of dry toluene. The flask is heated gently until the 25 mL of toluene have distilled into the graduated tube. The water distilled with the toluene separates to the bottom of the tube where the volume is recorded as mL or the weight as mg or percent.

To determine the sediment in heavy oil, extra heavy oil, and tar sand bitumen or in a heavy oil, extra heavy oil, and tar sand bitumen product, the method involves solvent extraction using a Soxhlet extractor (Speight, 2015a).

The Karl Fischer titration method (ASTM D1744), the Karl Fischer titration method (ASTM D377), and the colorimetric Karl Fischer titration method (ASTM D4298) still find wide application in many laboratories for the determination of water in liquid fuels, specifically the water content of aviation fuels.

The higher the bottom sediment and water content, the higher sludge and deposit formation rates can be expected in the stored crude oil.

5. Asphaltene Constituents and Instability-Incompatibility

An early hypothesis of the physical structure of crude oil (Pfeiffer and Saal, 1940) indicated that asphaltene constituents are the centers of micelles formed by adsorption or even by absorption of the part of the maltenes, that is, resin material, onto the surfaces or into the interiors of the asphaltene particles. Thus, most of those substances with greater molecular weight and with the most pronounced aromatic nature are situated closest to the nucleus and are surrounded by lower-boiling constituents of a less aromatic nature. The transition of the intermicellar (dispersed or oil) phase is gradual and almost continuous and is influenced by the makeup of the asphaltene fraction, which is composed of neutral, acidic, and basic constituents (Speight, 1994, 2014), and this is why the asphaltene fraction cannot be adequately represented by a so-called average structure. Various separation procedures applied to the asphaltene fraction have confirmed the complex chemical and functional nature of the asphaltene fraction (Fig. 5.3; Francisco and Speight, 1984; Speight, 1994, 2014). Since asphaltene constituents are incompatible with the oil fraction, asphaltene dispersion is attributable to the interactions between resin constituents (polar aromatic constituents) and the asphaltene constituents that each carries interacting chemical-physical sites and functional groups (Koots and Speight, 1975).

On this basis, the stability of crude oil can be represented by a three-phase system in which the asphaltene constituents, the aromatic constituents (including the resin constituents), and the saturate constituents are delicately balanced (Speight, 1994, 2014). Various factors, such as oxidation, can have an adverse effect on the system, leading to instability/incompatibility as a result of changing the polarity and bonding arrangements of the species in crude oil.

The evolution of this concept is perhaps the single most responsible agent for the perception that asphaltene constituents in crude oil, heavy oil, extra heavy oil, and tar sand bitumen crude oil exist in the manner of a colloidal system. Indices have been derived to describe this behavior. One strong factor in favor of such a system is the inability of most crude oils to disperse the asphaltene fraction in the absence of the resin fraction (Koots and Speight, 1975).

In heavy oil, extra heavy oil, and tar sand bitumen, the asphaltene constituents and the resin constituents have strong interactions to the extent that the asphaltene constituents are immiscible (insoluble/incompatible) with the

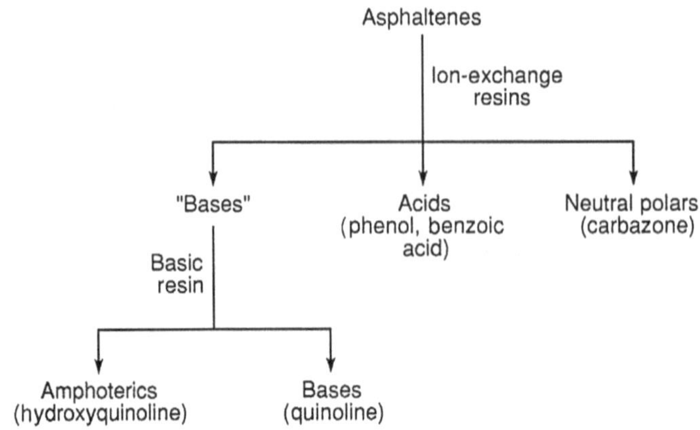

FIG. 5.3 Fractionation of the asphaltene fraction on the basis of functionality (Francisco and Speight, 1984). Simple examples of the functional types are given in parentheses.

remaining constituents in the absence of the resin constituents (Koots and Speight, 1975). And there appears to be points of structural similarity (for a crude oil) between the asphaltene and the resin constituents, thereby setting the stage for *a more than is generally appreciated* complex relationship but confirming the hypothesis that heavy oil, extra heavy oil, and tar sand bitumen is a continuum of chemical species, including the asphaltene constituents (Speight, 2014, 2015a).

This sets the stage for the incompatibility of the asphaltene constituents in any operation in which the asphaltene or resin constituents are physically or chemically altered. Disturbance of the asphaltene-resin relationships can be the stimulation by which, for example, some or all of the asphaltene constituents form a separate *insoluble* phase leading to such phenomena as coke formation (in thermal processes) or asphalt instability during use.

The problems with the asphaltene constituents have increased due to the need to extract even the heaviest crude oils and the trend to extract large amounts of lower-boiling fractions out of crude oil by among other methods cracking and visbreaking. There is a variety of issues that arise due to flocculation and/or sedimentation of asphaltene constituents and reacted asphaltene constituents (Table 5.2).

TABLE 5.2 Examples of Issue Arising From the Presence of Asphaltene Constituents in a Feedstock Such as Heavy Oil, Extra Heavy Oil, and Tar Sand Bitumen

Process	Issue
Oil recovery	Wellbore plugging and pipeline deposition
Visbreaking process	Degraded asphaltene constituents are more aromatic (loss of aliphatic chains) and less soluble
Cracking process	Degraded asphaltene constituents are more aromatic (loss of aliphatic chains) and less soluble
Emulsion formation	A sufficiently high degree of water contamination can lead to emulsions formation
	Asphaltene constituents are responsible for the undesired stabilization of emulsions, since asphaltene constituents are highly polar and surface active
Preheating	Preheating of heavy oil, extra heavy oil, and tar sand bitumen prior to combustion encourages reaction and precipitation of the reacted asphaltene constituents leading to coking
Combustion	A high content (>6% w/w) of asphaltene constituents in the results in ignition delay and poor combustion further leading to boiler fouling, poor heat transfer, stack solid emission, and corrosion problems

Continued

TABLE 5.2 Examples of Issue Arising From the Presence of Asphaltene Constituents in a Feedstock Such as Heavy Oil, Extra Heavy Oil, and Tar Sand Bitumen—cont'd

Process	Issue
Blending	The change of medium during mixing may cause destabilization of asphaltene constituents and destabilization of any reacted asphaltene constituents
Storage	Sedimentation and plugging can occur due to oxidation of the asphaltene constituents
	Increased polarity may cause asphaltene aggregation, and sludge/sediment formation can furthermore be accelerated if there is bacterial infestation in the heavy oil, extra heavy oil, and tar sand bitumen

Thus, the deposition of asphaltene constituents is the consequence of instability of the heavy oil, extra heavy oil, and tar sand bitumen or the heavy oil, extra heavy oil, and tar sand bitumen product.

The instability/incompatibility of heavy oil, extra heavy oil, and tar sand bitumen products is a precursor to either the formation of degradation products or the occurrence of undesirable changes in the properties of the heavy oil, extra heavy oil, and tar sand bitumen. Individually, the components of a product may be stable and in compliance with specifications, but their blend may exhibit poor stability properties, making them unfit for use.

Incompatibility in the distillate products is important to consumers and the producer. Distillate products that are made from the refining process based on straight-run distillation show very little incompatibility problems. However, at present and in the future, problems for refiners will continue to increase as the quality of the available crude decreases worldwide and more reliance is placed upon heavy oil, extra heavy oil, and tar sand bitumen as refinery feedstocks.

Incompatibility in heavy oil, extra heavy oil, and tar sand bitumen products can be linked to the presence of several different deleterious heteroatomic compound classes (Mushrush and Speight, 1995). The incompatibility observed in a blend is dependent on the blending stocks employed in its production. The lower-boiling cycle oils produced during catalytic process contain the unstable heteroatomic species that are responsible for the observed deterioration in some fuel products. The solution is to use straight-run distillate product but with a deficiency in supply, an option is to use chemical additives that overcome the incompatibilities of the variant chemical composition of the product known as light cycle oil.

In general, the reaction sequence for sediment formation can be envisaged as being dependent upon the most reactive of the various heteroatomic species that are present in heavy oil, extra heavy oil, and tar sand bitumen and their respective products (Pedley et al., 1986, 1988, 1989). The key reaction in many incompatibility processes is the generation of the hydroperoxide species from dissolved oxygen. Once the hydroperoxide concentration starts to increase, macromolecular incompatibility precursors form in the heavy oil, extra heavy oil, and tar sand bitumen or their respective products. Acid- or base-catalyzed condensation reactions then rapidly increase both the polarity, incorporation of heteroatoms, and the molecular weight.

When the product is transferred to a storage tank or some other holding tank, incompatibility can occur by the free-radical hydroperoxide-induced polymerization of active olefins. This is a relatively slow reaction, because the observed increase in hydroperoxide concentration is dependent on the dissolved oxygen content (Mayo and Lan, 1987; Mushrush and Speight, 1995).

The third incompatibility mechanism involves degradation when the heavy oil, extra heavy oil, and tar sand bitumen or the products are stored for prolonged periods, as might occur during stockpiling of fuel for military use (Goetzinger et al., 1983; Hazlett and Hall, 1985; Stavinoha and Westbrook, 1980). This incompatibility process involves (i) the buildup of hydroperoxide moieties after the gum reactions, (ii) a free-radical reaction with the various organo-sulfur compounds present that can be oxidized to sulfonic acids, and (iii) reactions such as condensations between organo-sulfur and nitrogen compounds and esterification reactions.

Examples of such occurrences are (i) asphaltene separation that occurs when the paraffinic nature of the liquid medium increases during thermal processes (Chapter 8, during catalytic processes (Chapter 9), during hydrotreating processes (Chapter 10), and during hydrocracking processes (Chapter 11); (ii) solid separation that occurs when there is a drop in temperature or the aromaticity of the liquid medium increases; (iii) sludge/sediment formation in a reactor that occurs when the solvent characteristics of the liquid medium change so that asphaltic or wax materials separate, (iv) coke formation that occurs at high temperatures and commences when the solvent power of the liquid phase is not sufficient to maintain the coke precursors in solution, and (v) sludge/sediment formation, in the liquid products, that occurs because of the interplay of several chemical and physical factors (Mushrush and Speight, 1995).

In mixtures as complex as heavy oil, extra heavy oil, and tar sand bitumen, production options and refining options can only be generalized because of difficulties in analyzing not only the products but also the feedstock and the intricate and complex nature of the molecules that make up the feedstock. Moreover, the incompatibility of feedstocks and products is an ongoing issue during refining. The occurrence of sediment during thermal operations and from products during storage reduces the efficiency of a variety of processes.

The complex nature of these species can be taken several steps further when it is acknowledged that the structures of the asphaltene constituents are multidimensional in space. Furthermore, the asphaltene constituents appear to play a

TABLE 5.3 Standard ASTM Methods for Asphaltene Precipitation

Method	Precipitant	Volume Precipitant Per Gram of Sample
ASTM D893	*n*-Pentane	10 mL
ASTM D2006	*n*-Pentane	50 mL
ASTM D2007	*n*-Pentane	10 mL
ASTM D3279	*n*-Heptane	100 mL
ASTM D4124	*n*-Heptane	100 mL

ASTM, 2017. Annual Book of Standards, ASTM International, West Conshohocken, PA.

major role in interactions that lead to incompatibility in feedstocks, particularly during production and refining operations. Thus, a key to decreasing incompatibility during production and refinery operations is to develop an understanding of the asphaltene constituents, which are the principal precursors to sludge and other sediments. Indeed, the presence of asphaltene constituents in feedstocks is of considerable concern to those processes that center around the use of a catalyst, which, due to incompatibility, can lead to the deposition of metals and carbonaceous residues onto the catalyst (Speight, 2014).

The asphaltene fraction is, by definition, a solubility class that is precipitated from heavy oil, extra heavy oil, and tar sand bitumen by the addition of an excess of a low-boiling liquid hydrocarbon such as heptane (Table 5.3). However, the nomenclature of the nonvolatile constituents of crude oil (i.e., the asphaltene constituents; the resin constituents; and, to some extent, part of the oil fraction insofar as nonvolatile oils do occur in heavy oil, extra heavy oil, and tar sand bitumen) is an operational aid and is not based on chemical or structural features (Fig. 5.1).

5.1 Composition

The asphaltene fraction of crude oil, heavy oil, extra heavy oil, and tar sand bitumen is a solubility class and, as such, is composed of many different types of molecular species (Speight, 2014). Therefore, there is no one description that can be used to define the asphaltene fraction or the asphaltene constituents.

Analytic data from various techniques and methods can only produce average values; such data do show that the asphaltene fraction contains the most aromatic and most polar species originally present in the crude oil. However, caution must be used in the application of average data to structural studies, where the formula for an average molecule might be derived. An average molecule that might be derived from the data may be very far from the truth and may only explain asphaltene behavior with difficulty and with much imagination.

5.2 Ultimate (Elemental) Composition

The elemental compositions of asphaltene fraction vary over only a narrow range, corresponding to H/C ratios of 1.15 ± 0.05%, although values outside of this range are often found (Speight, 2014). Notable variations do occur in the proportions of the heteroelements, in particular, in the proportions of oxygen and sulfur. On the other hand, the nitrogen content of the asphaltene fraction has a somewhat lesser degree of variation. This is perhaps not surprising, since exposing asphaltene constituents to atmospheric oxygen can substantially alter the oxygen content, and exposing a crude oil to elemental sulfur, or even to sulfur-containing minerals, can result in sulfur uptake.

5.3 Fractional Composition

The asphaltene fraction can be fractionated by the use of a variety of techniques (Francisco and Speight, 1984; Speight, 2014). Of specific interest is the observation that when asphaltene constituents are fractionated on the basis of aromaticity and polarity, it appears that the more aromatic species contain higher amounts of nitrogen (Speight, 1984, 1994, 2014). This suggests that the nitrogen species are located predominantly in aromatic systems. However, more important, the separation of the asphaltene constituents into a variety of functional (and polar) types (Francisco and Speight, 1984) has confirmed the complexity of the asphaltene fraction. High-performance liquid chromatography (HPLC) has also confirmed the diversity of the structural and functional types in the asphaltene fraction to the extent that different HPLC profiles can be expected for different asphaltene fractions.

In summary, fractionation data lend support to, and reinforce, the concept that asphaltene constituents are complex mixtures of molecular sizes and various functional types and that carbenes and carboids can also fit into this concept (Speight, 2014).

5.4 Instability and Incompatibility

The asphaltene constituents are the most troublesome fraction of heavy oil, extra heavy oil, and tar sand bitumen (as well as the crude oils that contain an asphaltene fraction) due to the ability of these constituents to precipitate as solids and subsequently deposit with changing pressure, temperature, and oil composition (Speight, 1994, 2014). This may affect and cause severe problems in both upstream and downstream operations leading to fouling in all cases (Speight, 2015b). Separated asphaltene constituents (as well as in the case of thermal processes) and separated reacted asphaltene constituents can deposit on reservoir rock (as well as in the pores of the reservoir rock leading to possible blocking of flow, particularly in the near wellbore region), in pipelines, in heat exchangers, and at the bottom of a distillation column, thereby affecting efficiency of production and refining processes and leading to equipment damage. Furthermore, it is not only the amount of precipitated asphaltene constituents that is important but also the compatibility of the asphaltene constituents with the remainder of the oil (the maltene fraction, that is, the deasphaltened oil fraction) that affects rheological properties of an oil (Koots and Speight, 1975; Speight, 1994, 2014; Laux et al., 1997; Andersen and Speight, 1999; Bearsley et al., 2004).

However, to explain the behavior of asphaltene constituents, it is necessary to distinguish between (i) precipitation, (ii) aggregation, and (iii) deposition. Precipitation is the process when the asphaltene constituents become a separate phase from the crude oil and remain suspended in the liquid phase. The precipitated asphaltene constituents associate (aggregation) and form larger particles (flocs). The asphaltene aggregates are initially suspended in the crude oil after which the flocs may attach to and accumulate on various surfaces (deposition).

Once a heavy oil, extra heavy oil, or tar sand bitumen has been defined sufficiently well to indicate the preferential methods of recovery and refining, it is the nature of the recovery and refining processing operations that dictate the means to proceed. However, it is the nature of the feedstock and the products that dictate the potential for instability and/or incompatibility. Moreover, the potential for instability/incompatibility during thermal recovery operations and refining operations is real and must be addressed during processing.

Examples of such occurrences are (i) asphaltene separation that occurs when the paraffinic nature of the liquid medium increases, (ii) sludge/sediment formation in a reactor that occurs when the solvent characteristics of the liquid medium change so that asphaltene constituents (or reacted asphaltene constituents) materials separate, (iii) coke formation that occurs at high temperatures and commences when the solvent power of the liquid phase is not sufficient to maintain the coke precursors in solution, (iv) sludge/sediment formation in the thermal products that occurs because of the interplay of several chemical and physical factors.

The formation of solid sediments or coke during thermal recovery and during thermal refinery processes can be (more than likely) a major limitation to the process. Furthermore, the presence of these different types of solids shows that solubility controls solids formation. The tendency toward solid formation changes in response to the relative amounts of the lower-boiling fractions (such as gases and naphtha), middle distillates, and residues and also in response to their changing chemical composition as the process proceeds.

The prime mover in the formation of incompatible products during the processing of feedstocks containing asphaltene constituents is the nature of the primary thermal decomposition products. There have been several attempts to focus attention on the asphaltene constituents during hydroprocessing studies.

The focus has been on the macromolecular changes that occur by investigation of the changes to the generic fractions, that is, the asphaltene constituents, the resin constituents, and the other fractions that make up such a feedstock.

This option suggests that the overall pathway by which hydrotreating and hydrocracking of heavy oils and residua occurs involves a stepwise mechanism:

Asphaltene constituents → polar aromatics (resin-type components)

Polar aromatics → aromatics

Aromatics → saturates

A direct step from either the asphaltene constituents or the resin-type components to the saturates is not considered a predominant pathway for hydroprocessing.

The means by which asphaltene constituents are desulfurized, as one step of a hydrocracking operation, are also suggested as part of this process. This concept can then be taken one step further to show the dealkylation of the aromatic systems as a definitive step in the hydrocracking process (Speight, 2014). It is also likely that molecular species within the asphaltene fraction that contain nitrogen and other heteroatoms and have lower volatility than their hydrocarbon analogs are the prime movers in the production of coke (Speight, 2014).

When catalytic processes are employed either during an *in situ* recovery process or in a refinery, complex molecules (such as those that may be found in the original asphaltene fraction or those formed during the process) are not sufficiently mobile (or are too strongly adsorbed by the catalyst) to be saturated by the hydrogenation components and, hence, continue to condense and eventually degrade to coke. These deposits deactivate the catalyst sites and eventually interfere with the hydroprocess.

A very convenient means of understanding the influence of feedstock on the hydrocracking process is through a study of the hydrogen content (H/C atomic ratio) and molecular weight (carbon number) of the feedstocks and products (Speight, 2014). Such data show the extent to which the carbon number must be reduced and/or the relative amount of hydrogen that must be added to generate the desired lower-molecular-weight, hydrogenated products. In addition, it is also possible to use data for hydrogen in residuum processing, where the relative amount of hydrogen consumed in the process can be shown to be dependent upon the sulfur content of the feedstock.

In summary, and as these concepts show, characterization data can be used as an integral part of refinery operations. There is the need to understand the nature of the asphaltene constituents and polar constituents of feedstocks in more detail in order to be able to predict product yield, product distribution, and product quality. It is this latter item that will allow some measure of predictability in terms of the compatibility/incompatibility of the product mix. The characterization techniques currently at hand (Speight, 2001, 2015a) are an aid to accomplishing this goal.

6. Blending

Blending (sometime simply referred to as *mixing*) (Hemrajani, 2004) is involved in every step of the oil industry from exploration to marketing products. While drilling oil and gas wells, drilling fluid is applied. The fluid consists of a mixture of clay and a stabilized water-in-oil emulsion. The emulsion is prepared batchwise by dispersing water in oil in agitated tanks. The main functions of drilling mud include providing hydrostatic pressure to prevent formation fluids from entering into the wellbore, keeping the drill bit cool and clean during drilling, carrying drill cuttings out of the well, and suspending the drill cuttings while drilling is paused, and the drilling assembly is brought in and out of the hole.

Meanwhile, blending is important for product sampling in the pipeline transport. When crude oil is sampled to determine its water content before its custody is transferred to refineries, the water has to be uniformly dispersed across the cross section of pipes. Thus, a blending system has to be installed upstream of the sampler. Adequate blending should create a good dispersion but still allow water to easily settle in downstream storage tanks. Optimum mixing can add a high value.

Viscous crude oil has become an increasingly important source of hydrocarbons around the world. Transportation of these viscous crudes from the source to the refinery is a challenge because existing pipelines were designed for less viscous crudes. One available technology is to blend crude oil into water with an emulsifier to form high oil content oil-in-water emulsions with a dramatic decrease in viscosity and pressure drop.

Blending is also used to control sludge accumulation in crude oil storage tanks. Crude oil usually carries a certain amount of bottom sludge and water, and since this sludge is heavier than crude oil, it settles in storage vessels at terminals and refineries. Excessive sludge accumulation can occur in tanks with poor mixers and at low ambient temperatures. Once the sludge is settled on the tank floor, it hardens and cannot be removed by normal pumping. The tank has to be taken off-line and cleaned, which is hazardous, expensive, and time-consuming and requires sludge disposal.

In the refinery, or for shipment to the refinery, the usual practice is to blend crude oils of similar characteristics, although fluctuations in the properties of the individual components of the blend crude oils may cause significant variations in the properties of the blend over a period. In fact, it is during the blending operation that instability or incompatibility of the components of the blend may be clearly manifested. Homogeneity of the blend and stability and consistency are key requirements for optimal blending operations.

Blending several crude oils prior to refining can eliminate the frequent need to change the processing conditions that may be required to process each of the crude oils individually. However, incompatibility of different crude oils, which can occur if, for example, a paraffinic crude oil is blended with a heavy asphaltic oil, can cause sediment formation in the unrefined feedstocks or in the products, thereby complicating the refinery process (Speight, 2014).

Fouling as it pertains to crude oil is deposit formation, encrustation, deposition, scaling, scale formation, slagging, and sludge formation that has an adverse effect on operations (Speight, 2015b). It is the accumulation of unwanted material (such as asphaltene-type deposits) within a processing unit or on the solid surfaces of the unit to the detriment of function. For example, when it does occur during refinery operations, the major effects include (i) loss of heat transfer as indicated by charge outlet temperature decrease and pressure drop increase, (ii) blocked process pipes, (iii) underdeposit corrosion and pollution, and (iv) localized hot spots in reactors, all of which culminate in production losses and increased maintenance costs.

Although crude oil fouling is a complex process involving several parallel mechanisms, it is assumed that the organic-derived fouling takes place through the destabilization of the asphaltene constituents of the oil, their deposition at the tube wall and subsequent degradation to carbonaceous fouling layer. The foulant typically—precipitated asphaltene constituents or in the case of a thermal process modified asphaltene-type material—can plug well and distillation equipment and, in many cases, decrease or stop oil production and can further increase the cost of oil recovery processes. Therefore, a detailed knowledge of factors that affect the composition and physicochemical structure of crude oils is necessary.

Asphaltene precipitation during oil production and processing is a serious problem. Asphaltene constituents are sensitive to shearing forces and electrostatic interactions, and crude distillation and vacuum units often are affected by asphaltene fouling. Crude preheat train and vacuum bottom heat exchangers can plug and as a result require chemical or mechanical cleaning; otherwise, throughput has to be reduced, leading to a loss of production. Thus, selection and compatibility considerations—through laboratory testing (Speight, 2015b)—are a critical part of a successful blending operation (Watkinson, 1992; Asomaning and Watkinson, 2000).

References

Andersen, S.I., Speight, J.G., 1999. Thermodynamic models for asphaltene solubility and precipitation. J. Pet. Sci. Eng. 22 (1), 53–66.

Asomaning, S., Watkinson, A.P., 2000. Petroleum stability and heteroatom species effects in fouling of heat exchangers by asphaltenes. Heat Transfer Eng. 21, 10–16.

ASTM, 2017. Annual Book of Standards. ASTM International, West Conshohocken, PA.

Batts, B.D., Fathoni, A.Z., 1991. A literature review on fuel stability studies with particular emphasis on diesel oil. Energy Fuel 5, 2–21.

Bearsley, S., Forbes, A., Haverkamp, R.G., 2004. Direct observation of the asphaltene structure in paving-grade bitumen using confocal laser-scanning microscopy. J. Microsc. 215 (2), 149–155.

Browarzik, D., Laux, H., Rahimian, I., 1999. Asphaltene flocculation in crude oil systems. Fluid Phase Equilib. 154, 285–300.

Fitzer, E., Mueller, K., Schaefer, W., 1971. The chemistry of the pyrolytic conversion of organic compounds to carbon. Chem. Phys. Carbon 7, 237–383.

Francisco, M.A., Speight, J.G., 1984. Asphaltene characterization by a non-spectroscopic method. Prepr. Div Fuel. Chem. Am. Chem. Soc. 29 (1), 36.

Gary, J.G., Handwerk, G.E., Kaiser, M.J., 2007. Petroleum Refining: Technology and Economics, fifth ed. CRC Press, Taylor & Francis Group, Boca Raton, FL.

Goetzinger, J.W., Thompson, C.J., Brinkman, D.W., 1983. A review of storage stability characteristics of hydrocarbon fuels. US Department of Energy. Report No. DOE/BETC/IC-83-3.

Hardy, D.R., Wechter, M.A., 1990. Insoluble sediment formation in middle-distillate diesel fuel: the role of soluble macromolecular oxidatively reactive species. Energy Fuel 4, 270–274.

Hazlett, R.N., Hall, J.M., 1985. Ebert, L.B. (Ed.), Chemistry of Engine Combustion Deposits. Plenum Press, New York.

Hemrajani, R.R., 2004. Fluid mixing technology in the petroleum industry. In: Paul, E.L., Atiemo-Obeng, V.A., Kresta, S.M. (Eds.), Handbook of Industrial Mixing: Science and Practice. John Wiley & Sons, Hoboken, NJ.

Hiley, R.W., Pedley, L.F., 1986. Formation of insolubles during storage of naval fuels. In: Proceedings of the 2nd International Conference on Long-Term Storage Stabilities of Liquid Fuels, San Antonio, TX, July 29-August 1, pp. 570–584.

Hsu, C.S., Robinson, P.R. (Eds.), 2017. Handbook of Petroleum Technology. Springer International Publishing AG, Cham.

Koots, J.A., Speight, J.G., 1975. The relation of petroleum resins to asphaltenes. Fuel 54, 179–184.

Laux, H., Rahimian, I., Butz, T., 1997. Thermodynamics and mechanism of stabilization and precipitation of petroleum colloids. Fuel Process. Technol. 53 (1), 69–79.

Magaril, R.Z., Akensova, E.I., 1967. Mechanism of coke formation during the cracking of petroleum tars. Izvestia Vyssh Ucheb. Zaved. Neft Gas 10 (11), 134–136.

Magaril, R.Z., Akensova, E.I., 1968. Study of the mechanism of coke formation in the cracking of petroleum resins. Int. Chem. Eng. 8 (4), 727–729.

Magaril, R.Z., Aksenova, E.I., 1970a. Mechanism of coke formation in the thermal decomposition of asphaltenes. Khim. Tekhnol. Topl. Masel. 15 (7), 22–24.

Magaril, R.Z., Aksenova, E.I., 1970b. Kinetics and mechanism of coking asphaltenes. Khim. Izvestia Vyssh. Ucheb. Zaved. Neft Gaz. 13 (5), 47–53.

Magaril, R.Z., Aksenova, E.I., 1972. Coking kinetics and mechanism of asphaltenes. Khim. Kim Tekhnol. Tr. Tyumen Ind. Inst. 169–172.

Magaril, R.Z., Ramazaeva, L.F., 1969. Study of carbon formation in the thermal decomposition of asphaltenes in solution. Izvestia Vyssh. Ucheb. Zaved. Neft Gaz. 12 (1), 61–64.

Magaril, R.Z., Ramazaeva, L.F., Aksenova, E.I., 1970. Kinetics of coke formation in the thermal processing of petroleum. Khim. Tekhnol. Topl. Masel. 15 (3), 15–16.

Magaril, R.Z., Ramazaeva, L.F., Aksenova, E.I., 1971. Kinetics of the formation of coke in the thermal processing of crude oil. Int. Chem. Eng. 11 (2), 250–251.

Mayo, F.R., Lan, B.Y., 1987. Gum and deposit formation from jet. Turbine and diesel fuels at 100°C. Ind. Eng. Chem. Prod. Res. 26, 215.

Mushrush, G.W., Speight, J.G., 1995. Petroleum Products: Instability and Incompatibility. Taylor & Francis Publishers, Washington, DC.

Mushrush, G.W., Beal, E.J., Hazlett, R.N., Hardy, D.R., 1990. Macromolecular oxidatively reactive species. Energy Fuel 5, 258.

Mushrush, G.W., Hazlett, R.N., Pellenbarg, R.E., Hardy, D.R., 1991. Energy Fuel 5, 258.

Parkash, S., 2003. Refining Processes Handbook. Gulf Professional Publishing, Elsevier, Amsterdam.

Pedley, J.F., Hiley, R.W., Hancock, R.A., 1986. Storage stability of petroleum-derived diesel fuel. 1. Analysis of sediment produced during the ambient storage of diesel fuel. Fuel 66, 1646–1651.

Pedley, J.F., Hiley, R.W., Hancock, R.A., 1987. Storage stability of petroleum-derived diesel fuel. 2. Fuel 66, 1646–1651.

Pedley, J.F., Hiley, R.W., Hancock, R.A., 1988. Storage stability of petroleum-derived diesel fuel. 3. Identification of compounds involved in sediment formation. Fuel 67, 1124–1130.

Pedley, J.F., Hiley, R.W., Hancock, R.A., 1989. Storage stability of petroleum-derived diesel fuel. 4. Synthesis of sediment precursor compounds and simulation of sediment formation using model systems. Fuel, 27–31.

Pfeiffer, J.P., Saal, R.N., 1940. Asphaltic bitumen as a colloid system. Phys. Chem. 44, 139–149.

Por, N., 1992. Stability Properties of Petroleum Products. Israel Institute of Petroleum and Energy, Tel Aviv.

Power, A.J., Mathys, G.I., 1992. Characterization of distillate fuel sediment molecules: functional group derivatization. Fuel 71, 903–908.

Schucker, R.C., Keweshan, C.F., 1980. Reactivity of cold lake asphaltenes. Prepr. Div. Fuel Chem. Am. Chem. Soc. 25, 155.

Speight, J.G., 1984. The chemical nature of petroleum asphaltenes. In: Tissot, B. (Ed.), Characterization of Heavy Crude Oils and Petroleum Residues. Editions Technip, Paris.

Speight, J.G., 1990. The chemistry of the thermal degradation of petroleum asphaltenes. Acta Pet. Sin. 6 (1), 29.

Speight, J.G., 1992. A chemical and physical explanation of incompatibility during refining operations. In: Proceedings of the 4th International Conference on the Stability and Handling of Liquid Fuels. vol. 1. US Department of Energy, Washington, DC, p. 169.

Speight, J.G., 1994. Chemical and physical studies of petroleum asphaltenes. In: Yen, T.F., Chilingarian, G.V. (Eds.), Asphaltenes and Asphalts. I. Developments in Petroleum Science. In: vol. 40. Elsevier, Amsterdam. (Chapter 2).

Speight, J.G., 1998. Thermal chemistry of petroleum constituents. In: Speight, J.G. (Ed.), Petroleum Chemistry and Refining. Taylor & Francis, Washington, DC. (Chapter 5).

Speight, J.G., 2001. Handbook of Petroleum Analysis. John Wiley & Sons, New York.

Speight, J.G., 2012. Crude Oil Assay Database. Knovel. Elsevier, New York. Online version available at. http://www.knovel.com/web/portal/browse/display?_EXT_KNOVEL_DISPLAY_bookid=5485&VerticalID=0. Accessed 4 June 2018.

Speight, J.G., 2013a. Heavy Oil Production Processes. Gulf Professional Publishing, Elsevier, Oxford.

Speight, J.G., 2013b. Oil Sand Production Processes. Gulf Professional Publishing, Elsevier, Oxford.

Speight, J.G., 2013c. Heavy and Extra Heavy Oil Upgrading Technologies. Gulf Professional Publishing, Elsevier, Oxford.

Speight, J.G., 2013d. The Chemistry and Technology of Coal, third ed. CRC Press, Taylor & Francis Group, Boca Raton, FL.

Speight, J.G., 2014. The Chemistry and Technology of Petroleum, fifth ed. CRC Press, Taylor & Francis Group, Boca Raton, FL.

Speight, J.G., 2015a. Handbook of Petroleum Product Analysis, second ed. John Wiley & Sons, Hoboken, NJ.

Speight, J.G., 2015b. Fouling in Refineries. Gulf Professional Publishing, Elsevier, Oxford.

Speight, J.G., 2017. Handbook of Petroleum Refining. CRC Press, Taylor & Francis Group, Boca Raton, FL.

Speight, J.G., Moschopedis, S.E., 1979. The production of low-sulfur liquids and coke from athabasca bitumen. Fuel Process. Technol. 2, 295.

Speight, J.G., Wernick, D.L., Gould, K.A., Overfield, R.E., Rao, B.M.L., Savage, D.W., 1985. Molecular weights and association of asphaltenes: a critical review. Rev. Inst. Fr. Pétrol. 40, 27.

Stavinoha, L.L., Henry, C.P. (Eds.), 1981. Distillate Fuel Stability and Cleanliness. American Society for Testing and Materials, West Conshohocken, PA. Special Technical Publication No. 751.

Stavinoha, L.L., Westbrook, S.R., 1980. Accelerated stability test techniques for diesel fuels. US Department of Energy Report No. DOE/BC/10043-12.

Watkinson, A.P., 1992. Chemical reaction fouling of organic fluid. Chem. Eng. Technol. 15, 82–90.

Waxman, M.H., Deeds, C.T., Closmann, P.J., 1980. Thermal alterations of asphaltenes in peace river tars. In: Paper No. SPE 9510, Proceedings of the 55th SPE Annual Fall Technical Conference, Dalla, TX, September 21–24. Society of Petroleum Engineers, Richardson, TX, pp. 728–747.

Wiehe, I.A., 1992. A solvent-resid phase diagram for tracking resid conversion. Ind. Eng. Chem. Res. 31, 530–536.

Wiehe, I.A., 1993. A phase-separation kinetic model for coke formation. Ind. Eng. Chem. Res. 32, 2447–2454.

Wiehe, I.A., 1994. The pendant-core building block model of petroleum residua. Energy Fuel 8, 536–544.

Wiehe, I.A., 1995. Polygon mapping with two-dimensional solubility parameters. Ind. Eng. Chem. Res. 34, 661–673.

Further Reading

Speight, J.G., 2000. The Desulfurization of Heavy Oils and Residua, second ed. Marcel Dekker, New York.

Part II

Upgrading

Chapter 6

Upgrading During Recovery

1. Introduction

Crude oil upgrading is of major economic importance to many countries in the world. Heavy crude oil, extra heavy crude oil, and tar sand bitumen exist in large quantities in the western hemisphere but are difficult to produce and transport because of their high viscosity. Some crude oils contain compounds such as sulfur and/or heavy metals causing additional refining problems and costs. In situ upgrading could be a very beneficial process for leaving the unwanted elements in the reservoir and increasing API gravity.

Heavy oil, extra heavy oil, and tar sand bitumen (the primary organic component of tar sand) are often defined (loosely and incorrectly) in terms of API gravity. A more appropriate definition of tar sand bitumen, which sets it aside from heavy oil and conventional crude oil, is based on the definition offered by the United States government as the *extremely viscous hydrocarbon that is not recoverable in its natural state by conventional oil well production methods including currently used enhanced recovery techniques* (Chapter 1). By inference, conventional crude oil and higher-viscosity oil (recoverable by *conventional oil well production methods including currently used enhanced recovery techniques*) are different to tar sand bitumen. Be that as it may, some stage of production, conventional crude oil (in the later stages of recovery), and high-viscosity oil (in the earlier stages of recovery) may require the application of enhanced oil recovery methods (Speight, 2009).

Whatever the definition, heavy oil, extra heavy oil, and tar sand are poor-quality refinery feedstocks. In reservoirs that contain high-viscosity oil, it is often desirable to initiate *enhanced oil recovery* (EOR) operations as early as possible. In tar sand deposits that contain bitumen, it is often desirable to initiate *enhanced oil recovery* (EOR) operations as early as possible. This may mean considerably abbreviating conventional secondary recovery operations or bypassing them altogether.

Thermal floods using steam and controlled in situ combustion methods are also used. Thermal methods of recovery reduce the viscosity of the crude oil by heat so that it flows more easily into the production well (Bennion et al., 1978; Graue, 2001).

Heavy Oil Recovery and Upgrading. https://doi.org/10.1016/B978-0-12-813025-4.00006-4

In fact, it has been reported (Henderson and Weber, 1965) that subjecting a high-viscosity feedstock to mild thermal cracking conditions effects a permanent and significant reduction in the viscosity and specific gravity. Although all feedstocks are susceptible to cracking, the temperatures and times required usually exceed those required to crack higher-boiling feedstocks such as those found in the Athabasca tar sands.

As in any field in which primary recovery operations are followed by secondary or enhanced recovery operations and there is a change in product quality, such is also the case for tar sand recovery operations. Thus, product oils recovered by thermal stimulation of tar sand deposits show some improvement in properties over those of the bitumen in place, especially when the recovery process involves in situ combustion (Rafenomanantsoa et al., 1998). Thus, advanced techniques are usually variations of secondary methods with a goal of improving the *sweeping* action of the invading fluid. Moreover, upgrading of extra heavy crude has also been proposed by using a hydrogen donor under steam-injection conditions (Ovalles et al., 2003, 2015).

The technologies applied to oil recovery involve different concepts, some of which can cause changes to the oil during production. Technologies such as alkaline flooding, microemulsion (micellar/emulsion) flooding, polymer-augmented waterflooding, and carbon dioxide miscible/immiscible flooding do not require or cause any change to the oil. The steaming technologies may cause some steam distillation that can augment the process when the steam-distilled material moves with the steam front and acts as a solvent for oil ahead of the steam front. Again, there is no change to the oil although there may be favorable compositional changes to the oil insofar as lower-boiling fractions (such as gases and naphtha) are recovered and heavier materials remain in the reservoir.

In some cases, hydrogen addition must be used during upgrading in order to stabilize the upgraded bitumen. This means that the cost of partial upgrading is not much reduced as compared with full upgrading. Therefore, the only choice currently is no upgrading or full upgrading. Other goals could be to achieve breakthroughs in upgrading technologies such as methods that would use far less energy or such as gasification at 800°C (14570°F) that is far less than current commercial temperatures. The technology where changes do occur involves combustion of the oil in situ. The concept behind any combustion technology requires that the oil be partially combusted as fuel and that thermal decomposition occurs to other parts of the oil. This is sufficient to cause irreversible chemical and physical changes to the oil to the extent that the product is markedly different to the oil in place. Recognition of this phenomenon is essential before combustion technologies are applied to oil recovery.

Furthermore, in any field where primary production is followed by a secondary or enhanced recovery method, there is the potential for noticeable differences in properties between the fluids produced. Significant differences may render the product outside of the range of acceptability for the usual refining options and force a higher demand for thermal process (i.e., coking) units. Thus, overloading thermal process units will increase as the proportion of the high-viscosity oil in the refinery feedstock increases.

Enhanced oil recovery (Chapters 2 and 3) is the incremental ultimate recovery of oil that can be economically recovered from a crude oil reservoir over oil that can be economically recovered by conventional primary and secondary methods. The intent of enhanced oil recovery is to increase the effectiveness of oil removal from pores of the rock (displacement efficiency) and to increase the volume of rock contacted by injected fluids (sweep efficiency). In addition, enhanced oil recovery processes use *thermal-*, *chemical-*, or *fluid-phase behavior* effects to reduce or eliminate the capillary forces that trap oil within pores, to thin the oil or otherwise improve its mobility, or to alter the mobility of the displacing fluids. In some cases, the effects of gravity forces, which ordinarily cause vertical segregation of fluids of different densities, can be minimized or even used to advantage. The various processes differ considerably in the complexity, the physical mechanisms responsible for oil recovery, and the amount of experience that has been derived from field application. The degree to which the enhanced oil recovery methods are applicable in the future will depend on the development of improved process technology. It will also depend on improved understanding of fluid chemistry, phase behavior, and physical properties and on the accuracy of geology and reservoir engineering in characterizing the physical nature of individual reservoirs.

Variations of the enhanced oil recovery theme include the use of steam and solvents as the means of reducing interfacial tension. The solvent approach has had some success when applied to bitumen recovery from mined tar sand, but when applied to nonmined material, phenomenal losses of solvent and bitumen are always a major obstacle. This approach should not be rejected out of hand since a novel concept may arise that guarantees minimal (acceptable) losses of bitumen and solvent. In fact, *miscible fluid displacement* (*miscible displacement*) is a process in which an alcohol, refined hydrocarbon, condensed crude oil gas, carbon dioxide, liquefied natural gas, or even exhaust gas is injected into an oil reservoir, at pressure levels such that the injected gas or alcohol and reservoir oil are miscible; the process may include the concurrent, alternating, or subsequent injection of water.

The procedures for miscible displacement are the same in each case and involve the injection of a slug of solvent that is miscible with the reservoir oil followed by the injection of either a liquid or a gas to sweep up any remaining solvent. As the miscible *slug* of solvent becomes enriched with oil as it passes through the reservoir, the composition changes, thereby reducing the effective scavenging action. However, changes in the composition of the fluid—which is a delicately balance system when unchanged—can also lead to wax deposition and deposition of asphaltene constituents (Speight, 1994).

Precipitation and deposition of high-boiling constituents—such as asphaltene constituents—inside reservoirs, processing, and transportation facilities is a major concern in the crude oil industry. The probability of asphaltene precipitation and deposition occurring during any enhanced oil recovery operation and the subsequent effects on reservoir performance should be anticipated at earlier stages of any development project, especially when changes to the asphaltene structure can be anticipated (Waxman et al., 1980). Once the asphaltene deposition occurs, it causes severe permeability and porosity reduction and wettability alteration, changing relative permeability in the reservoir and, in severe cases, plugging the wellbore and surface facilities. Therefore, caution is advised.

Microscopic observations of the leading edge of the miscible phase have shown that the displacement takes place at the boundary between the oil and the displacing phase. The small amount of oil that is bypassed is entrained and dissolved in the rest of the slug of miscible fluids; mixing and diffusion occur to permit complete recovery of the remaining oil. If a second miscible fluid is used to displace the first, another zone of displacement and mixing follows. The distance between the leading edge of the miscible slug and the bulk of pure solvent increases with the distance traveled, as mixing and reservoir heterogeneity cause the solvent to be dispersed.

Other parameters affecting the miscible displacement process are reservoir length, injection rate, porosity, and permeability of reservoir matrix; size and mobility ratio of miscible phases; gravitational effects; and chemical reactions.

Thermal recovery methods (Chapter 3) have found most use when heavy oil, extra heavy oil, or tar sand bitumen has an extremely high viscosity under reservoir conditions. For example, most heavy oils are highly viscous, with a viscosity ranging from a thousand centipoises to a million centipoises or more at the reservoir conditions. In addition, oil viscosity is also a function of temperature and API gravity (Speight, 2000, 2009).

Thermal enhanced oil recovery processes (i.e., cyclic steam injection, steam flooding, and in situ combustion) add heat to the reservoir to reduce oil viscosity and/or to vaporize the oil. In both instances, the oil is made more mobile so that it can be more effectively driven to producing wells. In addition to adding heat, these processes provide a driving force (pressure) to move oil to producing wells.

In situ combustion has long been used as an enhanced oil recovery method, and the potential for upgrading the oil during the process has long been recognized (Rafenomanantsoa et al., 1998; Castanier and Brigham, 2003). In situ combustion involves the injection of an oxidizing gas (air or oxygen-enriched air) to generate heat by burning a portion of the oil. Most of the oil is driven toward the producers by a combination of gas drive (from the combustion gases) and steam and water drive. This process is also called fireflooding to describe the movement of the burning front inside the deposit. Based on the respective directions of front propagation and airflow, the process can be forward, when the combustion front advances in the same direction as the airflow, or reverse, when the front moves against the airflow. For tar sand bitumen, numerous field observations have shown upgrading of 2°–6° API for tar sand bitumen undergoing combustion (Ramey et al., 1992; Lim et al., 2010). During in situ combustion of tar sand bitumen, temperatures of up to 700°C can be observed at the combustion front.

Forward combustion can be further characterized as *dry* when only air or enriched air is injected or *wet* when air and water are coinjected (Chapter 3). In the process, air is injected in the target formation for a short time, usually a few days to a few weeks, and the oil in the formation is ignited. Ignition can be induced using downhole gas burners, electric heaters, and/or injection of pyrophoric agents (not recommended) or steam. In some cases, autoignition occurs when the deposit temperature is fairly high and the oil reasonably reactive. This often happens to be the case for California heavy oil and extra heavy oil.

After ignition, the combustion front is propagated by a continuous flow of air. As the front progresses into the deposit, several zones can be found between the injector and the producer as a result of heat, mass transport, and chemical reactions occurring in the process. The burned zone is the volume already burned. This zone is filled with air and may contain small amounts of residual unburned organic solids. As it has been subjected to high temperatures, mineral alterations are possible. Because of the continuous airflow from the injector to the burned zone, temperature increases from injected air temperature at the injector to near-combustion-front temperature near the combustion front. There is no oil left in this zone.

The combustion front is the highest-temperature zone. It is very thin, often no more than several inches thick. It is in that region that oxygen combines with the fuel and high-temperature oxidation occurs. The products of the burning reactions are water and carbon oxides. The fuel is often misnamed coke. In fact, it is not pure carbon but a hydrocarbon with H/C atomic ratios ranging from about 1 to 2.0. This fuel is formed in the thermal cracking zone just ahead of the front and is the product of cracking and pyrolysis, which is deposited on the rock matrix. The amount of fuel burned is an important parameter because it determines how much air must be injected to burn a certain volume of deposit.

Chemical reactions are of two main categories: (i) oxidation, which occurs in the presence of oxygen, and (ii) pyrolysis, which is caused mainly by elevated temperatures. In general, at low temperature, oxygen combines with the oil to form oxidized hydrocarbon derivatives such as peroxides, alcohols, or ketones. This generally increases the oil viscosity but could increase oil reactivity at higher temperature. When oxygen contacts the oil at higher temperature, combustion occurs resulting in the production of water and carbon oxides. Of all the reactions that can occur during in situ combustion, only low-temperature oxidation can increase the viscosity of the oil. If the process is conducted judiciously, low-temperature oxidations are minimized because most of the oxygen injected is consumed at the burning front.

In situ combustion may make a comeback with a new concept—the toe-to-heel air injection process (THAI process)—which uses combinations of vertical injection wells and horizontal producer wells, arranged in a direct or staggered line drive (Greaves and Xia, 2004a,b; Xia and Greaves, 2006). Thus, the process is based on the geometry of horizontal wells that may solve the problems that have plagued conventional in situ combustion (Al-Marshed et al., 2015; Hart et al., 2017; Hart and Wood, 2018). The well geometry enforces a short flow path so that any instability issues associated with conventional combustion are reduced or even eliminated.

Unlike conventional refinery processing, downhole upgrading involves implementing upgrading processes (such as catalytic processes) within the oil-bearing geologic formations. By this means, impurities contained in heavy crude oil, extra heavy crude oil, and tar sand bitumen have the potential to remain in the formation or can be conveniently separated from the oil during the production process. Application of this type of concept would provide an improved feedstock for transportation and for refining. In order for the successful production of improved quality oil by means of a downhole upgrading project, several processing steps are necessary, such as (i) anticipated placement of one or more catalysts into an appropriate downhole location, (ii) mobilization of the oil over the catalyst bed(s), and (iii) creation of processing conditions that are necessary to achieve the desirable reasonable degree of upgrading. However, downhole processing differs from surface processing in that brine; high steam partial pressures and low hydrogen partial pressure need to be accommodated in the reaction zone (Weissman et al., 1996; Weissman, 1997).

When applied to Athabasca tar sand bitumen, downhole upgrading of the bitumen was significant, with the API gravity of the product oil having an increase in the API gravity on the order of 8° compared with the original bitumen (i.e., original bitumen, 8° API, and product oil, 16° API). The produced oil viscosity was also dramatically reduced, and compositional analysis (Superfund Amendments and Reauthorization Act (SARA) analysis) was used to assess the quality of the produced oil, showing that the saturate fraction of the bitumen was increased from ~16% to 72% w/w. Thus, the THAI process could well have a wider range of applications than SAGD, but in any case, a detailed knowledge of the reservoir or deposit is essential (Speight, 2013). SAGD generally works best in relatively thick (130 ft) homogeneous pay zones.

The CAPRI process involves the addition of gravel-packed catalyst, as used in a conventional refinery, between the tubing and the horizontal wellbore. Results have shown that the technique improves the quality of bitumen on the order of 6°–8° API points on top of the THAI in situ upgrades—that is, the product oil has an API gravity on the order of 22°–24°. Based on these data, in situ upgrading could produce a product that has an API gravity in excess of the 22° API requirement for produced fluids that can be transported by pipeline without diluent that also represents a major saving in surface upgrading and refining costs.

In situ conversion, or underground refining, is a promising new technology to tap the extensive reservoirs of heavy oil and deposits of extra heavy oil and tar sand bitumen. The new technology (Gregoli and Rimmer, 2000; Gregoli et al., 2000; Ovalles et al., 2001) features the injection of high-temperature, high-quality steam and hot hydrogen into a formation containing high-boiling hydrocarbon derivatives to initiate conversion of the high-boiling hydrocarbon derivatives into lower-boiling hydrocarbon derivatives. In effect, the higher-boiling hydrocarbon derivatives undergo partial underground refining that converts them into a synthetic crude oil (or *syncrude*). The heavier portion of the syncrude is treated to provide the fuel and hydrogen required by the process, and the lower-boiling portion is marketed as a conventional crude oil.

Thus, belowground, superheated steam and hot hydrogen are injected into a heavy oil or tar sand bitumen formation, which simultaneously produces the heavy oil or bitumen and converts it in situ (i.e., within the formation) into syncrude. Aboveground, the heavier fraction of the syncrude is separated and treated on-site to produce the fuel and hydrogen required by the process, while the lower-boiling fraction is sent to a conventional refinery to be made into crude oil products (Gregoli et al., 2000; Gregoli and Rimmer, 2000).

The potential advantages of an in situ process for heavy oil, extra heavy oil, and tar sand bitumen include (i) leaving the coke-forming precursors in the ground, (ii) leaving the heavy metals in the ground, (iii) reducing sand handling, and (iv) bringing a partially upgraded product to the surface. The extent of the upgrading can, hopefully, be adjusted by adjusting the exposure of the heavy oil, extra heavy oil, and tar sand bitumen to thermal effects in the subsurface.

In the *modified in situ extraction* processes, combinations of in situ and mining techniques are used to access the reservoir. A portion of the reservoir rock must be removed to enable the application of the in situ extraction technology. The most common method is to enter the reservoir through a large-diameter vertical shaft, excavate horizontal drifts from the bottom of the shaft, and drill injection and production wells horizontally from the drifts. Thermal extraction processes are then applied through the wells. When the horizontal wells are drilled at or near the base of the tar sand reservoir, the injected heat rises from the injection wells through the reservoir, and the drainage of produced fluids to the production wells is assisted by gravity.

Generally, as opposed to heavy oil recovery and (Venezuelan) extra heavy oil recovery, bitumen recovery from a tar sand formation requires a higher degree of thermal stimulation because the bitumen, in its immobile state, is extremely difficult to move to a production well. Extreme processes are required, usually in the form of a degree of thermal conversion that produces free-flowing product oil that will flow to the well and reduce the resistance of the bitumen to flow.

The concept of downhole catalytic upgrading of high-viscosity oil using in situ combustion is not new as evidenced by interest in the concept and work performed over the past four decades (e.g., Reichert et al., 1989;

Mamora et al., 1993; Weissman and Kessler, 1996; Weissman, 1997; Rafenomanantsoa et al., 1998; Moore et al., 1999; Fan and Liu, 2002). The downhole catalytic upgrading process requires suitable placement of a catalyst bed in the reservoir, the flow of oil through the catalytic bed at a specified temperature and pressure, and finally drainage of the upgraded oil into the production well.

Upgrading of any refinery feedstock is of major economic importance. Heavy crude oils exist in large quantities in the western hemisphere but are difficult to produce and transport because of their high viscosity. Some crude oils contain compounds such as sulfur and/or heavy metals causing additional refining problems and costs. In situ upgrading could be a very beneficial process for leaving the unwanted elements in the reservoir and increasing API gravity.

There are two ways that are currently practiced in bringing heavy crude oil, extra heavy crude oil, and tar sand bitumen to market. The first method is to upgrade the material in the oil field and leave much of the material behind as coke and then pipeline the upgraded material out as synthetic crude. In this method, the crude is fractionated, and the residue is coked. The products of the coking operation and in some cases some of the residue are hydrotreated. The hydrotreated materials are recombined with the fractionated lower-boiling materials to form synthetic crude that is then transported to market in a pipeline.

The second method is to effect partial upgrading in situ as part of the recovery process or, at least, at the surface before transportation (Colyar, 2009). Partial or field upgrading of heavy oil, extra heavy oil, and tar sand bitumen produces transportable synthetic crude oil and eliminates the need for diluents for transportation to refiners. However, the option to produce an acceptable pipeline material would be an ideal solution in terms of improving the API gravity of the oil; there may be limitations (Motaghi et al., 2010; Fellows et al., 2017). For example, the amount of viscous oil production could be limited by the recovery process, and the upgraded products must be compatible with the original or partially changed viscous oil. If the products and the original (partially changed) viscous oil have limited (or no) compatibility, this would limit the amount of dilution and could even adversely affect the effectiveness of the recovery process.

A final option that is already in common practice is to use traditional crude that is located in the general area to dilute the nontraditional crude to produce an acceptable pipeline material. This option is workable but does not represent any form of upgrading of the viscous oil—the operative word is dilution. This option also suffers from the limitation due to the potential for incompatibility and instability of the blend. If the viscous oil and the conventional crude oil have limited compatibility, this would limit the amount of dilution and, consequently, limit the amount of viscous oil recovery. Since this option does not involve upgrading, it will not be discussed further here.

Although this improvement in properties that are brought about by the recovery technology may not appear to be too drastic, nevertheless, the changes (usually measured by comparison of the before-and-after API gravity) may be sufficient to have major economic advantages for refinery operators. Any incremental increase in the units of hydrogen/carbon ratio can save amounts of costly hydrogen during upgrading. The same principles are also operative for reduction in the nitrogen, sulfur, and oxygen content. However, the limitations of processing tar sand bitumen depend to a large extent on the amount of nonvolatile higher-molecular-weight constituents (asphaltene constituents and resin constituents), which also contain the majority of the heteroatoms (i.e., nitrogen, oxygen, sulfur, and metals such as nickel and vanadium). These constituents are responsible for high yields of thermal and catalytic coke.

In catalytic processes, the asphaltene and resin constituents (as well as related constituents formed during the process) are not sufficiently mobile (i.e., they are strongly adsorbed by the catalyst) and fail to be converted to useful products either in a thermal process or in a hydroprocess. The chemistry of the thermal reactions of some of these constituents dictates that certain reactions, once initiated, cannot be reversed and proceed to completion (Speight, 2014, 2017). Coke is the eventual product, and deposits of such carbonaceous material and metal-bearing carbonaceous products deactivate the catalyst sites and eventually interfere with the catalytic process.

Thus, it is the purpose of this section to (i) present an outline of the options for surface upgrading facilities and (ii) compare the concepts for upgrading during an in situ recovery process. Nevertheless, some consideration of both options is warranted here with the possibility that one or the other (or both) must become a reality in the not too distant future.

2. In Situ Upgrading

Fluids produced from a well are seldom pure crude oil: in fact, a variety of materials may be produced by oil wells in addition to liquid and gaseous hydrocarbon derivatives (Speight, 2009). In addition, it has been recorded that there are noticeable differences in properties between the fluids produced (Shah et al., 2010). The differences in elemental composition may not reflect these differences to any great extent; indeed, the elemental composition of tar sand bitumen from different sources varies very little within narrow limits; more significant differences will be evident from an inspection of the physical properties. One issue that arises from the physical property data is that such oils may be outside the range of acceptability for refining techniques other than thermal options. In addition, overloading of thermal process units will increase as the proportion of the tar sand bitumen in the refinery feedstock increases. Obviously, there is a need for more and more refineries to accept larger proportions of oil sand bitumen and extra heavy oil as the refinery feedstock and have the capability to process such materials.

Technologies such as alkaline flooding, microemulsion (micellar/emulsion) flooding, polymer-augmented waterflooding, and carbon dioxide miscible/immiscible flooding do not require or cause any change to the any tar sand bitumen that is recovered by these methods. The technologies that use steam may cause some steam distillation that can augment the process when the steam-distilled material moves with the steam front and acts as a solvent for oil ahead of the steam front with a tendency to cause upgrading of the oil (Northrop and Venkatesan, 1993; Sharpe et al., 1995; Jaiswal and Mamora, 2007; Shah et al., 2010). This allows the transport and production of the lower-boiling fractions of the oil leaving behind the higher-boiling material—as in refinery distillation (Speight, 2014)—which often contain the majority of the undesirable compounds, which may contain sulfur or metals. Again, there is no chemical change to the bitumen although there may be favorable compositional changes to the bitumen insofar as lower-boiling constituents are recovered and higher-boiling constituents (usually equate to coke formation during refining) remain in the deposit.

In situ upgrading essentially involves using the reservoir or the production well as the reactor in which the properties of the oil are changed prior to the oil arriving at the surface facility. In this manner, as much as possible, the undesirable components of the oil (nitrogen, sulfur, metal constituents, and coke formers) would remain in the formation, and the recovered oil would have lesser amount of these components and be more preferential refinery feedstocks (Fig. 6.1).

In situ upgrading technologies such as catalytic thermal conversion in the presence of hydrogen, hydrogen-donor solvents, and the use of microbial communities generating methane (biodegradation of C_2-C_5 hydrocarbon derivatives) all have environmental advantages because of their ability to trap most of the pollutants generated from the viscous oil in the reservoir well. Application of these technologies will also reduce the need for surface upgrading, which results in the cleaner production of lower-viscosity oil, which is more easily transported without the use of diluents.

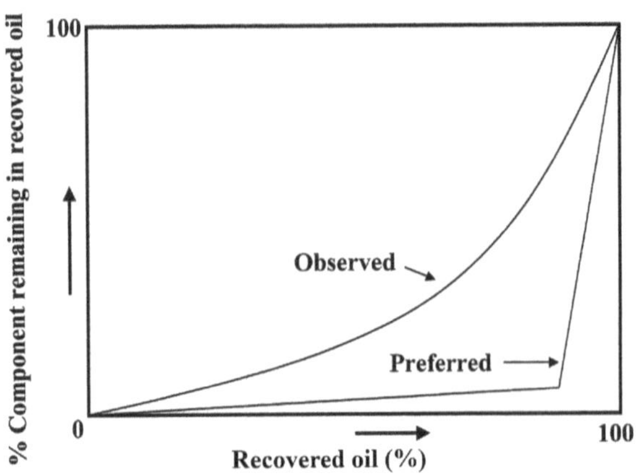

FIG. 6.1 Representation of the observed and preferred quality of the recovered oil.

Heavy oil, extra heavy oil, and bitumen are unconventional oil resources that are characterized by high viscosities (i.e., resistance to flow) and high densities compared with conventional oil. Most heavy oil, extra heavy oil, and bitumen deposits are very shallow and close to the surface. They originated as conventional oil that formed in deep formations but migrated to the surface region where they were degraded by bacteria and by weathering and where the lowest-boiling hydrocarbon derivatives escaped. Heavy oil, extra heavy oil, and bitumen are deficient in hydrogen and have high carbon, sulfur, and heavy metal content. Hence, they require additional processing (upgrading) to become a suitable feedstock for a normal refinery.

Recent developments in *upgrading* of viscous oil (Ancheyta and Speight, 2007; Speight, 2014, 2017) indicate that the near future could see a reduction of the differential cost of upgrading viscous oil as refineries evolve beyond this century (Speight, 2011). These processes are based on a better understanding of asphaltene solubility effects at high temperatures, the incorporation of a catalyst that is chemically precipitated internally during the upgrading, and the improvement of hydrogen addition or carbon rejection.

The potential advantages of an in situ process for bitumen upgrading include (i) leaving the carbon-forming precursors in the ground, (ii) leaving the heavy metals in the ground, (iii) bringing a partially upgraded product to the surface, and (iv) in the case of tar sand bitumen upgrading a reduction in the amounts of sand. The extent of the upgrading can, hopefully, be adjusted by adjusting the exposure of the heavy oil, extra heavy oil, and tar sand bitumen on the underground thermal effects.

In situ upgrading can reduce the viscosity of viscous oil by cracking long hydrocarbon chains and can improve oil quality by reducing or removing asphaltene constituents and resin constituents. Asphaltene constituents may contain iron, nickel, and vanadium, which are damaging to refineries. Excess carbon, in the form of coke, may be left in the reservoir.

The upgraded oil flows more readily into the wellbore (increasing recovery factor), is easier to lift to the surface, and may eliminate the need for a diluent for pipeline transportation. Furthermore, in situ upgrading might eliminate the need for surface upgrading facilities, thus reducing capital investments. In a conventional thermal process (e.g., SAGD), the viscous oil is heated in situ, but it may cool after being produced to the surface. It then has to be reheated for upgrading. In situ upgrading may be more energy-efficient as well. However, the increased mobilization of viscous oil in the reservoir by partial upgrading is not a new idea and still has many hurdles to overcome before it can be considered close to commercial. The product will be less viscous than the original viscous oil in place, but some property changes such as high olefin content from cracking are not necessarily positive.

In situ technologies offer the potential for upgrading during recovery. For example, it is to be expected that the product oil from an in situ combustion process (involving heavy oil, extra heavy oil, or tar sand bitumen) will show (i) a decrease in viscosity and (ii) a decrease in density—an increase in API gravity—and changes in other properties because of the thermal cracking reactions that will occur and because of the formation of alkane products and lower-molecular-weight aromatic products. This latter occurrence also improves catalyst life and activity and reduces the metal content.

Generally, one form of in situ upgrading involves the injection of a catalyst. As this process also involves contacting and mixing of the catalyst with the bitumen, it would be applied as an intermediate or supplemental process. An alternate to catalytic upgrading is in situ deasphalting process. Here, a solvent such as propane is injected to drop out some of the asphaltenes and lower the bitumen viscosity. This could be a stand-alone process or an additional benefit to a solvent-based recovery process such as VAPEX. In situ upgrading is also an additional benefit to the combustion recovery process. As the combustion front moves through the reservoir, any lower-boiling constituents of the bitumen are distilled off, thermal cracking takes place, and the coke product is consumed as fuel. The resulting produced oil has a much lower viscosity than the original bitumen. Another mechanism for in situ upgrading is bioconversion or bacterial upgrading as part of the microbial enhanced oil recovery (MEOR) process (Chapter 2) (Shennan and Levi, 1987; Bryant et al., 1989; Banat, 1995; Bryant and Lindsey, 1996; Monticello and Haney, 1996; Khire and Khan, 1994a,b; Omajali et al., 2017). The process depends on microbes that can convert the bitumen to lower-viscosity oil or methane. These microbes may be naturally occurring or injected into the reservoir. As additional nutrients are normally required for this process to take place at reasonable rates, these must be injected. As it is a relatively slow process, it would most likely be applied as a final "cleanup" process.

However, it is unlikely that any one in situ recovery process will be applicable to all reservoirs and no single recovery process will be able to access all the bitumen in a given reservoir. To achieve maximum recovery, it will be necessary to apply a combination of different processes, for example, a steam-based recovery process followed by in situ combustion, followed by in situ upgrading, followed by bioconversion of the residual hydrocarbon derivatives. This type of sequential recovery will require careful planning to ensure that the optimum sequence and timing is applied.

On the other hand, in situ recovery processes may have the added benefit of *leaving* some of the more obnoxious constituents (from the processing objective) in the ground. Processes that offer the potential for partial upgrading during recovery are varied but usually follow from a surface process. Not that this be construed as an easy task, there are many disadvantages that arise from attempting in situ upgrading. Nevertheless, there are three main approaches for heating the reservoir: (i) steam distillation; (ii) mild thermal cracking, such as the visbreaking process; and (iii) partial combustion.

2.1 Steam Distillation

Steam-injection pressures are limited during recovery of viscous oil because most viscous oil deposits are relatively shallow. The maximum steam temperature is limited by the ideal gas law. For example, at a 1000 m depth, the formation pressure is ~1450 psi, which permits a steam temperature of ~300°C (570°F), which is too low to provide significant upgrading on a short timescale. In situ combustion is capable of much higher temperatures (~700°C, 1290°F), which has the potential for measurable upgrading. Electric heating (resistance, induction, or RF) should also be able to achieve the high temperatures required for in situ upgrading (Mut, 2005). But steam distillation then becomes an option.

The principal mechanisms responsible for oil recovery during viscous oil recovery by steam-based processes are (i) thermal expansion of the oil, (ii) viscosity reduction, and (iii) steam distillation. Steam distillation, once an ignored phenomenon, deserves recognition as the main mechanism that reduces the residual oil saturation behind the hot-water front during steam flood particularly for low-viscosity crude oil and medium-viscosity crude oil. Furthermore, steam distillation can also play an important role during viscous oil recovery by steam flooding.

Vaporization of the lower-boiling hydrocarbon derivatives is induced by increasing the system temperature, and it is reflected by the increase in the system pressure. Steam will evolve from the aqueous phase and strip some hydrocarbon component from the oleic phase. Since the pressure of the system is increased, the apparent bubble point of the water is consequently increased, which causes the higher-molecular-weight, higher-boiling hydrocarbon derivatives to vaporize at the elevated saturation temperatures causing displacement of the oil by steam displacement and steam distillation (Volek and Pryor, 1972; Wu and Brown, 1975; Wu and Elder, 1983; Sarathi et al., 1988; Sharpe et al., 1995). In fact, several basic crude oil properties can be used to predict steam distillation yields reasonably well. A correlation using oil viscosity in centistokes at 100°F (38°C) can be used to predict the steam distillation yield as can the API gravity (Wu and Elder, 1983).

To reduce viscosity of heavy crude oil, solvents frequently are used for dilution—this is one of the most efficient methods of pipeline transportation of viscous oil. Solvents also are injected into the reservoir for well cleaning, stimulation, and fracturing and, less frequently, for miscible displacement. Thus, one benefit resulting from the distillation of viscous oil during a steam flood is that the lower-boiling constituents can act as a solvent to lower the viscosity of the original oil. In the steam distillation process, the vaporized distillate mixes with the original oil ahead of the steam condensation zone. Depending on the quantity of lower-boiling hydrocarbon available to mix with the oil, viscosity can be substantially decreased—provided the distillate has a sufficient amount of aromatic constituents or naphthenic constituents to maintain the asphaltene constituents in solution (Speight and Mitchell, 1973; Shu, 1984; Speight, 2014).

It has also been shown that, during steam-propane distillation of oil, the steam/propane vapor strips the more volatile components from the viscous oil and encourages a decrease of the boiling point of these components. The steam enriched with the propane and lower-boiling components flow through the steam zone to the condensation front where both steam and lower-boiling hydrocarbon derivatives condense. The condensed hydrocarbon derivatives are miscible with the oil, reducing overall viscosity of liquid hydrocarbon. The steam/propane distillation begins when the total vapor pressure (steam and propane) in the presence of two immiscible liquids (water and oil exerting its own vapor pressure at the temperature of the system) equals the total pressure on the system. It is also possible that the role of propane is to reduce the boiling point of the lower-molecular-weight components. As a result, viscous oil will begin distilling at temperatures much lower than the normal boiling points of the constituents, and as a consequence, the distillation yield increases.

As promising as this may seem, the steam distillation yields are mainly dependent on the oil composition and may not correlate with crude API gravity. Changes in steam saturation pressure and temperature have insignificant effect on the yields; however, superheated steam significantly increases the yields for some crude oils.

2.2 Mild Thermal Cracking

Viscous oil constituents can be cracked into lower-boiling hydrocarbon derivatives at high enough temperatures and pressures. As a result of the pyrolysis of viscous oil, carbon-carbon bonds in the hydrocarbon chain are broken by heat; essentially, the vibrational energy exceeds the chemical energy in the carbon-carbon bonds. Pyrolysis occurs in the absence of oxygen or a catalyst, but steam may be present. For example, steam cracking and thermal cracking are done in refineries at temperatures at or above 800°C (1470°F). Such high temperatures are difficult to achieve in the reservoir. Pyrolysis can still occur at lower temperatures but at much, much slower rates.

For example, viscous oil produced under primary and fireflood conditions showed a gradual increase in density, viscosity, and other properties over time. This indicates fractionation during flow through the formation; there was also a decrease in viscosity, density, and other properties as a result of the products of thermal cracking of the oil and the formation of smaller straight-chain alkanes and small aromatic molecules (Reichert et al., 1989).

It is also conceivable that under the appropriate conditions (as yet undefined), electric heating (resistance, induction, or radio frequency) should also be able to achieve the high temperatures required for in situ upgrading (Cunha, 2005).

2.3 Partial Combustion

The combustion front is the highest-temperature zone. It is very thin, often no more than several inches thick. It is in that region that oxygen combines with the fuel and oxidation occurs (Verkoczy and Freitag, 1997). The products of the burning reactions are water and carbon oxides. The fuel is often misnamed coke. In fact, it is not pure carbon but a hydrocarbon with H/C atomic ratios ranging from about 1 to 2.0. This fuel is formed in the thermal cracking zone just ahead of the front and is the product of cracking and pyrolysis, which is deposited on the rock matrix. The amount of fuel burned is an important parameter because it determines how much air must be injected to burn a certain volume of reservoir.

In situ combustion has long been used as an enhanced oil recovery method. For many viscous oils, numerous field observations have shown upgrading of 2°–6° API for heavy oils undergoing combustion (Ramey et al., 1992). During in situ combustion of viscous oils, temperatures of up to 700°C can be observed at the combustion front.

Forward in situ combustion by itself is already an effective in situ upgrading method with field improvements in gravity reported to be as much as 6° API (Ramey et al., 1992).

Forward in situ combustion by itself is already an effective in situ upgrading method with improvements in API gravity by as much as 6°. Another work (Greaves and Xia, 2004a,b; Greaves et al., 2005; Greaves and Xia, 2008) has followed up on this potential, but direct application of these laboratory results to the field is difficult; such technology deserves further research. The higher-boiling constituents often contain the majority of the undesirable compounds, which may contain sulfur or metals. In the following, we will discuss some hybrid methods that might be used to further upgrade the oil in situ.

Another possible in situ upgrading technique involves a combination of solvent injection and combustion. Cyclic oil recovery has numerous advantages both technically and economically. It can also be easily optimized in a given oil reservoir. Cyclic injection of solvents, either gas or liquid, is followed by in situ combustion of a small part of the reservoir not only to increase the temperature near the well but also to clean the wellbore region of all the residues left by the solvents. Alternate slugs of solvent and air would be injected, and production would occur after each solvent slug injection and after each combustion period. The process could be repeated until an economic limit is reached. One important fact to note is that both solvent injection and in situ combustion have been proved to be effective in a variety of reservoirs; however, the combination of the two methods has never been tried.

The most significant effect will be the precipitation and/or deposition of high-molecular-weight constituents such as asphaltene constituents or wax constituents. The produced oil is expected to be slightly upgraded by the solvent cycle. Unlike the classic well-to-well in situ combustion, we would only try to improve near-wellbore conditions by burning the solid residues left after the solvent cycle. The benefits of using combustion at this stage are expected to include (i) productivity improvement through the removal of the high-boiling constituents left from the solvent cycle, (ii) possible deactivation of the clays near the wellbore due to the high temperature of the combustion, and (iii) reduced viscosity of the oil due to temperature increase.

However, while in situ combustion is a relatively inexpensive process, it has major drawbacks. The high temperatures in the presence of oxygen that are encountered when the process is applied cause coke formation and the production of olefins and oxygenated compounds such as phenols and ketones, which in turn cause major problems when the produced liquids are processed in refinery units. Commonly, the processing of products from thermal cracking is restricted to delayed or fluid coking because the hydrocarbon is degraded to a degree that precludes processing by other methods.

One concept that relates to upgrading during in situ recovery utilizes a unique combination of operations to achieve hydrovisbreaking in formations in which viscous oil and commonly encountered levels of formation permeability combine to limit fluid mobility (Graue, 2001). The benefits of introduction of hydrogen during in situ retorting offer much promise. The possible application of such methods for selective separation of the metal constituents is an obvious benefit. For example, partial oxidation in the presence of steam may produce hydrogen for immediate pickup and result in integrated recovery and significant upgrading.

Downhole upgrading of Athabasca tar sand bitumen has been investigated in a series of experiments using THAI (toe-to-heel air injection) process, which uses combinations of vertical injection wells and horizontal producer wells, arranged in a direct or staggered line drive.

It is possible that the THAI process will be effective down to about 6–20 ft thick, as is common in many Saskatchewan viscous oil pools (Greaves and Xia, 2004a,b).

The THAI process potentially allows the inclusion of a catalytic upgrading stage since it provides favorable operating temperatures at the production well as the combustion zone is "anchored" to the horizontal well (Xia et al., 2002). The cracking reactions happening in the mobile oil zone create precursor conditions for the CAPRI process. The reactants are usually water (steam) and combustion gases, which pass through the mobile oil zone and in contact with the catalyst layer around the horizontal production well. The catalyst reacts with partially upgraded THAI oil and further upgrades it.

Thus, adding a catalyst (such as iron) to a thermal process may enhance in situ upgrading, even at the lower temperatures for steam injection (Jiang et al., 2005). Laboratory experiments combining in situ combustion with a catalyst in a horizontal producing well produced significantly upgraded oil. Thermal cracking occurred in the combustion zone, and additional upgrading was achieved by catalytic cracking in the production well (Xia et al., 2002). The downhole catalytic upgrading produced a low-boiling oil, characterized by a low viscosity, that was readily converted into naphtha and kerosene fractions, with a higher conversion on an fluid catalytic cracking (FCC) basis than that obtained with normal virgin bitumen vacuum gas oil (Greaves and Xia, 2004a,b; Greaves et al., 2005).

Thus, one form of in situ viscous oil upgrading involves the injection of a catalyst—hence, the CAPRI process involves the addition of gravel-packed catalyst, as used in a conventional refinery, between the tubing and the horizontal wellbore. Test results have shown the technique to add 6–8° API points in addition to the THAI in situ upgrading. Based on these data, the combination could deliver in situ upgrading above the 22° API requirement for produced fluids that can be transported by pipeline without diluent, which also represents a major saving in surface upgrading and refining costs.

2.4 Solvent Deasphalting

The application of lower-boiling hydrocarbon solvents to reduce or eliminate natural gas for steam generation has received significant recent interest. These lower-boiling hydrocarbon derivatives also have a natural tendency to cause asphaltene constituents to separate (Mitchell and Speight, 1973; Speight, 2009), thereby offering promise of some in situ upgrading.

VAPEX is the most advanced process in this area. Physical and chemical separations into fractions might lead to segregated and more targeted process steps, including more efficiently targeted hydrogen addition. There may be some overlap here with demetallization.

Thus, an extension of solvent recovery is the combined use of solvents and thermal stimulation to achieve some degree of in situ upgrading. A factor is a degree of upgrading that may occur in new recovery methods, but a potentially major factor is the likely conversion of bitumen-based residues in the future for energy, power, and hydrogen at production or upgrading stages and the possible application of mild in situ field upgrading to reduce dependence on diluent for transport to distant refineries. The potential move to *less severe* primary upgrading will place more emphasis on *conversion* at the secondary stage and heteroatom removal. The desire to reduce overall hydrogen consumption will place emphasis on lower-boiling by-product production and targeted hydrogen addition to the synthetic crude oil product.

During a solvent-based process for the recovery of viscous oil, such as vapor-assisted petroleum extraction (VAPEX), a condensable solvent is injected into a viscous oil reservoir. Solvent dissolution of or into viscous oil and possible asphaltene precipitation drastically reduce its viscosity so that the diluted viscous oil can flow toward a production well. In the past, several physical modeling studies have shown that the produced viscous oil has much less amount of high-boiling constituents than the original viscous oil. This phenomenon is often referred to as in situ upgrading.

Typically, after a solvent is made in contact with viscous oil at a relatively high pressure for a sufficiently long time, the solvent-viscous oil system at equilibrium state can be roughly divided into three different layers. The top layer is a solvent-enriched liquid phase, the middle layer comprises viscous oil with the dissolved solvent, and the bottom layer mainly consists of high-boiling constituents. The solvent-viscous oil mixtures in these three layers show rather different chemical and physical properties, such as solvent concentration, carbon number distribution, and viscosity. The top layer has the highest concentrations of solvent and lower-boiling components and the lowest viscosity of the oil even after its dissolved solvent is flashed off. The viscous oil in the middle layer has similar carbon number distribution to the original viscous oil. The bottom layer has the lowest solvent concentration and the highest concentration of high-boiling constituents.

The viscous oil in the bottom layer after its dissolved solvent is flashed off has much higher viscosity than the original viscous oil. These experimental results indicate that in a solvent-based viscous oil recovery process, the solvent-viscous oil mixture in the top and middle layers can be recovered because of its lower viscosity, whereas the viscous oil in the bottom layer may be left behind in the viscous oil reservoir because of its higher viscosity. In this way, the produced viscous oil is in situ upgraded during the solvent-based viscous oil recovery process.

However, there has been little work done to determine reservoir damage or deposit damage caused by asphaltene separation and, in addition, the site specificity of each reservoir, and/or each deposit needs to be considered before any specific plane is made for the development of a prospect.

An extension of solvent recovery is the combined use of solvents and thermal stimulation to achieve some degree of in situ upgrading. The potential move to *less severe* primary upgrading will place more emphasis on *conversion* at the secondary stage and heteroatom removal.

2.5 Microbial Enhanced Oil Recovery

Microbial enhanced oil recovery is different from the more conventional enhanced oil recovery methods such as carbon dioxide injection, steam injection, chemical surfactant, and polymer flooding, in that it involves the use of live microorganisms to use the oil as a feedstock, thereby changing the properties of the oil. The process takes place by different mechanisms, such as reduction of oil-water interfacial tension and alteration of wettability by surfactant production, selective plugging by microorganisms and their metabolites, oil viscosity reduction by gas production or degradation or biotransformation of long-chain saturated hydrocarbon derivatives, and production of acids that improves absolute permeability by dissolving minerals in the rock. The microbial metabolic products include biosurfactants, biopolymers, acids, solvents, gases, and enzymes. The bacteria used for microbial enhanced oil recovery are typically hydrocarbon-utilizing, nonpathogenic, and naturally occurring in many reservoirs.

3. Partial Upgrading at the Surface

The influx of highly viscous feedstocks into the refinery system can offset the shortages of conventional crude oil, but there is also a need for increased refining capacity and the need for lower-viscosity feedstocks. While new residue processing capacity needs to be added to existing refineries (Speight, 2011, 2014), there is the need for simple primary upgrading system to make the oil acceptable to a pipeline—many viscous oils are too viscous to transport by pipeline and fall outside of the specification required by pipeline owners. Partial upgrading reduces the viscosity of viscous feedstocks (such as heavy oil, extra heavy oil, and tar sand bitumen) so that the product oil can flow through pipelines more easily without requiring blending with a diluent.

In the past several decades, producers have evaluated field upgrading (moderate upgrading) processes. The interest was driven by concern for the future supply of diluent (typically gas field condensate) used to reduce bitumen viscosity for pipeline transport to distant refineries. The issue may resolve itself in time: it is now anticipated that more bitumen will be fully upgraded and bitumen might increasingly be transported with synthetic crude. Consequently, there are currently no projects that plan to employ field upgrading for transport purposes. The kinds of processes that have been proposed for field upgrading typically involve solvent-based deasphalting or mild, precoking thermal processes, such as visbreaking.

Although not in situ upgrading in the strictest sense, partial upgrading of viscous oil, extra heavy oil, and tar sand bitumen at the surface is the ultimate step in the recovery process; partial upgrading at the surface offers a means of preparing the oil for transport to the refinery. It can, therefore, be classed as a preupgrading step. The option of partial upgrading at the surface involves recovery of the viscous crude and sufficient upgrading to make the crude transportable by pipeline. As a result of such upgrading, the properties of the viscous oil would be modified to meet the specification required by the pipeline company. This could be either through dilution with a suitable solvent (presumably aromatic solvent that would prevent the asphaltene constituents from separating as a separate phase) or through partial thermal upgrading.

Traditional processing of viscous feedstock—such as by coking processes or hydrocracking processes—is very expensive processes and requires large scale to be viable. Thus, adaptation of such processes to the wellhead is not always viable. Thus, the manner in which refineries convert viscous oil into low-boiling high-value products has become a major focus of operations with new concepts evolving into new processes (Speight, 2011, 2014). Even though they may not be classed as conversion processes per se, pretreatment processes for removing asphaltene constituents and metals, sulfur, and nitrogen constituents are also important and can play an important role in selecting a suitable wellhead processing option.

Conceivably, viscous oil could be upgraded at the wellhead and sent through a pipeline to a refinery for further upgrading. However, this is not to be construed that viscous oil upgrading will *always* involve a coking step as the primary upgrading step. Other options, including some presented elsewhere (Speight, 2014, 2017), could well become predominant methods for upgrading in the future.

3.1 Thermal Cracking Processes

Thermal cracking processes offer attractive methods of feedstock conversion at low operating pressure without requiring expensive catalysts. Currently, the widest operated residuum conversion processes are visbreaking, delayed coking, and fluid coking that are still attractive processes for refineries from an economic point of view (Dickenson et al., 1997). The visbreaking process offers wide potential because of the mild nature of the process (Radovanović and Speight, 2011; Speight, 2012).

Visbreaking (*viscosity reduction* or *viscosity breaking*) is a mild form of thermal cracking insofar as the thermal reactions are not allowed to proceed to completion and are interrupted by quenching. Process conditions range from 455°C to 510°C (850–950°F) at a short residence time and from 50 to 300 psi at the heating coil outlet. It is the short residence time that brings to visbreaking the concept of being a mild thermal reaction in contrast to, for example, the delayed coking process where residence times are much longer and the thermal reactions are allowed to proceed to completion. Liquid-phase cracking takes place under these low-severity conditions to produce some naphtha and material in the kerosene and gas-oil boiling range. The gas oil may be used as additional feed for catalytic cracking units or as heating oil.

Visbreaking and variants and the recently demonstrated ORMAT process are examples of bulk thermal processes that convert residues without progressing all the way to solid coke. These processes have significant potential integrated with deasphalting to produce residues of varying yields on bitumen to meet future alternative energy and hydrogen production needs.

Hydrovisbreaking, a noncatalytic process, is conducted under similar conditions to visbreaking and involves treatment with hydrogen under mild conditions (RAROP, 1991, p. 57; Gregoli et al., 2000; Gregoli and Rimmer, 2000). The presence of hydrogen leads to more stable products (lower flocculation threshold) than can be obtained with straight visbreaking, which means that higher conversions can be achieved, producing a lower-viscosity product.

A recent variant of the visbreaking process is the *Aquaconversion process* (Marzin et al., 1998), which is a catalytic visbreaking process that operates in the presence of steam. The visbreaking technology is limited in conversion level because of the stability of the resulting product; because one process requirement is that the product has to be stable, standard visbreaking allows only an ~2°–6° API upgrading of the viscous oil and only a limited viscosity reduction, which does not ensure its transport without external diluent. The process pushes this maximum conversion level within the stability specification by adding a homogeneous catalyst in the presence of steam.

Other variants of visbreaking that may find use at the wellhead could the Tervahl T process and the Tervahl H process.

In the *Tervahl T process* (LePage et al., 1987; RAROP, 1991, p. 25), the feedstock is heated to the desired temperature using the coil heater, and heat is recovered in the stabilization section and held for a specified residence time in the soaking drum. The soaking drum effluent is quenched and sent to a conventional stabilizer or fractionator where the products are separated into the desired streams. The gas produced from the process is used for fuel.

In the *Tervahl H process*, the feedstock and hydrogen-rich stream are heated using heat recovery techniques and fired heater and held in the soak drum as in the Tervahl T process. The gas and oil from the soaking drum effluent are mixed with recycle hydrogen and separated in the hot separator where the gas is cooled, passed through a separator, and recycled to the heater and soaking drum effluent. The liquids from the hot and cold separator are sent to the stabilizer section where purge gas and synthetic crude are separated. The gas is used as fuel, and the synthetic crude can now be transported or stored.

Although not truly a thermal cracking process, upgrading by the use of cavitation energy is claimed to reduce the viscosity of bitumen by breaking up the aggregated molecules in bitumen (Chapter 7). The aim of the process is to change or modify the asphaltene microstructures in heavy oil, extra heavy oil, and tar sand bitumen, which comprise the highest-molecular-weight fraction and contribute to the high viscosities observed in these viscous feedstocks.

In the process (Yen, 1997; Mohapatra and Kirpalani, 2016), the viscous feedstock is subjected to a range of sonication frequencies (20kHz–1.1MHz). The sonication frequency of 574kHz with 50% power input resulted in low asphaltene content and lower viscosity suitable for improved transportability. Furthermore, the sonication treatment of bitumen under different conditions of frequencies and acoustic power decreased the H/C ratio. These results showed higher content of aromatic hydrogen and lower content of aliphatic hydrogen in bitumen treated under different conditions of sonication frequencies and intensity. The lowered metal content can be attributed to the reduced asphaltene formation as a result of sonication treatment of bitumen.

In terms of the mechanism of the reaction, it is believed that the collapse of bubbles during cavitation can give tiny hot spots with temperatures over 3000°C (5430°F) that initiate reaction (Thompson and Doraiswamy, 1999). However, at low temperature of the bulk liquid, the rate of conversion of hydrocarbon derivatives is insignificant, and addition or polymerization reactions are favored (Cataldo, 2000). Like ionizing radiation, significant conversion would require operation at temperatures comparable with conventional thermal processes. However, a combination of cavitation, heating to temperatures known to cause thermal reactions, and heating under pressure may result in partial upgrading, but the contribution of cavitation against the background effects of thermal reactions at elevated pressure is not clear.

3.2 Solvent Processes

The main solvent process for quick and convenient surface upgrading using a solvent is *dilution*. By this means, viscous oil can be diluted sufficiently so that the blend meets the specifications for shipping by pipeline or other means of transportation where specification must be met.

Dilution of viscous oil has been studied using various solvents, in terms of aromaticity, chemical nature, and viscosity (Speight, 2009, 2014; Motaghi et al., 2010). Dilution with low-viscosity hydrocarbon derivatives, such as low-viscosity crude oil or naphtha, has shown that the viscosity reduction efficiency is controlled by the sole viscosity of the diluent and not by its aromaticity. However, while blending diluents of different chemical nature and polarity can enhance viscosity reduction, there is always the risk (especially with paraffin-based solvents) that separation of asphaltene constituents will occur (Mitchel and Speight, 1973; Speight, 1979, 2014).

Solvent deasphalting processes allow the removal of sulfur and nitrogen compounds and metallic constituents by balancing yield with the desired feedstock properties (Ditman, 1973). In the process, the feedstock is mixed with dilution solvent from the solvent accumulator and then cooled to the desired temperature before entering the extraction tower. Because of its high viscosity, the charge oil neither can be cooled easily to the required temperature nor will it mix readily with solvent in the extraction tower. By adding a relatively small portion of solvent upstream of the charge cooler (insufficient to cause phase separation), the viscosity problem is avoided.

The *choice of solvent* is vital to the flexibility and performance of the unit. The solvent must be suitable, not only for the extraction of the desired oil fraction but also for the control of the yield and/or the quality of the deasphalted oil at temperatures that are within the operating limits (Speight, 2014).

4. Epilogue

In the not too distant past and even now, the mature and well-established processes such as visbreaking, delayed coking, fluid coking, flexicoking, propane deasphalting, and butane deasphalting were deemed adequate for upgrading viscous feedstocks. More options are now being sought in order to increase process efficiency in terms of the yields of the desired products.

There is (or will be) an obvious future need for partial upgrading during or immediately after recovery. On the other hand, hydrogen addition must be used during upgrading in order to stabilize the upgraded viscous oil—which could mean that the cost of partial upgrading is not much reduced as compared with full upgrading.

To achieve partial upgrading during recovery requires a further sequential operation before a transportable produced. A multistep system is required to achieve the necessary aims of viscous oil recovery with partial upgrading. What this might be is currently unknown, but there are possibilities.

The future development of the industry will also need to address the eventual, but not sudden, adoption of alternative energies such as for fuel cells and even hydrogen. Many of the technologies that will meet some of those long-term needs include hydrocracking, gasification, and syngas conversion. Thus, while the upgrading of heavy oil, extra heavy oil, and tar sand bitumen feedstocks has been seen as a major cost in competing with or replacing declining conventional crude oil feedstocks, the very need for upgrading may lead to a resource more versatile in addressing long-term, radical changes in the energy economy.

The use of surface field plants for partial upgrading of viscous feedstocks will address the following trends: (i) the need to take advantage of some relatively minor upgrading at the recovery stage; (ii) the need to take advantage of the necessity to move to alternative energy and hydrogen sources, particularly internally generated residues, a trend with very large impact on main upgrader plant process selection; (iii) the need to pay attention to any major environmental concerns in an integrated way; and (iv) the need to meet future crude quality trends in current planning.

In summary, the technologies applied to oil recovery involve different concepts, some of which can cause changes to the oil during production.

Finally, the potential for instability and incompatibility of the oil constituents during production must never be forgotten. Instability and incompatibility usually occur as a result of the separation or precipitation of asphaltene constituents. This may be caused by well stimulation, such as acidizing, which involves a drastic shift in local chemical equilibriums, pH, and liberation of carbon dioxide. Some of these changes may also increase the concentration of some ions, such as iron, which will promote the formation of asphaltenic sediment.

It is the purpose of this chapter (i) to present an outline of the options for surface upgrading facilities and (ii) to compare the concepts with in situ upgrading, in view of the origin and development of new process concepts—of which the VAPEX process (Chapter 2) and the THAI process (Chapter 3) are examples.

References

Al-Marshed, A., Hart, A., Leeke, G., Greaves, M., Wood, J., 2015. Optimization of heavy oil upgrading using dispersed nanoparticulate iron oxide as a catalyst. Energy Fuel 29, 6306–6316.

Ancheyta, J., Speight, J.G., 2007. Hydroprocessing of Heavy Oils and Residua. CRC-Taylor & Francis Group, Boca Raton, FL.

Banat, I.M., 1995. Biosurfactant production and possible uses in microbial enhanced oil recovery and oil pollution remediation. Bioresour. Technol. 51, 1–12.

Bennion, D.W., Vorndran, L., Donnelly, J.K., Moore, R.G., 1978. Fireflooding changes in athabasca bitumen and water properties. In: Strausz, O.P., Lown, E.M. (Eds.), Oil Sand and Oil Shale Chemistry. Verlag Chemie, New York, pp. 79–100.

Bryant, R.S., Lindsey, R.P., 1996. World-wide applications of microbial technology for improving oil recovery. In: Proceedings of the SPE Symposium on Improved Oil Recovery. Society of Petroleum Engineers, Richardson, TX, pp. 27–134.

Bryant, R.S., Donaldson, E.C., Yen, T.F., Chilingarian, G.V., 1989. Microbial enhanced oil recovery. In: Donaldson, E.C., Chilingarian, G.V., Yen, T.F. (Eds.), Enhanced Oil Recovery II: Processes and Operations. Elsevier, Amsterdam, pp. 423–450.

Castanier, L.M., Brigham, W.E., 2003. Upgrading of crude oil via in situ combustion. J. Pet. Sci. Eng. 39 (1–2), 125–136.

Cataldo, F., 2000. Ultrasound-induced cracking and pyrolysis of some aromatic and naphthenic hydrocarbons. Ultrason. Sonochem. 7 (1), 35–43.

Colyar, J., 2009. Has the time for partial upgrading of heavy oil and bitumen arrived? Pet. Technol. Q. Q4, 43–56.

Cunha, L.B., 2005. Recent In Situ Oil Recovery Techniques for Heavy and Extra Heavy Oil Reserves. SPE 94986.

Dickenson, R.L., Biasca, F.E., Schulman, B.L., Johnson, H.E., 1997. Hydrocarb. Process. 76 (2), 57.

Ditman, J.G., 1973. Deasphalt to get feed for lubes. Hydrocarb. Process. 52 (5), 110–113.

Fan, H., Liu, Y., 2002. Downhole catalysts upgrades heavy oil. Oil Gas J. 100 (11), 60–62.

Fellows, G.K., Mansell, R., Schlenker, R., Winter, J., 2017. Public-interest benefit evaluation of partial-upgrading technology. School of Publicly Policy Research Papers, Januaryvol. 10 (1). University of Calgary, Calgary.https://www.policyschool.ca/wp-content/uploads/2017/03/PIB-Evaluation-Fellows-Mansell-Schlenker-Winter.pdf. Accessed 29 March 2018.

Graue, D.J., 2001. Upgrading and Recovery of Heavy Crude Oils and Natural Bitumens by In Situ Hydrovisbreaking. United States Patent 6,328,104, December 11.

Greaves, M., Xia, T.X., 2004a. Downhole catalytic process for upgrading heavy oil: produced oil properties and composition. J. Can. Pet. Technol. 43 (9), 25–30.

Greaves, M., Xia, T.X., 2004b. Downhole upgrading of wolf lake oil using THAI/CAPRI processes—tracer tests. Preprint Paper Am. Chem. Soc. Div. Petrol. Chem. 49 (1), 69–72.

Greaves, M., Xia, T.X., 2008. Stability of THAI process—theoretical and experimental observations. J. Can. Pet. Technol. 47, 65.

Greaves, M., Xia, T.X., Ayasse, C., 2005. Underground Upgrading of Heavy Oil Using THAI—Toe-to-Heel Air Injection. SPE 97728.

Gregoli, A.A., Rimmer, D.P., 2000. Production of Synthetic Crude Oil From Heavy Hydrocarbons Recovered by In Situ Hydrovisbreaking. United States Patent 6,016,868, January 25.

Gregoli, A.A., Rimmer, D.P., Graue, D.J., 2000. Upgrading and Recovery of Heavy Crude Oils and Natural Bitumen by In Situ Hydrovisbreaking. United States Patent 6,016,867, January 25.

Hart, A., Wood, J., 2018. In situ catalytic upgrading of heavy crude with capri: influence of hydrogen on catalyst pore plugging and deactivation due to coke. Energies 11, 636–653.

Hart, A., Wood, J., Greaves, M., 2017. in situ catalytic upgrading of heavy oil using a pelletized Ni-Mo/Al_2O_3 catalyst in the THAI process. J. Pet. Sci. Eng. 156, 958–965.

Henderson, J.H., Weber, L., 1965. Physical upgrading of heavy crude oils by the application of heat. J. Can. Pet. Technol. (October-November), 206–212.

Jaiswal, N.J., Mamora, D.D., 2007. Distillation effects in heavy-oil recovery under steam injection with hydrocarbon additives. In: Paper No. 110712. Proceedings of the SPE Annual Technical Conference and Exhibition, Anaheim, CA, November 11-14.

Jiang, S., Liu, X., Liu, Y., Zhong, L., 2005. In situ upgrading heavy oil by aquathermolytic treatment under steam injection conditions. SPE.

Khire, J.M., Khan, M.I., 1994a. Microbially enhanced oil recovery (MEOR). Part 1. Importance and mechanisms of microbial enhanced oil recovery. Enzyme Microb. Technol. 16, 170–172.

Khire, J.M., Khan, M.I., 1994b. Microbially enhanced oil recovery (MEOR). Part 2. Microbes and the subsurface environment for microbial enhanced oil recovery. Enzyme Microb. Technol. 16, 258–259.

LePage, J.F., Morel, F., Trassard, A.M., Bousquet, J., 1987. Preprints Div. Fuel Chem. 32, 470.

Lim, G., Ivory, J., Coates, R., 2010. System and Method for the Recovery of Hydrocarbons by In-Situ Combustion. United States Patent 7,740.062, June 22.

Mamora, D.D., Ramey Jr., H.J., Brigham, W.E., Castanier, L.M., 1993. Kinetics of in situ combustion. Report No. DOE.BC/14600-51, July, US Department of Energy, Washington, DC.

Marzin, R., Pereira, P., Zacarias, L., Rivas, L., McGrath, M., Thompson, G.J., 1998. Resid conversion through the aquaconversion technology—an economical and environmental solution. SPE Paper No. 1998.086, Society of Petroleum Engineers, Richardson, TX.

Mitchell, D.L., Speight, J.G., 1973. The solubility of asphaltenes in hydrocarbon solvents. Fuel 52, 149.

Mohapatra, D.P., Kirpalani, D.M., 2016. Bitumen heavy oil upgrading by cavitation processing: effect on asphaltene separation, rheology, and metal content. Appl. Petrochem. Res. 6, 107–115.

Monticello, D.J., Haney III, W.M., 1996. Biocatalytic Process for Reduction of Petroleum Viscosity. United States Patent 5,529,930, June 25.

Moore, R.G., Laureshen, C.J., Mehta, S.A., Ursenbach, M.G., Belgrave, J.D.M., Weissman, J.G., Kessler, R.V., 1999. A Downhole Catalytic Upgrading Process for Heavy Oil Using In Situ Combustion. J. Can. Pet. Technol. 38 (13), 96.

Motaghi, M., Saxena, P., Ravi, R., 2010. Partial upgrading of heavy oil reserves. Pet. Technol. Q. Q4, 55–64.

Mut, S., 2005. Advances in an in situ upgrading process for unconventional oils. In: AAPG International Conference and Exhibition, Paris, France, September 11–14.

Northrop, P.S., Venkatesan, V.N., 1993. Analytical steam distillation model for thermal enhanced oil recovery processes. Ind. Eng. Chem. Res. 32, 2039–2046.

Omajali, J.B., Hart, A., Walker, M., Wood, J., Macaskie, L.E., 2017. In-situ catalytic upgrading of heavy oil using dispersed bionanoparticles supported on gram-positive and gram-negative bacteria. Appl. Catal. B Environ. 203 (2017), 807–819.

Ovalles, C., Vallejos, C., Vasquez, T., Martinis, J., Perez-Perez, A., Cotte, E., Castellanos, L., Rodriguez, H., 2001. Extra-heavy crude oil downhole upgrading process using hydrogen donors under steam injection conditions. Paper No. 69692, In: Proceedings of the SPE International Thermal Operations and Heavy Oil Symposium, Porlamar, Margarita Island, Venezuela, March 12-14.

Ovalles, C., Rengel-Unda, P., Bruzual, J., Salazar, A., 2003. Upgrading of extra-heavy crude using hydrogen donor under steam injection conditions. Characterization by pyrolysis GC-MS of the asphaltenes and effects of a radical initiator. Preprints. Am. Chem. Soc. Div. Fuel Chem. 48 (1), 59–60.

Ovalles, C., Rivero, V., Salazar, A., 2015. Downhole upgrading of Orinoco basin extra-heavy crude oil using hydrogen donors under steam injection conditions. Effect of the presence of iron nanocatalysts. Catalysts 5, 286–297.

Radovanović, L., and Speight, J.G. 2011. Visbreaking: a technology of the future. Proceedings of the First International Conference—Process Technology and Environmental Protection (PTEP 2011), December 7. University of Novi Sad, Technical Faculty Mihajlo Pupin, Zrenjanin. Page 335-338.

Rafenomanantsoa, A., Nicole, D., Rubini, P., Lauer, J.C., 1998. NMR and FIMS structural analysis of the oil obtained from the pyrolysis of bemolanga tar-sand bitumen (Madagascar) according to a post combustion process. Fuel 72 (1/2), 33–41.

Ramey Jr., H.J., Stamp, V.V., Pebdani, F.N., 1992. Case history of South Belridge, California, in-situ combustion oil recovery. In: Paper No. SPE 24200, Proceedings of the 9th SPE/DOE EOR Symposium, Tulsa, Oklahoma, April 21-24.

RAROP, 1991. Noguchi (Chairman), T. (Ed.), RAROP Heavy Oil Processing Handbook. Research Association for Residual Oil Processing. Ministry of Trade and International Industry (MITI), Tokyo.

Reichert, C., Fuhr, B., Sawatzky, H., Lafleur, R., Verkoczy, B., Soveran, D., Jha, K., 1989. Pet. Sci. Technol. 7, 851–878.

Sarathi, P.S., Roark, D., Strycker, A.R., 1988. Light oil steamflooding: a laboratory study. In: Paper No. SPE 17447, Proceedings of the 58th Annual California Regional Meeting of the SPE, Bakersfield, CA, March 23-25, pp. 431–442.

Shah, A., Fishwick, R., Wood, J., Leeke, G., Rigby, S., Greaves, M., 2010. A review of novel techniques for heavy oil and bitumen extraction and upgrading. Energy Environ. Sci. 3, 700–714.

Sharpe, H.N., Richardson, W.C., Lolley, C.S., 1995. Representation of steam distillation and in situ upgrading processes in heavy oil simulation. In: SPE 30301, Proceedings of the SPE International Heavy Oil Symposium, Calgary, Alberta, Canada, June 19-21. Society of Petroleum Engineers, Richardson, TX.

Shennan, J.L., Levi, J.D., 1987. In situ microbial enhanced oil recovery. In: Kosaric, N., Cairns, W.L., Gray, N.C.C. (Eds.), Biosurfactants and Biotechnology. Marcel Dekker, New York, pp. 163–180.

Shu, W.R., 1984. A viscosity correlation for mixtures of heavy oil, bitumen, and petroleum fractions. Soc. Pet. Eng. J. 24 (3), 277–292.

Speight, J.G., 1979. Studies on Bitumen Fractionation: (a) Fractionation by a Cryoscopic Method (b) Effect of Solvent Type on Asphaltene Solubility. Information Series No. 84, Alberta Research Council, Edmonton.

Speight, J.G., 1994. Chemical and physical studies of petroleum asphaltenes. In: Yen, T.F., Chilingarian, G.V. (Eds.), Asphaltenes and Asphalts. I. Developments in Petroleum Science. In: vol. 40. Elsevier, Amsterdam (Chapter 2).

Speight, J.G., 2000. The Desulfurization of Oil Sand Bitumen and Residua, second ed. Marcel Dekker, New York.

Speight, J.G., 2009. Enhanced Recovery Methods for Heavy Oil and Tar Sands. Gulf Publishing Company, Houston, TX.

Speight, J.G., 2011. The Refinery of the Future. Oxford, Gulf Professional Publishing.

Speight, J.G., 2012. Visbreaking: a technology of the past and the future. Scientia Iranica, 2012C 19 (3), 569–573.

Speight, J.G., 2013. Oil Sand Production Processes. Gulf Professional Publishing, Elsevier, Oxford.

Speight, J.G., 2014. The Chemistry and Technology of Petroleum, fifth ed. CRC Press, Taylor and Francis Group, Boca Raton, FL.

Speight, J.G., 2017. Handbook of Petroleum Refining. CRC Press, Taylor and Francis Group, Boca Raton, FL.

Thompson, L.H., Doraiswamy, L.K., 1999. Sonochemistry: science and Engineering. Ind. Eng. Chem. Res. 38 (4), 1215–1249.

Verkoczy, B., Freitag, N.P., 1997. Oxidation of heavy oils and their SARA fractions—its role in modeling in-situ combustion. In: Proceedings of the 7th Petroleum Conference, South Saskatchewan Section Petroleum Society of the CIM, Regina, Saskatchewan, October 12-22.

Volek, C.W., Pryor, J.A., 1972. Steam distillation drive—Brea field, California. J. Pet. Technol. 24 (8), 899–906.

Waxman, M.H., Deeds, C.T., Closmann, P.J., 1980. Thermal alterations of asphaltenes in peace river tars. In: Paper No. SPE 9510, Proceedings of the 55th SPE Annual Fall Technical Conference, Dallas, TX, September 21-214. Society of Petroleum Engineers, Richardson, TX, pp. 728–747.

Weissman, J.G., 1997. Review of processes for downhole catalytic upgrading of heavy crude oil. Fuel Process. Technol. 50 (2–3), 199–213.

Weissman, J.G., Kessler, R.V., 1996. Downhole heavy crude oil hydroprocessing. Appl. Catal. A Gen. 140, 1–16.

Weissman, J.G., Kessler, R.V., Sawicki, R.A., Belgrave, J.D.M., Laureshen, C.J., Mehta, S.A., Moore, R.G., Ursenbach, M.G., 1996. Downhole catalytic upgrading of heavy crude oil. Energy Fuel 10, 883–889.

Wu, C.H., Brown, A., 1975. A laboratory study on steam distillation in porous media. In: Paper No. 5569, Proceedings of the 1975 SPE Annual Meeting, Dallas, TX, September 28-October 1.

Wu, C.H., Elder, R.B., 1983. Correlation of crude oil steam distillation yields with basic crude oil properties. Soc. Pet. Eng. J. (December), 937–945.

Xia, T.X., Greaves, M., 2006. In situ upgrading of athabasca tar sand bitumen using THAI. Chem. Eng. Res. Des. 84, 856–864.

Xia, T.X., Greaves, M., Werfilli, W.S., Rathbone, R.R., 2002. Downhole Conversion of Lloydminster Heavy Oil Using THAI-CAPRI Processes. SPE 78998.

Yen, T.F., 1997. Upgrading Through Cavitation and Surfactant. In: Proceedings of the 15th World Petroleum Congress, Forum 17. John Wiley & Sons, Hoboken, NJ.

Further Reading

Khan, M.R., Patmore, D.J., 1998. Speight, J.G. (Ed.), Petroleum Chemistry and Refining. Taylor & Francis, Washington, DC (Chapter 6).

Chapter 7

Introduction to Upgrading Heavy Feedstocks

1. Introduction

In a very general sense, crude oil refining can be traced back over 5000 years to the times when asphalt materials and oils were isolated from areas where natural seepage occurred (Abraham, 1945; Forbes, 1958; Hoiberg, 1960). Any treatment of the asphalt (such as hardening in the air prior to use) or of the oil (such as allowing for more volatile components to escape prior to use in lamps) may be considered to be refining under the general definition of refining. However, crude oil refining as it is now practiced is a very recent science, and many innovations evolved during the 20th century.

Briefly, crude oil refining is the separation of crude oil into fractions and the subsequent treating of these fractions to yield marketable products (Parkash, 2003; Gary et al., 2007; Speight, 2011, 2014a, 2015a,b, 2017; Hsu and Robinson, 2017). In fact, a refinery is essentially a group of manufacturing plants that vary in number with the variety of products produced (Fig. 7.1). However, in addition to the simplified schematic of a refinery, the refinery (for the present purposes) can actually be considered as two refineries—(i) a section for low-viscosity feedstocks and (ii) a section for high-viscosity feedstocks (Fig. 7.2).

In this way, the processes can be selected and products manufactured to give a balanced operation in which the refinery feedstock oil is converted into a variety of products in amounts that are in accord with the demand for each. For example, the manufacture of products from the lower-boiling portion of crude oil automatically produces a certain amount of higher-boiling components using distillation and various thermal processes (Tables 7.1–7.4). If the latter cannot be sold as, say, heavy fuel oil, these products will accumulate until refinery storage facilities are full. To prevent the occurrence of such a situation, the refinery must be flexible and be able to change operations as needed. This usually means more processes are required for refining the heavier feedstocks: (i) thermal processes to change an excess of heavy fuel oil into more gasoline

Heavy Oil Recovery and Upgrading. https://doi.org/10.1016/B978-0-12-813025-4.00007-6

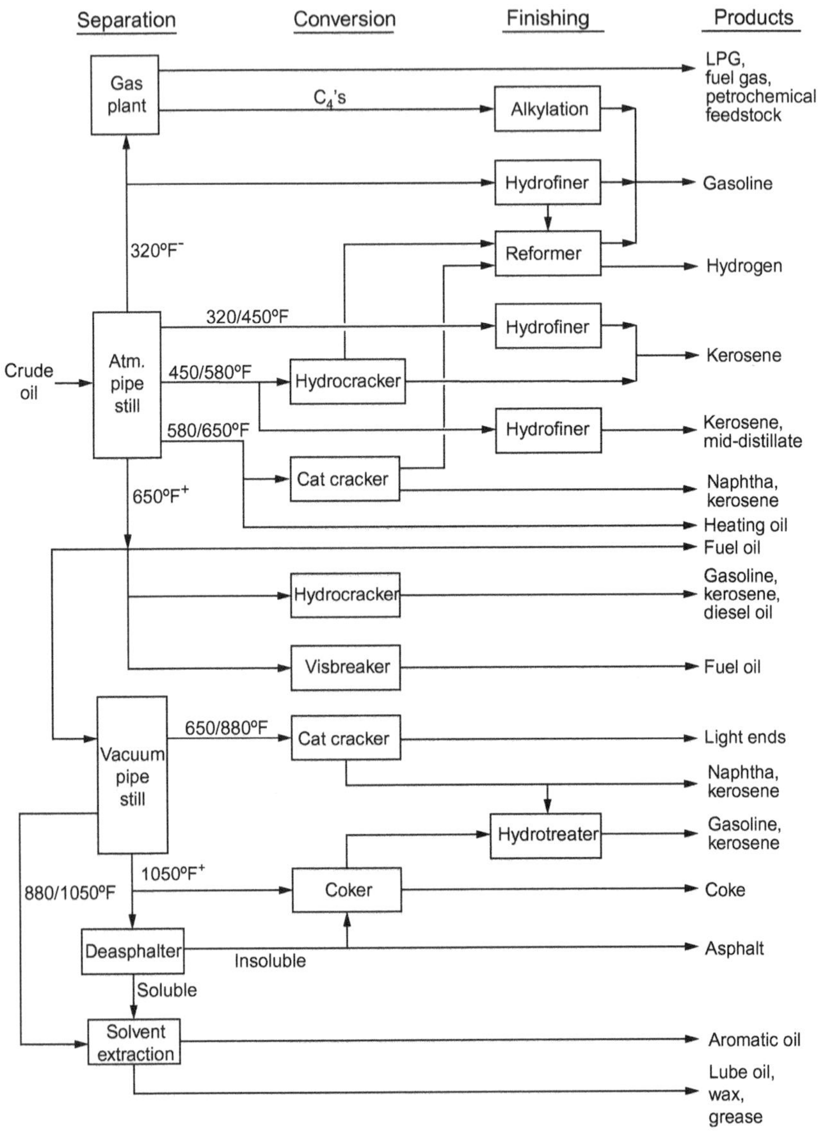

FIG. 7.1 Overview of a refinery showing placement of the various units.

with coke as the residual product or (ii) a vacuum distillation process to separate the viscous feedstock into lubricating oil stocks and asphalt.

The refining industry has been the subject of the four major forces that affect most industries and have hastened the development of new refining processes: (i) the demand for products such as gasoline, diesel, fuel oil, and jet fuel; (ii) feedstock supply, specifically the changing quality of refinery feedstock

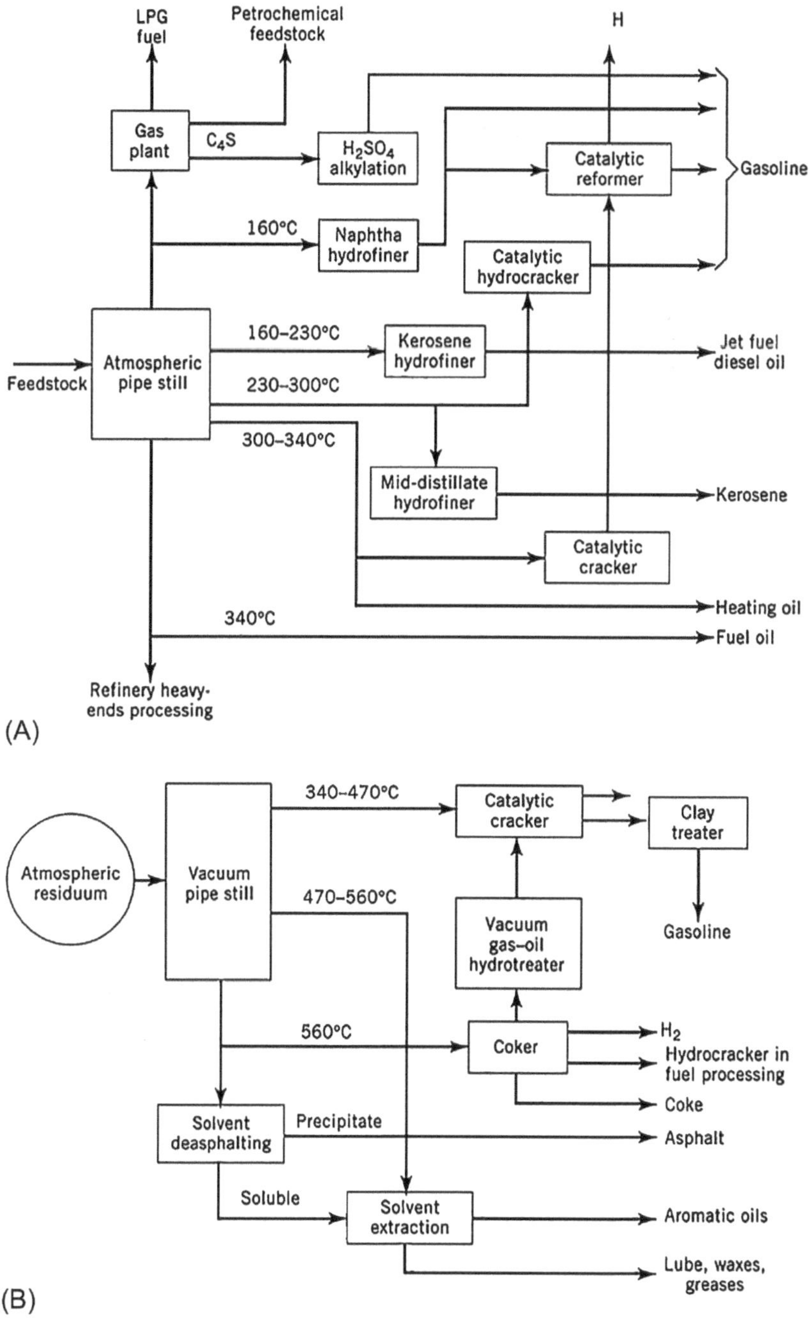

FIG. 7.2 General refinery operations: (A) light petroleum refining section and (B) heavy feedstock refining section.

TABLE 7.1 General: Boiling Fractions of Crude Oil

Boiling Range[a]		
Fraction	°C	°F
Light naphtha	−1–150	30–300
Gasoline	−1–180	30–355
Heavy naphtha	150–205	300–400
Kerosene	205–260	400–500
Light gas oil	260–315	400–600
Heavy gas oil	315–425	600–800
Lubricating oil	> 400	> 750
Vacuum gas oil	425–600	800–1100
Residuum	> 510	> 950

[a]*For convenience, boiling ranges are converted to the nearest 5°.*

TABLE 7.2 Comparison of Visbreaking With Delayed Coking and Fluid Coking

Visbreaking
Purpose: to reduce viscosity of fuel oil to acceptable levels
Conversion is not a prime purpose
Mild (470–495°C, 880–920°F) heating at pressures of 50–200 psi
Reactions quenched before going to completion
Low conversion (10%) to products boiling less than 220°C (430°F)
Heated coil or drum (soaker)
Delayed coking
Purpose: to produce maximum yields of distillate products
Moderate (480–515°C, 900–960°F) heating at pressures of 90 psi
Reactions allowed to proceed to completion
Complete conversion of the feedstock
Soak drums (845–900°F) used in pairs (one onstream and one offstream being decoked)
Coked until drum solid
Coke removed hydraulically from offstream drum
Coke yield: 20%–40% by weight (dependent upon feedstock)
Yield of distillate boiling below 220°C (430°F): ca. 30% (but feedstock-dependent)

TABLE 7.2 Comparison of Visbreaking With Delayed Coking and Fluid Coking—cont'd
Fluid coking
Purpose: to produce maximum yields of distillate products
Severe (480–565°C, 900–1050°F) heating at pressures of 10 psi
Reactions allowed to proceed to completion
Complete conversion of the feedstock
Oil contacts refractory coke
Bed fluidized with steam, heat dissipated throughout the fluid bed
Higher yields of light ends (<C_5) than delayed coking
Less coke make than delayed coking (for one particular feedstock)

TABLE 7.3 Summary of Catalytic Cracking Processes
Conditions
Solid acidic catalyst (silica-alumina, zeolite, etc.)
Temperature: 480–540°C (900–1000°F) (solid/vapor contact)
Pressure: 10–20 psi
Provisions needed for continuous catalyst replacement with heavier feedstocks (residua)
Catalyst may be regenerated or replaced
Feedstocks
Gas oils and residua
Residua pretreated to remove salts (metals)
Residua pretreated to remove high molecular weight (asphaltic constituents)
Products
Lower molecular weight than feedstock
Some gases (feedstock- and process-parameter-dependent)
Isoparaffins in product
Coke deposited on catalyst
Variations
Fixed bed
Moving bed
Fluidized bed

TABLE 7.4 Summary of Hydrocracking Process Operations

Conditions
Solid acid catalyst (silica-alumina with rare earth metals and various other options)
Temperature: 260–450°C (500–845°F) (solid/liquid contact)
Pressure: 1000–6000 psi hydrogen
Frequent catalyst renewal for heavier feedstocks
Gas oil: catalyst life up to 3 years
Heavy oil/tar sand bitumen: catalyst life less than 1 year
Feedstocks
Refractory (aromatic) streams
Coker oils
Cycle oils
Gas oils
Residua (as a full hydrocracking or hydrotreating option)
In some cases, asphaltic constituents (S, N, and metals) removed by deasphalting
Products
Lower-molecular-weight paraffins
Some methane, ethane, propane, and butane
Hydrocarbon distillates (full range depending on the feedstock)
Residual tar (recycle)
Contaminants (asphaltic constituents) deposited on the catalyst as coke or metals
Variations
Fixed bed (suitable for liquid feedstocks)
Ebullating bed (suitable for heavy feedstocks)

oil and geopolitics between different countries and the emergence of alternate feed supplies such as heavy oil, extra heavy oil, and tar sand bitumen; (iii) environmental regulations that include more stringent regulations in relation to sulfur in gasoline and diesel; and (iv) technology development such as new catalysts and processes (Table 7.5; Castañeda et al., 2014; Villasana et al., 2015).

With the onset of the 21st century and in spite of the availability of good-quality crude oil from tight formations (Chapter 1), refining technology continued to experience great innovation driven by the increasing supply of heavy

TABLE 7.5 Examples of Processes[a] for Upgrading Heavy Feedstocks

ABC process	Asphaltene cracking in the presence of hydrogen
Aquaconversion	A hydrovisbreaking technology that is a catalytic process that uses catalyst-activated transfer of hydrogen from water added to the feedstock in slurry mode. Reactions that lead to coke formation are suppressed, and there is no separation of asphaltene-type material
ART process	Efficient catalyst-feedstock contact
Asphalt coking technology (ASCOT) process	Combination of delayed coking and deep solvent deasphalting
BOC process	Temperature control and feedstock conversion limited to avoid coking
Cherry-P process	Feedstock slurred with coal
DSD process	Solvent chosen for yield of deasphaltened oil
Demex process	Less selective solvent than propane-based processes
ENI slurry technology	A process that is based on the slurry hydrotreatment of heavy feedstock at relatively low temperature in the presence of hydrogen and a dispersed catalyst, which is recycled to the slurry reactor via solvent deasphalting together with the asphaltene recycle
ET-II process	Feedstock mixed with high-boiling recycle oil
HOC process	Feedstock if first desulfurized (hydrotreated)
H-Oil process	An ebullated-bed process uses a single-stage, two-stage, or three-stage ebullated-bed reactors and can operate over a wide range of conversion levels. Adapted to process heavy feedstocks with high metal content and to desulfurize and demetallize feedstocks
HOT process	Steam-iron reaction to produce hydrogen in the cracker
HYCAR process	A noncatalytic process, based on visbreaking and involving treatment with hydrogen at mild conditions. Involves the use of three reactors:
	1. Visbreaking: this reactor carries out a moderate thermal cracking process in the presence of hydrogen 2. Hydrodemetallization: this reactor is to remove contaminants, particularly metals, prior to hydrocracking 3. Hydrocracking: desulfurization and denitrogenation take place in this reactor along with hydrocracking

Continued

TABLE 7.5 Examples of Processes for Upgrading Heavy Feedstocks—cont'd

Hycon process	Used to remove sulfur, metals, and asphaltene constituents and is typically operated in fixed-bed mode, but with increasing metal content in the feedstock, one or more moving-bed bunker reactors are added as the leading reactors for hydrodemetallization
Hyvahl-F and Hyvahl-S	These processes are used to hydrotreat heavy feedstocks and designed primarily for feedstock containing high concentrations of asphaltene, constituents, and metals
LC-Fining process	An ebullated-bed process is a hydrogenation process that can be operated for hydrodesulfurization, hydrodemetallization, and hydrocracking
MDS process	Solvent deasphalting and desulfurization for feedstock to catalytic cracker
Microcat-RC	A catalytic ebullated-bed hydroconversion process that operates at relatively moderate pressures and temperatures. The catalyst particles are dispersed uniformly throughout the feedstock, which results in less distance between particles and less time for a reactant molecule or intermediate to find an active catalyst site
MRH process	A hydrocracking process designed to upgrade heavy feedstocks containing large amounts of metals and asphaltene. The reactor is designed to maintain a mixed three-phase slurry of feedstock, fine powder catalyst, and hydrogen and to promote effective contact
OCR process	The process (which features onstream catalyst replacement) is a moving-bed reactor for hydroprocessing of heavy feedstocks that contain a significant amount of metals and operates in a countercurrent mode at high temperature and pressure
Residual oil supercritical extraction (ROSE) process	Deasphalting/deasphaltening depending upon the choice of solvent
RDS/VRDS	In both processes, desulfurization is used to remove sulfur and metallic constituents, while part of the feedstock is converted to lower-boiling products
Uniflex process	Uses a catalyst to promote hydrogenation and reduce coke formation, incorporates features of the former Canada Center for Mineral and Energy Technology (CANMET) process

[a]*Presented alphabetically rather than by preference.*

feedstocks of decreasing quality (such as heavy oil, extra heavy oil, and tar sand bitumen) and the fast increases in demand for clean and ultraclean vehicle fuels and petrochemical raw materials. This is continuing the movement from conventional methods of refining heavy feedstocks using (typically) coking technologies to more innovative processes (including hydrogen management) that will produce optimal amounts of liquid fuels from the feedstock and maintain emissions within environmental compliance (Davis and Patel, 2004; Penning, 2001; Speight, 2011).

In addition, the general trend throughout refining has been to produce more products from each barrel of crude oil and to process those products in different ways to meet the product specifications for use in modern engines. Overall, the demand for gasoline and diesel fuel has rapidly expanded, and demand has also developed for gas oils and fuels for domestic central heating and fuel oil for power generation, as well as for low-boiling distillates and other inputs, derived from refinery feedstocks, for the petrochemical industries.

Thus, upgrading heavy feedstocks has become a major economic incentive (Hedrick et al., 2006), and the feedstocks such as heavy oil, extra heavy oil, and tar sand bitumen exist in large quantities throughout the world but are difficult to produce and transport because of their high viscosity. Some refinery feedstocks contain compounds such as sulfur and/or heavy metals causing additional refining problems and costs. In situ upgrading could be a very beneficial process for leaving the unwanted elements in the reservoir and increasing API gravity.

As the need for the lower-boiling products developed, refinery feedstocks that could produce the desired quantities of the lower-boiling products became less available, and refineries had to introduce conversion processes to produce greater quantities of lower-boiling products from the higher-boiling fractions. The means by which a refinery operates in terms of producing the relevant products depends not only on the nature of the refinery feedstock but also on its configuration (i.e., the number of types of processes that are employed to produce the desired product slate), and the refinery configuration is, therefore, influenced by the specific demands of a market. Therefore, refineries need to be constantly adapted and upgraded to remain viable and responsive to ever-changing patterns of crude supply and product market demands. As a result, refineries have been introducing increasingly complex and expensive processes to gain higher yields of lower-boiling products from the heavy feedstocks.

Changes in the characteristics of the conventional refinery feedstock can be specified and will trigger changes in refinery configurations and corresponding investments. In the future, crude slate is expected to consist of higher proportions of heavier (more viscous) crude oils, sour (high-sulfur) crude oils, and extra light feedstocks such as natural gas liquids (NGLs). There will also be a shift toward viscous feedstocks such as Californian heavy oil, Venezuelan extra heavy oil, and Athabasca tar sand bitumen. These changes will require investment in upgrading either at field level to process tar sand bitumen and oil shale into synthetic crude at the refinery level (Speight, 2011).

In the not too distant past and even now, the mature and well-established processes such as visbreaking, delayed coking, fluid coking, flexicoking, propane deasphalting, and butane deasphalting were deemed adequate for upgrading heavy feedstocks. Moreover, more options are now being sought in order to increase process efficiency in terms of the yields of the desired products. Currently, there are currently four ways of bringing heavy feedstocks to market (Hedrick et al., 2006).

Partial upgrading is, in the context of this book, any combination of processing steps applied to heavy oil, extra have oil, or tar sand bitumen that prepares the oil to meet the specifications for pipeline transport or the next steps in conversion of the oil to products (Hart, 2014). For example, the approaches used for reducing the viscosity of heavy crude include heating, blending with light crude and with kerosene, and forming oil-in-water emulsions. Heating has a dramatic effect on the heavy crude viscosity but does not always achieve a practical level; consequently, blending the heavy crude with either light crude or kerosene requires substantial amounts of the diluent and raises the potential for instability of the blend or incompatibility of the blend components (Chapter 5).

Alternatively, emulsion formation can achieve a practical level of reduction that is achievable at 70%–75% v/v oil content, in the high-shear-rate range, and at 30–50°C (86–122°F), and providing these is no occurrence of instability of the emulsion (Mehrota, 1990; Khan, 1996; Al-Asmi et al., 1997; Urdahl et al., 1997; Zaki, 1997; Yaghi and Al-Bemani, 2002). In terms of pipeline transport of the oil, the addition of a diluent to the heavy oil, extra heavy oil, or tar sand bitumen prepares the oil to meet the specifications required by the pipeline company. In the true sense of the term, the addition of diluent is not an upgrading step insofar as the oil can be recovered unchanged by the removal of the diluent.

The heavy feedstocks add a new dimension of complicity of refining chemistry insofar as thermal decomposition processes, catalytic decomposition processes, pressure, and chemicals to rearrange the structures and bonding patterns of the high-molecular-weight feedstock constituents into different hydrocarbon products. In many cases, the chemical structures of the constituents of the products do not reflect the chemical structures of the constituents of the originally unprocessed feedstock. It is only in low-temperature solvent processes, such as deasphalting, that the original structures of the constituents are retained without change in the deasphalted oil and asphalt.

Understanding refining chemistry not only allows an explanation of the means by which these products can be formed from a feedstock but also offers the potential for predictability of the outcome, and the major processes by which these products are produced from refinery feedstock constituents involve thermal decomposition leading to many modern products. Even though refining chemistry might be represented by relatively simple equations, the chemistry of conversion process may be quite complex, and an understanding of the chemistry involved in the conversion of a refinery feedstock to a variety of products is essential to an understanding of refinery operations.

In fact, the complexity of the individual reactions occurring in an extremely complex mixture and the *interference* of the products with those from other components of the mixture is unpredictable. Or the *interference* of secondary and tertiary products with the course of a reaction and, hence, with the formation of primary products may also be cause for concern. Hence, caution is advised when applying the data from model compound studies to the behavior of crude oil, especially the molecularly complex viscous oils. These have few, if any, parallels in organic chemistry.

The yield and quality of refined crude oil products produced by any given oil refinery depend on the mixture of crude oil used as the refinery feedstock and the configuration of the refinery facilities. Light/sweet (low-sulfur) crude oil is generally more expensive and has inherent high yields of higher-value, low-boiling products such as naphtha, gasoline, jet fuel, kerosene, and diesel fuel. Viscous sour (high-sulfur) crude oil is generally less expensive and produces greater yields of lower-value, higher-boiling products that must be converted into lower-boiling products.

Finally, a comment about the chemistry of refining that has been dealt with in detail elsewhere (Speight, 2014a, 2017) will not be reproduced in this text.

2. Feedstock Quality

Over the past five decades, the feedstocks available to refineries have generally decreased in API gravity that has forced changes in refinery options for processing the heavier feedstocks (Parkash, 2003; Gary et al., 2007; Speight, 2011, 2014a, 2015a,b, 2017; Hsu and Robinson, 2017). There is, nevertheless, a major focus in refineries through a variety of conversion processes on the ways in which viscous feedstocks might be converted into low-boiling high-value products. Simultaneously, the changing refinery feedstock properties are reflected in changes such as an increase in asphaltene constituents and an increase in sulfur, metal, and nitrogen contents (Fig. 7.3). Pretreatment processes for removing those would also play an important role. However, the essential step required of refineries is the upgrading of viscous feedstocks, particularly residua (McKetta, 1992; Dickenson et al., 1997). In fact, the increasing supply of viscous crude oils is a matter of serious concern for the refining industry.

As a result of these changes, feedstock evaluation (always an important aspect of refining) is an even more important aspect of the prerefining examination of a refinery feedstock, especially the physical properties of the viscous feedstocks (Table 7.6). Evaluation, in this context, is the determination of the physical and chemical characteristics of crude oil, heavy oil, extra heavy oil, and tar sand bitumen since the yields and properties of products or factions produced from these feedstocks vary considerably and are dependent on the concentration of the various types of hydrocarbon derivatives and the amounts of the heteroatom compounds (i.e., molecular constituents containing nitrogen, oxygen, sulfur, and/or metals). Some types of feedstocks have economic

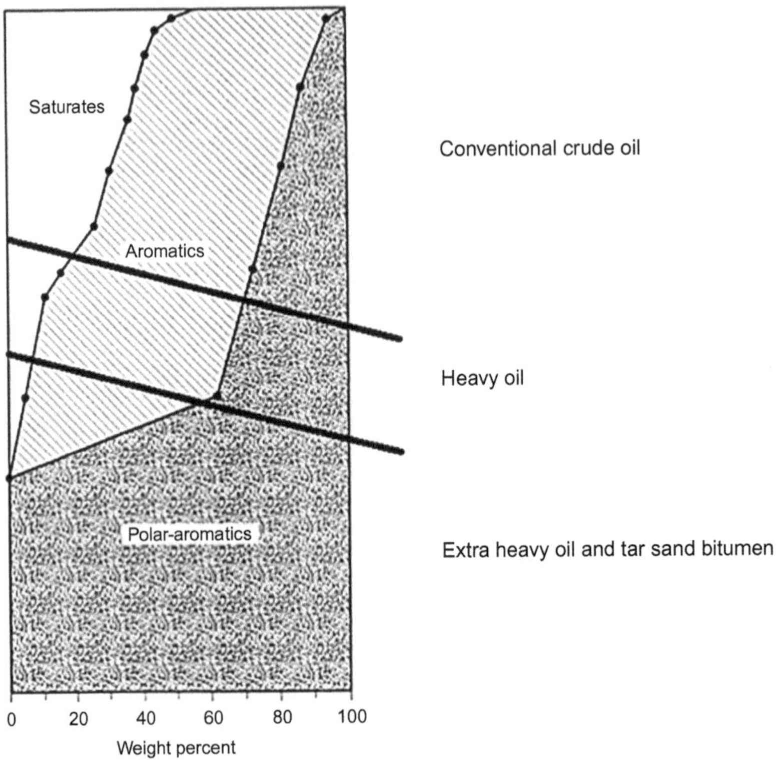

FIG. 7.3 Simplified representation of feedstock composition. The majority of the nitrogen-, sulfur-, and metal-containing constituents reside in the fraction designated as *polar aromatics*.

advantages as sources of fuels and lubricants with highly restrictive characteristics because they require less specialized processing than that needed for the production of the same products from many types of refinery feedstocks. Some feedstocks may contain unusually low concentrations of components that are desirable fuel or lubricant constituents, and the production of these products from such refinery feedstocks may not be economically feasible.

Since heavy oil, extra heavy oil, and tar sand bitumen exhibit a wide range of physical properties, it is not surprising the behavior of various feedstocks in these refinery operations is not simple. The atomic ratios from ultimate analysis (Speight, 2014a, 2015a) can give an indication of the nature of a feedstock and the generic hydrogen requirements to satisfy the refining chemistry (Speight, 2016), but it is not possible to predict with any degree of certainty how the feedstock will behave during refining. Any deductions made from such data are pure speculation and are open to much doubt.

In addition, the chemical composition of a feedstock is also an indicator of refining behavior (Parkash, 2003; Gary et al., 2007; Speight, 2014a, 2015a,b, 2017; Hsu and Robinson, 2017). Whether the composition is represented in

TABLE 7.6 Bitumen and Heavy Oil Properties

Source	Reservoir Temperature	Reservoir Permeability	Reservoir		Viscosity (cp)				
	°F	*md*	*Porosity*	API	*Reservoir Temp.*	Viscosity (cp)	Viscosity (cp)	Viscosity (cp)	Pour Point (°F)
Alberta									
Athabasca, Abasand	50			6.5				559 at 210°F	100
Athabasca, Bitumount				9.7				70 at 210°F	95
Athabasca, McMurray	54	2000	33	7	1,500,000				75
Athabasca, McMurray				8.9	1,011,600 at 59°F	265,300 at 77°F	4620 at 140°F	100 at 212°F	82
Athabasca, Mildred Lake	50			7.8				345	85
Carbonate	52			7	1,000,000			700 at 212°F	77
Cold Lake	55			11	100,000				50
Cold Lake Clearwater	55	1500	35	10.2	100,000			100 at 212°F	50
Lloydminster				15	3000			20 at 212°F	−8
Lloydminster	53			14		500 at 100°F			5
Peace River	60			9	100,000			100 at 212°F	30
Peace River	63	440	28	10	220,000				
Utah									

Continued

TABLE 7.6 Bitumen and Heavy Oil Properties—cont'd

Source	Reservoir Temperature	Reservoir Permeability	Reservoir	API	Viscosity (cp)	Viscosity (cp)	Viscosity (cp)	Viscosity (cp)	Pour Point (°F)
	°F	*md*	*Porosity*		*Reservoir Temp.*				
Asphalt Ridge	52	> 1000	27	12.9	1,000,000		20,000 at 140°F		95
Asphalt Ridge				9.1					126
Asphalt Ridge, Uinta County			10.2			16,910 at 149°F	539 at 225°F	113	
Asphalt Ridge NW	52			14.3	2,950,000	1,000,000 at 60°F			140
Asphalt Ridge NW	52	675	31	14	>1,000,000 at 60°F				140
Asphalt Ridge NW	50	651	26	14.2					140
Asphalt Ridge NW	< 6	2175	29.5	13	> 1,000,000 at 60°F			13 at 200°F	115
PR Spring	55	700	26	9	32,500,000	1,000,000 at 70°F	400,000 at 140°F		115
Sunnyside	50	800–1000	24	8	100,000				115
Tar Sand Triangle	50		22	5.5	1,300,000		700,000 140°F		95
Texas									
San Miguel, Saner Ranch	100		250–1000	2.1	2,000,000			2279 at 210°F	Solid

San Miguel, Street Ranch	95	<1000	27.5	−2	20,000,000		520,000 at 175°F 61,000 at 200°F	180	
California									
Casmalia			48						
Cat Canyon				12					
Brooks Sand	135	1400–5000	32	12	15,000				
S Sand	110	3450	37		12,000				
Coalinga Temblor	95	1000		12.0	3000			65 at 200°F	
Coalinga Etchegoin				11					
East Coalinga	95	1000		12	3000				
Kern River	86	2000	31		12,000			18 at 250°F	
Kern River	110			15.9		5047 at 80°F	302 at 140°F	44 at 200°F	
Midway-Sunset				11		786 at 100°F			55
Midway-Sunset	100			12		4000 at 100°F			45
Gore Sands	96	2000	22.7	13		6000 at 100°F		28 at 250°F	
Potter Sands	95	2000	35	11.6	9500			30 at 250°F	
Potter Sands	100	2000		11		4000 at 100°F			

Continued

TABLE 7.6 Bitumen and Heavy Oil Properties—cont'd

	Reservoir Temperature	Reservoir Permeability	Reservoir		Viscosity (cp)				
Source	°F	*md*	*Porosity*	API	*Reservoir Temp.*	Viscosity (cp)	Viscosity (cp)	Viscosity (cp)	Pour Point (°F)
Webster Sands	100	1300	28	14	1650				
Oxnard	122								
Yorba Linda	80		33	12	6400			11 at 300°F	
Lower Conglomerate	95	600	30	13–14	2000			8 at 300°F	
Olinda Fee No. 4	85	600		13		6400 at 100°F			
Olinda Fee Three 106				12.1					
Upper Conglomerate	85	600	30	12–13	6400			11 at 300°F	
Upper Conglomerate	80	4000	33	12					

terms of compound types or in terms of generic compound classes, it can enable the refiner to determine the nature of the reactions. Hence, chemical composition can play a large part in determining the nature of the products that arise from the refining operations. It can also play a role in determining the means by which a particular feedstock should be processed (Parkash, 2003; Gary et al., 2007; Speight, 2014a, 2017; Hsu and Robinson, 2017).

Therefore, the judicious choice of a refinery feedstock to produce any given product is just as important as the selection of the product for any given purpose. Thus, initial inspection of the nature of the feedstock will provide deductions related to the most logical means of refining. Indeed, careful evaluation of the feedstock from physical property data is a major part of the initial study of any heavy oil, extra heavy oil, and tar sand bitumen destined as a refinery feedstock. Proper interpretation of the data resulting from the inspection of crude oil requires an understanding of their significance.

Heavy oil, extra heavy oil, and tar sand bitumen exhibit a wide range of physical properties, and several relationships can be made between various physical properties (Speight, 2001). Whereas the properties such as viscosity, density, boiling point, and color of heavy oil, extra heavy oil, and tar sand bitumen may vary widely, the ultimate or elemental analysis varies, as already noted, over a narrow range for a large number of feedstock samples (Speight, 2014a, 2017). Of the data that are available, the proportions of the elements in crude oil vary only slightly over narrow limits:

Carbon 83.0%–87.0%
Hydrogen 10.0%–14.0%
Nitrogen 0.1%–2.0%
Oxygen 0.05%–1.5%
Sulfur 0.05%–6.0%
Metals (Ni and V) $<$1000 ppm

Nevertheless, there is a wide variation in physical properties from the less viscous (more mobile) crude oil at one extreme to the more viscous (asphaltic) crude oils at the other extreme. The majority of the more aromatic species and the heteroatoms occur in the higher-boiling fractions of feedstocks. The heavier feedstocks are relatively rich in these higher-boiling fractions (Speight, 2014a, 2015a). The carbon content is relatively constant, while the hydrogen and heteroatom contents are responsible for the major differences between heavy oil, extra heavy oil, and tar sand bitumen. Coupled with the changes brought about to the feedstock constituents by refinery operations, it is not surprising that heavy oil, extra heavy oil, and tar sand bitumen characterization is a monumental task.

Difficult-to-refine feedstocks, such as heavy oil, extra heavy oil, and tar sand bitumen, are characterized by low API gravity (high density) and high viscosity, high initial boiling point, high carbon residue, high nitrogen content, high sulfur content, and high metal content (Speight, 2013a, 2014a, 2017). In addition to these properties, the heavy feedstocks also have an increased molecular weight and reduced hydrogen content with a relatively low content of volatile saturated and aromatic constituents and a relatively high content of asphaltene and resin constituents that is accompanied by a high heteroatom (nitrogen, oxygen, sulfur, and metals) content. Thus, such feedstocks are of low volatility (especially tar sand bitumen and extra heavy oil) when compared with conventional crude oil (Tables 7.7 and 7.8) and not typically subject to distillation unless contained in the refinery feedstock as a stable blend with other refinery feedstocks (Chapter 5).

Upgrading viscous feedstocks began with the introduction of desulfurization processes (Speight, 2000, 2014a, 2017; Ancheyta and Speight, 2007; Hsu and Robinson, 2017). In the early days, the goal was desulfurization, but in later years, the processes were adapted to a 10%–30% partial conversion operation, as intended to achieve desulfurization and obtain low-boiling fractions simultaneously, by increasing severity in operating conditions. Refinery evolution has seen the introduction of a variety of cracking processes based on *thermal cracking*, *catalytic cracking*, and *hydroconversion*. Those processes are different from one another in cracking method, cracked product patterns, and product properties and will be employed in refineries according to their respective features. Thus, refining viscous feedstocks has become a major issue in modern refinery practice, and several process configurations have evolved to accommodate the viscous feedstocks (Table 1.1; RAROP, 1991; Shih and Oballa, 1991; Yoshida et al., 1997; Gary et al., 2007; Speight, 2014a, 2017; Hsu and Robinson, 2017).

Technologies for upgrading viscous feedstocks can be broadly divided into *carbon rejection* and *hydrogen addition* processes (Fig. 7.4). *Carbon rejection* redistributes hydrogen among the various components, resulting in fractions with increased H/C atomic ratios and fractions with lower H/C atomic ratios. On the other hand, *hydrogen addition* processes involve the reaction of refinery feedstocks with an external source of hydrogen and result in an overall increase in H/C ratio (Speight, 1994, 2011, 2014a, 2017).

The criteria to select one of the routes as an upgrading option depend on several factors that must be analyzed in detail when it comes to consider a project of this nature. For example, the technology of *hydrogen addition* produces a high yield of products with a commercial value larger than that of the *carbon rejection* technology but requires a larger investment and more natural gas availability to produce the amounts of hydrogen and steam required for these processes.

New processes for the conversion of heavy oil, extra heavy oil, and tar sand bitumen feedstocks will probably be used in concert with visbreaking with some degree of hydroprocessing as a primary conversion step. Other processes may replace or augment the deasphalting units in many refineries. Depending on the properties, an option for heavy oil, extra heavy oil, and tar sand bitumen, like the early option for tar sand bitumen, is to subject the feedstock to either delayed coking or fluid coking as the *primary upgrading* step with some prior distillation to remove a valuable gas-oil fraction (Fig. 7.5; Speight, 2009, 2011, 2014a, 2017).

TABLE 7.7 Distillation Profile of Leduc Crude Oil (Alberta, Canada) and Selected Properties of the Fractions (Charbonnier et al., 1969; Draper et al., 1977)

	Boiling Range			wt% Cumulative	Specific Gravity	API Gravity	Sulfur wt%	Carbon Residue (Conradson)
Whole Crude Oil	°C	°F	wt%	*100.0*	*0.828*	*39.4*	*0.4*	*1.5*
Fraction[a]								
1	0–50	0–122	2.6	2.6	0.650	86.2		
2	50–75	122–167	3.0	5.6	0.674	78.4		
3	75–100	167–212	5.2	10.8	0.716	66.1		
4	100–125	212–257	6.6	17.4	0.744	58.7		
5	125–150	257–302	6.3	23.7	0.763	54.0		
6	150–175	302–347	5.5	29.2	0.783	49.2		
7	175–200	347–392	5.3	34.5	0.797	46.0		
8	200–225	392–437	5.0	39.5	0.812	42.8		
9	225–250	437–482	4.7	44.2	0.823	40.4		
10	250–275	482–527	6.6	50.8	0.837	37.6		
11	< 200	< 392	5.4	56.2	0.852	34.6		
12	200–225	392–437	4.9	61.1	0.861	32.8		
13	225–250	437–482	5.2	66.3	0.875	30.2		
14	250–275	482–527	2.8	69.1	0.883	28.8		
15	275–300	527–572	6.7	75.4	0.892	27.0		
Residuum	> 300	> 572	22.6	98.4	0.929	20.8		6.6
Distillation loss				1.6				

[a]*Distillation at 765 mm Hg then at 40 mm Hg for fractions 11–15.*

TABLE 7.8 Distillation Profile of Athabasca Bitumen (Alberta, Canada) and Selected Properties of the Fractions (Charbonnier et al., 1969)

Feedstock	Boiling Range			wt% Cumulative	Specific Gravity	API Gravity	Sulfur wt%	Carbon Residue (Conradson)
Whole Bitumen	°C	°F	wt%	*100.0*	*1.030*	*5.9*	*5.8*	*19.6*
Fraction[a]								
1	0–50	0–122	0.0	0.0				
2	50–75	122–167	0.0	0.0				
3	75–100	167–212	0.0	0.0				
4	100–125	212–257	0.0	0.0				
5	125–150	257–302	0.9	0.9				
6	150–175	302–347	0.8	1.7	0.809	43.4		
7	175–200	347–392	1.1	2.8	0.823	40.4		
8	200–225	392–437	1.1	3.9	0.848	35.4		
9	225–250	437–482	4.1	8.0	0.866	31.8		
10	250–275	482–527	11.9	19.9	0.867	31.7		
11	< 200	< 392	1.6	21.5	0.878	29.7		
12	200–225	392–437	3.2	24.7	0.929	20.8		
13	225–250	437–482	6.1	30.8	0.947	17.9		
14	250–275	482–527	6.4	37.2	0.958	16.2		
15	275–300	527–572	10.6	47.8	0.972	14.1		
Residuum	> 300	> 572	49.5	97.3				39.6

[a]*Distillation at 762 mm Hg and then at 40 mm Hg for fractions 11–15.*

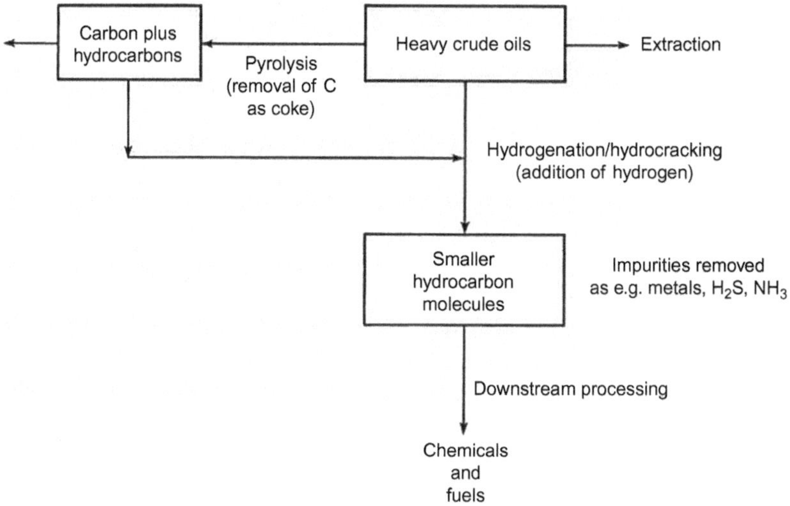

FIG. 7.4 Process options for refining heavy feedstocks.

After primary upgrading, the product streams are hydrotreated (referred to as secondary upgrading) and combined to form a *synthetic crude oil* (Table 7.9) that is shipped to a conventional refinery for further processing to liquid fuels. The synthetic crude oil is a blend of naphtha, distillate, and gas-oil range materials, with no residuum (1050°F+, 565°C+ material) (Speight, 2013c, 2014a, 2017).

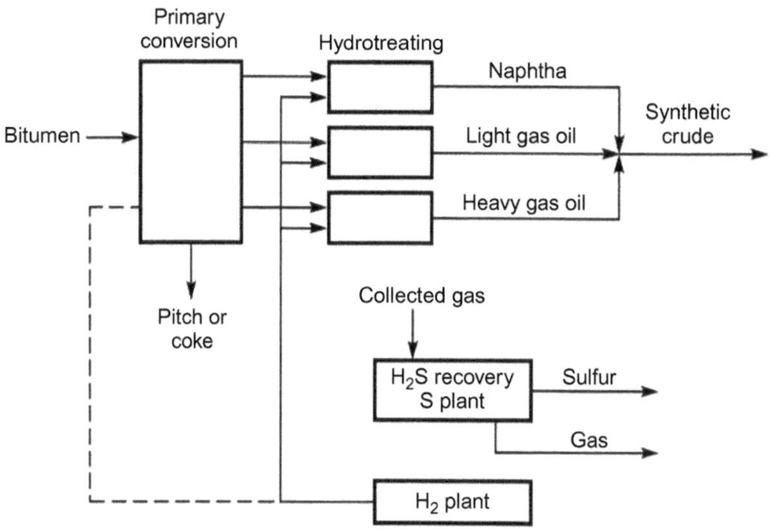

FIG. 7.5 Processing sequence for tar sand bitumen.

TABLE 7.9 Properties of Synthetic Crude Oil Produced Form Athabasca Bitumen

Property		Bitumen	Synthetic Crude Oil
Gravity, ° API		8	32
Sulfur, % w/w		4.8	0.2
Nitrogen, % w/w		0.4	0.1
Viscosity, centipoise at 100°F		500,000	10
Distillation profile, % w/w			
°C	°F		
0	30	0	5
30	85	0	30
220	430	1	60
345	650	17	90
550	1020	45	100

However, there is not one single heavy oil, extra heavy oil, and tar sand bitumen upgrading solution that will fit all refineries or because of the inherent differences, all viscous feedstocks. The properties of the viscous feedstock, existing refinery configuration, and desired product slate all can have a significant effect on the final configuration. Furthermore, a proper evaluation however is not a simple undertaking for an existing refinery. The evaluation starts with an accurate understanding of the nature of the feedstock along with corresponding conversion chemistry that needs to be assessed. Once the options have been defined, development of the optimal configuration for refining the incoming feedstocks can be designed. There is also recognition that in situ upgrading (upgrading during recovery operations) could be a very beneficial process for leaving the unwanted elements in the reservoir and increasing API gravity (Chapter 6).

However, the increased mobilization of viscous oil in the reservoir by partial upgrading is not a new idea and still has many hurdles to overcome before it can be considered close to commercial. The product will be less viscous than the heavy oil, extra heavy oil, and tar sand bitumen in place, but some property changes such as high olefin content from cracking are not necessarily positive. In summary, there are three main approaches for heating the reservoir: (i) steam distillation, (ii) mild thermal cracking (visbreaking), and (iii) partial combustion (Chapter 3) (Speight, 2009, 2013a).

Nevertheless, there is (or will be) an obvious future need for partial upgrading during or immediately after recovery. On the other hand, hydrogen addition must be used during upgrading in order to stabilize the upgraded heavy oil, extra heavy oil, and tar sand bitumen—which could mean that the cost of partial upgrading is not much reduced as compared with full upgrading. Therefore, the only choice currently is no upgrading or full upgrading. Other goals could be to achieve breakthroughs in upgrading technologies—such as nonthermal coking methods that would use far less energy or such as gasification at 800°C (1470°F), which is far less than current commercial temperatures. The technology where changes do occur involves combustion of the oil in situ. The concept of any combustion technology requires that the oil be partially combusted and that thermal decomposition occur to other parts of the oil. This is sufficient to cause irreversible chemical and physical changes to the oil to the extent that the product is markedly different to the oil in place. Recognition of this phenomenon is essential before combustion technologies are applied to oil recovery.

Although this improvement in properties may not appear to be too drastic, nevertheless, it usually is sufficient to have major advantages for refinery operators. Any incremental increase in the units of hydrogen/carbon ratio can save amounts of costly hydrogen during upgrading. The same principles are also operative for reductions in the nitrogen, sulfur, and oxygen contents. This latter occurrence also improves catalyst life and activity and reduces the metal content.

In short, in situ recovery processes (although less efficient in terms of bitumen recovery relative to mining operations) may have the added benefit of *leaving* some of the more obnoxious constituents (from the processing objective) in the ground. Processes that offer the potential for partial upgrading during recovery are varied but usually follow a surface process. Not that this be construed as an easy task, there are many disadvantages that arise from attempting in situ upgrading.

Finally, there is not a single in situ recovery process that will be applicable to all reservoirs, and no single recovery process will be able to access all the heavy oil, extra heavy oil, and tar sand bitumen in a given reservoir. To achieve maximum recovery, it will be necessary to apply a combination of different processes, for example, a steam-based recovery process followed by in situ combustion, followed by in situ upgrading, followed by bioconversion of the residual hydrocarbon derivatives. This type of sequential recovery will require careful planning to ensure that the optimum sequence and timing is applied.

To achieve partial upgrading during recovery requires a further sequential operation before a transportable produced. A multistep system is required to achieve the necessary aims of heavy oil, extra heavy oil, and tar sand bitumen recovery with partial upgrading. What this might be is currently unknown, but there are possibilities. Thus, a major decision at the time of recovery of heavy oil, extra heavy oil, and tar sand bitumen is to acknowledge the practical (or impractical) aspects of upgrading during recovery, partial upgrading at the surface, or full upgrading in a conversion refinery. For the purposes of this text only, upgrading in a conversion refinery has been assumed as the means of upgrading one or all of the three viscous feedstocks.

Furthermore, in hydroconversion processes such as LC-Fining and H-Oil where the NiMo or CoMo catalysts supported on alumina are used, it is crucial that the pore network of the support offers the maximum accessibility to asphaltene constituents. In order to produce lower-boiling products, the asphaltene constituents should be able to diffuse into the pore channels and then reach the active sites of the catalyst. The first active site that is accessible to the asphaltene constituents is the pore mouth, where plugging by metal deposition and/or coke deposition occurs. Hence, reaction rates of asphaltene conversion and demetallation could be controlled by the pore diffusion of reacting substances.

During hydroprocessing, the asphaltene constituents undergo a multitude of reactions involving both cracking and hydrogenation, which change their initial structure and relationship with the other constituents of the feedstocks (Speight, 1994, 2014a,b,c). Thus, below 350°C (650°F)—which is arbitrarily selected for convenience here as the cracking temperature—where cracking is minimal or certainly at a low level, the reaction chemistry is dominated by hydrogenation reactions, whereas above 350°C (650°F), cracking reactions play a greater role, thereby causing a rupture of alkyl side chains (dealkylation of aromatic moieties) and cracking of naphthene derivatives. The transition between these two reactions is more predominant within and above the temperature range 370–390°C (695–735°F), but it must be remembered that the so-called cracking temperature or the cracking range is feedstock-dependent.

Thus, with the complexity of the asphaltene fraction and the complex chemical structures of the individual constituents, any modification in the constitutional groups is very likely to alter the solubility/compatibility, and precipitation may occur within the catalyst pore system. At higher processing temperatures, cracking becomes dominant, yielding reacted asphaltene constituents with lower solubility and a higher tendency for incompatibility. As these reactions progress, the chemical structure of maltenes tends toward a higher degree of aliphatic character, and incompatibility (deposition) of the reacted asphaltene constituents is assured.

The problems related to the behavior of the asphaltene constituents can quickly reduce the benefits of processing a viscous feedstock. For example, incompatibility and/or extensive fouling in the feedstock preheat train may occur (Speight, 2015b) resulting in decreased energy efficiency, increased emissions of carbon dioxide, and increased frequency at which heat exchangers need to be cleaned. In a worst-case scenario, crude throughput may be reduced leading to significant financial losses.

Thus, heavy oil, extra heavy oil, and tar sand bitumen, while offering initial pricing advantages, may have composition problems that can cause severe problems at the refinery, harming infrastructure, yield, and profitability. Before refining, there is the need for comprehensive evaluations of these feedstocks, giving the potential buyer and seller the needed data to make informed decisions regarding fair pricing and the suitability of a particular crude oil as a feedstock for the refinery. This will assist the refiner to manage the ever-changing crude oil quality input to a refinery—including quality and quantity requirements and situations, crude oil variations, contractual specifications, and risks associated with such viscous feedstocks.

Finally, the limitations of processing these viscous feedstocks depend to a large extent on the tendency for coke formation and the deposition of metals and coke on the catalyst due to the higher molecular weight (low volatility) and heteroatom content. However, the essential step required of refineries is the upgrading of viscous feedstocks (McKetta, 1992; Dickenson et al., 1997). Also, there is one aspect of upgrading that cannot be ignored, and it relates to the compatibility/incompatibility of the refinery feedstocks. Blending feedstocks that are incompatible (Chapter 5) can lead to extensive fouling and processing difficulties due to unstable asphaltene constituents (Speight, 2014a,b, 2015b, 2017).

In fact, the increasing supply of viscous refinery feedstocks is a matter of serious concern for the refining industry. In order to satisfy the changing pattern of product demand, significant investments in refining conversion processes will be necessary to profitably utilize these viscous refinery feedstocks. The most efficient and economical solution to this problem will depend to a large extent on individual country and company situations.

3. Refinery Configuration

Crude oil refineries were originally designed and operated to run within a narrow range of crude oil feedstock and to produce a relatively fixed slate of crude oil products. Since the 1970s, refiners had to increase their flexibility in order to adapt to a more volatile environment. Several possible paths may be used by refiners to increase their flexibility within existing refineries. Examples of these paths are change in the severity of operating rules of some process units by varying the range of inputs used, thus achieving a slight change in output. Alternatively, refiners can install new processes, and this alternate scenario offers the greatest flexibility but is limited by the constraint of strict complementary of the new units with the rest of the existing plant and involves a higher risk than the previous ones. It is not surprising that many refiners decide to modify existing processes.

To convert a feedstock into the desired products in an economically feasible and environmentally acceptable manner, refinery processes for crude oil are generally divided into three categories: (i) separation processes, of which distillation is the prime example; (ii) conversion processes, of which coking and catalytic cracking are prime examples; and (iii) finishing processes, of which hydrotreating to remove sulfur is a prime example (Parkash, 2003; Gary et al., 2007; Speight, 2014a, 2017; Hsu and Robinson, 2017).

However, the configuration of refineries may vary from refinery to refinery. Some refineries may be more oriented toward the production of gasoline (large reforming and/or catalytic cracking), whereas the configuration of other refineries may be more oriented toward the production of middle distillates such as jet fuel and gas oil.

3.1 Topping Refinery and Hydroskimming Refinery

The simplest refinery configuration is the *topping refinery* (Table 7.10 and Fig. 7.6), which is designed to prepare feedstocks for petrochemical

TABLE 7.10 General Description of Refinery Types

Refinery Type	Processes	Name	Complexity	Complexity
Topping	Distillation	Skimming	Low	1
Hydroskimming	Distillation	Hydroskimming	Moderate	3
	Reforming			
	Hydrotreating			
Conversion	Distillation	Cracking	High	6
	Fluid catalytic cracking			
	Hydrocracking			
	Reforming			
	Alkylation			
	Hydrotreating			
Deep conversion	Distillation	Coking	Very high	10
	Coking			
	Fluid catalytic cracking			
	Hydrocracking			
	Reforming			
	Alkylation			
	Hydrotreating			

Indicates refinery complexity on an arbitrary scale of 1–10, with 10 being the most complex.

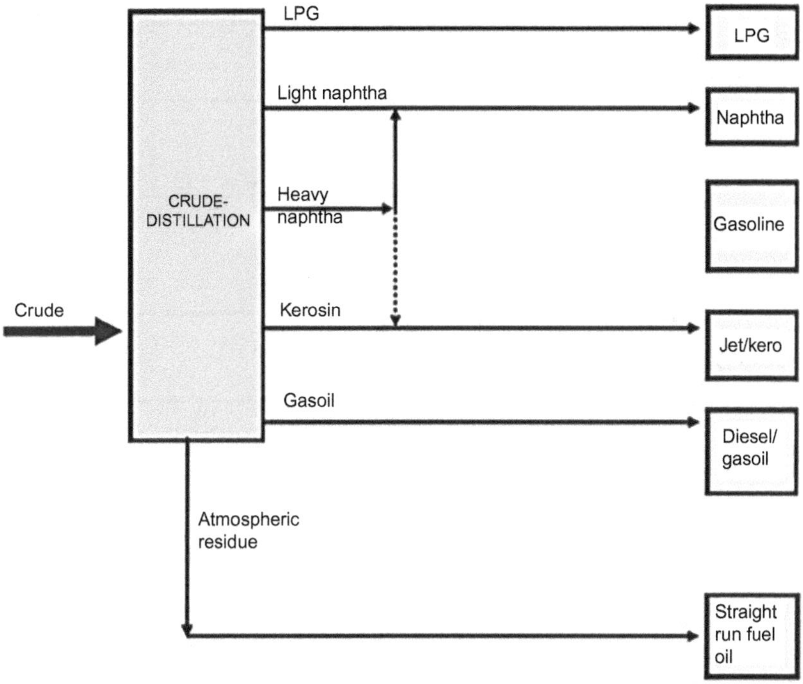

FIG. 7.6 A topping refinery.

manufacture or for the production of industrial fuels in remote oil production areas. The topping refinery consists of tankage, a distillation unit, recovery facilities for gases and low-boiling hydrocarbon derivatives, and necessary utility systems (steam, power, and water treatment plants).

A topping refinery (sometimes referred to as a modular mini refinery) is best utilized in emerging economies and in remote locations where naphtha and kerosene are needed. The local crude oil is typically the lowest cost feedstock because the transportation costs are minimized. A topping refinery with a viscous heavy crude and low API gravity will produce more fuel oil and less naphtha and kerosene. A crude oil with a high API gravity will produce less fuel oil and more naphtha and kerosene. Additionally, the sulfur content of the refinery feedstock determines refinery cost being as low-sulfur crudes may not require hydrotreaters.

Thus, a topping refinery will produce substantial quantities of unfinished products and is highly dependent on local markets, but the addition of hydrotreating and reforming units to this basic configuration results in a more flexible *hydroskimming refinery* (Table 7.10 and Fig. 7.7), which can also produce desulfurized distillate fuels and high-octane gasoline. These refineries may produce up to half of their output as residual fuel oil, and they face increasing market loss as the demand for low-sulfur (even no-sulfur) and high-sulfur fuel oil increases.

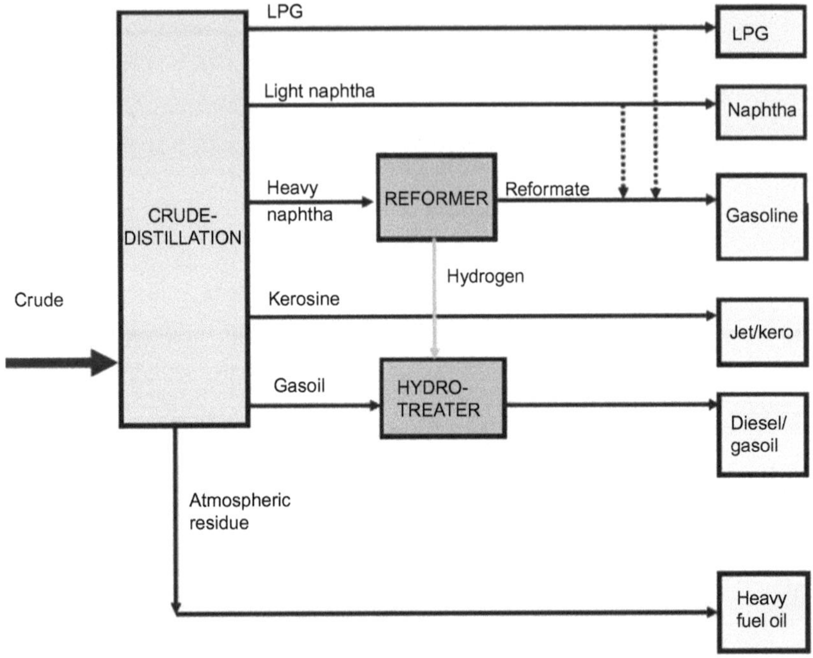

FIG. 7.7 A hydroskimming refinery.

A hydroskimming refinery is a refinery equipped with an atmospheric distillation unit, a naphtha reforming unit, and the necessary treating processes. A hydroskimming refinery is therefore more complex than a *topping refinery* (which just separates the crude into its constituent petroleum products by atmospheric distillation and produces naphtha). However, a hydroskimming refinery produces a surplus of fuel with a relatively unattractive price and demand.

The ability to switch between light and heavy crudes means that one crude may require a larger naphtha hydrotreater, a larger naphtha reformer, and a larger kerosene hydrotreater, whereas the other may not.

3.2 Conversion Refinery

The most versatile refinery configuration is known as the *conversion refinery*, of which the refinery based on coking technology (Table 7.10 and Fig. 7.8) and the refinery based on catalytic cracking (Fig. 7.9) are examples. In this context, the term *conversion* is the difference in the amount of unconverted oil between the feedstock and in the product(s) divided by the amount of unconverted oil in the feedstock.

Most refineries therefore add vacuum distillation and catalytic cracking, which adds one more level of complexity by reducing fuel oil by conversion

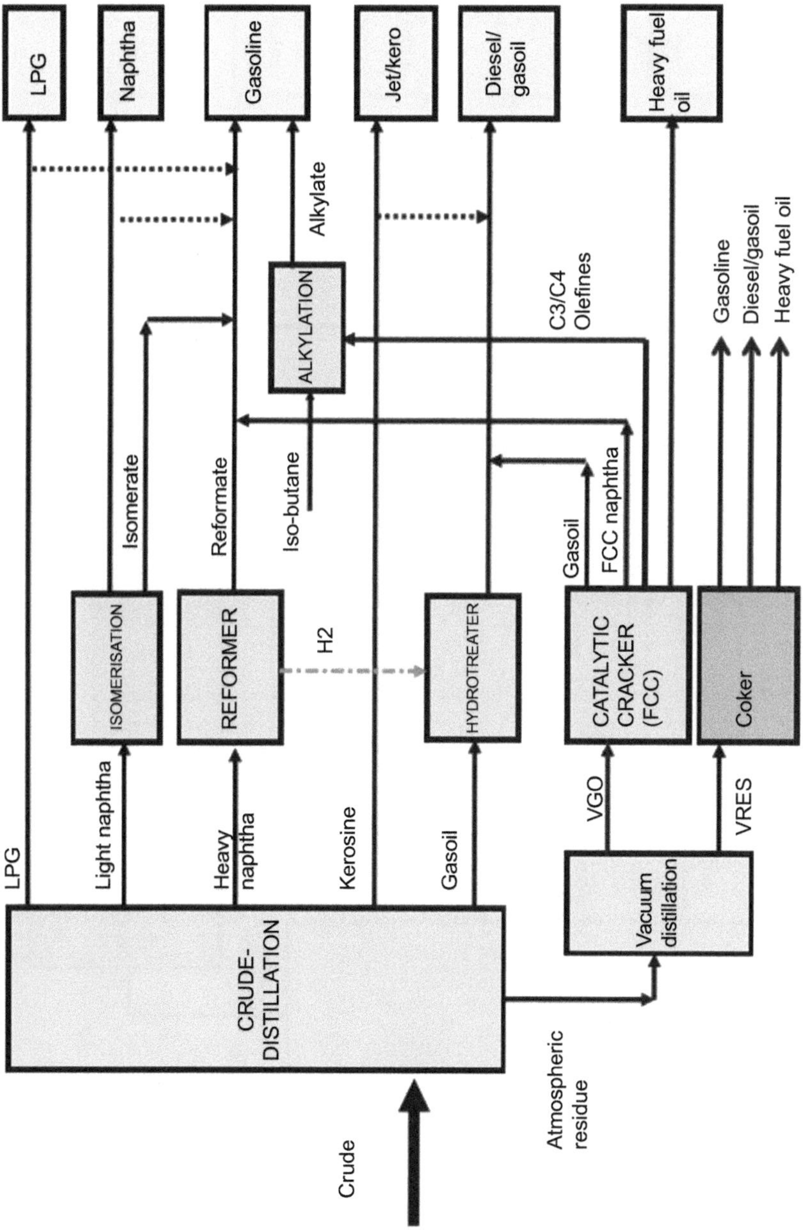

FIG. 7.8 A coking refinery.

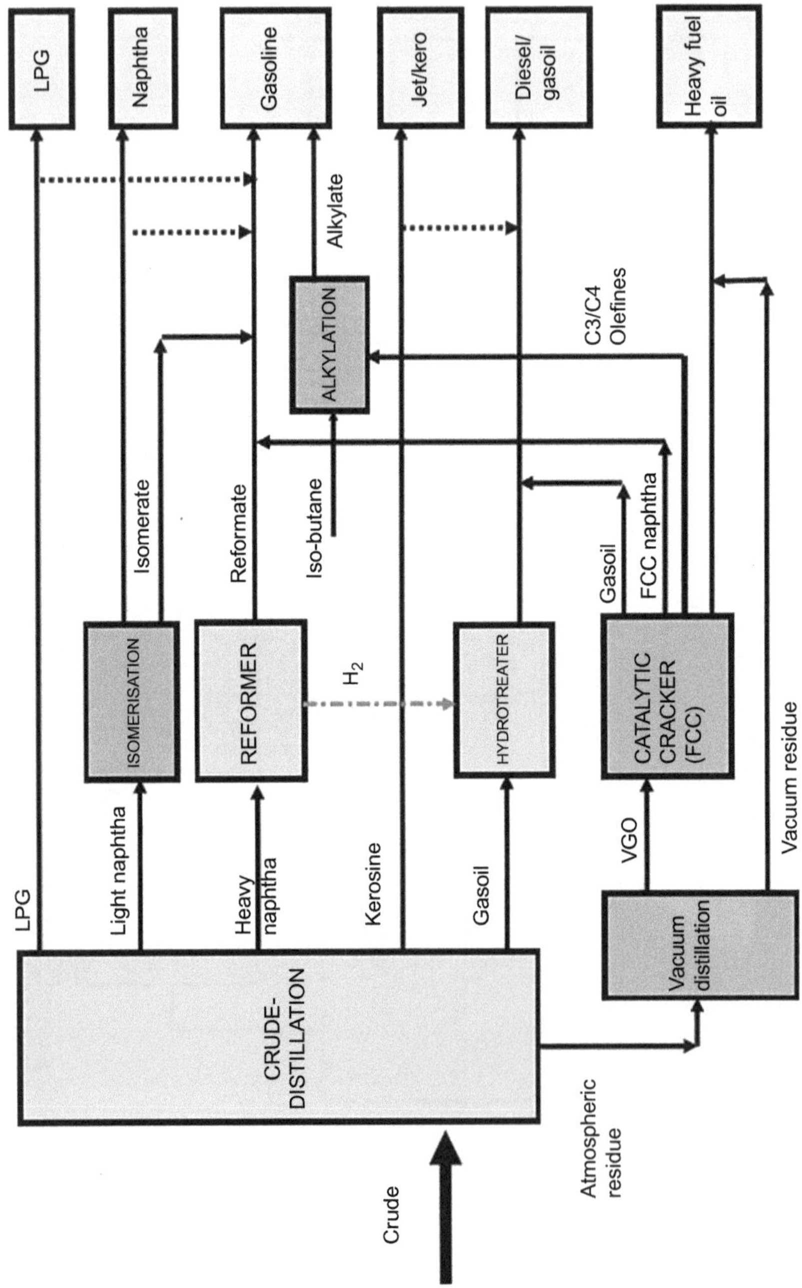

FIG. 7.9 A catalytic cracking refinery.

to low-boiling distillates and middle distillates. A coking refinery adds further complexity to the cracking refinery by high conversion of fuel oil into distillates and petroleum coke. Thus, a conversion refinery incorporates all the basic units found in both the topping and hydroskimming refineries, but it also features gas-oil conversion plants such as catalytic cracking and hydrocracking units; olefin conversion plants such as alkylation or polymerization units; and, frequently, coking units for sharply reducing or eliminating the production of residual fuels.

A modern conversion refinery may produce two-thirds of the product output as unleaded gasoline, with the balance distributed between liquefied petroleum gas, jet fuel, diesel fuel, and a small quantity of coke. Many such refineries also incorporate solvent extraction processes for manufacturing lubricants and petrochemical units with which to recover propylene, benzene, toluene, and xylenes for further processing into polymers.

Finally, the yields and quality of refined crude oil products produced by any given refinery depend on the mixture of crude oil used as feedstock and the configuration of the refinery facilities (Speight, 2013a, 2014a).

4. Upgrading

The influx of heavy oil, extra heavy oil, and tar sand bitumen into the refinery system can offset the shortages of conventional crude oil, but there is also the recognition of the need for increased refining capacity in terms of the units that can process these feedstocks (McGrath and Houde, 1999; Al-Marshed et al., 2015). New viscous feedstock upgrading capacity needs to be incorporated into existing refineries, or it could be built in separate, stand-alone upgrading facilities. If the oil is too viscous to transport by pipeline and/or there is the need for heat or energy at the production site, heavy oil, extra heavy oil, and tar sand bitumen upgrading in the field is attractive and may avoid extensive modifications of existing refineries (Hart, 2014). Traditional processes of viscous feedstocks, such as coking or hydrocracking, are expensive processes that require processing on the large scale to be viable. Thus, heavy oil, extra heavy oil, and tar sand bitumen refining is now in a significant transition period as the industry moves further into the 21st century and the demand for crude oil products has shown a sharp growth in recent decades (Speight, 2011).

In order to satisfy the changing pattern of product demand, significant investments in refining conversion processes will be necessary to profitably utilize viscous feedstocks. The most efficient and economical solution to this problem will depend to a large extent on individual refinery situations and the process temperature and pressure (Fig. 7.10) and feedstock conversion (Fig. 7.11). However, the most promising technologies will likely involve current (or modified) processes that are applied to the conversion of vacuum residua. But technologies are needed that will take the feedstock beyond current limits and, at the same time, reduce the amount of coke and other nonessential products (Fig. 7.1). Such a goal may require the use of two or more technologies

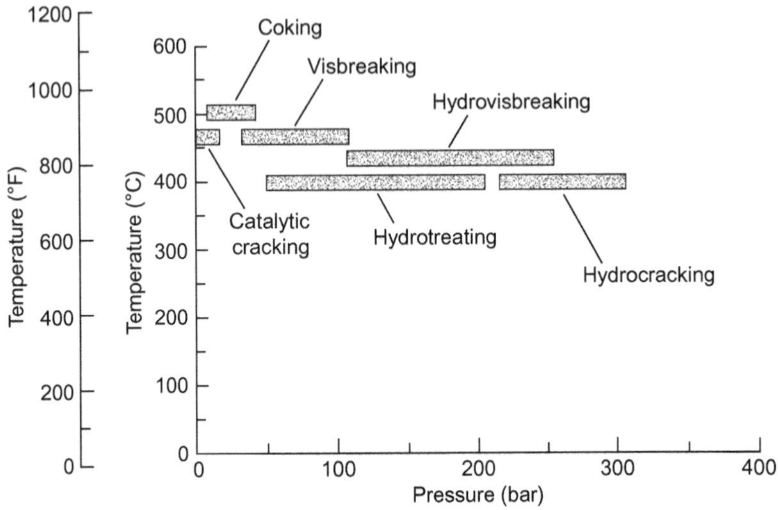

FIG. 7.10 Range of temperature and pressure for various processes.

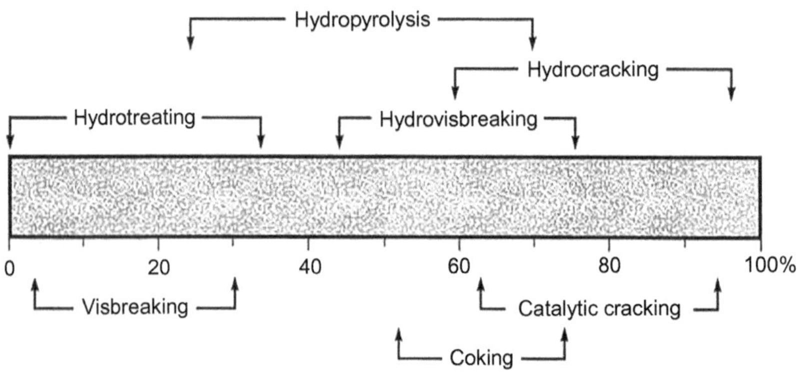

FIG. 7.11 Range for feedstock conversion for various processes.

in series rather than an attempt to develop a whole new one-stop conversion technology. Such is the nature of the modern refinery.

The manner in which refineries convert heavy oil, extra heavy oil, and tar sand bitumen into low-boiling high-value products has become a major focus of operations with new concepts, evolving into new processes (Khan and Patmore, 1998; Speight, 2000, 2014a, 2017). Even though they may not be classed as conversion processes per se, pretreatment processes for removing asphaltene constituents, metals, sulfur, and nitrogen constituents are also important and can play an important role.

New processes for the conversion of heavy oil, extra heavy oil, and tar sand bitumen will probably be used perhaps not in place of but in conjunction with visbreaking and coking options with some degree of hydroprocessing as a primary conversion step. In addition, other processes may replace or, more likely, augment the deasphalting units in many refineries. An exception, which may become the rule, is the upgrading of bitumen from tar sands (Speight, 1990, 2000, 2014a, 2017). The bitumen is subjected to either delayed coking or fluid coking as the *primary upgrading* step without prior distillation or topping. After primary upgrading, the product streams are hydrotreated and combined to form a synthetic crude oil that is shipped to a conventional refinery for further processing (Speight, 2013c, 2014a, 2017).

To upgrade heavy and extra heavy crude oils, two basic technological routes are known: (i) carbon rejection and (ii) hydrogen addition (Parkash, 2003; Gary et al., 2007; Speight, 2014a, 2017; Hsu and Robinson, 2017).

The upgrading process applied to heavy oil, extra heavy oil, and tar sand bitumen is based on the use of delayed coking as the principal technology, but the high coke production makes it necessary to analyze and develop a very serious consideration of other upgrading options, different from carbon rejection, such as hydrogen addition and other new technological upgrading proposals. The technology of hydrogen addition is a great advance, but one of the more important problems to solve in the hydrocarbon upgrading program is the ecological issue. Those hydrocarbon derivatives contain high concentrations of sulfur and metals, particularly nickel and vanadium. The processing implies a coke and sulfur generation greater than that of the light crude oil and medium crude oil and a larger amount of effluents and emissions to the atmosphere. Also, a high level of consciousness must be developed about the environmental impact of the extraction, transportation, upgrading, processing, and consumption of hydrocarbon derivatives with these characteristics, and technological solutions to preserve the ecological and environmental equilibrium of the affected area must be proposed and implemented as well.

In order to satisfy the changing patterns of feedstock slate and product demand, significant investments in refining conversion processes will be necessary to profitably utilize viscous feedstocks. The most efficient and economical solution to this problem will depend to a large extent on individual refinery situations, and this may require the use of two or more technologies in series rather than an attempt to develop a new one-stop technology for bitumen conversion (Speight, 2011).

The manner in which refineries convert tar sand bitumen into low-boiling high-value products has become a major focus of operations with new concepts evolving into new processes (Khan and Patmore, 1998; Speight, 2000).

For example, refineries can have any one or a conjunction of several configurations, but the refinery of the future will of necessity be required to be a *conversion refinery*, which incorporates all the basic building blocks found in both the *topping refinery* and the *hydroskimming refinery*, but it also features gas-oil conversion plants such as catalytic cracking and hydrocracking units; olefin conversion plants such as alkylation or polymerization units; and, frequently, coking units for sharply reducing or eliminating the production of residual fuels. Many such refineries also incorporate solvent extraction processes for manufacturing lubricants and petrochemical units with which to recover high-purity petrochemical feedstocks and petrochemical products.

Conceivably, tar sand bitumen could be upgraded in the same manner and, depending upon the upgrading facility, upgraded further for sales. However, this is not to be construed that bitumen upgrading will *always* involve a coking step as the primary upgrading step. Other options, including some presented elsewhere (Speight, 2014a,b,c), could well become predominant methods for upgrading in the future.

There are several proved technologies used worldwide in refineries and upgrading complexes for heavy oil, extra heavy oil, and tar sand bitumen that are described in the following chapters: (i) thermal cracking processes (Chapter 8), catalytic cracking (Chapter 9), (iii) hydrotreating (Chapter 10), (iv) hydrocracking (Chapter 11), (v) solvent treatment (Chapter 12), (vi) gasification (Chapter 13), and (vii) bioconversion (Chapter 14). These aspects of viscous feedstock conversion will not be discussed any further at this point but are the main focus of later chapters.

5. Options

Modern conversion refineries may produce two-thirds of their output as unleaded gasoline, with the balance distributed between liquefied petroleum gas, jet fuel, diesel fuel, and a small quantity of coke. Many such refineries also incorporate solvent extraction processes for manufacturing lubricants and petrochemical units with which to recover propylene, benzene, toluene, and xylenes for further processing into polymers. Also, many refineries offer options that can be accepted for upgrading viscous feedstocks.

The first option is to upgrade the viscous feedstock (partially or fully by the application of the appropriate processing steps) at or near to the wellhead, leaving much of the asphaltene-type material behind as coke, and the upgraded material will then be sent by pipeline as synthetic crude oil. With this method, the oil is fractionated, and the residue is coked—the products of the coking operation—and in some cases, some of the residue may also be hydrotreated in a field unit. The hydrotreated materials are recombined with the fractionated low-boiling materials to form synthetic crude that is then transported to market in a pipeline.

Examples of this type of processing can be seen in the current Canadian tar sand operations around Fort McMurray in Alberta, Canada (Parkash, 2003; Gary et al., 2007; Speight, 2011, 2014a, 2015a,b, 2017; Hsu and Robinson, 2017). The first option can be made more workable by the presence of abundant supplies of natural gas in the area and local electric power source. In addition, the presence of a scavenger-type catalyst in a relatively mild thermal or hydrotreating wellhead process can also assure the removal of the major coke formers as proto-coke—proto-coke is the material formed in the initial stages of the thermal process and has also been referred to as reacted asphaltene constituents (Speight and Moschopedis, 1979; Speight, 2014a). With the installation of suitable flue gas cleaning options, the proto-coke can be sued as process fuel. This concept can be extended to include a partial upgrading facility in which the viscous feedstock is upgraded to a level that make sit suitable for transportation (Motaghi et al., 2010a, 2010b, 2010c; Speight, 2013a, 2014a, 2017).

The second option is to build upgrading facilities at an established transportation terminal with abundant gas and electric resources. The liquid products from a coking operation can be hydrotreated and mixed back with the virgin materials. A pipeline from the upgrader to the oil field transports cutter stock to the oil field in sufficient quantity to produce pipeline-acceptable crude from the viscous feedstocks. There are several examples of this kind of facility located in the Jose, Venezuela, area that enable the production of extra heavy oil from the Orinoco River Basin (Venezuela).

A third option is to use conventional crude oil that is located in the general area to blend with the viscous feedstock to produce an acceptable pipeline material (Hart, 2014). While a seemingly viable option on paper, this option has a number of limitations. For example, the viscous feedstock production could be limited by the amount of conventional crude oil that is available for dilution. Another problem is the potential for incompatibility of the viscous feedstock and the conventional crude oil (Chapter 5) that would limit (perhaps even prohibit) the amount of dilution and the amount of viscous feedstock produced (Speight, 2011, 2014a).

The fourth and not always the final option—there may be other unforeseen options lurking out there that are specific to a particular area—is related to the established transportation terminal where a substantial oil field is located far from other fields, from power or from natural gas. This option includes building a reverse pipeline from a refinery to the oil field and a crude pipeline.

A fifth option, which often lurks in the sidelines, is the concept of the gasification refinery (Chapter 13) that could well be the most popular option in the future refinery (Chapter 16). Installation of a gasification unit would offer a technology not currently available in a nongasification refinery operation. The refinery would produce synthesis gas (from the viscous feedstock) from which liquid fuels would be manufactured using Fischer-Tropsch synthesis technology (Chadeesingh, 2011; Speight, 2011, 2013a,b, 2014c; Luque and Speight, 2015).

Briefly, the gasifier differs from a combustor in that the amount of air or oxygen available inside the gasifier is carefully controlled so that only a relatively small portion of the fuel burns completely. The partial oxidation process provides the heat, and rather than combustion, most of the carbon-containing feedstock is chemically broken apart by the heat and pressure applied in the gasifier resulting in the chemical reactions that produce synthesis gas.

Synthesis gas (syngas) is the name given to a gas mixture that contains varying amounts of carbon monoxide and hydrogen generated by the gasification of a carbon-containing fuel to a gaseous product with a heating value. The composition of the synthesis gas will vary because of dependence upon the conditions in the gasifier and the type of feedstock. Examples include the gasification of coal or viscous feedstocks (Speight, 2011). Synthesis gas is used as a source of hydrogen or as an intermediate in producing hydrocarbon derivatives using the Fischer-Tropsch synthesis (Chadeesingh, 2011). Viscous feedstock and biomass cogasification is therefore one of the most technically and economically convincing energy provision possibilities for a potentially carbon-neutral economy.

As a final note on this issue, viscous feedstock refining by gasification (Chapter 13) could well be the answer to viscous feedstock upgrading. The integration of gasification technology into a refinery offers alternate processing options for viscous feedstocks. Thus, the refinery of the future will have a gasification section devoted to the conversion of viscous feedstocks (including blends with coal and/or biomass) to produce Fischer-Tropsch hydrocarbon derivatives. Many refineries already have gasification capabilities, but the trend will increase to the point (over the next two decades) where nearly all refineries will feel the need to construct a gasification section to handle viscous feedstocks.

Since a refinery is a group of integrated manufacturing plants (Fig. 7.1) that are selected to give a balanced production of salable products in amounts that are in accord with the demand for each, it is necessary to prevent the accumulation of nonsalable products; the refinery must be flexible and be able to change operations as needed. The complexity of heavy oil, extra heavy oil, and tar sand bitumen is emphasized insofar as the actual amounts of the products vary significantly from one crude oil to another (Speight, 2014a, 2017).

Cavitation processing has been proposed as an environmentally friendly (low-temperature) alternative to solvent dilution or heat treatment of bitumen and other viscous oils to reduce viscosity and, hence, improve transportability (Yen, 1997; Fomitchev-Zamilov, 2014; Mohapatra and Kirpalani, 2016; Kirpalani and Mohapatra, 2017). Over the range of sonication frequencies (20 kHz–1.1 MHz), the sonication frequency of 574 kHz with 50% power input resulted in low asphaltene content and lower viscosity suitable for improved transportability. Furthermore, comparisons of different conditions of sonication frequencies and power inputs were carried out to investigate the effect of ultrasound on properties of asphaltene (elemental analysis and metal content). The sonication treatment of bitumen under different conditions of frequencies and acoustic power decreased the H/C ratio. These results showed higher content of aromatic hydrogen and lower content of aliphatic hydrogen in bitumen treated under different conditions of sonication frequencies and intensity. The lowered metal content can be attributed to the reduced asphaltene formation as a result of sonication treatment of bitumen.

No matter what option is selected for viscous feedstock upgrading, there remains the need for hydrogen at some point of the refinery scenario in order to (i) facilitate the conversion of the viscous feedstock to lower-boiling product while reducing the potential yield of coke and (ii) to assure that the primary products can be treated to meet sales specification.

6. Hydrogen Production

Throughout the previous sections of this chapter, there have been several acknowledgments of a very important property of crude oil and crude oil products. And that is the hydrogen content or the use of hydrogen during refining in hydrotreating processes, such as desulfurization, and in hydroconversion processes, such as hydrocracking. Although the hydrogen recycle gas may contain up to 40% by volume of other gases (usually hydrocarbon derivatives), hydrotreater catalyst life is a strong function of hydrogen partial pressure. While this aspect is presented in more detail elsewhere (Chapter 15), it is worthy of comment in the context of this chapter.

A critical issue facing the modern refineries is the changing landscape in processing viscous feedstock into refined transportation fuels under an environment of increasingly more stringent clean fuel regulations; decreasing fuel oil demand; and increasingly viscous, more sour crude supply. Hydrogen network optimization is at the forefront of world refinery options to address clean fuel trends, to meet growing transportation fuel demands, and to continue to make a profit from their crudes (Speight, 2016). A key element of a hydrogen network analysis in a refinery involves the capture of hydrogen in its fuel streams and extending its flexibility and processing options. Thus, innovative hydrogen network optimization will be a critical factor influencing refinery future operating flexibility and profitability in a shifting world of crude feedstock supplies and ultralow-sulfur (ULS) gasoline and diesel fuel.

As hydrogen use has become more widespread in refineries, hydrogen production has moved from the status of a high-tech specialty operation to an integral feature of most refineries. This has been made necessary by the increase in hydrotreating and hydrocracking, including the treatment of progressively heavier feedstocks. In fact, the use of hydrogen in thermal processes is perhaps the single most significant advance in refining technology during the 20th century. The continued increase in hydrogen demand over the last several decades is a result of the conversion of crude oil to match changes in product slate and the supply of viscous, high-sulfur oil and in order to make lower-boiling, cleaner, and more salable products. There are also many reasons other than product quality for using hydrogen in processes adding to the need to add hydrogen at relevant stages of the refining process and, most importantly, according to the availability of hydrogen.

With the increasing need for *clean* fuels, the production of hydrogen for refining purposes requires a major effort by refiners. In fact, the trend to increase the number of hydrogenation (*hydrocracking* and/or *hydrotreating*) processes in refineries coupled with the need to process the heavier oils, which require substantial quantities of hydrogen for upgrading because of the increased use of hydrogen in hydrocracking processes, has resulted in vastly increased demands for this gas. The hydrogen demands can be estimated to a very rough approximation using API gravity and the extent of the reaction, particularly the hydrodesulfurization reaction (Speight, 2000, 2014a, 2017; Ancheyta and Speight, 2007). But accurate estimation requires equivalent process parameters and a thorough understanding of the nature of each process. Thus, as hydrogen production grows, a better understanding of the capabilities and requirements of a hydrogen plant becomes ever more important to overall refinery operations as a means of making the best use of hydrogen supplies in the refinery.

References

Abraham, H., 1945. Asphalts and Allied Substances. vol. I. Van Nostrand Scientific Publishers, New York.

Al-Asmi, K., Benayoune, M., Kezar, L., 1997. Flow behavior of heavy-crude water mixtures. Pet. Sci. Technol. 15, 647–665.

Al-Marshed, A., Hart, A., Leeke, G., Greaves, M., Wood, J., 2015. Optimization of heavy oil upgrading using dispersed nanoparticulate iron oxide as a catalyst. Energy Fuel 29, 6306–6316.

Ancheyta, J., Speight, J.G., 2007. Hydroprocessing Heavy Oils and Heavy Feedstocks. CRC Press, Taylor & Francis Group, Boca Raton, FL.

Castañeda, L.C., Muñoz, J.A.D., Ancheyta, J., 2014. Current situation of emerging technologies for upgrading of heavy oils. Catal. Today 220–222, 248–273.

Chadeesingh, R., 2011. The Fischer-Tropsch process. In: Speight, J.G. (Ed.), The Biofuels Handbook. The Royal Society of Chemistry, London, pp. 476–517 (Part 3, Chapter 5).

Charbonnier, R.P., Draper, R.G., Harper, W.H., Yates, A., 1969. Information circular IC 232. Analyses and Characteristics of Oil Samples From Alberta. Fuels Research Center, Mines Branch, Department of Energy Mines and Resources, Ottawa.

Davis, R.A., Patel, N.M., 2004. Refinery hydrogen management. Pet. Technol. Q. (Spring), 29–35.

Dickenson, R.L., Biasca, F.E., Schulman, B.L., Johnson, H.E., 1997. Refiner options for converting and utilizing heavy fuel oil. Hydrocarb. Process. 76 (2), 57–62.

Draper, R.G., Kowalchuk, E., Noel, G., Furimsky, E., 1977. Analyses and characteristics of crude oil samples performed between 1969 to 1976. Report ERP/ERL 77-59 (TR)Energy Research Program, Energy Research Laboratories, Canada Center for Mineral and Energy Technology (CANMET), Energy Mines and Resources Canada, Ottawa.

Fomitchev-Zamilov, M., 2014. Athabasca bitumen upgrading with hydrodynamic cavitation. In: Proceedings of the CIM Conference, Vancouver, British Columbia, Canada. Canadian Institute of Mining and Metallurgy, Westmounthttps://www.researchgate.net/publication/262638029_Athabasca_Bitumen_Upgrading_with_Hydrodynamic_Cavitation. Accessed 25 July 2018.

Forbes, R.J., 1958. A History of Technology, V. Oxford University Press, Oxford.

Gary, J.H., Handwerk, G.E., Kaiser, M.J., 2007. Petroleum Refining: Technology and Economics, fifth ed. CRC Press, Taylor & Francis Group, Boca Raton, FL.

Hart, A., 2014. A review of technologies for transporting heavy crude oil and bitumen via pipelines. J. Pet. Explor. Prod. Technol. 4, 327–336.

Hedrick, B.W., Seibert, K.D., Crewe, C., 2006. A new approach to heavy oil and bitumen upgrading. Report No. AM-06-29UOP LLC, Des Plaines, IL.

Hoiberg, A.J., 1960. Bituminous Materials: Asphalts, Tars and Pitches, I & II. Interscience Publishers, New York.

Hsu, C.S., Robinson, P.R. (Eds.), 2017. Handbook of Petroleum Technology. Springer International Publishing AG, Cham.

Khan, M.R., 1996. Rheological properties of heavy oils and heavy oil emulsions. Energy Sources 18, 385.

Khan, M.R., Patmore, D.J., 1998. Heavy oil upgrading processes. In: Speight, J.G. (Ed.), Petroleum Chemistry and Refining. Taylor & Francis, Washington, DC (Chapter 6).

Kirpalani, D.M., Mohapatra, D.P., 2017. Towards the development of cavitation technology for upgrading bitumen: viscosity change and chemical cavitation yield measurements. Pet. Sci. 14, 404–411.

Luque, R., Speight, J.G. (Eds.), 2015. Gasification for Synthetic Fuel Production: Fundamentals, Processes, and Applications. Woodhead Publishing, Elsevier, Cambridge.

McGrath, M.J., and Houde, E.J. 1999. Proceedings of the AIChE Spring Meeting, Houston, TX, March 14-18.

McKetta, J.J. (Ed.), 1992. Petroleum Processing Handbook. Marcel Dekker, New York.

Mehrota, A.K., 1990. Modelling the effect of temperature, pressure, and composition on the viscosity of crude oil mixtures. Ind. Eng. Chem. 29, 1574.

Mohapatra, D.P., Kirpalani, D.M., 2016. Bitumen heavy oil upgrading by cavitation processing: effect on asphaltene separation, rheology, and metal content. Appl. Petrochem. Res. 6, 107–115.

Motaghi, M., Shree, K., Krishnamurthy, S., 2010a. Consider new methods for bottom of the barrel processing, part 1: advanced methods use molecule management to upgrade heavy ends. Hydrocarb. Process. 88 (2), 35–38.

Motaghi, M., Shree, K., Krishnamurthy, S., 2010b. Consider new methods for bottom of the barrel processing, part 2: new methods of molecule management dictate the best economics when upgrading residuum. Hydrocarb. Process. 88 (3), 55–58.

Motaghi, M., Saxena, P., Ravi, R., 2010c. Partial upgrading of heavy oil reserves. Pet. Technol. Q. Q4, 55–64.

Parkash, S., 2003. Refining Processes Handbook. Gulf Professional Publishing, Elsevier, Amsterdam.

Penning, R.T., 2001. Petroleum refining: a look at the future. Hydrocarb. Process. 80 (2), 45–46.

RAROP, 1991. Noguchi (Chairman), T. (Ed.), RAROP Heavy Oil Processing Handbook. Research Association for Residual Oil Processing. Ministry of Trade and International Industry (MITI), Tokyo.

Shih, S.S., Oballa, M.C. (Eds.), 1991. Tar sand upgrading technology.Symposium Series No. 282. American Institute for Chemical Engineers. NY.

Speight, J.G., 1990. In: Speight, J.G. (Ed.), Fuel Science and Technology Handbook. Marcel Dekker, New York (Chapters 12-16).

Speight, J.G., 1994. Chemical and physical studies of petroleum asphaltene constituents. In: Yen, T.F., Chilingarian, G.V. (Eds.), Asphaltene constituents and Asphalts. I. Developments in Petroleum Science. In: vol. 40. Elsevier, Amsterdam (Chapter 2).

Speight, J.G., 2000. The Desulfurization of Heavy Oils and Residua, second ed. Marcel Dekker, New York.

Speight, J.G., 2001. Handbook of Petroleum Analysis. John Wiley & Sons, Hoboken, NJ.

Speight, J.G., 2009. Enhanced Recovery Methods for Heavy Oil and Tar Sands. Gulf Publishing Company, Houston.

Speight, J.G., 2011. The Refinery of the Future. Gulf Professional Publishing, Elsevier, Oxford.

Speight, J.G., 2013a. Oil Sand Production Processes. Gulf Professional Publishing, Elsevier, Oxford.

Speight, J.G., 2013b. The Chemistry and Technology of Coal, third ed. CRC Press, Taylor & Francis Group, Boca Raton, FL.

Speight, J.G., 2013c. Heavy and Extra Heavy Oil Upgrading Technologies. Gulf Professional Publishing, Elsevier, Oxford.

Speight, J.G., 2014a. The Chemistry and Technology of Petroleum, fifth ed. CRC Press, Taylor & Francis Group, Boca Raton, FL.

Speight, J.G., 2014b. High Acid Crudes. Gulf Professional Publishing, Elsevier, Oxford.

Speight, J.G., 2014c. Gasification of Unconventional Feedstocks. Gulf Professional Publishing, Elsevier, Oxford.

Speight, J.G., 2015a. Handbook of Petroleum Product Analysis, second ed. John Wiley & Sons, Hoboken, NJ.

Speight, J.G., 2015b. Fouling in Refineries. Gulf Professional Publishing, Elsevier, Oxford.

Speight, J.G. 2016. Hydrogen in refineries. In: Hydrogen Science and Engineering: Materials, Processes, Systems, and Technology. D. Stolten and B. Emonts (Editors). Wiley-VCH Verlag GmbH & Co, Weinheim, Germany. (Chapter 1). Page 3-18. 2016.

Speight, J.G., 2017. Handbook of Petroleum Refining. CRC Press, Taylor & Francis Group, Boca Raton, FL.

Speight, J.G., Moschopedis, S.E., 1979. The production of low-sulfur liquids and coke from Athabasca bitumen. Fuel Process. Technol. 2, 295.

Urdahl, O., Fredheim, A.O., Loken, K.R., 1997. Viscosity measurements of water-in-oil emulsions under flowing conditions: a theoretical and practical approach. Colloids Surf. 123–124, 623.

Villasana, Y., Luis-Luis, M.A., Méndez, F.J., Labrador, H., Brito, J.L., 2015. Upgrading and hydrotreating heavy oils and residua. In: Sharma, U.C., Prasad, R., Sivakumar, S. (Eds.), Energy Science and Technology. Volume 3: Oil and Natural Gas. Studium Press, Houston, TX (Chapter 9).

Yaghi, B.M., Al-Bemani, A., 2002. Heavy crude oil viscosity reduction for pipeline transportation. Energy Sources 24 (2), 93–102.

Yen, T.F., 1997. Upgrading through cavitation and surfactant. In: Proceedings of the 15th World Petroleum Congress, Forum 17. John Wiley & Sons, Hoboken, NJ.

Yoshida, R., Miyazawa, M., Ishiguro, H., Itoh, S., Haraguchi, K., Nagesh, H., Narita, H., Yoshida, T., Maekawa, Y., Mitarai, Y., 1997. Chemical structure changes in cold lake oil-sand bitumen and catalytic activities during catalytic hydrotreatment. Fuel Process. Technol. 51, 195–203.

Zaki, N.N., 1997. Surfactant stabilized crude oil-in-water emulsions for pipeline transportation of viscous crude oils. Colloids Surf. 125, 19.

Chapter 8

Thermal Cracking Processes

1. Introduction

Distillation has remained a major refinery process and a process to which almost every crude oil that enters the refinery is subjected (Speight, 2014, 2017). However, not all crude oils yield the same distillation products nor are all oils amenable to the distillation process as the first step in refining. In fact, the nature of the refinery feedstock—especially heavy oil, extra heavy oil, and tar sand bitumen—dictates the processes that may be required for refining, and balancing product yield with demand is a necessary part of refinery operations (Parkash, 2003; Gary et al., 2007; Speight, 2014, 2017; Hsu and Robinson, 2017).

However, the balancing of product yield and market demand, without the manufacture of large quantities of fractions having low commercial value, has long required processes for the conversion of hydrocarbon derivatives of one molecular weight range and/or structure into some other molecular weight range and/or structure. Basic processes for this are still the so-called cracking processes in which relatively high-boiling constituents are cracked, that is, thermally decomposed into lower-molecular-weight, smaller, lower-boiling molecules, although reforming alkylation, polymerization, and hydrogen-refining processes have wide applications in making premium-quality products (Parkash, 2003; Gary et al., 2007; Hsu and Robinson, 2017; Speight, 2014, 2017).

After 1910 and the conclusion of World War I, the demand for automotive (and other) fuels began to outstrip the market requirements for kerosene, and refiners, needing to stay abreast of the market pull, were pressed to develop new technologies to increase gasoline yields. There being finite amounts of straight-run distillate fuels in viscous feedstocks, refiners had, of necessity, the urgency to develop processes to produce additional amounts of these fuels. The conversion of coal and oil shale to liquid through the agency of cracking had been known for centuries, and the production of various distillate fuels from crude oil through thermal methods had been known since at least the inception of Greek fire in earlier centuries.

Heavy Oil Recovery and Upgrading. https://doi.org/10.1016/B978-0-12-813025-4.00008-8

The discovery that higher-molecular-weight (higher-boiling) materials could be decomposed to lower-molecular-weight (lower-boiling) products was used to increase the production of kerosene and was called *cracking distillation*. In the process, a batch of crude oil was heated until most of the kerosene was distilled from it and the overhead material became dark in color. At this point, the still fires were lowered, the rate of distillation decreased, and the viscous oil was held in the hot zone, during which time some of the high-molecular-weight hydrocarbon derivatives were decomposed and rearranged into lower-molecular-weight products. After a suitable time, the still fires were increased, and distillation continued in the normal way. The overhead product, however, was a lower-boiling fraction suitable for kerosene instead of the higher-boiling fraction that would otherwise have been produced. Thus, it was not surprising that such technologies were adapted for the fledgling crude oil industry.

Thus, thermal cracking is a high-temperature process (in some cases, pressure is also applied) in which the constituents of petroleum are decomposed under the iinfluence of heat to break molecular bonds and yield products having a lower molecular weight than the feedstock. The process is one of several cracking methods used in the petroleum industry to process crude oil and other petroleum products for commercial use, and the major process variables are (i) feedstock type, (ii) temperature, (iii) pressure, and (iv) residence time in the hot zone, which need to be considered to maximize the yield of distillates and minimize the yield of coke (Parkash, 2003; Gary et al., 2007; Speight, 2014, 2017; Hsu and Robinson, 2017).

The earliest processes, which involved thermal cracking, consisted of heating heavier oils (for which there was a low market requirement) in pressurized reactors and thereby cracking or splitting their large molecules into the smaller ones that form the low-boiling, more valuable fractions such as gasoline, kerosene, and low-boiling industrial fuels. Gasoline manufactured by thermal cracking processes performed better in automobile engines than gasoline derived from straight distillation of crude oil. The development of more powerful aircraft engines in the late 1930s gave rise to a need to increase the combustion characteristics of gasoline to improve engine performance. Thus, during World War II and the late 1940s, improved refining processes involving the use of catalysts led to further improvements in the quality of transportation fuels and further increased their supply.

The 1950s and 1960s brought a large-scale demand for jet fuel and high-quality lubricating oils. The continuing increase in demand for crude oil products also heightened the need to process a wider variety of crude oils into high-quality products. Catalytic reforming of naphtha (Parkash, 2003; Gary et al., 2007; Speight, 2014, 2017; Hsu and Robinson, 2017) replaced the earlier thermal reforming process and became the leading process for upgrading fuel qualities to meet the needs of higher-compression engines. Hydrocracking, a catalytic cracking process conducted in the presence of hydrogen (Parkash, 2003;

Gary et al., 2007; Speight, 2014, 2017; Hsu and Robinson, 2017), was developed to be a versatile manufacturing process for increasing the yields of either gasoline or jet fuels.

In the early stages of thermal cracking process development, processes were generally classified as either liquid phase, high pressure (350–1500 psi), and low temperature (400–510°C, 750–950°F) or vapor phase, low pressure (<200 psi), and high temperature (540–650°C, 1000–1200°F). In reality, the processes were mixed phase with no process really being entirely liquid or vapor phase, but the classification (like many classifications of crude oil and related areas) was still used as a matter of convenience (Parkash, 2003; Gary et al., 2007; Hsu and Robinson, 2017; Speight, 2014).

The early processes were classified as liquid-phase processes and had the following advantages over vapor-phase processes: (i) large yields of gasoline of moderate octane number, (ii) low gas yields, (iii) ability to use a wide variety of feedstocks, (iv) long cycle time due to low coke formation, and (v) flexibility and ease of control. However, the vapor-phase processes had the advantages of operation at lower pressures and the production of a higher-octane gasoline due to the increased production of olefin derivatives and low-boiling aromatic derivatives. However, there were many disadvantages that curtailed the development of vapor-phase processes: (i) temperatures were required that the steel alloys available at the time could not tolerate, (ii) there were high gas yields and resulting losses since the gases were normally not recovered, and (iii) there was a high production of olefinic compounds that created naphtha with poor stability (increased tendency to form undesirable gum) (Mushrush and Speight, 1995; Speight, 2014) that required subsequent treating of the gasoline to stabilize it against gum formation. The vapor-phase processes were not considered suitable for the production of large quantities of gasoline but did find application in petrochemical manufacture due to the high concentration of olefin derivatives produced.

It is generally recognized that the most important part of any refinery, after the distillation units, is the gasoline (and liquid fuels) manufacturing facilities; other facilities are added to manufacture additional products as indicated by technical feasibility and economic gain. More equipment is used in the manufacture of gasoline, the equipment is more elaborate, and the processes are more complex than for any other product. Among the processes that have been used for liquid fuel production are thermal cracking, catalytic cracking, thermal reforming, catalytic reforming, polymerization, alkylation, coking, and distillation of fractions directly from crude oil (Fig. 8.1). Each of these processes may be carried out in a number of ways, which differ in details of operation, essential equipment, or both (Parkash, 2003; Gary et al., 2007; Hsu and Robinson, 2017; Speight, 2014).

Thermal processes are essentially processes that decompose, rearrange, or combine hydrocarbon molecules by the application of heat. The major variables involved are feedstock type, time, temperature, and pressure and, as such, are

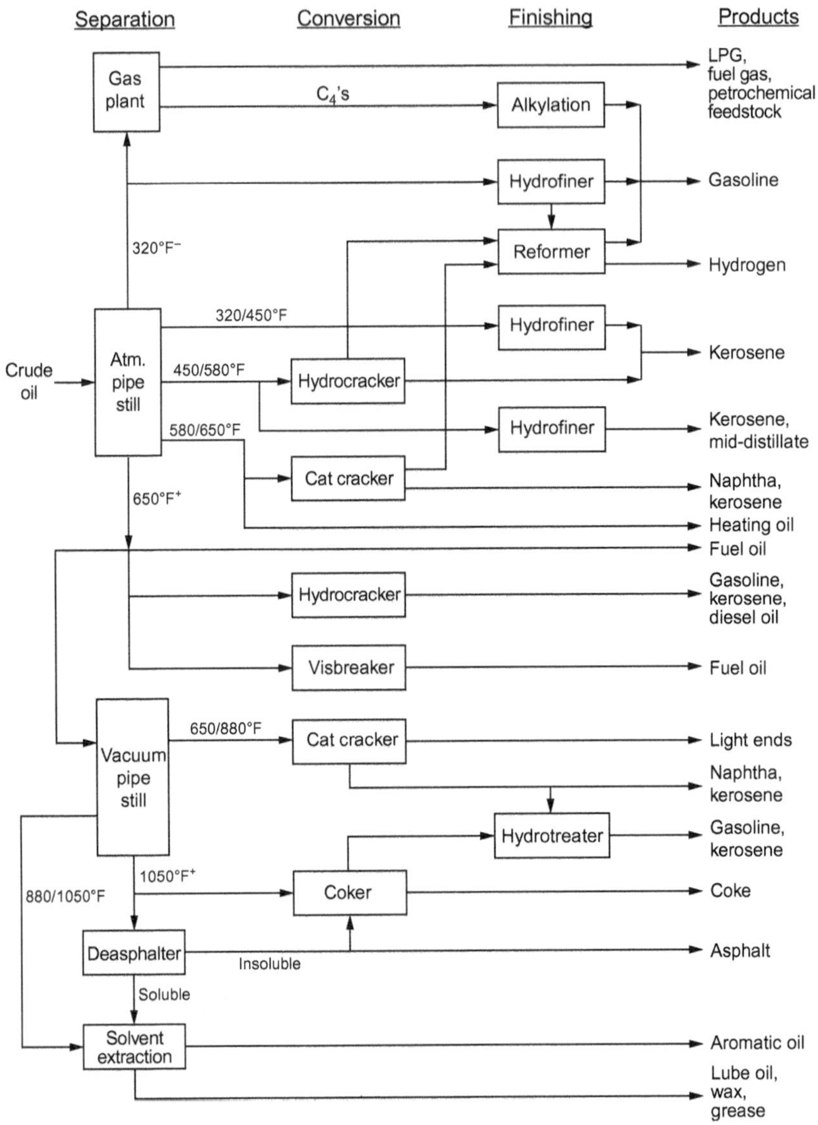

FIG. 8.1 Schematic representation of a refinery showing placement of the various thermal cracking units.

usually considered in promoting cracking (thermal decomposition) of the heavier molecules to lower-molecular-weight products and in minimizing coke formation. Thus, one of the earliest processes used in the crude oil industry is the noncatalytic conversion of higher-boiling crude oil stocks into lower-boiling products, known as *thermal cracking*.

The thermal decomposition (cracking) of high-molecular-weight hydrocarbon derivatives to lower molecular weight and normally more valuable hydrocarbon derivatives has long been practiced in the crude oil refining industry. Although catalytic cracking has generally replaced thermal cracking, noncatalytic cracking processes using high temperature to achieve the decomposition are still in operation. In several cases, thermal cracking processes to produce specific desired products or to dispose of specific undesirable charge streams are being operated or installed. The purpose of this chapter is to provide basic information to assist the practicing engineer/crude oil refiner to (i) determine if a particular thermal cracking process would be suitable for a specific application and could fit into the overall operation, (ii) develop a basic design for a thermal cracking process, and (iii) operate an existing or proposed process.

Conventional *thermal cracking* is the thermal decomposition, under pressure, of high-molecular-weight constituents (higher molecular weight and higher boiling than gasoline constituents) to form lower-molecular-weight (and lower-boiling) species. Thus, the thermal cracking process is designed to produce gasoline from higher-boiling charge stocks, and any unconverted or mildly cracked charge components (compounds that have been partially decomposed but are still higher boiling than gasoline) are usually recycled to extinction to maximize gasoline production. A moderate quantity of low-boiling hydrocarbon gases is also formed. As thermal cracking proceeds, reactive unsaturated molecules are formed that continue to react and can ultimately create higher-molecular-weight species that are relatively hydrogen-deficient and readily form coke. Thus, they cannot be recycled without excessive coke formation and are therefore removed from the system as cycle fuel oil.

When crude oil fractions are heated to temperatures over 350°C (660°F), the rates of the thermal cracking cause thermal decomposition to proceed at significant rates (Speight, 2000, 2014). Thermal decomposition does not require the addition of catalyst, and therefore, this approach is the oldest technology available for residue conversion. The severity of thermal processing determines the conversion and the product characteristics. Thermal treatment of residues ranges from mild treatment for reduction of viscosity to *ultrapyrolysis* (high-temperature cracking at very short residence time) for complete conversion to olefin derivatives and low-boiling products (light ends). The higher the temperature, the shorter the time required to achieve a given conversion but, in many cases, with a change in the chemistry of the reaction. The severity of the process conditions is the combination of reaction time and temperature to achieve a given conversion.

Thermal reactions, however, can give rise to a variety of different reactions so that selectivity for a given product changes with temperature and pressure. The mild- and high-severity processes are frequently used for processing of heavy oil, extra heavy oil, and tar sand bitumen, while conditions similar to *ultrapyrolysis* (high temperature and very short residence time) are only used commercially for cracking ethane, propane, butane, and low-boiling distillate feeds to produce ethylene and higher-molecular-weight olefin derivatives. Sufficiently, high temperatures convert oils entirely to gases and coke; cracking conditions are controlled to produce as much as possible of the desired product, which is usually gasoline but may be cracked gases for petrochemicals or a lower-viscosity oil for use as a fuel oil. The feedstock, or cracking stock, may be almost any fraction obtained from crude oil, but the greatest amount of cracking is carried out on gas oils, a term that refers to the portion of crude oil boiling between the fuel oils (kerosene and/or stove oil) and the residuum. Heavy oil, extra heavy oil, and tar sand bitumen are also cracked, but the processes are somewhat different from those used for gas oils.

Thus, thermal conversion processes are designed to increase the yield of lower-boiling products obtainable from crude oil either directly (by means of the production of gasoline components from higher-boiling feedstocks) or indirectly (by production of olefin derivatives and the like, which are precursors of the gasoline components). These processes may also be characterized by the physical state (liquid and/or vapor phase) in which the decomposition occurs. The state depends on the nature of the feedstock and conditions of pressure and temperature.

From the chemical viewpoint, the products of cracking are very different from those obtained directly from crude oil—the products are nonindigenous to crude oil because they are created from crude oil by application of an external force (heat). When a twelve-carbon atom hydrocarbon typical of straight-run gas oil is cracked or broken into two parts, one may be a six-carbon paraffin hydrocarbon and the other a six-carbon olefin hydrocarbon:

$$CH_3(CH_2)_{10}CH_3 \rightarrow CH_3(CH_2)_4CH_3 + CH_2 = CH(CH_2)_3CH_3$$

The paraffin may be the same as is found in straight-run (distilled) gasoline, but the olefin is new. Furthermore, the paraffin has an octane number approaching zero, but the olefin has an octane number approaching 100. Hence, naphtha formed by cracking *(cracked gasoline)* has a higher-octane number than straight-run gasoline. In addition to a large variety of olefin derivatives, cracking produces high-octane aromatic and branched-chain hydrocarbon derivatives in higher proportions than are found in straight-run gasoline. Diolefin derivatives are produced but in relatively small amounts; they are undesirable in gasoline because they readily combine to form gum. The overall complexity of such a reaction is illustrated by the following equations in which the products are subject to the position of bond scission within the starting molecule:

$$CH_3(CH_2)_{10}CH_3 \rightarrow CH_3(CH_2)_8CH_3 + CH_2 = CH_2$$

$$CH_3(CH_2)_{10}CH_3 \rightarrow CH_3(CH_2)_7CH_3 + CH_2 = CHCH_3$$

$$CH_3(CH_2)_{10}CH_3 \rightarrow CH_3(CH_2)_6CH_3 + CH_2 = CHCH_2CH_3$$

$$CH_3(CH_2)_{10}CH_3 \rightarrow CH_3(CH_2)_5CH_3 + CH_2 = CH(CH_2)_2CH_3$$

$$CH_3(CH_2)_{10}CH_3 \rightarrow CH_3(CH_2)_4CH_3 + CH_2 = CH(CH_2)_3CH_3$$

$$CH_3(CH_2)_{10}CH_3 \rightarrow CH_3(CH_2)_3CH_3 + CH_2 = CH(CH_2)_4CH_3$$

$$CH_3(CH_2)_{10}CH_3 \rightarrow CH_3(CH_2)_2CH_3 + CH_2 = CH(CH_2)_5CH_3$$

$$CH_3(CH_2)_{10}CH_3 \rightarrow CH_3CH_2CH_3 + CH_2 = CH(CH_2)_6CH_3$$

$$CH_3(CH_2)_{10}CH_3 \rightarrow CH_3CH_3 + CH_2 = CH(CH_2)_7CH_3$$

$$CH_3(CH_2)_{10}CH_3 \rightarrow CH_4 + CH_2 = CH(CH_2)_8CH_3$$

Furthermore, the primary products (unless the reaction conditions are monitored carefully) will react further to yield secondary, tertiary, and even quaternary products.

The hydrocarbon derivatives with the least thermal stability are the paraffin derivatives, and the olefin derivatives produced by the cracking of paraffin derivatives are also reactive. Cycloparaffin derivatives (naphthene derivatives) are less easily cracked, their stability depending mainly on any side chains present, but ring splitting may occur, and dehydrogenation can lead to the formation of unsaturated naphthene derivatives and aromatic derivatives. Aromatic derivatives are the most stable (*refractory*) hydrocarbon derivatives, the stability depending on the length and stability of side chains. Very severe thermal cracking of high-molecular-weight constituents can result to the production of excessive amounts of coke.

The higher-boiling oils produced by cracking are low-viscosity gas oil, high-viscosity gas oil, and a residual oil, which in the case of thermal cracking is usually (erroneously) called tar and in the case of catalytic cracking is called *cracked fractionator bottoms*. The residual oil may be used as heavy fuel oil, and gas oils from catalytic cracking are suitable as domestic and industrial fuel oils or as diesel fuels if blended with straight-run gas oils. Gas oils from thermal cracking must be mixed with straight-run (distilled) gas oils before they become suitable for domestic fuel oils and diesel fuels.

The gas oils produced by cracking are an important source of gasoline, and in a once-through cracking operation, all of the cracked material is separated into products and may be used as such. However, the cracked gas oils are more resistant to cracking (more refractory) than straight-run gas oils but can still be cracked to produce more gasoline. This is done in a recycling operation in which the cracked gas oil is combined with fresh feed for another trip through the cracking unit. The operation may be repeated until the cracked gas oil is almost completely decomposed (*cracking to extinction*) by recycling (*recycling to extinction*) the higher-boiling product, but it is more usual to withdraw part of the cracked gas oil from the system according to the need for fuel oils. The extent to which recycling is carried out affects the amount or yield of cracked gasoline resulting from the process.

The gases formed by cracking are particularly important because of their chemical properties and their quantity. Only relatively small amounts of paraffinic gases are obtained from crude oil, and these are chemically inactive. Cracking produces both paraffinic gases (e.g., propane, C_3H_8) and olefinic gases (e.g., propene, C_3H_6); the latter are used in the refinery as the feed for polymerization plants where high-octane polymer gasoline is made. In some refineries, the gases are used to make alkylate, a high-octane component for aviation gasoline and for motor gasoline. In particular, the cracked gases are the starting points for many petrochemicals (Speight, 2014).

The importance of solvents in coking has been recognized for many years (Langer et al., 1961, 1962), but their effects have often been ascribed to hydrogen-donor reactions rather than phase behavior. The separation of the phases depends on the solvent characteristics of the liquid. Addition of aromatic solvents will suppress phase separation, while paraffin derivatives will enhance separation. Microscopic examination of coke particles often shows evidence for the presence of a *mesophase*, spherical domains that exhibit the anisotropic optical characteristics of liquid crystal. This phenomenon is consistent with the formation of a second liquid phase; the mesophase liquid is denser than the rest of the hydrocarbon, has a higher surface tension, and likely wets metal surfaces better than the rest of the liquid phase. The mesophase characteristic of coke diminishes as the liquid phase becomes more compatible with the aromatic material (Speight, 1990, 2000).

Thermal cracking processes for the conversion of higher-boiling feedstocks to naphtha and kerosene are an obsolete process, since the antiknock requirement of modern automobile engines has outstripped the ability of the thermal cracking process to supply an economical source of high-quality fuel. New units are rarely installed, but a few refineries still operate thermal cracking units built in previous decades.

Feedstock properties such as carbon residue (potential coke formation); sulfur content (hydrogen needs for desulfurization); metallic constituents (catalyst rejuvenation); nitrogen content (catalyst rejuvenation); naphthenic or paraffinic character through use of a characterization factor or similar indicator (potential for cracking in different ways to give different products); and, to a lesser extent, asphaltene content (coke formation) since this last parameter is related to several of the previous parameters. *Process parameters* such as time-temperature-pressure relationships (distillate yield and coke yield), feedstock recycle ratio (distillate yield and coke yield plus overall conversion), and coke formation (the lack of liquid production when liquids are the preferred products). *Equipment parameters* such as batch operation, semicontinuous operation, or continuous operation (residence time and contact with the catalyst, if any); coke removal; and unit capacity that also dictates residence time. However, it is not the purpose of this text to present the detail of these three categories, but they should be borne in mind when considering and deciding upon the potential utility of any process presented throughout this and subsequent chapters.

In summary, the cracking of crude oil constituents can be visualized as a series of simple thermal conversions (Parkash, 2003; Gary et al., 2007;

Speight, 2014, 2017; Hsu and Robinson, 2017). The reactions involve the formation of transient highly reactive species that may react further in several ways to produce the observed product slate (Germain, 1969; Speight, 2000, 2014). Thus, even though chemistry and physics can be used to explain feedstock reactivity, the main objective of feedstock evaluation (Parkash, 2003; Gary et al., 2007; Speight, 2014, 2015a,b, 2017, Hsu and Robinson, 2017) is to allow a degree of predictability of feedstock behavior in thermal processes (Speight, 2000, 2014). And in such instances, chemical principles must be combined with engineering principles to understand feedstock processability and predictability of feedstock behavior. In the simplest sense, process planning can be built on an understanding of the following three parameter groups: (i) feedstock properties, (ii) process parameters, and (iii) equipment parameters (Speight, 2015a).

2. History

As the demand for gasoline increased with the onset of automobile sales, the issue of how to produce more gasoline from less crude oil was solved in 1913 by the incorporation of cracking units into refinery operations in which fractions higher boiling than gasoline were converted into gasoline by thermal decomposition.

The origins of cracking are unknown. There are records that illustrate the use of naphtha in *Greek fire* almost 2000 years ago (Speight, 2014), but whether the naphtha was produced naturally by distillation or by cracking distillation is not clear. Cracking was used commercially in the production of oils from coal and oil shale before the beginning of the modern crude oil industry. From this, the discovery that the higher-boiling materials from crude oil could be decomposed to lower-molecular-weight products was used to increase the production of kerosene and was called *cracking distillation* (Kobe and McKetta, 1958).

The precise origins of the modern version of cracking distillation, as applied in the modern crude oil industry, are also unknown. However, it is essential to recognize that the production of volatile product by the destructive distillation of wood and coal was known for many years, if not decades or centuries, before the birth of the modern crude oil industry. Indeed, the production of *spirits of fire* (i.e., naphtha, the flammable constituent of *Greek fire*) was known from early times. The occurrence of bitumen at Hit (Mesopotamia) that was used as mastic by the Assyrians was further developed for use in warfare through the production of naphtha by destructive distillation.

At the beginning of the 20th century, the yields of gasoline and kerosene fractions were usually markedly increased by means of *cracking distillation*, but the technique was not entirely suitable for gasoline production. As the need for gasoline arose, the necessity of prolonging the cracking process became apparent and led to a process known as *pressure cracking*, which is a batch operation in which feedstock was heated to ~425°C (800°F) in stills (shell stills) especially reinforced to operate at pressures as high as 95 psi for 24 h.

Distillation was then started, and during the next 48–72 h, a low-boiling distillate was obtained, which was treated with sulfuric acid to remove unstable gum-forming components (olefin derivatives and diolefin derivatives) and then redistilled to produce a naphtha (cracked gasoline, boiling range < 205°C, < 400°F) and residual fuel oil (Stephens and Spencer, 1956).

The Burton cracking process for the large-scale production of naphtha (cracked gasoline) was first used in 1912. The process employed batch distillation in horizontal shell stills and operated at ~400°C (ca. 750°F) and 75–95 psi and was the first successful method of converting higher-boiling feedstocks to gasoline. However, batch heating gas oil was considered inefficient, and during the years 1914–22, a number of successful continuous cracking processes were developed. In these processes, gas oil was continuously pumped through a unit that heated the gas oil to the required temperature and held it for a time under pressure, and then, the cracked product was discharged into a distillation unit for separation into gases, gasoline, gas oil, and cracked residuum (often called *tar*).

The *tube-and-tank cracking process* is typical of the early continuous cracking processes. Gas oil, preheated by exchange with the hot products of cracking, was pumped into the cracking coil (up to several hundred feet long) that lined the inner walls of a furnace where oil or gas burners raised the temperature of the gas oil to 425°C (800°F). The hot gas oil passed from the cracking coil to a reaction chamber (soaker) where the gas oil was held under these temperature and pressure conditions until the cracking reactions are completed. The cracking reactions formed coke, which over the course of several days filled the soaker. The gas-oil stream was then switched to a second soaker, and drilling operations similar to those used in drilling an oil well cleaned out the first soaker. The cracked material (other than coke) left the onstream soaker to enter an evaporator (tar separator) maintained under a much lower pressure than the soaker, where, because of the lower pressure, all the cracked material except the tar became vaporized. The vapor left the top of the separator, where it was distilled into separate fractions: gases, gasoline, and gas oil. The tar that was deposited in the separator was pumped out for use as asphalt or as a heavy fuel oil.

Shortly thereafter, in 1921, a more advanced thermal cracking process that operated at 750–860°F (400–460°C) was developed (*Dubbs process*). In the process, a reduced crude (such as an atmospheric residuum or a toped crude oil) was the feedstock, and the process also employed the concept of recycling in which the gas oil was combined with fresh (viscous) feedstock for further cracking. In a typical application of conventional thermal cracking (Fig. 8.2), the feedstock (reduced crude, i.e., residuum or flashed crude oil) is preheated by direct exchange with the cracked products in the fractionating columns. Cracked gasoline and middle distillate fractions were removed from the upper section of the column. Low-boiling and high-boiling distillate fractions were removed from the lower section and were pumped to separate heaters. Higher temperatures were used to crack the more refractory low-boiling distillate

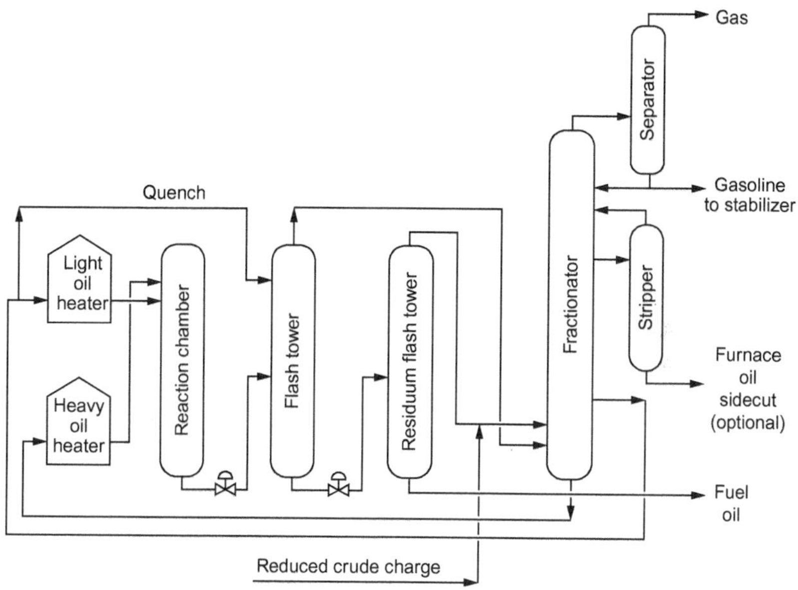

FIG. 8.2 A thermal cracking unit.

fraction. The streams from the heaters were combined and sent to a soaking chamber, where additional time is provided to complete the cracking reactions. The cracked products were then separated in a low-pressure flash chamber where a heavy fuel oil is removed as the nonvolatile product. The remaining cracked products were sent to the fractionating columns.

As refining technology evolved throughout the 20th century, the feedstocks for cracking processes became the residuum or gas oil from a distillation unit. In addition, the residual oil produced as the end product of distillation processes and even some of the higher-boiling crude oil constituents often contain substantial amounts of asphaltic materials, which preclude use of the residuum as fuel oils or lubricating stocks (Speight, 2000, 2014, 2017; Parkash, 2003; Gary et al., 2007; Hsu and Robinson, 2017).

However, subjecting the heavy oil, extra heavy oil, and tar sand bitumen directly to thermal processes has become economically advantageous since, on the one hand, the end result is the production of lower-boiling products but, on the other hand, the asphaltene constituents and the resin constituents that are concentrated in heavy oil, extra heavy oil, and tar sand bitumen are precursors to high yields of thermal coke (i.e., coke formed in noncatalytic processes) (Parkash, 2003; Gary et al., 2007; Speight, 2014, 2017; Hsu and Robinson, 2017). Although new thermal cracking units are now under development for heavy oil, extra heavy oil, and tar sand bitumen (Speight, 2008, 2014), processes that can be regarded as having evolved from the original concept of thermal cracking are visbreaking and the various coking processes (Speight, 2014).

In the modern refinery, thermal cracking processes offer attractive methods of conversion of viscous feedstocks because they enable low operating pressure, while involving high operating temperature, without requiring expensive catalysts. Currently, the most widely operated viscous feedstock conversion processes are visbreaking and delayed coking. And these are still attractive processes for refineries from an economic point of view and will continue to be so well into the 21st century (Dickenson et al., 1997; Gary et al., 2007; Speight, 2011).

The majority of regular thermal cracking processes use temperatures of 455–540°C (850–1005°F) and pressures of 100–1000 psi. This approach is the oldest technology available for residue conversion, and the severity of thermal processing determines the conversion and the product characteristics. Thermal treatment of viscous feedstocks ranges from mild treatment for reduction of viscosity to *ultrapyrolysis* (high-temperature cracking at very short residence time) for better conversion to overhead products. A higher temperature requires a shorter time to achieve a given conversion, but in many cases, there is *a change in the chemistry of the reaction*, so merely raising the temperature does not necessarily accomplish the same goals in terms of product slate and product yields.

Low pressures (<100 psi) and temperatures in excess of 500°C (930°F) tend to produce lower-molecular-weight hydrocarbon derivatives than those produced at higher pressures (400–1000 psi) and at temperatures below 500°C (930°F). The reaction time is also important; low-boiling feeds (gas oils) and recycle oils require longer reaction times than the readily cracked high-boiling residues. Mild cracking conditions (defined here as a low conversion per cycle) favor a high yield of gasoline components with low gas production and decreased coke production, but the gasoline quality is not high, whereas more severe conditions give increased gas production and increased coke production and reduced gasoline yield (but of higher quality). With limited conversion per cycle, the heavier residues must be recycled. However, the recycled oils become increasingly refractory upon repeated cracking, and if they are not required as a fuel oil stock, they may be subjected to a coking operation to increase gasoline yield or refined by means of a hydrogen process.

Thus, thermal cracking processes offer attractive methods of feedstock conversion at low operating pressure without requiring expensive catalysts. Currently, the widest operated residuum conversion processes are visbreaking, delayed coking, and fluid coking that are still attractive processes for refineries that process viscous feedstocks such as heavy oil, extra heavy oil, and tar sand bitumen. Currently, the widest operated residuum conversion processes are visbreaking and delayed coking, and other processes that have also received some attention for bitumen upgrading include partial upgrading (a form of thermal deasphalting), flexicoking, Eureka process, and various hydrocracking processes (Parkash, 2003; Gary et al., 2007; Speight, 2014, 2017; Hsu and Robinson, 2017).

3. Visbreaking

Visbreaking (*viscosity reduction* or *viscosity breaking*), a mild form of thermal cracking, was developed in the late 1930s to produce more desirable and valuable products (Parkash, 2003; Gary et al., 2007; Joshi et al., 2008; Stell et al., 2009a,b; Carrillo and Corredor, 2013; Speight, 2014, 2017). The processes that can be regarded as having evolved from the original concept of thermal cracking are visbreaking and the various coking processes (Table 8.1). The process is a relatively mild, liquid-phase thermal cracking process used to convert high-viscosity feedstocks to lower-viscosity fractions suitable for use in heavy fuel oil. This ultimately results in less production of fuel oil since less cutter stock (low-viscosity diluent) is required for blending to meet fuel oil viscosity specifications. The cutter stock no longer required in fuel oil may then be used in more valuable products. A secondary benefit from the visbreaking operation is the production of gas oil and gasoline streams that usually have higher product values than the visbreaker charge. Visbreaking produces a small quantity of low-boiling hydrocarbon gases and a larger amount of gasoline and remains a process of promise for viscous feedstocks (Stark and Falkler, 2008; Stark et al., 2008).

The process can also be used as the first step in upgrading viscous feedstocks (Schucker, 2003). In such a process, the viscous feedstock is first thermally cracked using visbreaking or hydrovisbreaking technology to produce a product that is lower in molecular weight and boiling point than the feed. The product is then deasphalted using an alkane solvent at a solvent-to-feed ratio of <2 wherein separation of solvent and deasphalted oil from the asphaltenes is achieved through the use of a two-stage membrane separation system in which the second stage is a centrifugal membrane.

Visbreaking, unlike conventional thermal cracking, typically does not employ a recycle stream. Conditions are too mild to crack a gas-oil recycle stream, and the unconverted viscous feedstock stream, if recycled, would cause excessive heater coking. The boiling range of the product viscous feedstock stream is extended by visbreaking so that low-boiling gas oil and high-boiling gas oil can be fractionated from the product viscous feedstock stream, if desired. In some present applications, the high-boiling gas-oil stream is recycled and cracked to extinction in a separate higher-temperature heater with the production of products that are lower boiling than the original feedstock (Ballard et al., 1992; Parkash, 2003; Negin and Van Tine, 2004; Gary et al., 2007; Speight, 2014, 2017; Hsu and Robinson, 2017). Low residence times are required to avoid polymerization and coking reactions, although additives can help to suppress coke deposits on the tubes of the furnace.

Visbreaking conditions range from 455°C to 510°C (850–950°F) at a short residence time and from 50 to 300 psi at the heating coil outlet. It is the short residence time that brings to visbreaking the concept of being a mild thermal reaction in contrast to, for example, the delayed coking process where residence

TABLE 8.1 Summary of the Various Thermal Cracking Processes

Thermal cracking
Prime purpose: conversion
Semicontinuous process
High conversion
Process configuration: various, depending on feedstock
Visbreaking
Prime purpose: viscosity reduction
Low conversion (10%) to products boiling <220°C (430°F)
Mild (470–495°C; 880–920°F) heating at pressures of 50–200 psi
Thermal reactions quenched before going to completion
Heated coil or drum (soaker)
Delayed coking
Prime purpose: conversion to distillates
Complete conversion of the feedstock
Moderate—short residence time in hot zone (480–515°C; 900–960°F)
Pressures on the order of 90 psi
Thermal reactions allowed to proceed to completion out if hot zone
Soak drums (845–900°F) used in pairs (one onstream and one offstream for decoking)
Coke yield: 20%–40% w/w (dependent upon feedstock)
Fluid coking
Prime purpose: conversion to distillates
Complete conversion of the feedstock
Severe—longer residence time in hot zone (480–565°C; 900–1050°F)
Pressure on the order of 10 psi
Thermal reactions allowed to proceed to completion in hot zone
Coke bed fluidized with steam; heat dissipated throughout the fluid bed
Higher yields of hydrocarbon gases ($<C_5$) than delayed coking
Less coke yield than delayed coking (for one particular feedstock)

times are much longer and the thermal reactions are allowed to proceed to completion. The visbreaking process uses a quench operation to terminate the thermal reactions. Liquid-phase cracking takes place under these low-severity conditions to produce some naphtha and material in the kerosene and gas-oil boiling range. The gas oil may be used as additional feed for catalytic cracking units or as heating oil.

Two visbreaking processes are commercially available (Figs. 8.3–8.5); the first process is the coil, or furnace, type, and the second process is the soaker type: (i) the coal visbreaking process and (ii) the soaker visbreaking process, respectively. The *coil visbreaking process* (Fig. 8.4) achieves conversion by high-temperature cracking within a dedicated soaking coil in the furnace. With conversion primarily achieved as a result of temperature and residence time, coil visbreaking is described as a high-temperature, short-residence-time route. The main advantage of the coil-type design is the two-zone fired heater that provides better control of the material being heated, and with the coil-type design, decoking of the heater tubes is accomplished more easily by the use of steam-air decoking. In the alternative *soaker visbreaking process* (Fig. 8.5), the bulk of the cracking reaction occurs not in the furnace but in a drum located after the furnace (the *soaker*) in which the heated feedstock is held at an elevated temperature for a predetermined period of time to allow cracking to occur before being quenched and then passed to a fractionator. In soaker visbreaking, lower temperatures are used than in coil visbreaking. Consequently, the soaker visbreaking process is described as a low-temperature, high-residence-time route. Product quality and yields from the coil and soaker drum design are

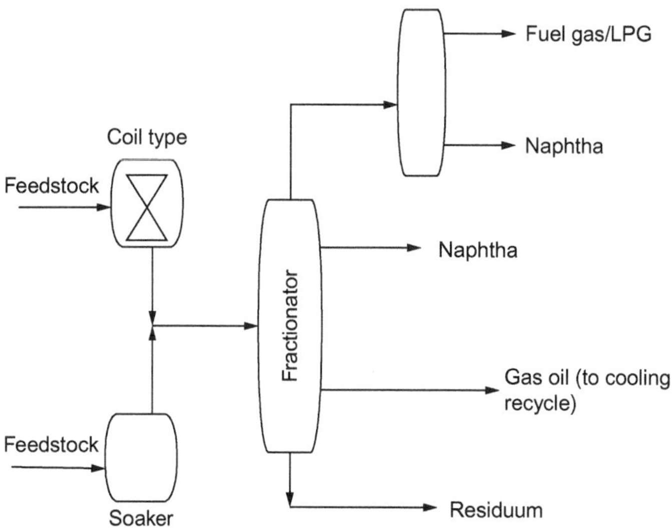

FIG. 8.3 General representation of a visbreaking unit.

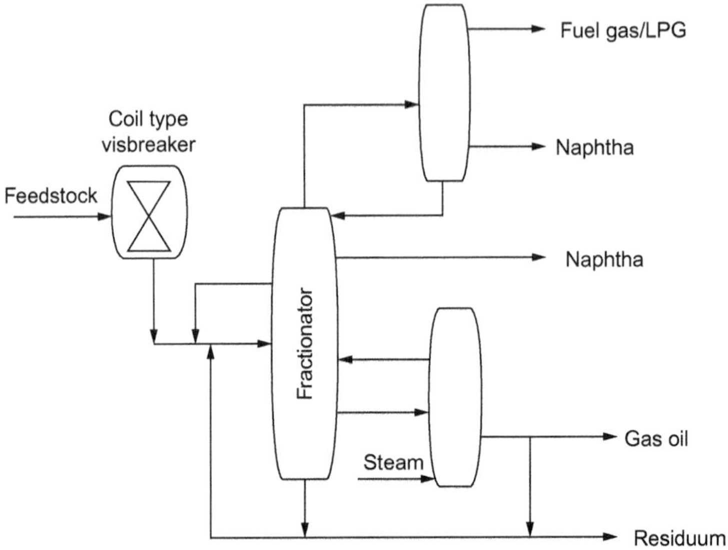

FIG. 8.4 A coil-type visbreaker.

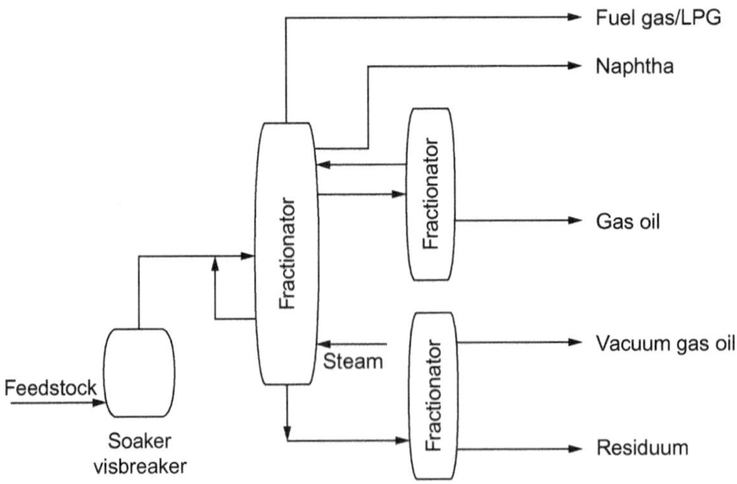

FIG. 8.5 A soaker-type visbreaker.

essentially the same at a specified severity being independent of visbreaker configuration. By providing the residence time required to achieve the desired reaction, the soaker drum design allows the heater to operate at a lower outlet temperature (thereby saving fuel), but there are disadvantages.

In the process, the viscous feedstock is passed through a furnace where it is heated to a temperature of 480°C (895°F) under an outlet pressure of ~100 psi. The cracked products are then passed into a flash-distillation chamber. The overhead material from this chamber is then fractionated to produce naphtha and low-boiling gas oil. The liquid products from the flash chamber are cooled with a gas-oil flux and then sent to a vacuum fractionator. This yields a high-boiling gas-oil distillate and a residuum of reduced viscosity—a 5%–10% v/v conversion of residuum to naphtha is usually sufficient to afford at least an approximate fivefold reduction in viscosity. Reduction in viscosity is also accompanied by a reduction in the pour point. An alternative option is to use lower furnace temperatures and longer times, achieved by installing a soaking drum between the furnace and the fractionator. The disadvantage of this approach is the need to remove coke from the soaking drum.

The higher heater outlet temperature specified for a coil visbreaker is an important advantage of coil visbreaking. The higher heater outlet temperature is used to recover significantly higher quantities of visbroken gas oil. This capability cannot be achieved with a soaker visbreaker without the addition of a vacuum flasher. In terms of product yield, there is little difference between the two options (soaker visbreaker compared with coil visbreaker approaches). However, each offers significant advantages in particular situations. For example, the cracking reaction forms coke as a by-product. In coil visbreaking, this lays down in the tubes of the furnace and will eventually lead to fouling or blocking of the tubes. The lower temperatures used in the soaker approach mean that these units use less fuel. In cases where a refinery buys fuel to support process operations, any savings in fuel consumption could be extremely valuable. In such cases, soaker visbreaking may be advantageous. In fact, most of the existing visbreaker are the soaker type, which utilize a soaker drum in conjunction with a fired heater to achieve conversion and which reduces the temperature required to achieve conversion while producing a stable residue product, thereby increasing the heater run length and reducing the frequency of unit shut down for heater decoking.

Decoking is accomplished by a high-pressure water jet. First, the top and bottom heads of the coke drum are removed after which a hole is drilled in the coke from the top to the bottom of the vessel and a rotating stem is lowered through the hole, spraying a water jet sideways. The high-pressure jet cuts the coke into lumps, which fall out the bottom of the drum for subsequent loading into trucks or railcars for shipment to customers. Typically, coke drums operate on fixed cycles that depend upon the feedstock and the coking parameters. Cokers produce no liquid residue but yield up to 30% coke by weight. Much of the low-sulfur product is employed to produce electrodes for the electrolytic smelting of aluminum.

The main disadvantage is the decoking operation of the heater and soaker drum, and although decoking requirements of the soaker drum design are not as frequent as those of the coil-type design, the soaker design requires more equipment for coke removal and handling. The customary practice of removing coke from a drum is to cut it out with high-pressure water, thereby producing a significant amount of coke-laden water that needs to be handled, filtered, and then recycled for use again.

The Shell soaker visbreaking process is suitable for the production of fuel oil by residuum (atmospheric residuum, vacuum residuum, or solvent deasphalter bottoms) viscosity reduction with maximum production of distillates. The basic configuration of the process includes the heater, soaker, and fractionator and more recently a vacuum flasher to recover more distillate products (Parkash, 2003; Gary et al., 2007; Speight, 2014, 2017; Hsu and Robinson, 2017). The cut point of the viscous gas-oil stream taken from the vacuum flasher is ~520°C (970°F). In the process, the feedstock is preheated before entering the visbreaker heater, where the residue is heated to the required cracking temperature. Heater effluent is sent to the soaker drum where most of the thermal cracking and viscosity reduction takes place under controlled conditions. Soaker drum effluent is flashed and then quenched in the fractionator, and the flashed vapors are fractionated into gas, naphtha, gas-oil, and visbreaker residue. The visbreaker residue is steam stripped in the bottom of the fractionator and pumped through the cooling circuit for further processing. Visbreaker gas oil, which is recovered as a sidestream, is steam stripped, cooled, and sent for further processing. As expected, product yields are dependent on feed type and product specifications. The viscous gas-oil stream for the visbreaker can be used as feedstock for a thermal distillate cracking unit or for a catalytic cracker for the production of lower-boiling distillate products.

Other variations of visbreaking technology include the Tervahl T and Tervahl H processes. The Tervahl T alternative (Fig. 8.6) includes only the thermal section to produce a synthetic crude oil with better transportability by having reduced viscosity and greater stability. The Tervahl H alternative adds hydrogen that also increases the extent of the desulfurization and decreases the carbon residue of the product. The Aquaconversion process (Fig. 8.7) is a hydrovisbreaking process that uses catalyst-activated transfer of hydrogen from water added to the feedstock. Reactions that lead to coke formation are suppressed, and there is no separation of asphaltene-type material (Marzin et al., 1998).

Atmospheric and vacuum residua are the usual feedstocks to a visbreaker although heavy oil, extra heavy oil, and tar sand bitumen are also acceptable feedstocks. The viscous feedstocks will typically achieve a conversion to gas, naphtha, and gas oil in the order of 10–50% w/w, depending on the severity and feedstock characteristics. The conversion of the viscous feedstock to distillate (low-boiling products) is commonly used as a measurement of the severity of the visbreaking operation, and the conversion is determined as the amount of 345°C$^+$ (650°F$^+$) material present in the atmospheric residuum or the 482°C$^+$ (900°F$^+$) material present in the vacuum residuum that is converted (*visbroken*) into lower-boiling components.

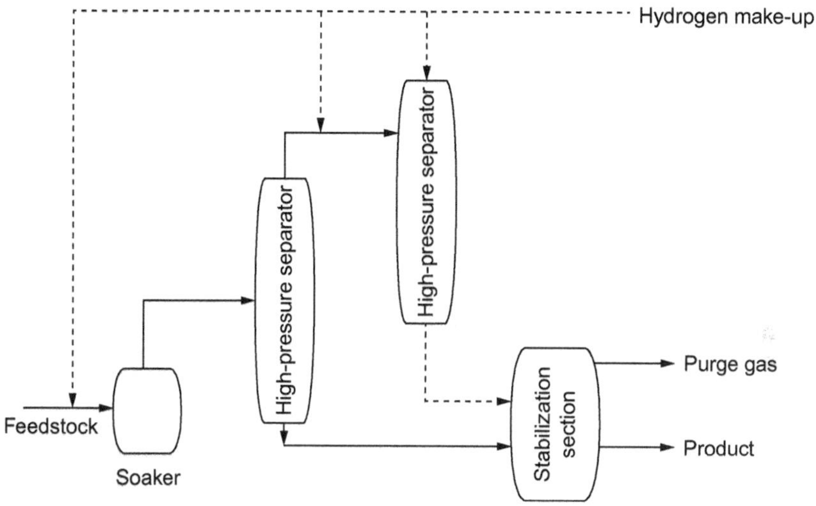

FIG. 8.6 The Tervahl T and Tervahl H process configurations.

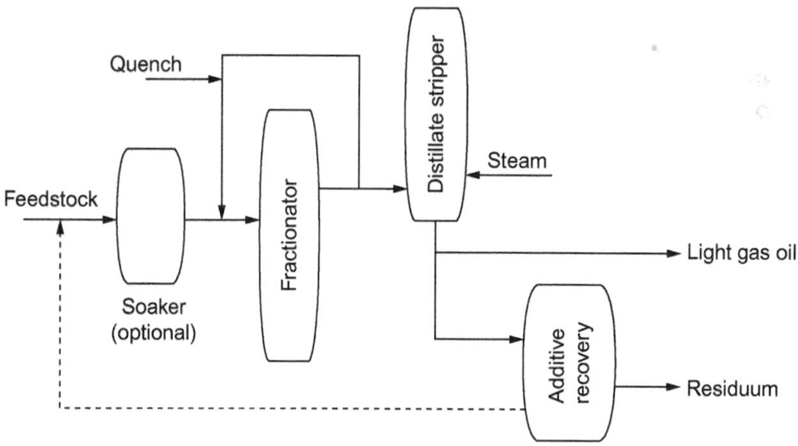

FIG. 8.7 The aquaconversion process.

The extent of feedstock conversion is limited by a number of feedstock characteristics, such as asphaltene content that varies with the type of feedstock and, hence, the type of residuum and, more particularly, carbon residue (Parkash, 2003; Gary et al., 2007; Speight, 2014, 2017; Hsu and Robinson, 2017). In very general terms, paraffinic feedstocks will have a low heptane-asphaltene content (0%–8% by weight), whereas naphthenic feedstock will have a much higher heptane-asphaltene content (10%–20% by weight) with the mixed crude oils having intermediate values. Of course, when the heptane asphaltenes are concentrated in the nonvolatile portion (through distillation), the proportions of the asphaltenes will be much higher. Thus, feedstocks with a high heptane-asphaltene content will result in an overall lower conversion than feedstocks with a lower heptane-asphaltene content while maintaining production of a stable fuel oil from the visbreaker bottoms. Minimizing the sodium content to almost a negligible amount and minimizing the Conradson carbon weight percent will result in longer cycle run lengths.

In addition, variations in feedstock quality will impact the level of conversion obtained at a specific severity. For example, for a given feedstock, as the severity is increased, the viscosity of the 205°C$^+$ (400°F$^+$) visbroken residue (often referred to as *visbroken tar* or *visbreaker tar*) initially decreases and then, at higher severity levels, increases dramatically, indicating the formation of coke precursors and their initial phase separation as sediment. The point at which this viscosity reversal occurs differs from feedstock to feedstock but can be estimated from the amount of low-molecular-weight hydrocarbon gases ($\leq C_3$) (Negin and Van Tine, 2004).

Thus, a viscous feedstock, such as a crude oil residuum or tar sand bitumen, is passed through a furnace where it is heated to a temperature of 480°C (895°F) under an outlet pressure of ~100 psi. The heating coils in the furnace are arranged to provide a soaking section of low heat density, where the charge remains until the visbreaking reactions are completed. The cracked products are then passed into a flash-distillation chamber. It is advisable to maintain the flash zone temperature as low as possible to minimize the potential for coking. Under fixed flashing conditions, increasing the yield of the residuum will reduce this temperature.

The overhead material from this chamber is then fractionated to produce a low-quality gasoline as an overhead product and low-boiling gas oil as bottoms. The liquid products from the flash chamber are cooled with a gas-oil flux and then sent to a vacuum fractionator. This yields a heavy gas-oil distillate and a residuum of reduced viscosity. A quench oil may also be used to terminate the reactions and will also influence the temperature of the flash zone.

A 5%–10% v/v conversion of viscous feedstocks to naphtha is usually sufficient to afford at least an approximate fivefold reduction in viscosity. Reduction in viscosity is also accompanied by a reduction in the pour point. However, the reduction in viscosity of nonvolatile portion tends to reach a limiting value with conversion, although the total product viscosity can continue to decrease. The minimum viscosity of the unconverted residue can lie outside the range of allowable conversion if sediment begins to form. When shipment of the visbreaker product by pipeline is the process objective, the addition of a diluent such as gas condensate can be used to achieve a further reduction in viscosity.

Conversion of heavy oil, extra heavy oil, and tar sand bitumen in visbreaking follows first-order reaction kinetics.

Briefly, fouling (a deposit buildup in refinery processes that impedes heat transfer and/or reduces throughput) is the leading cause of diminished efficiency and productivity in refineries. The energy lost due to this inefficiency must be supplied by burning additional fuel or reducing feed. While most fouling is caused by the deposition of heavier hydrocarbon species coming directly from the crude oil, a small undetermined percentage is related to corrosion and scale deposits, either actively participating as loose corrosion products or by scale acting as a substrate for hydrocarbon deposition. Fouling will also occur in the drum of a soaker visbreaker, though the lower temperatures used in the soaker drum lead to fouling at a much slower rate. Coil visbreaking units therefore require frequent decoking. Soaker drums require far less frequent attention, but them being taken out of service normally requires a complete halt to the operation.

Thus, the severity of the visbreaking operation is generally limited by the stability of the visbroken product—generally a prelude to the onset of fouling. If overcracking occurs, the resulting fuel oil may form excessive deposits in storage or when used as a fuel in a furnace. The visbreaking correlations presented are based on operating to levels where the fuel oil quality will be limited by this test. This severity level is well within the operating limits that would be imposed by excessive coke formation in properly designed visbreaking furnaces.

The main limitation of the visbreaking process and for that matter all thermal processes is that the products can be unstable. Thermal cracking at low pressure gives olefin derivatives, particularly in the naphtha fraction. These olefin derivatives give a very unstable product, which tends to undergo secondary reactions to form gum and intractable nonvolatile material. Product stability of the visbreaker residue is a main concern in selecting the severity of the visbreaker operating conditions. Severity, or the degree of conversion, can cause phase separation of the fuel oil even after cutter stock blending. Increasing visbreaking severity and percent conversion will initially lead to a reduction in the visbroken fuel oil viscosity. However, visbroken fuel oil stability will decrease as the level of severity—and hence conversion—is increased beyond a certain point, dependent on feedstock characteristics.

The instability of the visbroken fuel oil is related to the asphaltene constituents and their thermal stability in the residuum. Asphaltenes are high-molecular-weight nonvolatile compounds that can be classified according to their solubility in various solvents. The asphaltene constituents can be thermally altered during visbreaking operations. In addition, during visbreaking, some of the high-molecular-weight constituents, including some of the asphaltene constituents, are converted to lower-boiling and medium-boiling paraffinic components, some of which are removed from the residuum. The asphaltenes and thermally altered, being unchanged, are thus concentrated in the product residuum (that may contain new paraffinic material) and if the extent of the visbreaking reaction is too high, the asphaltene constituents or altered asphaltene constituents will phase tend to precipitate in the product fuel oil, creating an unstable fuel oil.

A common method of measuring the amount of asphaltenes in crude oil is by the addition of a low-boiling liquid hydrocarbon such as *n*-pentane (ASTM D893, 2015; ASTM D, 2007) or *n*-heptane (ASTM D3279, 2015; ASTM D4124, 2015) by which treatment the asphaltenes separate as a solid (Speight, 2000, 2014). Since the amount of asphaltenes in the visbreaking unit charge residuum may limit the severity of the visbreaking operations, the *n*-pentane insoluble content or *n*-heptane insoluble content of the feedstock is used as the correlating parameter in various visbreaking correlations. However, these correlations can be visbreaker and feedstock-dependent, and application from one unit to another and one feedstock to another may be misleading.

Sulfur in the visbroken residuum can also be an issue since the sulfur content of the visbreaker residuum is often higher (~0.5% w/w or greater) than the sulfur in the feedstock. Therefore, it can be difficult to meet the commercial sulfur specifications of the refinery product residual fuel oil, and blending with low-sulfur cutter stocks may be required.

Visbreaking, like thermal cracking, is a *first-order reaction*. However, due to the visbreaking severity limits imposed by fuel oil instability, operating conditions do not approach the level where secondary reactions, polymerization and condensation, occur to any significant extent. The first-order reaction rate equation altered to fit the visbreaking reaction is

$$K = (1/t)(\ln 100/X_1)$$

where K is the first-order reaction velocity constant, 1/s; t is the time at thermal conversion conditions, s; and $X_1 = 900°\text{F}+$ visbroken residuum yield, volume %. The thermal conversion reactions are generally assumed to start at 425°C (800°F) although some visbreaking occurs below this temperature. Therefore, the residence time in the 425–450°C (800–865°F) reaction zone should be 613 s.

The central piece of *equipment* in any thermal process and visbreaking is, with no exception, the heater. The heater must be adequate to efficiently supply the heat required to accomplish the desired degree of thermal conversion. A continually increasing temperature gradient designed to give most of the temperature increase in the front part of the heater tubes with only a slow rate of increase near the outlet is preferred. Precision control of time and temperature is usually not critical in the processes covered in this article. Usually, all that is required is to design to some target temperature range and then adjust actual operations to achieve the desired cracking. In the higher-temperature processes (e.g., ethylene manufacture), temperature control does become of prime importance due to equilibrium considerations.

Equipment design to minimize coke formation is of importance. The excessive production of coke adversely affects the thermal cracking process in the following ways: (i) reduces heat transfer rates, (ii) increases pressure drops,

(iii) creates overheating, (iv) reduces run time, and (v) requires the expense of removing the coke from the equipment. In addition, the metallurgy of the equipment, specifically the heater tubes and pumps, in the high-temperature, corrosive environments must be adequate to prevent expensive destruction and replacement of equipment. In the early days of thermal cracking, the metallurgy of the heater tubes was not of sufficient quality to permit extended periods of high temperatures. Modern improvements in the quality of steel have extended the durability of the thermal cracking equipment.

The advances in heater design have reached a point where very efficient furnace and heating tube arrangements can be built that give the refiner the desired thermal cracking operation. The practice of the refiner is generally to set the specifications the heater is expected to meet for the specific application and have a heater manufacturer prepare a suitable design. Proper *tube size* selection depends upon minimizing pressure drop while obtaining good turbulence for proper heat transfer that is also dependent upon the charge rate that ultimately affects the residence time and, therefore, the extent of the conversion.

The charge stock *liquid velocity* should be sufficient to provide enough turbulence to ensure a good rate of heat transfer and to minimize coking. A minimum linear cold 15.6°C (60°F) velocity on the order of 5 ft./s for a 100% liquid charge rate is considered to be sufficient. The maximum velocity would be limited to ~10 ft./s due to excessive pressure drop. The velocities at the higher cracking temperatures would, of course, be greater due to the partial vaporization of the charge.

Most of the *heat supplied* to the charge stock is radiant heat. The convection section of the heater is used primarily to supply preheat to the charge prior to the main heating in the radiant section. The *heat transfer rate* in the convection section will range from 3000 to 10,000 Btu/ft^2 of tube outside area per hour with an average rate of 5000 Btu/ft^2/h. The heating rates in the radiant section will range from 8000 to 20,000 Btu/ft^2/h depending upon the charge stock, with heavier oil generally requiring the lower heating rate.

The heating tube *outlet temperature* will depend upon the charge stock being processed and the degree of thermal conversion required. The outlet temperature will vary from a minimum of 425°C (800°F) for visbreaking to a maximum of 595°C (1100°F) for thermal reforming. The combustion chamber temperature will range from 650°C to 870°C (1200–1600°F) at a point ~1 ft. below the radiant tubes. Flue gas temperatures are usually high (425–595°C, 800–1100°F) particularly since the viscous feedstock is usually entering the heater at a high temperature from a fractionating tower. An exception to the charge entering at a high temperature would be when charging gasoline to a thermal reformer. However, since thermal reforming requires high temperatures, flue gas temperatures will also be high.

Since it is desirable to maintain different temperature increase rates throughout the charge heating, that is, rapid increase at the beginning of the heating coil and a lower rate near the outlet, zone temperature control within the furnace is desired. A three-zone furnace is preferred with the first zone giving the greatest rate of temperature increase and the last zone the least.

Coke formation limits the operation of the heater, and techniques should be employed to minimize coke formation in the heater tubes. Coking occurs on the walls of the tubes, particularly where turbulence is low and temperature is high. Maintaining sufficient turbulence assists in limiting coke formation. Baffles within the tubes are sometimes used, but water injection into the charge stream is the preferred method. Water is usually injected at the inlet although water also may be injected at additional points along the heater tubes. The water, in addition to providing turbulence in the heater tubes as it is vaporized to steam, also provides a means to control temperature. The optimum initial point of water injection into the heater tubes is at the point of incipient cracking where coke would start to form. An advantage to this injection point is the elimination of the additional pressure drop that would have been created by the presence of water between the heater inlet and the point of incipient cracking.

The preferred method to remove coke (decoke) from the heater tubes is to burn off the coke using a steam-air mixture. The heater tubes, therefore, should be capable of withstanding temperatures up to 760°C (1400°F) at low pressures for limited time periods. The heater tubes along with the tube supports should be designed to handle the thermal expansion extremes that would be encountered. Mechanical means, such as drills, can also be used to remove coke, but most modern heaters use the steam-air combustion technique. Parallel heaters may be employed so that one can be decoked while permitting cracking to proceed in the other heater(s).

The metallurgy of thermal cracking units is variable although alloy steel tubes of 7%–9% chromium are usually satisfactory to resist sulfur corrosion in thermal cracking heaters. If the hydrogen sulfide content of the cracked products exceeds 0.1 mol% in the cracking zone, a higher alloy steel may be required. Stabilized stainless steel, such as Type 321 or 347, would be suitable in this case. Other alloys, such as the Inconel or Incoloy alloys, could also be used. Seamless tubes with welded return bends are now normally used in heaters. Flanged return bends were used in earlier thermal cracking units to facilitate cleaning. However, the use of steam air to burn out the coke essentially eliminates the need for flanged fittings that, in turn, reduces the possibility of dangerous leaks.

A useful tool to aid in the design and operation of thermal cracking units is the soaking volume factor (SVF). This factor combines time, temperature, and pressure of thermal cracking operations into a single numerical value. The SVF is defined as the *equivalent* coil volume in cubic feet per daily barrel of charge (fresh plus recycle) if the cracking reaction had occurred at 800°F and 750 psi:

$$\mathrm{SVF}_{750\mathrm{psi}/800\mathrm{F}} = 1/FRK_{\mathrm{p}}\mathrm{d}V$$

where $\mathrm{SVF}_{750\mathrm{psi}/800\mathrm{F}}$ is the SVF at base reaction conditions of 750 psi gauge pressure and 800°F, cubic feet of coil volume per total charge throughput in barrels per day; F is the charge (fresh plus recycle) throughput rate, barrels per day; R is the ratio of reaction velocity constant at temperature Y and reaction velocity constant at 800°F, K_T/K_{800}; K_p is the pressure correction factor for pressures other than 750 psi gauge; and dV is the incremental coil volume, cubic feet.

When an additional soaking drum is used, the SVF for the soaking drum should be added to the coil SVF. The SVF for the drum may be determined from

$$\mathrm{SVF_D} = \mathrm{DV}/F(K_{\mathrm{TD}})(K_{\mathrm{p}})$$

$\mathrm{SVF_D}$ is the SVF of the drum. DV is the volume of drum, ft^3. F is the charge (fresh plus recycle) throughput rate, bbl/day. K_{TD} is the reaction velocity constant for the mean drum temperature, and K_p is the pressure correction factor for the mean drum pressure.

The SVF will range from 0.03 for visbreaking of viscous feedstocks to ~1.2 for low-boiling gas-oil cracking. The SVF is a numerical expression of cracking rate and thus can be correlated with product yield, and quality SVF may also be translated into cracking coils and still volumes of known dimensions under design conditions of temperature and pressure.

A cracking unit seldom operates very long at design conditions. Charge stock quality changes, desired product yields and qualities change, or additional capacity is required. These changes require a SVF that is different than the design SVF. The SVF may be varied by (i) varying pressure at constant temperature and feed rate; (ii) varying temperature at constant feed rate, the pressure gradient varying with the effect upon cracking rate and fluid density in the cracking coil; and (iii) varying the soaking volume at constant temperature and pressure by varying heater feed rate and/or varying the number of tubes in the section above 425°C (800°F).

With the advent of higher firing rate and better efficiency heaters, the use of external *soaking drums* to provide additional reaction times is of less importance in thermal cracking operations. In modern units, the coil in the heater is usually sufficient to provide the temperature-time relationships required. A possible exception would be the case where it is desirable to crack a considerable amount of viscous nonvolatile material. The temperature required probably could not be successfully obtained in a heater coil without excessive coking. A reaction chamber (soaking drum) is employed where the hotter, cleaner low-boiling gas oil is used to supply heat to the heavier dirty oil stream in a soaking drum. A low-temperature low-boiling gas-oil stream is also frequently used to wet the walls of the soaking drum to minimize coking. Parallel soakers could be used to allow one to be decoked, while the other is used for the cracking operations.

The *pumps* used in thermal cracking operations must be capable of operation for extended periods handling a high-temperature (above 230°C, 450°F, and up to 345°C, 650°F) corrosive liquid. In addition, since coke particles are formed in thermal cracking, the pumps must be able to withstand the potential erosion of the metal parts by the coke particles. In the early days of thermal cracking, reciprocating pumps were commonly used, but in later units, centrifugal pumps have been used. A preferred centrifugal pump would be of the coke-crushing type or may have open impellers with case wear plates substituted for the front rings. The metal should be 12% chromium steel alloy or a higher alloy if serious corrosion is potential.

Heat exchangers should be constructed to provide easy cleaning since high temperatures and coke particles can create extensive fouling of the exchangers. The downstream processing equipment (*flash drums*, *separators*, and *fractionating towers*) is typically of standard design, and no special design specifications are required other than minimizing potential coke buildup. This can be accomplished by designing the equipment, so there would be no significant holdup or dormant spots in the process equipment where coke could accumulate.

In thermal cracking operations, there is a considerable amount of *excess heat that* cannot be economically utilized within the cracking unit itself. When a thermal cracking unit is being considered, it is desirable to construct the unit in conjunction with some other unit, such as a crude still, which could utilize the excess heat to preheat the crude oil charge. Alternately, the excess heat could be used in steam generation facilities.

Visbreaking may be the most underestimated and/or undervalued process in a refinery. The process may find rejuvenated use not only for viscous feedstock (including tar sand bitumen) but also for biofeedstocks. This visbreaking process possesses sufficient hardware flexibility to accommodate feedstock blending (crude oil feedstocks and biofeedstocks) and of the unit as well as a high measure of reliability and predictive operations/maintenance, thereby minimizing unplanned shutdowns.

The severity of visbreaker operation is generally limited by the stability requirement of the product and the extent of fouling and coke laydown in the visbreaker heater (Speight, 2015a). The former requirement means that the stability of the residue must be sufficient to ensure that the finished fuel resulting from blending with diluents (that are less aromatic than the residue) is stable and that asphaltene flocculation does not occur. Where the residue is converted to an emulsion, blend stability is improved, and severity/conversion can be increased, subject to acceptable levels of heater fouling and coke deposition (Miles, 2009). Operational modifications, such as increasing steam injection or recycling high-boiling distillates from the visbreaker fractionator, may help mitigate coking tendency and enhance yield, while some relatively low-cost options to increase heater capacity might be implemented in certain instances.

In terms of processing biofeedstocks, many biofeedstocks have a high oxygen content and high mineral content, which could (even when blended) disqualify the use of the biomaterial as a feedstock to a hydroprocessing unit. Refiners are very wary of high-oxygen and high-mineral feedstocks because of the increased hydrogen requirements (hydrogen is an expensive refinery commodity) to remove the oxygen from the hydrocarbon products with the appearance of the additional hydrogen as water. However, blending a biofeedstock with a resid as feedstock to a visbreaking unit to produce additional fuel products is a concept that could pay dividends and provide refineries with a source of fuels to supplement crude oil feedstocks. In the visbreaker, the feedstock is converted to overhead (volatile products) and coke (if the unit is operated beyond the typical operating point or coke-forming threshold) (Speight, 2014). The majority of the nitrogen, sulfur, and minerals appear in the coke. Oxygen often appears in the volatile product as water and carbon dioxide, unfortunately removing valuable hydrogen from the internal hydrogen management system.

Alternatively, another option is the preparation of a feedstock that is acceptable to a refinery. In particular, any process that reduces the mineral matter in the biofeedstock and reduces the oxygen content in the biofeed would be a benefit.

This can be accomplished by one or two preliminary treatment steps (such as the visbreaking process) in which the feedstock is demineralized and the oxygen constituents are removed as overhead (volatile) material giving the potential for then production of a fraction rich in oxygen functions that may be of some use to the chemical industry. Such a process might have to be established at a biofeedstock production site, unless the refinery has the means by which to accommodate the feedstock in an already existing unit.

In a manner similar to the visbreaking process where the biofeedstock is blended with a residuum, the biofeedstock alone would be heated in a visbreaker-type reactor (at a lower temperature than the conventional visbreaking temperature) to the point where hydrocarbon derivatives (or alcohols) are evolved and coke starts to form. As the coke forms, the mineral matter is deposited with the coke, and the oxygen constituents are deoxygenized leaving a (predominantly) hydrocarbon product as a liquid that will ensure easy separation from the coke and mineral matter.

As an addendum to the description of visbreaking, *hydrovisbreaking*, a noncatalytic process, is conducted under similar conditions to visbreaking and involves treatment with hydrogen under mild conditions (Speight, 2014, 2017). The presence of hydrogen leads to more stable products (lower flocculation threshold) than can be obtained with straight visbreaking, which means that higher conversions can be achieved, producing a lower-viscosity product. In addition, the HYCAR process is composed fundamentally of three parts: (i) visbreaking, (ii) hydrodemetallization, and (iii) hydrocracking. In the visbreaking section, the viscous feedstock is subjected to moderate thermal cracking while no coke formation is induced. The visbroken oil is fed to the demetallization reactor in the presence of catalysts, which provides sufficient pore for diffusion and adsorption of high-molecular-weight constituents. The product from this second stage proceeds to the hydrocracking reactor, where desulfurization and denitrogenation take place along with hydrocracking.

4. Coking

Coking is a thermal process for the continuous conversion of heavy oil, extra heavy oil, and tar sand bitumen into lower-boiling products. The feedstock can be atmospheric residuum, vacuum residuum, or cracked residuum, and the products are gases, naphtha, fuel oil, gas oil, and coke. Coking processes generally utilize longer reaction times than thermal cracking processes. To accomplish this, drums or chambers (reaction vessels) are employed, but it is necessary to use two or more such vessels so that coke removal can be accomplished in those vessels not onstream without interrupting the semicontinuous nature of the process. Gas oil may be the major product of a coking operation and serves primarily as a feedstock for catalytic cracking units. The coke can be used as fuel, but processing for specialty uses, such as electrode manufacture, production of chemicals, and metallurgical coke, is also possible. For these latter uses, the coke may require treatment to remove sulfur and metal impurities—calcined crude oil coke can be used for making anodes for aluminum manufacture and a variety of carbon or graphite products such as brushes for electric equipment.

Thus, coking is a thermal cracking-type operation used to convert low-grade feedstocks to coke, gas, and distillates. Two types of crude oil coking processes are presently operating: (i) delayed coking, which uses multiple coking chambers to permit continuous feed processing wherein one drum is making coke and one drum is being decoked; and (ii) fluid coking, which is a fully continuous process where product coke can be withdrawn as a fluidized solid.

Crude oil residua obtained from the vacuum distillation tower as a nonvolatile (bottoms) fraction, heavy oil, extra heavy oil, and tar sand bitumen are the usual feedstocks to coking units. Atmospheric tower bottoms (sometimes referred to as *long residua*) may be charged to coking units, but it is generally not always acceptable to thermally degrade any gas-oil fraction contained in the heavy feedstock. Other feedstocks to coking units are deasphalter bottoms (often referred to as *deasphalter pitch*) and tar sand bitumen and cracked residua (thermal tars). The products are gases, naphtha, fuel oil, gas oil, and coke. The gas oil may be the major product of a coking operation and serves primarily as a feedstock for catalytic cracking units. The coke obtained is usually used as fuel, but processing marketing for specialty uses, such as electrode manufacture, production of chemicals, and metallurgical coke, is also possible and increases the value of the coke. For these uses, the coke may require treatment to remove sulfur and metal impurities. Furthermore, the increasing attention paid to reducing atmospheric pollution has also served to direct some attention to coking, since the process not only concentrates such pollutants as feedstock sulfur in the coke but also usually yields products that can be conveniently subjected to desulfurization processes.

Coking processes have the virtue of eliminating the residue fraction of the feed, at the cost of forming a solid carbonaceous product. The yield of coke in a given coking process tends to be proportional to the carbon residue content of the feed (measured as the Conradson carbon residue) (Speight, 2014, 2015b). The formation of large quantities of coke is a severe drawback unless the coke can be put to use. Calcined crude oil coke can be used for making anodes for aluminum manufacture and a variety of carbon or graphite products such as brushes for electric equipment. These applications, however, require a coke that is low in mineral matter and sulfur.

If the feedstock produces a high-sulfur, high-ash, and high-vanadium coke, one option for the use of the coke is combustion of the coke to produce process steam (and large quantities of sulfur dioxide unless the coke is first gasified or the combustion gases are scrubbed). Another option is stockpiling. For some feedstocks, particularly from heavy oil, extra heavy oil, and tar sand bitumen, the combination of poor coke properties for anode use, limits on sulfur dioxide emissions, and loss of liquid product volume have tended to relegate coking processes to a strictly secondary role in any new upgrading facilities.

4.1 Delayed Coking

Delayed coking is a severe form of thermal cracking process that falls in the temperature range of 450–470°C (840–890°F). It is one of the technologies with greater use to upgrade high-molecular-weight viscous fractions, since almost 30% of this type of installations uses this process. It started to be used in the 1920s, with the necessity to increase the yield of low-boiling oil products, and there since it has been the preferred option at many refineries, because it permits to handle heavier fractions containing impurities. It is a semicontinuous process based on the alternate use of the drums in filling, coking, and emptying cycles, due to which viscous hydrocarbon derivatives with low commercial value, coming mainly from the bottom of atmospheric distillation towers or vacuum columns, are converted into lower-molecular-weight hydrocarbon derivatives of higher value and a solid by-product known as coke, whose value will depend on its properties, such as sulfur and metal content. Top vapors of drums are taken to the separating column, where they are split into wet gas, nonstabilized naphtha, low-boiling gas oil, high-boiling gas oil, and viscous recycle oil. Wet gas is processed in gas plants, while liquids must be hydrotreated to reduce its reactivity and improve their quality. Once it has been cooled, coke is hydraulically cut and then handled for disposal, whether mine storage (Canadian case) or commercialization (Venezuelan case). Product yield is in the range of 65%–70%, whereas the balance stays as coke.

In the process, the cracking reactions are given sufficient (extended or delayed) time to proceed to completion in coke drums that are specially designed to accumulate the coke and not in the heater tubes, which otherwise would have led to the premature shutdown of the unit. Sufficient heat is introduced in the heater tubes for complete destructive distillation (with simultaneous removal of distillate), but the reduction to coke does not occur unless and until the residue enters the coke drum. Heating is achieved in the furnace to initiate cracking, and the actual reactions are complemented and completed in the huge and tall coke drums.

Delayed coking is the oldest, most widely used process and has changed very little since the process was first brought onstream ~80 years ago. It is a semicontinuous (semibatch) process in which the heated charge is transferred to large coking (or soaking) drums that provide the long residence time needed to allow the cracking reactions to proceed to completion (McKinney, 1992; Feintuch and Negin, 2004). The process is widely used for treating viscous feedstocks and is particularly attractive when the green coke produced can be sold for anode or graphitic carbon manufacture or when there is no market for fuel oils. The process uses long reaction times in the liquid phase to convert the residue fraction of the feed to gases, distillates, and coke. The condensation reactions that give rise to the highly aromatic coke product also tend to retain sulfur, nitrogen and metals, so that the coke is enriched in these elements relative to the feed.

In the process (Fig. 8.8), the feedstock is charged to the fractionator and subsequently charged with an amount of recycle material (usually ~10%, but as much as 25%, of the total feedstock) from the coker fractionator through a preheater and then to one of a pair of coke drums; the heater outlet temperature varies from 480°C to 515°C (895–960°F) to produce the various products. The cracked products leave the drum as overheads to the fractionator, and coke deposits form on the inner surface of the drum. The majority of the sulfur originally in the feedstock remains in the coke (Table 8.2), but this is dependent upon the chemical character of the sulfur in the feedstock insofar as ring sulfur will have a higher retention in the coke than nonring sulfur. A pair of coke

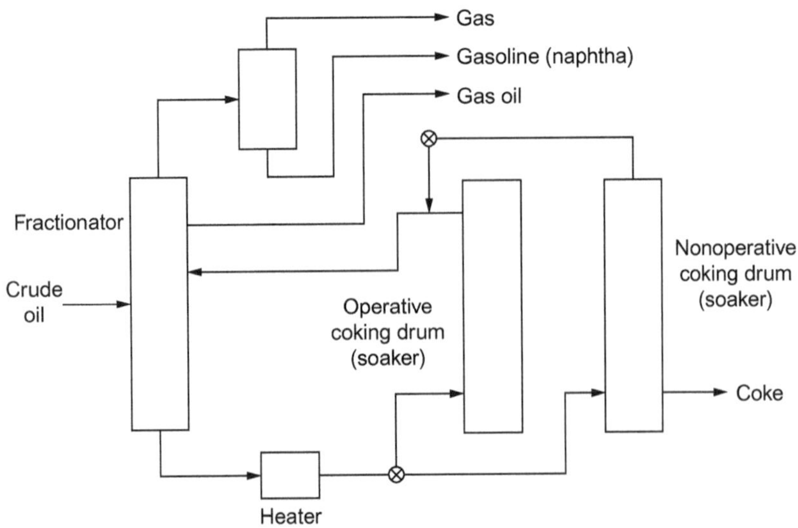

FIG. 8.8 The delayed coking process.

TABLE 8.2 Relationship of Feedstock Sulfur to Coke Sulfur

API Gravity	Sulfur	Sulfur in Coke	$\%S_{coke}/\%S_{feedstock}$
2.5	3.5	6.5	1.83
4.5	4.5	7.0	1.55
6	5.4	10.8	2.01
8	5.3	5.0	1.06
19	0.6	0.6	1.00

drums is used so that while one drum is onstream, the other is being cleaned allowing continuous processing, and the drum operation cycle is typically 48 h. The temperature in the coke drum ranges from 415°C to 450°C (780–840°F) at pressures from 15 to 90 psi (103–620 kPa) (Parkash, 2003; Gary et al., 2007; Speight, 2014, 2017; Hsu and Robinson, 2017).

Fractionators separate the overhead products from the coke drum to fuel gas (low-molecular-weight gases up to and including ethane), propane and propylene ($CH_3CH_2CH_3$ and $CH_3CH{=}CH_2$), butane-butene ($CH_3CH_2CH_2CH_3$ and $CH_3CH_2CH{=}CH_2$), naphtha, low-boiling gas oil, and high-boiling gas oil. Yields and product quality vary widely due to the broad range of feedstock used for coking units (Parkash, 2003; Gary et al., 2007; Speight, 2014, 2017; Hsu and Robinson, 2017).

Coker naphtha typically has a boiling range up to 220°C (430°F), is olefinic, and must be upgraded by hydrogen processing for the removal of olefin derivatives and sulfur. They are then used conventionally for reforming to gasoline or as chemical feedstocks. Middle distillates, boiling in the range of 220–360°C (430–680°F), are also hydrogen treated for improved storage stability, sulfur removal, and nitrogen reduction. They can then be used for either diesel or burner fuels or thermally processed to lower-boiling naphtha. The gas oil boiling up to ~510°C (950°F) end point may be charged to a fluid catalytic cracking unit immediately or after hydrogen upgrading when low sulfur is a requirement.

The coke drums are on a 48 h cycle (Parkash, 2003; Gary et al., 2007; Speight, 2014, 2017; Hsu and Robinson, 2017) in which the coke drum is usually onstream for ~24 h before becoming filled with porous coke after which time the coke is removed by the following procedure: (i) The coke deposit is cooled with water; (ii) one of the heads of the coking drum is removed to permit the drilling of a hole through the center of the deposit; and (iii) a hydraulic cutting device, which uses multiple high-pressure water jets, is inserted into the hole, and the wet coke is removed from the drum. Typically, 24 h is required to complete the cleaning operation and to prepare the coke drum for subsequent use onstream (Parkash, 2003; Gary et al., 2007; Speight, 2014, 2017; Hsu and Robinson, 2017).

A well-designed delayed coker will have an operating efficiency of better than 95%, although delayed coking units are generally scheduled for shutdown for cleaning and repairs on a 12–18 month schedule, depending on the most economical cycle for the refinery. In terms of process efficiency, the feedstock *heater* and the *coke drums* are the most critical parts of the delayed coking process. The function of the heater or furnace is to preheat the charge quickly, to avoid preliminary decomposition, to the required temperature. Since coking is endothermic, the furnace outlet temperature must be ~55°C (100°F) higher than the coke drum temperature to provide the necessary process heat. The heater run length is a function of coke laydown in heater tubes, and careful design is necessary to avoid premature shutdown with cycle lengths preferably at least 1 year. When the charge stock is derived from crude distillation, double desalting is desirable since salt deposits will shorten heater cycles.

The heater for a delayed coking unit does not require as broad an operating range as a thermal cracking or visbreaking heater where both contact time and temperature can be varied to achieve the desired level of conversion. The coker heater must reach a fixed outlet temperature for the required coke drum temperatures. Thus, the coker heater requires a short residence time, high radiant heat flux, and good control of heat distribution.

The function of the coke drum is to provide the residence time required for the coking reactions to proceed to completion and to accumulate the coke. In sizing coke drums, a superficial vapor velocity in the range of 0.3–0.5 ft./s is used, and coke drums with heights of 97 ft. (30 m) have been constructed and approach a practical limit for hydraulic coke cutting. Drum diameters up to 26 ft. (8 m) have been commonly used, and larger drums are feasible for efficient processing. Various types of level detectors are used to permit drum filling to within 7–8 ft. (2–2.5 m) of the upper tangent line of the drum monitor coke height in the drum during onstream service.

Hydraulic cutters are used to remove coke from the drum, and the first step is to bore a vertical pilot hole through the coke after which cutting heads with horizontally directed nozzles and then undercut the coke and drop it out of the bottom of the drum. Hydraulic pressures in the range of 3000–3600 psi are used in the 26 ft. diameter coking drums.

In regard to the process parameters and product yields, an increase in the coking temperature (i) decreases coke production, (ii) increases liquid yield, and (iii) increases gas-oil end point. On the other hand, increasing pressure and/or recycle ratio (i) increases gas yield, (ii) increases coke yield, (iii) decreases liquid yield, and (iv) decreases gas-oil end point. As an example, increasing the pressure from the currently designed 15–35 psi causes the higher-boiling products to remain in the hot zone longer causing further decomposition and an increase in the yield of the naphtha fraction, a decrease in the yield of the middle distillate-gas-oil fraction, and an increase in the yield of coke.

In the past, many delayed coking units were designed to provide complete conversion of atmospheric residue to naphtha, kerosene, and other low-boiling products. However, some units have been designed to minimize coke yield and produce heavy coker gas oil (HCGO) that is catalytically upgraded. The yield slate for a delayed coker can be varied to meet a refiner's objectives through the selection of operating parameters. Furthermore, delayed coking has an increasingly important role to play in the integration of modem crude oil refineries because of the inherent flexibility of the process to handle even the heaviest of residues. The flexibility of operation inherent in delayed coking permits refiners to process a wide variety of crude oil feedstocks including high-sulfur viscous feedstocks.

Low-pressure coking is a process designed for a once-through, low-pressure operation. The process is similar to delayed coking except that recycling is not usually practiced and the coke chamber operating conditions are 435°C (815°F), 25 psi. Excessive coking is inhibited by the addition of water to the feedstock in order to quench and restrict further reactions of the reactive intermediates.

High-temperature coking is a semicontinuous process designed to convert viscous feedstocks to gas oil and coke as the primary products. In the process, the feedstock is transported to the heater (370°C, 700°F, and 30 psi) and finally to the coking unit, where temperatures may be as high as 980–1095°C (1800–2000°F). Volatile materials are fractionated, and after the cycle is complete, coke is collected for sulfur removal before storage.

Delayed coking is likely to remain the workhorse of thermal cracking processes for the foreseeable future. Online spalling, decoking techniques have been developed, based on successful, similar application on delayed coker heaters. Coker heater operation (in delayed coking units and in visbreaker units) is improved by online spalling of the coker heater pipes by the addition of off-line pipe that is added to the online coker heater pipes. When an online pipe is to be spalled, flow is diverted to the off-line pipe allowing for full operation of the coker heater. In another embodiment, a thermal transfer resistant zone plate is movably mounted in the radiant section of the coker heater. By moving the zone plate from an operating position to a spalling position and adjusting the temperature of the plurality of burners, the temperature of the pipes in the zone of the heater radiant section to be spalled can be lowered, while the temperature in the remaining zones of the heater radiant section are fully operational.

The delayed coking process will remain a preferred residue upgrading option because of its ability to handle the heaviest, contaminated crudes. Globally, approximately one-third of installed residue upgrading plant is by delayed coking. Although a mature process, in recent years, many developments have taken place, including (i) development of automated coke drum unheeding devices, allowing the operator to carry out the decoking procedure safely from a remote location; (ii) understanding of process parameters affecting yields, coker product qualities, and coke qualities (e.g., shot coke); and (iii) design and operation of major equipment items, in particular coke drums (allowing shorter coking cycles) and the delayed coker heater (online spalling/decoking and minimization of coking in furnace tubes).

Considering the need for expanding viscous feedstock processing (Parkash, 2003; Gary et al., 2007; Speight, 2014, 2017; Hsu and Robinson, 2017), there will be a need to incorporate operational flexibility into the original design of the delayed coker and make a few key equipment choices with long-term goals in mind. In recent projects and licensing proposals, refineries are also incorporating unique long-term rationales into their designs (Wodnik and Hughes, 2005). For example, some locations intentionally leave plot space and specify design criteria around the coker to allow for easy placement of an additional pair of coke drums with minimal debottlenecking of existing assets. This decision is based upon future plans to construct either another sour crude train or other projects to make more coker feedstock available from existing refinery units. These units utilize the benefit of being online and generating earnings to help pay for the future expansion projects. Incorporating distillate recycle in the processing scheme or at least designing the coker so that recycle technology can be added later at minimal cost is another design strategy possibility.

In terms of the chemistry of the process, coke formation begins only after an accumulation of considerable amount of asphaltene-type constituent that are formed as part of the thermal process. The rate of coke formation in a given case is determined by the rate of increase of the asphaltenes in the cracked residue. The process of formation of a new solid phase is made up of the precipitation of asphaltene constituents from the saturated solution and their subsequent condensation. In addition, the time for the inception of coke formation coincides with the time of maximum yield of asphaltene constituents (Magaril and Aksenova, 1968; Magaril et al., 1971).

4.2 Fluid Coking

Throughout the history of the refining industry, with only short-term exceptions, there has been a considerable economic driving force for upgrading heavy oil, extra heavy oil, and tar sand bitumen. This has led to the development of processes to reduce unwanted yield of nonvolatile materials including processes such as thermal cracking, visbreaking, delayed coking, vacuum distillation, and deasphalting.

Fluid coking is a continuous coking process that converts viscous feedstocks, such as vacuum residua, deasphalter bottoms, and tar sand bitumen. In the reactor, the feedstock is thermally cracked into low-boiling products and coke. The product yield of the process is in the range of 70%–75% per load weight, whereas the balance is coke. The process is also valuable for the conversion of heavy oil, extra heavy oil, and tar sand bitumen to distillates (Parkash, 2003; Gary et al., 2007; Speight, 2014, 2017; Hsu and Robinson, 2017). In the process, the hot feedstock is sprayed onto a fluidized bed of hot, fine coke particles, which permits the coking reactions to be conducted at higher temperatures and shorter contact times than can be employed in delayed coking. These conditions result in decreased yields of coke; greater quantities of more valuable liquid product are recovered in the fluid coking process.

As a brief history, in the late 1940s and early 1950s, there was a large incentive to develop a continuous process to convert viscous feedstocks into lower-boiling products. During this period, fluid coking using the principle of fluidized solids was developed, and contact coking, using the principle of a moving solid bed, was also developed, and the first commercial fluid coker went onstream in late 1954. During the late 1960s, environmental considerations indicated that, in many areas, it would no longer be possible to utilize high-sulfur coke as a boiler fuel. This and other environmental considerations resulted in the development of flexicoking to convert the coke product from a fluid coker to clean fuel. Flexicoking is a carbon rejection process developed as a modification to fluid coking, adding a step for coke gasifying to produce flexigas, a coke gas with a low-heating power (80–100 Btu/ft^3), reducing considerably the coke production. The product yield is similar to that of fluid coking, but coke production is reduced from 24 to 4%, converting it into flexigas. The first commercial flexicoking unit went onstream in 1976.

Fluid coking (Fig. 8.9) is a continuous process that uses the fluidized solid technique to convert viscous feedstocks such as vacuum residua, cracked residua, heavy oil, extra heavy oil, and tar sand bitumen to more valuable products (Roundtree, 1997; Parkash, 2003; Gary et al., 2007; Speight, 2014, 2017; Hsu and Robinson, 2017). This coking process allows improvement in the yield of distillates by reducing the residence time of the cracked vapors and also allows simplified handling of the coke product. Heat for the process is supplied by partial combustion of the coke with the remaining coke being drawn as product. The new coke is deposited in a thin fresh layer (ca. 0.005 mm, 5 μm) on the outside surface of the circulating coke particle, giving an onion skin effect.

The equipment for the fluid coking process is similar to that used in fluid catalytic cracking (Parkash, 2003; Gary et al., 2007; Speight, 2014, 2017; Hsu and Robinson, 2017) and follows comparable design concepts except that the fluidized coke solids replace catalyst. Small particles of coke made in the process circulate in a fluidized state between the vessels and are the heat transfer medium, and thus, the process requires no high-temperature preheat furnace.

Fluid coking uses two vessels, a reactor and a burner; coke particles are circulated between these to transfer heat (generated by burning a portion of the coke) to the reactor (Fig. 8.9; Blaser, 1992; Parkash, 2003; Gary et al., 2007; Speight, 2014, 2017; Hsu and Robinson, 2017). The reactor holds a bed of fluidized coke particles, and steam is introduced at the bottom of the reactor to fluidize the bed. The feed coming from the bottom of a vacuum tower at, for example, 260–370°C (500–700°F) is injected directly into the reactor. The temperature in the coking vessel ranges from 480 to 565°C (900–1050°F), with short residence times of the order of 15–30 s, and the pressure is substantially

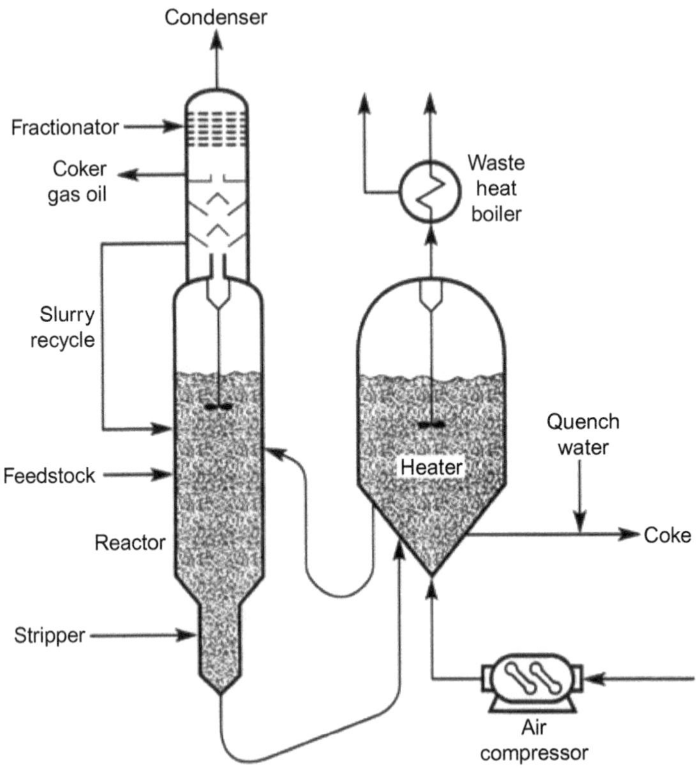

FIG. 8.9 The fluid coking process.

atmospheric, so the incoming feed is partly vaporized and partly deposited on the fluidized coke particles. The material on the particle surface then cracks and vaporizes, leaving a residue that dries to form coke. The vapor products pass through cyclones that remove most of the entrained coke.

Vapor products leave the bed and pass through cyclones that are necessary for the removal of the entrained coke. The cyclones discharge the vapor into the bottom of a scrubber, and any coke dust remaining after passage through the cyclones is scrubbed out with a pump-around stream, and the products are cooled to condense the viscous tar-like product. The resulting slurry is recycled to the reactor. The scrubber overhead vapors are sent to a fractionator where they are separated into wet gas, naphtha, and various gas-oil fractions. The wet gas is compressed and fractionated into the desired components.

In the reactor, the coke particles flow down through the vessel into the stripping zone. The stripped coke then flows down a standpipe and through a slide valve that controls the reactor bed level. A riser carries the *cold coke* to the burner. Air is introduced to the burner to burn part of the coke to provide reactor heat. The hot coke from the burner flows down a standpipe through a slide valve that controls coke flow and thus the reactor bed temperature. A riser carries the hot coke to the top of the reactor bed. Combustion products from the burner bed pass through two stages of cyclones to recover coke fines and return them to the burner bed.

Coke is withdrawn from the burner to keep the solid inventory constant. To aid in keeping the coke from becoming too coarse, large particles are selectively removed as product in a quench elutriator drum and coke fines are returned to the burner. The product coke is quenched with water in the quench elutriator drum and pneumatically transported to storage. A simple jet attrition system in the reactor provides additional seed coke to maintain a constant particle size within the system.

Due to the higher thermal cracking severity used in fluid coking compared with delayed coking, the products are somewhat more olefinic than the products from delayed coking. In general, products are handled for upgrading in a comparable manner from both coking processes. Delayed coking and fluid coking have been the processes of choice for primary conversion of Athabasca bitumen to liquid products for more than four decades (Fig. 8.10; Speight, 1990, 2000, 2014, 2017, Parkash, 2003; Gary et al., 2007; Hsu and Robinson, 2017). Both processes are termed the *primary conversion* processes for the tar sand plants in Fort McMurray, Alberta, Canada. The unstable liquid product streams are hydrotreated before recombining to the synthetic crude oil.

Coke, being a product of the process, must be withdrawn from the system to keep the solids inventory from increasing. The net coke produced is removed from the burner bed through a quench elutriator drum, where water is added

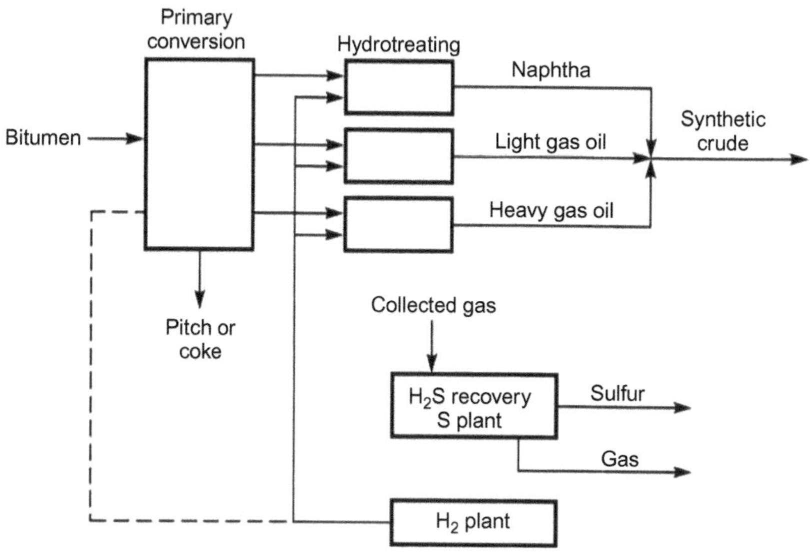

FIG. 8.10 Processing sequence for tar sand bitumen.

for cooling and cooled coke is withdrawn and sent to storage. During the course of the coking reaction the particles tend to grow in size. The size of the coke particles remaining in the system is controlled by a grinding system within the reactor.

The coke product from the fluidized process is a laminated sphere with an average particle size of 0.17–0.22 mm (170–220 μm), readily handled by fluid transport techniques. It is much harder and denser than delayed coke and in general is not as desirable for manufacturing formed products.

The yields of products are determined by the feed properties, the temperature of the fluid bed, and the residence time in the bed. The lower limit on operating temperature is set by the behavior of the fluidized coke particles. If the conversion to coke and light ends is too slow, the coke particles agglomerate in the reactor, a condition known as *bogging*. The use of a fluidized bed reduces the residence time of the vapor-phase products in comparison with delayed coking, which in turn reduces cracking reactions. The yield of coke is thereby reduced, and the yield of gas-oil and olefin derivatives increased. An increase of 5°C (9°F) in the operating temperature of the fluid-bed reactor typically increases gas yield by 1% w/w and naphtha by ~1% w/w.

The disadvantage of burning the coke to generate process heat is that sulfur from the coke is liberated as sulfur dioxide (SO_2). The gas stream from the coke burner also contains carbon monoxide (CO), carbon dioxide (CO_2), and nitrogen (N_2). An alternate approach is to use a coke gasifier to convert the carbonaceous solids to a mixture of carbon monoxide (CO), carbon dioxide (CO_2), and hydrogen (H_2).

The liquid products from the coker can, following cleanup via commercially available gas-oil hydrodesulfurization technology (Parkash, 2003; Gary et al., 2007; Speight, 2014, 2017; Hsu and Robinson, 2017), provide large quantities of low-sulfur fuel (<0.2% by weight sulfur). The incentive for fluid coking or flexicoking increases relative to alternate processing, such as direct hydroprocessing, as feedstock quality (Conradson carbon, metals, sulfur, nitrogen, etc.) decreases. Changes in yields and product quality result from a change from a low cut-point, high reactor temperature operation to a high cut-point operation with a lower reactor temperature (Parkash, 2003; Gary et al., 2007; Speight, 2014, 2017; Hsu and Robinson, 2017).

Fluid coke is used in electrodes for aluminum manufacture, in silicon carbide manufacture, in ore sintering operations, and as fuel. The coke from a feedstock containing a large amount of contaminants may not be suitable for these uses, from either a product contamination or environmental standpoint. The flexicoking process overcomes this problem by converting part of the gross coke to a gas that can be burned in process furnaces and boilers. The coke fines from a flexicoking unit contain most of the metals in the feedstock and may be suitable for metal recovery.

The fluid coking processes can be used to produce a high yield of low-sulfur fuel oil and to completely eliminate residual fuel oil and nonvolatile residua from the refinery product slate (Parkash, 2003; Gary et al., 2007; Speight, 2014, 2017; Hsu and Robinson, 2017). The different distributions are obtained by varying the fluid coker/flexicoker operating conditions and changing the downstream processing of the coker reactor products. In fact, there are many *process variations* that can be used to adapt the process to particular refining situations. Once-through or partial recycle coking can be used where there is a small market for heavy fuel oil or where a quantity of high-sulfur material can be blended into the fuel oil pool.

In reference to the process parameters, the *reactor temperature* is normally set at 510–540°C (950–1000°F). Low temperature favors high liquid yields and reduces the unsaturation of the gas but increases the reactor holdup requirements. The burner temperature is normally 55–110°C (100–200°F) above the reactor temperature. Regulating the amount of coke sent to the reactor from the burner controls the reactor temperature. Burner temperature is controlled by the air rate to the burner.

Low pressure provides maximum gas-oil recycle cut point, minimizes steam requirements, and reduces air blower horsepower. Reactor pressure normally adjusts to the gas compressor suction pressure but is higher due to the pressure drop through the piping, the condenser, the fractionation tower, and the reactor cyclone. The unit pressure balance required for coke circulation is normally controlled at a fixed differential pressure relative to the reactor that sets the burner pressure. *Reactor coke level* is controlled by the cold coke slide valve on the transfer line from the reactor to the burner, and *burner coke level* is controlled by the coke withdrawal rate.

In all coking processes, *product yields* are a function of feed properties, the severity of the operation, and recycle cut point (Parkash, 2003; Gary et al., 2007; Speight, 2014, 2017; Hsu and Robinson, 2017). Severity is a function of time and temperature since low severity and high gas-oil cut point favor high liquid yields, whereas high severity and low gas-oil cut point increase coke and gas yields. Data from these sources indicate that the gross coke yield is directly related to the API gravity of the feedstock (Table 8.3), which in turn is related to the carbon residue (Speight, 2014, 2015a,b, 2017). Coke quality and gas quality are also important (Parkash, 2003; Gary et al., 2007; Speight, 2014, 2017; Hsu and Robinson, 2017).

In most cases, high liquid yield and minimum coke and gas yields are required, and in theory, two cracking rates should be considered. The first is the rate at which the liquid cracks and vaporizes after initially laying down on the coke particles that determines the reactor holdup. The second is that vapor-phase cracking determines the distribution of the products between gas, naphtha, and gas oil. The vapor residence time can be determined from the reactor volume and the volume flow of hydrocarbon vapor and steam and can be divided into time in the fluid bed and time in the disperse phase. The former is a function of the coke holdup or hourly space velocity that is normally expressed as reciprocal hours. For maximum liquid yield, the secondary

TABLE 8.3 Relationship of Coke Yield to API Gravity

	Yield (% w/w)	
API Gravity	*Delayed Coke*	*Fluid Coke*
26	8	3
16	18	11
10	27	17
6	35	23
4	42	29

cracking time should be kept at a minimum, and thus, it is normally desirable to design the unit for the maximum hourly space velocity.

The maximum rate at which feed can be injected into a fluid coker is limited by a condition known as *bogging*. The conditions required to avoid a bogging are the following: (i) The feedstock must be uniformly distributed over the entire surface of the heat transfer medium; (ii) the layer of feed material on the particles should not be too great, and the thickness of the sticky plastic layer depends on the specific flow rate of feedstock, its coking factor, and the recirculation rate of the heat transfer medium; (iii) the bed temperature and the initial temperature of the heat transfer medium should be sufficiently high that the first stage of the process is completed in a short time; and (iv) the heat transfer medium should not consist of particles that are too fine. The heat reserve of the granules should be sufficient to cover the entire energy requirements in connection with heating the feedstock, supplying the energy for the endothermic cracking reaction, and evaporating the decomposition products. If the feed injection rate exceeds the vaporization rate for an extended period of time, the thickness of the tacky oil film on the particles will increase until the particles rapidly agglomerate, causing the bed to lose fluidity. When fluidization is lost, the heat transfer rate is greatly reduced, further aggravating the condition. Coke circulation cannot be maintained due to the loss of reactor fluidization.

For comparative purposes because of the similarity of the processes, there are some notable differences between the operation of a fluid coker and a fluid catalytic cracking unit (FCCU), and some of these differences tend to make the fluid coker easier to operate. The fluid coker heat balance is very easy to maintain, as there is always an excess of carbon to burn, whereas a fluid catalytic cracking unit has a sensitive interaction between heat balance and intensity balance and therefore between carbon burned and carbon produced, which complicates control, especially during operating changes, startup, and shutdown.

In addition, recovery from upsets caused by the loss of utilities such as steam and air is normally easier and faster with a fluid coker than with a fluid catalytic cracking unit. The fluid coker normally operates well at low feed rates, and turndown to low rate is normally limited by the ability of the tower to maintain fractionation of the products. The fluid coker properly can operate at any feed rate that will provide enough coke to heat balance.

However, the fluid coker has some inherent features that can create problems if proper precautions are not followed. The viscous residuum can become a near-solid phase if the lines are not properly heat traced and insulated. Low reactor temperature results in reactor bogging. If the particle size of the circulating coke is not properly controlled, the size can grow to the point that coke circulation problems are encountered. The feed nozzles must be maintained and occasionally cleaned to prevent poor feed distribution followed by excessive agglomerate formation. Control of the reactor bed level is critical since an excessively high bed level will flood the reactor cyclone and allow coke to be carried to the scrubber where it will plug the lines.

Along similar lines to the fluid coking process, the *rapid thermal processing* (RTP, now the heavy-to-light upgrading technology (HTL)) process was developed by Ivanhoe Energy Inc. in the 1980s. The process uses a circulating transport bed of hot sand to rapidly induce thermal cracking of the viscous feedstock in the absence of air to produce a low-boiling synthetic crude oil (Veith, 2007; Koshka et al., 2008; Silverman et al., 2011).

4.3 Flexicoking

Flexicoking is a direct descendent of fluid coking (Fig. 8.11) and uses the same configuration as the fluid coker but includes a gasification section in which excess coke can be gasified to produce refinery fuel gas (Roundtree, 1997; Marano, 2003; Parkash, 2003; Gary et al., 2007; Speight, 2014, 2017; Hsu and Robinson, 2017). The flexicoking process was designed during the late 1960s and the 1970s as a means by which excess coke could be reduced in view of the gradual incursion of the heavier feedstocks in refinery operations. Such feedstocks are notorious for producing high yields of coke (>15% by weight) in thermal and catalytic operations.

In the process, the viscous feedstock enters the scrubber for direct-contact heat exchange with the overhead product vapors from the reactor. The higher-boiling products (> 525°C/> 975°F) present in the overhead condense in the scrubber and return to the reactor as a recycle stream with fresh feedstock. Lower-boiling overhead constituents in the scrubber go to a conventional fractionator and also to light ends recovery. The feedstock is thermally cracked in the reactor fluidized bed to a range of gas and liquid products and coke. The coke inventory is maintained by circulating the bed coke from the reactor to the heater via the cold coke transfer line. In the heater, the coke is heated by the gasifier products and circulated back to the reactor via the hot coke transfer line to supply the heat that sustains the thermal cracking process.

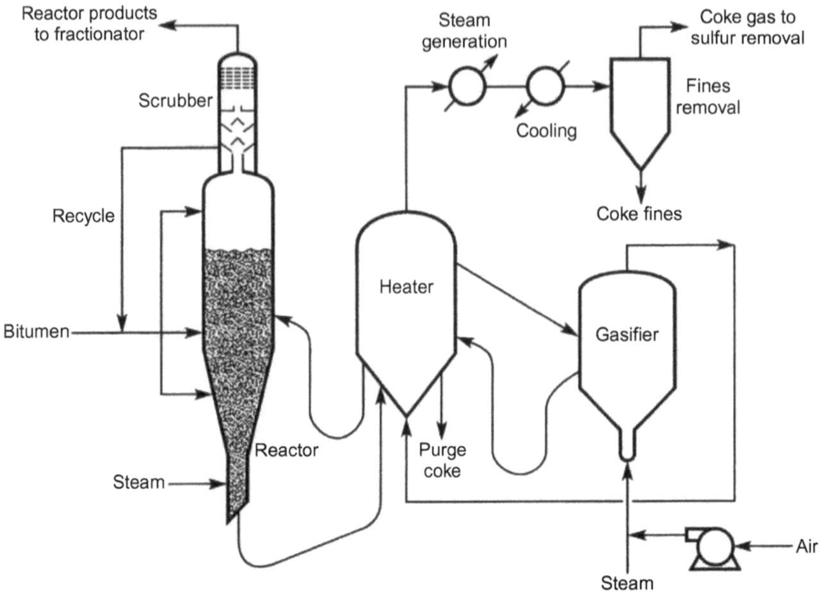

FIG. 8.11 The flexicoking process.

Excess coke is converted to a low-heating value gas in a fluid-bed gasifier with steam and air. The air is supplied to the gasifier to maintain temperatures of 830–1000°C (1525–1830°F) but is insufficient to burn all of the coke. The gasifier products, consisting of a gas and coke mixture, return to the heater to heat up the coke. The gas exits the heater overhead and goes to steam generation, to dry/wet particulate removal, and to desulfurization. The clean flexigas is then ready for use as fuel in refinery boilers and furnaces and/or for steam and power generation. Approximately 95% w/w of the coke generated in the reactor is converted in the process. Only a small amount of product coke is collected as fines from the flexigas and purged from the heater to extract feed metals.

A typical gas product, after removal of hydrogen sulfide, contains carbon monoxide (CO, 18%), carbon dioxide (CO_2, 10%), hydrogen (H_2, 15%), nitrogen (N_2, 51%), water (H_2O, 5%), and methane (CH_4, 1%). The heater is located between the reactor and the gasifier, and it serves to transfer heat between the two vessels. The heater temperature is controlled by the rate of coke circulation between the heater and the gasifier. Adjusting the air rate to the gasifier controls the unit inventory of coke, and the gasifier temperature is controlled by steam injection to the gasifier.

Yields of liquid products from flexicoking are the same as from fluid coking, because the coking reactor is unaltered. The main drawback of gasification is the requirement for a large additional reactor, especially if high conversion of the coke is required. Units are designed to gasify 60%–97% w/w of the coke from the reactor. Even with the gasifier, the product coke will contain more sulfur than the feed, which limits the attractiveness of even the most advance of coking processes.

The flexicoking process produces a clean fuel gas with a heating value of ~90 Btu/ft^3 or higher. The coke gasification can be controlled to burn ~95% of the coke to maximize production of coke gas or at a reduced level to produce both gas and a coke that has been desulfurized by ~65%. This flexibility permits adjustment for coke market conditions over a considerable range of feedstock properties. Fluid coke is currently being used in power plant boilers.

Fluid coking and flexicoking are versatile processes that are applicable to a wide range of viscous feedstocks and provide a variety of products. The feedstock should have a carbon residue in excess of 5% w/w, and there is no upper limit on the carbon residue. Suitable feedstocks include vacuum residua, asphalt, and visbreaker residuum (visbreaker tar), heavy oil, extra heavy oil, and tar sand bitumen.

5. Other Processes

Typically, in terms of upgrading tar sand bitumen, the feedstock was processed (on the surface) by delayed coking or fluid coking to produce a synthetic crude oil (Fig. 8.10; Speight, 2013, 2014, 2017). In a variation of the *bitumen upgrading process*, bitumen from the separation plant is first sent to an atmospheric distillation tower where the resulting products are naphtha, low-boiling gas oil, high-boiling gas oil, and residuum (NRCAN, 2015). Each of the naphtha, low-boiling gas-oil, and high-boiling gas-oil streams is then sent to the appropriate (and separate) hydrotreater to remove sulfur and nitrogen by adding hydrogen. The resulting products from each of the separate hydrotreaters are then combined to produce synthetic crude oil. The residuum from the atmospheric distillation tower is further separated into two streams, viscous vacuum gas oil and vacuum residue. The viscous vacuum gas oil is mixed with the gas oil from the atmospheric distillation tower to undergo hydrotreating. The vacuum residuum is further upgraded either in a hydroconverter or in a coker to produce low-boiling liquid products and by-product coke (in the coker). The overhead product from the hydroconverter or coker is further separated into naphtha, low-boiling gas oil and high-boiling gas oil by distillation. Each of these three streams are mixed with the corresponding liquid product from the atmospheric distillation tower and then sent to the separate hydrotreaters to remove sulfur and nitrogen by adding hydrogen. The final products from the upgrading facility are synthetic crude oil and coke.

The *decarbonizing* thermal process is designed to minimize coke and gasoline yields but, at the same time, to produce maximum yields of gas oil and should not be confused with the propane decarbonizing process, which is a deasphalting process. The thermal decarbonizing process is essentially the same as the delayed coking process, but lower temperatures and pressures are employed.

For example, pressures range from 10 to 25 psi, heater outlet temperatures may be 485°C (905°F), and coke drum temperatures may be of the order of 415°C (780°F). Decarbonizing in this sense of the term should not be confused with *propane decarbonizing*, which is essentially a solvent deasphalting process (Chapter 12; Parkash, 2003; Gary et al., 2007; Speight, 2014, 2017; Hsu and Robinson, 2017).

High-temperature coking is a semicontinuous thermal conversion process designed for high-melting feedstocks that yields coke and gas oil as the primary products. The coke may be treated to remove sulfur to produce a low-sulfur coke ($\leq$ 5%), even though the feedstock contained as much as 5% w/w sulfur. In the process, the feedstock is transported to the pitch accumulator, then to the heater (370°C (700°F) and 30 psi), and finally to the coke oven, where temperatures may be as high as 980–1095°C (1800–2000°F). Volatile materials are fractionated, and after the cycle is complete, coke is collected for sulfur removal and quenching before storage.

Mixed-phase cracking (also called *liquid-phase cracking*) is a continuous thermal decomposition process for the conversion of viscous feedstocks to products boiling in the gasoline range. The process generally employs rapid heating of the feedstock (kerosene, gas oil, reduced crude, or even whole crude), after which it is passed to a reaction chamber and then to a separator where the vapors are cooled. Overhead products from the flash chamber are fractionated to gasoline components and recycle stock, and flash chamber bottoms are withdrawn as a heavy fuel oil. Coke formation, which may be considerable at the process temperatures (400–480°C, 750–900°F), is minimized by the use of pressures in excess of 350 psi.

Vapor-phase cracking is a high-temperature (545–595°C, 1000–1100°F), low-pressure (<50 psi) thermal conversion process that favors dehydrogenation of feedstock (gaseous hydrocarbon derivatives to gas oils) components to olefin derivatives and aromatic derivatives. Coke is often deposited in heater tubes, causing shutdowns. Relatively large reactors are required for these units.

Selective thermal cracking is a thermal conversion process that utilizes different conditions depending on the nature of the feedstock. For example, a heavy oil, extra heavy oil, and tar sand bitumen may be cracked at 494–515°C (920–960°F) and 300–500 psi; a low-boiling gas oil may be cracked at 510–530°C (950–990°F) and 500–700 psi (Fig. 8.12). Each feedstock has its own particular characteristics that dictate the optimum conditions of temperature and pressure for maximum yields of the products. These factors are utilized in selective combination of cracking units in which the more refractory feedstocks are cracked for longer periods of time or at higher temperatures than the less stable feedstocks, which are cracked at lower temperatures.

The process eliminates the accumulation of stable low-boiling material in the recycle stock and also minimizes coke formation from high-temperature cracking of the higher-boiling material. The end result is the production of fairly high yields of gasoline, middle distillates, and olefin gases.

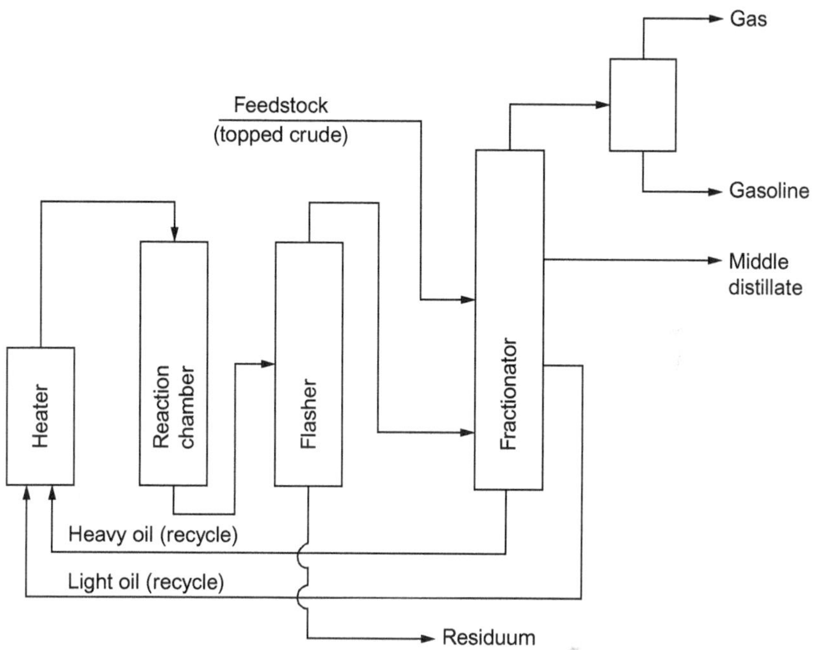

FIG. 8.12 Selective thermal cracking.

The thermal cracking of naphtha involves the upgrading of low-octane fractions of catalytic naphtha to higher-quality material. The process is designed, in fact, to upgrade the heavier portions of naphtha, which contain virgin feedstock, and to remove naphthene derivatives and paraffin derivatives. Some high-molecular-weight aromatic derivatives are produced by condensation reactions, and substantial quantities of olefin derivatives occur in the product streams.

6. Options for Heavy Feedstocks

The limitations of processing the more complex difficult-to-convert heavy oil, extra heavy oil, and tar bitumen depend to a large extent on the amount of non-volatile higher-molecular-weight constituents, which also contain the majority of the heteroatoms (i.e., nitrogen, oxygen, sulfur, and metals such as nickel and vanadium) (Chapter 1). The chemistry of the thermal reactions of some of these constituents dictates that certain reactions, once initiated, cannot be reversed and proceed to completion and coke is the eventual product (Parkash, 2003; Gary et al., 2007; Speight, 2014, 2017; Hsu and Robinson, 2017).

Upgrading viscous feedstocks, such as crude oil residua that are similar in character to heavy oil, extra heavy oil, and tar sand bitumen, began with the introduction of desulfurization processes that were designed to reduce the sulfur content of the feedstock and products therefrom. In the early days of such upgrading, the goal was desulfurization, but in later years, the processes were adapted to a 10%–30% partial conversion operation, as intended to achieve desulfurization and obtain low-boiling fractions simultaneously, by increasing severity in operating conditions.

Although new thermal cracking units are now under development for viscous feedstocks (Speight, 2011), processes that can be regarded as having evolved from the original concept of thermal cracking are visbreaking and the various coking processes.

In summary, there is a need to improve conversion of viscous feedstocks, and part of the future growth will be at or near recovery sites at heavy crude oil reservoirs, extra heavy crude oil deposits, and tar sand bitumen deposits in order to improve the quality to ease transportation and open markets for crudes of otherwise marginal value. Thus, refinery evolution has seen the introduction of a variety of viscous feedstock cracking processes (some use catalysts and are, of necessity, included here). These processes are different from one another in cracking method, cracked product patterns, and product properties and will be employed in refineries according to their respective features.

It is the purpose of this chapter to present these processes in relation to their use in modern refineries and the information that should be borne in mind when considering and deciding upon the potential utility of any process presented throughout this and subsequent chapters. In addition, the importance of solvents to mitigate coke formation has been recognized for many years (e.g., Langer et al., 1961), but their effects have often been ascribed to hydrogen-donor reactions rather than phase behavior. The separation of the phases depends on the solvent characteristics of the liquid. Addition of aromatic solvents will suppress phase separation, while paraffin-based solvents will enhance separation.

6.1 Aquaconversion Process

The *Aquaconversion process* (Fig. 8.7) consists of an improved version of the viscosity reduction process to decrease both the viscosity and the density of viscous feedstocks, achieving a conversion much larger than that of the conventional visbreaking. The novel catalytic system allows hydrogen from the water to be transferred to the resid when operated at the conditions normally used for the visbreaking process. This hydrogen incorporation is much lower than that obtained when using a deep hydroconversion process under high hydrogen partial pressure. Nevertheless, it is high enough to saturate the free radicals, formed within the thermal process, which would normally lead to polymerization reactions that form large asphaltenes and cause stability problems. With this hydrogen incorporation, a higher conversion level can be reached, thus enabling higher API and viscosity improvements to be achieved while maintaining product stability.

Put simply, the Aquaconversion process is a hydrovisbreaking technology that uses catalyst-activated transfer of hydrogen from water added to the feedstock. Reactions that lead to coke formation are suppressed, and there is no separation of asphaltene-type material (Marzin et al., 1998; Pereira et al., 1998, 2001; Parkash, 2003; Gary et al., 2007; Speight, 2014, 2017; Hsu and Robinson, 2017). The important aspect of the Aquaconversion technology is that it does not produce any solid by-product such as coke, nor requires any hydrogen source or high-pressure equipment. In addition, the Aquaconversion process can be implanted in the production area, and thus, the need for external diluent and its transport over large distances is eliminated.

Typically, visbreaking technology is limited in conversion level because of the stability of the resulting product. The Aquaconversion process extends the maximum conversion level within the stability specification by adding a homogeneous catalyst in the presence of steam.

Low-boiling distillates from the raw crude can be used as diluent for both the production and desalting processes. Also some catalyst processes have been developed such as catalytic aquathermolysis, which is used widely for upgrading viscous oil. In this process to maximize the upgrading effect, the suitable catalysts should be chosen.

In order to achieve the desired result, the catalyst precursor is dispersed in the feedstock by reaction with its polar components. The polar nature of the resulting mixture allows the catalyst to migrate toward the feedstock's more aromatic, multiring components. The catalyst, which is produced and activated by heat in the presence of water, then catalyzes the dissociation of water into hydrogen and oxygen radicals. As thermal cleavage of the carbon-carbon bonds progresses, hydrocarbon free radicals would begin to be formed. Unlike the typical visbreaking reaction sequence, where these materials react to eventually form asphaltene-type products, the Aquaconversion reaction mechanism reduces the tendency for the formation of high-molecular-weight products. The catalyst also accomplishes dealkylation of the alkyl aromatic structures to form smaller aromatic derivatives, hydrogen, and carbon dioxide (Yoshida et al., 1997). In addition, oxygen from the water saturates some hydrocarbon free radicals to form carbon oxides, mostly carbon dioxide because the thermodynamic equilibrium favors its formation at the Aquaconversion temperature. The entire reaction sequence can effectively terminate several undesirable asphaltene polymerization reactions and results in a catalytic steam visbroken product with lower asphaltene content than conventional visbreaking (Marzin et al., 1998).

6.2 Asphalt Coking Technology Process

The asphalt coking technology (ASCOT) process is a residual oil upgrading process that integrates the delayed coking process and the deep solvent deasphalting process (low-energy deasphalting, LEDA) (Bonilla, 1985; Bonilla and Elliott, 1987). Removing the deasphalted oil fraction prior to application of the delayed coking process has two benefits: (i) In the coking process, this fraction is thermally cracked to extinction, degrading this material as an FCC feedstock, and (ii) thermally cracking this material to extinction results in conversion of a significant portion to coke.

In the process, the feedstock is brought to the desired extraction temperature (50–230°C, 120–445°F, at 300–500 psig) and then sent to the extractor where solvent (straight-run naphtha and coker naphtha; solvent-to-oil ratio = 4:1 to 13:1) flows upward, extracting soluble material from the down-flowing feedstock. The solvent-deasphalted phase leaves the top of the extractor and flows to the solvent recovery system where the solvent is separated from the deasphalted oil and recycled to the extractor. The deasphalted oil is sent to the delayed coker (heater outlet temperature, 480–510°C, 900–950°F, at 15–35 psig, and a recycle ratio of 0.05:0.25 on fresh feedstock) where it is combined with the high-boiling coker gas oil from the coker fractionator and sent to the coker gas-oil stripper where low-boiling hydrocarbon derivatives are stripped off and returned to the fractionator.

The stripped deasphalted oil/high-boiling coker gas-oil mixture is removed from the bottom of the stripper and used to provide heat to the naphtha stabilizer-reboiler before being sent to battery limits as a cracking stock. The raffinate phase containing the asphalt and some solvent flows at a controlled rate from the bottom of the extractor and is charged directly to the coking section.

The solvent contained in the asphalt and deasphalted oil is condensed in the fractionator overhead condensers, where it can be recovered and used as lean oil for a propane/butane recovery in the absorber, eliminating the need to recirculate lean oil from the naphtha stabilizer. The solvent introduced in the coker heater and coke drums results in a significant reduction in the partial pressure of asphalt feed, compared with a regular delayed coking unit. The low asphalt partial pressure results in low coke and high liquid yields in the coking reaction.

With the ASCOT process, there is a significant reduction in by-product fuel as compared with either solvent deasphalting or delayed coking, and the process can be tailored to process a specific quantity or process to a specific quality of cracking stock (Parkash, 2003; Gary et al., 2007; Speight, 2014, 2017; Hsu and Robinson, 2017).

6.3 Cherry-P Process

The comprehensive heavy ends reforming refinery (Cherry-P) process is a process for the conversion of viscous crude oil or residuum into distillate and a cracked residuum (Ueda, 1976, 1978). In this process, the principal aim is to upgrade viscous crude oil residues at conditions between those of conventional visbreaking and delayed coking. Although coal is added to the feedstock, it is not intended to be a coprocessing feedstock, but the coal is intended to act as a scavenger to prevent the buildup of coke on the reactor wall. The use of scavengers in the process is projected to increase (Stark and Falkler, 2008; Stark et al., 2008; Speight, 2014). The superior surface activity of the asphaltene constituents will facilitate removal by adsorption on different surfaces (Speight, 1994, 2014, 2015a,b).

In the process, the feedstock is mixed with coal powder in a slurry mixing vessel (without a catalyst or hydrogen), heated in the furnace, and fed to the reactor where the feedstock undergoes thermal cracking reactions for several hours at a temperature higher than 400–450°C (750–840°F) and under pressure (70–290 psi) with a residence time on the order of 1–5 h. Gas and distillate from the reactor are sent to a fractionator, and the cracked residuum residue is extracted out of the system after distilling low-boiling fractions by the flash drum and vacuum flasher to adjust its softening point. Distillable product yields of 44% by weight on total feed are reported (Parkash, 2003; Gary et al., 2007; Speight, 2014, 2017; Hsu and Robinson, 2017). Since this yield is obtained when using anthracite, the proportion that is derived from the coal is likely to be very low and unlikely to cause compatibility reactions in downstream reactors due to the presence of phenols and other polar species (Speight, 1990, 2014).

6.4 Continuous Coking Process

A new coking process that can accept a viscous feedstock and continuously discharge vapor and dry crude oil coke particles has been developed and is designed to. A new coking process is designed to accept viscous feedstocks (including vacuum residua) from a conventional residuum heater in a coking plant and continuously discharges volatiles and dry coke particles (Sullivan, 2011).

In the process, the reactor/devolatilizer receives heavy oil, extra heavy oil, and tar sand bitumen from a fired heater at a temperature of 450–535°C (850–1000°F), similar to the heater discharge temperature from the heater in a delayed coke plant. The feed flows directly into one end of the horizontal reactor/devolatilizer. A kneading/mixing action inside the reactor/devolatilizer causes new surfaces of the residuum mass to be continuously exposed to the gas phase several times per second. This creation of new liquid-gas interface enhances the rapid mass transfer of volatiles into the gas phase similar to what happens with thin films. The volatile products are then rapidly cooled in a quench vessel to retard possible degradation by secondary cracking. With the rapid reduction of volatile content in the residuum mass, the carbonization reaction rates are accelerated, enabling continuous and rapid production of solid coke particles. The rapid release of volatiles also reduces the opportunity for the volatiles to react with the hot residuum and become locked in the coke. In order to reduce secondary cracking of the volatiles after they are released into the vapor phase, the vapor flows immediately (0.1–0.2 s) into the quench section after leaving the reactor. In the quench section, the temperature of the vapor is reduced to 51–80°C (125–175°F) by countercurrent flow against gas-oil (or residuum feedstock) liquid droplets in a low-pressure-drop vapor-liquid contactor. The gas-oil quench liquid also scrubs entrained solids out of the vapor stream and reduces the potential for downstream fouling.

The process also causes the carbonization reactions to proceed more rapidly, and it produces uniform composition and uniform size of coke particles that have a low volatile content. The process uses a kneading and mixing action to continuously expose new resid surface to the vapor space and causes a more complete removal of volatiles from the produced crude oil coke. Not only are more valuable volatiles recovered, but also the volatiles are likely to be richer in middle distillates. As a result of kneading/mixing action by the reactor/devolatilizer, new surfaces of the residuum mass are continuously exposed to the gas phase, enhancing the rapid mass transfer of volatiles into the gas phase. The volatiles are then rapidly cooled to retard degradation. With the rapid reduction of volatiles content in the resid mass, the carbonization reaction rates are accelerated, enabling continuous and rapid production of solid crude oil coke particles. The short contact time of the volatiles with the hot residuum minimizes thermal degradation of volatiles.

Concurrently with the carbonization reactions and the formation of coke, some cracking of side chains off the larger molecules likely occurs. These smaller, low-boiling molecules produced from cracking reactions join the population of the indigenous volatiles. Some volatiles may be generated even after the solid coke is formed. In the delayed coking process, many of these late-forming volatiles remain trapped in the coke. The process promotes the release of these late-forming cracked volatiles, allowing them to escape into the gas phase by breaking the solid coke into small particles.

In addition to the recovery of additional and more valuable volatiles, there are other benefits of the new process compared with the delayed coking process. The consumption of utilities is less because no steam or water is required. Since there is no quenching, energy from the hot coke is recovered. The process is continuous so is never opened to the atmosphere. There is no cutting procedure as in the delayed coking process where high-pressure water is used to cut the coke out of the drums. Therefore, no volatiles and no coke particles are released into the atmosphere.

6.5 ET-II Process

The ET-II process is a thermal cracking process for the production of distillates and cracked residuum for use as metallurgical coke and is designed to accommodate feedstocks such as heavy oil, extra heavy oil, and tar sand bitumen (Kuwahara, 1987). The distillate (*cracked oil*) is suitable as a feedstock to hydrocracker and fluid catalytic cracking. The basic technology of the ET-II process is derived from that of the original Eureka process.

In the process, the feedstock is heated up to 350°C (660°F) by passage through the preheater and fed into the bottom of the fractionator, where it is mixed with recycle oil, and the high-boiling fraction of the cracked oil. The ratio of recycle oil to feedstock is within the range 0.1%–0.3% by weight. The feedstock mixed with recycle oil is then pumped out and fed into the cracking heater, where the temperature is raised to approximately 490–495°C (915–925°F), and the outflow is fed to the stirred-tank reactor where it is subjected to further thermal cracking. Both cracking and condensation reactions take place in the reactor.

The cracked oil and gas products, together with steam from the top of the reactor, are introduced into the fractionator where the oil is separated into two fractions, *cracked light oil* and *vacuum gas oil and pitch* (Parkash, 2003; Gary et al., 2007; Speight, 2014, 2017; Hsu and Robinson, 2017).

The heat required for the cracking reaction is brought in by the effluent itself from the cracking heater and by the superheated steam, which is heated in the convection section of the cracking heater and blown into the reactor bottom. The superheated steam reduces the partial pressure of the hydrocarbon derivatives in the reactor and accelerates the stripping of volatile components from the cracked residuum. This residual product is discharged through a transfer pump and transferred to a cooling drum, where the thermal cracking reaction is terminated by quenching with a water spray after which it is sent to the pitch water slurry preparation unit.

6.6 Eureka Process

The Eureka process is a thermal cracking process to produce a cracked oil and aromatic residuum from nonvolatile residual materials (Aiba et al., 1981; Parkash, 2003; Gary et al., 2007; Ohba et al., 2008; AlHumaidan et al., 2013a,b; Speight, 2014, 2017; Hsu and Robinson, 2017). The cracking reactions occur under lower cracked oil partial pressure by introducing steam into the reactor. The unconverted cracked residuum (pitch) in the reactor behaves as a homogeneous system that provides stable and trouble-free operating conditions. The cracked oil is further hydrotreated, cracked, and/or hydrocracked to produce marketable fuels, and the cracked residuum is utilized as a boiler fuel or as a gasification (partial oxidation) feedstock for hydrogen production or synthesis gas production (Chapter 13).

In this process, the viscous feedstock is fed to the preheater and then enters the bottom of the fractionator, where it is mixed with the recycle oil. The mixture is then fed to the reactor system that consists of a pair of reactors operating alternately. In the reactor, thermal cracking reaction occurs in the presence of superheated steam that is injected to strip the cracked products out of the reactor and supply a part of heat required for cracking reaction. At the end of the reaction, the bottom product is quenched. The oil and gas products (and steam) pass from the top of the reactor to the lower section of the fractionator, where a small amount of entrained material is removed by a wash operation. The upper section is an ordinary fractionator, where the heavier fraction of cracked oil is drawn as a sidestream. The process bottoms (pitch) can be used as boiler fuel, as partial oxidation feedstock for producing hydrogen and carbon monoxide, and as binder pitch for manufacturing metallurgical coke (Parkash, 2003; Gary et al., 2007; Speight, 2014, 2017; Hsu and Robinson, 2017).

The process reactions proceed at lower cracked oil partial pressure by injecting steam into the reactor, keeping crude oil pitch in a homogeneous liquid state, and unlike a conventional delayed coker, a higher cracked oil yield can be obtained. A wide range of heavy oil, extra heavy oil, and tar sand bitumen can be used as feedstock, such as atmospheric and vacuum residue of crude oils, various cracked residues, asphalt products from solvent deasphalting, and native asphalt. After hydrotreating, the cracked oil is used as feedstock for a fluid catalytic cracker or hydrocracker.

The original Eureka process uses two batch reactors, while the newer ET-II and high conversion soaker cracking (HSC) processes both employ continuous reactors.

6.7 FTC Process

The fluid thermal cracking (FTC) process is a heavy oil, extra heavy oil, and tar sand bitumen upgrading process in which the feedstock is thermally cracked to produce distillate and coke (Miyauchi et al., 1981; Miyauchi et al., 1987; Parkash, 2003; Gary et al., 2007; Speight, 2014, 2017; Hsu and Robinson, 2017).

The feedstock, mixed with recycle stock from the fractionator, is injected into the cracker and is immediately absorbed into the pores of the particles by capillary force and is subjected to thermal cracking. In consequence, the surface of the noncatalytic particles is kept dry, and good fluidity is maintained allowing a good yield of and selectivity for middle distillate products. Hydrogen-containing gas from the fractionator is used for the fluidization in the cracker. Excessive coke caused by the metals accumulated on the particle is suppressed under the presence of hydrogen. The particles with deposited coke from the cracker are sent to the gasifier, where the coke is gasified and converted into carbon monoxide (CO), hydrogen (H_2), carbon dioxide (CO_2), and hydrogen sulfide (H_2S) with steam and air. Regenerated hot particles are returned to the cracker.

6.8 HSC Process

The high conversion soaker cracking (HSC) process is a cracking process designed for moderate conversion, higher than visbreaking but lower than coking (Watari et al., 1987; Washimi, 1989; Parkash, 2003; Gary et al., 2007; Speight, 2014, 2017; Hsu and Robinson, 2017). The process is an advanced continuous thermal cracking technology with a proprietary soaking drum, featuring a wide range of conversion levels between visbreaking and coking while producing pumpable liquid residue at process temperature. A broad range of viscous feedstocks such as heavy crude, long and short residue with high contents of sulfur and heavy metals, and even visbroken residue can be charged to the HSC process. The cracked distillates from the HSC process are mostly low-boiling gas oil and high-boiling gas oil with fewer unsaturated compounds than coker distillates. The high-boiling gas-oil fraction serves as the feedstock to the fluid catalytic cracking unit, and the cracked residue can be used as the fuel for boiler at the power station. The process uses no hydrogen, no catalyst, and no high-pressure equipment. The process economics is benefited by low investment cost and low utility consumptions due to its simple process scheme as visbreaking process. The process features less gas make and a higher yield of distillate compared with other thermal cracking processes. The process can be used to convert a wide range of feedstocks with high sulfur and metal content, including heavy oil, extra heavy oil, and tar sand bitumen. As a note of interest, the HSC process employs continuous reactors, whereas the original Eureka process (*q.v.*) uses two batch reactors.

In the process, the preheated feedstock enters the bottom of the fractionator, where it is mixed with the recycle oil. The mixture is pumped up to the charge heater and fed to the soaking drum (ca. atmospheric pressure, steam injection at the top and bottom), where sufficient residence time is provided to complete the thermal cracking. In the soaking drum, the feedstock and some product flow downward passing through a number of perforated plates while steam with cracked gas and distillate vapors flow through the perforated plates countercurrently.

The volatile products from the soaking drum enter the fractionator where the distillates are fractionated into desired product oil streams, including a high-boiling gas-oil fraction. The cracked gas product is compressed and used as refinery fuel gas after sweetening. The cracked oil product after hydrotreating is used as fluid catalytic cracking or hydrocracker feedstock. The residuum is suitable for use as boiler fuel, road asphalt, binder for the coking industry, and a feedstock for partial oxidation.

6.9 Selective Cracking Process

Selective cracking is a thermal conversion process that utilizes different conditions depending on feedstock composition. For example, a feedstock such as heavy oil, extra heavy oil, and tar sand bitumen may be cracked at 494–515°C (920–960°F) and 300–500 psi. A lower-boiling gas oil may be cracked at 510–530°C (950–990°F) and 500–700 psi (Moschopedis et al., 1998; Parkash, 2003; Gary et al., 2007; Speight, 2014, 2017; Hsu and Robinson, 2017).

The process eliminates the accumulation of stable low-boiling material in the recycle stock and also minimizes coke formation from high-temperature cracking of the higher-boiling material. The end result is the production of fairly high yields of gasoline, middle distillates, and olefin gases.

6.10 Shell Thermal Cracking Process

The Shell thermal distillate cracking unit is based on the principle of converting high-boiling feedstocks to lower-boiling products (Parkash, 2003; Gary et al., 2007; Speight, 2014, 2017; Hsu and Robinson, 2017). Thermal cracking of heavy oil, extra heavy oil, and tar sand bitumen takes place in the liquid phase in a furnace, at elevated pressure and temperature, and the products are residuum and distillate products.

In the process, the feedstock (e.g., the gas-oil fraction from the atmospheric distillation unit or vacuum gas oil from the vacuum distillation unit) is sent to a surge drum. Liquid from this drum is pumped to the distillate heater, which typically operates at a pressure of ~490°C (915°C) and 290 psi. Under these conditions, the cracking reactions take place in the liquid phase. Fluid from the distillate heater is then routed to the combi tower where separation is achieved between residue, gas-oil, and lower-boiling products. In addition, a gas-oil fraction is taken from the combi tower, returned to the surge drum, and then recycled through the distillate heater. The bottom product of the combi tower is routed to a vacuum flasher where high-boiling gas oil is recovered from the residuum stream and routed back to the distillate heater.

The vacuum flashed residuum from the vacuum flasher can be routed to fuel oil blending or can be used internally as refinery fuel. The recycling of high-boiling gas oil from both the vacuum flasher and the combi tower to the distillate heater means that all of the high-boiling gas oil is converted to lower-boiling products. Low-boiling gas oil from the combi tower is first stripped and is then routed to a hydrotreater. Alternatively, the low-boiling gas oil can be used as cutter stock. The feedstock and product requirements of the thermal distillate cracking process are flexible, and the process has the capability to optimize conversion through adjustment of the recycle rate of the high-boiling gas-oil fraction.

The *Shell deep thermal gas-oil process* is a combination of the *Shell deep thermal conversion process* and the *Shell thermal gas-oil process*. In this alternative high conversion scheme, the heavy gas oil (HGO) from the atmospheric distillation unit and the vacuum gas oil (VGO) from the vacuum flasher or vacuum distillation unit are cracked in a distillate thermal cracking heater into lower-boiling range gas oil.

A related process, the *deep thermal conversion* (DTC) process (Fig. 8.13) offers a bridge between visbreaking and coking and provides maximum distillate yields by applying deep thermal conversion followed by vacuum flashing of the products. In the process, the heated vacuum residuum is charged to the heater and from there to the soaker where conversion occurs. The products are then led to an atmospheric fractionator to produce gases, naphtha, kerosene, and gas oil. The fractionator residuum is sent to a vacuum flasher that recovers additional gas oil and distillate. The next steps for the coke are dependent on its potential use, and it may be isolated as *liquid* coke (pitch, cracked residuum) or

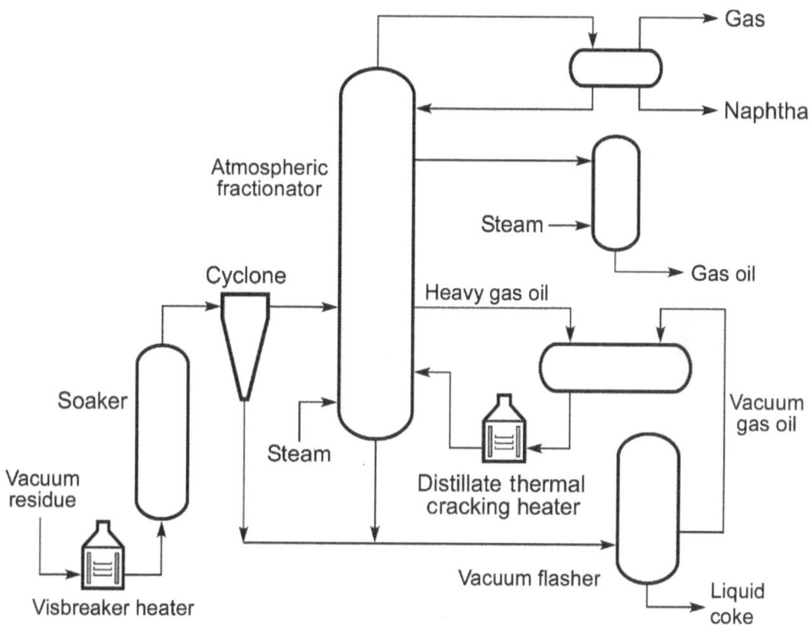

FIG. 8.13 The deep thermal conversion (DTC) process.

solid coke. The process yields a maximum of distillates from viscous feedstocks (such as vacuum residua, heavy oil, extra heavy oil, and tar sand bitumen) and by vacuum flashing the cracked residue. The liquid coke, not suitable for blending to commercial fuel, is used for specialty products, gasification, and/or combustion, for example, to generate power and/or hydrogen.

6.11 Tervahl-T Process

The Tervahl T process offers options that allow the process to accommodate differences in the feedstock and the desired sale of products. In the process (LePage et al., 1987; Peries et al., 1988; Parkash, 2003; Gary et al., 2007; Speight, 2014, 2017; Hsu and Robinson, 2017), the feedstock is heated to the desired temperature using the coil heater, and heat is recovered in the stabilization section and held for a specified residence time in the soaking drum. The soaking drum effluent is quenched and sent to a conventional stabilizer or fractionator where the products are separated into the desired streams (Parkash, 2003; Gary et al., 2007; Speight, 2014, 2017; Hsu and Robinson, 2017). The gas produced from the process is used for fuel.

In the related Tervahl H process (a hydrogenation process but covered here for convenient comparison with the Tervahl T process), the feedstock and hydrogen-rich stream are heated using heat recovery techniques and fired heater and held in the soak drum as in the Tervahl T process. The gas and oil from the soaking drum effluent are mixed with recycle hydrogen and separated in the hot separator where the gas is cooled, passed through a separator, and recycled to the heater and soaking drum effluent. The liquids from the hot and cold separator are sent to the stabilizer section where purge gas and synthetic crude are separated. The gas is used as fuel, and the synthetic crude can now be transported or stored.

References

Aiba, T., Kaji, H., Suzuki, T., Wakamatsu, T., 1981. The Eureka process. Chem. Eng. Progr. (February), 37.

AlHumaidan, F., Hauser, A., Al-Rabiah, H., Lababidi, H., Bouresli, R., 2013a. Studies on thermal cracking behavior of vacuum residues in Eureka process. Fuel 109, 635–646.

AlHumaidan, F., Haitham, M.S., Lababidi, H., Al-Rabiah, H., 2013b. Thermal cracking kinetics of Kuwaiti vacuum residues in Eureka process. Fuel 109, 923–931.

ASTM D, 2007. Standard test method for characteristic groups in rubber extender and processing oils and other petroleum-derived oils by the clay-gel absorption chromatographic method. In: Annual Book of Standards. ASTM International, West Conshohocken, PA, p. 2015.

ASTM D3279, 2015. Standard test method for n-heptane Insolubles. In: Annual Book of Standards. ASTM International, West Conshohocken, PA.

ASTM D4124, 2015. Standard test method for separation of asphalt into four fractions. In: Annual Book of Standards. ASTM International, West Conshohocken, PA.

ASTM D893, 2015. Standard test method for insolubles in used lubricating oils. In: Annual Book of Standards. ASTM International, West Conshohocken, PA.

Ballard, W.P., Cottington, G.I., Cooper, T.A., 1992. Cracking, thermal. In: McKetta, J.J. (Ed.), Crude Oil Processing Handbook. Marcel Dekker, New York, p. 309.

Blaser, D.E., 1992. Coking, crude oil (fluid). In: McKetta, J.J. (Ed.), Petroleum Processing Handbook. Marcel Dekker, New York, p. 255.

Bonilla, J., 1985. Energy Prog. 5 (4), 239–244.

Bonilla, J., Elliott, J.D., 1987. Asphalt Coking Method. United States Patent 4,686,027, August 11.

Carrillo, J.A., Corredor, L.M., 2013. Heavy crude oil upgrading: Jazmin crude. Adv. Chem. Eng. Sci. 3, 46–55.

Dickenson, R.L., Biasca, F.E., Schulman, B.L., Johnson, H.E., 1997. Hydrocarb. Process. 76 (2), 57.

Feintuch, H.M., Negin, K.M., 2004. FW delayed-coking process. In: Meyers, R.A. (Ed.), Handbook of Petroleum Refining Processes. second ed. McGraw-Hill, New York, p. 33 (Chapter 12).

Gary, J.G., Handwerk, G.E., Kaiser, M.J., 2007. Petroleum Refining: Technology and Economics, fifth ed. CRC Press, Taylor & Francis Group, Boca Raton, FL.

Germain, J.E., 1969. Catalytic Conversion of Hydrocarbons. Academic Press, New York.

Hsu, C.S., Robinson, P.R. (Eds.), 2017. Handbook of Petroleum Technology. Springer International Publishing AG, Cham.

Joshi, J.B., Pandit, A.B., Kataria, K.L., Kulkarni, R.P., Sawarkar, A.N., Tandon, D., Ram, Y., Kumar, M.M., 2008. Ind. Eng. Chem. Res. 47, 8960–8988.

Kobe, K.A., McKetta, J.J., 1958. Advances in Petroleum Chemistry and Refining. Interscience, New York.

Koshka, E., Kuhach, J., Veith, E., 2008. Improving Athabasca bitumen development economics through integration with HTL upgrading. In: Proceedings of the World Heavy Oil Congress, Edmonton, Alberta, Canada, Alberta Department of Energy, Edmonton, Alberta, Canada, March.

Kuwahara, I., 1987. The ET-II process. Koagaku Kogaku 51, 1.

Langer, A.W., Stewart, J., Thompson, C.E., White, H.T., Hill, R.M., 1961. Thermal hydrogenation of crude residua. Ind. Eng. Chem. 53, 27–30.

Langer, A.W., Stewart, J., Thompson, C.E., White, H.T., Hill, R.M., 1962. Hydrogen donor diluent visbreaking of residua. Ind. Eng. Chem. Process Design Dev. 1, 309–312.

LePage, J.F., Morel, F., Trassard, A.M., Bousquet, J., 1987. Preprint. Div. Fuel Chem. 32, 470.

Magaril, R.Z., Aksenova, E.I., 1968. Study of the mechanism of coke formation in the cracking of petroleum resins. Int. Chem. Eng. 8, 727–729.

Magaril, R.Z., Ramazaeva, L.F., Aksenova, E.I., 1971. Kinetics of the formation of coke in thermal processing of crude oil. Int. Chem. Eng. 11 (2), 250–251.

Marano, J.J., 2003. Refinery technology profiles: gasification and supporting technologies. In: Report Prepared for the United States Department of Energy, National Energy Technology Laboratory. United States Energy Information Administration, Washington, DC. June.

Marzin, R., Pereira, P., McGrath, M.J., Feintuch, H.M., Thompson, G., 1998. A new option for residue conversion and heavy oil upgrading. Oil Gas J. 97 (44), 79.

McKinney, J.D., 1992. Coking, petroleum (delayed and fluid). In: McKetta, J.J. (Ed.), Petroleum Processing Handbook. Marcel Dekker, New York, p. 245.

Miles, J., 2009. Maximizing distillate yields and refinery economics—an alternative solution to conventional fuel oil production or residue conversion. In: Proceedings of the Session A, 14th Annual Meeting—European Refining Technology Conference, November 11.

Miyauchi, T., Furusaki, S., Morooka, Y., 1981. Upgrading resid. In: Advances in Chemical Engineering. Academic Press, New York (Chapter 11).

Miyauchi, T., Tsutsui, T., Nozaki, Y., 1987. A new fluid thermal cracking process for upgrading resid. In: Paper 65B, Proceedings of the Spring National Meeting, American Institute of Chemical Engineers, Houston, March 29.

Moschopedis, S.E., Ozum, B., Speight, J.G., 1998. Upgrading heavy oils. Rev. Process Chem. Eng. 1 (3), 201–259.

Mushrush, G.W., Speight, J.G., 1995. Petroleum Products: Instability and Incompatibility. Taylor & Francis Publishers, Philadelphia, PA.

Negin, K.M., Van Tine, F.M., 2004. FW/UOP visbreaking process. In: Meyers, R.A. (Ed.), Handbook of Petroleum Refining Processes. second ed. McGraw-Hill, New York, p. 3 (Chapter 12).

NRCAN, 2015. Upgrading and Refining Process Development. National Research Council of Canada, Ottawa.http://www.nrcan.gc.ca/energy/oil-sands/upgrading/5879. [(Accessed 15 August 2015)].

Ohba, T., Shibutani, I., Watari, R., Inomata, J., Nagata, H., 2008. The advanced EUREKA process: environment friendly thermal cracking process. In: Paper No. WPC-19-2856, Proceedings of the 19th World Crude Oil Congress, Madrid, Spain, June 29–July 3.

Parkash, S., 2003. Refining Processes Handbook. Gulf Professional Publishing, Elsevier, Amsterdam.

Pereira, P., Marzin, R., McGrath, M. and Thompson, G.J. 1998. Proceedings of the 17th World Energy Congress, Houston, TX.

Pereira, P., Flores, C., Zbinden, H., Guitian, J., Solari, R.B., Feintuch, H., Gillis, D., 2001. Oil Gas J. May 14.

Peries, J.P., Quignard, A., Farjon, C., Laborde, M., 1988. Thermal and catalytic ASVAHL processes under hydrogen pressure for converting heavy crudes and conventional residues. Rev. Inst. Fr. Pétrol. 43 (6), 847–853.

Roundtree, E.M., 1997. Fluid coking. In: Meyers, R.A. (Ed.), Handbook of Petroleum Refining Processes. second ed. McGraw-Hill, New York, p. 1 (Chapter 12).

Schucker, R.C., 2003. Heavy Oil Upgrading Process. United States Patent 6,524,469, February 25.

Silverman, M.A., Pavel, S.K., Hillerman, M.D., 2011. HTL heavy oil upgrading: a key solution for heavy oil upstream and midstream operations. In: Paper No. WHOC11–419, Proceedings of the World Heavy Oil Congress, Edmonton, Alberta, Canada.

Speight, J.G., 1990. Tar sands. In: Speight, J.G. (Ed.), Fuel Science and Technology Handbook. Marcel Dekker, New York (Chapters 12–16).

Speight, J.G., 1994. Chemical and physical studies of petroleum asphaltenes. In: Yen, T.F., Chilingarian, G.V. (Eds.), Asphaltenes and Asphalts. I. Developments in Petroleum Science. In: vol. 40. Elsevier, Amsterdam (Chapter 2).

Speight, J.G., 2000. The Desulfurization of Heavy Oils and Residua, second ed. Marcel Dekker, New York.

Speight, J.G., 2008. Synthetic Fuels Handbook: Properties, Processes, and Performance. McGraw-Hill, New York.

Speight, J.G., 2011. The Refinery of the Future. Gulf Professional Publishing, Elsevier, Oxford.

Speight, J.G., 2013. Heavy and Extra-Heavy Oil Upgrading Technologies. Gulf Professional Publishing, Elsevier, Oxford.

Speight, J.G., 2014. The Chemistry and Technology of Petroleum, fifth ed. CRC Press, Taylor & Francis Group, Boca Raton, FL.

Speight, J.G., 2015a. Fouling in Refineries. Gulf Professional Publishing, Elsevier, Oxford.

Speight, J.G., 2015b. Handbook of Petroleum Product Analysis, second ed. John Wiley & Sons, Hoboken, NJ.

Speight, J.G., 2017. Handbook of Petroleum Refining. CRC Press, Taylor & Francis Group, Boca Raton, FL.

Stark, J.L., Falkler, T., 2008. Method for Improving Liquid Yield During Thermal Cracking of Hydrocarbons. United States Patent 7,425,259 September 16, 2008.

Stark, J.L., Falkler, T., Weers, J.J., Zetlmeisl, M.J., 2008. Method for Improving Liquid Yield During Thermal Cracking of Hydrocarbons. United States Patent 7,416,654 August 26.

Stell, R.C., Balinsky, G.J., McCoy, J.N., Keusenkothen, P.F., 2009a. Process and Apparatus for Cracking Hydrocarbon Feedstock Containing Resid. United States Patent 7,588,737, September 15.

Stell, R.C., Dinicolantonio, A.R., Frye, J.M., Spicer, D.B., McCoy, J.N., Strack, R.D., 2009b. Process for Steam Cracking Heavy Hydrocarbon Feedstocks. United States Patent 7,578,929, August 25.

Stephens, M.M., Spencer, O.F., 1956. Petroleum Refining Processes. Penn State University Press, University Park, PA.

Sullivan, D.W., 2011. New continuous coking process. In: Proceedings of the 14th Topical Symposium on Refinery Processing, AIChE Spring Meeting and Global Congress on Process Safety, Chicago, IL, March 13-17.

Ueda, K., 1976. J. Jpn. Pet. Inst. 19 (5), 417.

Ueda, H., 1978. J. Fuel Soc. Jpn. 57, 963.

Veith, E.J., 2007. Performance of heavy-to-light-crude-oil upgrading process. In: Proceedings of the SPE International Oil Conference and Exhibition, Veracruz, Mexico, June 27–30, Society of Crude oil Engineers, Richardson, TX.

Washimi, K., 1989. Hydrocarb. Process. 68 (9), 69.

Watari, R., Shoji, Y., Ishikawa, T., Hirotani, H., and Takeuchi, T. 1987. Annual Meeting, National Petroleum Refiners Association, San Antonio, TX, Paper AM-87-43.

Wodnik, R., Hughes, G.C., 2005. Delayed coking advances. Pet. Technol. Q. Q4, 1–6.

Yoshida, R., Miyazawa, M., Ishiguro, H., Itoh, S., Haraguchi, K., Nagaishi, H., Narita, H., Yoshida, T., Maekawa, Y., Mitarai, Y., 1997. Chemical structure changes in cold lake oil-sand bitumen and catalytic activities during catalytic hydrotreatment. Fuel Process. Technol. 51, 195–203.

Chapter 9

Catalytic Cracking Processes

1. Introduction

Catalytic cracking is a conversion process that can be applied to a variety of feedstocks ranging from gas oil to oil and residuum (Parkash, 2003; Gary et al., 2007; Speight, 2014, 2017; Hsu and Robinson, 2017). The concept of catalytic cracking is basically the same as thermal cracking, but it differs by the use of a catalyst that is not (in theory) consumed in the process, and it is one of several practical applications used in a refinery that employ a catalyst to improve process efficiency and product slate. Catalytic cracking is different to thermal cracking insofar as a catalyst is used for the catalytic process. The mechanism of catalytic cracking is also different insofar as the mechanism of the thermal process involves free-radical intermediates, whereas the mechanism of the catalytic process involves ionic intermediates (Table 9.1). However, there has been the claim that, in some cases, the feedstock constituents decompose by thermolysis before the species may come into contact with the catalyst.

Catalytic cracking is widely used to convert viscous feedstocks—such as heavy oil, extra heavy oil, and tar sand bitumen—into more valuable naphtha (a blend stock for gasoline manufacture) and other low-boiling products. As the demand for gasoline increased, catalytic cracking replaced thermal cracking with the evolution of catalytic cracking. Fluid catalytic cracking (FCC) refers to the behavior of the catalyst during this process insofar as the fine, powdered catalyst (typically zeolites, which have a particle size on the order of 70 μm) takes on the properties of a fluid when it is mixed with the vaporized feed. Fluidized catalyst particles circulate continuously between the reaction zone and the regeneration zone.

In terms of process parameters, catalytic cracking is typically performed at temperatures ranging from 485 to 540°C (900–1000°F) and pressures up to 100 psi. Feedstocks for the process have typically been the atmospheric and/or the vacuum gas-oil fraction of crude oil (Table 9.2), but the focus is shifting to gas-oil-residuum blends, gas-oil-viscous feedstock blends, and gas-oil-bitumen blends (Shidhaye et al., 2015). In some cases, heavy oil, extra heavy oil, and tar sand bitumen had been blended with the minimum amount of gas oil (added as a flux) as the feedstock to catalytic cracking units. In the process,

Heavy Oil Recovery and Upgrading. https://doi.org/10.1016/B978-0-12-813025-4.00009-X

TABLE 9.1 Comparison of Thermal Cracking and Catalytic Cracking

Thermal cracking

- No catalyst
- Free-radical reaction mechanism
- Moderate yields of naphtha and other distillates
- Gas yields are feedstock-dependent
- Low-to-moderate product selectivity
- Low-octane naphtha
- Low-to-moderate yield of C_4 olefins
- • Low-to-moderate yields of aromatic derivatives

Catalytic cracking

- Uses a catalyst
- Ionic reaction mechanism
- High yields of naphtha and other distillates
- Lower gas yields than the thermal process
- High product selectivity
- Low yields of *n*-alkanes—high-octane naphtha
- Chain branching and high yield of C_4 olefins
- • High yields of aromatic derivatives

TABLE 9.2 Preferred Composition of the FCC Feedstock

Test	Data Range
Gravity, API	19.5–23.0
Density at 15°C, kg/L	0.9153–0.9366
Distillation (D 1160),°C (°F)	
Initial boiling point	275 (525)
End point	345 (655)
Flash point	116–143 (240–290)
Pour point	17–38 (60–100)
Viscosity at 50°C (122°F), cSt	20–50
Sulfur, % w/w	1.1–1.4
Carbon residue, % w/w	0.1–0.5
Aniline point	73–79 (163–174)
Asphaltene content, % w/w	< 2
Nitrogen, ppm	1200–1700
Basic nitrogen, ppm	400–600
Vanadium, ppm	< 0.05
Nickel, ppm	< 0.10

the feedstock enters the unit at temperatures on the order of 485–540°C (900–1000°F), and the circulating catalyst provides heat from the regeneration zone to the oil feed. Carbon (coke) on the catalyst is burned off in the regenerator, raising the catalyst temperature to 620–735°C (1150–1350°F), before the catalyst returns to the reactor.

The preferred feedstock to a fluid catalytic cracking unit has been and continues to be the portion of the crude oil that has an initial boiling point (at atmospheric pressure) of ~275°C (525°F) up to the initial boiling point of the atmospheric residuum (345°C, 655°F) (Table 9.2). On occasion, the vacuum gas oil (boiling range, 345–510 or 345–565°C, 655–950 or 655–1050°F) may also be used as feedstock to the fluid catalytic cracking unit. However, the changing slate of refinery feedstocks has caused this to change. Currently, the feedstocks for catalytic cracking can be any one (or blends) of the following: (i) straight-run gas oil, (ii) vacuum gas oil, (iii) atmospheric residuum, and (iv) vacuum residuum, with special emphasis on the heavier feedstocks such as heavy oil, extra heavy oil, and tar sand bitumen (Lifschultz, 2005; Ross et al., 2005; Speight, 2014). If blends of the above feedstocks are employed, compatibility of the constituents of the blends (i.e., no phase separation) must be assured, or excessive coke (and metals) will be laid down onto the catalyst.

In addition, there are several *pretreatment options* for the feedstocks that offer process benefits, and these are (i) deasphalting to prevent excessive coking on catalyst surfaces; (ii) demetallization, that is, removal of nickel, vanadium, and iron to prevent catalyst deactivation; (iii) the use of a short residence time as a means of preparing the feedstock; (iv) hydrotreating or mild hydrocracking to prevent excessive coking in the fluid catalytic cracking unit; (v) blending with an aromatic gas-oil type to prevent phase separation; and (vi) staged partial conversion (Birch and Ulivieri, 2000; Speight, 2000, 2014, 2017; Patel et al., 2002, 2004; Gary et al., 2007; Dziabala et al., 2011; Hsu and Robinson, 2017). Hydrotreating the feedstock to the fluid catalytic cracker improves the yield and quality of naphtha (Table 9.3) and reduces the sulfur oxide (SO_x) emissions from the catalytic cracker unit (Sayles and Bailor, 2005). Refineries wishing to process heavier crude oil may only have the option to desulfurize the resulting high-sulfur naphtha produced in the process.

On a global basis, the effect of declining crude quality (Speight, 2011a, 2014) may be looked upon as influencing the fluid catalytic cracking feedstock quality and amount. However, this will be a secondary factor compared with the changes required in the refined product slate (Benazzi, and Cameron, 2005). In addition to the heavier viscous crude oils—as a blend or as a hydrotreated feedstock—the production of synthetic crude oil from tar sand bitumen will increase dramatically in the next decade (Patel, 2007; Speight, 2008, 2009, 2011a). For example, the synthetic crude oil from Canadian tar sand sources is projected to increase to 3.0 million barrels per day by 2015. With Canadian reserves in excess of 170 billion barrels (170×10^9 bbl) of viable oil, economic forecasts predict that tar sand deposits will continue to be a significant crude source

TABLE 9.3 Feedstock and Product Data for the Fluid Catalytic Process With and Without Feedstock Hydrotreating

	Without Feedstock Hydrotreating	With Feedstock Hydrotreating
Feedstock (> 370°C, > 700°F)		
API	15.1	20.1
Sulfur, % w/w	3.3	0.5
Nitrogen, % w/w	0.2	0.1
Carbon residue, % w/w	8.9	4.9
Nickel + vanadium, ppm	51.0	7.0
Products		
Naphtha (C5–221°C, C5–430°F), % v/v	50.6	58.0
Light cycle oil (221–360°C, 430–680°F), % v/v	21.4	18.2
Residuum (>360°C, > 680°F), % w/w	9.7	7.2
Coke, % w/w	10.3	7.0

(and, hence, feedstock to the catalytic cracking unit) for the foreseeable future (Schiller, 2011). With the increasing focus to reduce sulfur content in fuels, the role of *desulfurization* in the refinery becomes more and more important. Currently, the process of choice is the hydrotreater in which the unfinished fuel is hydrotreated to remove sulfur from the fuel. Hydrotreating of the feedstock to the catalytic cracking unit can increase conversion by 8%–12% v/v and with most feedstocks (Salazar-Sotelo et al., 2004). Thus, it will be possible to reduce the sulfur content of the naphtha/gasoline product to levels low enough to meet the future low-sulfur gasoline pool specifications.

Finally, the use of biofeedstocks (such as animal fats, vegetable oils, cellulosic materials, and lignin) in the fluid catalytic cracking unit will be used to increase the yield of light cycle oil and will also provide high-quality products in terms of cetane number (Speight, 2008). Practical implementation in a refinery will be accompanied by blending with vacuum gas oil or resid (Speight, 2011a). From a strategic point of view, refiners should not try to compete with biofuel producers, but rather try to use renewable feedstocks in traditional crude oil refining processes and make products that are compatible with conventional hydrocarbon fuels (Speight, 2008, 2011a).

Furthermore, fluid catalytic cracking technology represents one of the most expanded processes producing the precursors to liquid fuels (naphtha and kerosene) and automobile fuels from gas-oil distillates and from viscous feedstocks. A key factor is the use of active, stable, and selective (*tailor-made*) catalysts to convert specific feedstocks (especially viscous feedstocks) into desired products. Thus, the refinery process (Fig. 9.1) can be applied to a variety

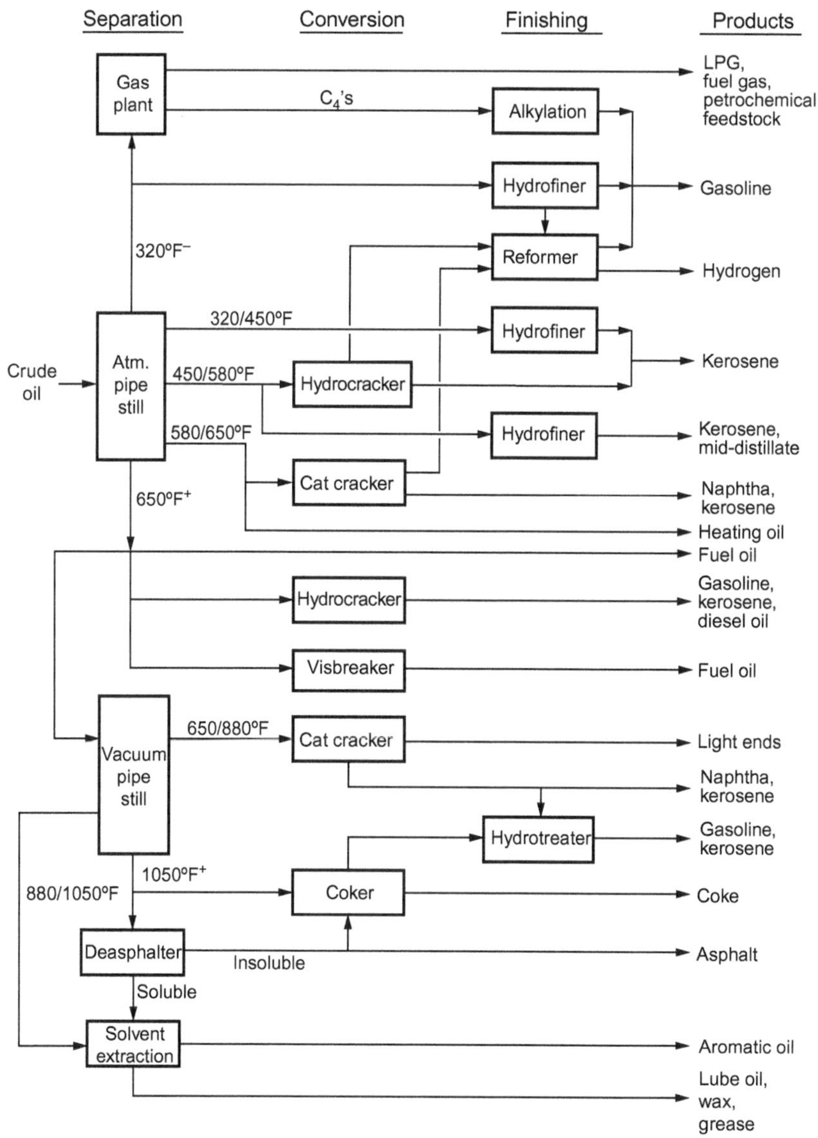

FIG. 9.1 Generalized refinery layout showing relative placement of the catalytic cracking units.

of feedstocks ranging from gas oil to heavy oil, extra heavy oil, and tar sand bitumen. It is one of several practical applications used in a refinery that employ a catalyst to improve process efficiency (Parkash, 2003; Gary et al., 2007; Speight, 2014, 2017; Hsu and Robinson, 2017). The original incentive to develop cracking processes arose from the need to increase gasoline supplies, and since cracking could virtually double the volume of naphtha from a barrel of crude oil, the purpose of cracking was wholly justified.

In the 1930s, thermal cracking units produced approximately half the total naphtha manufactured, the octane number of which was ~70 compared with ~60 for straight-run naphtha. These were usually blended together with light ends and sometimes with polymer gasoline and reformatted to form a gasoline base stock with an octane number of ~65. The addition of *tetraethyl lead* (*ethyl fluid*) increased the octane number to ~70 for the *regular-grade* gasoline and 80 for *premium-grade* gasoline. The thermal reforming and polymerization processes that were developed during the 1930s could be expected to further increase the octane number of gasoline to some extent, but something new was needed to break the octane barrier that threatened to stop the development of more powerful automobile engines. In 1936, a new cracking process opened the way to higher-octane gasoline; this process was catalytic cracking. Since that time, the use of catalysts in the crude oil industry has spread to other processes (Bradley et al., 1989; Parkash, 2003; Gary et al., 2007; Speight, 2014, 2017; Hsu and Robinson, 2017).

Catalytic cracking is basically the same as thermal cracking, but it differs by the use of a catalyst that is not (in theory) consumed in the process (Parkash, 2003; Gary et al., 2007; Speight, 2014, 2017; Hsu and Robinson, 2017). The catalyst directs the course of the cracking reactions to produce more of the desired products that can be used for the production of better-quality gasoline and other liquid fuels (Avidan and Krambeck, 1990; Parkash, 2003; Gary et al., 2007; Speight, 2014, 2017; Hsu and Robinson, 2017).

Catalytic cracking has a number of advantages over thermal cracking. The naphtha produced by catalytic cracking has a higher octane number and consists largely of *iso*paraffin derivatives and aromatic derivatives. The *iso*paraffin derivatives and aromatic hydrocarbon derivatives have high octane numbers and greater chemical stability than monoolefin derivatives and diolefin derivatives. The olefin derivatives and diolefin derivatives are present in much greater quantities in thermally cracked naphtha. Furthermore, olefin derivatives (e.g., $RCH{=}CH_2$ where $R{=}H$ or an alkyl group) and smaller quantities of methane (CH_4) and ethane ($CH_3{\cdot}CH_3$) are produced by catalytic cracking and are suitable for petrochemical use (Parkash, 2003; Gary et al., 2007; Speight, 2014, 2017; Hsu and Robinson, 2017). Sulfur compounds are changed in such a way that the sulfur content of naphtha produced by catalytic cracking is lower than the sulfur content of naphtha produced by thermal cracking. Catalytic cracking produces less residuum and more of the useful gas-oil constituents than thermal cracking. Finally, the process has considerable flexibility, permitting the manufacture of both automobile gasoline and aviation gasoline and a variation in the gas-oil production to meet changes in the fuel oil market.

Catalytic cracking in the usual commercial process involves contacting a feedstock (usually a gas-oil fraction) with a catalyst under suitable conditions of temperature, pressure, and residence time. By this means, a substantial part (>50%) of the feedstock is converted into naphtha and lower-boiling products, usually in a single-pass operation. However, during the cracking reaction, carbonaceous material is deposited on the catalyst, which markedly reduces its activity, and removal of the deposit is very necessary. The carbonaceous deposit arises from the presence of high-molecular-weight polar species in the feedstock (Parkash, 2003; Gary et al., 2007; Speight, 2014, 2017; Hsu and Robinson, 2017). Removal of the deposit from the catalyst is usually accomplished by burning in the presence of air until catalyst activity is reestablished.

The reactions that occur during catalytic cracking are complex (Germain, 1969), but there is a measure of predictability now that catalyst activity is better understood. The major catalytic cracking reaction exhibited by paraffin derivatives is carbon-carbon bond scission into a lighter paraffin and olefin. Bond rupture occurs at certain definite locations on the paraffin molecule, rather than randomly as in thermal cracking. For example, paraffin derivatives tend to crack toward the center of the molecule, the long chains cracking in several places simultaneously. Normal paraffin derivatives usually crack at carbon-carbon bonds or still nearer the center of the molecule. On the other hand, isoparaffin derivatives tend to rupture between carbon atoms alpha and beta to a tertiary carbon. In either case, catalytic cracking tends to yield products containing three or four carbon atoms rather than the one- or two-carbon-atom molecules produced in thermal cracking.

As in thermal cracking (Chapter 8), high-molecular-weight constituents usually crack more readily than small molecules, unless there has been some recycle, and the constituents of the recycle stream have become more refractory and are less liable to decompose. Paraffin derivatives having more than six carbon atoms may also undergo rearrangement of their carbon skeletons before cracking, and a minor amount of dehydrocyclization also occurs, yielding aromatic derivatives and hydrogen.

Olefin derivatives are the most reactive class of hydrocarbon derivatives in catalytic cracking and tend to crack from 1000 to 10,000 times faster than in thermal processes. Severe cracking conditions destroy olefin derivatives almost completely, except for those in the low-boiling naphtha and gaseous hydrocarbon range, and as in the catalytic cracking of paraffin derivatives, *iso*-olefin derivatives crack more readily than *n*-olefin derivatives. The olefin derivatives tend to undergo rapid isomerization and yield mixtures with an equilibrium distribution of double-bond positions. In addition, the chain-branching isomerization of olefin derivatives is fairly rapid and often reaches equilibrium. These branched-chain olefin derivatives can then undergo hydrogen transfer reactions with naphthene derivatives and other hydrocarbon derivatives. Other olefin reactions include polymerization and condensation to yield aromatic molecules, which in turn may be the precursors of coke formation.

In catalytic cracking, the cycloparaffin (naphthene) species crack more readily than paraffin derivatives but not as readily as olefin derivatives. Naphthene cracking occurs by ring rupture and by rupture of alkyl chains to yield olefin derivatives and paraffin derivatives, but the formation of methane and the C2 hydrocarbon derivatives (ethane, $CH_3 \cdot CH_3$; ethylene, $CH_2{=}CH_2$; and acetylene, $CH{\equiv}CH$) is relatively minor.

Aromatic hydrocarbon derivatives exhibit wide variations in their susceptibility to catalytic cracking. The benzene ring is relatively inert, and condensed ring compounds, such as naphthalene, anthracene, and phenanthrene, crack very slowly. When these aromatic derivatives crack, a substantial part of their *conversion* is reflected in the amount of coke deposited on the catalyst. Alkylbenzenes with attached groups of C2 or larger primarily form benzene and the corresponding olefin derivatives, and heat sensitivity increases as the size of the alkyl group increases.

The several processes currently employed commercially for catalytic cracking differ mainly in the method of catalyst handling, although there is an overlap with regard to catalyst type and the nature of the products. The catalyst, which may be an activated natural or synthetic material, is employed in bead, pellet, or microspherical form and can be used as a *fixed-bed*, *moving-bed*, or *fluid-bed* configurations.

The fixed-bed process was the first to be used commercially and uses a static bed of catalyst in several reactors, which allows a continuous flow of feedstock to be maintained. Thus, the cycle of operations consists of (i) flow of feedstock through the catalyst bed, (ii) discontinuance of feedstock flow and removal of coke from the catalyst by burning, and (iii) insertion of the reactor onstream. On the other hand, the *moving-bed process* uses a reaction vessel in which cracking takes place and a kiln in which the spent catalyst is regenerated; catalyst movement between the vessels is provided by various means.

The *fluid-bed process* differs from the fixed-bed and moving-bed processes insofar as the powdered catalyst is circulated essentially as a fluid with the feedstock (Sadeghbeigi, 1995). The several fluid catalytic cracking processes in use differ primarily in mechanical design. Side-by-side reactor-regenerator configuration or a configuration where the reactor is either above or below the regenerator is the main mechanical variation. From a flow standpoint, all fluid catalytic cracking processes contact the feedstock and any recycle streams with the finely divided catalyst in the reactor.

Feedstocks may range from naphtha fractions (included in normal heavier feedstocks for upgrading) to an atmospheric residuum (*reduced crude*). Feedstock preparation (to remove *metallic constituents* and *high-molecular-weight nonvolatile materials*) is usually carried out through the application of any one of several other processes: coking, propane deasphalting, furfural extraction, vacuum distillation, viscosity breaking, thermal cracking, and hydrodesulfurization (Speight, 2000).

The major process variables are temperature, pressure, catalyst-oil ratio (ratio of the weight of catalyst entering the reactor per hour to the weight of oil charged per hour), and space velocity (weight or volume of the oil charged per hour per weight or volume of catalyst in the reaction zone). Wide flexibility in product distribution and quality is possible through the control of these variables along with the extent of internal recycling, which is necessary. Increased conversion can be obtained by applying higher temperature or higher pressure. Alternatively, lower space velocity and higher catalyst-oil ratio will also contribute to an increased conversion.

When cracking is conducted in a single stage, the more reactive hydrocarbon derivatives may be cracked, with a high conversion to gas and coke, in the reaction time necessary for reasonable conversion of the more refractory hydrocarbon derivatives. However, in a two-stage process, gas and naphtha from a short-reaction-time, high-temperature cracking operation are separated before the main cracking reactions take place in a second-stage reactor. For the short time of the first stage, a flow line or vertical riser may act as the reactor, and some conversion is effected with minimal coke formation. Cracked gases are separated and fractionated; the catalyst and residue, together with recycle oil from a second-stage fractionator, pass to the main reactor for further cracking. The products of this second-stage reaction are gas, naphtha and gas-oil streams, and recycle oil.

Most fluid catalytic cracking units are operated to maximize conversion to naphtha and liquefied petroleum gas (Speight, 2011a, 2014). In the current context, the catalyst, which may be an activated natural or a synthetic material, is employed in bead, pellet, or microspherical form in any one (or all) of the several available or *fluidized-bed* (*fluid-bed*) configurations, which differ primarily in mechanical design (Sadeghbeigi, 1995; Hunt, 1997; Johnson and Niccum, 1997; Ladwig, 1997; Hemler and Smith, 2004; Speight 2011a, 2014). In addition, as the worldwide consumption of fuels has increased, product demand pattern has continued to shift toward distillate fuels such as gasoline, diesel, and kerosene-jet fuel with varying demands for the more viscous fuels using processes such as the Houdry fixed-bed catalytic cracking process (Fig. 9.2; Parkash, 2003; Ross et al., 2005; Gary et al., 2007; Speight, 2014, 2017; Hsu and Robinson, 2017). On the other hand, the octane number of the naphtha is also enhanced by overcracking the middle-boiling-point fraction with low octane number. This technique is more effective in the case of that octane number enhancement in fluid catalytic cracking naphtha and if an increase in propylene yield has a priority over naphtha production (Buchanan et al., 1996; Imhof et al., 2005).

The last 60 years have seen substantial advances in the development of catalytic processes. This has involved not only rapid advances in the chemistry and physics of the catalysts themselves but also major engineering advances in reactor design, for example, the evolution of the design of the catalyst beds from *fixed beds* to *moving beds* to *fluidized beds*. Catalyst chemistry and physics and bed design have allowed major improvements in process efficiency and

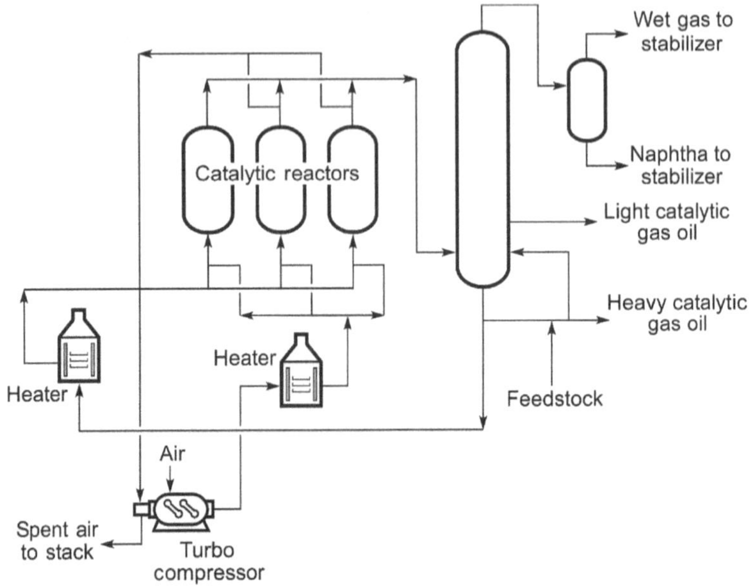

FIG. 9.2 The Houdry fixed-bed catalytic cracking process.

product yields (Sadeghbeigi, 1995). Most importantly, in terms of catalyst use, the most important concerns of the crude oil refining industry in the near future are as follows: (i) meeting the growing market of cleaner fuels, (ii) the gradual substitution of scarce light low-sulfur refinery feedstocks by the heavier high-sulfur feedstocks, (iii) the decreasing demand for heavy fuel oil, and (iv) the need to update processing operations (Parkash, 2003; Swaty, 2005; Gary et al., 2007; Gembicki et al., 2007; Bridjanian and Khadem Samimi, 2011; Letzsch, 2011; Speight, 2011a, 2014, 2017).

As the trend toward processing heavy oil, extra heavy oil, and tar sand bitumen increases (Patel, 2007; Speight, 2011a), evolving environmental mandates require lower levels of sulfur in the final fuel product and a reduction in emissions of sulfur dioxide (EPA, 2010). Reducing the sulfur concentration requires not only more efficient process options and specialized catalysts but also special options required to process heavy oil, extra heavy oil, and tar sand bitumen (Gembicki et al., 2007; Patel, 2007; Runyan, 2007). Furthermore, stricter environmental regulations are on the horizon. That venerable, almost revered, Bunker fuel oil was, in the past, released to markets that served as an outlet for a large percentage of the organic sulfur in the refinery feedstock.

2. Fixed-Bed Processes

Although fixed-bed catalytic cracking units have been phased out of existence, they represented an outstanding chemical engineering commercial development by incorporating a fully automatic instrumentation system, which provided a short-time reactor/purge/regeneration cycle, a novel molten salt heat transfer system, and a flue gas expander for recovering power to drive the regeneration air compressor. Historically, the Houdry fixed-bed process, which went onstream in June 1936, was the first of the modern catalytic cracking processes. Only the McAfee batch process that employed a metal halide catalyst but has long since lost any commercial significance preceded it.

In a *fixed-bed process*, the catalyst in the form of small lumps or pellets is made up of layers or beds in several (four or more) catalyst-containing drums called converters (Parkash, 2003; Gary et al., 2007; Speight, 2014, 2017; Hsu and Robinson, 2017). Feedstock vaporized at ~450°C (840°F) and <7–15 psi pressure is passed through one of the converters where the cracking reactions take place. After a short time, deposition of coke on the catalyst renders the catalyst ineffective, and using a synchronized valve system, the feedstock stream was passed into a converter, while the catalyst in the first converter was regenerated by carefully burning the coke deposits with air. After ~10 min, the catalyst is ready to go onstream again.

The requirement of complete vaporization necessarily limited feeds to those with a low-boiling range, and higher-boiling feedstock constituents are retained in a separator before the feedstock is passed into the bottom of the upflow fixed-bed reactors. The catalyst consisted of a pelletized natural silica-alumina catalyst and was held in reactors or *cases* ~11 ft. (3.4 m) in diameter and 38 ft. (11.6 m) length for a 15,000 bbl/day unit. Cracked products are passed through the preheat exchanger, condensed, and fractionated in a conventional manner. The reactors are operated at ~30 psi and 480°C (900°F).

The heat of reaction and some of the required feedstock circulating a molten salt through vertical tubes are distributed through the reactor beds. The reaction cycle of an individual reactor was ~10 min, after which the feedstock was automatically switched to a new reactor that had been regenerated. The reactor was purged with steam for ~5 min and then isolated by an automatic cycle timer. Regeneration air was introduced under close control, and carbon was burned off at a rate at which the recirculating salt stream could control the bed temperature. This stream comprised a mixture of potassium nitrate (KNO_3) and sodium nitrate ($NaNO_3$), which melts at 140°C (284°F) and was cooled in the reactors through which feedstock was being processed. The regeneration cycle lasted ~10 min. The regenerated bed was then purged of oxygen and automatically cut back into cracking service. There were three to six reactors in a unit. Naphtha yields diminished over the life of the catalysts (18 months) from 52% to 42% by volume, based on fresh feedstock.

Equilibrium was never reached in this cyclic process. The gas-oil conversion, that is, the amount of feedstock converted to lighter components, was high at the start of a reaction cycle and progressively diminished as the carbon deposit accumulated on the catalyst until regeneration was required. Multiple parallel reactors were used to approach a steady-state process. However, the resulting process flows were still far from steady state. The reaction bed temperature varied widely during reaction and regeneration periods, and the temperature differential within the bed during each cycle was considerable.

Fixed-bed catalytic cracking units have now generally been replaced by moving-bed or fluid-bed processes.

3. Moving-Bed Processes

The fixed-bed process had obvious capacity and mechanical limitations that needed improvement, and such improvement, thus, was replaced by a moving-bed process in which the hot salt systems were eliminated. The catalyst was lifted to the top of the reactor system and flowed by gravity down through the process vessels. The plants were generally limited in size to units processing up to ~30,000; these units have been essentially replaced by larger fluid-solid units.

In the moving-bed processes, the catalyst is a pelletized form [~0.125 in (3 mm)-diameter beads that flow by gravity from the top of the unit through a seal zone to the reactor that operates at ~10 psi 455–495°C (850–925°F)]. The catalyst then flows down through another seal and countercurrent through a stripping zone to the regenerator or kiln that operates at a pressure that is close to atmospheric. In early moving-bed units, built around 1943, bucket elevators were used to lift the catalyst to the top of the structure. In later units, built ~1949, a pneumatic lift was used. This pneumatic lift permitted higher catalyst circulation rates, which in turn permitted injection of all liquid feedstocks and feedstocks that had a higher-boiling range. A primary airstream was used to convey the catalyst. A secondary airstream was injected through an annulus into which the catalyst could flow. Varying the secondary air rate varied the circulation rate.

The lift pipe is tapered to a larger diameter at the top and minimizes erosion and catalyst attrition at the top. This taper is also designed so that the total collapse of circulation will not occur instantaneously when a specific concentration or velocity of solids, below which particles tend to drop out of the flowing gas stream, is experienced. The taper can be designed so that this potential separation of solids is preceded by a pressure instability that can alert the operators to take corrective action.

The *Airlift Thermofor Catalytic Cracking (Socony Airlift TCC) process* (Parkash, 2003; Gary et al., 2007; Speight, 2014, 2017; Hsu and Robinson, 2017) is a moving-bed, reactor-over-generator continuous process for the conversion of high-boiling gas oil into lower-boiling high-quality naphtha and middle distillate fuel oil. Feedstock preparation may consist of flashing in a tar separator to obtain vapor feed, and the tar separator bottoms may be sent to a vacuum tower from which the liquid feedstock is produced.

The gas-oil vapor-liquid flows downward through the reactor concurrently with the regenerated synthetic bead catalyst. The catalyst is purged by steam at the base of the reactor and gravitates into the kiln, or regeneration is accomplished by the use of air injected into the kiln. Approximately 70% of the carbon on the catalyst is burned in the upper kiln burning zone and the remainder in the bottom burning zone. Regenerated, cooled catalyst enters the lift pot, where low-pressure air transports it to the surge hopper above the reactor for reuse.

The *Houdriflow catalytic cracking process* (Parkash, 2003; Gary et al., 2007; Speight, 2014, 2017; Hsu and Robinson, 2017) is a continuous, moving-bed process employing an integrated single vessel for the reactor and regenerator kiln. The charge stock, sweet or sour, can be any fraction of the crude boiling between naphtha and soft asphalt. The catalyst is transported from the bottom of the unit to the top in a gas lift employing compressed flue gas and steam. The reactor feedstock and catalyst pass concurrently through the reactor zone to a disengager section, in which vapors are separated and directed to a conventional fractionation system. The spent catalyst, which has been steam purged of residual oil, flows to the kiln for regeneration, after which steam and flue gas are used to transport the catalyst to the reactor.

Houdresid catalytic cracking process (Fig. 9.3) is a process that uses a variation of the continuously moving catalyst bed designed to obtain high yields of high-octane naphtha and light distillate from reduced crude charge (Parkash, 2003; Gary et al., 2007; Speight, 2014, 2017; Hsu and Robinson, 2017). Residuum cuts ranging from crude tower bottoms to vacuum bottoms, including residua high in sulfur or nitrogen, can be employed as the feedstock, and the catalyst is synthetic or natural (Alvarenga Baptista et al., 2010a,b). Although the equipment employed is similar in many respects to that used in Houdriflow units, novel process features modify or eliminate the adverse effects and catalyst and product selectivity usually resulting when heavy metals—iron, nickel, copper, and vanadium—are present in the fuel. The Houdresid catalytic reactor

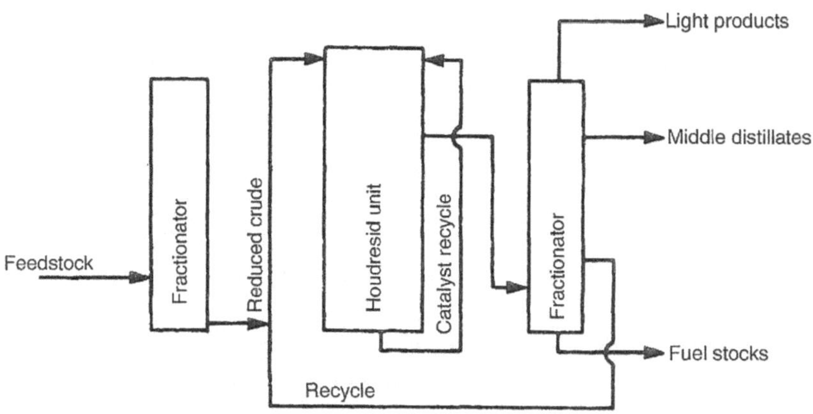

FIG. 9.3 The Houdresid catalytic cracking process.

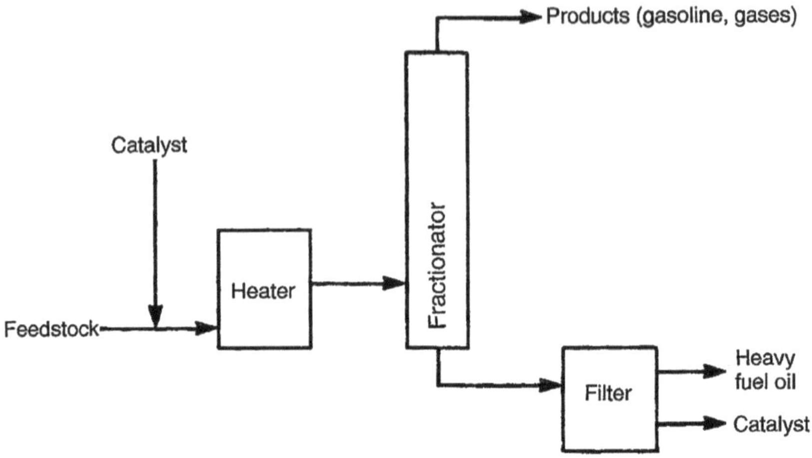

FIG. 9.4 The suspensoid catalytic cracking process.

and catalyst-regenerating kiln are contained in a single vessel. Fresh feedstock plus recycled gas oil are charged to the top of the unit in a partially vaporized state and mixed with steam.

The *Suspensoid catalytic cracking process* (Fig. 9.4) was developed from the thermal cracking process carried out in tube and tank units (Parkash, 2003; Gary et al., 2007; Speight, 2014, 2017; Hsu and Robinson, 2017). Small amounts of powdered catalyst or a mixture with the feedstock are pumped through a cracking coil furnace. Cracking temperatures are 550–610°C (1025–1130°F), with pressures of 200–500psi. After leaving the furnace, the cracked material enters a tar separator where the catalyst and tar are left behind. The cracked vapors enter a bubble tower where they are separated into two parts, gas oil and pressure distillate. The latter is separated into naphtha and gases. The spent catalyst is filtered from the tar, which is used as an industrial fuel oil. The process is actually a compromise between catalytic and thermal cracking. The main effect of the catalyst is to allow a higher cracking temperature and to assist mechanically in keeping coke from accumulating on the walls of the tubes. The normal catalyst employed is spent clay obtained from the contact filtration of lubricating oils (2–10lb./bbl of feed).

4. Fluid-Bed Processes

The application of fluidized solid techniques to catalytic cracking resulted in a major process breakthrough. It was possible to transfer all of the regeneration heat to the reaction zone. Much larger units could be built, and higher-boiling feedstocks could be processed. In fact, the improvement in catalysts and unit configurations has permitted the catalytic cracking of higher-boiling (poorer-quality) feedstocks such as residua. Presently, there are a number of processes that allow catalytic cracking of viscous feedstocks (Parkash, 2003; Gary et al., 2007; Speight, 2014, 2017).

The first fluid catalytic cracking units were Model I upflow units in which the catalyst flowed up through the reaction and regeneration zones in a riser type of flow regime. Originally, the Model I unit was designed to feedstock from a reduced crude to a vaporizer furnace where all of the gas oil was vaporized and fed, as vapor, to the reactor. The nonvolatile residuum (bottoms) bypassed the cracking section. The original Model I upflow design (1941) was superseded by the Model II downflow design (1944) followed by the Model III (1947) balanced-pressure design with the later introduction of the Model IV low-elevation design (Parkash, 2003; Gary et al., 2007; Speight, 2014, 2017; Hsu and Robinson, 2017).

Of the catalytic cracking process concepts, the *fluid-bed catalytic cracking process* (Fig. 9.5) is the most widely used process and is characterized by the use of a finely powdered catalyst that is moved through the processing unit (Parkash, 2003; Gary et al., 2007; Speight, 2014, 2017; Hsu and Robinson, 2017). The catalyst particles are of such a size that when *aerated* with air or hydrocarbon vapor, the catalyst behaves like a liquid and can be moved through pipes. Thus, vaporized feedstock and fluidized catalyst flow together into a reaction chamber where the catalyst, still dispersed in the hydrocarbon vapors, forms beds in the reaction chamber and the cracking reactions take place. The cracked vapors pass through cyclones located in the top of the reaction chamber, and the catalyst powder is thrown out of the vapors by centrifugal force. The cracked vapors then enter the bubble towers where fractionation into gases, naphtha, low-boiling gas oil, and high-boiling gas oil is achieved.

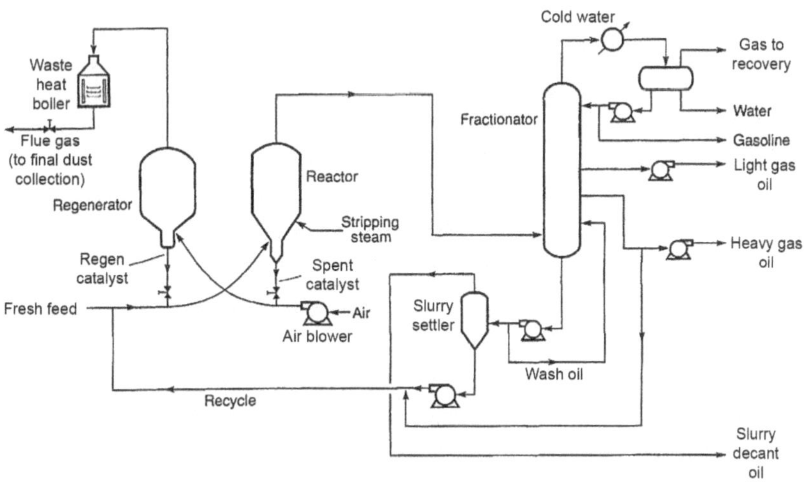

FIG. 9.5 The fluid-bed catalytic cracking process.

Since the catalyst in the reactor becomes contaminated with coke, the catalyst is continuously withdrawn from the bottom of the reactor and lifted by means of a stream of air into a regenerator where the coke is removed by controlled burning. The regenerated catalyst then flows to the fresh feedstock line, where the heat in the catalyst is sufficient to vaporize the fresh feedstock before it reaches the reactor, where the temperature is ~510°C (950°F).

The *Model IV fluid-bed catalytic cracking unit* (Parkash, 2003; Gary et al., 2007; Speight, 2014, 2017; Hsu and Robinson, 2017) involves a process in which the catalyst is transferred between the reactor and regenerator by means of U bends and the catalyst flow rate can be varied in relation to the amount of air injected into the spent-catalyst U bend. Regeneration air, other than that used to control circulation, enters the regenerator through a grid, and the reactor and regenerator are mounted side by side.

The *Orthoflow fluid-bed catalytic cracking process* (Parkash, 2003; Gary et al., 2007; Speight, 2014, 2017; Hsu and Robinson, 2017) uses the unitary vessel design, which provides a straight-line flow of catalyst and thereby minimizes the erosion encountered in pipe bends. Commercial Orthoflow designs are of three types: models A and C, with the regenerator beneath the reactor, and model B, with the regenerator above the reactor. In all cases, the catalyst-stripping section is located between the reactor and the regenerator. All designs employ the heat-balanced principle incorporating fresh feed-recycle feedstock cracking.

The *Universal Oil Products (UOP) fluid-bed catalytic cracking process* (Parkash, 2003; Gary et al., 2007; Speight, 2014, 2017; Hsu and Robinson, 2017) is adaptable to the needs of both large and small refineries. The major distinguishing features of the process are (i) elimination of the air riser with its attendant large expansion joints; (ii) elimination of considerable structural steel supports; and (iii) reduction in regenerator and in air-line size through the use of a mild pressure operation, on the order of 15–18 psi. The UOP process is also designed to produce low-molecular-weight olefin derivatives (for alkylation, polymerization, etherification, or petrochemicals), liquefied crude oil gas, high-octane naphtha, distillates, and fuel oils.

In the process, a side-by-side reactor/regenerator configuration and a patented preacceleration zone are used to condition the regenerated catalyst before feedstock injection—the riser terminates in a vortex separation system. A high-efficiency stripper then separates the remaining hydrocarbon derivatives from the catalyst, which is then reactivated in a combustor-style regenerator. The reactor zone features a short-contact-time riser, a state-of-the-art riser termination device for quick separation of catalyst and vapor (with high hydrocarbon containment technology), and a portion of the stripped (carbonized) catalyst from the reactor that is blended with the hot regenerated catalyst in a proprietary mixing chamber for delivery to the riser. Additionally, the recycling of cooler that partially spent catalyst back to the base of the riser lowers the reactor inlet temperature, which results in a reduction of undesirable thermally produced products, including dry gas. The ability to vary the carbonized/regenerated catalyst ratio provides considerable flexibility to handle changes in feedstock quality and enables a real-time switch between gasoline, olefin derivatives, or distillate operating modes. For heavier feedstocks, a two-stage regenerator is used—in the first stage, the bulk of the carbon is burned from the catalyst, forming a mixture of carbon monoxide and carbon dioxide. In the second stage, the remaining coke is burned from the catalyst resulting in only low levels of carbon on the regenerated catalyst. A catalyst cooler is located between the stages.

The Shell two-stage fluid-bed catalytic cracking process (Fig. 9.6) was devised to permit greater flexibility in shifting product distribution when dictated by demand. Thus, feedstock is first contacted with cracking catalyst in a riser reactor, that is, a pipe in which fluidized catalyst and vaporized oil flow concurrently upward, and the total contact time in this first stage is of the order of seconds. High temperatures 470–565°C (875–1050°F) are employed to reduce undesirable coke deposits on catalyst without destruction of naphtha by secondary cracking. Other operating conditions in the first stage are a pressure of 16psi and a catalyst-oil ratio of 3:1 to 50:1, and volume conversion ranges between 20% and 70% have been recorded (Parkash, 2003; Gary et al., 2007; Speight, 2014, 2017; Hsu and Robinson, 2017).

All or part of the unconverted or partially converted gas-oil product from the first stage is then cracked further in the second-stage fluid-bed reactor.

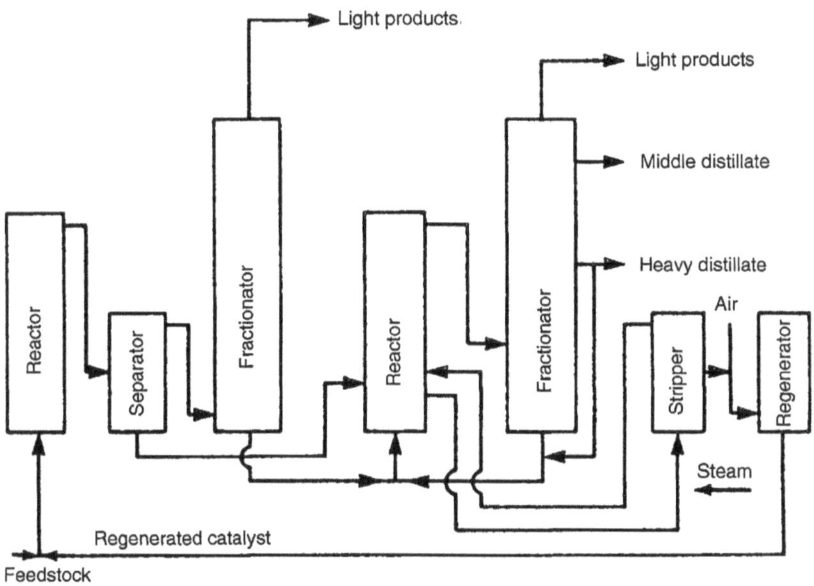

FIG. 9.6 A two-stage fluid catalytic cracking process.

Operating conditions are 480–540°C (900–1000°F) and 16psi with a catalyst-oil ratio of 2:1 to 12:1. Conversion in the second stage varies between 15% and 70%, with an overall conversion range of 50%–80%.

The residuum cracking unit (M. W. Kellogg Company/Phillips Petroleum Company) offers conversions up to 85% *w*/w of atmospheric residua or equivalent feedstocks (Parkash, 2003; Gary et al., 2007; Speight, 2014, 2017; Hsu and Robinson, 2017). The unit is similar to the Orthoflow C unit, but there are some differences that enhance performance on residua. The catalyst flows from the regenerator to a plug valve that controls the flow to hold the reactor temperature. Steam is injected upstream of the feedstock point to accelerate the catalyst and disperse it so as to avoid high rates of coke formation at the feedstock point. The feedstock, atomized with steam, is then injected into this stream through a multiple nozzle arrangement. The flow rates are adjusted to control the contact time in the riser since the effects of metal poisoning on yields are claimed to be largely a function of the time that the catalyst and oil are in contact. Passing the mix through a rough cut cyclone stops the reaction.

The *Gulf residuum process* consists of cracking a residuum that has been previously hydrotreated to low-sulfur and metal levels. In this case, high conversions are obtained, but coke yield and hydrogen yield are kept at conventional levels by keeping metals on catalyst low.

Other processes include the *deep catalytic cracking process* (DCC process) that is a fluidized catalytic process for selectively cracking a variety of feedstocks to low-molecular-weight olefin derivatives (Fig. 9.7; Chapin and Letzsch, 1994; Parkash, 2003; Gary et al., 2007; Speight, 2014, 2017; Hsu and Robinson, 2017). A traditional reactor/regenerator unit design is employed with a catalyst having physical properties much like those of a catalyst for a fluid catalytic cracking unit. The unit may be operated in two operational modes, maximum propylene (Type I) or maximum isoolefin derivatives (Type II), and each mode utilizes a unique catalyst and reaction conditions. The Type I unit uses both riser and bed cracking at relatively severe reactor conditions, while the Type II unit uses only riser cracking like a modern fluid catalytic cracking unit at milder conditions. The overall flow scheme of the process is similar to that of a conventional fluid catalytic cracking unit, but changes in the areas of catalyst development, process variable selection and severity, and gas plant design enable the production of higher yields of olefin derivatives than the conventional fluid catalytic cracking processes. The products are low-molecular-weight olefin derivatives, high-octane naphtha, light cycle oil, dry gas, and coke. Propylene yields over 24% *w*/w are achievable with paraffin feedstocks.

There is also the *flexicracking process* (Fig. 9.8) that is designed for the conversion of gas oils, residua, and deasphalted oils to distillates (Draemel, 1992; Parkash, 2003; Gary et al., 2007; Speight, 2014, 2017; Hsu and Robinson, 2017).

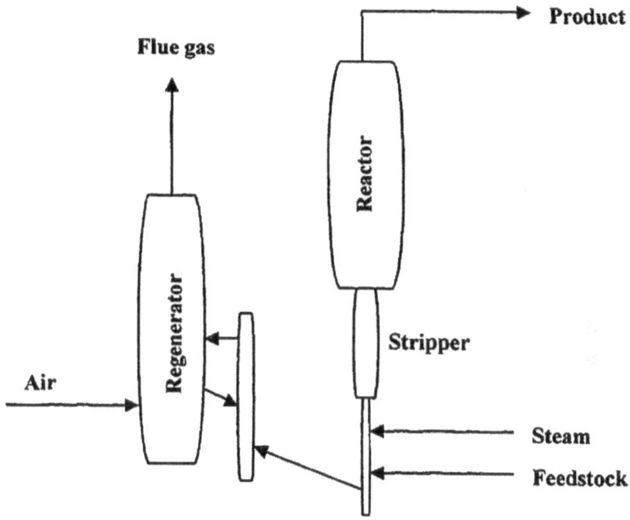

FIG. 9.7 The deep catalytic cracking process.

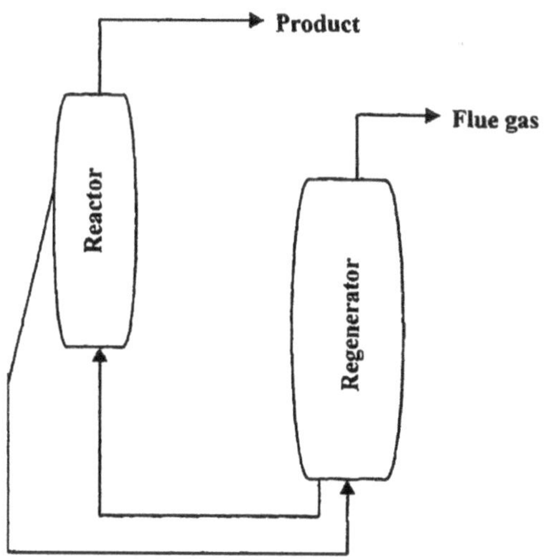

FIG. 9.8 The flexicracking process.

5. Process Variables

5.1 Feedstock Quality

Generally, the ability of any single unit to accommodate wide variations in feedstock is an issue related to the flexibility of the process (Navarro et al., 2015). Initially, catalytic cracking units were designed to process gas-oil feedstocks, but many units have been modified successfully, and new units are designed, to handle more complex feedstocks and feedstock blends containing residua.

Vacuum gas oil (ibp, 315–345°C, 600–650°F; fbp, 510–565°C, 950–1050°F), as produced by vacuum flashing or vacuum distillation, is the usual feedstock with final boiling point being limited by the carbon-forming constituents (measured by the Conradson carbon residue) or metal content since both properties have adverse effects on cracking characteristics. The vacuum residua (565°C$^+$, 1050°F$^+$) are occasionally included in cat cracker feedstock when the units (residuum cat crackers) are capable of handling such materials (Parkash, 2003; Gary et al., 2007; Speight, 2014, 2017; Hsu and Robinson, 2017). In such cases, if the residua are either relatively low in terms of carbon-forming constituents and metals (as a residuum from a waxy crude), the effects of these properties are relatively small. Many units also recycle slurry oil (455°C$^+$, 850°F$^+$) and a high-boiling cycle oil stream. Gas oils from thermal cracking or coking processes (Chapter 8), gas oils from hydrotreating processes (Chapter 10), and gas-oil fractions from deasphalting processes (Chapter 12) are often included in feedstocks for catalytic cracking units (Parkash, 2003; Gary et al., 2007; Speight, 2014, 2017).

The general feedstock quality effects can be indicated by characterization factor, *K*:

$$K = (\text{MABP})^{1/3}/\text{specific gravity at } 60^{\circ}\text{F}/60^{\circ}\text{F}$$

MABP is the mean-average boiling point expressed in degrees Rankin (°R = °F + 460). However, a single parameter such as this can only reflect general trends, and even then, the accuracy and meaningful nature of the data may be very questionable (Parkash, 2003; Gary et al., 2007; Speight, 2014, 2017; Hsu and Robinson, 2017). Generally speaking, coke yield increases as the characterization factor (K) decreases or as the feedstock becomes less paraffinic, and the API gravity decreases and as the conversion increases (Maples, 2000; Parkash, 2003; Gary et al., 2007; Speight, 2014, 2017; Hsu and Robinson, 2017). With straight-run gas oils, naphtha yield increases as the characterization factor decreases (i.e., the paraffin character of the oil decreases), but the opposite effect is obtained with cracked stocks or cycle oils.

Either molecular weight, or average boiling point, or feedstock boiling range is an important feedstock characteristic in determining cat cracking yields and product quality. In general, for straight-run fractions, thermal sensitivity (increased thermal decomposition or cracking) increases as molecular weight increases; coke and naphtha production (at constant processing conditions) also increase with the heavier feedstocks, but, not to become too enthusiastic approximately the word *increase*, coke yield also increases (Parkash, 2003; Gary et al., 2007; Speight, 2014, 2017; Hsu and Robinson, 2017).

However, the characterization factor and feedstock boiling range are generally insufficient to characterize a feedstock for any purpose other than approximate comparisons. A more detailed description of the feedstock is needed to reflect and predict the variations in feedstock composition and cracking behavior (Speight, 2014, 2000).

Irrespective of the source of the high-boiling feedstocks, a number of issues typically arise when these materials are processed in a fluid catalytic cracking unit, although the magnitude of the problem can vary substantially.

Viscous feedstocks have high levels of contaminants (Table 9.4) that will affect the process and must be removed. Examples are the carbon-forming constituents that yield high levels of (Conradson) carbon residue, and the overall coke production (as carbon on the catalyst) is high. Burning this coke requires additional regeneration air that might be constraint that limits the capacity of the

TABLE 9.4 Feedstock Contaminants That Affect Catalytic Cracking Processes[a]

Contaminants	Effect on Catalyst	Mitigation	Process
Sulfur	Catalyst fouling	Hydrodesulfurization	Hydroprocessing
	Deactivation of active sites		
Nitrogen	Adsorption of basic nitrogen	Hydrodemetallization	Hydroprocessing
	Destruction of active sites		
Metals	Fouling of active sites	Demetallization	Demet, Met-X
	Fouling of pores		
Particulate matter	Deactivation of active sites	Filter/pretreatment	Clay filtration/ guard bed
	Pore plugging		
Coke precursors	Formation of coke	Remove asphaltene constituents	Mild hydrocracking/ hydroprocessing
	Catalyst fouling	Remove resin constituents	
	Deactivation of active sites		
	Pore plugging		

[a]*Also applicable to hydrocracking processes.*

unit. Metals in the viscous feedstocks also deposit (almost quantitatively) on the catalyst where two significant effects are caused. First, the deposited metals can accelerate certain metal-catalyzed dehydrogenation reactions, thereby contributing to light gas (hydrogen) production and to the formation of additional coke. A second and more damaging effect is the situation in which the deposition of the metals causes a decline in catalyst activity because of the limited access to the active catalytic sites. This latter effect is normally controlled by catalyst makeup practices (adding and withdrawing catalyst).

The amount of sulfur and nitrogen in the products, waste streams, and flue gas generally increases when high-boiling feedstocks are processed because these feedstock components typically have higher sulfur and nitrogen contents than gas oil. However, in the case of nitrogen, the issue is not only one of higher nitrogen levels in the products but also (because of the feedstock nitrogen that is basic in character) catalyst poisoning that reduces the useful activity of the catalyst.

Heat-balance control may be the most immediate and troublesome aspect of processing high-boiling feedstocks. As the contaminant carbon increases, the first response is usually to increase in regenerator temperature. Adjustments in operating parameters can be made to assist in this control, but eventually, a point will be reached for heavier feedstocks when the regenerator temperature is too high for good catalytic performance. At this point, some external heat removal from the regenerator is required and would necessitate a mechanical modification like a catalyst cooler.

For the last two decades, demetallized oil (produced by the extraction of a vacuum-tower bottom stream using a light paraffinic solvent) has been included as a component of the feedstock in fluid catalytic cracking units. Modern solvent-extraction processes, such as the Demex process, provide a higher demetallized oil yield that is possible in the propane deasphalting process that has been used to prepare fluid catalytic cracker feedstock and demetallized feedstocks for other processes. Consequently, the demetallized oil is more heavily contaminated. In general, demetallized oils are still good cracking stocks, but most feedstocks can be further improved by hydrotreating to reduce contaminant levels and to increase their hydrogen content, thereby becoming a more presentable and process-friendly feedstock.

In many cases, atmospheric residua have been added as a blended component to feedstocks for existing fluid catalytic cracking units as a means of converting the highest-boiling constituents of crude oil. In fact, in some cases, the atmospheric residuum has ranged from a relatively low proportion of the total feedstock all the way to a situation in which it represents the entire feedstock to the unit. To improve the handling of these high-boiling feedstocks, several units have been revamped to upgrade them from their original gas-oil designs, whereas other units have taken a stepwise approach to residuum processing whereby modifications to the operating conditions and processing techniques are made as more experience is gained in the processing of residua.

5.2 Feedstock Preheating

In a heat-balanced commercial operation, increasing the temperature of the feedstock to a cracking reactor reduces the heat that must be supplied by the combustion of the coked catalyst in the regenerator. Feedstock preheating is usually supplied by heat exchange with hot product streams, a fired preheater, or both. When the feedstock rate, the recycle rate, and the reactor temperature are held constant as the feedstock preheat is increased, the following changes in operation result: (i) The catalyst/oil ratio (catalyst circulation rate) is decreased to hold the reactor temperature constant; (ii) conversion and all conversion-related yields, including coke, decline due to the decrease in catalyst/oil ratio and severity; (iii) the regenerator temperature will usually increase; although the total heat released in the regenerator and the air required by the regenerator are reduced by the lower coke yield, the lower catalyst circulation usually overrides this effect and results in an increase in regenerator temperature; and (iv) as a result of the lower catalyst circulation rate, residence time in the stripper and overall stripper efficiency are increased, liquid recovery is increased, and a corresponding decrease in coke usually results. Advantage is usually taken of these feedstock preheat effects, including the reduced air requirement, by increasing the total feedstock rate until coke production again requires all of the available air (Parkash, 2003; Gary et al., 2007; Speight, 2014, 2017; Hsu and Robinson, 2017).

5.3 Feedstock Pressure

Catalytic cracker pressures are generally set slightly above atmospheric by balancing the yield and quality debits of high pressure plus increased regeneration air compression costs against improved era king and regeneration kinetics; the lower cost of smaller vessels; plus, in some cases, power recovery from the regenerator stack gases. Representative yield and product effects show, at the same conversion level, that coke and naphtha yields are increased marginally and light gas yields are reduced at the higher pressure (Parkash, 2003; Gary et al., 2007; Speight, 2014, 2017; Hsu and Robinson, 2017).

Pressure levels in commercial units are generally in the range of 15–35 psi. Lowering the partial pressure of the reacting gases with steam will improve yields somewhat, but the major beneficial effect of feedstock injection steam is that it atomizes the feedstock to small droplets that will vaporize and react quickly. If feedstock is not atomized, it will soak into the catalyst and possibly crack to a higher coke make.

Both pressure and partial pressure of the feedstock, or steam/feedstock ratio, are generally established in the design of a commercial unit and thus are usually not available as independent variables over any significant range. However, in some units, injector steam is varied over a narrow range to balance carbon make with regeneration carbon burn-off.

5.4 Feedstock Conversion

All of the independent variables in catalytic cracking have a significant effect on conversion that is truly a dependent variable but can be shown as a function of API gravity and a variety of other functions (Maples, 2000). The detailed effects of changing conversion depend on which the conversion is changed, that is, by temperature, space velocity, catalyst/oil ratio, and catalyst activity. Increasing conversion increases yields of naphtha and all light products up to a conversion level of 60%–80% by volume in most cases. At this high conversion level, secondary reactions become sufficient to cause a decrease in the yields of olefin derivatives and naphtha. However, the point at which this occurs is the feedstock, operating conditions, catalyst activity, and other parameters.

5.5 Reactor Temperature

The principal effects of increasing reactor temperature at constant conversion are to decrease naphtha yield and coke yield and to increase yields of methane (CH_4), ethane (C_2H_6), propane ($CH_3CH_2CH_3$), and total butane (C_4H_{10}) yield; yields of the pentanes (C_5H_{12}) and higher-molecular-weight paraffin decrease, while olefin yields are increased.

The effect of reactor temperature on a commercial unit is, of course, considerably more complicated as variables other than temperature must be changed to maintain heat balance. For example, in order to increase the temperature of a reactor at constant fresh feedstock rate, the interrelated changes of recycle rate, space velocity, and feedstock preheat are required to maintain heat balance on the unit by increasing circulation rate and coke yield. The combined effects of higher reactor temperature and higher conversion resulted in the following additional changes: (i) yields of butanes and propane are increased, (ii) naphtha yield is increased, and (iii) the yield of light catalytic cycle oil is decreased. Thus, the effects of an increase in reactor temperature on an operating unit reflect not only the effects of temperature per se but also the effects of several concomitant changes such as increased conversion and increased catalyst/oil ratio.

5.6 Recycle Rate

With most feedstocks and catalyst, naphtha yield increases with increasing conversion up to a point, passes through a maximum, and then decreases. This phenomenon (*overcracking*) is due to the increased thermal stability (*refractory character*) of the unconverted feedstock as conversion increases and the destruction of naphtha through secondary reactions, primarily cracking of olefin derivatives. The onset of secondary reactions and the subsequent leveling off or decrease in naphtha yield can be avoided by recycling a portion of the reactor product, usually a fractionator product with a boiling points on the order of 345–455°C (650–850°F). Other tests have shown the following effects of increasing recycle rate when space velocity was simultaneously adjusted to maintain conversion constant: (i) The naphtha yield increased significantly; (ii) the coke yield decreases appreciably; (iii) there is a decrease in the yield of dry gas components, propylene, and propane; (iv) the yield of butane decreased, while the yield of butylene increased; and (v) the yields of light catalytic cycle oil and clarified oil increased, but the yield of catalytic cycle oil is decreased.

With the introduction of high-activity zeolite catalysts, it was found that in once-through cracking operations with no recycle, the maximum in naphtha yield was located at much higher conversions. In effect, the higher-activity catalysts were allowing higher conversions to be obtained at severity levels that significantly reduced the extent of secondary reactions (*overcracking*). Thus, on many units employing zeolite catalysts, recycle has been eliminated or reduced to less than 15% of the fresh feedstock rate.

5.7 Space Velocity

The role of space velocity as an independent variable arises from its relation to *catalyst contact time* or *catalyst residence time*. Thus,

$$\Theta = 60/(\text{WHSV} \times \text{CO})$$

In the equation, Θ is catalyst residence time in minutes, WHSV is the weight hourly space velocity on a total weight basis, and C/O is the catalyst/oil weight ratio.

The catalyst/oil ratio is a dependent variable so that catalyst time becomes directly related to the weight hourly space velocity. When catalyst contact time is low, secondary reactions are minimized; thus, naphtha yield is improved, and light and coke yields are decreased. In dense-bed units, the holdup of catalyst in the reactor can be controlled within limits, usually by a slide valve in the spent-catalyst standpipe, and feedstock rate can also be varied within limits. Thus, there is usually some freedom to increase space velocity and reduce catalyst residence time. In a riser-type reactor, holdup and feedstock rate are not independent, and space velocity is not a meaningful term. Nevertheless, with both dense-bed and riser-type reactors, contact times are usually minimized to improve selectivity. An important step in this direction was the introduction of high-activity zeolite catalyst (Junaid et al., 2010). These catalysts require short contact times for optimum performance and have generally moved cracking operations in the direction of minimum holdup in dense-bed reactors or replacement of dense-bed reactors with short-contact-time riser reactors.

Strict comparisons of short-contact-time riser cracking versus the longer-contact-time dense-bed mode of operations are generally not available due to differences in cat activity, carbon content of the regenerated catalyst, or factors other than contact time but inherent in the two modes of operation. However, in general, improvements in catalyst activity have resulted in the need for less catalyst in the reaction zone.

Many units are designed with only riser cracking, that is, no dense-bed catalyst reactor cracking occurs, and all cracking is done in the catalyst/oil transfer lines leading into the reactor cyclone vessel. However, in some of these instances, the reactor temperature must be increased to the 550–565°C (1020–1050°F) range in order to increase the intensity of cracking conditions to achieve the desired conversion level. This is because not enough catalyst can be held in the riser zone, since the length of the riser is determined by the configuration and elevation of the major vessels in the unit. Alternatively, superactive catalysts can be used to achieve the desired conversion in the riser.

Significant selectivity disadvantages have not been shown if a dispersed catalyst phase or even a very small dense bed is provided downstream of the transfer line riser cracking zone. In this case, the cracking reaction can be run at a lower temperature, say 510°C (950°F), which will reduce light gas make and increase naphtha yield when compared with the higher-temperature (550–565°C, 1020–1050°F) operation.

5.8 Catalyst Activity

Catalyst activity as an independent variable is governed by the capability of the unit to control the carbon content of the spent catalyst and the quantity and quality of fresh catalyst that can be continuously added to the unit. The carbon content of the regenerated catalyst is generally maintained at the lowest practical level to obtain the selectivity benefits of low carbon on the catalyst. Thus, catalyst addition is, in effect, the principal determinant of catalyst activity.

The deliberate withdrawal of catalyst over and above the inherent loss rate through regenerator stack losses and decant or clarified oil, if fly, and a corresponding increase in fresh catalyst addition rate are generally not practiced as a means of increasing the activity level of the circulating catalyst. If a higher activity is needed, the addition of a higher-activity fresh catalyst to the minimum makeup rate to maintain inventory is usually the more economical route. The general effects of increasing activity are to permit a reduction in severity and thus reduce the extent of secondary cracking reactions. Higher activity typically results in more naphtha and less coke. In other cases, higher-activity catalysts are employed to increase the feedstock rate at essentially constant conversion and constant coke production so that the coke burning or regenerator air compression capacities are fully utilized.

5.9 Catalyst/Oil Ratio

The dependent variable catalyst/oil ratio is established by the unit heat balance and coke make that in turn are influenced by almost every independent variable. Since catalyst/oil ratio changes are accompanied by one or more shifts in other variables, the effects of catalyst/oil ratio are generally associated with other effects. A basic relation, however, in all catalyst/oil ratio shifts is the effect on conversion and carbon yield. At constant space velocity and temperature, increasing catalyst/oil ratio increases conversion (Parkash, 2003; Gary et al., 2007; Speight, 2014, 2017; Hsu and Robinson, 2017). In addition to increasing conversion, higher catalyst/oil ratios generally increase coke yield at a constant conversion. This increase in coke is related to the hydrocarbon derivatives entrapped in the pores of the catalyst and carried through the stripper to the regenerator. Thus, this portion of the catalyst/oil ratio effect is highly variable and depends not only on the catalyst/oil ratio change but also on the catalyst porosity and stripper conditions. The following changes, in addition to increased coke yield, accompany an increase in the catalyst/oil ratio in the range of 5–20 at constant conversion, reactor temperature, and catalyst activity: (i) decreased hydrogen yield; (ii) decreased yields of methane, ethane, propane, and butane; and (iii) little effect on the naphtha yield or octane number (Parkash, 2003; Gary et al., 2007; Speight, 2014, 2017; Hsu and Robinson, 2017).

The catalyst-oil ratio (v/v) ranges from 5:1 to 30:1 for the different processes, although most processes are operated to 10:1. By comparison, the catalyst-oil volume ratio for moving-bed processes may be substantially lower than 10:1. In a traditional fluid catalytic cracking unit, increasing the catalyst-oil ratio to increase conversion also increases the coke yield and catalyst circulation to the regenerator. A high catalyst-oil ratio is necessary to maintain high reaction temperature by transferring enough heat from regenerator to reactor in commercial units. It is possible, using a specific type of catalyst (containing low-acid-density zeolite), to suppress hydrogen transfer and maximize olefin production (Soni et al., 2009; Fujiyama et al., 2010).

The UOP RxCat technology provides the ability to increase both conversion and selectivity by recycling a portion of the carbonized catalyst back to the base of the reactor riser. The carbonized catalyst circulated from the stripper back to the base of the riser is effectively at the same temperature as the reactor. Since the recycle catalyst adds no heat to the system, the recycle is heat-balance neutral. For the first time, the catalyst circulation up to the riser can be varied independently from the catalyst circulation rate to the regenerator and is decoupled from the unit heat balance (Wolschlag and Couch, 2010b; Wolschlag, 2011). When process conditions are changed so that an increase in the catalyst/oil ratio occurs, an increase in conversion is also typically observed. By increasing the catalyst/oil ratio, the effects of operating at high reaction temperature (thermal cracking) are minimized. High catalyst/oil ratio maintains heat balance, thereby achieving high reaction temperature. It also increases conversion and maximizes low-molecular-weight olefin derivative production (Maadhah et al., 2008).

5.10 Regenerator Temperature

Catalyst circulation, coke yield, and feedstock preheating are the principal determinants of regenerator temperature that is generally allowed to respond as a dependent variable within limits. Mechanical or structural specifications in the regenerator section generally limit regenerator temperature to a maximum value specific to each unit. However, in some cases, the maximum temperature may be set by catalyst stability. In either case, if regenerator temperature is too high, it can be reduced by decreasing feedstock preheat; catalyst circulation is then increased to hold a constant reactor temperature, and this increased catalyst circulation will carry more heat from the regenerator and lower the regenerator temperature. The sequence of events is actually more complicated as the shift in catalyst/oil ratio, and to a lesser extent, the shift in carbon content of the regenerated catalyst will change coke make and the heat release in the regenerator.

5.11 Regenerator Air Rate

The amount of air required for regeneration depends primarily on coke production. Regenerators have been operated with only a slight excess of air leaving the dense phase. With less air, carbon content of the spent catalyst increases, and a reduction in coke yield is required of air; burning of carbon monoxide (CO) to carbon dioxide (CO_2) above the dense bed will occur. This *after burning* must be controlled as extremely high temperature in the absence of the heat sink provided by the catalyst in the dense phase.

In terms of catalyst regeneration, higher-stability catalysts are available, and regenerator temperatures can be increased by 38–65°C (100–150°F) up to the 720–745°C (1325–1370°F) range without significant thermal damage to the catalyst. At these higher temperatures, the oxidation of carbon monoxide to carbon dioxide is greatly accelerated, and the regenerator can be designed to absorb the heat of combustion in the catalyst under controlled conditions. The carbon burning rate is improved at the higher temperatures, and there usually are selectivity benefits associated with the lower carbon on regenerated catalyst; this high-temperature technique usually results in carbon on regenerated catalyst levels of 0.05% by weight or less. In addition, the use of catalysts containing promoters for the oxidation of carbon monoxide to carbon dioxide produces a major effect of high-temperature regeneration, that is, low-carbon-monoxide-content regenerator stack gases and the resultant regenerator conditions may result in a lower-carbon-content regenerated catalyst.

5.12 Process Design

Process design improvement will continue to focus on (i) modification of existing units, (ii) commercialized technology changes, and (iii) new directions in processing technology to allow processing of a variety of feedstocks (Chen, 2006; Freel and Graham, 2011).

5.12.1 Modifications for Existing Units

The feedstock injection system is by far the most critical breakthrough of reactor design. In addition, a higher regenerator temperature is used to achieve more complete catalyst regeneration. The typical riser top temperature is on the order of 510–565°C (950–1050°F), but typical regenerated catalyst temperature is much higher—on the order of 675–760°C (1250–1400°F). Feedstock injection reduces thermal cracking reactions by cooling off the lower riser quickly through rapid mixing and vaporization of the feed.

As the fluid catalytic cracking feedstock moves to the more viscous feedstocks, vaporization of the feedstock becomes more difficult. However, the newest generation of side-entry fluid catalytic cracking feedstock nozzles generates more uniform feedstock distribution (and more rapid mixing) as a result of better control of homogeneity of two-phase flow and atomization at the nozzle exit using two-phase choke flow. Some older fluid catalytic cracking units still retain the original feedstock injection system located at the bottom of the riser (bottom-entry nozzles). A new generation of feedstock injection technology uses a similar side-entry atomization mechanism. For catalyst circulation, the bottom-entry nozzles have the advantage of reducing pressure drop through the riser. This system also enables longer riser residence time if riser height is limited.

Current feedstock nozzles optimize the temperature profile in the riser and substantially reduce dry gas yield, thereby increasing naphtha yield. These results are in line with the expectation that better feedstock injection design reduces thermal cracking reactions, which are the primary source for dry gas. Process/hardware technologies to improve light cycle oil yield include proper feedstock injection systems and risers/reaction zone designs as claimed by Petrobras, Shell, and Sinopec in their latest commercial processes.

The fluid catalytic cracking riser is known for its shortcomings of density and velocity variations, but the newest riser internal technology minimizes these shortcomings and promotes ideal plug flow. Improved riser reaction termination technology sharpens the termination of reactions by the combination of the unique design of primary stripper cyclones and close-coupled secondary cyclones and designs to reduce coke formation (Hedrick and Palmas, 2011). Due to the development of highly active zeolite fluid catalytic cracking catalyst, the reaction time has been shortened significantly to a few seconds in the modern riser reactor. Since catalytic cracking reactions can only occur after the vaporization of the liquid hydrocarbon feedstock, mixing and feedstock vaporization must take place in the riser as quickly as possible; otherwise, thermal cracking reactions will dominate (Chen, 2004). An efficient product separator suppresses side reactions (oligomerization and hydrogenation of low-molecular-weight olefin derivatives) and coke formation accelerated by condensation.

The high yields of products generated in the highly selective reactor riser environment must be preserved in the rest of the reaction system. Improvements in the design of riser separation or termination systems focus on the rapid disengagement of catalyst from the cracked products in a highly contained system. Product vapors are quickly directed to the fractionation system for thermal quench and recovery (Ross et al., 2005; Fujiyama et al., 2011). The design improves both the separation system to reduce dry gas and the stripping system to reduce coke. In addition, the pressure drop is extremely low in order to limit dip-leg immersion requirements to seal the positive pressure separator and so that the capacity of the unit will not be limited. The dip-leg size and flux are optimized to minimize gas entrainment with the catalyst and even allow for stripping within the dip legs.

Prospective techniques for fluid catalytic cracking control technology fall into four general categories: (i) hydrodesulfurization, (ii) catalyst additives, (iii) scrubbing, and (iv) chemical reaction such as selective noncatalytic reduction (SNCR) and selective catalytic reduction (SCR) (Bouziden et al., 2002; Couch et al., 2004). Also, hydrotreating of the fluid catalytic cracking feedstock will decrease feedstock sulfur and nitrogen, thereby decreasing sulfur emissions from the unit, as well as nitrogen in the coke on catalyst, and consequently nitrogen oxide (NOx) emissions from the regenerator.

5.12.2 Commercial Technology Changes

Dramatic yield improvements have been demonstrated commercially using the latest fluid catalytic cracking process technology advances. Riser termination designs, in particular, have received much attention, for new units and for revamped units. Significant postriser cracking occurs in commercial fluid catalytic cracking units resulting in substantial production of dry gas and other lower-value products (Avidan and Krambeck, 1990). Furthermore, the major process licensors have developed advanced riser termination systems to minimize postriser cracking (Draemel, 1992; Wrench and Glascow, 1992; Upson and Wegerer, 1993; Long et al., 1993), and many units have implemented these in both new unit and revamp applications. In addition, some refiners have implemented their own "in-house" designs for the same purpose.

Due to the complexity and diversity of existing fluid catalytic cracking units and new unit design differences, there are many variations of these systems such as (i) closed cyclones, (ii) close-coupled cyclones, (iii) direct-connected cyclones, (iv) coupled cyclones, (v) high-containment systems, and (vi) short-contact-time systems. There are differences in the specific designs, and some may be more appropriate for specific unit configurations than others, but all serve the same fundamental purpose of reducing the undesirable postriser reactions.

Proper catalyst selection is essential to realizing the maximum potential benefit from these hardware improvements. Both fresh catalyst properties and catalyst management policy are important variables. These include the use of (i) high-activity catalyst for maximum conversion or throughput,

(ii) high-matrix catalyst for reducing slurry yield, and (iii) metal-tolerant catalyst for processing resid feed. In all cases, improved yields are achieved when incorporating catalyst change effects, which may not have been possible without the advanced hardware, due to unit constraints.

The downer is a gas-solid cocurrent downflow reactor, which has the potential to overcome the drawback of a conventional upflow reactor (or a riser) caused by back-mixing of catalyst. In the downer, gas and solid catalysts move downward together with the assist of the gravity; this can avoid the back-mixing of catalyst in the reactor.

The operation of the downer is affected by various key parameters including (i) recycled catalyst flow rate, (ii) superficial gas velocity, (iii) spent-catalyst flow rate, and (iv) carbon content on the spent catalyst. The parameters that affect the temperature of the downer regenerator should be carefully selected as they have the most significant effect to a regeneration process. High regeneration temperature could deactivate the catalyst permanently, but low-temperature operation lowers the regeneration performance (Chuachuensuk et al., 2010).

A downflow reactor system has been adopted for the high-severity fluid catalytic cracking (HS-FCC) process. The downer permits higher catalyst/oil ratios because the lifting of catalyst by vaporized feedstock is not required. As with most reactor designs involving competing reactions and secondary product degradation, there is a concern over catalyst-feedstock contacting, back-mixing, and control of the reaction time and temperature. The downflow reactor would ensure plug flow without back-mixing (Maadhah et al., 2008).

The development of a highly active zeolite catalyst has led to the reaction time that has being decreased to a few seconds in the riser reactor. The short contact time (short residence time, < 0.5 s) of feedstock and product hydrocarbon derivatives in the downer should be favorable for minimizing thermal cracking. Undesirable secondary reactions such as polymerization reactions and hydrogen transfer reactions, which consume olefin derivatives, are suppressed. In order to attain the short residence time, the catalyst and the products have to be mixed and dispersed at the reactor inlet and separated immediately at the reactor outlet. For this purpose, a high-efficiency product separator has been developed capable of suppressing side reactions (oligomerization and hydrogenation of low-molecular-weight olefin derivatives) and coke formation accelerated by condensation (Nishida and Fujiyama, 2000). The short-contact-time reactor affords (i) minimal back-mixing and erosion, (ii) efficient catalyst/oil contacting, (iii) reduced hydrogen transfer, and (iv) high yield selectivity.

The overcracking of naphtha to gases is minimized by reducing the contact time between catalyst and hydrocarbon products. Addition of ZSM-5 additive enhances the octane number in fluid catalytic cracking naphtha by overcracking of naphtha fraction. On the other hand, the octane number enhancement is achieved by overcracking of the middle-boiling-point fraction with low octane number (Buchanan et al., 1996). Short-contact-time riser cracking is inherently more flexible than a typical fluid catalytic cracking unit because the product slate can be easily adjusted to maximize propylene, maximize naphtha, or produce combinations such as propylene plus ethylene or propylene plus naphtha (Jakkula and Niemi, 1997). This process flexibility is a key variable in maximizing profitability in a given market scenario.

The UOP millisecond catalytic cracking process (MSCC process) involves injection of the feedstock perpendicular to a downflowing steam of catalyst (Schnaith et al., 1998; Harding et al., 2001). The basic MSCC reactor configuration consists of an injection zone, a central dilute phase disengaging zone, the lower dense phase collection zone, and an upper inertial separation zone. The short contact time combined with the low-volume reaction zone reduces secondary cracking reactions and produces more naphtha and less coke compared with conventional fluid catalytic cracking. Another benefit is that the low coke yield allows heavier feedstocks (conventionally high-coke-make feedstocks) to be processed.

Improvements in the design of riser separation or termination systems focus on the rapid disengagement of catalyst from the cracked products in a highly contained system. Product vapors are quickly directed to the fractionation system for thermal quench and recovery (Ross et al., 2005). Using a two-riser system, a viscous feedstock can be treated by the use of a two-riser system (two-stage riser fluid catalytic cracking, TSRFCC). The spent catalyst from the other of the two risers is fed to the inlet of the first riser to produce relatively mild cracking conditions. Improved total naphtha plus distillate yields are achieved, and the novel two-riser system facilitates heat balancing of the system (Krambeck and Pereira, 1986; Shan et al., 2003).

5.12.3 New Directions

Continued research and technology developments for fluid catalytic cracking will focus on (i) widening the boiling range of the feedstock that can be processed in the unit, (ii) maximizing diesel yield and low-molecular-weight olefin yield, and (iii) providing operational flexibility to allow the unit to take advantage of favorable market opportunities.

A new hydrocarbon conversion process, *high-severity fluid catalytic cracking* (HS-FCC), has been developed to maximize propylene production in oil refineries (Fujiyama et al., 2005; Maadhah et al., 2008). The yield of propylene was maximized using a combination of three factors: (i) catalyst properties, (ii) reaction conditions, and (iii) reactor design. Optimization of reaction conditions and catalyst development found that high reaction temperature accelerated catalytic cracking rather than hydrogen transfer. As a result, the olefin/paraffin ratio of the product was higher at high reaction temperatures.

The special features of this process include (i) rapid feedstock vaporization, (ii) downflow reactor, (iii) high severity, (iv) short contact time, and (v) high catalyst/oil ratio. Since the fluid catalytic cracking process involves successive reactions, the desired products such as olefin derivatives and naphtha are considered intermediate products. A suppression of back-mixing by using the downer reactor is the key to achieving maximum yield of these intermediates. Compared with conventional fluid catalytic cracking processes, the HS-FCC has modifications in the reactor/regenerator and stripper sections (Fujiyama, 1999; Ino and Ikeda, 1999; Fujiyama et al., 2000; Nishida and Fujiyama, 2000; Ino et al., 2003).

Recent enhancements made to resid fluid catalytic cracking units permit feeding significant amounts of viscous crude oil while simultaneously improving yields and service factors. Traditional technology has been modified in key areas including (i) catalyst design to accommodate higher metal feedstock and to minimize the amount of coke formed on the catalyst, (ii) feedstock injection, (iii) riser pipe design and catalyst/oil product separation to minimize overcracking, (iv) regenerator design improvements to handle high coke output and avoid damage to catalyst structure, and (v) overall reactor and regenerator design concepts.

These developments have allowed fluid catalytic cracking units to substantially increase residue processing capabilities, and substantial portions of refinery residua are processes (as blends with gas oils) in fluidized-bed units, thereby increasing naphtha and diesel production. This will not only continue but also increase in the future.

6. Catalysts

Commercial synthetic catalysts are amorphous and contain more silica that is called for the preceding formulas; they are generally composed of 10%–15% alumina (Al_2O_3) and 85%–90% silica (SiO_2). The natural materials, montmorillonite, a nonswelling bentonite, and halloysite, are hydrosilicates of aluminum, with a well-defined crystal structure and approximate composition of Al_2O_3 $4Si_2O{\cdot}xH_2O$. Some of the newer catalysts contain up to 25% alumina and are reputed to have a longer active life.

However, cracking occurs over many types of catalytic materials, and cracking catalysts can differ markedly in both activity to promote the cracking reaction and in the quality of the products obtained from cracking the feedstocks (Gates et al., 1979; Wojciechowski and Corma, 1986; Stiles and Koch, 1995; Cybulski and Moulijn, 1998; Occelli and O'Connor, 1998; Domokos et al., 2010). Activity can be related directly to the total number of active (acid) sites per unit weight of catalyst and also to the acidic strength of these sites. Differences in activity and acidity regulate the extent of various secondary reactions occurring and thus the product quality differences. The acidic sites are considered to be Lewis- or Brønsted-type acid sites, but there is much controversy as to which type of site predominates.

Briefly and by way of a historical introduction, the first acid catalyst, tested for cracking of the viscous crude oil fraction, was aluminum chloride

($AlCl_3$), but the problems with catalyst manipulation, corrosion, and waste treatment or disposal put the use of this catalyst at a serious disadvantage. In the 1940s, silica-alumina catalysts were created and showed great improvement over the use of catalytic clay minerals and natural clay catalysts. After natural aluminosilicate minerals were found to be adequate to the task, synthetic aluminosilicates were prepared and showed enhanced cracking properties. Both types (natural and synthetic) of aluminosilicates were known for the presence of Lewis acid sites. The early synthetic amorphous aluminosilicate catalysts contained ~13% w/w alumina (Al_2O_3), which was boosted to 25% w/w alumina in the mid-1950s. In the original fixed-bed process, activated bentonite was used (probably in the form of pellets). For the Thermofor catalytic cracking unit, the catalysts were of spherical shape (diameter, approximately 1–2 mm). In 1948, the first spray-dried catalyst was introduced, and the microspherical particles (50–100 μm) were produced with the similar particle size distribution as the ground catalysts. However, the spherical particles showed both improved fluidization properties and significant reduction of attrition losses.

The most significant advance came in 1962 when zeolite catalysts (particularly zeolite Y at that time) were incorporated into the silica-alumina structures after which advances in catalysts have produced the greatest overall performance of fluid catalytic cracking units over the last 50 years. The presence of the zeolites has resulted in the presence of strong Brønsted acid sites with very easily accessible Lewis acid sites also being present. These new types of catalysts possess the properties required of a successful catalyst: activity, stability, selectivity, correct pore size, resistance to fouling, and low cost. The first commercial zeolite catalysts were introduced in 1964, and zeolite catalysts remain in use in modern refineries.

From this point on, catalyst development has proceeded at a rapid rate, and the development of active and stable catalysts (typically acid catalysts) has paralleled equipment design and development. The ultimate goal is the development of catalysts that are resistant to the obnoxious constituents of viscous feedstocks. Generally, the philosophy of the catalyst preparation for fluid catalytic cracking units is to have weak acid centers in macroporous part of catalyst particles to insure precracking of viscous feedstock constituents to lower-molecular-weight producers that enter to the mesopores with stronger acidity. Cracking in mesopores leads to even low-molecular-weight products that can enter the zeolite micropores and crack over strongest zeolite acid centers into the desired products, typically naphtha constituents.

6.1 Catalyst Types

The first cracking catalysts were acid-leached *montmorillonite clays*. The acid leach was to remove various metal impurities, principally iron, copper, and nickel that could exert adverse effects on the cracking performance of the catalyst. The catalysts are first used in fixed-bed and moving-bed reactor systems in the form of shaped pellets. Later, with the development of the fluid catalytic cracking process, the clay catalysts were made in the form of a ground, sized powder. Clay catalysts are relatively inexpensive and have been used extensively for many years.

The desire to have catalysts that were uniform in composition and catalytic performance led to the development of *synthetic catalysts*. The first synthetic cracking catalyst consisting of 87% silica (SiO_2) and 13% alumina (Al_2O_3) was used in pellet form and used in fixed-bed units in 1940. Catalysts of this composition were ground and sized for use in fluid catalytic cracking units. In 1944, catalysts in the form of beads approximately 2.5–5.0mm diameter were introduced and comprised ~90% silica and 10% alumina and were extremely durable. One version of these catalysts contained a minor amount of chromia (Cr_2O_3) to act as an oxidation promoter.

Neither silica (SiO_2) nor alumina (Al_2O_3) alone is effective in promoting catalytic cracking reactions. In fact, they (and also activated carbon) promote hydrocarbon decompositions of the thermal type. Mixtures of anhydrous silica and alumina ($SiO_2 \cdot Al_2O_3$) or anhydrous silica with hydrated alumina ($2SiO_2 \cdot 2Al_2O_3 \cdot 6H_2O$) are also essentially noneffective. A catalyst having appreciable cracking activity is obtained only when prepared from hydrous oxides followed by partial dehydration (*calcining*). The small amount of water remaining is necessary for proper functioning.

The catalysts are porous and highly adsorptive, and their performance is affected markedly by the method of preparation. Two catalysts that are chemically identical but have pores of different size and distribution may have different activity, selectivity, temperature coefficient of reaction rate, and response to poisons. The intrinsic chemistry and catalytic action of a surface may be independent of pore size, but small pores appear to produce different effects because of the manner and time in which hydrocarbon vapors are transported into and out of the interstices.

In addition to synthetic catalysts comprising silica-alumina, other combinations of *mixed oxides* were found to be catalytically active and were developed during the 1940s. These systems included silica (SiO_2), magnesia (MgO), silica-zirconia (SiO_2-ZrO), silica-alumina-magnesia, silica-alumina-zirconia, and alumina-boria (Al_2O_3-B_2O_3). Of these, only silica-magnesia was used in commercial units but operating difficulties developed with the regeneration of the catalyst that at the time demanded a switch to another catalyst. Further improvements in silica-magnesia catalysts have since been made. High yields of desirable products are obtained with hydrated aluminum silicates. These may be either activated (acid-treated natural clays of the bentonite type) or synthesized silica-alumina or silica-magnesia preparations. Both the natural and the synthetic catalysts can be used as pellets or beads and also in the form of powder; in either case, replacements are necessary because of attrition and gradual loss of efficiency (DeCroocq, 1984; Le Page et al., 1987).

During the period 1940–62, the cracking catalysts used most widely commercially were the aforementioned acid-leached clays and silica-alumina. The latter was made in two versions; *low alumina* (~13% Al_2O_3) and *high alumina* (~25% Al_2O_3) contents. High-alumina-content catalysts showed a higher equilibrium activity level and surface area.

During the 1958–60 period, *semisynthetic catalysts* of silica-alumina catalyst were used in which approximately 25%–35% kaolin was dispersed throughout the silica-alumina gel. These catalysts could be offered at a lower price and therefore were disposable, but they were marked by a lower catalytic activity and greater stack losses because of increased attrition rates. One virtue of the semisynthetic catalysts was that a lesser amount of adsorbed, unconverted, high-molecular-weight products on the catalyst was carried over to the stripper zone and regenerator. This resulted in a higher yield of more valuable products and also smoother operation of the regenerator as local hot spots were minimized.

The catalysts must be stable to physical impact loading and thermal shocks and must withstand the action of carbon dioxide, air, nitrogen compounds, and steam. They should also be resistant to sulfur compounds; the synthetic catalysts and certain selected clays appear to be better in this regard than average untreated natural catalysts.

Commercially used cracking catalysts are *insulator catalysts* possessing strong protonic (acidic) properties. They function as catalyst by altering the cracking process mechanisms through an alternative mechanism involving *chemisorption* by *proton donation* and *desorption*, resulting in cracked oil and theoretically restored catalyst. Thus, it is not surprising that all cracking catalysts are poisoned by proton-accepting vanadium.

The catalyst-oil volume ratios range from 5:1 to 30:1 for the different processes, although most processes are operated to 10:1. However, for moving-bed processes, the catalyst-oil volume ratios may be substantially lower than 10:1.

Crystalline *zeolite catalysts* having molecular sieve properties were introduced as selective adsorbents in the 1955–59 period. In a relatively short time period, all of the cracking catalyst manufacturers were offering their versions of zeolite catalysts to refiners. The intrinsically higher activity of the crystalline zeolites vis-à-vis conventional amorphous silica-alumina catalysts coupled with the much higher yields of naphtha and decreased coke and light ends yields served to revitalize research and development in the mature refinery process of catalytic cracking.

A number of *zeolite catalysts* have been mentioned as having catalytic cracking properties, such as synthetic faujasite (X- and Y-types), offretite, mordenite, and erionite. Of these, the faujasites have been most widely used commercially. While faujasite is synthesized in the sodium form, base exchange removes the sodium with other metal ions that, for cracking catalysts, include magnesium, calcium, rare earths (mixed or individual), and ammonium. In particular, mixed rare earths alone or in combination with ammonium ions have been the most commonly used forms of faujasite in cracking catalyst formulations. Empirically, X-type faujasite has a stoichiometric formula of Na_2O $A1_2O_3$ $2.5SiO_2$ and Y-type faujasite Na_2O $A1_2O_3$ 4.8SiO2. Slight variations in the silica/alumina ($SiO_2/A1_2O_3$) ratio exist for each of the types. Rare earth-exchanged Y-type faujasite retains much of its crystallinity after steaming at 825°C (1520°F) with steam for 12h. The rare earth forms X-faujasite, which is thermally stable in dry air, and will lose its crystallinity at these temperatures in the presence of steam.

6.2 Catalyst Manufacture

While each manufacturer has developed proprietary procedures for making silica-alumina catalyst, the general procedure consists of (i) the gelling of dilute sodium silicate solution ($Na_2O{\cdot}3.25SiO_2{\cdot}xH_2O$) by the addition of an acid (H_2SO_4 and CO_2) or an acid salt such as aluminum sulfate, (ii) aging the hydrogel under controlled conditions, (iii) adding the prescribed amount of alumina as aluminum sulfate and/or sodium aluminate, (iv) adjusting the pH of the mixture, and (v) filtering the composite mixture. After filtering, the filter cake can either be (i) washed free of extraneous soluble salts by a succession of slurrying and filtration steps and spray-dried or (ii) spray-dried and then washed free of extraneous soluble salts before flash-drying the finished catalyst.

There are a number of critical areas in the preparative processes that affect the physical and catalytic properties of the finished catalyst. Principal among them is the concentration and temperature of the initial sodium silicate solution. The amount of acid added to effect gelation, the length of time of aging the gel, the method and conditions of adding the aluminum salt to the gel, and its incorporation therein. Under a given set of conditions, the product catalyst is quite reproducible in both physical properties and catalytic performance.

During the period 1940–62, a number of improvements were made in silica-alumina catalyst manufacture. These included continuous production lines versus batch-type operation, introduction of spray drying to eliminate grinding and sizing of the catalyst while reducing catalyst losses as fines, improving catalyst stability by controlling pore volume, and improved wash procedures to remove extraneous salts from high-alumina-content catalysts to improve equilibrium catalyst performance.

Zeolite cracking catalysts are made by dispersing or imbedding the crystals in a matrix. The matrix is generally amorphous silica-alumina gel and may also contain finely divided clay. The zeolite content of the composite catalyst is generally in the range of 5%–16% by weight. If clay (e.g., kaolin) is used in the matrix, it is present in an amount of 25%–45% by weight, the remainder being the silica-alumina hydrogel *glue* that binds the composite together. The zeolite may be preexchanged to the desired metal form and calcined to lock the exchangeable metal ions into position before compositing with the other ingredients. In an alternate scheme, sodium-form zeolite is composited with the other components, washed, and then treated with a dilute salt solution of the desired metal ions before the final drying step.

As stated above, the matrix generally consists of silica-alumina, but several catalysts have been commercialized that contain (i) silica-magnesia and kaolin and (ii) synthetic montmorillonite-mica and/or kaolin as the matrix for faujasite.

6.3 Catalyst Selectivity

In the catalytic cracking process, the most abundant products are those having three, four, and five carbon atoms. On a weight basis, the four-carbon-atom fraction is the largest. The differences between the catalysts of the mixed oxide type lie in the relative action toward promoting the individual reaction types included in the overall cracking operation. For example, silica-magnesia catalyst under a given set of cracking conditions will give a higher conversion to cracked products than silica-alumina catalyst. However, the products from a silica-magnesia (SiO_2-MgO) catalyst have a higher average molecular weight, hence a lower volatility, lesser amounts of highly branched/acyclic isomers, but more olefin derivatives among the naphtha boiling range products (C_4—220°C, 430°F) than the products from a silica-alumina catalyst. With these changes in composition, the naphtha from cracking with a silica-magnesia catalyst is of lower octane number.

These differences between catalysts may also be described as differences in the intensity of the action at the individual active catalytic centers. That is, a catalyst such as silica-alumina would give greater intensity of reaction than silica-magnesia as observed from the nature and yields of the individual cracked products and the automobile gasoline octane number. Titration of these two catalysts shows silica-alumina to have a lower acid titer than silica-magnesia, but the acid strength of the sites is higher.

While each of the individual component parts in these catalysts is essentially nonacidic, when mixed together properly, they give rise to a titratable acidity as described above. Many of the secondary reactions occurring in the cracking process may also be promoted with strong mineral acids, such as concentrated sulfuric and phosphoric acids, aluminum halides, hydrogen fluoride, and hydrogen fluoride-boron trifluoride (BF_3) mixtures. This parallelism lends support to the concept of the active catalytic site as being acidic. Zeolites have a much higher active site density (titer) than the amorphous mixed oxides, which may account in large part for their extremely high cracking propensity. In addition, these materials strongly promote complex hydrogen transfer reactions among the primary products so that the recovered cracked products have a much lower olefin and higher paraffin content than are obtained with the amorphous mixed oxide catalysts. This hydrogen transfer propensity of zeolites to saturate primary cracked product olefin derivatives to paraffin derivatives minimizes the reaction of polymerizing the olefin derivatives to form a coke deposit, thus accounting in part for the much lower coke yields with zeolite catalysts than with amorphous catalysts.

Activity of the catalyst varies with faujasite content as does the selectivity of the catalyst to coke and naphtha. As the faujasite content drops below 5% by weight, the catalyst starts to show some of the cracking properties of the matrix, while for zeolite contents of 10% by weight or higher, very little change in selectivity patterns is noted. The various ion-exchanged forms of the faujasite can result in slightly different cracking properties; for example, using high cerium content, mixed rare earths improve carbon burning rates in the regenerator, the use of H-form of faujasite improves selectivity to propane-pentane fractions, and the use of a minor amount of copper form faujasite increases low-molecular-weight olefin yield and naphtha octanes.

6.4 Catalyst Deactivation

A cracking catalyst should maintain its cracking activity with little change in product selectivity as it ages in a unit. A number of factors contribute to degrade the catalyst: (i) the combination of high temperature, steam partial pressure, and time; (ii) impurities present in the fresh catalyst; and (iii) impurities picked up by the catalyst from the feedstock while in use. Under normal operating conditions, the catalyst experiences temperatures of 480–515°C (900–960°F) in the reactor and steam stripper zones and temperatures of 620–720°C (1150–1325°F) and higher in the regenerator accompanied by a substantial partial pressure of steam. With mixed oxide amorphous gel catalysts, the plastic nature of the gel is such that the surface area and pore volume decrease rather sharply in the first few days of use and then at a slow inexorable rate thereafter. This plastic flow also results in a loss in the number and strength of the active catalytic sites.

Zeolite catalysts comprising both amorphous gel and crystalline zeolite degrade from instability of the gel, as stated above, and also from loss in crystallinity. The latter also results from the combined effects of time, temperature, and steam partial pressure. When crystallinity is lost, the amorphous residue is relatively low in activity, approximating that of the amorphous gel matrix. The rate of degradation of the amorphous gel component may not be the same as that of the zeolite crystals; for example, the gel may degrade rapidly and through thermoplastic flow effectively coat the crystals and interfere with the diffusion of hydrocarbon derivatives to the catalytic sites in the zeolite. Catalyst manufacturers try to combine high stability in the matrix with high-stability zeolite crystals in making zeolite catalysts.

Residual impurities in freshly manufactured catalysts are principally sodium and sulfate. These result from the use of sodium silicate and aluminum sulfate in making the silica-alumina gel matrix and subsequent washing of the composite catalyst with ammonium sulfate to remove sodium. Generally, the sodium content of the amorphous gel is <01% w/w (as Na_2O) and sulfate <0.5% w/w.

With zeolite catalysts, the residual sodium may be primarily associated with the zeolite, so that sodium levels may range from approximately 0.2%–0.80% for the composite catalyst. Sulfate levels in zeolite catalysts are still <0.5%. An excessive amount of sodium reacts with the silica in the matrix under regenerator operating conditions and serves as a flux to increase the rate of surface area and pore volume loss. Sodium faujasite is not as hydrothermally stable as other metal-exchanged (e.g., mixed rare earths) forms of faujasite. It is most desirable to reduce the sodium content of the faujasite component to <5.0% by weight (as Na_2O) with rare earths or with mixtures of rare earths and ammonium ions.

Finally, catalysts can degrade as a result of impurities picked up from the feedstock being processed. These impurities are sodium, nickel, vanadium, iron, and copper. Sodium as laid down on the catalyst not only acts to neutralize active acid sites, reducing catalyst activity, but also acts as a flux to accelerate matrix degradation. Freshly deposited metals are effective *poisons* to cracking catalysts because of the loss of active surface area by metal deposition (Otterstedt et al., 1986; Parkash, 2003; Gary et al., 2007; Speight, 2014, 2017; Hsu and Robinson, 2017). Zeolite catalysts are less responsive to metal contaminants than amorphous gel catalysts. Hence, equilibrium catalysts can tolerate low levels of these metals so long as they have enough time to become buried. A sudden deposition of fresh metals can cause adverse effects on unit performance. Metal levels on equilibrium catalysts reflect the metal content of the feeds being processed; typical ranges are 200–1200 ppm V, 150–500 ppm Ni, and 5–45 ppm Cu. Sodium levels are in the range of 0.25%–0.8% by weight (as Na_2O).

6.5 Catalyst Stripping

Catalyst leaving the reaction zone is fluidized with reactor product vapors that must be removed and recovered with the reactor product. In order to accomplish this, the catalyst is passed into a stripping zone where most of the hydrocarbon is displaced with steam.

Stripping is generally done in a countercurrent contact zone where shed baffles or contactors are provided to insure equal vapor flow up through the stripper and efficient contacting. Stripping can be accomplished in a dilute catalyst phase. Generally, a dense phase is used, but with lighter feeds or higher reactor temperature and high conversion operations, a significant portion of the contacting can be done in a dilute phase.

The amount of hydrocarbon carried to the regenerator is dependent upon the amount of stripping steam used per pound of catalyst and the pressure and temperature at which the stripper operates.

Probes at the stripper outlet have been used to measure the composition of the hydrocarbon vapors leaving the stripper. When expressed as percent of coke burned in the regenerator, the strippable hydrocarbon is only 2%–5%. Very poor stripping is shown when the hydrogen content of the regenerator coke is on the order of 10% w/w or higher. Good stripping is shown by 6%–9% by weight hydrogen levels.

The proper level of stripping is found in many operating units by reducing the stripping steam until there is a noticeable effect or rise in regenerator temperature. Steam is then marginally increased above this rate. In some units, stripping steam is used as a control variable to control the carbon burning rate or differential temperature between the regenerator bed and cyclone inlets.

In summary, the catalytic cracking unit is an extremely dynamic unit, primarily because there are three major process flow streams (the catalyst, hydrocarbon, and regeneration air), all of which interact with each other. Problems can arise in the equipment and flowing streams that are sometimes difficult to diagnose because of the complex affects they can create (Parkash, 2003; Gary et al., 2007; Speight, 2014, 2017; Hsu and Robinson, 2017).

6.6 Catalyst Treatment

The latest technique developed by the refining industry to increase naphtha yield and quality is to treat the catalysts from the cracking units to remove metal poisons that accumulate on the catalyst. Nickel, vanadium, iron, and copper compounds contained in catalytic cracking feedstocks are deposited on the catalyst during the cracking operation, thereby adversely affecting both catalyst activity and selectivity. Increased catalyst metal contents affect catalytic cracking yields by increasing coke formation, decreasing naphtha and butane and butylene production, and increasing hydrogen production.

The recent commercial development and adoption of cracking catalyst-treating processes definitely improve the overall catalytic cracking process economics.

6.6.1 Demet

In the *Demet process*, a cracking catalyst is subjected to two pretreatment steps (Fig. 9.9). The first step effects vanadium removal and the second nickel removal, to prepare the metals on the catalyst for chemical conversion to compounds (chemical treatment step) that can be readily removed through water washing (catalyst wash step). The treatment steps include the use of a sulfurous compound followed by chlorination with an anhydrous chlorinating agent (e.g., chlorine gas) and washing with an aqueous solution of a chelating agent such as citric acid ($HO_2CCH_2C(OH)(CO_2H)CH_2CO_2H$ or 2-hydroxy-1,2,3-propanetricarboxylic acid). The catalyst is then dried and further treated before returning to the cracking unit (Parkash, 2003; Gary et al., 2007; Speight, 2014, 2017; Hsu and Robinson, 2017).

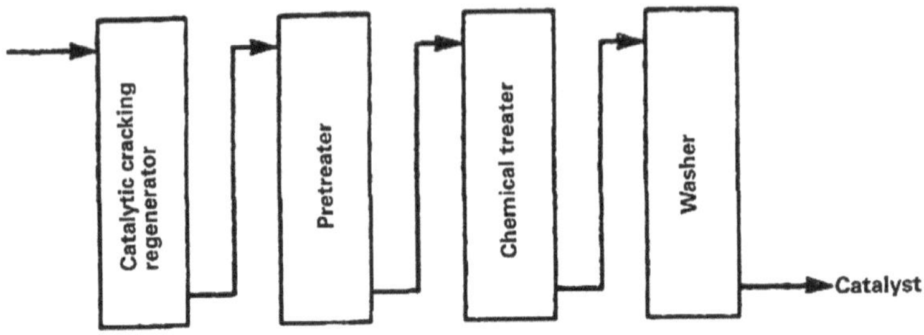

FIG. 9.9 The Demet process.

6.6.2 *Met-X*

The *Met-X process* consists of cooling, mixing, ion-exchange separation, filtration, and resin regeneration. Moist catalyst from the filter is dispersed in oil and returned to the cracking reactor in a slurry (Fig. 9.10; Parkash, 2003; Gary et al., 2007; Speight, 2014, 2017; Hsu and Robinson, 2017). On a continuous basis, the catalyst from a cracking unit is cooled and then transported to a stirred reactor and mixed with an ion-exchange resin (introduced as slurry). The catalyst-resin slurry then flows to an elutriator for separation. The catalyst slurry is taken overhead to a filter, and the wet filter cake is slurried with oil and pumped into the catalytic cracker feedstock system. The resin leaves the bottom of the elutriator and is regenerated before returning to the reactor.

6.7 Recent Advances

Advances in fluid catalytic cracking catalysts have concentrated on modifying zeolite Y for improved coke selectivity, higher cracking activity, and greater stability through manipulation of extra-framework aluminum or through the generation of mesoporous nature of the zeolite crystals. Extra-framework aluminum is introduced either by steaming or via ion exchange. The development of improved fluid catalytic cracking catalysts includes modifying a single crystal structure to achieve multiple catalytic objectives (Degnan, 2000).

In order to meet demands for a higher-octane product, an increasing number of refiners have switched to using ultrastable Y (USY) zeolite-type catalysts. These catalysts, which feature zeolites with a reduced unit cell size, are a proved means of increasing the octane number of fluid catalytic cracking naphtha (which is used for gasoline production in the blending operation). However, there are trade-offs associated with using a reduced unit cell-size zeolite—one of the more important properties being affected is catalyst activity. In addition, the use of ZSM-5 (a pentasil zeolite concentrate that typically incorporates up to 25%–40% pentasil) enhances the octane number in fluid catalytic cracking naphtha by overcracking of naphtha/gasoline fraction. This reaction mainly occurs on the stronger acid site of ZSM-5 than that of the USY catalyst.

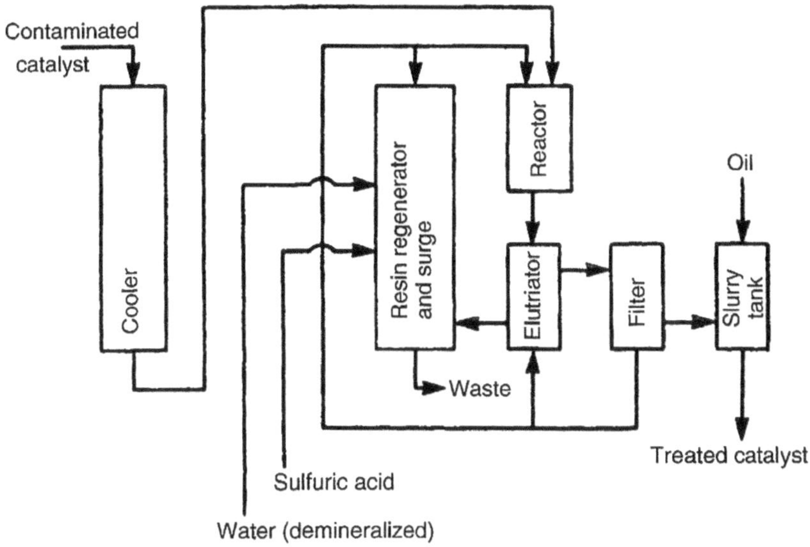

FIG. 9.10 The Met-X process.

Comparing with the result with typical fluid catalytic cracking catalyst, the yield of liquefied petroleum gas is increased by overcracking, and the octane number is enhanced for the mixed catalyst, although the yield of fluid catalytic cracking naphtha is decreased. The increase of liquefied petroleum gas yield is due to the production of propylene in the case of mixed catalyst (Buchanan et al., 1996; Imhof et al., 2005; Li et al., 2007).

6.7.1 Matrix, Binder, Zeolite

The *matrix* in the fluid catalytic cracking catalyst is often considered to be that part of the catalyst other than the zeolites—it may or may not have catalytic activity. Very often, it does possess sufficient catalytic activity toward some components of the feedstock, and in this case, they are described by the term *active matrix*. The matrix consists of two main components—firstly, a synthetic component like amorphous silica, silica-alumina gel, or silica-magnesia gel, which serve as the binder and also exhibit catalytic properties. The other component is natural or chemically modified clay, such as kaolinite, halloysite, or montmorillonite. The clays provide mechanical stability.

The functions of the *fillers* and *binders* incorporated into the fluid catalytic cracking catalysts are often similar to those performed by the matrix. Sometimes, additional fillers like kaolin may be provided for physical integrity and as a more efficient fluidizing medium. The binder performs the all-important function of holding the catalyst, the matrix, and the filler glued together. This is especially important when the catalyst contains a higher amount of zeolites. The filler and binder minimize the production of catalyst fines in the reactor-regenerator system and help to control or mitigate catalyst losses.

6.7.2 *Additives*

In addition to cracking catalyst described above, a series of *additives* have been developed that catalyze or otherwise alter the primary catalyst's activity/selectivity or act as pollution control agents. Additives are most often prepared in microspherical form to be compatible with the primary catalysts and are available separately in compositions that (i) enhance naphtha octane and low-molecular-weight olefin formation, (ii) selectively crack heavy cycle oil, (iii) passivate vanadium and nickel present in many viscous feedstocks, (iv) oxidize coke to carbon dioxide, and (v) reduce sulfur dioxide emissions. Both vanadium and nickel deposits on the cracking catalyst are extremely deleterious when present in excess of 3000 ppm on the catalyst. Formulation changes to the catalyst can improve tolerance to vanadium and nickel, but the use of additives that specifically passivate either metal is often preferred.

The ability of small amounts of ZSM-5 added to the fluid catalytic cracking unit to improve naphtha octane number while producing more low-molecular-weight olefin derivatives has prompted a substantial amount of process and catalyst research into zeolite-based fluid catalytic cracking additives. Significant advances have been made in stabilizing ZSM-5 to harsh fluid catalytic cracking regenerator conditions that, in turn, have led to reductions in the level of ZSM-5 needed to achieve desired uplifts and wider use of the less expensive additives (Degnan, 2000).

6.7.3 *Metal Traps*

The performance of fluid catalytic cracking catalysts for processing viscous feedstocks is often determined by the tolerance of the catalyst to metal contaminants. Different approaches are used to manage contaminant metals such as riser pipe design (Hedrick and Palmas, 2011). Metal traps have been introduced into catalysts in order to protect the zeolite from poisonous metals that have accumulated in large amounts as a result of lowering the amount of catalyst use and in order to give the catalyst metal tolerance so that the formation of hydrogen and coke due to the dehydrogenation over poisonous metals would be reduced.

As a result, catalyst manufacturers have developed more stable zeolites and a series of vanadium traps to increase the ability of the zeolite to handle vanadium. These traps are based on barium (Ba), titanium (Ti), rare earth elements, and other elements. Some are more effective than others, but the basic idea is the same, that is, to keep the vanadium away from the zeolite by binding to the surface of an inactive particle (Harding et al., 2001).

6.7.4 Low Rare Earths

Rare earth elements inhibit alumina removal from a zeolite; a higher concentration of acid sites will be found in a rare earth-exchanged catalyst. This improves both the activity and the hydrothermal stability of the catalyst. On average, these sites are weaker and in closer proximity to each other than those found in a more highly dealuminated catalyst characterized by lower unit cell-size measurements.

The addition of rare earth into the zeolite inhibits the degree of unit cell-size shrinkage during equilibration in the regenerator. Steam in the fluid catalytic cracking regenerator removes active acidic alumina from the zeolite. Rare earth inhibits the extraction of aluminum from the zeolite's structure (alumina removal) that in turn increases the equilibrium unit cell size for fluid catalytic cracking catalysts. Since reducing the equilibrium unit cell size of a fluid catalytic cracking catalyst has the effect of improving octane, adding rare earth decreases the octane.

Because rare earth oxides promote hydrogen transfer, the yield of C_3 and C_4 olefin derivatives in the liquefied petroleum gas fraction will be lower. The resulting reduction in the total liquefied petroleum gas yield results in a reduction in the wet gas yield, which can have a major effect on plant operations, as compressor capacity is often the limiting factor for fluid catalytic cracking unit throughput.

6.7.5 Catalysts for Olefin Production

The market for propene derivatives and butene derivatives is cyclical and depends on the local demand for plastic precursors and alkylate feed. There are several key determinants of low-molecular-weight olefin selectivity (Harding et al., 2001; Soni et al., 2009). In addition to increasing reactor temperature and using ZSM-5, catalysts can be designed with lower hydrogen transfer (to reduce the conversion of isobutylene to isobutane) and moderate matrix activity (to increase the C4 olefin content of the product).

The addition of ZSM-5 is the single most important method to increase the yields of low-molecular-weight olefin derivatives at the expense of naphtha. Catalyst producers have developed methods to increase the concentration of the ZSM-5 in their additives in order to avoid dilution effects at high levels of additive. The breakthrough technologies in this area involve the stabilization of the ZSM-5 to hydrothermal deactivation at higher concentrations in the additive.

Current commercial catalysts range from 10% to 25% ZSM-5, although evolving techniques are expected to allow much higher levels in an additive particle. Several recent efforts have also been made to change the selectivity of ZSM-5 by increasing the Si/Al ratio to increase the ratio of butene to propene. However, this approach reduces the overall activity quite significantly.

6.7.6 Catalysts for Jet and Diesel Production

In markets dominated by fluid catalytic cracking-based refineries, the need to increase distillate production has taken on a new dimension, posing interesting challenges while presenting some unique opportunities (Yung and Pouwels, 2008). Much attention on the possibility of tighter specifications for aromatic derivatives, density, boiling range, and cetane number can be expected. New technologies are being developed to address these challenges. For example, a commercialized moderate-pressure hydrocracking (MPHC) suite of technologies can provide an economic solution to desulfurize high-boiling gas oil with options to convert part of the feedstock into valuable higher-quality distillate products and to maximize the production of Jet A1 and diesel products (Degnan, 2000; Hilbert et al., 2008).

The catalysts are large pore, highly dealuminated zeolite Y and improved base metal combinations that have been tailored to operate well at low hydrogen partial pressures. Processes have been developed around USY-based catalysts for partial conversion of vacuum gas oils, cracked gas oils, deasphalted oil, and fluid catalytic cracking light cycle oil. The processes comprise a dual-catalyst system consisting of an amorphous hydrotreating catalyst (normally NiMo/Al_2O_3) and a metal-containing USY-based hydrocracking catalyst. Single-stage, single-pass conversions are typically in the range of 30%–60% w/w. The process requirements are similar to those used in vacuum gas-oil hydrodesulfurization, which has led to several catalytic hydrodesulfurization revamps (Degnan, 2000).

For cracking catalysts, there is usually a trade-off between high activity and high middle distillate selectivity. The objective for new cracking catalyst development is often to increase activity while maintaining selectivity or vice versa. This can be achieved by altering the catalyst's acidic function and/or the hydrogenation function, as is illustrated by the UOP Unicracking process (Ackelson, 2004; Abdo et al., 2008).

In addition to control and optimization of process conditions in the Unicracking process, catalyst design principles when applied to this problem take advantage of improved molecular-scale definition of feedstock composition, appropriate selection of catalytic materials, and enhanced knowledge of the impact of specific process conditions on catalyst performance to deliver the needed level of activity and selectivity to meet cycle length and product yields (Motaghi et al., 2010).

In the early 1990s, Zeolyst introduced Z-603, a zeolite catalyst with high activity and also good middle distillate selectivity. The improved performance of the catalyst versus other zeolite and amorphous silica-alumina catalysts was the result of two major developments—a new dealuminated zeolite Y used in conjunction with an amorphous silica-alumina catalysts powder and an improved hydrogenation function. This catalyst has been used successfully at Shell's Godorf (Germany) refinery and at Alliance Refining Company

(Rayong, Thailand) refinery. Because of its higher activity and stability, Z-603 has extended the cycle length beyond the capability of amorphous silica-alumina catalysts with only a small reduction in middle distillate yield relative to the amorphous silica-alumina catalysts (Huve et al., 2004).

Continuing development efforts have led to a new range of cracking catalysts (e.g., Z-623 and Z-613) with improved middle distillate yields/activities. Z-623 is significantly more active than the Z-603 catalyst with very little middle distillate yield loss. The benefits of the Z-623 catalyst have been demonstrated in a number of commercial cases. At Alliance Refining Company, the catalyst has made it possible to increase the hydrocracker's cycle length from 2 years to >3 years. At Shell's Godorf refinery, the catalyst was a key component of the hydrocracker revamp that has produced a 15% *v*/v increase in throughput of feedstock without any cycle length reduction (Huve et al., 2004).

6.7.7 New Directions

With the increasing focus to reduce sulfur content in fuels, the role of *desulfurization* in the refinery becomes more and more important. Currently, the process of choice is the hydrotreater to remove the sulfur from the fuel. Because of the increased attention for fuel desulfurization, various new process concepts are being developed with various claims of efficiency and effectiveness.

Resid fluid catalytic cracking is an important component in the upgrading of heavy oil, extra heavy oil, and tar sand bitumen, with unit profitability depending upon the extent to which viscous hydrocarbon derivatives in the feedstock are cracked into valuable products. The product slate, in turn, depends upon the feedstock characteristics, the catalyst, the hardware, and the operating conditions. Exemplifying a trend toward heavier feedstocks, the majority of the fluid-bed catalytic cracking units scheduled to start up before 2015 are expected to process vacuum gas-oil and/or or residuum feedstocks.

Product recycling and multiple reaction sections are the prevalent technology trends. Improving feedstock injectors, riser termination and catalyst separation devices, strippers, and regenerator components are revamp options for existing units. As feedstocks get heavier, the trend toward a higher stripper residence time and, consequently, increased mass transfer between entrained hydrocarbon derivatives and steam will continue. Moreover, the role of the regenerator continues to evolve because of reduction requirements in the carbon dioxide emissions. The need for the development of flue gas treating systems points to the need for continued innovation in that area.

The use of biofeedstocks (such as animal fats and vegetable oils) in the fluid catalytic cracking will be used to increase the yield of light cycle oil and will also provide high-quality products. Practical implementation in a refinery will, more than likely, be accompanied by blending with vacuum gas oil or residuum (Speight, 2011a,b).

Oils derived from biofeedstock oils are generally best upgraded by HZSM-5 or ZSM-5, as these zeolite catalysts promote high yields of liquid products and propylene. Unfortunately, these feeds tend to coke easily, and high acid numbers and undesirable by-products such as water and carbon are and will continue to be additional challenges. Waxy feeds obtained from biomass gasification followed by Fischer-Tropsch synthesis to hydrocarbon liquids and waxes (biomass to liquids) are especially suited for increasing light cycle oil production in the catalytic cracking unit, due to the high paraffinic character, low sulfur content, and low aromatic derivative yield of the feed. A major disadvantage for biomass-to-liquid products is the intrinsically low coke yield that can disrupt the heat balance of the catalytic cracking unit.

The major developmental opportunity area in the next decade may well be at the refinery-petrochemical interface. Processes that maximize olefin derivatives and aromatic derivatives while integrating clean fuel production will continue to look attractive to refiners who are seeing their fuel product margins come under increasing competitive pressure.

Furthermore, in order to alleviate growing concerns over energy security and global climate change, the fluid catalytic cracking unit will take on two additional roles: (i) acceptance of biomass feedstocks and (ii) reduction in carbon dioxide emissions. With committed efforts by refiners and technology developers, the process will continue to demonstrate adaptability amid changing market requirements through the installation of dual-radius feedstock distributors and spent-catalyst distributors (Wolschlag and Couch, 2010a).

Methods to reduce emissions from the fluid catalytic cracking unit and other refinery operations lend themselves to two conflicting categories: those that hinder the formation of carbon dioxide and those that aim to produce a pure carbon dioxide stream for capture. Improving unit energy efficiency, reducing coke yields, and shifting vacuum gas oil to the hydrocracking unit belong to the former category, while carbon-capture methods typically comprise the latter (Spoor, 2008). As refiners continue to develop a comprehensive strategy to reduce carbon dioxide emissions, a balance between these two categories must be found. This happy medium will be influenced greatly by the fluid catalytic cracking unit heat balance, which determines the amount of coke burned in the regenerator in present unit designs.

7. Options for Heavy Feedstocks

The fluid catalytic cracking process, using vacuum gas oil as the feedstock, was introduced into refineries in the 1930s. In recent years, because of a trend for low-boiling products, most refineries perform the operation by partially blending residua into vacuum gas oil. However, conventional fluid catalytic cracking processes have limits in residuum processing, so residuum fluid catalytic cracking processes have lately been employed one after another. Because the residuum fluid catalytic cracking process enables efficient gasoline production directly from residua, it will play the most important role as a residuum cracking process, along with a residuum hydrotreating process.

The processes described below are the evolutionary offspring of the fluid catalytic cracking and the residuum catalytic cracking processes. Some of these newer processes use catalysts with different silica/alumina ratios as acid support of metals such as Mo, Co, Ni, and W. In general, the first catalyst used to remove metals from oils was the conventional hydrodesulfurization (HDS) catalyst. Diverse natural minerals are also used as raw material for elaborating catalysts addressed to the upgrading of viscous fractions. Among these minerals are clay minerals; manganese nodules; bauxite activated with vanadium (V), nickel (Ni), chromium (Cr), iron (Fe), and cobalt (Co); and the high-iron-oxide-content iron laterites, sepiolite minerals, and mineral nickel and transition metal sulfides supported on silica and alumina. Other kinds of catalysts, such as vanadium sulfide, are generated in situ, possibly in colloidal states.

In the past decades, in spite of the difficulty of handling viscous feedstocks, residuum fluidized catalytic cracking (RFCC) has evolved to become a well-established approach for converting a significant portion of the heavier fractions of the crude barrel into a high-octane gasoline blending component. Residuum fluidized catalytic cracking, which is an extension of conventional fluid catalytic cracking technology for applications involving the conversion of highly contaminated residua, has been commercially proved on feedstocks ranging from gas-oil-residuum blends to atmospheric residua and blends of atmospheric and vacuum residua. In addition to high naphtha yields, the residuum fluidized catalytic cracking unit also produces gaseous, distillate, and fuel oil-range products.

The product quality from the residuum fluidized catalytic cracker is directly affected by its feedstock quality. In particular and unlike hydrotreating, the residuum fluidized catalytic cracking redistributes sulfur among the various products, but does not remove sulfur from the products unless, of course, one discounts the sulfur that is retained by a coke formed on the catalyst. Consequently, tightening product specifications have forced refiners to hydrotreat some or all of the products from the resid cat cracking unit. Similarly, in the future, the emissions of sulfur oxides (SOx) from a resid cat cracker may become more of an obstacle for residue conversion projects. For these reasons, a point can be reached where the economic operability of the unit can be sufficient to justify hydrotreating the feedstock to the cat cracker.

As an integrated conversion block, residue hydrotreating and residuum fluidized catalytic cracking complement each other and can offset many of the inherent deficiencies related to residue conversion.

7.1 Asphalt Residual Treating Process

The asphalt residual treating process (ART process) is a process for increasing the production of transportation fuels and reduces heavy fuel oil production, without hydrocracking (Bartholic, 1981a,b; Bartholic and Haseltine, 1981; Bartholic, 1989; Parkash, 2003; Gary et al., 2007; Speight, 2014, 2017). The process is a flexible selective vaporization process that can be used for removing essentially all of the metals and substantial proportion of the carbon residue, nitrogen, and sulfur compounds from the viscous feedstock (Bartholic and Haseltine, 1981). In addition, the process can be considered as an efficient carbon rejection process followed by catalytic hydrogenation.

In the process, the preheated feedstock (which may be whole crude, atmospheric residuum, vacuum residuum, or tar sand bitumen) is injected into a stream of fluidized, hot catalyst (trade name, ArtCat). The unit configuration is similar to that of a riser fluid catalytic cracking unit where complete mixing of the feedstock with the catalyst is achieved in the contactor, which is operated within a pressure-temperature envelope to ensure selective vaporization. The vapor and the contactor effluent are quickly and efficiently separated from each other, and entrained hydrocarbon derivatives are stripped from the contaminant (containing spent solid) in the stripping section. The contactor vapor effluent and vapor from the stripping section are combined and rapidly quenched in a quench drum to minimize product degradation. The cooled products are then transported to a conventional fractionator that is similar to that found in a fluid catalytic cracking unit. Spent solid from the stripping section is transported to the combustor bottom zone for carbon burn-off.

The process is claimed to require lower hydrogen consumption that is claimed in comparison with direct hydrocracking processes (Dennis, 1989; Suchanek and Moore, 1986).

Contact of the feedstock with the fluidizable catalyst in the short-residence-time contactor causes the lower-boiling components of the feedstock to vaporize, and asphaltene constituents (high-molecular-weight compounds) are cracked to yield lower-boiling compounds and coke. The metals present and some of the sulfur and the nitrogen compounds in the nonvolatile constituents are retained on the catalyst. At the exit of the contacting zone, the oil vapors are separated from the catalyst and are rapidly quenched to minimize thermal cracking of the products. The catalyst, which holds metals, sulfur, nitrogen, and coke, is transferred to the regenerator where the combustible portion is oxidized and removed. Regenerated contact material, bearing metals but very little coke, exits the regenerator and passes to the contactor for further removal of contaminants from the charge stock.

In the combustor, coke is burned from the spent solid that is then separated from combustion gas in the surge vessel. The surge vessel circulates regenerated catalyst streams to the contactor inlet for feedstock vaporization and to the combustor bottom zone for premixing. The components of the combustion gases include carbon dioxide (CO_2), nitrogen (N_2), oxygen (O_2), sulfur oxides (SO_X), and nitrogen oxides (NO_X) that are released from the catalyst with the combustion of the coke in the combustor. The concentration of sulfur oxides in the combustion gas requires treatment for their removal.

7.2 Residue Fluid Catalytic Cracking Process

The residue fluid catalytic cracking process (also known as the HOC process) is a version of the fluid catalytic cracking process that has been adapted to the conversion of residua that contain high amounts of metal and asphaltene constituents). Depending on quality and product objectives, feedstocks with vanadium plus nickel content of 5–30 ppm and a carbon residue on the order of 5%–10% w/w can be processed without feedstock pretreatment. The process, when coupled with hydrodesulfurization, provides a particularly effective means for meeting demands for naphtha or diesel fuel blending stock and low-molecular-weight olefin derivatives. The asphaltene constituents are converted to coke (which deposits on the catalyst) and distillate from the cracked fragments. The catalyst is regenerated by combusting its carbon deposits, and the heat generated from the combustion of carbon deposits is used to produce high-pressure steam.

In the process, a viscous feedstock is desulfurized, and the nonvolatile fraction from the hydrodesulfurization unit is charged to the residuum fluid catalytic cracking unit. The reaction system is an external vertical riser terminating in a closed cyclone system. Dispersion steam in amounts higher than that used for gas oils is used to assist in the vaporization of any volatile constituents of viscous feedstocks. A two-stage stripper is utilized to remove hydrocarbon derivatives from the catalyst. Hot catalyst flows at low velocity in dense phase through the catalyst cooler and returns to the regenerator. Regenerated catalyst flows to the bottom of the riser to meet the feed. The coke deposited on the catalyst is burned off in the regenerator along with the coke formed during the cracking of the gas-oil fraction. If the feedstock contains high proportions of metals, control of the metals on the catalyst requires excessive amounts of catalyst withdrawal and fresh catalyst addition. This problem can be addressed by feedstock pretreatment.

Regenerator bed temperatures are on the order of 730°C (1300°F), and feedstock introduction systems are designed for efficient mixing of oil and catalyst and rapid quenching of catalyst temperature to the equilibrium mix temperature. The reaction system is an external vertical riser providing very low contact times and terminating in the riser cyclones for rapid separation of catalyst and vapors.

The high amounts of coke produced in cracking of residua can cause extreme temperatures and excessive catalyst deactivation. Steam coils located within the regenerator bed and/or external catalyst coolers remove the excess heat produced by the high coke yields. When the metal content and carbon residual data are in excess of 30 ppm and 10% by weight, respectively, the feedstock may require pretreatment.

The feedstocks for the process are rated on the basis of carbon residue and content of metals: thus, *good-quality feedstocks* have <5% by weight carbon residue and <10 ppm metals. *Medium-quality feedstocks* have >5% but

$<10\%$ by weight carbon residue and >10 but <30 ppm metals. *Poor-quality feedstocks* have >10 but $<20\%$ by weight carbon residue and >30 but less than <150 ppm metals. Finally, *bad-quality feedstocks* have $>20\%$ by weight carbon residue and >150 ppm metals. One might question the value of this rating of the feedstocks for the HOC process since these feedstock ratings can apply to virtually many fluid catalytic cracking processes.

Overall, the process is similar to the fluid catalytic cracking process in configuration, catalysts, and product handling but differs from gas-oil cracking in that the heat release, due to burning the coke produced from the asphaltenes in the charge, is considerably greater. In addition, the much higher content of feedstock metals—particularly nickel and vanadium—requires special consideration in catalyst development and in operation. With the need to convert residual fuels to naphtha and middle distillates, the installation of new viscous feedstock cracking units and the conversion of fluid catalytic cracking units to handle residua have become a refining necessity.

Special catalysts are required for heavy oil, extra heavy oil, and tar sand bitumen cracking units because of the required specifications for activity and selectivity. Some of the catalysts have a high zeolite content, or they have pore structures that avoid trapping large molecules and causing coke production. Poisons such as sodium and vanadium accelerate the deactivation rate of catalyst, and high amounts of sodium are usually avoided by double desalting of the crude oil. While customized catalysts can improve heavy oil, extra heavy oil, and tar sand bitumen cracking operations, optimization can also take place by closer process control. For example, high cracking activity, if used correctly, can override the adverse dehydrogenation activity of the metals. Thus, low contact time in risers, along with rapid and efficient separation of catalyst and oil vapors, is preferred in reduced crude cracking. Also, heavy catalytic gas oil or slurry oil recycle produces higher yields of coke and higher heat release than it removes through vaporization and therefore should be minimized or eliminated. In addition to reducing coke yield, lower regenerator temperatures can be realized by direct heat removal or lower heat generation.

7.3 Heavy Oil Treating Process

The heavy oil treating process (HOT process) is a catalytic cracking process for upgrading viscous feedstocks such as topped crude oils, vacuum residua, and solvent deasphalted bottoms using a fluidized bed of iron ore particles (Parkash, 2003; Gary et al., 2007; Speight, 2014, 2017; Hsu and Robinson, 2017).

The main section of the process consists of three fluidized reactors, and separate reactions take place in each reactor (*cracker*, *regenerator*, and *desulfurizer*):

$$Fe_3O_4 + \text{asphaltene constituents} \rightarrow \text{coke}/Fe_3O_4 + \text{Oil} + \text{Gas (in the } \textit{cracker}\text{)}$$

$$3FeO + H_2O \rightarrow Fe_3O_4 + H_2 \text{ (in the } \textit{cracker}\text{)}$$

$$Coke/Fe_3O_4 + O_2 \rightarrow 3FeO + CO + CO_2 \text{ (in the } \textit{regenerator}\text{)}$$

$$FeO + SO_2 + 3CO \rightarrow FeS + 3CO_2 \text{ (in the } \textit{regenerator}\text{)}$$

$$3FeS + 5O_2 \rightarrow Fe_3O_4 + 3SO_2 \text{ (in the } \textit{desulfurizer}\text{)}$$

In the *cracker*, feedstock cracking and the steam-iron reaction take place simultaneously under the conditions similar usual to thermal cracking. Any unconverted feedstock is recycled to the cracker from the bottom of the scrubber. The scrubber effluent is separated into hydrogen gas, liquefied petroleum gas, and liquid products that can be upgraded by conventional technologies to priority products.

In the *regenerator*, coke deposited on the catalyst is partially burned to form carbon monoxide in order to reduce iron tetroxide and to act as a heat supply. In the *desulfurizer*, sulfur in the solid catalyst is removed and recovered as molten sulfur in the final recovery stage.

7.4 R2R Process

The R2R process is a fluid catalytic cracking process for the conversion of viscous feedstocks (Heinrich and Mauleon, 1994; Inai, 1994; Parkash, 2003; Gary et al., 2007; Speight, 2014, 2017). In the process, the feedstock is vaporized upon contacting hot regenerated catalyst at the base of the riser and lifts the catalyst into the reactor vessel separation chamber where rapid disengagement of the hydrocarbon vapors from the catalyst is accomplished by both a special solid separator and cyclones. The bulk of the cracking reactions takes place at the moment of contact and continues as the catalyst and hydrocarbon derivatives travel up the riser. The reaction products, along with a minute amount of entrained catalyst, then flow to the fractionation column. The stripped spent catalyst, deactivated with coke, flows into the Number 1 regenerator.

Partially regenerated catalyst is pneumatically transferred via an air riser to the Number 2 regenerator, where the remaining carbon is completely burned in a dryer atmosphere. This regenerator is designed to minimize catalyst inventory and residence time at high temperature while optimizing the coke burning rate. Flue gases pass through external cyclones to a waste heat recovery system. Regenerated catalyst flows into a withdrawal well, and after, stabilization is charged back to the oil riser.

7.5 Reduced Crude Oil Conversion Process

Another role of the *residuum fluid catalytic cracking process* is to generate high-quality gasoline blending stock and petrochemical feedstock. Olefin derivatives (propene, the isomeric butene derivatives, and the isomeric pentene derivatives) serve as feedstock for alkylating processes, for polymer gasoline, and for additives for reformulated gasoline.

In the reduced crude oil conversion process (RCC process), the clean regenerated catalyst enters the bottom of the reactor riser where it contacts low-boiling hydrocarbon *lift gas* that accelerates the catalyst up the riser prior to feedstock injection (Parkash, 2003; Gary et al., 2007; Speight, 2014, 2017). At the top of the lift gas zone, the feedstock is injected through a series of nozzles located around the circumference of the reactor riser. The catalyst/oil disengaging system is designed to separate the catalyst from the reaction products and then rapidly remove the reaction products from the reactor vessel. Spent catalyst from reaction zone is first steam stripped, to remove adsorbed hydrocarbon, and then routed to the regenerator. In the regenerator, all of the carbonaceous deposits are removed from the catalyst by combustion, restoring the catalyst to an active state with a very low carbon content. The catalyst is then returned to the bottom of the reactor riser at a controlled rate to achieve the desired conversion and selectivity to the primary products.

7.6 Shell FCC Process

The Shell FCC process is designed to maximize the production of distillates from residua (Parkash, 2003; Gary et al., 2007; Speight, 2014, 2017; Hsu and Robinson, 2017). In the process, the preheated feedstock (vacuum gas oil and atmospheric residuum) is mixed with the hot regenerated catalyst. After reaction in a riser, volatile materials and catalyst are separated after which the spent catalyst is immediately stripped of entrained and adsorbed hydrocarbon derivatives in a very effective multistage stripper. The stripped catalyst gravitates through a short standpipe into a single vessel, simple, reliable, and yet efficient catalyst regenerator. Regenerative flue gas passes via a cyclone/swirl tube combination to a power recovery turbine. From the expander turbine, the heat in the flue gas is further recovered in a waste heat boiler. Depending on the environmental conservation requirements, a de-NO_xing, de-SO_xing, and particulate emission control device can be included in the flue gas train.

There is a claim that feedstock pretreatment of bitumen (by hydrogenation) prior to fluid catalytic cracking (or for that matter any catalytic cracking process) can result in enhanced yield of naphtha. It is suggested that mild hydrotreating be carried out upstream of a fluid catalytic cracking unit to provide an increase in yield and quality of distillate products (Long et al., 1993; Parkash, 2003; Gary et al., 2007; Speight, 2014, 2017). This is in keeping with earlier work (Speight and Moschopedis, 1979) where mild hydrotreating of bitumen was reported to produce low-sulfur liquids that would be amenable to further catalytic processing.

7.7 S&W Fluid Catalytic Cracking Process

The S&W FCC process is also designed to maximize the production of distillates from residua (Parkash, 2003; Gary et al., 2007; Speight, 2014, 2017; Hsu and Robinson, 2017). In the process, the feedstock is injected into a stabilized, upward flowing catalyst stream whereupon the feedstock-steam-catalyst mixture travels up the riser and is separated by a high-efficiency inertial separator. The product vapor goes overhead to the main fractionator.

The spent catalyst is immediately stripped in a staged, baffled stripper to minimize hydrocarbon carryover to the regenerator system. The first regenerator (650–700°C, 1200–1290°F) burns 50%–70% of the coke in an incomplete carbon monoxide combustion mode running countercurrently. This relatively mild, partial regeneration step minimizes the significant contribution of hydrothermal catalyst deactivation. The remaining coke is burned in the second regenerator (ca. 775°C, 1425°F) with an extremely low steam content. Hot clean catalyst enters a withdrawal well that stabilizes its fluid qualities prior to being returned to the reaction system.

8. Other Options

Catalytic cracking is widely used in the crude oil refining industry to convert viscous feedstocks into more valuable naphtha and lighter products. As the demand for higher-octane gasoline has increased, catalytic cracking has replaced thermal cracking. Two of the most intensive and commonly used catalytic cracking processes in crude oil refining are fluid catalytic cracking and hydrocracking. In the fluid catalytic cracking process, the fine, powdery catalyst (typically zeolites, which have an average particle size of ~70 μm) takes on the properties of a fluid when it is mixed with the vaporized feed. Fluidized catalyst circulates continuously between the reaction zone and the regeneration zone.

Several process innovations have been introduced in the form of varying process options, some using piggyback techniques (where one process works in close conjunction with another process; please see above); there are other options that have not yet been introduced or even invented but may well fit into the refinery of the future.

In the fluid catalytic cracker, the major developments are in integration with sulfur removal to produce low-sulfur naphtha without octane loss (Babich and Moulijn, 2003; Rama Rao et al., 2011). This development will build on the development of new catalysts (see above). Furthermore, recent enhancements made to RFCC units to permit feeding significant amounts of viscous crude oils while simultaneously improving yields and service factors have focused on improved feedstock injection and dispersion, reduced contact time of products and catalyst, improved separation of products and catalyst, and regenerator heat removal. Traditional technology has been modified in key areas including (i) catalyst design to accommodate higher metal feedstock and to minimize the amount of coke formed on the catalyst, (ii) feedstock injection, (iii) riser pipe design and catalyst/oil product separation to minimize overcracking, (iv) regenerator design improvements to handle high coke output and avoid damage to catalyst structure, and (v) overall reactor/regenerator design concepts.

Furthermore, power recovery turbines can be installed on catalytic cracking units to produce power from the pressure of the off-gases of the catalytic cracking.

Hydrotreating of the feedstock to the catalytic cracking unit can increase conversion by 8%–12% v/v, and with most feeds, it will be possible to reduce the sulfur content of the gasoline/naphtha product to levels low enough to meet the future low-sulfur gasoline pool specifications.

The major developments in desulfurization have three main routes such as advanced hydrotreating (new catalysts and catalytic distillation) and reactive adsorption (such as the use of metal oxides that will chemically abstract sulfur) (Babich and Moulijn (2003)). In such concepts, already onstream in some refineries, the number of units performing chemical desulfurization will increase in the future.

There are also proposals (some of which have been put into practice) to bypass the atmospheric and vacuum distillation units by feeding crude oil directly into a thermal cracking process, which would provide sufficient flexibility to supply a varying need of products with net energy savings.

Light cycle oil from catalytic cracking units will be increased by modifying feedstock composition, introducing improved catalysts and additives, and modifying operating conditions (such as recycle ratio, temperature, and catalyst/oil ratio). The addition of an active alumina matrix is a common feature to help refiners increase light cycle oil production when cracking viscous feedstocks. A comprehensive survey of patent literature in the report found the use of an inorganic additive to occur more than once in catalyst formulations, and metal-doped anionic clays and amorphous silicoaluminophosphates (SAPO) are cited, among other inventions. There is some overlap in catalysts tailored for the production of light cycle oil and catalysts tailored for residuum feedstocks.

Complete process technologies for residuum catalytic cracking units are the most comprehensive approach to improve resid processing operations but present, as might be anticipated, the most expensive.

Process/hardware technologies to improve light cycle oil yield from the fluid catalytic cracking unit will include improved feedstock injection systems and riser pipe and reaction zone designs. In fact, over the next two decades, fluid catalytic cracking units will take on two additional roles—a user of biomass feedstocks and to reduce carbon dioxide emissions—to alleviate growing concerns over energy security and global warming.

The implementation of biofeedstock processing techniques in crude oil refineries can result in a competitive advantage for both refiners and society at large. First, the processes provide refineries with alternative feeds that are renewable and could be lower in cost than crude oil. Second, they can reduce the costs of producing fuels and chemicals from biofeedstocks by utilizing the existing production and distribution systems for crude oil-based products and avoiding the establishment of parallel systems. Finally, the use of biofeedstocks provides a production base for fuels and chemicals that is less threatened by changes in government policies toward fossil fuel feeds and renewable energies.

Biofeedstocks that are able to be processed in the fluid catalytic cracking unit can be categorized as biomass-derived oils (both lignocellulosic materials and free carbohydrates) or triglycerides and their free fatty acids. The operating conditions and catalysts used for each type of feedstock to achieve a desired product slate vary, and each feedstock comes with inherent advantages and disadvantages. Most of the research work completed to date has been performed on relatively pure biofeedstocks as opposed to blends of bio-based materials with traditional catalytic cracker feedstocks. Practical implementation in a refinery will, more than likely, be accompanied by blending with vacuum gas oil or residuum. In fact, biomass constituents can be blended with the feedstock and fed to a fluid catalytic cracking unit (Speight, 2011a,b). The acidity of the oil (caused by the presence of oxygen functions) can be reduced by means of a moderate thermal treatment at temperatures in the range of 320–420°C (610–790°F).

In terms of processability, triglycerides are the best suited biofeedstock for the catalytic cracker. These materials generally produce high-quality diesel and high-octane naphtha and are low in sulfur.

Finally, the use of microwaves to generate heat in a uniform and controlled fashion is well known, and microwave technology with catalysts is another technology that has been applied recently in upgrading of viscous feedstock (Mutyala et al., 2010; Lam et al., 2012), for example, catalytic hydroconversion of residue from coal liquefaction by using microwave irradiation with a Ni catalyst (Wang et al., 2008) and microwave-assisted desulfurization of sour crude oil using iron powder as catalyst (Leadbeater and Khan, 2008). Also, upgrading of Athabasca bitumen with microwave-assisted catalytic pyrolysis was carried out in one study. Silicon carbide is used for pyrolysis, and nickel and molybdenum nanoparticles as catalysts are used to enhance the pyrolysis performance. The results of the work suggest that microwave heating with nanoparticle catalyst can be a useful tool for the upgrading of viscous crudes such as bitumen because of rapid heating and energy efficiency (Jeon et al., 2012).

9. Coke Formation

Feedstock properties that are most important to consider when processing residue in a fluid catalytic cracking unit are as follows: (i) asphaltene constituents and content, which can cause deactivation of downstream catalyst systems; (ii) vanadium, which is the controlling parameter setting the fluid catalytic cracking unit catalyst makeup rates; (iii) carbon residue, which is the major factor affecting coke burning and catalyst cooling requirements; and (iv) hydrogen content, which impacts the fluid catalytic cracking unit conversion and yield selectivity. Even if blended in small concentrations into feedstocks for fluid catalytic cracking units, heavy oil, extra heavy oil, and tar sand bitumen contain higher concentrations of metals and coke-forming precursors than would be suitable for processing in a fluid catalytic cracking unit because of the impact of the contaminants on the catalyst makeup rate and the yields of the products. Therefore, before processing heavy oil, extra heavy oil, and tar sand bitumen in a catalytic cracking unit, these feedstocks may first require some form of pretreatment (such as mild coking to remove the contaminants in the first-formed coke or the use of a scavenger such as calcium oxide or zinc oxide to remove the contaminants), vacuum distillation (where applicable), mild hydrotreating, or solvent deasphalting to reduce the amount of metals and the coke-forming constituents.

The formation of coke deposits has been observed in virtually every unit in operation, and the deposits can be very thick with thicknesses up to 4 ft., which have been reported (McPherson, 1984). Coke has been observed to form where condensation of hydrocarbon vapors occurs. The reactor walls and plenum offer a colder surface where hydrocarbon derivatives can condense. Higher-boiling constituents in the feedstock may be very close to their dew point, and they will readily condense and form coke nucleation sites on even slightly cooler surfaces.

Nonvaporized feedstock droplets readily collect to form coke precursors on any available surface since the high-boiling feedstock constituents do not vaporize at the mixing zone of the riser. Thus, it is not surprising that residuum processing makes this problem even worse. Low-residence-time cracking also contributes to coke deposits since there is less time for heat to transfer to feedstock droplets and vaporize them. This is an observation in line with the increase in coking when short-contact-time *riser crackers* (*q.v.*) were replacing the longer-residence-time fluid-bed reactors.

Higher-boiling feedstocks that have high aromaticity result in higher yields of coke. Furthermore, polynuclear aromatic derivatives and aromatic derivatives containing heteroatoms (i.e., nitrogen, oxygen, and sulfur) are more facile coke makers than simpler aromatic derivatives. Feedstock quality alone is not a foolproof method of predicting where coking will occur. However, it is known that feedstock hydrotreaters rarely have coking problems. The hydrotreating step mitigates the effect of the coke formers, and coke formation is diminished.

It is evident that significant postriser cracking occurs in commercial catalytic cracking units resulting in substantial production of dry gas and other low-valued products (Avidan and Krambeck, 1990). There are two mechanisms by which this postriser cracking occurs, thermal and dilute phase catalytic cracking.

Thermal cracking results from extended residence times of hydrocarbon vapors in the reactor disengaging area and leads to high dry gas yields via nonselective free-radical cracking mechanisms. On the other hand, dilute phase catalytic cracking results from extended contact between catalyst and hydrocarbon vapors downstream of the riser. While much of this undesirable cracking was eliminated in the transition from bed to riser cracking, there is still a substantial amount of nonselective cracking occurring in the dilute phase due to the significant catalyst holdup.

Many catalytic cracking units are equipped with advanced riser termination systems to minimize postriser cracking (Long et al., 1993). However, due to the complexity and diversity of catalytic cracking units, there are many variations of these systems such as closed cyclones, and many designs are specific to the unit configuration, but all serve the same fundamental purpose of reducing the undesirable postriser reactions. Furthermore, there are many options for taking advantage of reduced postriser cracking to improve yields. A combination of higher reactor temperature, higher cat/oil ratio, higher feedstock rate, and/or poorer-quality feedstock is typically employed. Catalyst modification is also appropriate, and typical catalyst objectives such as low coke and dry gas selectivity are reduced in importance due to the process changes, while other features such as activity stability and bottom cracking selectivity become more important for the new unit constraints.

Certain catalyst types seem to increase coke deposit formation. For example, these catalysts (some rare earth zeolites) tend to form aromatic derivatives from naphthene derivatives as a result of secondary hydrogen transfer reactions, and the catalysts contribute to coke formation indirectly because the products that they produce have a greater tendency to be coke precursors. In addition, high-zeolite-content, low-surface-area cracking catalysts are less efficient at heavy oil, extra heavy oil, and tar sand bitumen cracking than many amorphous catalysts because the nonzeolite catalysts contained a matrix that was better able to crack viscous feedstocks and convert the coke precursors. The active matrix of some modern catalysts serves the same function.

Once coke is formed, it is a matter of where it will appear. Coke deposits are most often found in the reactor (or disengager), transfer line, and slurry circuit and cause major problems in some units such as increased pressure drops, when a layer of coke reduces the flow through a pipe, or plugging, when chunks of coke spall off and block the flow completely. Deposited coke is commonly observed in the reactor as a black deposit on the surface of the cyclone barrels, reactor dome, and walls. Coke is also often deposited on the cyclone barrels 180° away from the inlet. Coking within the cyclones can be potentially very troublesome since any coke spalls going down into the dip leg could restrict catalyst flow or jam the flapper valve. Either situation reduces cyclone efficiency and can increase catalyst losses from the reactor. Coke formation also occurs at nozzles that can increase the nozzle pressure drop. It is possible for steam or instrument nozzles to be plugged completely, a serious problem in the case of unit instrumentation.

Coking in the transfer line between the reactor and main fractionator is also common, especially at the elbow where it enters the fractionator. Transfer line coking causes pressure drop and spalling and can lead to reduced throughput. Furthermore, any coke in the transfer line that spalls off can pass through the fractionator into the circulating slurry system where it is likely to plug up exchangers, resulting in lower slurry circulation rates and reduced heat removal. Pressure balance is obviously affected if the reactor has to be run at higher pressures to compensate for transfer line coking. On units where circulation is limited by low slide valve differentials, coke laydown may then indirectly reduce catalyst circulation. The risk of a flow reversal is also increased. In units with reactor grids, coking increases grid pressure drop, which can directly affect the catalyst circulation rate.

Shutdowns and start-ups can aggravate problems due to coking. The thermal cycling leads to differential expansion and contraction between the coke and the metal wall that will often cause the coke to spall in large pieces. Another hazard during shutdowns is the possibility of an internal fire when the unit is opened up to the atmosphere. Proper shutdown procedures that ensure that the internals have sufficiently cooled before air enters the reactor will eliminate this problem. In fact, the only defense against having coke plugging problems during start-up is to thoroughly clean the unit during the turnaround and remove all the coke. If strainers are on the line(s), they will have to be cleaned frequently.

The two basic principles to minimize coking are to avoid dead spots and prevent heat losses. An example of minimizing *dead spots* is using purge steam to sweep out stagnant areas in the disengager system. The steam prevents collection of high-boiling condensable products in the cooler regions. Steam also provides a reduced partial pressure or steam distillation effect on the high-boiling constituents and causes enhanced vaporization at lower temperatures. Steam for purging should preferably be superheated since medium-pressure low-velocity steam in small pipes with high heat losses is likely to be very wet at the point of injection and will cause more problems. *Cold spots* are often caused by heat loss through the walls in which case increased thermal resistance might help reduce coking. The transfer line, being a common source of coke deposits, should be as heavily insulated as possible, provided that stress-related problems have been taken into consideration.

In some cases, changing catalyst type or the use of an additive (*q.v.*) can alleviate coking problems. The catalyst types that appear to result in the least coke formation (not delta coke or catalytic coke) contain low or zero earth zeolites with moderate matrix activities. Eliminating high-boiling recycle streams can lead to reduced coke formation. Since clarified oil is a desirable feedstock to make needle coke in a coker, then it must also be a potential coke maker in the disengager.

One of the trends in recent years has been to improve product yields by means of better feedstock atomization. The ultimate objective is to produce an oil droplet small enough so that a single particle of catalyst will have sufficient energy to vaporize it. This has the double benefit of improving cracking selectivity and reducing the number of liquid droplets that can collect to form coke nucleation sites.

References

Abdo, S.F., Thakkar, V., Ackelson, D.B., Wang, L., Rossi, R.J., 2008. Maximize diesel with UOP enhanced two-stage Unicracking™ technology. In: Proceedings of the 18th Annual Saudi-Japan Symposium, Dhahran, Saudi Arabia.

Ackelson, 2004. UOP unicracking process for hydrocracking. In: Meyers, R.A. (Ed.), Handbook of Petroleum Refining Processes. McGraw-Hill, New York. (Chapter 7.2).

Alvarenga Baptista, C.M., de, L., Cerqueira, H.S., Sandes, E.F., 2010a. Process for Fluid Catalytic Cracking of Hydrocarbon Feedstocks With High Levels of Basic Nitrogen. United States Patent 7,736,491, June 15.

Alvarenga Baptista, C.M., De, L., Moreira, E.M., Cerqueira, H.S., 2010b. Process for Fluid Catalytic Cracking of Hydrocarbon Feedstocks With High Levels of Basic Nitrogen. United States Patent 7,744,745, June 29.

Avidan, A.A., Krambeck, F.J., 1990. FCC closed cyclone system eliminates post riser cracking. In: Proceedings of the Annual Meeting, National Petrochemical and Refiners Association.

Babich, I.V., Moulijn, J.A., 2003. Science and technology of novel processes for deep desulfurization of oil refinery streams: a review. Fuel 82, 607–631.

Bartholic, D.B., 1981a. Preparation of FCC Charge From Residual Fractions. United States Patent 4,243,514, January 6.

Bartholic, D.B., 1981b. Upgrading Petroleum and Residual Fractions Thereof. United States Patent 4,263,128, April 21.

Bartholic, D.B., 1989. Process for Upgrading Tar Sand Bitumen. United States Patent 4,804,459, February 14.

Bartholic, D.B., Haseltine, R.P., 1981. Oil Gas J. 79 (45), 242.

Benazzi, E., Cameron, C., 2005. Future refinements. In: Fundamentals of the Global Oil and Gas Industry.pp. 111–113.

Birch, C.H., Ulivieri, R., 2000. ULS Gasoline and Diesel Refining Study. Pervin and Gertz, Houston, TX.

Bouziden, G., Gentile, K., Kunz, R.G., 2002. Selective catalytic reduction of NOx from fluid catalytic cracking case study: BP whiting refinery. In: Paper ENV-03-128, NPRA Meeting. National Environmental and Safety Conference, New Orleans, LA, April 23–24.

Bradley, S.A., Gattuso, M.J., Bertolacini, R.J., 1989. Characterization and catalyst development. Symposium Series No. 411, American Chemical Society, Washington, DC.

Bridjanian, H., Khadem Samimi, A., 2011. Bottom of the barrel, an important challenge of the petroleum refining industry. Crude Oil Coal 53 (1), 13–21. Available Fromwww.vurup.sk/pc.

Buchanan, J.S., Santiesteban, J.G., Haag, W.O., 1996. Mechanistic considerations in acid-catalyzed cracking of olefins. J. Catal. 158, 279–287.

Chapin, L., Letzsch, W., 1994. Deep catalytic cracking maximize olefin production. Paper No. AM-94-43, Proceedings of the NPRA Annual Meeting, Mar. 20–22.

Chen, Y.-M., 2004. Recent advances in FCC technology. In: Proceedings of the 2004 AIChE Annual Meeting, Austin, TX, November 7–12.

Chen, Y.-M., 2006. Recent advances in FCC technology. Powder Technol. 163, 2–8.

Chuachuensuk, A., Paengjuntuek, W., Kheawhom, S., and Arpornwichanopa, A. 2010. Proceedings of the 20th European Symposium on Computer Aided Process Engineering—ESCAPE20. S. Pierucci and G. Buzzi Ferraris (Editors). Elsevier, Amsterdam.

Couch, K.A., Siebert, K.D., Van Opdorp, P.J., 2004. Controlling FCC yield and emissions. In: Proceedings of the NPRA Annual Meeting, March.

Cybulski, A., Moulijn, J.A. (Eds.), 1998. Structured Catalysts and Reactors. Marcel Dekker, New York.

DeCroocq, D., 1984. Catalytic Cracking of Heavy Petroleum Fraction. Editions Technip, Paris.

Degnan, T.F., 2000. Applications of zeolites in petroleum refining. Top. Catal. 13, 349–356.

Dennis, G.E., 1989. Use of small scale ART processing for wellhead viscosity reduction and upgrading of heavy oils. In: Proceedings of the 4th UNITAR/UNDP International Conference on Heavy Crude and Tar Sands, August 7–12, Edmonton, Alberta, Canadavol. 5. pp. 89–102.

Domokos, L., Jongkind, H., Stork, W.H.J., Van Den Tol-Kershof, J.M.H., 2010. Catalyst Composition, Its Preparation and Use. United States Patent 7,749,937 July 6 2010.

Draemel, D.C., 1992. Flexicracking IIIR—ER&E's latest cat cracking design. Proceedings of the JPI Petroleum Refining Conference, Japanese Petroleum Institute, Tokyo, Japan.

Dziabala, B., Thakkar, V.P., Abdo, S.F., 2011. Combination of Mild Hydrotreating and Hydrocracking for Making Low Sulfur Diesel and High Octane Naphtha. United States Patent 8,066,867. November 29.

EPA, 2010. Available and Emerging Technologies for Reducing Greenhouse Gas Emissions From the Petroleum Refining Industry. Sector Policies and Programs Division, Office of Air Quality Planning and Standards, United States Environmental Protection Agency, Research Triangle Park, NC.

Freel, B., Graham, R.G., 2011. Products Produced From Rapid Thermal Processing of Heavy Hydrocarbon Feedstocks. United States Patent 8,062,503, November 22.

Fujiyama, Y., 1999. Process for Fluid Catalytic Cracking of Oils. United States Patent 5,904,837, May 18.

Fujiyama, Y., Adachi, M., Okuhara, T., Yamamoto, S., 2000. Process for Fluid Catalytic Cracking of Heavy Fraction Oils. United States Patent 6,045,690, April 4.

Fujiyama, Y., Redhwi, H., Aitani, A., Saeed, R., Dean, C., 2005. Demonstration plant for new FCC technology yields increased propylene. Oil Gas J. (September 26), 62–67.

Fujiyama, Y., Al-Tayyar, M.H., Dean, C.F., Aitani, A., Redhwi, H.H., Tsutsui, T., Mizuta, K., 2010. Development of high severity FCC process for maximizing propylene production—catalyst development and optimization of reaction conditions. J. Jpn. Crude Oil Inst. 53 (6), 336–341.

Fujiyama, Y., Okuhara, T., Uchiura, A., 2011. Method of Designing Gas-Solid Separator. United States Patent 8,070,846, December 6.

Gary, J.G., Handwerk, G.E., Kaiser, M.J., 2007. Crude Oil Refining: Technology and Economics, fifth ed. CRC Press, Taylor & Francis Group, Boca Raton, FL.

Gates, B.C., Katzer, J.R., Schuit, G.C.A., 1979. Chemistry of Catalytic Processes. McGraw-Hill, New York.

Gembicki, V.A., Cowan, T.M., Brierley, G.R., 2007. Update processing operations to handle heavy feedstocks. Hydrocarb. Process. 86 (2), 41–53.

Germain, G.E., 1969. Catalytic Conversion of Hydrocarbons. Academic Press, New York.

Harding, R.H., Peters, A.W., Nee, J.R.D., 2001. New developments in FCC catalyst technology. Appl. Catal. A Gen. 221, 389–396.

Hedrick, B.W., Palmas, P., 2011. Process for Contacting High Contaminated Feedstocks With Catalyst in an FCC Unit. United States Patent 8,062.506, November 22.

Heinrich, G., Mauleon, J.-L., 1994. The R2R process: 21st century FCC technology. Rev. Inst. Fr. Pétrol. 49 (5), 509–520.

Hemler, C.L., Smith, L.F., 2004. UOP fluid catalytic cracking process. In: Meyers, R.A. (Ed.), Handbook of Petroleum Refining Processes. McGraw-Hill, New York (Chapter 3.3).

Hilbert, T.L., Chitnis, G.K., Umansky, B.S., Kamienski, P.W., Patel, V., Subramanian, A., 2008. Consider new technology to produce clean diesel. Hydrocarb. Process. 87 (2), 47–56.

Hsu, C.S., Robinson, P.R. (Eds.), 2017. Handbook of Petroleum Technology. Springer International Publishing AG, Cham.

Hunt, D.A., 1997. Meyers, R.A. (Ed.), Handbook of Petroleum Refining Processes. McGraw-Hill, New York (Chapter 3.5).

Huve, L.G., Creyghton, E.J., Ouwehand, C., van Veen, J.A.R., Hanna, A., 2004. New Catalyst Technologies Expand Hydrocrackers' Flexibility and Contributions. Report No. CRI424/0704Criterion Catalysts and Technologies, Shell Global Solutions International BV, Amsterdam.

Imhof, P., Rautiainen, E.P.H., Gonzalez, J.A., 2005. Maximize propylene yields. Hydrocarb. Process. 84 (9), 109–114.

Inai, K., 1994. Operation results of the R2R process. Rev. Inst. Fr. Pétrol. 49 (5), 521–527.

Ino, T., Ikeda, S., 1999. Process for Fluid Catalytic Cracking of Heavy Fraction Oil. United States Patent 5,951,850, September 14.

Ino, T., Okuhahra, T., Abul-Hamayel, M., Aitani, A., Maghrabi, A., 2003. Fluid Catalytic Cracking Process for Heavy Oil. United States Patent No. 6,656,346, December 2.

Jakkula, J.H., Niemi, V.M., 1997. Short contact time catalytic cracking process: results from bench scale unit. Paper 29241. Proceedings of the 15th World Crude Oil Congress, Beijing, China, October 12–17.

Jeon, S.G., Kwak, N.S., Rho, N.S., Ko, C.H., Na, J.-G., Yi, K.B., Park, S.B., 2012. Catalytic pyrolysis of Athabasca bitumen in H2 atmosphere using microwave irradiation. Chem. Eng. Res. Des. 90, 1292–1296.

Johnson, T.E., Niccum, P.K., 1997. Meyers, R.A. (Ed.), Handbook of Petroleum Refining Processes. McGraw-Hill, New York (Chapter 3.2).

Junaid, A.S.M., Wang, W., Street, C., Rahman, M., Gersbach, M., Zhou, S., McCaffrey, W., Kuznicki, S.M., 2010. Viscosity reduction and upgrading of Athabasca oil sands bitumen by natural zeolite cracking. Int. J. Mater. Metall. Eng. 4 (9), 609–614.

Krambeck, F.J., Pereira, C.J., 1986. FCC Processing Scheme With Multiple Risers. United States Patent 4,606,810, August 19.

Ladwig, P.K., 1997. Meyers, R.A. (Ed.), Handbook of Petroleum Refining Processes. McGraw-Hill, New York (Chapter 3.1).

Lam, S.S., Russell, A.D., Lee, C.L., Chase, H.A., 2012. Microwave-heated pyrolysis of waste automotive engine oil: influence of operation parameters on the yield, composition, and fuel properties of pyrolysis oil. Fuel 92, 327–339.

Le Page, J.F., Cosyns, J., Courty, P., Freund, E., Franck, J.P., Jacquin, Y., Juguin, B., Marcilly, C., Martino, G., Miguel, J., Montarnal, R., Sugier, A., von Landeghem, H., 1987. Applied Heterogeneous Catalysis. Editions Technip, Paris.

Leadbeater, N.E., Khan, M.R., 2008. Microwave-promoted desulfurization of heavy oil and a review of recent advances on process technologies for upgrading of heavy and sulfur containing crude oil. Energy Fuel 22, 1836–1839.

Letzsch, W., 2011. Innovation drives new catalyst development. Catal. Hydrocarb. Process. C93–C94.

Li, X., Li, C., Zhang, J., Yang, C., Shan, H., 2007. Effects of temperature and catalyst to oil weight ratio on the catalytic conversion of heavy oil to propylene using ZSM-5 and USY catalysts. J. Nat. Gas Chem. 16 (1), 92–99.

Lifschultz, 2005. Oil refiner's gathering storm: help is on the way. Hydrocarb. Process. 84 (9), 59–62.

Long, S.L., Johnson, A.R., Dharia, D., 1993. Advances in residual oil FCC. In: Paper No. AM-93-50, Proceedings of the Annual Meeting, National Petrochemical and Refiners Association.

Maadhah, A.G., Fujiyama, Y., Redhwi, H., Abul-Hamayel, M., Aitani, A., Saeed, M., Dean, C., 2008. A new catalytic cracking process to maximize refinery propylene. Arab. J. Sci. Eng. 33 (1B), 17–28.

Maples, R.E., 2000. Petroleum Refinery Process Economics, second ed. PennWell Corporation, Tulsa, OK.

McPherson, L.J., 1984. Causes of FCC Reactor Coke Deposits Identified. Oil Gas J. 139September 10.

Motaghi, M., Shree, K., Krishnamurthy, S., 2010. Consider new methods for bottom of the barrel processing—part 1. Hydrocarb. Process. 89 (2), 35–40.

Mutyala, S., Fairbridge, C., Jocelyn Paré, J.R., Bélanger, J.M.R., Ng, S., Hawkins, R., 2010. Microwave applications to oil sands and petroleum: a review. Fuel Process. Technol. 91, 127–135.

Navarro, U., Ni, M., Orlicki, D.F., 2015. FCC 101: how to estimate product yields cost-effectively and improve operations. Hydrocarb. Process. 94 (2), 41–52.

Nishida, S., Fujiyama, Y., 2000. Separation Device. United States Patent 6,146,597. November 14.

Occelli, M.L., O'Connor, P., 1998. Fluid Cracking Catalysts. Marcel Dekker, New York.

Otterstedt, J.E., Gevert, S.B., Jaras, S.G., Menon, P.G., 1986. Processing heavy oils. Appl. Catal. 22, 159–179.

Parkash, S., 2003. Refining Processes Handbook. Gulf Professional Publishing, Elsevier, Amsterdam.

Patel, S., 2007. Canadian oil sands—opportunities, technologies, and challenges. Hydrocarb. Process. 86 (2), 65–74.

Patel, R., Zeuthen, P., Schaldemose, M., 2002. Advanced FCC feed pretreatment technology and catalysts improves FCC profitability. Proceedings of the NPRA Annual Meeting, San Antonio, March 2002, National Petrochemical and Refiners Association, Washington, DC.

Patel, R., Moore, H., Hamari, B., 2004. FCC hydrotreater revamp for low-sulfur gasoline. In: Proceedings of the NPRA Annual Meeting, San Antonio, TX, March, National Petrochemical and Refiners Association, Washington, DC.

Rama Rao, M., Soni, D., Siele, G.M., Bhattacharyya, D., 2011. Convert bottom-of-the-barrel and diesel into light olefins. Hydrocarb. Process. 90 (2), 46–49.

Ross, J., Roux, R., Gauthier, T., Anderson, L.R., 2005. Fine-tune FCC operations for changing fuels market. Hydrocarb. Process. 84 (9), 65–73.

Runyan, J., 2007. Is bottomless-barrel refining possible. Hydrocarb. Process. 86 (9), 81–92.

Sadeghbeigi, R., 1995. Fluid Catalytic Cracking: Design, Operation, and Troubleshooting of FCC Facilities. Gulf Publishing Company, Houston, TX.

Salazar-Sotelo, D., Maya-Yescas, R., Mariaca-Domínguez, E., Rodríguez-Salomón, S., Aguilera-López, M., 2004. Effect of hydrotreating FCC feedstock on product distribution. Catal. Today 98 (1–2), 273–280.

Sayles, S., Bailor, J., 2005. Upgrade FCC hydrotreating. Hydrocarb. Process. 84 (9), 87–90.

Schiller, R., 2011. Effect of synthetic crude feedstocks on FCC yield. Refin. Oper. 2 (4), 1–2.

Schnaith, M.W., Sexson, A., Tru, D., Bartholic, D.B., Lee, Y.K., Yoo, I.S., Kang, H.S., 1998. Oil Gas J. 96 (25), 53.

Shan, H.H., Zhao, W., He, C.Z., Zhang, J.F., Yang, C.H., 2003. Maximum FCC diesel yield with TSRFCC technology. Preprints. Div. Fuel Chem. Amer. Chem. Soc. 48 (2), 710–711.

Shidhaye, H., Kukade, S., Kumar, P., Rao, P.V.C., Choudary, N.V., 2015. Improve FCC margins by processing more vacuum resid in feed. Hydrocarb. Process. 94 (12), 35–38.

Soni, D., Rama Rao, M., Saidulu, G., Bhattacharyya, D., Satheesh, V.K., 2009. Catalytic cracking process enhances production of olefins. Pet. Technol. Q. 14 (Q4), 95–100.

Speight, J.G., 2000. The Desulfurization of Heavy Oils and Residua, second ed. Marcel Dekker, New York.

Speight, J.G., 2008. Synthetic Fuels Handbook: Properties, Processes, and Performance. McGraw-Hill, New York.

Speight, J.G., 2009. Enhanced Recovery Methods for Heavy Oil and Tar Sands. Gulf Publishing Company, Houston, TX.

Speight, J.G., 2011a. The Refinery of the Future. Gulf Professional Publishing, Elsevier, Oxford.

Speight, J.G. (Ed.), 2011b. Biofuels Handbook. Royal Society of Chemistry, London.

Speight, J.G., 2014. The Chemistry and Technology of Petroleum, fifth ed. CRC Press, Taylor & Francis Group, Boca Raton, FL.

Speight, J.G., 2017. Handbook of Petroleum Refining. CRC Press, Taylor & Francis Group, Boca Raton, FL.

Speight, J.G., Moschopedis, S.E., 1979. The production of low-sulfur liquids and coke from Athabasca bitumen. Fuel Process. Technol. 2, 295.

Spoor, R.M., 2008. Low-carbon refinery: dream or reality. Hydrocarb. Process. 87 (11), 113–117.

Stiles, A.B., Koch, T.A., 1995. Catalyst Manufacture. Marcel Dekker, New York.

Suchanek, A.J., Moore, A.S., 1986. Modern residue upgrading by ART. Proceedings of the NPRA National Meeting, March 23–25. National Petrochemical and Refiners Association, Washington, DC.

Swaty, T.E., 2005. Global refinery industry trends: the present and the future. Hydrocarb. Process. 84 (9), 35–46.

Upson, L.L., Wegerer, D.E., 1993. Rapid disengager techniques in riser design. Proceedings of the 3rd International Symposium on Advances in Fluid Catalytic Cracking, Am. Chem. Soc., August.

Wang, T.X., Zong, Z.M., Zhang, V.W., Wei, Y.B., Zhao, W., Li, B.M., Wei, X.Y., 2008. Microwave-assisted hydroconversions of demineralized coal liquefaction residues from Shenfu and Shengli coals. Fuel 87, 498–507.

Wojciechowski, B.W., Corma, A., 1986. Catalytic Cracking: Catalysts, Chemistry, and Kinetics. Marcel Dekker, New York.

Wolschlag, L.M., 2011. Innovations developed using sophisticated engineering tools. Proceedings of the AIChE 2011 Regional Process Technology Conference, American Institute of Chemical Engineers.

Wolschlag, L.M., Couch, K.A., 2010a. Upgrade FCC performance. Hydrocarb. Process. 89 (9), 57–65.

Wolschlag, L.M., Couch, K.A., 2010b. UOP FCC Innovations Developed Using Sophisticated Engineering Tools. Report No. AM-10-109UOP LLC, Des Plaines, IL.

Wrench, R.E., Glascow, P.E., 1992. FCC hardware options for the modern cat cracker. AIChE Symp. Ser. 29, 1992.

Yung, K.Y., Pouwels, A.C., 2008. Fluid catalytic cracking—a diesel producing machine. Hydrocarb. Process. 87 (2), 79–83.

Further Reading

Hudec, P. n.d. FCC catalyst—key element in refinery technology. Proceedings of the 45th International Petroleum Conference, Bratislava, Slovak Republic, June 13.

Chapter 10

Upgrading by Hydrotreating

1. Introduction

Hydroprocessing (covers the process terms *hydrotreating* and *hydrocracking*) is a refining technology in which a feedstock is treated with hydrogen at temperature and under pressure that can affect the product slate in refineries by strategic placement (Fig. 10.1). In fact, the use of hydrogen in thermal processes is perhaps the single most significant advance in refining technology during the 20th century (Scherzer and Gruia, 1996; Dolbear, 1998).

Catalytic hydrotreating, also referred to as *hydroprocessing* or *hydrodesulfurization*, commonly appears in multiple locations in a refinery (Speight, 2000, 2014; Parkash, 2003; Ancheyta and Speight, 2007; Gary et al., 2007; Hsu and Robinson, 2017). Furthermore, *hydrotreating*, a term often used synonymously with *hydrodesulfurization* (HDS), is a catalytic refining process widely used to remove sulfur from crude oil products such as naphtha, kerosene, and fuel oil (Speight, 2000, 2014, 2017; Gary et al., 2007; Hsu and Robinson, 2017). The objective of the hydrotreating process is to remove sulfur and other unwanted compounds, for example, unsaturated hydrocarbon derivatives and nitrogen from refinery process streams. Hydrodesulfurization processes typically include facilities for the capture and removal of the resulting hydrogen sulfide (H_2S) gas, which is subsequently converted into by-product elemental sulfur or sulfuric acid.

In contrast to the hydrocracking process (Chapter 11), hydrotreating is a process in which thermal decomposition is usually minimized (Figs. 10.2 and 10.3; Parkash, 2003; Gary et al., 2007; Speight, 2014, 2017; Hsu and Robinson, 2017). Hydrotreating (nondestructive hydrogenation) is generally used for the purpose of improving product quality without appreciable alteration of the boiling range. Mild processing conditions are employed so that only the more unstable materials are attacked. Thus, nitrogen, sulfur, and oxygen compounds undergo hydrogenolysis to split out ammonia, hydrogen sulfide, and water, respectively. Olefin derivatives are saturated, and unstable compounds, such as diolefin derivatives, which might lead to the formation of gums or insoluble materials, are converted to more stable compounds. Heavy metals present in the feedstock are also usually removed during hydrogen processing.

Heavy Oil Recovery and Upgrading. https://doi.org/10.1016/B978-0-12-813025-4.00010-6

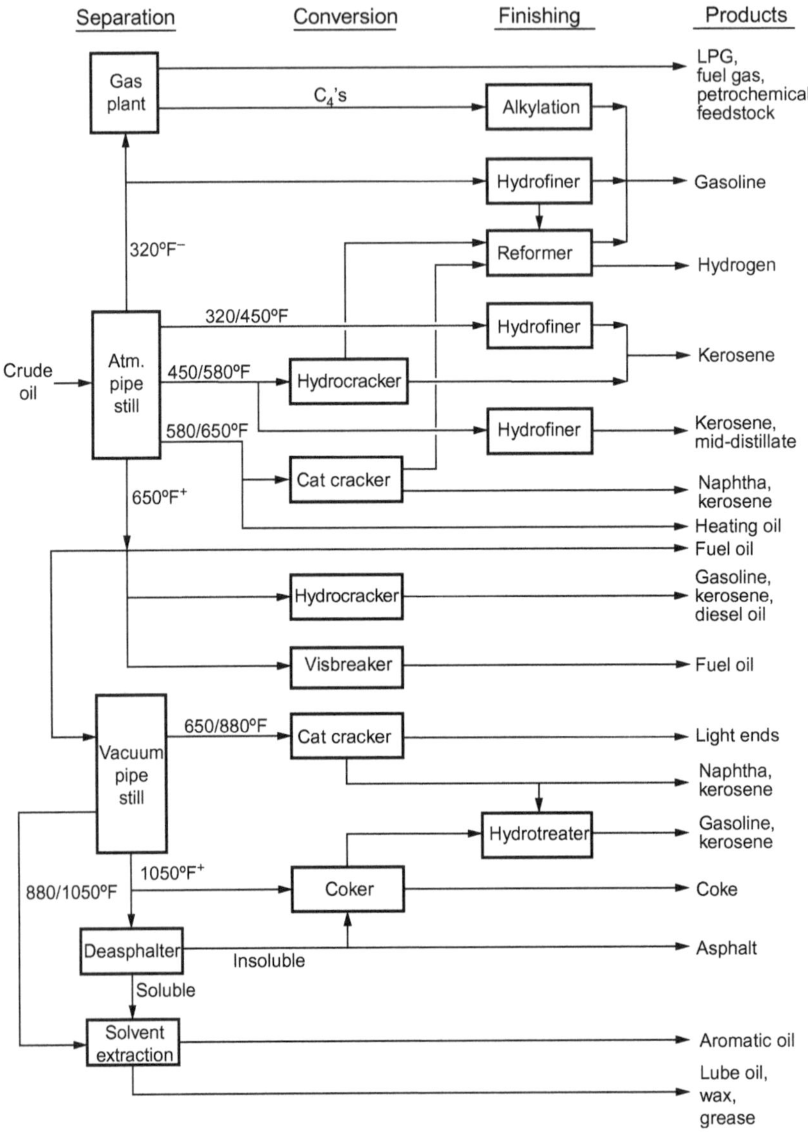

FIG. 10.1 Generalized refinery layout showing relative placement of the hydrotreating (hydrofining) units.

The major differences between hydrotreating and hydrocracking are the time at which the feedstock remains at reaction temperature and the extent of the decomposition of the nonheteroatom constituents and products. The upper limits of hydrotreating conditions may overlap with the lower limits of hydrocracking conditions. And where the reaction conditions overlap,

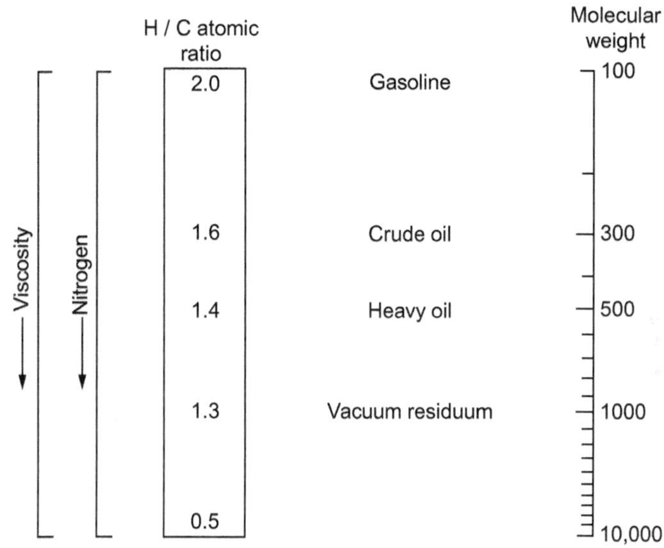

FIG. 10.2 Atomic hydrogen/carbon ratios of various feedstocks.

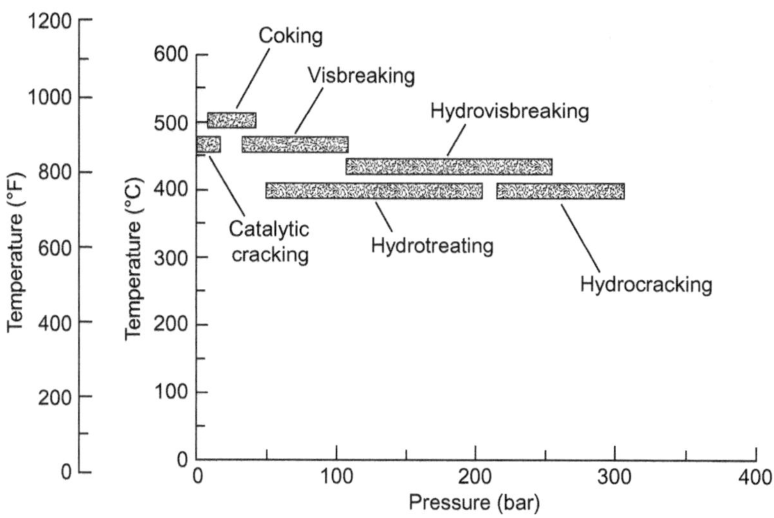

FIG. 10.3 Temperature and pressure ranges for various processes.

feedstocks to be hydrotreated will generally be exposed to the reactor temperature for shorter periods; hence, the reason why hydrotreating conditions may be referred to as (relatively) mild.

A growing dependence on high-heteroatom viscous oils and residua has emerged as a result of continuing decreasing availability of conventional crude oil through the worldwide depletion of crude oil reserves. Thus, the ever growing tendency to convert as much as possible of lower-grade feedstocks to liquid products is causing an increase in the total sulfur content in penultimate products. Refiners must, therefore, continue to remove substantial portions of sulfur from the lighter products so that the final products meet specifications. However, residua, heavy oil, extra heavy oil, and tar sand bitumen pose a particularly difficult problem in terms of the propensity of these feedstocks to form coke and to shorten catalyst life (Speight, 2000; Ancheyta and Speight, 2007).

Hydrotreating and desulfurization are processes for treatment of crude oil products to ensure that they meet performance specifications. During the hydrotreating and desulfurization processes, several chemical conversions occur: (i) Olefin derivatives are converted to alkanes, (ii) aromatic derivatives are converted to cycloalkanes and alkanes, and (iii) sulfur is removed (Dolbear, 1998; Parkash, 2003; Gary et al., 2007; Hsu and Robinson, 2017; Speight, 2014, 2017). With the influx of the heavier feedstocks into refineries, hydroprocessing will assume a greater role in the refinery of the future (Speight, 2011). In addition to being recognized as a chemical hydrogenation process, the process uses the principle that the presence of hydrogen during a thermal reaction of a crude oil feedstock terminates many of the coke-forming reactions and enhances the yields of the lower-boiling components, such as naphtha and kerosene, and middle-distillate fractions.

Hydrotreating processes for heavy oil, extra heavy oil, and tar sand bitumen have three definite roles: (i) desulfurization to supply low-sulfur fuel oils, (ii) pretreatment of the feedstock for fluid catalytic cracking processes, and (iii) hydrocracking to produce feedstocks for fluid catalytic cracking processes. There are several commercial technologies worldwide that compete with thermal cracking in the heavy oil, extra heavy oil, and tar sand bitumen, and examples are LC-Fining, HDH Plus, H-Oil, and HyCon, which are based on hydrogen addition in order to increase production and quality of liquid products and reduce the coke generation in the process.

The hydrotreating processes are becoming more popular as pretreating processes where the main goal is to remove sulfur, metal, and asphaltene contents from residua and other viscous feedstocks to a desired level and, at the same time, maintain hydrogen consumption at acceptable levels that are dictated, for example, by the degree of desulfurization and/or hydrocracking (Speight, 2000, 2014, 2017). Hydrocracking process options for residua are also seeing an increase in use but have been considered in the past hydrogen sinks (being somewhat uncontrollable and therefore wasteful in hydrogen use) and are, therefore, seemingly less popular than the hydrotreating processes. Hopefully, this attitude is changing as more hydroconversion processes are being introduced for viscous feedstock conversion.

In general, viscous feedstocks such as heavy oil, extra heavy oil, and tar sand bitumen consist of high proportions of undesirable contaminants such as asphaltene constituents, metal-containing constituents (vanadium, nickel, and iron), sulfur-containing constituents, and nitrogen-containing constituents.

Catalytic hydrotreatment followed by fluid catalytic cracking of the hydrotreated feedstock is conceived as the most efficient way of maximizing the quantity of light oil products from viscous feedstocks in modern refineries (Lee et al., 1996; Kim and Hanson, 1998; Villasana et al., 2015). The hydrotreating process is capable of removing heteroatoms (metals, sulfur, and nitrogen) and simultaneously converting asphaltene constituents to products that are relatively easier to convert than the original asphaltene constituents.

Catalysts such as Co-Mo, Ni-Mo, or Ni-W supported on alumina have been conventionally used in hydrotreating processes for hydrodesulfurization (HDS), hydrodenitrogenation (HDN), and hydrodemetallization (HDM). Although such conventional supported catalysts have very high initial activity, they do suffer from rapid deactivation, particularly when applied to viscous feedstocks such as heavy oil, extra heavy oil, and tar sand bitumen that typically have a high content of asphaltene constituents and metal-containing constituents. Furthermore, a large proportion of the sulfur and metals in viscous feedstocks is contained in high-molecular-weight hydrocarbon fractions such as the resin fraction and the asphaltene fraction.

However, viscous feedstock upgrading using fixed-bed technologies have some limitations at high conversion level due to sediment formation or asphaltene precipitation, but when working at moderate severity, sediment formation is minimized, and fixed-bed hydroprocessing becomes more attractive than other technologies (Khan and Patmore, 1998; Parkash, 2003; Gary et al., 2007; Speight, 2014, 2017; Hsu and Robinson, 2017). The catalyst designs for viscous feedstock upgrading are forced toward the large pore catalyst, which can retain an optimum quantity of metal and have good level of sulfur and nitrogen removals. It is also evident that the steady improvements in catalysts have led to longer run lengths and lower operating temperatures. However, the opinions vary about the catalyst use, for example, nature of catalyst; definitely, the existing catalyst may not work as effectively as required. In this respect, some thought that ring-opening (acidic) catalyst technologies were already well advanced, while others thought that better selectivity for hydrogen addition and hydrocracking (e.g., less gas make) was needed by using acid-base nature of catalyst (Rana et al., 2008).

The hydrotreating process uses the principle that the presence of hydrogen during *mild* thermal treatment reaction of a viscous feedstock removes the heteroatoms and metals (Hunter et al., 2010). On the other hand, the presence of hydrogen during cracking (*hydrocracking*) terminates many of the coke-forming reactions and enhances the yields of the lower-boiling components, such as naphtha, kerosene, and jet fuel. However, hydrocracking (Chapter 11) should not be regarded as a competitor for catalytic cracking (Chapter 9). Catalytic cracking units normally use gas oil distillates as feedstocks, whereas hydrocracking feedstock usually consists of refractive gas oils derived from cracking and coking operations. Hydrocracking is a supplement to, rather than a replacement for, catalytic cracking.

In hydrotreating, the feedstock is reacted with hydrogen at elevated temperatures, in the range of 300–450°C (570–840°F), and elevated pressures, in the range of 120–2200 psi, under the presence of a hydrogenation catalyst typically cobalt-molybdenum (Co-Mo) or nickel-molybdenum (Ni-Mo) on gamma-alumina (γ-Al_2O_3). The catalysts are produced in the oxide forms and are sulfided before their use in the process. During hydrotreating, the heteroatoms are removed in the form of hydrogen sulfide (H_2S), ammonia (NH_3), and water (H_2O); simultaneously, metal species such as vanadium (V) and nickel (Ni) are also removed by hydrogenation reactions.

Attention must also be given to the coke mitigation aspects. For example, in the hydrogen addition options, particular attention must be given to hydrogen management, thereby promoting asphaltene fragmentation to lighter products rather than to the production of coke. The presence of a material with good solvating power, to diminish the possibility of coke formation, is preferred. In this respect, it is worth noting the reappearance of donor solvent processing of viscous oils (Vernon et al., 1984), which has its roots in the older hydrogen donor diluent visbreaking process (Carlson et al., 1958; Langer et al., 1961, 1962; Menoufy et al., 2014).

By hydrotreating, hydrogen is also added to the feedstock, which in the case of the viscous feedstock reduces the tendency for coke formation during subsequent thermal or catalytic cracking processes. Fouling of hydrotreating catalysts by metal deposits and by coke depositions becomes unavoidable, which may cause expensive plant shutdowns for the replacement of the catalyst. Catalyst reactivation and replacement of poisoned catalyst by the fresh catalyst are important elements of reactor design.

The distinguishing feature of the hydroprocesses is that, although the composition of the feedstock is relatively unknown and a variety of reactions may occur simultaneously, the final product may actually meet all the required specifications for its particular use. Thus, the purposes of refinery hydroprocesses are as follows: (i) to improve existing products, (ii) to enable the products to meet market specifications; (iii) to develop new products or even new uses for existing products; (iv) to convert inferior or low-grade materials into valuable products; and (v) to transform near-solid heavy oil, extra heavy oil, and tar sand bitumen into liquid fuels.

There is a rough correlation between the quality of crude oil products and their hydrogen content (Dolbear, 1998). It so happens that desirable aviation gasoline, kerosene, diesel fuel, and lubricating oil are made up of hydrocarbon derivatives containing high proportions of hydrogen (Fig. 10.2). In addition, it is usually possible to convert olefin derivatives and higher-molecular-weight constituents to paraffin derivatives and monocyclic hydrocarbon derivatives by hydrogen addition processes. These facts have for many years encouraged attempts to employ hydrogenation for refining operations; despite considerable technical success, such processes were not economically possible until low-priced hydrogen became available as a result of the rise of reforming, which converts naphthene derivatives to aromatic derivatives with the release of hydrogen.

As already noted, hydroprocesses for the conversion of feedstocks may be classified as *nondestructive* and *destructive*. Nondestructive hydrogenation is characterized by the removal of heteroatom constituents as the hydrogenated analogs:

$$RSR^1 + H_2 \rightarrow RH + R^1H + H_2S$$

$$2RN(R_2)R^1 + 3H_2 \rightarrow 2RH + 2R^1H + 2R^2H + 2NH_3$$

$$ROR^1 + H_2 \rightarrow RH + R^1H + H_2O$$

In addition, the saturation of olefin derivatives in the products from thermal processes also occurs:

$$RCH = CHR^1 + H_2 \rightarrow R - CH_2CH_2 - R^1$$

Aromatic constituents may also be saturated to produce cycloaliphatic derivatives (naphthene derivatives) by this treatment:

$$C_6H_6 + 3H_2 \rightarrow C_6H_{12}$$

Benzene → cyclohexane

Any metals (usually nickel and vanadium) present in the feedstock are usually removed during hydrogen processing as a cost to the hydrogen consumption and not, of course, as their hydrogen analogs but by deposition on the catalyst through changes in the chemical properties of the metal-containing constituents by the high temperatures and the presence of hydrogen. These two process parameters progressively affect the ability of the organic structures or retain the metals within the organic matrix and deposition ensues.

Thus, hydrotreating of distillates may be defined simply as the removal of sulfur, nitrogen, and oxygen compounds as well as olefinic compounds by selective hydrogenation. The hydrotreating catalysts are usually cobalt plus molybdenum (CoS-MoS) or nickel plus molybdenum (NiS-MoS) sulfides impregnated on an alumina (Al_2O_3) base. The hydrotreating operating conditions 1000–2000psi hydrogen and ~370°C (700°F) are such that appreciable hydrogenation of aromatic derivatives does not occur. The desulfurization reactions are invariably accompanied by small amounts of hydrogenation and hydrocracking, the extent of which depends on the nature of the feedstock and the severity of desulfurization. In summary, hydrotreating (nondestructive hydrogenation) is generally used for the purpose of improving product quality without appreciable alteration of the boiling range. Mild processing conditions are employed so that only the more unstable materials are attacked.

Destructive hydrogenation (*hydrogenolysis* or *hydrocracking*) is characterized by the cleavage of carbon-carbon linkages accompanied by hydrogen saturation of the fragments to produce lower-boiling products:

$$RCH_2CH_2R^1 + H_2 \rightarrow RCH_3 + CH_3R^1$$

Such treatment requires processing thermal regimes that are similar to those used in catalytic cracking (Chapter 9) and the use of high hydrogen pressures to minimize the reactions that lead to coke formation.

The major differences between *hydrotreating* and *hydrocracking* are the time at which the feedstock remains at reaction temperature and the extent of the decomposition of the nonheteroatom constituents and products. The upper limits of hydrotreating conditions may overlap with the lower limits of hydrocracking conditions (Fig. 10.3) with a lower overall conversion for hydrotreating (Fig. 10.4). And where the reaction conditions overlap, feedstocks to be hydrotreated will generally be exposed to the reactor temperature for shorter periods, hence the reason why hydrotreating conditions may be referred to as *mild*. All is relative.

Unsaturated compounds, such as olefin derivatives, are not indigenous to viscous feedstocks and produced during cracking processes and need to be removed from product streams because of the tendency of unsaturated compounds and heteroatomic polar compounds to form gum and sediment (Speight, 2014). On the other hand, aromatic compounds are indigenous to viscous feedstocks and some may be formed during cracking reactions. The most likely explanation is that the aromatic compounds present in product streams are related to the aromatic compounds originally present in viscous feedstocks but now having shorter alkyl side chains. Thus, in addition to olefin derivatives, product streams will contain a range of aromatic compounds that should be removed to enable many of the product streams to meet product specifications.

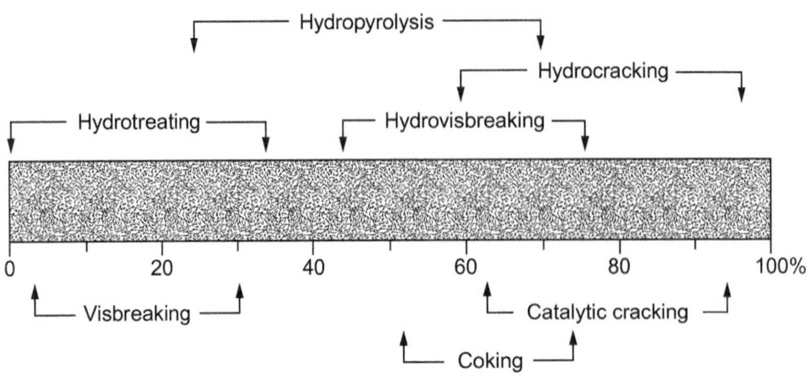

FIG. 10.4 Feedstock conversion in various processes.

Of the aromatic constituents, the polycyclic aromatic derivatives are first partially hydrogenated before cracking of the aromatic nucleus takes place. The sulfur and nitrogen atoms are converted to hydrogen sulfide and ammonia, but a more important role of the hydrogenation is probably to hydrogenate the coke precursors rapidly and prevent their conversion to coke.

One of the problems in the processing of high-sulfur and high-nitrogen feeds is the large quantity of hydrogen sulfide and ammonia that are produced. Substantial removal of both compounds from the recycle gas can be achieved by the injection of water in which, under the high-pressure conditions employed, both hydrogen sulfide and ammonia are very soluble compared with hydrogen and hydrocarbon gases. The solution is processed in a separate unit for the recovery of anhydrous ammonia and hydrogen sulfide.

Hydrotreating is carried out by charging the feedstock to the reactor, together with a portion of all the hydrogen produced in the catalytic reformer. Suitable catalysts are tungsten-nickel sulfide, cobalt-molybdenum-alumina, nickel oxide-silica-alumina, and platinum-alumina. Most processes employ cobalt-molybdenum catalysts, which generally contain ~10% molybdenum oxide and <1% cobalt oxide supported on alumina. The temperatures employed are in the range of 300–345°C (570–850°F), and the hydrogen pressures are ~500–1000 psi.

The reaction generally takes place in the vapor phase but, depending on the application, may be a mixed-phase reaction. The reaction products are cooled in a heat exchanger and led to a high-pressure separator where hydrogen gas is separated for recycling. Liquid products from the high-pressure separator flow to a low-pressure separator (stabilizer) where dissolved light gases are removed. The product may then be fed to a reforming or cracking unit if desired.

With the influx of viscous feedstocks in refineries, hydrotreating has taken on a new role insofar as hydrotreating units are usually placed upstream of units where catalyst deactivation may occur from feedstock impurities or to lower impurities in finished products, like jet fuel or diesel fuel. A large refinery may have five or more hydrotreaters, and three types of hydrotreaters are typically found in all refineries: (i) the naphtha hydrotreater, which pretreats feedstock to the reformer; (ii) the kerosene hydrotreater, also referred to as the *middle-distillate hydrotreater*, which is used to treat middle distillates from the atmospheric crude tower; and (iii) the gas oil hydrotreater, also referred to as the *diesel hydrotreater*, which treats gas oil from the atmospheric crude tower or pretreats vacuum gas oil entering a cracking unit. It is this hydrotreater that may also be used to treat blends of gas oil with heavier feedstocks.

In the process, the feedstock to the hydrotreater is mixed with hydrogen-rich gas before entering a fixed-bed reactor. In the presence of a catalyst, hydrogen reacts with the oil feedstock to produce hydrogen sulfide, ammonia, saturated hydrocarbon derivatives, and other free metals. The metals remain on the surface of the catalyst, and other products leave the reactor with the oil-hydrogen stream. Oil is separated from the hydrogen-rich gas stream, and any remaining light ends (C_4 and lighter) are removed in the stripper. The gas stream is treated to remove hydrogen sulfide, and then, it is recycled to the reactor (Parkash, 2003; Gary et al., 2007; Speight, 2014, 2017; Hsu and Robinson, 2017).

Most hydrotreating reactions are carried out below 425°C (800°F) to minimize cracking. Product streams vary considerably depending on (i) the characteristics of the feedstock, (ii) the catalyst, and (iii) the process parameters. The predominant reaction type is hydrodesulfurization, although many reactions take place in hydrotreating including denitrogenation, deoxygenation, hydrogenation, and hydrocracking. Almost all hydrotreating reactions are exothermic, and depending on the specific conditions, a temperature rise through the reactor of 3–11°C (5–20°F) is usually observed.

Generally, it is more economical to hydrotreat high-sulfur feedstocks before catalytic cracking than to hydrotreat the products from catalytic cracking. The advantages are the following: (i) The products require less finishing; (ii) sulfur is removed from the catalytic cracking feedstock, and corrosion is reduced in the cracking unit; (iii) carbon formation during cracking is reduced, and higher conversions result; and (iv) the catalytic cracking quality of the gas oil fraction is improved.

One of the chief problems with the processing of heavy oil, extra heavy oil, and tar sand bitumen is the deposition of metals, in particular vanadium, on the catalyst. It is not possible to remove vanadium from the catalyst, which must therefore be replaced when deactivated, and the time taken for catalyst replacement can significantly reduce the unit time efficiency. Fixed-bed catalysts tend to plug owing to solids in the feedstock or carbon deposits when processing heavy oil, extra heavy oil, and tar sand bitumen feedstocks. As mentioned previously, the highly exothermic reaction at high conversion gives difficult reactor design problems in heat removal and temperature control.

It is the physical and chemical composition of a feedstock that plays a large part not only in determining the nature of the products that arise from refining operations but also in determining the precise manner by which a particular feedstock should be processed. Furthermore, it is apparent that the conversion of heavy oil, extra heavy oil, and tar sand bitumen requires new lines of thought to develop suitable processing scenarios. Indeed, the use of thermal (*carbon rejection*) processes and of hydrothermal (*hydrogen addition*) processes, which were inherent in the refineries designed to process lighter feedstocks, has been a particular cause for concern. This has brought about the evolution of processing schemes that accommodate the heavier feedstocks. As a point of reference, an example of the former option is the delayed coking process in which the feedstock is converted to overhead with the concurrent deposition of coke, for example, that used by Suncor, Inc., at their tar sand plant (Speight, 2013, 2014).

The hydrogen addition concept is illustrated, in part, by the hydrotreating processes in which hydrogen is used in an attempt to remove the heteroatoms and metals and to *stabilize* the reactive fragments produced by the low degree of hydrocracking, thereby decreasing their potential for recombination to heavier

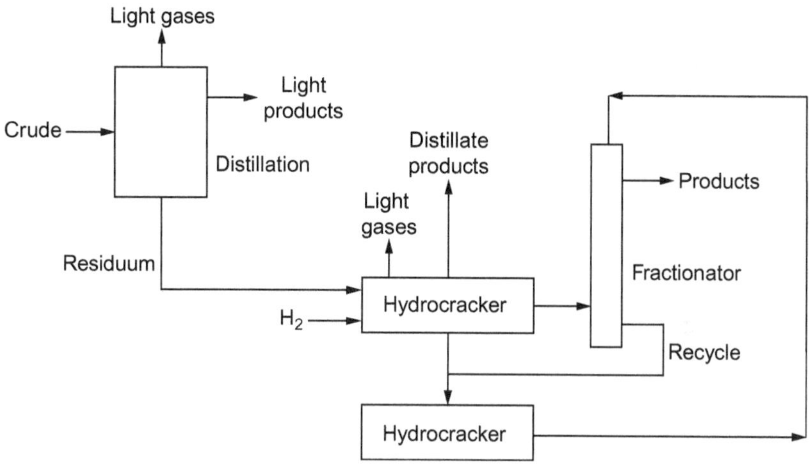

FIG. 10.5 A single-stage and two-stage hydroprocessing configuration.

products and ultimately to coke. The choice of processing schemes for a given hydrotreating application depends upon the nature of the feedstock and the product requirements. The process can be simply illustrated as a single-stage or as a two-stage operation (Fig. 10.5; Parkash, 2003; Gary et al., 2007; Speight, 2014, 2017; Hsu and Robinson, 2017).

The refining industry often employs two-stage processes in which the feedstock undergoes both hydrotreating and hydrocracking. In the first, or pretreating, stage, the main purpose is the conversion of nitrogen compounds in the feedstock to hydrocarbon derivatives and to ammonia by hydrogenation and mild hydrocracking. Typical conditions are 340–390°C (650–740°F), 150–2500 psi, and a catalyst contact time of 0.5–1.5 h. Up to 1.5% w/w hydrogen is absorbed, partly by conversion of the nitrogen compounds but chiefly by aromatic compounds that are hydrogenated. Thus, the single-stage process can be used to facilitate hydrotreating, and the two-stage process may then be used for hydrocracking primarily to produce distillates from high-boiling feedstocks. Both processes use an extinction-recycling technique to maximize the yields of the desired product. Significant conversion of viscous feedstocks can be accomplished by including the second stage and hydrocracking at high severity. For some applications, the products boiling up to 340°C (650°F) can be blended to give the desired final product.

Product distribution and quality vary considerably depending upon the nature of the feedstock constituents and on the process. Different process configurations will produce variations in the product slate from any one particular feedstock, and the *feedstock recycle* option adds another dimension to variations in product slate (Parkash, 2003; Gary et al., 2007; Speight, 2014, 2017; Hsu and Robinson, 2017).

In modern refineries, hydrotreating is one of several process options that can be applied to the production of liquid fuels from the heavier feedstocks (Speight, 2000). A most important aspect of the modern refinery operation is the desired product slate, which dictates the matching of a process with any particular feedstock to overcome differences in feedstock composition. Hydrogen consumption is also a parameter that varies with feedstock composition (Parkash, 2003; Gary et al., 2007; Speight, 2014, 2017; Hsu and Robinson, 2017), thereby indicating the need for a thorough understanding of the feedstock constituents if the process is to be employed to maximum efficiency.

A convenient means of understanding the influence of feedstock on the hydrotreating process is through a study of the hydrogen content (H/C atomic ratio) and molecular weight (carbon number) of the various feedstocks or products (Parkash, 2003; Gary et al., 2007; Speight, 2014, 2017; Hsu and Robinson, 2017). These data show the carbon number and/or the relative amount of hydrogen must be added to generate the desired heteroatom removal. In addition, it is also possible to use data for hydrogen usage in processing heavy oil, extra heavy oil, and tar sand bitumen, where the relative amount of hydrogen consumed in the process can be shown to be dependent upon the sulfur content of the feedstock (Speight, 2000, 2014, 2017; Parkash, 2003; Ancheyta and Speight, 2007; Gary et al., 2007; Hsu and Robinson, 2017).

The commercial processes for treating or finishing viscous feedstocks with hydrogen all operate in essentially the same manner with similar parameters. The feedstock is heated and passed with hydrogen gas through a tower or reactor filled with catalyst pellets. The reactor is maintained at a temperature of 260–425°C (500–800°F) at pressures from 100 to 1000 psi, depending on the particular process, the nature of the feedstock, and the degree of hydrogenation required. After leaving the reactor, excess hydrogen is separated from the treated product and recycled through the reactor after the removal of hydrogen sulfide. The liquid product is passed into a stripping tower, where steam removes dissolved hydrogen and hydrogen sulfide, and after cooling the product is run to finished product storage or, in the case of feedstock preparation, pumped to the next processing unit.

It is most important to reduce the nitrogen content of the product oil to <0.001% w/w (10 ppm). This stage is usually carried out with a bifunctional catalyst containing hydrogenation promoters, for example, nickel and tungsten or molybdenum sulfides, on an acid support, such as silica-alumina. The metal sulfides hydrogenate aromatic derivatives and nitrogen compounds and prevent deposition of carbonaceous deposits; the acid support accelerates nitrogen removal as ammonia by breaking carbon-nitrogen bonds. The catalyst is generally used as 1/8 by 1/8 in (0.32 by 0.32 cm) or 1/16 by 1/8 in (0.16 by 0.32 cm) pellets, formed by extrusion.

Most of the hydrocracking is accomplished in the second stage, which resembles the first but uses a different catalyst. Ammonia and some naphtha are usually removed from the first-stage product, and then the remaining oil, which is low in nitrogen compounds, is passed over the second-stage catalyst. Again, typical conditions are 300–370°C (600–700°F), 1500–2500 psi hydrogen pressure, and 0.5–1.5 h contact time; 1%–1.5% by weight hydrogen may be absorbed. Conversion to naphtha or jet fuel is seldom complete in one contact with the catalyst, so the lighter oils are removed by distillation of the products and the heavier, higher-boiling product combined with fresh feedstock and recycled over the catalyst until it is completely converted.

2. Process Chemistry and Parameters

At this point and in the context of this chapter, it is pertinent that there be a discussion of the techniques for desulfurization and concurrent demetallization of various feedstocks. Thus, desulfurization is the removal of sulfur or sulfur compounds from viscous feedstocks, and demetallization is the removal of metals or metal-containing constituent (such as porphyrins) compounds from viscous feedstocks.

The problems encountered in hydrotreating any viscous feedstock (especially heavy oil, extra heavy oil, and tar sand bitumen) can be equated to the amount of higher-boiling constituents (Speight, 2000; Ancheyta and Speight, 2007). Moreover, metalloporphyrin derivatives are present in significant quantities, and these are precursors of metal deposits under hydroprocessing conditions that cause permanent deactivation. Also, nitrogen-containing compounds can poison acid sites and diminish hydrocracking reacting speed (Parkash, 2003; Ancheyta and Speight, 2007; Gary et al., 2007; Speight, 2014, 2017; Hsu and Robinson, 2017). In addition, hydroprocessing reactions occur in active sites of catalysts; therefore, an appropriate pore size distribution is required to guarantee the access of the molecules to the active site. The loss of active sites is associated to pore blocking, poisoning, sintering, and metal and coke deposits. In order to overcome these complications, some solutions imply the application of more severe conditions when heavier feedstocks are treated.

Processing these feedstocks not only is a matter of applying know-how derived from refining *conventional* crude oils that often used hydrogen-to-carbon (H/C) atomic ratios as the main criterion for determining process options but also requires knowledge of several other properties (Parkash, 2003; Gary et al., 2007; Speight, 2014, 2017; Hsu and Robinson, 2017). The materials are complex not only in terms of the carbon number and boiling point ranges but also because a large part of this *envelope* falls into a range about which very little is known about model compounds (Speight, 2000, 2014, 2017; Parkash, 2003; Ancheyta and Speight, 2007; Gary et al., 2007; Hsu and Robinson, 2017). It is also established that the majority of the higher-molecular-weight materials produce coke (with some liquids), while the majority of the lower-molecular-weight constituents produce liquids (with some coke). It is to the latter trend that hydrotreating is aimed.

Thus, there is the potential for the application of more efficient conversion processes to viscous feedstock refining. Hydrocracking is probably the most versatile of crude oil refining processes because of its applicability to a wide range of feedstocks. In fact, hydrocracking can be applied to the conversion of the heavier feedstocks, and there are a variety of processes that are designed specifically for this particular use (Parkash, 2003; Gary et al., 2007; Speight, 2014, 2017; Hsu and Robinson, 2017). It is worthy of note at this point that there is often very little effort (it is often difficult) to differentiate between a hydrotreating process and a hydrocracking process (Chapter 11; Parkash, 2003; Gary et al., 2007; Speight, 2014, 2017; Hsu and Robinson, 2017).

In actual practice, the reactions that are used to chemically define the processes, that is, cracking and subsequent hydrogenation of the fragments, hydrogenation of unsaturated material, and hydrodesulfurization and hydrodenitrogenation (Ancheyta and Speight, 2007), can all occur (or be encouraged to occur) in any one particular process. Thus, hydrocracking will, in all likelihood, be accompanied by hydrodesulfurization, thereby producing not only low-boiling products but also low-boiling products that are low in sulfur.

Moreover, it is the physical and chemical composition of a feedstock that plays a large part not only in determining the nature of the products that arise from refining operations but also in determining the precise manner by which a particular feedstock should be processed. Furthermore, it is apparent that the conversion of heavy oil, extra heavy oil, and tar sand bitumen requires new lines of thought to develop suitable processing scenarios. Indeed, the use of thermal (*carbon rejection*) and hydrothermal (*hydrogen addition*) processes that were inherent in the refineries designed to process lighter feedstocks has been a particular cause for concern and has brought about the evolution of processing schemes that will accommodate the heavier feedstocks. However, processes based upon carbon rejection are not chemically efficient since they degrade usable portions of the feedstock to coke.

Thus, there is the potential for the application of more efficient conversion processes to viscous feedstock refining. Hydroprocessing, that is, hydrotreating (in the present context), is probably the most versatile of viscous feedstock refining processes because of its applicability to a wide range of feedstocks. In fact, hydrotreating can be applied to the removal of heteroatoms from heavier feedstocks as a first step in preparing the feedstock for other process options.

The sulfur content and metal content of the feedstock vary with viscous feedstock type (Speight, 2000; Ancheyta and Speight, 2007) and may or may not present a problem to the refiner, depending on the amount of metals present and the downstream processing required. Vanadium (V) and nickel (Ni), the primary metals found in viscous feedstocks, can range from <1 part per million (ppm by weight) in some viscous feedstocks to as high as 1100 ppm vanadium and 85 ppm nickel for Boscan (Venezuela) heavy oil. Vanadium is usually present in higher concentrations than nickel for Middle East crude oils and Venezuelan crude oils. However, for many US crude oils, particularly those from

California, the nickel content may be higher than the vanadium content. Other metals, such as sodium and iron, are also found in quantities up to 100 and 60 ppm, respectively, though usually much lower.

A number of methods are available for segregating metals from viscous feedstocks (especially heavy oil, extra heavy oil, and tar sand bitumen). For example, deasphalting removes metals insofar as they appear in the separated asphalt (Speight, 2000), and coking processes cause the metals to concentrate in the coke (Speight, 2000); there is a similar segregation of sulfur, but it is not as dramatic as the metal segregation (Kim and Hanson, 1998; Speight, 2000).

Hydrotreating processes, in particular the hydrodesulfurization of residua, are catalytic processes. Hydrocarbon feedstock and hydrogen are passed through a catalyst bed at elevated temperatures and pressures. Some of the sulfur atoms attached to hydrocarbon molecules react with hydrogen on the surface of the catalyst to form hydrogen sulfide (H_2S), and thermodynamic equilibrium calculations show that these reactions can be driven to almost 100% completion (Speight, 2000).

The operating conditions in distillate hydrodesulfurization are dependent upon the stock to be charged and the desired degree of desulfurization or quality improvement. Kerosene and light gas oils are generally processed at mild severity and high throughput, whereas light catalytic cycle oils and thermal distillates require slightly more severe conditions. Higher-boiling distillates, such as vacuum gas oils and lube oil extracts, require the most severe conditions (Parkash, 2003; Gary et al., 2007; Speight, 2014, 2017; Hsu and Robinson, 2017). In terms of the kinetics, the presence of a complex mixture of sulfur-bearing compounds results in an apparent reaction order between first and second. First-order behavior has been shown if either liquid holdup or effective catalyst wetting is accounted for.

In reality, no single processing scheme is the best choice for heavy oil, extra heavy oil, and tar sand bitumen. Refiners must consider the potential of proved processes, evaluate the promise of newer ones, and choose based on the situation. The best selection will always depend on the feedstock properties, the market for products, and financial and environmental consideration. Although there are no simple solutions, the available established processes and the growing number of new ones under development offer some reasonable choices. The issue then becomes how to most effectively handle the asphaltene fraction of the feedstock at the most reasonable cost. Solutions to this processing issue can be separated into two broad categories: (a) conversion of asphaltenes into another, salable product and (b) use of the asphaltenes by concentration into a marketable, or useable, product such as asphalt.

Also, the principal variables affecting the required severity in hydrotreating are as follows: (i) process chemistry, (ii) process configuration, (iii) reactor type, (iv) feedstock properties, (v) catalyst type, and (vi) process temperature and space velocity.

2.1 Process Chemistry

Hydrodesulfurization and demetallization occur simultaneously on the active sites within the catalyst pore structure. The important effect of hydrogen is the minimization of coking reactions. If the hydrogen pressure is too low for the required duty at any position within the reaction system, premature aging of the remaining portion of catalyst will be encountered. In addition, the effect of hydrogen pressure on desulfurization varies with feedstock boiling range. For a given feedstock, there exists a threshold level above which hydrogen pressure is beneficial to the desired desulfurization reaction. Below this level, desulfurization drops off rapidly as hydrogen pressure is reduced.

Sulfur and nitrogen occurring in heavy oil, extra heavy oil, and tar sand bitumen are converted to hydrogen sulfide and ammonia in the catalytic reactor, and these gases are scrubbed out of the reactor effluent gas stream. The metals in the feedstock are deposited on the catalyst in the form of metal sulfides, and cracking of the feedstock to distillate produces a laydown of carbonaceous material on the catalyst; both events poison the catalyst and activity or selectivity suffers. The deposition of carbonaceous material is a fast reaction that soon equilibrates to a particular carbon level and is controlled by hydrogen partial pressure within the reactors. On the other hand, metal deposition is a slow reaction that is directly proportional to the amount of feedstock passed over the catalyst.

The life of a catalyst used to hydrotreat heavy oil, extra heavy oil, and tar sand bitumen is dependent on the rate of carbon deposition and the rate at which organometallic compounds decompose and form metal sulfides on the surface. Several different metal complexes exist in the asphaltene fraction of the residuum, and an explicit reaction mechanism of decomposition that would be a perfect fit for all of the compounds is not possible. However, in general terms, the reaction can be described as hydrogen (A) dissolved in the feedstock contacting an organometallic compound (B) at the surface of the hydrotreating catalyst and producing a metal sulfide (C) and a hydrocarbon (D):

$$A + B \rightarrow C + D$$

Different rates of reaction may occur with various types and concentrations of metallic compounds. For example, a medium-metal-content feedstock will generally have a lower rate of demetallization compared with high-metal-content feedstock. And although individual organometallic compounds decompose according to both first- and second-order rate expressions, for reactor design, a second-order rate expression is applicable to the decomposition of residuum as a whole.

Finally, it has been recognized over the past three decades that desulfurization of gas oil or viscous feedstock that is subsequently fed to fluid catalytic cracking units allows the feedstock to be catalytically cracked without excessive sulfur oxide emissions in the regenerator flue gas. There are additional advantages such as (i) higher conversions of the feedstock, (ii) higher yields of naphtha, (iii) less sulfur in the naphtha and other distillates, (iv) lower yields of coke, and (v) reduced consumption of the cracking catalyst consumption.

Most of the metals in the catalytic cracking feedstock are (i) removed during desulfurization, eliminating a principal source of cracking catalyst deactivation or (ii), more preferably for the viscous feedstock, removed by the use of a guard-bed (demetallization reactor) or mild thermal treatment in which the metals are separated as part of the initially formed coke.

2.2 Process Configuration

All hydrodesulfurization processes react hydrogen with a hydrocarbon feedstock to produce hydrogen sulfide and a desulfurized hydrocarbon product (Speight, 2000, 2014, 2017; Parkash, 2003; Ancheyta and Speight, 2007; Gary et al., 2007; Hsu and Robinson, 2017). The feedstock is preheated and mixed with hot recycle gas containing hydrogen, and the mixture is passed over the catalyst in the reactor section at temperatures between 290 and 445°C (550–850°F) and pressures between 150 and 3000 psig. The reactor effluent is then cooked by heat exchange, and desulfurized liquid hydrocarbon product and recycle gas are separated at essentially the same pressure as used in the reactor. The recycle gas is then scrubbed and/or purged of the hydrogen sulfide and light hydrocarbon gases, mixed with fresh hydrogen makeup, and preheated prior to mixing with hot hydrocarbon feedstock.

The recycle gas scheme is used in the hydrodesulfurization process to minimize physical losses of expensive hydrogen. Hydrodesulfurization reactions require a high hydrogen partial pressure in the gas phase to maintain high desulfurization reaction rates and to suppress carbon laydown (catalyst deactivation). The high hydrogen partial pressure is maintained by supplying hydrogen to the reactors at several times the chemical hydrogen consumption rate. The majority of the unreacted hydrogen is cooled to remove hydrocarbon derivatives, recovered in the separator, and recycled for further utilization. Hydrogen is physically lost in the process by solubility in the desulfurized liquid hydrocarbon product and from losses during the scrubbing or purging of hydrogen sulfide and light hydrocarbon gases from the recycle gas.

2.3 Process Reactors

2.3.1 Downflow Fixed-Bed Reactor

The reactor design commonly used in hydrodesulfurization of distillates is the fixed-bed reactor design in which the feedstock enters at the top of the reactor and the product leaves at the bottom of the reactor (Parkash, 2003; Gary et al., 2007; Speight, 2014, 2017; Hsu and Robinson, 2017). The catalyst remains in a stationary position (fixed-bed) with hydrogen and a viscous feedstock passing in a downflow direction through the catalyst bed. The hydrodesulfurization reaction is exothermic, and the temperature rises from the inlet to the outlet of each catalyst bed. With a high hydrogen consumption and subsequent large temperature rise, the reaction mixture can be quenched with cold recycled gas at intermediate points in the reactor system. This is achieved by dividing the catalyst charge into a series of catalyst beds, and the effluent from each catalyst bed is quenched to the inlet temperature of the next catalyst bed.

The extent of desulfurization is controlled by raising the inlet temperature to each catalyst bed to maintain constant catalyst activity over the course of the process. Fixed-bed reactors are mathematically modeled as plug-flow reactors with very little back mixing in the catalyst beds. The first catalyst bed is poisoned with vanadium and nickel at the inlet to the bed and may be a cheaper catalyst (*guard bed*). As the catalyst is poisoned in the front of the bed, the temperature exotherm moves down the bed, and the activity of the entire catalyst charge declines, thus requiring a raise in the reactor temperature over the course of the process sequence. After catalyst regeneration, the reactors are opened and inspected, and the high-metal-content catalyst layer at the inlet to the first bed may be discarded and replaced with fresh catalyst. The catalyst loses activity after a series of regenerations, and consequently, after a series of regenerations, it is necessary to replace the complete catalyst charge. In the case of very-high-metal-content feedstocks (such as heavy oil, extra heavy oil, and tar sand bitumen), it is often necessary to replace the entire catalyst charge rather than to regenerate it. This is due to the fact that the metal contaminants cannot be removed by economical means during rapid regeneration, and the metals have been reported to interfere with the combustion of carbon and sulfur, catalyzing the conversion of sulfur dioxide (SO_2) to sulfate (SO_4^{2-}) that has a permanent poisoning effect on the catalyst.

Fixed-bed hydrodesulfurization units are generally used for distillate hydrodesulfurization and may also be used for residuum hydrodesulfurization but require special precautions in processing. The residuum must undergo two-stage electrostatic desalting so that salt deposits do not plug the inlet to the first catalyst bed, and the residuum must be low in vanadium and nickel content to avoid plugging the beds with metal deposits. Hence, the need for a guard bed in residuum hydrodesulfurization reactors.

During the operation of a fixed-bed reactor, contaminants entering with fresh feedstock are filtered out and fill the voids between catalyst particles in the bed. The buildup of contaminants in the bed can result in the channeling of reactants through the bed and reducing the hydrodesulfurization efficiency. As the flow pattern becomes distorted or restricted, the pressure drop throughout the catalyst bed increases. If the pressure drop becomes high enough, physical damage to the reactor internals can result. When high pressure drops are observed throughout any portion of the reactor, the unit is shut down, and the catalyst bed is skimmed and refilled.

With fixed-bed reactors, a balance must be reached between reaction rate and pressure drop across the catalyst bed. As catalyst particle size is decreased, the desulfurization reaction rate increases but so does the pressure drop across the catalyst bed. Expanded-bed reactors do not have this limitation, and small 1/32 in (0.8 mm) extrudate catalysts or fine catalysts may be used without increasing the pressure drop.

2.3.2 Upflow Expanded-Bed Reactor

Expanded-bed reactors are applicable to distillates but are commercially used for very viscous, high metals and/or dirty feedstocks having extraneous fine solid material (Lee et al., 1996). These reactors operate in such a way that the catalyst is in an expanded state so that the extraneous solids pass through the catalyst bed without plugging. They are isothermal, which conveniently handles the high-temperature exotherms associated with high hydrogen consumptions. Since the catalyst is in an expanded state of motion, it is possible to treat the catalyst as a fluid and to withdraw and add catalyst during operation.

Expanded beds of catalyst are referred to as particulate fluidized insofar as the feedstock and hydrogen flow upward through an expanded bed of catalyst with each catalyst particle in independent motion (Parkash, 2003; Gary et al., 2007; Speight, 2014, 2017; Hsu and Robinson, 2017). Thus, the catalyst migrates throughout the entire reactor bed. Expanded-bed reactors are mathematically modeled as back-mix reactors with the entire catalyst bed at one uniform temperature. Spent catalyst may be withdrawn and replaced with fresh catalyst on a daily basis. Daily catalyst addition and withdrawal eliminate the need for costly shutdowns to change out catalyst and also result in a constant equilibrium catalyst activity and product quality. The catalyst is withdrawn daily with a vanadium, nickel, and carbon content that is representative on a macroscale of what is found throughout the entire reactor. On a microscale, individual catalyst particles have ages from that of fresh catalyst to as old as the initial catalyst charge to the unit, but the catalyst particles of each age group are so well dispersed in the reactor that the reactor contents appear uniform.

In the unit, the feedstock and hydrogen recycle gas enter the bottom of the reactor, pass up through the expanded catalyst bed, and leave from the top of the reactor (Parkash, 2003; Gary et al., 2007; Speight, 2014, 2017; Hsu and Robinson, 2017). Commercial expanded-bed reactors normally operate with 1/32 in (0.8 mm) extrudate catalysts that provide a higher rate of desulfurization than the larger catalyst particles used in fixed-bed reactors. With extrudate catalysts of this size, the upward liquid velocity based on fresh feedstock is not sufficient to keep the catalyst particles in an expanded state. Therefore, for each part of the fresh feed, several parts of product oil are taken from the top of the reactor, recycled internally through a large vertical pipe to the bottom of the reactor, and pumped back up through the expanded catalyst bed. The amount of catalyst bed expansion is controlled by the recycle of product oil back up through the catalyst bed.

The expansion and turbulence of gas and oil passing upward through the expanded catalyst bed are sufficient to cause almost complete random motion in the bed (particulate fluidized). This effect produces the isothermal operation. It also causes almost complete back-mixing. Consequently, in order to effect near complete sulfur removal (over 75%), it is necessary to operate with two or more reactors in series. The ability to operate at a single temperature throughout the reactor or reactors and to operate at a selected optimum temperature rather than an increasing temperature from the start to the end of the run results in more effective use of the reactor and catalyst contents. When all these factors are put together, that is, the use of a smaller catalyst particle size, isothermal, fixed temperature throughout run, back-mixing, daily catalyst addition, and constant product quality, the reactor size required for an expanded bed is often smaller than that required for a fixed bed to achieve the same product goals. This is generally true when the feeds have high initial boiling points and/or the hydrogen consumption is very high.

2.3.3 Demetallization Reactor

The demetallization reactor (guard-bed reactor) is used when feedstocks that have relatively high metal contents (>300 ppm) substantially increase catalyst consumption because the metals poison the catalyst, thereby requiring frequent catalyst replacement. The usual desulfurization catalysts are relatively expensive for these consumption rates, but there are catalysts that are relatively inexpensive and can be used in the first reactor to remove a large percentage of the metals. Subsequent reactors downstream of the first reactor would use normal hydrodesulfurization catalysts. Since the catalyst materials are proprietary, it is not possible to identify them here. However, it is understood that such catalysts contain little or no metal promoters, that is, nickel, cobalt, and molybdenum. Metal removal on the order of 90% has been observed with these materials.

Thus, one method of controlling demetallization is to employ separate smaller *guard reactors* just ahead of the fixed-bed hydrodesulfurization reactor section. The preheated feedstock and hydrogen pass through the guard reactors that are filled with an appropriate catalyst for demetallization that is often the same as the catalyst used in the hydrodesulfurization section. The advantage of this system is that it enables replacement of the most contaminated catalyst (*guard bed*), where pressure drop is highest, without having to replace the entire inventory or shut down the unit. The feedstock is alternated between guard reactors, while catalyst in the idle guard reactor is being replaced.

When the expanded-bed design is used, the first reactor could employ a low-cost catalyst (5% of the cost of Co/Mo catalyst) to remove the metals, and subsequent reactors can use the more selective hydrodesulfurization catalyst (Parkash, 2003; Gary et al., 2007; Speight, 2014, 2017; Hsu and Robinson, 2017). The demetallization catalyst can be added continuously without taking the reactor out of service, and the spent demetallization catalyst can be loaded to >30% vanadium, which makes it a valuable source of vanadium.

2.4 Feedstock Effects

The composition of the various feedstocks under consideration in this text (i.e. heavy oil, extra heavy oil, and tar sand bitumen) may, at first sight, seem to be of minor importance when the problem of the hydrotreating (hydrodesulfurization) of viscous feedstocks comes under consideration. However, consideration of the variation in process conditions for different feedstocks presents some indication of the problems that may be encountered where the feedstocks are less well defined. Molecular composition is as important as molecular weight (or boiling range). Such is the nature of the problem when dealing with various residua and viscous oils that are (to say the least) unknown in terms of their chemical composition. In fact, the complexity of heavy oil, extra heavy oil, and tar sand bitumen has allowed little more than speculation as to the molecular structure of the constituents.

One of the major drawbacks to defining the influence of the feedstock on the process is that the research with respect to feedstocks has been fragmented. A conventional hydrotreating catalyst is used, and the results obtained are only valid for the process parameters, including the reactor system. More rigorous correlation is required, and there is a need to determine the optimum temperature for each type of sulfur compound. In order to obtain a useful model, the intrinsic kinetics of the reaction for a given catalyst should also be known. In addition, other factors that influence the desulfurization process such as (i) catalyst inhibition or deactivation by hydrogen sulfide, (ii) effect of nitrogen compounds, and (iii) the effects of various solvents should also be included in order to obtain a comprehensive model that is independent of the feedstock. The efficacy of other catalytic systems on the various viscous feedstocks also needs to be evaluated.

The character of the *feedstock properties*, especially the feedstock boiling range, has a definite effect on the ultimate design of the desulfurization unit and process flow and the reactor type. In agreement, there is a definite relationship between the percent by weight sulfur in the feedstock and the hydrogen requirements (Maples, 2000; Speight, 2000; Ancheyta and Speight, 2007). Of particular importance is the content of asphaltene constituents and resin constituents.

High amounts of asphaltene constituents and resin constituents require a high hydrogen partial pressure and may actually limit the maximum level of hydrodesulfurization, or final traces of sulfur in the residuum may only be eliminated under extremely severe reaction conditions where hydrocracking is the predominant reaction in the process. High asphaltene and resin contents are also responsible for high viscosity, which may increase the resistance to mass transfer of the reactants to the catalyst surface and thus decrease the rate of sulfur removal. In an instance such as this, it may be deemed satisfactory by the refiner to simply blend the viscous feedstock with a lighter feedstock, thereby alleviating the problem for that particular process. Similarly, a high asphalt content of a feedstock usually indicates a high sulfur content of that particular feedstock, which will require the higher partial pressure of hydrogen and the more severe reaction conditions indicated in the earlier part of this section. The sulfur content of a feedstock may also be expressed in terms of the asphaltene content alone, there seemingly being a relationship between the asphaltenes and the asphaltene sulfur.

One suggested method of facilitating hydrodesulfurization of viscous oils and residua is to remove completely, or partially, all of the asphaltic from the stock (Parkash, 2003; Gary et al., 2007; Speight, 2014, 2017; Hsu and Robinson, 2017). In the first instance, the asphaltic constituents can be completely and conveniently removed from a viscous feedstock by treatment with liquid propane under pressure whereupon the asphaltenes and the resins are deposited as viscous, thick, black asphalt leaving the lower-molecular-weight oil fraction in the liquid phase. Alternatively, use of other hydrocarbon solvents allows adjustment of the yield of precipitate in the form of partial removal of the asphaltenes and resins, thereby retaining some, or all, of the resins in the stock to serve as potential liquid-forming materials.

In addition, the reaction rate constant in the kinetic relationships decreases rapidly with increasing average boiling point in the kerosene and light gas oil range but much more slowly in the high-boiling gas oil range. This is attributed to the difficulty in removing sulfur from ring structures present in the entire high-boiling gas oil range.

The hydrodesulfurization of light (low-boiling) distillate (naphtha) is one of the more common catalytic hydrodesulfurization processes since it is usually used as a pretreatment of such feedstocks prior to catalytic reforming. Hydrodesulfurization of such feedstocks is required because sulfur compounds poison the precious-metal catalysts used in reforming, and desulfurization can be achieved under relatively mild conditions and is near quantitative. If the feedstock arises from a cracking operation, hydrodesulfurization will be accompanied by some degree of saturation resulting in increased hydrogen consumption.

The hydrodesulfurization of low-boiling (naphtha) feedstocks is usually a gas-phase reaction and may employ the catalyst in fixed beds, and (with all of the reactants in the gaseous phase) only minimal diffusion problems are encountered within the catalyst pore system. It is, however, important that the feedstock be completely volatile before entering the reactor as there may be the possibility of pressure variations (leading to less satisfactory results) if some of the feedstock enters the reactor in the liquid phase and is vaporized within the reactor.

In applications of this type, the sulfur content of the feedstock may vary from 100 ppm to 1%, and the necessary degree of desulfurization to be effected by the treatment may vary from as little as 50% to >99%. If the sulfur content of the feedstock is particularly low, it will be necessary to presulfide the catalyst. For example, if the feedstock only has 100–200 ppm sulfur, several days may be required to sulfide the catalyst as an integral part of the desulfurization process even with complete reaction of all of the feedstock sulfur to, say, cobalt and molybdenum (catalyst) sulfides. In such a case, presulfiding can be conveniently achieved by the addition of sulfur compounds to the feedstock or by the addition of hydrogen sulfide to the hydrogen.

Under the relatively mild processing conditions used for the hydrodesulfurization of the viscous feedstock that is difficult to achieve, process conditions may dictate that only part of the feedstock is actually in the vapor phase and that sufficient liquid phase is maintained in the catalyst bed to carry the larger molecular constituents of the feedstock through the bed. If the amount of liquid phase is insufficient for this purpose, molecular stagnation (leading to carbon deposition on the catalyst) will occur.

Hydrodesulfurization of middle distillates causes a more marked change in the specific gravity of the feedstock, and the amount of low-boiling material is much more significant when compared with the naphtha-type feedstock. In addition, the somewhat more severe reaction conditions (leading to a designated degree of hydrocracking) also lead to an overall increase in hydrogen consumption when middle distillates are employed as feedstocks in place of the naphtha.

High-boiling distillates, such as the atmospheric and vacuum gas oils, are not usually produced as a refinery product but merely serve as feedstocks to other processes for conversion to lower-boiling materials. For example, gas oils can be desulfurized to remove >80% of the sulfur originally in the gas oil with some conversion of the gas oil to lower-boiling materials (Parkash, 2003; Gary et al., 2007; Speight, 2014, 2017; Hsu and Robinson, 2017). The treated gas oil (which has a reduced carbon residue and lower sulfur and nitrogen contents relative to the untreated material) can then be converted to lower-boiling products in, say, a catalytic cracker where an improved catalyst life and volumetric yield may be noted.

In summary, the hydrodesulfurization of the viscous feedstock (especially heavy oil, extra heavy oil, and tar sand bitumen) can be achieved using a variety of processes. One major disadvantage of this type of feedstock is that the catalyst does become poisoned by metal contaminants in the feedstock because the relatively high amounts (relatively to the contaminants in distillate feedstocks) of these contaminants will be present. Thus, the catalyst may be regenerated, and onstream times between catalyst regeneration (while varying with the process conditions and application) may be of the order of 1 year or less (Parkash, 2003; Gary et al., 2007; Speight, 2014, 2017; Hsu and Robinson, 2017).

While many of the conventional design criteria in distillate desulfurization (hydrogen partial pressure, degree of desulfurization, and gas circulation rates) must be considered in residuum desulfurization, an additional important criterion is the effect of *metals accumulation* on the catalyst. The effective life of a particular catalyst will vary depending on its pore structure and total pore volume. It is also dependent upon the particular feedstock being processed and the operating conditions employed. As mentioned previously, many of the competitive processes use different catalyst characteristics that have been tailored to achieve the objectives of most concern to the individual licensor. Thus, some processes will remove more metals from the feedstock, while others will reject metals to a greater extent.

In general terms, catalysts that show better selectivity for metals removal will also hold more total metals before they become inoperable for the required desulfurization duty. This holding capacity for metals has been defined as a *saturation level* that increases with decreasing size for a given catalyst. However, the selectivity for demetallization over desulfurization reactions also increases with decreasing size. The combination of these effects results in an optimum particle size to maximize the cycle life.

2.5 Catalysts

The selection of the catalysts must take into consideration the properties of the feedstock to be hydroprocessed. Differences in feedstock composition influence the choice of processing schemes. The heteroatom content and molecular weight distribution are important aspects of feedstock properties, and in addition, for viscous feedstocks, the amount of asphaltene constituents and metals in the feedstock should be determined (Speight, 2015). In fact, based on the variation in feedstock properties, a catalyst or catalytic system suitable for hydroprocessing the various feedstocks does not exist, and the catalyst must be designed according to the feedstock properties and process parameters.

Hydrodesulfurization catalysts typically consist of metals impregnated on a porous alumina support (Furimsky, 1998; Ancheyta and Speight, 2007). Almost all of the surface area is found in the pores of the alumina (200–300 m^2/g), and the metals are dispersed in a thin layer over the entire alumina surface within the pores. This type of catalyst does display a huge catalytic surface for a small weight of catalyst. Cobalt (Co), molybdenum (Mo), and nickel (Ni) are the most commonly used metals for desulfurization catalysts. The catalysts are manufactured with the metals in an oxide state. In the active form, they are the sulfide state, which is obtained by sulfiding the catalyst either prior to use or with the feedstock during actual use. Any catalyst that exhibits hydrogenation activity will catalyze hydrodesulfurization to some extent. However, the group VIB metals (chromium, molybdenum, and tungsten) are particularly active for desulfurization, especially when promoted with metals from the iron group (iron, cobalt, and nickel).

Cobalt-molybdenum catalysts are by far the most popular choice for desulfurization. Nickel-molybdenum catalysts are often chosen instead of cobalt-molybdenum catalysts when higher activity for the saturation of polynuclear aromatic compounds or nitrogen removal is required or when more refractory sulfur compounds such as those in cracked feedstocks must be desulfurized. In some applications, nickel-cobalt-molybdenum (Ni-Co-Mo) catalysts appear to offer a useful balance of hydrotreating activity. Nickel-tungsten (Ni-W) is usually chosen only when very high activity for aromatic derivative saturation is required along with activity for sulfur and nitrogen removal. There are several different compositions for available catalysts (Parkash, 2003; Gary et al., 2007; Speight, 2014, 2017; Hsu and Robinson, 2017).

Cobalt-molybdenum (Co/Mo) and nickel-molybdenum (Ni/Mo) catalysts resist poisoning and are the most universally applied catalysts for hydrodesulfurization of everything from naphtha to heavy oil, extra heavy oil, and tar sand bitumen. In addition, cobalt-molybdenum (Co/Mo) and nickel-molybdenum (Ni/Mo) catalysts promote both demetallization and desulfurization. The vanadium deposition rate at a given desulfurization level is a function of the pore structure of the alumina support and the types of metals on the support whereas a catalyst support having small pores preferentially removes sulfur with a low degree of demetallization (Parkash, 2003; Gary et al., 2007; Speight, 2014, 2017; Hsu and Robinson, 2017).

Catalyst life depends on the charge stock properties and the degree of desulfurization desired. The only permanent poisons to the catalyst are metals in the feedstock that deposit on the catalyst, usually quantitatively, causing permanent deactivation as they accumulate. However, this is usually of little concern except when deasphalted oils are used as feedstocks since most distillate feedstocks contain low amounts of metals. Nitrogen compounds are a temporary poison to the catalyst, but there is essentially no effect on catalyst aging except that caused by a higher temperature requirement to achieve the desired desulfurization. Hydrogen sulfide can be a temporary poison in the reactor gas, and recycle gas scrubbing is employed to counteract this condition.

Providing that pressure drop buildup is avoided, cycles of 1 year or more and ultimate catalyst life of 3 years or more can be expected. The catalyst employed can be regenerated by normal steam-air or recycle combustion gas-air procedures. The catalyst is restored to near fresh activity by regeneration during the early part of its ultimate life. However, permanent deactivation of the catalyst occurs slowly during usage and repeated regenerations, so replacement becomes necessary.

The general *catalyst types* used in residuum desulfurization are combinations of metal oxides on alumina (Al_2O_3) or silica-stabilized alumina (SiO_2-Al_2O_3) supports. Molybdenum always seems to be one of the metals, with cobalt and/or nickel being used in combination with the molybdenum in many cases. The supports are usually tailored to the process objectives since different support can be made to accomplish particular goals of a specific unit. For example, smaller pored catalysts will tend to remove less metals than larger pored catalysts and be more active for desulfurization reactions. However, the metal-holding capacity of the small-pored catalyst will also be less than the large-pored catalyst, which results in a sacrifice of catalyst life. The better catalyst selection would depend on the combination of these two characteristics that will allow the best life activity relationship for a given application. In some cases, critical combinations of large and small pore sizes are used to arrive at the best catalyst for a given feedstock and operating conditions. The objectives of the process unit are also important since metal removal would be more critical in an application to produce heavy oil, extra heavy oil, and tar sand bitumen feedstock for catalytic cracking versus an application to produce heavy oil, extra heavy oil, and tar sand bitumen fuel.

Catalyst size and shape are also important factors in residue desulfurization processes. Smaller size contributes to improved desulfurization and demetallization, but pressure drop considerations increase in importance.

A gradual loss of *catalyst activity* occurs during normal operation of the residue process. Therefore, a gradual increase in catalyst temperature is required through the cycle to maintain the desired product sulfur content. This loss in activity is caused by deposition of coke and metals (from nickel and vanadium in the feedstock) on the catalyst surface and in the catalyst pores. Ultimate *catalyst life* is directly related to the total metal tolerance of the catalyst, which is a function of particle size, shape, and pore size and volume. Metals deposited cause permanent deactivation of the catalyst and preclude the restoration of catalyst activity by normal regeneration procedures. Spent catalysts are either discarded or returned to reclaimers for the recovery of the various metals of value. The amount of coke deposited on the catalyst depends primarily on hydrogen partial pressure but is also influenced by the asphaltene content of the feedstock. Higher hydrogen pressure decreases coking, while higher asphaltene content increases coking.

Besides the effect of hydrogen partial pressure on *catalyst aging*, maintenance of adequate amounts of hydrogen within the system is required. Circulation rates and purity requirements are set to avoid a shortage of hydrogen anywhere in the reaction system in order to prevent undesirable side reactions.

2.6 Temperature and Space Velocity

Temperature and *space velocity* are very important variables that influence the operation of the process. In a given system, a reduction in space velocity without an appropriate reduction in temperature will result in feedstock overtreating. This will lead to irreversible premature aging of the catalyst by virtue of increased coking and incremental metal deposition.

Since *catalyst temperature* is increased over the length of an operating cycle, both yields of lighter materials and properties of the remaining feedstock are affected. The magnitude of the variations will depend on the catalyst selected for the operation and the operating conditions employed (Parkash, 2003; Gary et al., 2007; Speight, 2014, 2017; Hsu and Robinson, 2017). The product properties affected to the greatest extent are viscosity and pour point and, with changes in distillate yields, indicate that cracking reactions are increasing as the run progresses (due to temperature increases that lead to a greater degree of thermal cracking).

A higher *reaction temperature* increases the rate of desulfurization at constant feedstock rate, and the start-of-run temperature is set by the design desulfurization level, space velocity, and hydrogen partial pressure. The capability to increase temperature as the catalyst deactivates is built into the most process or unit designs. Temperatures of 415°C (780°F) and above result in excessive coking reactions and higher than normal catalyst aging rates. Therefore, units are designed to avoid the use of such temperatures for any significant part of the life cycle.

As the *space velocity* is increased, desulfurization is decreased, but increasing the hydrogen partial pressure and/or the reactor temperature can offset the detrimental effect of increasing space velocity.

In actual practice, the reactions that are used to chemically define the processes, that is, hydrodesulfurization, hydrodenitrogenation, and hydrocracking reactions (Parkash, 2003; Gary et al., 2007; Speight, 2014, 2017; Hsu and Robinson, 2017), can occur (or be encouraged to occur) in any one particular process. Thus, hydrodesulfurization may be accompanied, in all likelihood, by a degree of hydrocracking as determined by the refiner, thereby producing not only products that are low in sulfur but also low-boiling products. Thus, the choice of processing schemes for a given hydroprocess depends upon the nature of the feedstock and the product requirements (Parkash, 2003; Ancheyta and Speight, 2007; Gary et al., 2007; Speight, 2014, 2017; Hsu and Robinson, 2017).

The single-stage process can be used to produce naphtha but is more often used to produce middle distillate from vacuum gas oil and may be used to remove the heteroatoms from heavy oil, extra heavy oil, and tar sand bitumen with a specified degree of hydrocracking. The two-stage process was developed primarily to produce high yields of naphtha from straight-run gas oils, and the first stage may actually be a purification step to remove sulfur-containing and nitrogen-containing organic materials. Both processes use an extinction/recycle technique to maximize the yields of the desired product. Significant conversion of viscous feedstocks can be accomplished by hydrocracking at high severity. For some applications, the products boiling up to 340°C (650°F) can be blended to give the desired final product.

In reality, no single bottom-of-the-barrel processing scheme is always the best choice.

The hydrodesulfurization process variables (Speight, 2000) usually require some modification to accommodate the various feedstocks that are submitted for this particular aspect of refinery processing (Parkash, 2003; Gary et al., 2007; Speight, 2014, 2017; Hsu and Robinson, 2017). The main point of this section is to outline the hydrotreating process with particular reference to the heavier oils and heavy oil, extra heavy oil, and tar sand bitumen. However, some reference to the lighter feedstocks is warranted. This will serve as a base point to indicate the necessary requirements for heavy oil, extra heavy oil, and tar sand bitumen and residuum hydrodesulfurization.

One particular aspect of the hydrotreating process that needs careful monitoring, with respect to feedstock type, is the exothermic nature of the reaction.

The heat of the reaction is proportional to the hydrogen consumption, and with the more saturated lower-boiling feedstocks where hydrocracking may be virtually eliminated, the overall heat production during the reaction may be small, leading to a more controllable temperature profile. However, with viscous feedstocks where hydrogen consumption is appreciable (either by virtue of the hydrocracking that is necessary to produce a usable product or by virtue of the extensive hydrodesulfurization that must occur), it may be desirable to provide internal cooling of the reactor. This can be accomplished by introducing cold recycle gas to the catalyst bed to compensate for excessive heat. One other generalization may apply to the lower-boiling feedstocks in the hydrodesulfurization process. The process may actually have very little effect on the properties of the feedstock (assuming that hydrocracking reactions are negligible)—removal of sulfur will cause some drop in specific gravity that could give rise to volume recoveries approaching (or even above) 100%. Furthermore, with the assumption that cracking reactions are minimal, there may be a slight lowering of the boiling range due to sulfur removal from the feedstock constituents. However, the production of lighter fractions is usually small and may only amount to some 1%–5% by weight of the products boiling below the initial boiling point of the feedstock.

One consideration for viscous feedstocks is that it may be more economical to hydrotreat and desulfurize high-sulfur feedstocks before catalytic cracking than to hydrotreat the products from catalytic cracking. This approach (DeCroocq, 1984; Speight, 2000, 2014) has the potential for several advantages such as the following: (i) The products require less finishing; (ii) sulfur is removed from the catalytic cracking feedstock, and corrosion is reduced in the cracking unit; (iii) coke formation is reduced; (iv) higher feedstock conversions; and (v) the potential for better-quality products. The downside is that many of the heavier feedstocks act as hydrogen sinks in terms of their ability to interact with the expensive hydrogen. A balance of the economic advantages/disadvantages must be struck on an individual feedstock basis.

In terms of the feedstock composition, it must be recognized that when catalytic processes are employed for viscous feedstocks, complex molecules (such as those that may be found in the original asphaltene fraction) or those formed during the process are not sufficiently mobile. They are also too strongly adsorbed by the catalyst to be saturated by the hydrogenation component and, hence, continue to react and eventually degrade to coke. These deposits deactivate the catalyst sites and eventually interfere with the hydroprocess by causing a decrease in the relative rate of hydrodesulfurization (Parkash, 2003; Gary et al., 2007; Speight, 2014, 2017; Hsu and Robinson, 2017).

Viscous feedstocks, such as heavy oil, extra heavy oil, and tar sand bitumen, require more severe hydrodesulfurization conditions to produce low-sulfur liquid product streams that can then, as is often now desired, be employed as feedstocks for other refining operations. Hydrodesulfurization of the heavier feedstocks is normally accompanied by a high degree of hydrocracking, and thus, the process conditions required to achieve 70%–90% desulfurization will also effect substantial conversion of the feedstock to lower-boiling products. In addition, the extent of hydrodesulfurization reaction of viscous feedstocks is dependent upon the temperature, and the reaction rate increases with increase in temperature.

In contrast to the lighter feedstocks that may be subjected to the hydrodesulfurization operation, the process catalysts are usually susceptible to poisoning by nitrogen (and oxygen) compounds and metallic salts (in addition to the various sulfur-compound types) that tend to be concentrated in heavy oil, extra heavy oil, and tar sand bitumen or exist as an integral part of the heavy oil, extra heavy oil, and tar sand bitumen matrix. Thus, any processing sequence devised for hydrodesulfurization of heavy oil, extra heavy oil, and tar sand bitumen must be capable of accommodating the constituents that adversely affect the ability of the catalyst to function in the most efficient manner possible.

The conditions employed for the hydrodesulfurization of the heavier feedstocks may be similar to those applied to the hydrodesulfurization of gas oil fractions but with the tendency to increased pressures. However, carbon deposition on and metal contamination of the catalyst is much greater when heavy oil, extra heavy oil, and tar sand bitumen are employed as feedstocks, and unless a low level of desulfurization is acceptable, frequent catalyst regeneration is necessary.

A wide choice of commercial processes is available for the catalytic hydrodesulfurization of heavy oil, extra heavy oil, and tar sand bitumen. The suitability of any particular process depends not only upon the nature of the feedstock but also on the degree of desulfurization that is required. There is also a dependence on the relative amounts of the lower-boiling products that are to be produced as feedstocks for further refining and generation of liquid fuels.

There is, however, one aspect of feedstock properties that has not yet been discussed fully and that is feedstock composition. This particular aspect of the nature of the feedstock is, in fact, related to the previous section where the influence of various feedstock types on the hydrodesulfurization process was noted, but it is especially relevant when heavy oil, extra heavy oil, and tar sand bitumen from various sources are to be desulfurized.

Heavy oil, extra heavy oil, tar sand bitumen, and other viscous feedstocks contain impurities other than sulfur, nitrogen, and oxygen, and the most troublesome of these impurities are the organometallic compounds of nickel and vanadium. The metal content of a residuum can vary from several parts per million to >1000 ppm, and there does seem to be a more-than-chance relationship between the metal content of a feedstock and its physical properties (Reynolds, 1998; Parkash, 2003; Gary et al., 2007; Speight, 2014, 2017; Hsu and Robinson, 2017). In the hydrodesulfurization of the heavier feedstocks, the metals (nickel plus vanadium) are an important factor since large amounts (over 150 ppm) will cause rapid deterioration of the catalyst. The free metals, or the sulfides, deposit on the surface of the catalyst and within the pores of the catalyst, thereby poisoning the catalyst by making the more active catalyst sites inaccessible to the feedstock and the hydrogen. This results in frequent replacement of an expensive process commodity unless there are adequate means by which the catalyst can be regenerated.

The problem of metal deposition on the hydrodesulfurization catalysts has generally been addressed using any one of three methods: (i) suppressing deposition of the metals on the catalyst, (ii) development of a catalyst that will accept the metals and can tolerate high levels of metals without marked reduction in the hydrodesulfurization capabilities of the catalyst, and (iii) removal of the metal contaminants before the hydrodesulfurization step. The first two methods involve a careful and deliberate choice of the process catalyst and the operating conditions. However, these methods may only be viable for feedstocks with <150 ppm total metals since the decrease in catalyst activity is directly proportional to the metal content of the feedstock. There are, however, catalysts that can tolerate substantial proportions of metals within their porous structure before the desulfurizing capability drops to an unsatisfactory level. Unfortunately, data on such catalysts are extremely limited because of their proprietary nature and details are not always available, but tolerance levels for metals that are equivalent to 15%–65% by weight of the catalyst have been quoted.

The third method may be especially applicable to feedstocks with a high metal content and requires a separate demetallization step just prior to the hydrodesulfurization reactor by the use of a guard-bed reactor. Such a step might involve passage of the feedstock through a demetallization chamber that contains a catalyst with a high selectivity for metals but whose activity for sulfur removal is low. Nevertheless, demetallization applied as a separate process can be used to generate low-metal-content feedstocks and will allow a more active and stable desulfurization system so that a high degree of desulfurization can be achieved on high-metal-content feedstocks with an acceptable duration of operation (Parkash, 2003; Gary et al., 2007; Speight, 2014, 2017; Hsu and Robinson, 2017).

3. Hydrotreating Heavy Feedstocks

The objectives of upgrading viscous feedstocks (such as heavy oil, extra heavy oil, and tar sand bitumen) are as follows: (i) reduction of metals (such as nickel, vanadium, and iron); (ii) reduction in the sulfur content; (iii) reduction in the amount of coke formers in the feedstock; (iv) reduction in nitrogen; and, last but certainly not least, (v) the conversion of the asphaltene and resin constituents into lower-molecular-weight easier-to-refine molecular species (such as naphtha, middle distillate, and gas oil) in order to produce higher-value feedstocks for heavy oil, extra heavy oil, and tar sand bitumen conversion units.

Advances made in hydrotreating processes have made the utilization of heavier feedstocks almost a common practice for many refineries. Upgrading processes can be used for the upgrading of atmospheric (650°F$^+$ or 345°C$^+$) and vacuum (1050°F$^+$ or 565°C$^+$) heavy oil, extra heavy oil, and tar sand bitumen. However, there is no hydroprocessing process that can universally be applicable to upgrade all viscous feedstocks. As a result, several hydroprocessing processes are developed for different commercial applications; many other processes are in their development stages.

Two routes exist for residue upgrading: (i) carbon rejection, such as coking processes, and (ii) hydrogen addition, such as hydrotreating processes that use a fixed-bed unit or an ebullated-bed unit (Parkash, 2003; Gary et al., 2007; Speight, 2014, 2017; Hsu and Robinson, 2017). The hydrogen addition route is a more expensive option relative to a carbon rejection option but results in a significantly higher yield of liquid products.

Hydroprocessing units with fixed-bed reactors must be shut down to remove the spent catalyst when catalyst activity declines below an acceptable level (due to the accumulation of coke, metals, and other contaminants). There are a few types of hydroprocessing reactors with moving or ebullating catalyst beds. In ebullated-bed hydroprocessing, the catalyst within the reactor bed is not fixed. In such a process, the hydrocarbon feedstock stream enters the bottom of the reactor and flows upward through the catalyst. The catalyst is kept in suspension by the pressure of the fluid feed. Ebullating-bed reactors are capable of converting the most problematic feeds, such as atmospheric resids, vacuum resids, heavy oil, extra heavy oil, and tar sand bitumen feedstocks (all of which have a high content of asphaltenes, metals, sulfur, and sediments), to lower-boiling, more valuable products while simultaneously removing contaminants. The function of the catalyst is to remove contaminants such as sulfur and nitrogen heteroatoms, which accelerate the deactivation of the catalyst, while cracking (converting) the feedstock to lighter products. Because ebullating-bed reactors perform both hydrotreating and hydrocracking functions (Chapter 12), they are also referred to as dual-purpose reactors (Parkash, 2003; Gary et al., 2007; Speight, 2014, 2017; Hsu and Robinson, 2017).

Both the fixed-bed processes and the ebullated-bed processes require a catalyst system with a pore size distribution to match the changing molecular structure of the feedstock constituents. The catalyst can be designed for high metal uptake capacity and moderate sulfur conversion to be applied in the front-end reactor when processing high-metal-containing feedstocks ($>$ 70 ppm vanadium). On the other hand, the catalyst may be designed for moderate metal removal capacity but higher activity for sulfur and conversion of the coke precursors that is applied in front-end reactors when processing feedstocks with a lower metal content ($<$ 70 ppm vanadium) or in middle reactors when processing high-metal-containing feedstocks. Catalysts with a high propensity for sulfur, removal of coke precursors, and removal of nitrogen are applied in the middle and/or tail-end reactors.

3.1 Processes

In refining viscous feedstocks, hydrodesulfurization (HDS) processes and hydrodemetallization (HDM) processes are used to reduce or eliminate poisoning of sophisticated and expensive catalysts that are used in the downstream refining steps (i.e., fluid catalytic cracking, FCC, reforming, and hydrotreating). The hydrodemetallization process is a pretreatment process for the viscous feedstock by which metals and part of heteroatom contaminates are removed along with conversion of the residue to a lighter fractions. The hydrodemetallization process uses low-cost catalysts either in a fixed-bed or moving-bed reactor operating at moderate temperatures and pressures (580–2900 psi) and at relatively high liquid hourly space velocity (LHSV). Guard-bed catalysts are often used also.

Processes for the direct desulfurization of heavy oil, extra heavy oil, and tar sand bitumen have a similar flow to distillate hydrodesulfurization but with distinguishing features such as the catalyst compositions and shapes employed (Parkash, 2003; Gary et al., 2007; Speight, 2014, 2017; Hsu and Robinson, 2017). Examples of such processes are the RDS/VRDS process for heavy oil, extra heavy oil, and tar sand bitumen and the residfining process (a derivative of the hydrofining process) (Parkash, 2003; Gary et al., 2007; Speight, 2014, 2017; Hsu and Robinson, 2017).

3.1.1 Resid Desulfurization and Vacuum Resid Desulfurization Process

The resid desulfurization process (RDS process) and the vacuum resid desulfurization process (VRDS process) are designed to remove sulfur, nitrogen, asphaltene, and metal contaminants from heavy oil, extra heavy oil, and tar sand bitumen and are also capable of accepting whole viscous feedstocks or topped viscous feedstocks as feedstocks to the refining unit(s). The major product of the processes is a low-sulfur fuel oil, and the amount of naphtha and middle distillates is maintained at a minimum to conserve hydrogen. The basic elements of each process are similar and consist of a once-through operation of the feedstock coming into contact with hydrogen and the catalyst in a downflow reactor that is designed to maintain activity and selectivity in the presence of deposited metals. Moderate temperatures and pressures are employed to reduce the incidence of hydrocracking and, hence, minimize production of low-boiling distillates (Parkash, 2003; Gary et al., 2007; Speight, 2014, 2017; Hsu and Robinson, 2017). The combination of a desulfurization step and a vacuum residuum desulfurizer (VRDS) is often seen as an attractive alternate to the atmospheric residuum desulfurizer (RDS) because the combination route uses less hydrogen for a similar investment cost. Both the RDS and VRDS processes can be coupled with other processes (such as delayed coking, fluid catalytic cracking, and solvent deasphalting) to achieve the most optimum refining performance.

3.1.2 Residfining Process

The residfining process is a catalytic fixed-bed process for the desulfurization and demetallization of viscous feedstocks (Parkash, 2003; Gary et al., 2007; Speight, 2014, 2017; Hsu and Robinson, 2017). The process can also be used to pretreat heavy oil, extra heavy oil, and tar sand bitumen to suitably low contaminant levels prior to catalytic cracking. In the process, liquid feedstock to the unit is filtered, pumped to pressure, preheated, and combined with treat gas prior to entering the reactors. A guard reactor would typically be employed to prevent plugging/fouling of the main reactors with provisions to periodically remove the guard while keeping the main reactors on-line. The temperature rise associated with the exothermic reactions is controlled utilizing either a gas or liquid quench. A train of separators is employed to separate the gas and liquid products. The recycle gas is scrubbed to remove ammonia (NH_3) and hydrogen sulfide (H_2S). It is then combined with fresh makeup hydrogen before being reheated and recombined with fresh feed. The liquid product is sent to a fractionator where the product is fractionated.

The different catalysts allow other minor differences in operating conditions and peripheral equipment. Primary differences include the use of higher-purity hydrogen makeup gas (usually 95% or greater), inclusion of filtration equipment in most cases, and facilities to upgrade the off-gases to maintain higher concentration of hydrogen in the recycle gas. Most of the processes utilize downflow operation over fixed-bed catalyst systems, but exceptions to this are the H-Oil and LC-Fining processes (which are predominantly conversion processes) that employ upflow designs and ebullating catalyst systems with continuous catalyst removal capability and the Shell process (a conversion process) that may involve the use of a *bunker flow* reactor ahead of the main reactors to allow periodic changeover of catalyst (Speight, 2000, 2014, 2017; Parkash, 2003; Ancheyta and Speight, 2007; Gary et al., 2007; Hsu and Robinson, 2017).

3.1.3 Other Processes

The primary objective in most of the residue desulfurization processes is to remove sulfur with minimum consumption of hydrogen. Substantial percentages of nitrogen, oxygen, and metals are also removed from the feedstock. However, complete elimination of other reactions is not feasible, and in addition, hydrocracking, thermal cracking, and aromatic saturation reactions occur to some extent. Certain processes, that is, H-Oil process using a single-stage or a two-stage reactor and the LC-Fining process using an expanded-bed reactor, can be designed to accomplish greater amounts of hydrocracking to yield larger quantities of lighter distillates at the expense of desulfurization (Speight, 2000, 2014, 2017; Parkash, 2003; Ancheyta and Speight, 2007; Gary et al., 2007; Hsu and Robinson, 2017).

Removal of nitrogen is much more difficult than removal of sulfur. For example, nitrogen removal may be only ~25%–30% when sulfur removal is at a 75%–80% level. Metals are removed from the feedstock in substantial quantities and are mainly deposited on the catalyst surface and exist as metal sulfides at processing conditions. As these deposits accumulate, the catalyst pores eventually become blocked and inaccessible; thus, catalyst activity is lost.

Desulfurization of heavy oil, extra heavy oil, and tar sand bitumen is considerably more difficult than desulfurization of distillates (including vacuum gas oil) because many more contaminants are present and very large, complex molecules are involved. The most difficult portion of feedstock in residue desulfurization is the asphaltene fraction that forms coke readily, and it is essential that these large molecules be prevented from condensing with each other to form coke, which deactivates the catalyst. This is accomplished by the selection of proper catalysts, use of adequate hydrogen partial pressure, and assuring intimate contact of the hydrogen-rich gases and oil molecules in the process design.

4. Other Options

Environmental concerns and newly enacted rules and regulations intend to ensure that viscous feedstock products are expected to meet lower and lower limits on contaminants, such as sulfur and nitrogen. New regulations require the removal of sulfur compositions from liquid hydrocarbon derivatives, such as those used in gasoline, diesel fuel, and other transportation fuels.

In the hydrodesulfurization process, high temperatures and pressures may be required to obtain the desired low levels of sulfur. High-temperature processing of olefinic naphtha, however, may result in a lower-grade fuel due to the saturation of olefin derivatives leading to an octane loss. Low-octane naphtha will, more than likely, require additional refining, isomerization, blending, and the like to produce higher-quality fuels suitable for use in gasoline. Such extra processing adds additional cost, expense, and complexity to the process and may result in other undesirable changes in the products. As a result, the future will see processes that are more chemically precise in hydrotreating and will offer higher efficiency and conversion to selected products on the base that is not always practice currently.

Five of the most common approaches to upgrading hydrotreaters for clean-fuel production (in order of increasing capital cost) are currently and will continue (at least for the next two decades) (i) developing higher-activity and more resilient catalysts, (ii) replacing reactor internals for increased efficiency, (iii) adding reactor capacity to accommodate viscous feedstocks and increase naphtha and kerosene production for gasoline and diesel fuel products, (iv) increasing hydrogen partial pressure, and (v) creating process design and hardware that are more specialized and focus on process schemes that effectively reduce hydrogen consumption.

Finally, hydrotreating of heavy oil, extra heavy oil, and tar sand bitumen requires considerably different catalysts and process flows, depending on the specific operation so that efficient hydroconversion through uniform distribution of liquid, hydrogen-rich gas, and catalyst across the reactor is assured. There will also be automated demetallization of fixed-bed systems and more units that operate as ebullating-bed hydrocrackers (Chapter 11).

Finally, severe hydrotreating requires high-purity (>99% v/v) hydrogen, while less severe hydrotreating can employ lower-purity (<90% v/v) hydrogen. Refiners will continue to optimize hydrogen use by cascading hydrogen through the refinery. High-purity hydrogen will continue to be used only where required, and the lower-purity hydrogen purged from these applications is used for services that do not require high-purity hydrogen.

4.1 Catalyst Technology

Conventional hydroprocessing catalysts are generally in the form of a carrier of a refractory oxide material on which hydrogenation metals are deposited, the choice and amount of each component being determined by the end use. Refractory oxide materials usual in the art are amorphous or crystalline forms of alumina, silica, and combinations thereof. These oxide materials can have some intrinsic catalytic activity but often only provide the support on which active metal compounds are held. Generally, the thermal stability, low surface area, and poor mechanical strength have all hindered the commercial exploitation of certain metal oxide-supported catalyst systems. The intrinsic activity of hydrogenation metals-on-catalyst is superior to alumina-based catalysts. Catalyst synthesis will attempt to harness the intrinsic activity of various metals and remedy the deficiencies that currently plague low metal loading and thermal instability by using mixed oxides.

Use of ultra-deep desulfurization of liquid hydrocarbon fuels such as gasoline, diesel fuel, and jet fuel to satisfy new environmental regulations and fuel cell applications is receiving increased attention and will continue to receive attention. Conventional hydrodesulfurization (HDS) technology is difficult and costly to use to remove sulfur compounds from liquid hydrocarbon fuels to levels suitable to match environmental regulations. Several nonhydrodesulfurization-based desulfurization technologies for use with liquid fuels have been initiated. These technologies include (i) *biodesulfurization* (Ranson and Rivas, 2008) and (ii) adsorptive desulfurization (Song et al., 2010). Both of these technologies will continue to receive attention and have a high potential for incorporating into future refineries. In fact, the biodesulfurization technology is likely to see application during microbial enhanced oil recovery processes.

The current trend in hydroprocessing is the treatment of viscous sour feeds that contain compounds such as sulfur, nitrogen, aromatic derivatives, iron, and other undesirable components. These compounds pose significant problems with catalyst poisoning; however, developments are keeping pace with increased demand. In light of growing demand for ultra-low-sulfur diesel (ULSD), light cycle oil hydrotreating is receiving much attention. Feeds such as these are typically high in heavy metals, which will require additional unit modifications and/or the installation of guard beds/reactors.

In fact, the Topsøe ultra-low-sulfur diesel process yields a maximum of distillates by applying deep thermal conversion of the vacuum residue feedstock and by vacuum flashing the cracked residue (Egebjeng et al., 2011). The process is a hydrotreating process that combines a high-activity catalyst and state-of-the-art reactor internals and can be applied over a very wide range of reactor pressures.

In addition, hydrotreating feedstocks prior to sending the feedstock to the fluid catalytic cracking units is another important focus and will continue to be important or even increase in importance. Many fluid catalytic cracking unit incorporate pretreaters (in the form of hydrotreating the feedstocks or guard beds/reactors) to meet the sulfur requirements of the naphtha (and gasoline) products. Installation of such reactor units will necessarily increase as viscous feedstocks are incorporated into gas oils (fed to the fluid catalytic cracking unit) or become the sole feedstock for the catalytic cracking unit. Proved technology is available to remove sulfur, metal, and asphaltene content while converting an important part of the feedstock to lighter quality products. This technology will improve, and operations will be varied to upgrade heavy oil, extra heavy oil, tar sand bitumen, deasphalter bottoms, and biofeedstocks.

Thus, with the increasing focus to reduce sulfur content in fuels, the role of *desulfurization* in the refinery becomes more and more important. Currently, the process of choice is the hydrotreater, in which hydrogen is added to the fuel to remove the sulfur from the fuel. Some hydrogen may be lost to reduce the octane number of the fuel, which is undesirable. Because of the increased attention for fuel desulfurization, various new process concepts are being developed with various claims of efficiency and effectiveness.

The major developments in three main routes to desulfurization are advanced hydrotreating (new catalysts, catalytic distillation, and processing at mild conditions), reactive adsorption (type of adsorbent used and process design), and oxidative desulfurization (catalyst and process design).

The demand for low-sulfur transportation fuels requires that refiners evaluate the many different options for reaching the target. Selection of catalyst types is one of the important decisions and depends on (i) the feedstock, (ii) the operation conditions, (iii) the availability of hydrogen, and (iv) the desired properties of the product that may play a role (Tippet et al., 1999).

Besides the issues related to the legislative drive for removing sulfur, the refiners will be faced with a growing demand for diesel fuel, which may be met by producing fewer low-value products such as heating oil. This can be done by converting high-molecular-weight fractions by hydrocracking or mild hydrocracking (hydrotreating) processes, or one may adopt upgrading processes, for example, for light cycle oil. The latter alternative will require innovative technology that not only removes refractory sulfur species in the presence of high amounts of nitrogen but also performs a certain degree of ring opening to reach a reasonable product cetane number. The innovative might well involve (in part) the use of guard beds or reactor or the use of scavengers (such as calcium oxide, alone or supported on the catalyst) that will remove sulfur and coke formers during the hydrotreating process.

Catalyst development will accelerate, including catalysts used for pretreating feedstocks to the fluid catalytic cracking unit that will eliminate much of the need for naphtha treating (Topsøe et al., 2004). The catalysts will have multiple functions, such as (i) optimized hydrodesulfurization and (ii) minimized hydrogen consumption. The addition of metals such as iron, tungsten, niobium, boron, and phosphorus to catalyst compositions and the use of unsupported nanoparticles will increase and begin (if not already beginning) commercial utilization.

The challenges that the refining industry is facing in regard to hydrodesulfurization processes in the next two decades requite for major developments within hydroprocessing catalyst technology (Lautenschlager Moro, 2003). Areas such as (i) catalyst supports, (ii) catalyst morphology, and (iii) reaction pathways will continue to provide new opportunities for the development of improved commercial hydrotreating catalysts. Breakthrough technology is essential if catalysts are to be developed that are able to exhibit high activity to produce the desire products.

Briefly, the two desulfurization processes used for fuel purification (desulfurization) are (i) sweetening and (ii) hydrotreating. Sweetening is effective only against mercaptans, which are the predominant species in low-boiling naphtha. Hydrotreating is effective against all sulfur species and is more widely used.

In the sweetening process, a light naphtha stream is washed with amine to remove hydrogen sulfide and then reacted with caustic, which promotes the conversion of mercaptans to disulfides.

$$R-SH \rightarrow RSSR$$

The disulfides can subsequently be extracted and removed in what is referred to as extractive sweetening.

In the hydrotreating process, the feedstock is reacted with hydrogen, in the presence of a solid catalyst. The hydrogen removes sulfur by conversion to hydrogen sulfide, which is subsequently separated and removed from the reacted stream. As the reaction is favored by both temperature and pressure, hydrotreaters are typically designed and operated at ~370°C (700°F) and 1000–2000 psi hydrogen. The lower ends of the ranges typically apply to naphtha desulfurization, while gas oil desulfurization requires a more severe operation.

Hydrogen is provided in the form of treating gas at a purity that is typically around on the order of 90% by volume although gas with as little as 60% by volume hydrogen is reputed to be used. Hydrogen is produced by catalytic reformers or hydrogen generation units (Parkash, 2003; Gary et al., 2007; Speight, 2014, 2017; Hsu and Robinson, 2017) and distributed to the hydrotreaters through a refinery-wide network.

In a hydrotreating unit, feedstock and treating gas are combined and brought to the reaction temperature and pressure, prior to entering the reactor. The reactor is a vessel preloaded with solid catalyst, which promotes the reaction. The catalyst is slowly deactivated by the continuous exposure to high temperatures and by the formation of a coke layer on its surface. Refineries have to shut down the units periodically and regenerate or replace the catalyst.

The severity of operation of an existing unit can be increased by increasing the reaction temperature, but there is a negative impact on the catalyst life. The severity of operation can also be increased by increasing the catalyst volume of the unit. In this case, the typical solution is to add a second reactor identical to the existing one, doubling the reactor volume. The pressure of an existing unit cannot be changed to increase its severity, because the pressure is related to material of construction and thickness of metal surfaces. If higher pressure is required, the typical solution is to install a new unit and use the existing one for a less severe service.

One new technology is the use of *adsorption by metal oxides* in which the oxides react either by physical adsorption or by chemical adsorption insofar as adsorption followed by chemical reaction is promoted. The major distinction of this type of process from conventional hydrotreating is that the sulfur in the sulfur-containing compounds adsorbs to the catalyst after the feedstock-hydrogen mixture interacts with the catalyst. The catalyst does need to be regenerated constantly.

Another option involves *sulfur oxidization* in which a feedstock and water emulsion is reacted with hydrogen peroxide (or another oxidizer) to convert the sulfur in sulfur-containing compounds to sulfones. The sulfones are separated from the hydrocarbon derivatives for postprocessing. The major advantages of this new technology include low reactor temperatures and pressures, short residence time, no emissions, and no hydrogen requirement. The technology preferentially treats dibenzothiophene derivatives, one of streams that are most difficult to desulfurize.

One way to add to the supply of ultra-low-sulfur fuels is to turn to a nonoil-based diesel fuel. The *Fischer-Tropsch process*, for example, can be used to convert natural gas to a synthetic, sulfur-free diesel fuel. Commercial viability of gas-to-liquid projects depends (in addition to capital costs) on the market for viscous feedstock products and possible price premiums for gas-to-liquid fuels as well as the value of any by-products.

A second way to avoid desulfurization is with *biodiesel* made from vegetable oil or animal fats. Although other processes are available, most biodiesel is made with a base-catalyzed reaction. In the process, fat or oil is reacted with an alcohol, such as methanol, in the presence of a catalyst to produce glycerin and methyl esters or biodiesel. The methanol is charged in excess to assist in quick conversion and recovered for reuse. The catalyst, usually sodium or potassium hydroxide, is mixed with the methanol. Biodiesel is a strong solvent and can dissolve paint and deposits left in fuel lines by crude oil-based diesel, sometimes leading to engine problems. Biodiesel also freezes at a higher temperature than crude oil-based diesel.

4.2 Bioconversion

Refiners are being continually challenged to produce products with ever-decreasing levels of sulfur. At the same time, the supplies of light, low-sulfur crude oil that favor distillate production are limited and even decreasing. Generally, the sulfur content of crude oil continues to rise with the accompanying decrease in API gravity and an increase in the proportion of nonvolatile material in the feedstock. These factors require the viscous feedstocks to be processed more severely to produce naphtha and kerosene for use in transportation fuels. Thus, many refineries are now configured for maximum naphtha production that also includes increasingly processing highly aromatic distillate by-products, such as light cycle oil, for the additional feedstock to produce more distillate.

At this point, it is pertinent that a brief review of the potential methods for fuel polishing should be introduced. Biocatalyst desulfurization of the constituents of viscous feedstocks is one of a number of possible modes of applying biologically based processing to the needs of the refining industry in terms of processing and spill cleanup (McFarland et al., 1998; Setti et al., 1999). In addition, *Mycobacterium goodie* has been found to desulfurize benzothiophene (Li et al., 2005). The desulfurization product was identified as α-hydroxystyrene. This strain appeared to have the ability to remove organic sulfur from a broad range of sulfur species in naphtha. When straight-run naphtha containing various organic sulfur compounds was treated with immobilized cells of *M. goodie* for 24h at 40°C (104°F), the total sulfur content significantly decreased, from 227 to 71 ppm at 40°C. Furthermore, when immobilized cells were incubated at 40°C (104°F) with *M. goodie*, the sulfur content of the naphtha decreased from 275 to 54ppm in two consecutive reactions.

A dibenzothiophene-degrading bacterial strain, *Nocardia* sp., was able to convert dibenzothiophene to 2-hydroxybiphenyl as the end metabolite through a sulfur-specific pathway (Chang et al., 1998). Other organic sulfur compounds, such as thiophene derivatives, thiazole derivatives, sulfides, and disulfides were also desulfurized by *Nocardia* sp. When a sample in which dibenzothiophene was dissolved in hexadecane and treated with growing cells, the dibenzothiophene was desulfurized in ~80h.

The soil-isolated strain microbe identified as *R. erythropolis* can efficiently desulfurize benzonaphthothiophene (Yu et al., 2006). The desulfurization product was α-hydroxy-ß-phenyl-naphthalene. Resting cells were able to desulfurize diesel oil (total organic sulfur, 259 ppm) after hydrodesulfurization, and the sulfur content of diesel oil was reduced by 94.5% after 24 h at 30°C (86°F). Biodesulfurization of viscous feedstocks was also investigated, and after 72 h at 30°C (86°F), 62.3% of the total sulfur content in Fushun crude oil (initial total sulfur content, 3210 ppm) and 47.2% of the sulfur in Sudanese crude oil (initial total sulfur, 1237 ppm) were removed (see also Abbad-Andaloussi et al., 2003).

The recovery of heavy crude oil, facilitated by microorganisms, was suggested in the 1920s and received growing interest in the 1980s as microbial enhanced oil recovery. However, such projects have been slowed to get under way although in situ biosurfactant and biopolymer applications continue to garner interest (Van Hamme et al., 2003). In fact, studies have been carried out on biological methods of removing heavy metals such as nickel and vanadium from viscous feedstock distillate fractions, coal-derived liquid shale, bitumen, and synthetic fuels. However, further characterization on the biochemical mechanisms and bioprocessing issues involved in viscous feedstock upgrading are required in order to develop reliable biological processes.

For upgrading options, the use of microbes has to show a competitive advantage of enzyme over the tried-and-true chemical methods prevalent in the industry. Currently, the range of reactions using microbes is large but is usually related to production of bioactive compounds or precursors. But the door is not closed, and the issues of biodesulfurization and bioupgrading remain open for the challenge of bulk viscous feedstock processing. These drawbacks limit the applicability of this technology to specialty chemicals and steer it away from bulk viscous feedstock processing.

Biodesulfurization is, therefore, another technology to remove sulfur from the feedstock. However, several factors may limit the application of this technology, however. Many ancillary processes novel to viscous feedstock refining would be needed, including a biocatalyst fermenter to regenerate the bacteria. The process is also sensitive to environmental conditions such as sterilization, temperature, and residence time of the biocatalyst. Finally, the process requires the existing hydrotreater to continue in operation to provide a lower sulfur feedstock to the unit and is more costly than conventional hydrotreating. Nevertheless, the limiting factors should not stop the investigations of the concept, and work should be continued with success in mind.

Once the concept has been proved on the scale that a refiner would require, the successful microbial technology will most probably involve a genetically modified bacterial strain for (i) upgrading distillates and other viscous feedstock fractions in refineries; (ii) upgrading viscous feedstocks upstream; and (iii) dealing with environmental problems that ace industry, especially in areas related to spillage of viscous feedstocks and products. These developments are part of a wider trend to use bioprocessing to make products and do many of the tasks that are accomplished currently by conventional chemical processing. If commercialized for refineries, however, biologically based approaches will be at scales and with economic impacts beyond anything previously seen in industry.

In addition, the successful biodesulfurization process will, most likely, be based on naturally occurring aerobic bacteria that can remove organically bound sulfur in heterocyclic compounds without degrading the fuel value of the hydrocarbon matrix. Because of the susceptibility of bacteria to heat, the process will need to operate at temperatures and pressures close to ambient and also use air to promote sulfur removal from the feedstock.

4.3 Biofeedstocks

There are also biofeedstock issues that will become relevant in the refinery of the future, not the least of which will be the incorporation of biomass or first-generation biofeedstocks (i.e., feedstocks produced from biomass) into refinery hydrotreaters. For example, there are already reports of (i) refining extracted bio-oils being combined with refinery streams for hydroprocessing, (ii) hydrogenation of animal fats to produce a high-cetane diesel-range product, and (iii) hydrogenation of palm oil. However, there are issues related to quality of hydrotreated vegetable oils in terms of high paraffin content, low filter plugging points, and low density that must and will be resolved. Resolution of such issues will lead to recommendations on the means by which biofeedstocks can be (or will be) incorporated into existing hydrotreating units based on process efficiency and economic considerations.

References

Abbad-Andaloussi, S., Warzywoda, M., Monot, F., 2003. Rev. Inst. Fr. Pétrol. 58 (4), 505–513.

Feedstock evaluation and composition. Ancheyta, J., Speight, J.G. (Eds.), 2007. Hydroprocessing of Heavy Oils and Residua. CRC Press, Taylor & Francis Group, Boca Raton, FL.

Carlson, C.S., Langer, A.E., Stewart, J., Hill, R.M., 1958. Ind. Eng. Chem. 50, 1067.

Chang, J.H., Rhee, S.K., Chang, Y.K., Chang, H.N., 1998. Biotechnol. Prog. 14 (6), 851–855.

DeCroocq, D., 1984. Catalytic Cracking of Heavy Crude Oil Hydrocarbons. Editions Technip, Paris.

Dolbear, G.E., 1998. Hydrocracking: reactions, catalysts, and processes. In: Speight, J.G. (Ed.), Crude Oil Chemistry and Refining. Taylor & Francis, Washington, DC. (Chapter 7).

Egebjeng, R., Knudsen, K., Grennfelt, E., 2011. Bigger is better: industrial-scale production of renewable diesel. In: Proceedings of the NPRA Annual Meeting, San Antonio, TX, March, National Petrochemical & Refiners Association, Washington, DC.

Furimsky, E., 1998. Selection of catalysts and reactors for hydroprocessing. Appl. Catal. A Gen. 171, 177–206.

Gary, J.G., Handwerk, G.E., Kaiser, M.J., 2007. Crude Oil Refining: Technology and Economics, fifth ed. CRC Press, Taylor & Francis Group, Boca Raton, FL.

Hsu, C.S., Robinson, P.R. (Eds.), 2017. Handbook of Petroleum Technology. Springer International Publishing AG, Cham.

Hunter, M.G., Vivas, A.H., Jensen, L.S., Low, G.G., 2010. Partial Conversion Hydrocracking Process and Apparatus. United States Patent 7,763,218, July 27.

Khan, M.R., Patmore, D.J., 1998. Heavy oil upgrading process. In: Speight, J.G. (Ed.), Crude Oil Chemistry and Refining. Taylor & Francis, Washington, DC. (Chapter 6).

Kim, J.-W., Hanson, F.V., 1998. Catalytic hydrotreatment of PR spring bitumen over an HDM catalyst. Energy 23 (3), 221–229.

Langer, A.W., Stewart, J., Thompson, C.E., White, H.T., Hill, R.M., 1961. Ind. Eng. Chem. 53, 27.

Langer, A.W., Stewart, J., Thompson, C.E., White, H.T., Hill, R.M., 1962. Ind. Eng. Chem. Proc. Design and Development 1, 309.

Lautenschlager Moro, L.F., 2003. Process technology in the crude oil refining industry current situation and future trends. Comput. Chem. Eng. 27, 1303–1305.

Lee, D.K., Yoon, W.Y., Woo, S.I., 1996. Hydrotreatment of an atmospheric residual oil over the dispersed cobalt and molybdenum catalysts in a carbon expanded-bed reactor. Fuel 75 (10), 1186–1192.

Li, F., Xu, P., Feng, J., Meng, L., Zheng, Y., Luo, L., Ma, C., 2005. Appl. Environ. Microbiol. 71 (1), 276–281.

Maples, R.E., 2000. Crude Oil Refinery Process Economics, second ed. PennWell Corporation, Tulsa, OK.

McFarland, B.L., Boron, D.J., Deever, W., Meyer, J.A., Johnson, A.R., Atlas, R.M., 1998. Crit. Rev. Microbiol. 24, 99–147.

Menoufy, M.F., Ahmed, H.S., Bentiha, M.A., Sayed, M.A., 2014. A comparative study on hydrocracking and hydrovisbreaking combination for heavy vacuum residue conversion. Fuel 119, 106–110.

Parkash, S., 2003. Refining Processes Handbook. Gulf Professional Publishing, Elsevier, Amsterdam.

Rana, M.S., Ancheyta, J., Maity, S.K., Marroquin, 2008. Comparison between refinery processes for heavy oil upgrading: a future fuel demand. Int. J. Oil, Gas and Coal Technology 1 (3), 250–282.

Ranson, I., Rivas, C.M., 2008. Biodesulfurization of Hydrocarbons. United States Patent 7,338,795, March 4.

Reynolds, J.G., 1998. Metals and heteroatoms in heavy crude oils. In: Speight, J.G. (Ed.), Crude Oil Chemistry and Refining. Taylor & Francis Publishers, Washington, DC. (Chapter 3).

Scherzer, J., Gruia, A.J., 1996. Hydrocracking Science and Technology. Marcel Dekker, New York.

Setti, L., Farinelli, P., Di Martino, S., Frassinetti, S., Lanzarini, G., Pifferia, P.G., 1999. Appl. Microbiol. Biotechnol. 52, 111–117.

Song, C., Ma, X., Watanabe, S., Sun, F., 2010. Oxidatively Regenerable Adsorbents for Sulfur Removal. United States Patent 7,731,837, June 8.

Speight, J.G., 2000. The Desulfurization of Heavy Oils and Residua, second ed. Marcel Dekker, New York.

Speight, J.G., 2011. The Refinery of the Future. Gulf Professional Publishing, Elsevier, Oxford.

Speight, J.G., 2013. Heavy and Extra Heavy Oil Upgrading Technologies. Gulf Professional Publishing, Elsevier, Oxford.

Speight, J.G., 2014. The Chemistry and Technology of Petroleum, fifth ed. CRC Press, Taylor & Francis Group, Boca Raton, FL.

Speight, J.G., 2015. Handbook of Petroleum Product Analysis, second ed. John Wiley & Sons, Hoboken, NJ.

Speight, J.G., 2017. Handbook of Petroleum Refining. CRC Press, Taylor & Francs Group, Boca Raton, FL.

Tippet, T., Knudsen, K.G., Cooper, B.C. 1999. Proceedings of the NPRA Annual Meeting, Paper AM-99-06.

Topsøe, H., Egeberg, R.G., Knudsen, K.G., 2004. Preprint. Div. Fuel Chem. Am. Chem. Soc. 49 (2), 568.

Van Hamme, J.D., Singh, A., Ward, O.P., 2003. Microbiol. Mol. Biol. Rev. 67 (4), 503–549.

Vernon, L.W., Jacobs, F.E., Bauman, R.F., 1984. Process for Converting Petroleum Residuals. United States Patent 4,425,224, January 10.

Villasana, Y., Luis-Luis, M., Méndez, F.J., Labrador, H., Brito, J.L., 2015. Upgrading and hydrotreating of heavy oils and residua. In: Sharma, U.C., Prasad, R., Sivakumar, S. (Eds.), Energy Science and Technology Volume 3—Oil and Natural Gas, pp. 304–328. (Chapter 9).

Yu, B., Xu, P., Shi, Q., Ma, C., 2006. Appl. Environ. Microbiol. 72, 54–58.

Further Reading

Speight, J.G., 2012. Visbreaking: a technology of the past and the future. Sci. Iran. 19 (3), 569–573.

Chapter 11

Upgrading by Hydrocracking

1. Introduction

Hydrocracking is a refining technology that, like hydrotreating (Chapter 10), also falls under the general process umbrella of *hydroprocessing*. The outcome is the conversion of a variety of feedstocks to a range of products, and units to accomplish this goal can be found at various points in a refinery as opposed to hydrotreating (Chapter 12), which was designed primarily as a hydrodesulfurization process (Fig. 11.1) (Parkash, 2003; Gary et al., 2007; Speight, 2014, 2017; Hsu and Robinson, 2017).

Hydrocracking is a more recent process development compared with the older thermal cracking, visbreaking, and coking. The major differences between *hydrocracking* and *hydrotreating* are the time at which the feedstock remains at reaction temperature and the extent of the decomposition of the nonheteroatom constituents and products. The lower limits of hydrocracking conditions may overlap with the upper limits of hydrotreating conditions. Where the reaction conditions overlap, feedstocks to be hydrocracked will generally be exposed to the reactor temperature for longer periods, hence the reason why hydrocracking conditions may be referred to as (relatively) severe.

The use of hydrogen in thermal processes is perhaps the single most significant advancement in refining technology during the 20th century (Scherzer and Gruia, 1996; Dolbear, 1998; Parkash, 2003; Rana et al., 2007; Gary et al., 2007; Speight, 2014, 2017; Hsu and Robinson, 2017). The ability of refiners to cope with the renewed trend toward distillate production from heavier feedstocks with low atomic hydrogen/carbon ratios has created a renewed interest in hydrocracking. Without the required conversion units, heavier crude oils produce in lower yields of naphtha and middle distillate. To maintain current naphtha and middle distillate production levels, additional conversion capacity is required because of the differential in the amount of distillates produced from light crude oil and the distillate products produced from heavier feedstocks (Parkash, 2003; Gary et al., 2007; Speight, 2014, 2017; Hsu and Robinson, 2017).

The concept of hydrocracking allows the refiner to produce products having a lower molecular weight with a higher hydrogen content and a lower yield of coke. In summary, hydrocracking facilities add flexibility to refinery processing

Heavy Oil Recovery and Upgrading. https://doi.org/10.1016/B978-0-12-813025-4.00011-8

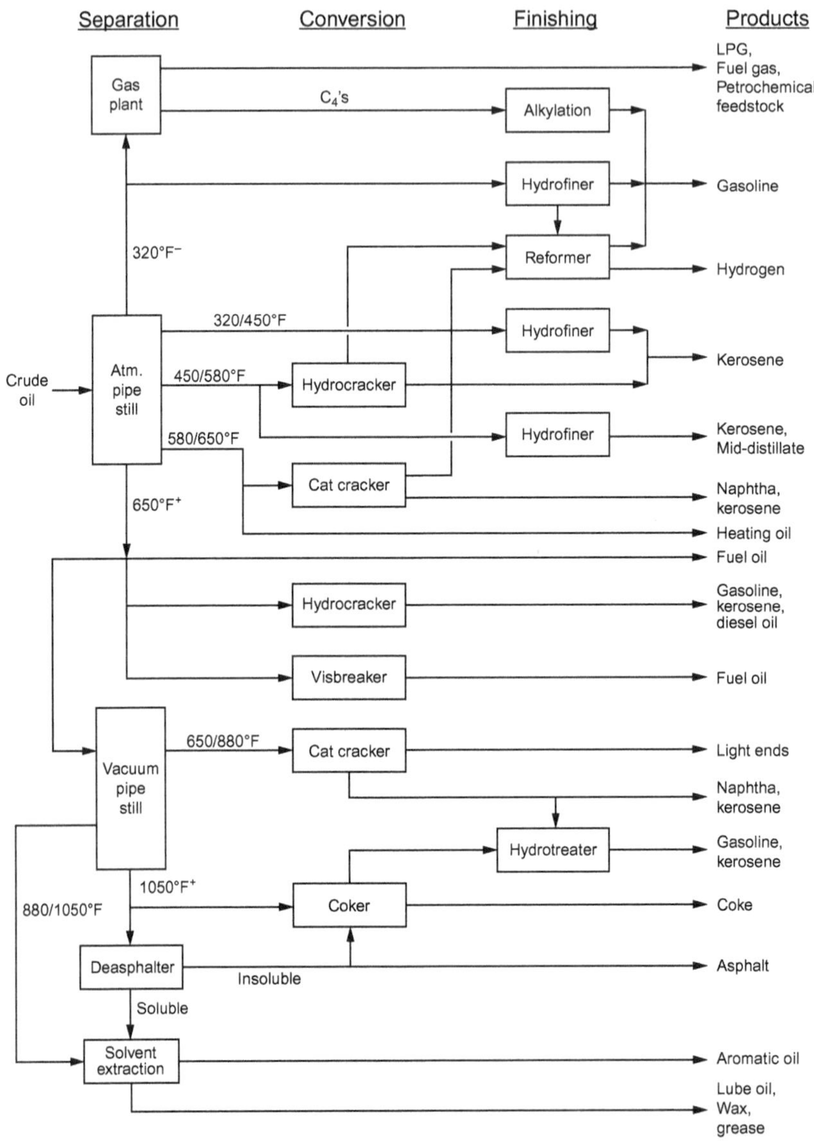

FIG. 11.1 Generalized refinery layout showing relative placement of the hydrocracking units.

and to the product slate. Hydrocracking is more severe than hydrotreating (Chapters 5 and 10) there being the intent, in hydrocracking processes, to convert the feedstock to lower-boiling products rather than to treat the feedstock for heteroatom and metal removal only. Process parameters emphasize the relatively severe nature of the hydrocracking process (Parkash, 2003; Gary et al., 2007; Speight, 2014, 2017; Hsu and Robinson, 2017).

The older hydrogenolysis type of hydrocracking practiced in Europe during and after World War II used tungsten sulfide (WS_2) or molybdenum sulfide (MoS) as catalysts. These processes required high reaction temperatures and operating pressures, sometimes in excess of approximately 3000 psi for continuous operation. The modern hydrocracking processes were initially developed for converting refractory feedstocks to naphtha and jet fuel; process and catalyst improvements and modifications have made it possible to yield products from gases and naphtha to furnace oils and catalytic cracking feedstocks.

Hydrocracking is an extremely versatile process that can be utilized in many different ways, and one of the advantages of hydrocracking is its ability to break down high-boiling aromatic stocks produced by catalytic cracking or coking. To take full advantage of hydrocracking, the process must be integrated in the refinery with other process units (Fig. 11.1). In naphtha production, for example, the hydrocracker product must be further processed in a catalytic reformer as it has a high naphthene content and relatively low octane number. The high naphthene content makes the hydrocracker naphtha an excellent feedstock for catalytic reforming, and good yields of high-octane-number gasoline can be obtained.

If high-molecular-weight crude oil fractions are *pyrolyzed*, that is, if no hydrogenation occurs, progressive cracking and condensation reactions generally lead to the final products. These products are usually (i) gaseous and low-boiling liquid compounds of high hydrogen content; (ii) liquid material of intermediate molecule weight with a hydrogen-carbon atomic ratio differing more or less from that of the original feedstock, depending on the method of operation; and (iii) material of high molecular weight, such as coke, possessing a lower hydrogen-carbon atomic ratio than the starting material. Highly aromatic or refractory recycle stocks or gas oils that contain varying proportions of highly condensed aromatic structures (e.g., naphthalene and phenanthrene) usually crack, in the absence of hydrogen, to yield intractable residues and coke.

An essential difference between *pyrolysis* (thermal decomposition, usually in the absence of any added agent) and *hydrogenolysis* (thermal decomposition in the presence of hydrogen and a catalyst or a hydrogen-donating solvent) of crude oil is that in pyrolysis, a certain amount of polymerized heavier products, like cracked residuum and coke, is always formed along with the light products, such as gas and naphtha. During hydrogenolysis (*destructive hydrogenation*), polymerization may be partly or even entirely prevented so that only light products are formed. The prevention of coke formation usually results in an increased distillate (e.g., naphtha) yield. The condensed type of molecule, such as naphthalene or phenanthrene, is one that is closely associated with the formation of coke, but in an atmosphere of hydrogen and in contact with catalysts, these condensed molecules are converted into lower-molecular-weight saturated compounds that boil within the gasoline range.

The mechanism of hydrocracking is basically similar to that of catalytic cracking but with concurrent hydrogenation. The catalyst assists in the production of carbonium ions via olefin derivative intermediates, and these intermediates are quickly hydrogenated under the high hydrogen partial pressures employed in hydrocracking. The rapid hydrogenation prevents adsorption of olefin derivatives on the catalyst and, hence, prevents their subsequent dehydrogenation, which ultimately leads to coke formation so that long onstream times can be obtained without the necessity of catalyst regeneration.

One of the most important reactions in hydrocracking is the partial hydrogenation of polycyclic aromatic derivatives followed by rupture of the saturated rings to form substituted monocyclic aromatic derivatives. The side chains may then be split off to give isoparaffins. It is desirable to avoid excessive hydrogenation activity of the catalyst so that the monocyclic aromatic derivatives become hydrogenated to naphthene derivatives; furthermore, repeated hydrogenation leads to loss in octane number, which increases the catalytic reforming required to process the hydrocracked naphtha.

Side chains of three or four carbon atoms are easily removed from an aromatic ring during catalytic cracking, but the reaction of aromatic rings with shorter side chains appears to be quite different. For example, hydrocracking single-ring aromatic derivatives containing four or more methyl groups produces largely isobutane and benzene. It may be that successive isomerization of the feedstock molecule adsorbed on the catalyst occurs until a four-carbon side chain is formed, which then breaks off to yield isobutane and benzene. Overall, coke formation is very low in hydrocracking since the secondary reactions and the formation of the precursors to coke are suppressed as the hydrogen pressure is increased.

Like many refinery processes, the problems encountered in hydrocracking viscous feedstocks can be directly equated to the amount of complex, higher-boiling constituents that may require pretreatment (Speight, 2000, 2014; Moulton and Erwin, 2005; Stratiev and Petkov, 2009; Bridjanian and Khadem Samimi, 2011). Processing these feedstocks is not merely a matter of applying know-how derived from refining *conventional* crude oils but requires a knowledge of composition (Parkash, 2003; Gary et al., 2007; Speight, 2014, 2017; Hsu and Robinson, 2017).

It is the physical and chemical composition of a feedstock that plays a large part not only in determining the nature of the products that arise from refining operations but also in determining the precise manner by which a particular feedstock should be processed. Furthermore, it is apparent that the conversion of viscous feedstocks requires new lines of thought to develop suitable processing scenarios (Babich and Moulijn, 2003). Indeed, the use of thermal (*carbon rejection*) processes and of hydrothermal (*hydrogen addition*) processes, which were inherent in the refineries designed to process lower-boiling feedstocks, has been a particular cause for concern. This has brought about the evolution of processing schemes that accommodate the heavier feedstocks (Khan and Patmore, 1998; Speight, 2014). As a point of reference, an example of the former option is the delayed coking process in which the feedstock is converted to overhead with the concurrent deposition of coke, for example, that used by Suncor, Inc., at their oil sand plant (Speight, 1990, 2014).

The hydrogen addition concept is illustrated by the hydrocracking process in which hydrogen is used in an attempt to *stabilize* the reactive fragments produced during the cracking, thereby decreasing their potential for recombination to heavier products and ultimately to coke. The choice of processing schemes for a given hydrocracking application depends upon the nature of the feedstock and the product requirements. The process can be simply illustrated as a single-stage or as a two-stage operation (Fig. 11.2) (Parkash, 2003; Gary et al., 2007; Hansen et al., 2010; Speight, 2014, 2017; Hsu and Robinson, 2017).

The single-stage process can be used to produce naphtha but is more often used to produce middle distillate from vacuum gas oil. The single-stage process may contain two reactors but without separation. On the other hand, in the two-stage processes, the undesirable products (such as hydrogen sulfide and ammonia) of the first stage are eliminated before the second stage. The most common reactor design for viscous feedstock hydroprocessing is the downflow and fixed-bed reactor. The single stage with recycle is a commonly used configuration. In this process option, the uncracked residual oil from the bottom of reaction product fractionation tower is recycled back into the single reactor for further cracking. However, for single-stage hydrocracking of viscous feedstocks, it is advisable that the feedstock should first be hydrotreated to remove ammonia and hydrogen sulfide or the catalyst used in the single reactor must be capable of both hydrotreating and hydrocracking.

The two-stage process was developed primarily to produce high yields of naphtha from straight-run gas oil, and the first stage may actually be a purification step to remove sulfur-containing (and nitrogen-containing) organic materials. In terms of sulfur removal, it appears that nonasphaltene sulfur

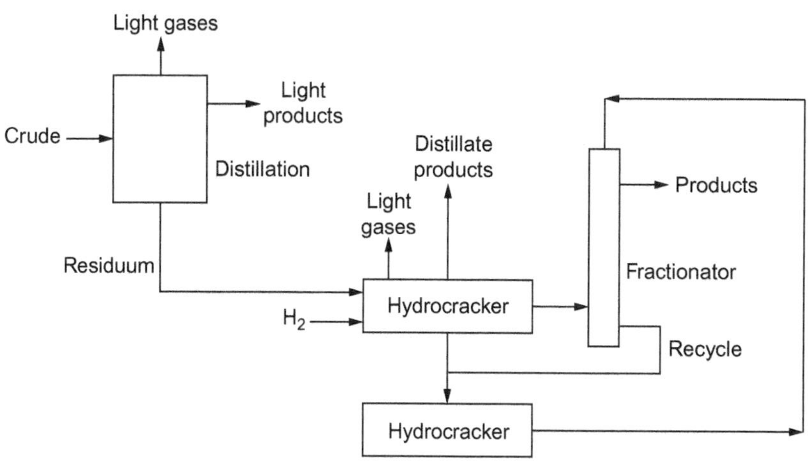

FIG. 11.2 A single-stage and two-stage hydrocracking unit.

may be removed before the more refractory asphaltene sulfur, thereby requiring thorough desulfurization (Speight, 2000, 2014; Ancheyta and Speight, 2007). This is a good reason for processes to use an extinction-recycling technique to maximize desulfurization and the yields of the desired product. Significant conversion of viscous feedstocks can be accomplished by hydrocracking at high severity. For some applications, the products boiling up to 340°C (650°F) can be blended to give the desired final product.

The two-stage hydrocracker process configuration uses two reactors, and the residual product from the bottom of reaction product fractionation tower is recycled back into the second reactor for further cracking. Since the first-stage reactor accomplishes both hydrotreating and hydrocracking, the second-stage reactor feedstock is virtually free of ammonia and hydrogen sulfide. This permits the use of high-performance noble metal (palladium and platinum) catalysts that are susceptible to poisoning by sulfur or nitrogen compounds. The process is best suited for large units and for processing difficult, high-nitrogen feedstocks. Almost all of the unconverted bottoms are recycled, and conversion levels of 95%–99% can be achieved. The two-stage hydrocracker is typically installed as a stand-alone unit and does not involve integration with any other units. In any multibed hydrocracking reactor, particulates can accumulate and affect the top-bed catalyst performance, even with sophisticated automatic backwash feedstock filters, which are considered necessary for all vacuum gas oil. Furthermore, the two-stage hydrocracker process configuration is best suited for large units and for processing difficult, high-nitrogen feedstocks and offers higher conversion of feedstock through two distinct reactor stages with intermediate fractionation. This helps to maximize the yield of high-quality products (kerosene and low-boiling gas oil). Other configurations—such as the single-stage design (once-through and recycle mode)—can be used to produce base oils for lubricating oil blending or as feedstocks for fluid catalytic cracking units and feedstocks for ethylene cracker feedstocks (Speight, 2000, 2014, 2017; Parkash, 2003; Ancheyta and Speight, 2007; Gary et al., 2007; Hsu and Robinson, 2017).

Hydrocracking is similar to catalytic cracking, with hydrogenation superimposed and with the reactions taking place either simultaneously or sequentially. While whole families of catalysts are required depending on feedstock available and the desired product slate or product character, the number of process stages is also important to catalyst choice. Generally, one of three options is utilized by the refinery. Thus, depending on the feedstock being processed and the type of plant design employed (*single stage* or *two stage*), flexibility can be provided to vary product distribution among the following principal end products.

Fundamentally, the trend toward lower API gravity feedstocks is related to an increase in the hydrogen/carbon atomic ratio of crude oils because of the higher content of residuum (Speight, 2000, 2014, 2017; Parkash, 2003; Ancheyta and Speight, 2007; Gary et al., 2007; Hsu and Robinson, 2017). This can be overcome by upgrading methods that lower this ratio by adding hydrogen, rejecting carbon, or using a combination of both methods. Though several technologies exist to upgrade viscous feedstocks, selection of the optimum process units is very much dependent on each refiner's needs and goals, with the market pull being the prime motivator. Furthermore, processing option systems to *dig deeper into the barrel* by converting more of the higher-boiling materials to distillable products should be not only cost-effective and reliable but also flexible. Hydrocracking adds that flexibility and offers the refiner a process that can handle varying feeds and operate under diverse process conditions. Utilizing different types of catalysts can modify the product slate produced, but reactor design and number of processing stages play a role in this flexibility.

Finally, a word about conversion measures for upgrading processes. Such measures are necessary for *any conversion process* but more particularly for hydrocracking processes where hydrogen management is an integral, and essential, part of process design.

The objective of any upgrading process is to convert viscous feedstock into marketable products by reducing their heteroatom (nitrogen, oxygen, and sulfur) and metal contents modifying their asphaltenic structures (reducing coke precursors) and converting the high-molecular-weight polar species large molecules into lower-molecular-weight and lower-boiling hydrocarbon products. Upgrading processes are evaluated on the basis of liquid yield (i.e., naphtha, distillate, and gas oil), heteroatom removal efficiency, feedstock conversion (FC), carbon mobilization (CM), and hydrogen utilization (HU), along with other process characteristics. Definition of RC, CM, and HU is as follows:

$$\mathrm{FC} = (\mathrm{Feedstock_{IN}} - \mathrm{Feedstock_{OUT}})/\mathrm{Feedstock_{IN}} \times 100$$

$$\mathrm{CM} = \mathrm{Carbon_{LIQUIDS}}/\mathrm{Carbon_{FEEDSTOCK}} \times 100$$

$$\mathrm{HU} = \mathrm{Hydrogen_{LIQUIDS}}/\mathrm{Hydrogen_{FEEDSTOCK}} \times 100$$

High carbon mobilization ($\mathrm{CM} < 100\%$) and high hydrogen utilization (HU) correspond to high feedstock conversion (RC) processes involving hydrogen addition such as hydrocracking. Since hydrogen is added, hydrogen utilization can be $>100\%$. These tasks can be achieved by using thermal and/or catalytic processes. Low carbon mobilization and low hydrogen utilization correspond to low feedstock conversion such as coking (carbon rejection) processes (Chapter 8). Maximum efficiency from an upgrading process can be obtained by maximizing the liquid yield and its quality by minimizing the gas (C_1–C_4) yield, simultaneously. Under these operating conditions, the hydrogen consumption would be the most efficient, that is, hydrogen is consumed to increase the liquid yield and its quality. Several process optimization models can be formulated if the reaction kinetics is known.

2. Process Chemistry and Parameters

Hydrocracking has become an indispensable processing technology to modern crude oil refining and petrochemical industry due to its flexibility to feedstocks and product scheme and high-quality products. Particularly, high-quality naphtha, jet fuel, diesel, and lube base oil can be produced through this technology. The hydrocracker provides a better balance of naphtha and distillates, improves gasoline yield and octane quality, and can supplement the fluid catalytic cracker to upgrade viscous feedstocks. In the hydrocracker, light fuel oil is converted into lower-boiling products under a high hydrogen pressure and over a hot catalyst bed—the main products are naphtha, jet fuel, and diesel oil.

In terms of process parameters and variables, the focus is typically on (i) process flow, (ii) feedstock properties, (iii) reaction temperature, (iv) hydrogen partial pressure, (v) gas recycle, and (vi) coke formation. For the viscous feedstocks, which will increase in amounts in terms of hydrocracking feedstocks, reactor designs focus on online catalyst addition and withdrawal.

2.1 Chemistry

The objective of the hydrocracking process is to convert higher-molecular-weight feedstocks into high-quality, lower-molecular-weight products such as naphtha and kerosene from which liquid fuels can be produced. The choice of processing schemes for a given hydrocracking application depends upon the nature of the feedstock and the product requirements (Rashid, 2007; Speight, 2011a, 2014). A two-stage process is typically employed. In the first (pretreat) step, polyaromatic compounds are saturated, and organic nitrogen and sulfur are converted to ammonia and hydrogen sulfide. The organic nitrogen contained in the feedstock would otherwise inhibit the activity of the cracking catalyst. In the second (cracking) step, higher-molecular-weight hydrocarbon molecules are preferentially cracked over an acidic metal-containing hydrocracking catalyst. The product yields and product properties are determined by the feedstock, the cracking catalyst selectivity, and the process conditions.

A particular feature of the hydrocracking process, as compared with its alternatives, is its flexibility with respect to product production and the relatively high quality of the products. On the whole, hydrocracking can handle a wider range of feedstock than catalytic cracking, although the latter process has seen some recent catalyst developments that narrowed the gap. There are also examples where hydrocracking is complementary rather than alternative to the other conversion process; an example, cycle oils, which cannot be recycled to extinction in the catalytic cracker, can be processed in the hydrocracker.

2.2 Process Configuration

The simplest form of the hydrocracking process is the *single-stage* process in which the layout of the reactor section generally resembles that of hydrotreating unit. This configuration finds application in cases where only a moderate degree of conversion (say 60% or less) is required. It may well apply to process where the feedstock is pretreated prior to introduction to a fluid catalytic cracking unit. Another form of hydrocracking process for heavier feedstocks is a *two-stage* operation (Speight, 2000, 2014; Ancheyta and Speight, 2007). Generally, the first stage of the two-stage unit resembles a *single-stage once-through* (SSOT) unit. This configuration uses only one reactor, and any uncracked nonvolatile material from the bottom of the distillation tower is not recycled for further cracking. For single-stage hydrocracking, either the feedstock must first be hydrotreated to remove ammonia and hydrogen sulfide or the catalyst used in the single reactor must be capable of both hydrotreating and hydrocracking. This flow scheme has been very popular since it can be used to maximize the yield of transportation fuels and is an attempt to combat the adverse effect of ammonia and nitrogen compounds on catalyst activity. Similarly, the *series-flow* version of the multistage hydrocracker has also been developed. The two-stage flow scheme has been very popular since it maximizes the yield of transportation fuels and has the flexibility to produce naphtha and kerosene to meet seasonal swings in demand for fuels.

Thus, the two-stage hydrocracker consists of two reactor stages together with a product distillation section. The choice of catalyst in each reaction stage depends on the product slate required and the character of the feedstock. In general, however, the first-stage catalyst is designed to remove nitrogen and high-molecular-weight aromatic derivatives from raw crude oil stocks. The second-stage catalyst carries out a selective hydrocracking reaction on the cleaner oil produced in the first stage. Both reactor stages have similar process flow schemes. The feedstock is combined with a preheated mixture of makeup hydrogen and hydrogen-rich recycled gas and heated to reactor inlet temperature via a feed-effluent exchanger and a reactor charge heater. The reactor charge heater design philosophy is based on many years of safe operation with such two-phase furnaces. The feed-effluent exchangers take advantage of special high-pressure exchanger design features developed by Chevron engineers to give leak-free end closures. From the charge heater, the partially vaporized feedstock enters to the top of the reactor. The catalyst is loaded in separate beds in the reactor with facilities between the beds for quenching the reaction mix and ensuring good flow distribution through the catalyst.

The reactor effluent is cooled through a variety of heat exchangers including the feed-effluent exchanger and one or more air coolers. Deaerated condensate is injected into the first-stage reactor effluent before the final air cooler in order to remove ammonia and some of the hydrogen sulfide. This prevents solid ammonium bisulfide from depositing in the system. A body of expertise in the field of material selection for hydrocracker cooling trains is quite important for proper design.

The reactor effluent leaving the air cooler is separated into hydrogen-rich recycled gas, a sour water stream, and a hydrocarbon liquid stream in the high-pressure separator. The sour water effluent stream is often then sent to a plant for ammonia recovery and for purification, so water can be recycled back to the hydrocracker. The hydrocarbon-rich stream is pressure reduced and fed to the distillation section after light products are flashed off in a low-pressure separator. The hydrogen-rich gas stream from the high-pressure separator is recycled back to the reactor feedstock by using a recycle compressor. Sometimes with sour feeds, the first-stage recycled gas is scrubbed with an amine system to remove hydrogen. If the sulfur content of the feedstock is high, this option can improve the performance of the catalyst and result in less costly materials of construction.

The distillation section consists of a hydrogen sulfide (H_2S) stripper and a recycle splitter. This latter column separates the product into the desired cuts. The column bottom stream is recycled back to the second-stage feed. The recycle cut point is changed depending on the light products needed. It can be as low as 160°C (320°F) if naphtha production is maximized (for aromatic derivatives) or as high as 380°C (720°F) if a low pour point diesel is needed. Between these two extremes, a recycle cut point of 260–285°C (500–550°F) results in high yields of high smoke point, low freeze point jet fuel.

In the *series-flow* configuration, the principal difference is the elimination of first-stage cooling and gas/liquid separation and the ammonia removal step. The effluent from the first stage is mixed with more recycled gas and routed direct to the inlet of the second reactor. In contrast with the amorphous catalyst of the two-stage process, the second reactor in series flow generally has a zeolite catalyst, based on crystalline silica-alumina. AS in the two-stage process, material not converted to the product boiling range is recycled from the fractionation section.

A *single-stage once-through* (SSOT) unit resembles the first stage of the two-stage configuration. This type of hydrocracker usually requires the least capital investment. The feedstock is not completely converted to lower-boiling products.

A *single-stage recycle* (SSREC) unit converts a viscous feedstock completely into lower-boiling products with a flow scheme resembling the second stage of the two-stage plant. Such a unit maximizes the yield of naphtha, jet fuel, or diesel depending on the recycle cut point used in the distillation section. This type of unit is more economical than the more complex two-stage unit when plant design capacity is less than approximately 10,000–15,000 bbl/day. Commercial SSREC plants have operated to produce low pour point diesel fuel from waxy Middle Fast vacuum gas oils.

Building on the theme of *one-* or *two-stage* hydrocracking, the *once-through partial conversion* (OTPC) concept evolved. This concept offers the means to convert vacuum gas oil feedstock into high-quality gasoline, jet fuel, and diesel products by a partial conversion operation. The advantage is lower initial capital investment and also lower utility consumption than a plant designed for total conversion. Because total conversion of the higher-molecular-weight compounds in the feedstock is not required, once-through hydrocracking can be carried out at lower temperatures and in most cases at lower hydrogen partial pressures than in recycle hydrocracking, where total conversion of the feedstock is normally an objective.

The purpose of mild hydrocracking is to convert vacuum gas oil to low-sulfur distillates at operating conditions consistent with those for hydrotreating equipment (Parkash, 2003; Gary et al., 2007; Speight, 2014, 2017; Hsu and Robinson, 2017). Full conversion of the feedstock does not occur in the mild hydrocracking process. Typically, the process yields conversions of 20%–60%. The products obtained through mild hydrocracking are high-quality, low-sulfur/nitrogen diesel and unconverted vacuum gas oil fractions. The vacuum gas oil fraction is desirable as fluid catalytic cracking feedstock due to its high hydrogen content and reduced sulfur and nitrogen levels. The product properties of the fractions depend on the feedstock characteristics and the process operating conditions.

Most often, mild hydrocracking units are redesigns of existing hydrotreating vacuum gas oil process units. The process employs a single reactor and operates on a once-through basis, designed to partially convert the vacuum gas oil into low-sulfur naphtha or distillate. The feedstock to a mild hydrocracking unit is mostly vacuum gas oil but can also be other viscous feedstock. Catalysts used in this type of unit are multipurpose in that they not only perform the hydrotreating functions of desulfurization and denitrogenation but also convert the fuel oil into lower-boiling middistillates. The catalysts are mildly acidic, usually consisting of cobalt or nickel oxide combined with molybdenum or tungsten oxide, supported on amorphous silica-alumina or mildly acidic zeolite.

The process operates under temperature conditions of 350–440°C (660–835°F) and pressures of 450–1500psi. The hydrogen partial pressure has the greatest effect on the mild hydrocracking process. Higher pressures result in higher reaction rates and increased catalyst stability. Lower pressures facilitate deactivation of the catalyst due to the fact that the reactive coke precursors are not hydrogenated quickly enough to prevent coke formation on the catalyst. Reactor pressure is dependent on the available pressure of the hydrogen gas, and to compensate for varying pressures, the reactor temperature can be adjusted to achieve similar results.

The *recycle hydrocracking unit* is designed to operate at hydrogen partial pressures from approximately 1200 to 2300psi depending on the type of feedstock being processed. Hydrogen partial pressure is set in the design in part not only depending on required catalyst cycle length but also to enable the catalyst to convert high-molecular-weight polynuclear aromatic and naphthene compounds that must be hydrogenated before they can be cracked. Hydrogen partial pressure also affects properties of the hydrocracked products that depend on hydrogen uptake, such as jet fuel aromatic derivative content and smoke point and diesel cetane number. In general, the higher the feedstock end point, the higher the required hydrogen partial pressure necessary to achieve satisfactory performance of the plant.

Once-through partial conversion hydrocracking of a given feedstock may be carried out at hydrogen partial pressures significantly lower than required for recycle total conversion hydrocracking. The potential higher catalyst deactivation rates experienced at lower hydrogen partial pressures can be offset by using higher activity catalysts and designing the plant for lower catalyst space velocities. Catalyst deactivation is also reduced by the elimination of the recycle stream. The lower capital cost resulting from the reduction in plant operating pressure is much more significant than the increase resulting from the possible additional catalyst requirement and larger volume reactors. Additional capital cost savings from once-through hydrocracking result from the reduced overall required hydraulic capacity of the plant for a given fresh feedstock rate as a result of the elimination of a recycle oil stream. Hydraulic capacity at the same fresh feedstock rate is 30%–40% lower for a once-through plant compared with one designed for recycle.

Utility savings for a once-through versus recycle operation arise from lower pumping and compression costs as a result of the lower design pressure possible and also lower hydrogen consumption. Additional savings are realized as a result of the lower oil and gas circulation rates required, since recycle of oil from the fractionator bottoms is not necessary. Lower capital investment and operating costs are obvious advantages of once-through hydrocracking compared with a recycle design. This type of operation may be adaptable for use in an existing gas oil hydrotreater or in a viscous feedstock desulfurization unit. The change from hydrotreating to hydrocracking service will require some modifications and capital expenditure, but in most cases, these changes will be minimal.

One disadvantage of once-through hydrocracking compared with a recycle operation is a somewhat reduced flexibility for varying the ratio of naphtha to middle distillate that is produced. A greater quantity of naphtha can be produced by increasing conversion and production of jet fuel plus diesel can also be increased. But selectivity for higher-boiling products is also a function of conversion. Selectivity decreases as once-through conversion increases. If conversion is increased too much, the yield of desired product will decrease, accompanied by an increase in light ends and gas production. Higher yields of naphtha or jet fuel plus diesel are possible from a recycle than from a once-through operation. However, the fact that unconverted oil is produced by the plant is not necessarily a disadvantage. The unconverted oil produced by once-through hydrocracking is a high-quality, low-sulfur and low-nitrogen material that is an excellent feedstock stock for a fluid catalytic cracking unit or ethylene pyrolysis furnace or a source of high viscosity index lube oil base stock. The properties of the oil are a function of the degree of conversion and other plant operating conditions.

Middle distillate products made by once-through hydrocracking are generally higher in aromatic derivative content of poorer burning quality than those produced by recycle hydrocracking. However, the quality is generally better than produced by catalytic cracking or from straight run. Middle distillate product quality improves as the degree of conversion increases and as the hydrogen partial pressure is increased.

The hydrocracking process employs high-activity catalysts that produce a significant yield of light products. Catalyst selectivity to middle distillate is a function of both the conversion level and operating temperature, with values in excess of 90% being reported in commercial operation. In addition to the increased hydrocracking activity of the catalyst, percentage desulfurization and denitrogenation at start-of-run conditions are also substantially increased. End of cycle is reached when product sulfur has risen to the level achieved in conventional vacuum gas oil hydrodesulfurization process.

An important consideration, however, is that commercial hydrocracking units are often limited by design constraints of an existing vacuum gas oil hydrotreating units. Thus, the proper choice of catalyst(s) is critical when searching for optimum performance. Typical commercial distillate hydrocracking (DHC) catalysts contain both the hydrogenation (metal) and cracking (acid sites) functions required for service in existing desulfurization units.

2.3 Process Reactors

Like hydrotreating units, many different hydrocracking unit designs are marketed; they all work along the same principle—all processes use the reaction of hydrogen with the hydrocarbon feedstock to produce hydrogen sulfide and a desulfurized hydrocarbon product (Parkash, 2003; Gary et al., 2007; Speight, 2014; Hsu and Robinson, 2017; Speight, 2017). This also invokes the concept of different process types (usually names after the types of reactor employed). Each reactor has its own particular merits and, providing the choice is made according to the feedstock, has an excellent chance of producing the desired products.

The reaction temperature is typically on the order of 290–445°C (550–850°F) with a hydrogen gas pressure at the upper end of the range 150–3000 psi—the higher temperature maximizes cracking reactions. In some designs, the feedstock is heated and then mixed with the hydrogen rather than the option of passing moderately heated feedstock (i.e., feedstock not at the full reactor temperature) and moderately heated hydrogen (i.e., hydrogen not at the full reactor temperature) into the reactor. The gas mixture is led over a catalyst bed of metal oxides (most often cobalt or molybdenum oxides on different metal carriers). The catalysts help the hydrogen to react with sulfur and nitrogen to form hydrogen sulfides (H2S) and ammonia. The reactor effluent is then cooled, and the oil feedstock and gas mixture are then separated in a stripper column. Part of the stripped gas may be recycled to the reactor. Generally, with few exceptions, hydrocracking reactors fall into the same groups as those used for hydrotreating: (i) downflow fixed-bed reactor, (ii) upflow expanded-bed reactor, and (iii) demetallization reactor.

2.3.1 General Aspects

Fixed-bed designs have suffered from (i) mechanical inadequacy when used for the heavier feedstocks and (ii) short catalyst lives—6 months or less—even though large catalyst volumes are used (LHSV typically of 0.5–1.5). Refiners will attempt to overcome these shortcomings by innovative designs, allowing better feedstock flow and catalyst utilization or online catalyst removal. For example, processes that focus on onstream catalyst replacement (OCR) in which a lead, moving-bed reactor is used to demetallized viscous feedstock ahead of the fixed-bed hydrocracking reactors will continue to be of interest and see further use.

The use of ebullating-bed technologies was first introduced in the 1960s in an attempt to overcome problems of catalyst aging and poor distribution in fixed-bed designs. Hydrogen and feedstock enter at the bottom of the reactor, thereby expanding the catalyst bed. Although catalyst performance can be kept constant because catalyst can be replaced online, the ebullition results in a back mixed reactor; therefore, desulfurization and hydroconversion are less than obtainable in a fixed-bed unit. Currently, in order to limit coking, most commercial ebullating-bed units operate in the 70%–85% desulfurization range and 50%–70% *v*/v, viscous feedstock conversion.

Development work will continue, and ebullating-bed units will see more use and have a greater impact of viscous feedstock conversion operations. Improvements such as (i) second-generation catalyst technology, which will allow higher conversion to a stable product; (ii) catalyst rejuvenation, which allows spent catalyst to be reused to a greater extent than current operations allow; and (iii) new reactor designs raising single train-size greater throughput.

Slurry-phase hydrocracking of viscous feedstocks and the latest development of dispersed catalysts present strong indications that such technologies will play a role in future refineries (Motaghi et al., 2011). Catalysts for slurry-phase hydrocracking of viscous feedstocks have undergone two development phases: (i) heterogeneous solid powder catalysts, which have low catalytic activity and will produce a large number of solid particles in bottom oil making the catalyst difficult to dispose and utilize, and (ii) homogeneous dispersed catalysts, which are divided into water-soluble dispersed catalysts and oil-soluble dispersed catalysts (Zhang et al., 2007). Dispersed catalysts are highly dispersed and have greater surface-area-to-volume ratio. Therefore, they show high catalytic activity and good performance. They are desirable catalysts for slurry-phase hydrocracking of viscous feedstocks and will be used more prominently in future hydrocracking operations (Bhattacharyya and Mezza, 2010; Bhattacharyya et al., 2011; Motaghi et al., 2011).

In spite of the numerous process design variations (Parkash, 2003; Gary et al., 2007; Speight, 2014, 2017; Hsu and Robinson, 2017), process design innovations and hardware innovations will continue. Although conventional (high-pressure) hydrocracking will still be used to address the need to produce more gasoline and diesel, moderate-pressure hydrocracking (where control of the reaction chemistry is more possible), is used to introduce a mild hydrocracking unit upstream of the fluid catalytic cracking unit to maintain that unit at full capacity. Alternatively, more refiners will turn to two-stage recycle (TSR) hydrocracking and reverse-staging configurations.

2.3.2 Design Improvements

Operating severity increases with heavier feedstocks, so catalyst loading and high-performance reactor internals are becoming even more important to get the most out of the catalyst. Furthermore, the importance of feedstock filtering and particulate and metal trapping must be included within the design of the technology.

Guard reactors are used in hydrocracking processes to protect catalysts in subsequent reactors, including precious metal hydrocracking catalysts, from contaminants in feedstocks that are not previously hydrotreated. If a hydrocracking unit is designed to accept feedstocks that have not been hydrotreated previously, a guard reactor precedes the first hydrocracking reactor in the process flow. The purpose of the guard reactor is to reduce the metal content of the feedstock to the hydrocracking units and to convert organic sulfur and nitrogen compounds to hydrogen sulfide and ammonia and to (Table 11.1).

Guard reactors also serve the purpose of reducing the metal content in the feedstock to the hydrocracking units. Catalysts used in guard reactors are usually modified hydrotreating catalysts such as Co-Mo on silica-alumina. Most of the metals in the feedstock will be deposited on the catalyst in the guard reactor, and there will be a substantial reduction in the coke yield, resulting from a feedstock that is low in metals and carbon-forming precursors (Parkash, 2003; Gary et al., 2007; Speight, 2014, 2017; Hsu and Robinson, 2017). Guard reactors are now (and will continue to be) recognized as being essential in the hydrocracking processes to protect catalysts in subsequent reactors from contaminants in feedstocks that are not previously hydrotreated.

Reactor internals are a critical part of any hydroprocessing technology package. Maximum catalytic performance can now be achieved through good vapor/liquid mixing and distribution across the catalyst bed. Good distribution results in maximum catalyst utilization from start to end of run, maximizing product selectivity. The reactor internals that are used in a hydrocracker can play an important role in determining its capital cost, as they can improve the volume available for catalyst and therefore can help to reduce the reactor size. Reactor internals also have a major influence on its performance as they can drive its onstream factor, utilization, cycle length, product yields, and even product quality. In fact, reactor internals should be custom designed and fabricated for each application to ensure maximum performance over the desired range of operating conditions—off-the-shelf units are not as popular as they were in the past. Consequently, reactor internals are being designed to (i) distribute gas and liquid uniformly, (ii) minimize thermal instabilities, and (iii) maximize reactor catalyst inventory and catalyst utilization. This is now recognized as being critical for applications with more stringent product specifications, such as ultra-low-sulfur diesel production.

Reactor temperature controls and safety instrumented systems are now used in the hydrocracking process found in many refineries. A significant change in feedstock flow rate can result in a temperature runaway due to rapid change of

TABLE 11.1 Feedstock Contaminants That Affect Hydrocracking Processes[a]

Contaminants	Effect on Catalyst	Mitigation	Process
Sulfur	Catalyst fouling	Hydrodesulfurization	Hydroprocessing
	Deactivation of active sites		
Nitrogen	Adsorption of basic nitrogen	Hydrodemetallization	Hydroprocessing
	Destruction of active sites		
Metals	Fouling of active sites	Demetallization	Demet, Met-X
	Fouling of pores		
Particulate matter	Deactivation of active sites	Filter/pretreatment	Clay filtration/ guard bed
	Pore plugging		
Coke precursors	Formation of coke	Remove asphaltene constituents	Mild hydrocracking/ hydroprocessing
	Catalyst fouling	Remove resin constituents	
	Deactivation of active sites		
	Pore plugging		

[a]*Also applicable to catalytic cracking processes.*

the hydrogen-to-hydrocarbon ratio. Significant changes in the feedstock flow rate are the result of failures in feedstock flow controllers and feedstock pumps. The temperature rise in this scenario is moderately fast, but recovery is possible through automatic and manual adjustments of quench rates and readjustment of feedstock flow rates. Excessive temperature of the reactor feedstock can also result in temperature runaway. Excessive temperature of the reactor feedstock is possible as the result of a failure of temperature control in the charge heater such that maximum firing occurs.

Due to the highly exothermic reactions, coupled with the high apparent activation energy (a measure of the sensitivity for the activity to temperature changes) in the second stage of hydrocracking, advanced control for the hydrocracker reactor has become essential for maintaining a reliable hydrocracking operation. Poor control in the reactor is still often reported—even with the implementation of the advanced control mechanism—due to poor reactor internals and the problematic overall control scheme implemented for the hydrocracking reactor.

The conventional control scheme using heater firing control has been found to be insufficient for the hydrocracker—attempting to maintain a constant reactor inlet temperature (a precursor for temperature runaway) with poor internals and deficient overall reactor control scheme leads to further operating problems. With a properly designed quench box and an improved reactor control scheme, the bed inlet temperature spread can be reduced to <3°C (5°F), and the temperature variation at the reactor inlet can be reduced to <1.5°C (3°F) with a low capital investment.

Reactor internals are a critical part of hydrocracking technology. Maximum catalytic performance can only be achieved through efficient vapor/liquid mixing and distribution across the catalyst bed.

Options have also arisen for taking advantage of reduced postriser cracking to improve product yields. A combination of higher reactor temperature, higher cat/oil ratio, higher feedstock rate, and/or poorer quality feedstock is typically employed. The types of modifications to the catalyst are also necessary (on a unit and feedstock basis) to complement these designs, particularly for revamp applications.

The concept of single-stage and double-stage hydroprocessing operations has long been recognized (Speight, 2011a, 2014) and, with the advent of heavier feedstocks as part of the reactor feed, is becoming a necessity. However, the performance of a hydroprocessing operation is determined not only by the number of stages and the catalyst loaded but also by the design of the reactor internals. Antifouling trays are employed to reduce pressure drop buildup and maximize unit run length. Increased efficiency and technologies to counteract fouling are particularly important when operators are processing increasingly difficult feedstocks, such as viscous feedstocks (Kunnas et al., 2010).

Most designs in commercial operation process a variety of feedstocks in multiple flow schemes. The higher-molecular-weight feedstock components in the blends cover coker gas oil, vacuum gas oil, and deasphalted oil. These feedstock blends are processed in hydrocracking units with various objectives and flow schemes including single-stage once-through and recycle as well as two-stage and separate hydrotreating flow schemes.

In a single-stage unit, the feedstock is first hydrotreated, and the reactor effluent goes through gas/liquid separation. The hydrocracking reactor effluent goes through gas/liquid separation and then to the fractionator. This configuration uses recycle for optimization of yield and processing severity. The flow scheme is designed to assure that high-quality product is produced in terms of ultra-low-sulfur diesel with a cetane index over 55.

In hydrocrackers that process vacuum gas oils or other feeds with similar boiling ranges, the typical once-through conversion exceeds 60% *w*/w. If the unconverted oil is recycled, the overall conversion can exceed 95% w/w. As with mild hydrocracking, the unconverted bottoms are high-value oils, which usually are sent to fluid catalytic cracking units, lube plants, or olefin derivative plants. For heavier feedstocks, conversions are much lower, especially in fixed-bed units.

The limitation in fixed-bed reactor is the catalyst bed poisoning with time. Catalyst life depends on the rate of deactivation by coke and metal deposits and sintering of the active phases. Information regarding the activity, selectivity, and deactivation of the individual catalyst is, therefore, highly desirable for optimizing reactor loading in the multiple catalyst systems. These parameters have to be optimum for hydroprocessing operation, which can be achieved by properly matching the type of reactor and catalyst, along with properties of viscous feeds. However, in ebullated-bed units, the conversion of 565°C+ (1050°F+) residue can exceed 60% *w*/w (Kressmann et al., 2000).

Many hydrocrackers in the refineries operate in mild hydrocracking mode. For these units, the main objectives are to obtain a certain minimum conversion and to meet specific product properties such as sulfur content, density, and cetane number. Typical pressures are in the 850–1560 psig range. Typical conversion is 10%–20% for lower pressure units and 30%–50% for higher pressure units. The demand for refined products has increased to the extent that refiners desire larger hydrocracking reactors that can operate at higher pressures with design conditions that are even more severe. New feedstock diets for refineries utilize more difficult to "crack" crudes; demand for reactors that can withstand higher temperatures (>450°C, >840°F) and higher hydrogen partial pressures likewise is increasing. Under such severe processing conditions, reactor vessels are constructed from low alloy chromium (Cr)-molybdenum (Mo) steel of various grades.

As an example of mild hydrocracking (and hydrotreating), the DOW Isotherming process provides a means to upgrade gas oil, deasphalted oil, and fluid catalytic cracking cycle oils. The products are low-sulfur, low-nitrogen fluid catalytic cracking feed, low-sulfur gasoline-kerosene-type products, and low-nitrogen fuels and/or downstream feedstocks. In the process, fresh feedstock, after heat exchange, is combined with recycle product and hydrogen in a mixer internal to the reactor. The liquid feedstock with soluble hydrogen is fed to the Isotherming reactor/bed one where partial desulfurization, removal of metals, or even mild hydrocracking occurs. The stream is resaturated with additional hydrogen in a second mixer and fed to the second Isotherming reactor/bed where further mild hydrocracking takes place. The treated oil from the second Isotherming bed may then be fed to additional Isotherming beds to achieve the desired level of conversion. Treated oil from the last bed is recycled back to the inlet of bed one. This recycle stream delivers recycled hydrogen to the reactors and also acts as a heat sink; thus, a nearly isothermal reactor operation is achieved.

Removal of sulfur and nitrogen from feedstocks is affected by the chemical composition of supported molybdate catalysts. Cobalt and nickel, when added to these catalysts, have a promoting effect on these reactions. However, the relative rates always follow the same trend; that is, the hydrodesulfurization is the fastest reaction followed by hydrodenitrogenation. The hydrodenitrogenation (HDN) reaction rates are dependent on the catalyst, the concentration of the reactive species (organic nitrogen and hydrogen), and reaction temperature. Ni/Mo catalysts generally have higher denitrogenation activity than the Co/Mo catalysts. The affinity of the organic nitrogen compounds for the catalytic sites has been reported as extremely high, and if the organic nitrogen concentration exceeds 20ppm, as much as 90% of the catalyst sites may be occupied.

Ammonia inhibits the hydrocracking catalyst activity, requiring higher operating temperatures to achieve target conversion, but this has been found to result in better liquid yields than would be the case if no ammonia was present. There is no interstage product separation in single-stage or series-flow operation. Two-stage hydrocrackers employ interstage product separation that removes hydrogen sulfide and ammonia, and the second-stage hydrocracking catalyst is exposed to lower levels of these gases. However, caution is needed as some two-stage hydrocracker designs do result in high levels of hydrogen sulfide in the second stage.

Aromatic derivative saturation can be achieved by the use of a pretreatment step. The main objectives for the pretreat catalyst are (i) the removal of organic nitrogen and sulfur from the feedstock to levels that allow the second-stage catalysts to better perform the hydrocracking function and (ii) the initiation of the sequence of hydrocracking reactions by saturation of the aromatic compounds in the feedstock. Pretreated catalysts must have adequate activity to achieve both objectives within the operating limits of the unit (hydrogen partial pressure, temperature, and LHSV).

A hydrocracking step is necessary for obtaining high-quality fuels from Fischer-Tropsch wax. Isomerization is an important reaction that takes place during the hydroconversion process. The amount and the type of the isomers in the produced fuels heavily influence both cold flow properties and cetane number (Gamba et al., 2010). The reaction temperature and the space velocity exhibit a considerable impact on the conversion and on the isoparaffin to n-paraffin ratio and on the boiling range of the hydrocracking product. At severe reactor conditions, increased hydrogenolysis activity of the base metal catalyst will be observed, resulting in an increased methane formation.

The quality of the diesel fuel product shows a strong dependence on the conversion achieved from hydrocracking. Cetane numbers of up to 80 can be recoded under low-conversion conditions due to lower isoparaffin to n-paraffin ratios of the hydrocracking products, whereas high-conversion conditions lead to an increase in the isoparaffin to n-paraffin ratio, which leads to improved cold flow properties with cold filter plugging points down to or below −27°C (−17°F) (Olschar et al., 2007).

The shortage of good source of lubricant base oil is creating a high demand for feedstocks and units that produce such materials. Generally, very high viscosity index base oils are produced by severe hydrocracking of vacuum distillate fractions in a fuel hydrocracker. Another route for very high viscosity index base oil production is hydroisomerization, in which slack wax, produced by solvent dewaxing, is converted to branched chain paraffins under hydroisomerization conditions with an appropriate catalyst. The conversion rate of the process is approximately 80–85.

Mild hydrocracking technology and catalysts enable a medium conversion of heavier feedstocks to lower-boiling products. The Topsøe mild hydrocracking technology portfolio includes staged partial conversion and back-end shift. Staged partial conversion is a new pretreatment technology designed to produce low-sulfur fluid catalytic cracking feedstock to allow ultra-low-sulfur diesel production without gasoline post treatment (Section 2.2). Back-end shift technology significantly reduces the distillation temperature by selective hydrocracking of the viscous hydrocarbon derivatives present in the back-end distillation with high diesel yields and moderate hydrogen consumption.

For the viscous feedstocks, which will increase in amount in terms of hydrocracking feedstocks, reactor designs will continue to focus on online catalyst addition and withdrawal. Fixed-bed designs have suffered from (i) mechanical inadequacy when used for the heavier feedstocks and (ii) short catalyst lives—6 months or less—even though large catalyst volumes are used (LHSV typically in the order of 0.5–1.5). Refiners will attempt to overcome these shortcomings by innovative designs, allowing better feedstock flow and catalyst utilization or online catalyst removal. For example, the onstream catalyst replacement (OCR) process, in which a lead, moving-bed reactor is used to demetallized viscous feedstock ahead of the fixed-bed hydrocracking reactors, has seen some success. But whether this will be adequate for continuous hydrocracking viscous feedstock remains a question. The OCR process enables refiners to process viscous feedstocks with >400 ppm metals (Ni + V) or to achieve deeper desulfurization. The life of the downstream catalyst is improved substantially, and problems from pressure drop buildup are reduced.

2.4 Feedstocks and Hydrogen Requirements

Processing heavier feedstocks poses many challenges to hydrocracking operations in terms of producing the desired products from various feedstocks. It is also important to consider various other aspects of the hydrocracking unit design to maximize catalyst utilization and pretreatment of the feedstock using a hydrotreating unit (Morel et al., 2009; Dziabala et al., 2011).

The hydrocracking process operating parameters (400–815°C, 750–1500°F, and 1000–2000 psi) depend on the nature of the feedstock and the relative rates of the two competing reactions, hydrogenation and cracking. Under these conditions, viscous aromatic feedstock is converted into lower-boiling products under a wide range of very high pressures and high temperatures. When the feedstock has a high paraffinic content, the primary function of hydrogen is to prevent the formation of polycyclic aromatic compounds. Hydrogenation also serves to convert sulfur and nitrogen compounds present in the feedstock to hydrogen sulfide and ammonia.

It is estimated that the refining industry will require >14 trillion cubic feet of hydrogen to meet processing requirements between 2010 and 2030 with the Asia Pacific region and the Middle East region representing approximately 40% of global requirements (Vauk et al., 2008). In any such scenario, hydrogen management is key (Patel et al., 2005; Luckwal and Mandal, 2009). The chemical hydrogen consumption for the production of ultra-low-sulfur diesel (ULSD) will be significantly higher than the consumption for low-sulfur diesel production (Morel et al., 2009). Several factors influence the amount of the increased hydrogen consumption including (i) catalyst selection for low-sulfur diesel versus ultra-low-sulfur diesel, (ii) the amount of cracked material boiling over 325°C (620°F) in the feedstock, (iii) unit pressure, and (iv) the flow rate (LHSV).

Many refineries chose to use Co/Mo catalyst in the production of low-sulfur diesel. The Co/Mo catalyst provided an effective route to sulfur removal in lower pressure units with the added benefit of lower hydrogen consumption since the Co/Mo catalysts have lower hydrogenation activity compared with the Ni/Mo catalyst. Moreover, the Co/Mo catalysts actually have higher activity for direct hydrodesulfurization and less organic nitrogen inhibition. Refiners with feedstock that contains a high amount of high-boiling cracked material currently using Co/Mo catalyst will experience a large increase in hydrogen consumption when switching to Ni/Mo catalysts for ULSD production.

Hydrogen consumption increases significantly when producing a 10 ppm sulfur product compared with production of a 500 ppm sulfur product. Increases in chemical hydrogen consumption from 40% to 100% can be expected. In addition, the higher consumption results in higher heat release. The increased heat release requires more quench gas (as much as 950 ft^3/bbl) to limit the temperature rise.

Because of the operating temperatures and the presence of hydrogen, the equipment must be capable of handling high-sulfur feedstocks and must be able to withstand the possibility of severe corrosion. When processing high-nitrogen feedstock, the ammonia and hydrogen sulfide form ammonium hydrosulfide (bisulfide), which causes serious corrosion at temperatures below the water dew point.

When hydrogen availability is depleted by new hydroprocessing demands, the refinery needs to be flexible, and alternative sources of hydrogen are needed (Vauk et al., 2008). Each refinery's configuration, crude slate, and new production requirements are different, so various hydrogen network optimization considerations must be considered (Patel et al., 2005; Luckwal and Mandal, 2009). Hydrogen network optimization is increasingly more important to maximizing higher-value transportation fuel production, minimizing clean fuel investments, and overall refinery profitability. Refiners must seek to unlock the hydrogen contained in refinery offgases, improve hydroprocessing network purities, and consider new hydrogen production options. The optimum solution for an overall refinery hydrogen network will continue to be complex and involve the screening of numerous options.

3. Catalysts

The hydrocracking process employs high-activity catalysts that produce a significant yield of light products. In addition to the increased hydrocracking activity of the catalyst, percentage desulfurization and denitrogenation at start-of-run conditions are also substantially increased. Thus, several types of catalysts are used in hydrocracking process, and the catalysts combine cracking activity and hydrogenation activity to achieve the conversion of specific feedstocks into desirable products. Hydrocracking reactions require a dual-function catalyst with high cracking and hydrogenation activities. The former activity (high cracking activity) is provided by an acidic support or by a clay support, whereas the hydrogenation activity is provided by metals on the support. The catalyst base, such as acid-treated clay, usually supplies the cracking function or alumina or silica-alumina that is used to support the hydrogenation function supplied by metals, such as nickel, tungsten, platinum, and palladium. These highly acid catalysts are very sensitive to nitrogen compounds in the feed, which break down the conditions of reaction to give ammonia and neutralize the acid sites. As many gas oils contain substantial amounts of nitrogen (up to approximately 2500ppm), a purification stage is frequently required. Denitrogenation and desulfurization can be carried out using cobalt-molybdenum or nickel-cobalt-molybdenum on alumina or silica-alumina.

Acid sites (crystalline zeolite, amorphous silica-alumina, and mixture of crystalline zeolite and amorphous oxides) provide cracking activity. Metals (noble metal such as palladium and platinum or nonnoble metal sulfides such as molybdenum, tungsten, cobalt, or nickel) provide hydrogenation-dehydrogenation activity. These metals catalyze the hydrogenation of feedstocks making them more reactive for cracking and heteroatom removal as well reducing the coke yield. Zeolite-based hydrocracking catalysts have the following advantages of (i) greater acidity resulting in greater cracking activity, (ii) better thermal/hydrothermal stability, (iii) better naphtha selectivity, (iv) better resistance to nitrogen and sulfur compounds, (v) low coke-forming tendency, and (vi) easy regenerability.

Palladium sulfide and promoted group VI sulfides (nickel-molybdenum or nickel-tungsten) provide the hydrogenation function. These active compositions saturate aromatic derivatives in the feed, saturate olefin derivatives formed in the cracking, and protect the catalysts from poisoning by coke. Zeolites or amorphous silica-alumina provides the cracking functions. The zeolites are usually type Y (faujasite) ion exchanged to replace sodium with hydrogen and make up 25%–50% of the catalysts. Pentasils (silicalite or ZSM-5) may be included in dewaxing catalysts.

Hydrocracking catalysts, such as nickel (5% by weight) on silica-alumina, work best on feedstocks that have been hydrofined to low nitrogen content and a low sulfur content. The nickel catalyst then operates well at 350–370°C (660–700°F) and a pressure of approximately 1500 psi to give good conversion of feedstock to lower-boiling liquid fractions with minimum saturation of single-ring aromatic derivatives and a high isoparaffin to *n*-paraffin ratio in the lower-molecular-weight paraffins.

The catalyst used in a single-stage process comprises a hydrogenation function in combination with a strong cracking function. Sulfided metals such as cobalt, molybdenum, and nickel provide the hydrogenation function. An acidic support, usually alumina, attends to the cracking function. Nitrogen compounds and ammonia produced by hydrogenation interfere with acidic activity of the catalyst. In the cases where high/full conversion is required, the reaction temperatures and run lengths of interest in conventional operation can no longer be adhered to. And moreover, conversion asymptotes out with increasing hydrogen pressure (Speight, 2014), so more hydrogen in the reactor is not the answer. In fact, it becomes necessary to switch to a different reactor bed system or to a multistage process, in which the cracking reaction mainly takes place in an additional reactor. The operating temperature influences reaction selectivity since the activation energy for hydrotreating reactions is much lower than the activating energy for the hydrocracking reaction. Therefore, raising the temperature in a viscous feedstock hydrotreater increases the extent of hydrocracking relative to hydrotreating, which also increases the hydrogen consumption.

Multistage reaction catalysts take advantage of staged reactions with different catalytic attributes much the same way that staged hydrotreating catalyst loading permits different reaction zones in a fixed-bed reactor vessel. The concept of staged reactions is not new to the refining industry, but its application to a circulating system such as the fluid catalytic cracking system is a step forward in catalyst technology. The multistage reaction catalyst platform uses existing catalyst technologies and combines two or more existing functionalities within the same catalyst particle. The location of the various stages can be specifically engineered to achieve maximum value for the refiner. This staging approach can be applied to allow processing of heavier feedstocks or to maximize specific product yields.

The transition metal sulfides such as molybdenum (Mo), cobalt (Co), and nickel (Ni) will remain and are still the industry favorites, because of their excellent hydrogenation, hydrodesulfurization (HDS), and hydrodenitrogenation (HDN) activities, as well as their availability and cost. However, greater attention will be paid to the size of the particles, pore volume and size distribution, pore diameter, and the shape of the particles to maximize utilization of the catalyst. Thus, the future challenge for refiners will be to obtain higher conversion of heavier and more refractive feedstocks and to make cleaner transportation fuels in larger quantities while reducing the emissions of greenhouse gases. One of the main constraints to meeting these objectives is that deeper conversion of heavier oil tends to result in reduced stability of the unconverted product, leading to higher fouling rates in the reactors and downstream equipment and thus reduced reliability of the process.

Moreover, the increasing importance of hydrodesulfurization (HDS) and hydrodenitrogenation (HDN) in crude oil processing in order to produce clean-burning fuels has led to a surge of research on the chemistry and engineering of heteroatom removal, with sulfur removal being the most prominent focus. Most of the earlier works are focused on (i) catalyst characterization by physical methods; (ii) low-pressure reaction studies of model compounds having relatively high reactivity; (iii) process development; or (iv) cobalt-molybdenum (Co-Mo) catalysts, nickel-molybdenum catalysts (Ni-Mo), or nickel-tungsten (Ni-W) catalysts supported on alumina, often doped by fluorine or phosphorus.

The need to develop catalysts that can carry out deep hydrodesulfurization and deep hydrodenitrogenation has become even more pressing in view of recent environmental regulations limiting the amount of sulfur and nitrogen emissions. The development of a new generation of catalysts to achieve this objective of low nitrogen content and a low sulfur content in the processing of different feedstocks presents an interesting challenge for catalyst development. Recent developments (from the catalyst and process aspects) have focused on improvement of (i) the desulfurization level, (ii) the denitrogenation level, (iii) the run length, and (iv) the conversion level of fixed-bed hydroprocessing units. These improvements include the development of a multiple-swing reactor system and the development of complex associations of guard-bed materials and catalysts with particle-size, activity, pore-size, and shape grading.

The deposition of coke and metals onto the catalyst diminishes the cracking activity of hydrocracking catalysts. Basic nitrogen plays a major role because of the susceptibility of such compounds for the catalyst and their predisposition to form coke (Speight, 2000, 2014). However, zeolite catalysts can operate in the presence of substantial concentrations of ammonia, in marked contrast to silica-alumina catalysts, which are strongly poisoned by ammonia. Similarly, sulfur-containing compounds in a feedstock adversely affect the noble metal hydrogenation component of hydrocracking catalysts. These compounds are hydrocracked to hydrogen sulfide, which converts the noble metal to the sulfide form. The extent of this conversion is a function of the hydrogen and hydrogen sulfide partial pressures.

Removal of sulfur from the feedstock results in a gradual increase in catalyst activity, returning almost to the original activity level. As with ammonia, the concentration of the hydrogen sulfide can be used to control precisely the activity of the catalyst. Nonnoble metal-loaded zeolite catalysts have an inherently different response to sulfur impurities since a minimum level of hydrogen sulfide is required to maintain the nickel-molybdenum and nickel-tungsten in the sulfide state.

Hydrodenitrogenation is more difficult to accomplish than hydrodesulfurization, but the relatively smaller amounts of nitrogen-containing compounds in conventional crude oil made this of little concern to refiners. However, the trend to heavier feedstocks in refinery operations, which are richer in nitrogen than the conventional feedstocks, has increased the awareness of refiners to the presence of nitrogen compounds in crude feedstocks. For the most part, however, hydrodesulfurization catalyst technology has been used to accomplish hydrodenitrogenation although such catalysts are not ideally suited to nitrogen removal. However, in recent years, the limitations of hydrodesulfurization catalysts when applied to hydrodenitrogenation have been recognized, and there has been the need to manufacture catalysts more specific to nitrogen removal.

The character of the hydrotreating processes is chemically very simple since they essentially involve removal of sulfur and nitrogen as hydrogen sulfide and ammonia, respectively:

$$S_{feedstock} + H_2 \rightarrow H_2S$$

$$2N_{feedstock} + 3H_2 \rightarrow 2NH_3$$

However, nitrogen is the most difficult contaminant to remove from feedstocks, and processing conditions are usually dictated by the requirements for nitrogen removal.

In general, any catalyst capable of participating in hydrogenation reactions may be used for hydrodesulfurization. The sulfides of hydrogenating metals are particularly used for hydrodesulfurization, and catalysts containing cobalt, molybdenum, nickel, and tungsten are widely used on a commercial basis. *Hydrotreating catalysts* are usually cobalt-molybdenum catalysts, and under the conditions whereby nitrogen removal is accomplished, desulfurization usually occurs and oxygen removal. Indeed, it is generally recognized that fullest activity of the hydrotreating catalyst is not reached until some interaction with the sulfur (from the feedstock) has occurred, with part of the catalyst metals converted to the sulfides. Too much interaction may of course lead to catalyst deactivation.

The reactions of hydrocracking require a dual-function catalyst with high cracking and hydrogenation activities. The catalyst base, such as acid-treated clay, usually supplies the cracking function or alumina or silica-alumina that is used to support the hydrogenation function supplied by metals, such as nickel, tungsten, platinum, and palladium. These highly acid catalysts are very sensitive to nitrogen compounds in the feed, which break down the conditions of reaction to give ammonia and neutralize the acid sites. As many viscous gas oils contain substantial amounts of nitrogen (up to approximately 2500 ppm), a purification stage is frequently required. Denitrogenation and desulfurization can be carried out using cobalt-molybdenum or nickel-cobalt-molybdenum on alumina or silica-alumina.

Hydrocracking catalysts typically contain separate hydrogenation and cracking functions. Palladium sulfide and promoted group VI sulfides (nickel-molybdenum or nickel-tungsten) provide the hydrogenation function. These active compositions saturate aromatic derivatives in the feed, saturate olefin derivatives formed in the cracking, and protect the catalysts from poisoning by coke. Zeolites or amorphous silica-alumina provides the cracking functions. The zeolites are usually type Y (faujasite) ion exchanged to replace sodium with hydrogen and make up 25%–50% of the catalysts. Pentasils (silicalite or ZSM-5) may be included in dewaxing catalysts.

Hydrocracking catalysts, such as nickel (5% by weight) on silica-alumina, work best on feedstocks that have been hydrofined to low nitrogen and sulfur levels. The nickel catalyst then operates well at 350–370°C (660–700°F) and a pressure of approximately 1500 psi to give good conversion of feedstock to lower-boiling liquid fractions with minimum saturation of single-ring aromatic derivatives and a high isoparaffin to *n*-paraffin ratio in the lower-molecular-weight paraffins.

The poisoning effect of nitrogen can be offset to a certain degree by operation at a higher temperature. However, the higher temperature tends to increase the production of material in the methane (CH_4) to butane (C_4H_{10}) range and decrease the operating stability of the catalyst so that it requires more frequent regeneration. Catalysts containing platinum or palladium (approximately 0.5% wet) on a zeolite base appear to be somewhat less sensitive to nitrogen than are nickel catalysts, and successful operation has been achieved with feedstocks containing 40 ppm nitrogen. This catalyst is also more tolerant of sulfur in the feed, which acts as a temporary poison, the catalyst recovering its activity when the sulfur content of the feedstock is reduced.

On such catalysts as nickel or tungsten sulfide on silica-alumina, isomerization does not appear to play any part in the reaction, as uncracked normal paraffins from the feedstock tend to retain their normal structure. Extensive splitting produces large amounts of low-molecular-weight (C3–C6) paraffins, and it appears that a primary reaction of paraffins is catalytic cracking followed by hydrogenation to form isoparaffins. With catalysts of higher hydrogenation activity, such as platinum on silica-alumina, direct isomerization occurs. The product distribution is also different, and the ratio of low- to intermediate-molecular-weight paraffins in the breakdown product is reduced.

In addition to the chemical nature of the catalyst, the physical structure of the catalyst is also important in determining the hydrogenation and cracking capabilities, particularly for viscous feedstocks. When gas oils and viscous feedstocks are used, the feedstock is present as liquids under the conditions of the reaction. Additional feedstock and the hydrogen must diffuse through this liquid before reaction can take place at the interior surfaces of the catalyst particle.

At high temperatures, reaction rates can be much higher than diffusion rates, and concentration gradients can develop within the catalyst particle. Therefore, the choice of catalyst porosity is an important parameter. When feedstocks are to be hydrocracked to liquefied crude oil gas and naphtha, pore diffusion effects are usually absent. High surface area (approximately 300 m^2/g) and low to moderate porosity (from 12 D pore diameter with crystalline acidic components to 50 D or more with amorphous materials) catalysts are used. With reactions involving high-molecular-weight feedstocks, pore diffusion can exert a large influence, and catalysts with large pore diameters are necessary for more efficient conversion.

Aromatic hydrogenation in crude oil refining may be carried out over supported metal or metal sulfide catalysts depending on the sulfur and nitrogen levels in the feedstock. For hydrorefining of feedstocks that contain appreciable concentrations of sulfur and nitrogen, sulfided nickel-molybdenum (Ni-Mo), nickel-tungsten (Ni-W), or cobalt-molybdenum (Co-Mo) on alumina (γ-A_2-O_3) catalysts is generally used, whereas supported noble metal catalysts have been used for sulfur- and nitrogen-free feedstocks. Catalysts containing noble metals on zeolite Y have been reported to be more sulfur-tolerant than those on other supports. Within the series of cobalt- or nickel-promoted group VI metal (Mo or W) sulfides supported on γ-Al_2O_3, the ranking for hydrogenation is

$$\mathrm{Ni-W > Ni-Mo > Co-Mo > Co-W}$$

Nickel-tungsten (Ni-W) and nickel-molybdenum (Ni-Mo) on Al_2O_3 catalysts are widely used to reduce sulfur, nitrogen, and aromatic derivative levels in crude oil fractions by hydrotreating.

Molybdenum sulfide (MoS_2), usually supported on alumina, is widely used in crude oil processes for hydrogenation reactions. It is a layered structure that can be made much more active by the addition of cobalt or nickel. When promoted with cobalt sulfide (CoS), making what is called *cobalt-moly* catalysts, it is widely used in hydrodesulfurization (HDS) processes. The nickel sulfide (NiS)-promoted version is used for hydrodenitrogenation (HDN) and hydrodesulfurization (HDS). The closely related tungsten compound (WS_2) is used in commercial hydrocracking catalysts. Other sulfides (iron sulfide, FeS; chromium sulfide, Cr_2S_3; and vanadium sulfide, V_2S_5) are also effective and used in some catalysts. A valuable alternative to the base metal sulfides is palladium sulfide (PdS). Although it is expensive, palladium sulfide forms the basis for several very active catalysts, while clay minerals that are also used as cracking catalysts for viscous feedstocks, especially for demetallization of viscous crude oil, are much cheaper.

The choice of hydrogenation catalyst depends on what the catalyst designer wishes to accomplish. In catalysts to make naphtha, for instance, vigorous cracking is needed to convert a large fraction of the feedstock to the kinds of molecules that will make a good gasoline blending stock. For this vigorous cracking, a vigorous hydrogenation component is needed. Since palladium is the most active catalyst for this, the extra expense is warranted. On the other hand, many refiners wish only to make acceptable diesel, a less demanding application. For this, the less expensive molybdenum sulfides are adequate.

The cracking reaction results from attack of a strong acid on a paraffinic chain to form a carbonium ion (carbocation, e.g., R^+) (Dolbear, 1998). Strong acids come in two fundamental types, Brønsted and Lewis acids. *Brønsted acids* are the familiar proton-containing acids; *Lewis acids* are a broader class including inorganic and organic species formed by positively charged centers. Both kinds have been identified on the surfaces of catalysts; sometimes, both kinds of sites occur on the same catalyst. The mixture of Brønsted and Lewis acids sometimes depends on the level of water in the system.

Examples of Brønsted acids are the familiar proton-containing species such as sulfuric acid (H_2SO_4). Acidity is provided by the very active hydrogen ion (H^+), which has a very high positive charge density. It seeks out centers of negative charge such as the pi electrons in aromatic centers. Such reactions are familiar to organic chemistry students, who are taught that bromination of aromatic derivatives takes place by attack of the bromonium ion (Br^+) on such a ring system. The proton in strong acid systems behaves in much the same way, adding to the pi electrons and then migrating to a site of high electron density on one of the carbon atoms. These acids all have high positive charge densities. Examples are aluminum chloride ($AlCl_3$) and the bromonium ion (Br^+). Such strong positive species have become known as Lewis acids. This class obviously includes proton acids, but the latter are usually designated Brønsted acids in honor of the Danish chemist J.N. Brønsted, who contributed greatly to the understanding of the thermodynamics of aqueous solutions.

In reactions with hydrocarbon derivatives, both Lewis and Brønsted acids can catalyze cracking reactions. For example, the proton in Brønsted acids can add to an olefin double bond to form a carbocation. Similarly, a Lewis acid can abstract a hydride from the corresponding paraffin to generate the same intermediate (Dolbear, 1998). Although these reactions are written to show identical intermediates in the two reactions, in real catalytic systems, the intermediates would be different. This is because the carbocations would probably be adsorbed on surface sites that would be different in the two kinds of catalysts.

Zeolites and amorphous silica-alumina provide the cracking function in hydrocracking catalysts. Both of these have similar chemistry at the molecular level, but the crystalline structure of the zeolites provides higher activities and controlled selectivity not found in the amorphous materials.

Chemists outside the catalyst field are often surprised that a solid can have strong acid properties. In fact, many solid materials have acid strength matching that of concentrated sulfuric acid. Some specific examples are (i) amorphous silica-alumina (SiO_2/Al_2O_3), (ii) zeolites, (iii) activated acid-leached clay minerals, (iv) aluminum chloride ($AlCl_3$) and many related metal chlorides, (v) amorphous silica magnesia compounds (SiO_2/MgO), (vi) chloride-promoted alumina ($Al_2O_3.Cl$), and (vii) phosphoric acid supported on silica gel (H_3PO_4/SiO_2). Each of these is applied in one or more commercial catalysts in the crude oil refining industry. For commercial hydrocracking catalysts, only zeolites and amorphous silica-alumina are used commercially.

Zeolites are highly porous crystals veined with submicroscopic channels that consist primarily of silicon, aluminum, and oxygen and host an assortment of other elements. The channels contain water (hence the bubbling at high temperatures), which can be eliminated by heating (combined with other treatments) without altering the crystal structure (Occelli and Robson, 1989). Typical naturally occurring zeolites include *analcite* (also called *analcime*) $Na(AlSi_2O_6)$, and *faujasite* $Na_2Ca(AlO_2)2(SiO_2)_4{\cdot}H_2O$ that is the structural analog of the synthetic *zeolite X* and *zeolite Y*. *Sodalite* $(Na_8[(Al_2O_2)_6(SiO_2)_6]Cl_2)$ contains the truncated octahedral structural unit known as the *sodalite cage* that is found in several zeolites. The corners of the faces of the cage are defined by either four or six Al/Si atoms, which are joined together through oxygen atoms. The zeolite structure is generated by joining *sodalite* cages through the four-Si/Al rings, so enclosing a cavity or *super cage* bounded by a cube of eight *sodalite* cages and readily accessible through the faces of that cube (*channels* or *pores*). Joining sodalite cages together through the six-Si/Al faces generates the structural frameworks of faujasite, zeolite X, and zeolite Y. In zeolites, the effective width of the pores is usually controlled by the nature of the cation (M^+ or M^{2+}).

Natural zeolites form hydrothermally (e.g., by the action of hot water on volcanic ash or lava), and synthetic zeolites can be made by mixing solutions of aluminates and silicates and maintaining the resulting gel at temperatures of 100°C (212°F) or higher for appropriate periods. *Zeolite A* can form at temperatures below 100°C (212°F), but most zeolite syntheses require hydrothermal conditions (typically 150°C/300°F at the appropriate pressure). The reaction mechanism appears to involve dissolution of the gel and precipitation as the crystalline zeolite and the identity of the zeolite produced depend on the composition of the solution. Aqueous alkali metal hydroxide solutions favor zeolites with relatively high aluminum contents, while the presence of organic molecules such as amines or alcohols favors highly siliceous zeolites such as *silicalite* or ZSM-5. Various tetra-alkyl ammonium cations favor the formation of certain specific zeolite structures and are known as *template* ions, although it should not be supposed that the channels and cages form simply by the wrapping of alumina-silica fragments around suitably shaped cations.

Zeolite catalysts have also found use in the refining industry during the last two decades. Like the silica-alumina catalysts, zeolites also consist of a framework of tetrahedra usually with a silicon atom or an aluminum atom at the center. The geometric characteristics of the zeolites are responsible for their special properties, which are particularly attractive to the refining industry (DeCroocq, 1984). Specific zeolite catalysts have shown up to 10,000 times more activity than the so-called conventional catalysts in specific cracking tests. The mordenite-type catalysts are particularly worthy of mention since they have shown up to 200 times greater activity for hexane cracking in the temperature range 360–400°C (680–750°F).

Other zeolite catalysts have also shown remarkable adaptability to the refining industry. For example, the resistance to deactivation of the type Y zeolite catalysts containing either noble or nonnoble metals is remarkable, and catalyst life of up to 7 years has been obtained commercially in processing gas oil in the Unicracking-JHC processes. Operating life depends on the nature of the feedstock, the severity of the operation, and the nature and extent of operational upsets. Gradual catalyst deactivation in commercial use is counteracted by incrementally raising the operating temperature to maintain the required conversion per pass. The more active a catalyst, the lower is the temperature required. When processing for naphtha, lower operating temperatures have the additional advantage that less of the feedstock is converted to isobutane.

Any given zeolite is distinguished from other zeolites by structural differences in its unit cell, which is a tetrahedral structure arranged in various combinations. Oxygen atoms establish the four vertices of each tetrahedron, which are bound to, and enclose, either a silicon (Si) or an aluminum (Al) atom. The vertex oxygen atoms are each shared by two tetrahedrons, so that every silicon atom or aluminum atom within the tetrahedral cage is bound to four neighboring caged atoms through an intervening oxygen. The number of aluminum atoms in a unit cell is always smaller than, or at most equal to, the number of silicon atoms because two aluminum atoms never share the same oxygen.

The aluminum is actually in the ionic form and can readily accommodate electrons donated from three of the bound oxygen atoms. The electron donated by the fourth oxygen imparts a negative, or anionic, charge to the aluminum atom. This negative charge is balanced by a cation from the alkali metal or the alkaline earth groups of the periodic table. Such cations are commonly sodium, potassium, calcium, or magnesium. These cations play a major role in many zeolite functions and help to attract polar molecules, such as water. However, the cations are not part of the zeolite framework and can be exchanged for other cations without any effect on crystal structure.

Zeolites provide the cracking function in many hydrocracking catalysts, as they do in fluid catalytic cracking catalysts. The zeolites are crystalline aluminosilicates, and in almost all commercial catalysts today, the zeolite used is faujasite. Pentasil zeolites, including silicalite and ZSM-5, are also used in some catalysts for their ability to crack long-chain paraffins selectively.

Typical levels are 25%–50% by wt zeolite in the catalysts, with the remainder being the hydrogenation component and a silica (SiO_2) or alumina (Al_2O_3) binder. Exact recipes are guarded as trade secrets.

Crystalline zeolite compounds provide a broad family of solid acid catalysts. The chemistry and structures of these solids are beyond the scope of this book. What is important here is that the zeolites are not acidic as crystallized. They must be converted to acidic forms by ion exchange processes. In the process of doing this conversion, the chemistry of the crystalline structure is often changed. This complication provides tools for controlling the catalytic properties, and much work has been done on understanding and applying these reactions as a way to make catalysts with higher activities and more desirable selectivity.

As an example, the zeolite faujasite crystallizes with the composition SiO_2 $(NaAlO_2)_x(H_2O)_y$. The ratio of silicon to aluminum, expressed here by the subscript x, can be varied in the crystallization from 1 to >10. What does not vary is the total number of silicon and aluminum atoms per unit cell, 192. For legal purposes to define certain composition of matter patents, zeolites with a ratio of 1:1.5 are called type X; those with ratio >1.5 are type Y.

Both silicon and aluminum in zeolites are found in tetrahedral oxide sites. The four oxides are shared with another silicon or aluminum (except that two aluminum ions are never found in adjacent, linked tetrahedral). Silicon with a plus four charge balances exactly half of the charge of the oxide ions it is linked to; since all of the oxygens are shared, silicon balances all of the charge around it and is electrically neutral. Aluminum, with three positive charges, leaves one charge unsatisfied. Sodium neutralizes this charge.

The sodium, as expected from its chemistry, is not linked to the oxides by covalent bonds as the silicon and aluminum are. The attraction is simply *ionic*, and sodium can be replaced by other cations by ion exchange processes. In extensive but rarely published experiments, virtually every metallic and organic cation has been exchanged into zeolites in studies by catalyst designers.

The most important ion exchanged for sodium is the proton. In the hydrogen ion form, faujasite zeolites are very strong acids, with strengths approaching that of oleum. Unfortunately, direct exchange using mineral acids such as hydrochloric acid is not practical. The acid tends to attack the silica-alumina network, in the same way that strong acids attack clays in the activation processes developed by Houdry. The technique adopted to avoid this problem is indirect exchange, beginning with exchange of ammonium ion for the sodium. When heated to a few hundred degrees, the ammonium decomposes, forming gaseous ammonia and leaving behind a proton:

$$R^- NH_4^+ \rightarrow R^- H^+ + NH_3 \uparrow$$

The step is accompanied by a variety of solid-state reactions that can change the zeolite structure in subtle but important ways. This chemistry and the related structural alterations have been described in many articles.

While zeolites provided a breakthrough that allowed catalytic hydrocracking to become commercially important, continued advances in the manufacture of amorphous silica-alumina made these materials competitive in certain kinds of applications. This was important, because patents controlled by Unocal and Exxon dominated the application of zeolites in this area. Developments in amorphous catalysts by Chevron and UOP allowed them to compete actively in this area.

Typical catalysts of this type contain 60–80 wt% of the silica-alumina, with the remainder being the hydrogenation component. The compositions of these catalysts are closely held secrets. Over the years, broad ranges of silica/alumina molar ratios have been used in various cracking applications, but silica is almost always in excess for high acidity and stability. A typical level might be 25 wt% alumina (Al_2O_3).

Amorphous silica-alumina is made by a variety of precipitation techniques. The whole class of materials traces its beginnings to silica gel technology, in which sodium silicate is acidified to precipitate the hydrous silica-alumina sulfate; sulfuric acid is used as some or all of the acid for this precipitation, and a mixed gel is formed. The properties of this gel, including acidity and porosity, can be varied by changing the recipe—concentrations, order of addition, pH, temperature, aging time, and the like. The gels are isolated by filtration and washed to remove sodium and other ions.

Careful control of the precipitation allows the pore-size distributions of amorphous materials to be controlled, but the distributions are still much broader than those in the zeolites. This limits the activity and selectivity. One effect of the reduced activity has been that these materials have been applied only in making middle distillates: diesel and turbine fuels. At higher process severities, the poor selectivity results in production of unacceptable amounts of methane (CH_4) to butane (C_4H_{10}) hydrocarbon derivatives.

Hydrocarbon derivatives, especially aromatic hydrocarbon derivatives, can react in the presence of strong acids to form coke. This coke is a complex polynuclear aromatic material that is low in hydrogen. Coke can deposit on the surface of a catalyst, blocking access to the active sites and reducing the activity of the catalyst. Coke poisoning is a major problem in fluid catalytic cracking catalysts, where coked catalysts are circulated to a fluidized bed combustor to be regenerated. In hydrocracking, coke deposition is virtually eliminated by the catalyst's hydrogenation function.

However, the product referred to as *coke* is not a single material. The first products deposited are tarry deposits that can, with time and temperature, continue to become more complex. In a well-designed hydrocracking system, the hydrogenation function adds hydrogen to the tarry deposits. This reduces the concentration of coke precursors on the surface. There is, however, a slow accumulation of coke that reduces activity over a one-to-two-year period. Refiners respond to this slow reduction in activity by raising the average temperature of the catalyst bed to maintain conversions. Eventually, however, an upper limit to the allowable temperature is reached, and the catalyst must be removed and regenerated.

Catalysts carrying coke deposits can be regenerated by burning off the accumulated coke. This is done by service in rotary or similar kilns rather than leaving catalysts in the hydrocracking reactor, where the reactions could damage the metals in the walls. Removing the catalysts also allows inspection and repair of the complex and expensive reactor internals, discussed below. Regeneration of a large catalyst charge can take weeks or months, so refiners may own two catalyst loads, one in the reactor and one regenerated and ready for reload.

The thermal reactions also convert the metal sulfide hydrogenation functions to oxides and may result in agglomeration. Excellent progress has been made since the 1970s in regenerating hydrocracking catalysts; similar regeneration of hydrotreating catalysts is widely practiced.

After combustion to remove the carbonaceous deposits, the catalysts are treated to disperse active metals. Vendor documents claim >95% recovery of activity and selectivity in these regenerations. Catalysts can undergo successive cycles of use and regeneration, providing long functional life with these expensive materials.

The type of catalyst used can influence the product slate obtained. For example, for a mild hydrocracking operation at constant temperature, the selectivity of the catalyst varies from approximately 65% to approximately 90% by volume. Indeed, several catalytic systems have now been developed with a group of catalysts specifically for mild hydrocracking operations. Depending on the type of catalyst, they may be run as a single catalyst or in conjunction with a hydrotreating catalyst. Insight into catalyst nanostructures is leading to the development of high-activity catalysts, which provide solutions and designs to meet many product specifications. In addition, such insights are leading to optimization of hydrocracker units with respect to yield structure, product properties, and throughput and onstream efficiency, resulting in improved refinery margins.

The development of new hydrocracking catalysts is very dependent on new or modified materials. Topsøe has found unique methods of preparing hydroprocessing catalysts and through an extensive understanding of the chemistry has demonstrated a high level of expertise in making catalyst carriers with a uniform distribution of acidic sites and hydrogenation metal sites. Many zeolite hydrocracking catalysts are now offered in the trilobe shape that reduces the diffusion path and decrease the pressure drop. The design significantly enhances the accessibility of the active catalyst sites and thus provides a substantial enhance of the catalyst activity.

In order to optimize overall unit performance, catalysts with pore-size distribution to match the changing molecular structure of the oil as it processes through the reactor system are necessary. They are applied in the front-end reactor when processing high metal-containing feedstocks (>70 ppm vanadium). Catalysts that exhibit the high activity for sulfur, coke precursors, and nitrogen conversion are applied in the middle and/or tail end reactors.

4. Hydrocracking Heavy Feedstocks

The goals of *heavy feedstock hydroconversion* are to convert feedstocks to low-sulfur liquid product oils or, in some cases, to pretreat feedstocks for fluid catalytic cracking processes. Some of the processes available for hydroprocessing viscous feedstocks are presented below. However, when applied to viscous feedstocks (such as heavy oil, extra heavy oil, and tar sand bitumen), the problems encountered can be directly equated to the amount of complex, higher-boiling constituents that may require pretreatment (Speight and Moschopedis, 1979; Speight, 2000; Parkash, 2003; Gary et al., 2007; Speight, 2014, 2017; Hsu and Robinson, 2017). Furthermore, the majority of the higher-molecular-weight materials produce high yield of (35%–60% by weight) coke. It is this trend of coke formation that hydrocracking offers some relief. Thus, the major goal of *heavy feedstock hydroconversion* is cracking of viscous feedstocks with desulfurization, metal removal, denitrogenation, and asphaltene conversion. However, asphaltene constituents and metal-containing constituents exert a strong deactivating influence on the catalyst that markedly decreases the hydrogenolysis rate of sulfur compounds, practically without having impact on coke formation. In addition, nitrogen-containing compounds are adsorbed on acid sites, blocking the sites and thereby lowering catalyst activity. Thus, during the hydrocracking of viscous feedstocks, preliminary feedstock hydrodesulfurization and demetallization over special catalyst are advantageous.

The processes that follow are available for conversion of viscous feedstocks to a variety of product slates and are listed in alphabetical order with no other preference in mind.

4.1 Asphaltenic Bottom Cracking Process

The asphaltenic bottom cracking (ABC) process can be used for distillate production, hydrodemetallization, asphaltene cracking, and moderate hydrodesulfurization as well as sufficient resistance to coke fouling and metal deposition using such feedstocks as heavy oil, extra heavy oil, tar sand bitumen, vacuum residua, thermally cracked residua, solvent-deasphalted bottoms, and bitumen with fixed catalyst beds (Kressmann et al., 1998; Parkash, 2003; Gary et al., 2007; Speight, 2014; Hsu and Robinson, 2017; Speight, 2017). The process can be combined with (i) solvent deasphalting for complete or partial conversion of the residuum; (ii) hydrodesulfurization to promote the conversion of residue, to treat feedstock with high metals, and to increase catalyst life; or (iii) hydrovisbreaking to attain high conversion of the viscous feedstock with favorable product stability.

In the process, the feedstock is pumped up to the reaction pressure and mixed with hydrogen. The mixture is heated to the reaction temperature in the charge heater after a heat exchange and fed to the reactor. In the reactor, hydrodemetallization and subsequent asphaltene cracking with moderate hydrodesulfurization take place simultaneously under conditions similar to residuum hydrodesulfurization. The reactor effluent gas is cooled, cleaned up, and recycled to the reactor section, while the separated liquid is distilled into distillate fractions and vacuum residue that is further separated by deasphalting into deasphalted oil and asphalt using butane or pentane (Chapter 12).

In case of the ABC-hydrodesulfurization catalyst combination, the ABC catalyst is placed upstream of the hydrodesulfurization catalyst and can be operated at a higher temperature than the hydrodesulfurization catalyst under conventional viscous feedstock hydrodesulfurization conditions. In the VisABC process, a soaking drum is provided after heater, when necessary. Hydrovisbroken oil is first stabilized by the ABC catalyst through hydrogenation of coke precursors and then desulfurized by the HDS catalyst.

4.2 Chevron RDS Process and VRDS Process

The RDS/VRDS process (like the Residfining process) is designed to hydrotreat vacuum gas oil, atmospheric residuum, vacuum residuum, and other viscous feedstocks to remove sulfur metallic constituents while part of the feedstock is converted to lower-boiling products. In the case of viscous feedstocks, the asphaltene content is reduced. The process consists of a once-through operation and is ideally suited to produce feedstocks for residuum fluid catalytic crackers or delayed coking units to achieve minimal production of nonvolatile products in a refinery.

The basic elements of each process are similar and consist of a once-through operation of the feedstock coming into contact with hydrogen and the catalyst in a downflow reactor that is designed to maintain activity and selectivity in the presence of deposited metals. Moderate temperatures and pressures are employed to reduce the incidence of hydrocracking and, hence, minimize production of low-boiling distillates. The combination of a desulfurization step and a viscous feedstock desulfurizer (VRDS) is often seen as an attractive alternate to the atmospheric residuum desulfurizer (RDS). In addition, either RDS option or the VRDS option can be coupled with other processes (such as delayed coking, fluid catalytic cracking, and solvent deasphalting) to achieve the most optimum refining performance.

The *Chevron deasphalted oil hydrotreating* process is designed to desulfurize viscous feedstocks that have had the asphaltene fraction removed by prior application of a deasphalting process. The principal product is a low-sulfur fuel oil that can be used as a blending stock or as a feedstock for a fluid catalytic unit. The process employs a downflow, fixed-bed reactor containing a highly selective catalyst that provides extensive desulfurization at low pressures with minimal cracking and, therefore, low consumption of hydrogen.

In the process, which is designed for viscous feedstocks including a wide range of viscous feedstocks, the feedstock and hydrogen are charged to the reactors in a once-through operation. The catalyst combination can be varied significantly according to feedstock properties to meet the required product qualities. Product separation is done by the hot separator, cold separator, and fractionator. Recycle hydrogen passes through a hydrogen sulfide absorber. The *onstream catalyst replacement* (OCR) reactor technology improves catalyst utilization and increases run length with high metals and viscous feedstocks.

This technology allows spent catalyst to be removed from one or more reactors and replaced with fresh while the reactors continue to operate normally. The novel use of upflow reactors in the onstream catalyst replacement technology provides increased tolerance of feedstock solids while maintaining low-pressure drop. A related technology (upflow reactor (UFR) technology) uses a multibed upflow reactor for minimum pressure drop in cases where onstream catalyst replacement is not necessary. Onstream catalyst replacement technology and upflow reactor technology are particularly well suited to revamp existing RDS/VRDS units for additional throughput or heavier feedstock. The products (residuum fluid catalytic cracking feedstock, coker feedstock, solvent deasphalter feedstock, low-sulfur fuel oil, and vacuum gas oil) are suitable for further upgrading by fluid catalytic cracking units or hydrocrackers for naphtha/middistillate manufacture. Middistillate material can be directly blended into low-sulfur diesel or further hydrotreated into ultra-low-sulfur diesel (ULSD). Thus, the process can be integrated with residuum fluid catalytic cracking units to minimize catalyst consumption, improve yields, and reduce sulfur content of fluid catalytic cracking products. RDS/VRDS also can be used to substantially improve the yields of downstream coking units and solvent deasphalting units.

4.3 ENI Slurry Phase Technology

The advent of slurry-phase hydrocracking into the refineries has caused much interest. This technology adopts high operating pressures and can achieve near-complete (if not, complete) conversion of the residuum while producing finished salable products (Motaghi et al., 2010). Slurry-phase hydrocracking can be used to convert viscous feedstock in the presence of hydrogen under severe process conditions—on the order of 450°C (840°F) and 2000–3000 psi. To prevent excessive coking, finely powdered additives are added to the feedstock. Inside the reactor, the feedstock/powder mixture behaves as a single phase due to the small size of the additive particles.

For example, in the ENI slurry-phase process that is based on an organic oleo-soluble molybdenum compound and the catalyst precursor, the catalyst is added to the feedstock before it enters the reactor (Bellussi et al., 2013). In the process, fresh feedstock is sent to the fresh feedstock heater and then mixed with the proprietary catalyst makeup and sent to the upflow slurry bubble column reactor. Hot hydrogen is also sent to the slurry reactor providing the thermocatalytic hydroconversion of the feedstock. The reactor effluent is collected in a hot high-pressure separator where a gas–vapor stream and a vacuum gas oil stream are separated. The stream is subsequently cooled and sent to a cold high-pressure separator to separate the gas stream, rich in hydrogen, and the hydrocarbon liquid stream. The liquid stream from the cold high-pressure separator is sent to the light distillate stabilizer from which low-boiling distillates are separated and sent to battery limit. The reaction occurs at 400–450°C (750–840°F) and at approximately 2200 psi with hydrogen fed from the bottom of the reactor. Under the reaction conditions, the catalyst precursor forms highly dispersed molybdenum sulfide (MoS_2) nanoparticles. The in situ formation of the catalyst preparation method enables the dispersion of molybdenum sulfide (MoS_2) mainly as single layers within the slurry reactor.

The unconverted nonvolatile fraction at the bottom of the vacuum distillation column, containing all of the catalyst, is recycled back to the reactor, and only a small part of the viscous fraction is purged (1%–3% *w*/w of the fresh feedstock) to avoid the accumulation of coke precursors and of Ni and V sulfides from the organometallic compounds contained in the feedstock. With the purge, a limited amount of molybdenum is also removed; therefore, an equivalent amount is fed continuously to the reactor to maintain concentration constant. The purge can be used as a fuel in the cement or steel industries. In order to facilitate the handling of the purge and its blending with other streams, the viscosity can be adjusted by adding a small amount of a low-value flow improver (such as vacuum gas oil). The purge can also be treated in a centrifugal decanter to recover the liquid fraction, which is recycled back to the reactor, and a solid product (cake) containing high-molecular-weight hydrocarbon derivatives, coke, and concentrated metal sulfides. The cake can be processed further to recover the metals (molybdenum, vanadium, and nickel). Since the largest part of the catalyst is not lost, but is recycled to the reaction section, the process can operate at a higher catalyst concentration than in the case with other slurry technologies.

The process is very flexible with regard to the feedstock and can accept feedstocks such as heavy crude, extra heavy crude, tar sand bitumen, refinery visbroken residua, and other nonvolatile residua. The typical overall performance of the process is (i) metal removal, >99%; (ii) Conradson carbon residue reduction, >97%; (iii) sulfur reduction, >85%; and (iv) nitrogen reduction, >40%. Furthermore, because of the recycling of unconverted products and the dispersed catalyst, the process has the ability to reach total conversion of the feedstock.

4.4 Gulf Resid Hydrodesulfurization Process

The Gulf Resid HDS process is a regenerative fixed-bed process to upgrade viscous feedstocks by catalytic hydrogenation to fuel oil or to high-quality catalytic charge stocks (Speight, 2000, 2014, 2017; Parkash, 2003; Ancheyta and Speight, 2007; Gary et al., 2007; Hsu and Robinson, 2017). Desulfurization and quality improvement are the primary purposes of the process, but if the operating conditions and catalysts are varied, light distillates can be produced, and the viscosity of the feedstock can be lowered. Long onstream cycles are maintained by reducing random hydrocracking reactions to a minimum, and whole crude oils, heavy oil, extra heavy oil, and tar sand bitumen may serve as feedstock. This process is suitable for the desulfurization of high-sulfur viscous feedstocks to produce low-sulfur fuel oils or catalytic cracking feedstocks.

In addition, the process can be used, through alternate design types, to upgrade high-sulfur crude oils or bitumen that are unsuited for the more conventional refining techniques.

The process has three basic variations—the Type II unit, the Type III unit, and the Type IV unit with the degree of desulfurization, and process severity, increasing from Type I to Type IV. Thus, liquid products from Type III and IV units can be used directly as catalytic cracker feedstocks and perform similarly to virgin gas oil fractions, whereas liquid products from the Type II unit usually need to be vacuum-flashed to provide a feedstock suitable for a catalytic cracker.

In the process, fresh, filtered feedstock is heated together with hydrogen and recycled gas and charged to the downflow reactor from which the liquid product goes to fractionation after flashing to produce the various product streams. Each process type is basically similar to its predecessor but will differ in the number of reactors. For example, modifications necessary to convert the Type II to the Type III process consist of the addition of a reactor and related equipment, while the Type III process can be modified to a Type IV process by the addition of a third reactor section. Types III and IV are especially pertinent to the problem of desulfurizing viscous feedstocks since they have the capability of producing extremely low-sulfur liquids from high-sulfur viscous feedstocks.

The catalyst is a metallic compound supported on pelleted alumina and may be regenerated in situ with air and steam or flue gas through a temperature cycle of 400–650°C (750–1200°F). Onstream cycles of 4–5 months can be obtained at desulfurization levels of 65%–75%, and catalyst life may be as long as 2 years.

4.5 H-G Hydrocracking Process

The H-G hydrocracking process may be designed with either a single- or a two-stage reactor system for conversion of gas oil to lower-boiling fractions (Speight, 2000, 2014, 2017; Parkash, 2003; Ancheyta and Speight, 2007; Gary et al., 2007; Hsu and Robinson, 2017). The feedstock is mixed with recycled gas oil, makeup hydrogen, and hydrogen-rich recycled gas and then heated and charged to the reactor. The reactor effluent is cooled and sent to a high-pressure separator, where hydrogen-rich gas is flashed off, scrubbed, and then recycled to the reactor. Separator liquid passes to a stabilizer for removal of butanes and lower-boiling products, and the bottoms are taken to a fractionator for separation; any unconverted material is recycled to the reactor.

4.6 H-Oil Process

The H-Oil process is a catalytic process that that uses a single-stage, two-stage, or three-stage ebullated-bed reactor in which, during the reaction, considerable hydrocracking takes place (Table 11.2; Fig. 11.3) (Speight, 2000, 2014, 2017; Parkash, 2003; Ancheyta and Speight, 2007; Gary et al., 2007; Hsu and

TABLE 11.2 General Process Parameters for Ebullated Bed (H-Oil and LC-Fining) Processes

Parameter	H-Oil	LC-Fining
Temperature (°C)	415–440	385–450
Temperature (°F)	780–825	725–840
Pressure (psi)	2440–3000	1015–2740
Hydrogen/bbl	1410	1350
Conversion	45–90	40–95
HDS	55–90	60–90
HDM	65–90	50–95

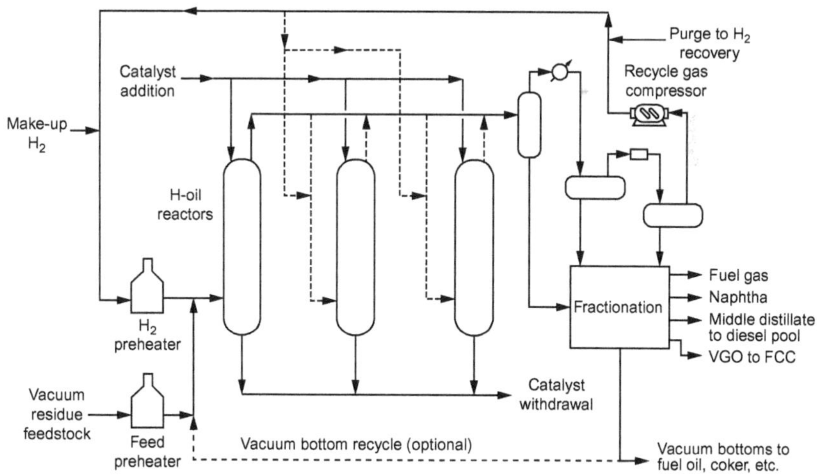

FIG. 11.3 The H-Oil process.

Robinson, 2017). The process is used to upgrade sulfur-containing viscous feedstocks to low-sulfur distillates, thereby reducing fuel oil yield. A modification of H-Oil called Hy-C cracking converts high-boiling distillates to middle distillates.

The process is designed for hydrogenation of viscous feedstocks in an ebullated-bed reactor to produce upgraded crude oil products (Colyar et al., 1998; Parkash, 2003; Gary et al., 2007; Speight, 2014; Hsu and Robinson, 2017; Speight, 2017). The process is able to convert all types of feedstocks to either distillate products or desulfurize and demetallize viscous feedstocks for feedstock to coking units or fluid catalytic cracking units, for production of low-sulfur fuel oil, or for production to asphalt blending.

A wide variety of process options can be used with the H-Oil process depending on the specific needs. In all cases, a catalytic ebullated-bed reactor system is used to provide an efficient hydroconversion. The system insures uniform distribution of liquid, hydrogen-rich gas and catalyst across the reactor. The ebullated-bed system operates under essentially isothermal conditions, exhibiting little temperature gradient across the bed (Kressmann et al., 2000). The heat of reaction is used to bring the feedstock oil and hydrogen up to reactor temperature.

In the process, feedstock (which may be combined with recycled residuum) and hydrogen are fed upward through the reactors as a liquid-gas mixture at a velocity such that catalyst is in continuous motion. A catalyst of small particle size can be used, giving efficient contact among gas, liquid, and solid with good mass and heat transfer. Part of the reactor effluent is recycled back through the reactors for temperature control and to maintain the requisite liquid velocity. The entire bed is held within a narrow temperature range, which provides essentially an isothermal operation with an exothermic process. Because of the movement of catalyst particles in the liquid-gas medium, deposition of tar and coke is minimized, and fine solids entrained in the feedstock do not lead to reactor plugging. The catalyst can also be added and withdrawn from the reactor without destroying the continuity of the process. The reactor effluent is cooled by exchange and separates into vapor and liquid. After scrubbing in a lean oil absorber, hydrogen is recycled, and the liquid product is either stored directly or fractionated before storage and blending.

A variation of this process (the HDH resid hydrocracking process) was originally developed for the upgrading of viscous feedstocks from the Orinoco Oil Belt, Venezuela. In the process, the viscous feedstock is slurried with a low-cost catalyst and fed into a series of upflow bubbling (slurry) reactors operating at 420–480°C temperature under hydrogen partial pressure. The reaction products are fractionated using a high-pressure, hot separator.

While ebullated-bed processes are continuous and produce higher levels of liquid fuels (no coke), it is not always possible to achieve complete viscous feedstock conversion, and the unit may still produce 20%–30% *v*/v of a high-boiling product (Motaghi et al., 2010). Ebullated beds have also been prone to high operating costs and have sometimes been plagued with poor operability. The quality of liquid products, although improved over coking, still requires secondary processing to produce clean fuels. The inability to achieve near-complete conversion requires further processing of unconverted resid. As a result, ebullated-bed technologies have not achieved huge deployment, which, when coupled with the high capital cost, makes them the least robust at low-oil price scenarios.

In terms of advancement of ebullated-bed technology, the original H-Oil process has evolved, during the past decade, into various configurations that have the potential to play major roles in viscous feedstock upgrading up to and beyond the year 2020. A new development in H-Oil process technology is interstage separation for a two-stage unit design (Kressmann et al., 2004). In this configuration, an additional vessel is fed the first-stage reactor effluent (mixed phase) and separates it into vapor and liquid products. The interstage liquid is fed to the second-stage reactor and the vapor to the overhead of the hot high-pressure separator located after the second-stage reactor. With interstage separation, off-loading of the first-stage reactor gas results in improved reaction kinetics in the second-stage reactor since the amount of gas holdup in the reactor is greatly reduced, and increasing liquid holdup enables greater conversion of the feedstock.

The H-Oil$_{DC}$ process (previously known as the T-Star process) is a specially engineered, ebullated-bed process for the treatment of vacuum gas oils. Because of the ability to replace the catalyst bed incrementally, the H-Oil$_{DC}$ reactor can operate indefinitely—typically, 4–5 years between turnarounds to coincide with the inspection and maintenance schedule for a fluid catalytic cracking unit. The difficult processing requirements that result from stricter environmental regulations and the processing of viscous feedstocks makes H-Oil$_{DC}$ a preferred choice for pretreatment of fluid catalytic cracker feedstocks.

The H-Oil$_{HCC}$ is a viscous crude conversion process that produces synthetic crude oil. The objective of the unit is to enable just enough conversion to reduce viscosity and increase stability so that the product can be readily transported to an upgrading center. Among the improvements made to the traditional H-Oil technology are the integration of an interstage separator between reactors in series and the application of cascade catalyst utilization.

4.7 HYCAR Process

Briefly, *hydrovisbreaking*, a noncatalytic process, is conducted under similar conditions to visbreaking and involves treatment with hydrogen under mild conditions. The presence of hydrogen leads to more stable products (lower *flocculation threshold*) than can be obtained with straight visbreaking, which means that higher conversions can be achieved, producing a lower viscosity product.

The HYCAR process is composed fundamentally of three parts: (i) visbreaking, (ii) hydrodemetallization, and (iii) hydrocracking. In the visbreaking section, the viscous feedstock (e.g., vacuum residuum, extra heavy oi, or tar sand bitumen) is subjected to moderate thermal cracking while no coke formation is induced. The visbreaker oil is fed to the demetallization reactor in the presence of catalysts, which provides sufficient pore for diffusion and adsorption of high-molecular-weight constituents. The product from this second stage proceeds to the hydrocracking reactor, where desulfurization and denitrogenation take place along with hydrocracking.

4.8 Hyvahl-F Process

The process is used to hydrotreat viscous feedstocks to convert the feedstock to naphtha and middle distillates (Peries et al., 1988; Billon et al., 1994; Parkash, 2003; Gary et al., 2007; Speight, 2014, 2017; Hsu and Robinson, 2017). The main features of this process are its dual catalyst system and its fixed-bed swing-reactor concept. The first catalyst has a high capacity for metals (to 100% by *w*/w new catalyst) and is used for both hydrodemetallization (HDM) and most of the conversion. This catalyst is resistant to fouling, coking, and plugging by asphaltene constituents (as well as by reacted asphaltene constituents) and shields the second catalyst from the same. Protected from metal poisons and deposition of coke-like products, the highly active second catalyst can carry out its deep hydrodesulfurization (HDS) and refining functions. Both catalyst systems use fixed beds that are more efficient than moving beds and are not subject to attrition problems. The swing-reactor design reserves two of the HDM reactors as guard reactors: one of them can be removed from service for catalyst reconditioning and put on standby while the rest of the unit continues to operate. >50% of the metals are removed from the feedstock in the guard reactors.

In the process, the preheated feedstock enters one of the two guard reactors where a large proportion of the nickel and vanadium is adsorbed and hydroconversion of the high-molecular-weight constituents commences. Meanwhile, the second guard reactor catalyst undergoes a reconditioning process and then is put on standby. From the guard reactors, the feedstock flows through a series of hydrodemetallization reactors that continue the metal removal and the conversion of viscous oil. The next processing stage, hydrodesulfurization, is where most of sulfur, some of the nitrogen, and metals are removed. A limited amount of conversion also takes place. From the final reactor, the gas phase is separated, hydrogen is recirculated to the reaction section, and the liquid products are sent to a conventional fractionation section for separation into naphtha, middle distillates, and heavier streams.

A related process—the Hyvahl-M process—employs countercurrent moving-bed reactors and is recommended for feedstocks containing substantial amounts of metals and asphaltene constituents.

4.9 IFP Hydrocracking Process

The process features a dual catalyst system: the first catalyst is a promoted nickel-molybdenum amorphous catalyst. It acts to remove sulfur and nitrogen and hydrogenate aromatic rings. The second catalyst is a zeolite that finishes the hydrogenation and promotes the hydrocracking reaction.

In the single-stage process, the first reactor effluent is sent directly to then second reactor, followed by the separation and fractionation steps (Speight, 2000, 2014, 2017; Parkash, 2003; Ancheyta and Speight, 2007; Gary et al.,

2007; Hsu and Robinson, 2017). The fractionator bottoms are recycled to the second reactor or sold. In the two-stage process, feedstock and hydrogen are heated and sent to the first reaction stage where conversion to products occurs (RAROP, 1991, p. 85). The reactor effluent phases are cooled and separated, and the hydrogen-rich gas is compressed and recycled. The liquid leaving the separator is fractionated, the middle distillates and lower-boiling streams (Parkash, 2003; Gary et al., 2007; Speight, 2014, 2017; Hsu and Robinson, 2017) are sent to storage, and the high-boiling stream is transferred to the second reactor section and then recycled back to the separator section.

4.10 Isocracking Process

The Isocracking process is a high-pressure, moderate-temperature conversion unit, and the designs include single-stage once-through, single-stage recycle, and two-stage recycle processes. A two-stage hydrocracking unit with intermediate distillation represents the most common process configuration for maximizing middle distillates. The process has been applied commercially in the full range of process flow schemes: single-stage, once-through liquid; single-stage, partial recycle of viscous feedstocks; single-stage recycle to extinction of the feedstock (100% conversion); and two-stage recycle to extinction of the feedstock (Speight, 2000, 2014, 2017; Parkash, 2003; Ancheyta and Speight, 2007; Gary et al., 2007; Hsu and Robinson, 2017). The preferred flow scheme will depend on the feedstock properties; the processing objectives; and, to some extent, the specified feedstock rate. The process uses multibed reactors, and in most applications, a number of catalysts are used in a reactor. The catalysts are dual function being a mixture of hydrous oxides (for cracking) and heavy metal sulfides (for hydrogenation). The catalysts are used in a layered system to optimize the processing of the feedstock that undergoes changes in its properties along the reaction pathway (Parkash, 2003; Gary et al., 2007; Speight, 2014, 2017; Hsu and Robinson, 2017).

In the process, the feedstock (typically a blend of various gas oils) is sent to the first stage of the hydrocracker and is severely hydrotreated. Most of the sulfur and nitrogen compounds are removed from the feedstock, and many of the aromatic derivatives are saturated. In addition, significant conversion to light products occurs in the first stage. The liquid product from the first stage is then sent to a common fractionation section where, to prevent overcracking, lower-boiling products are removed by distillation. The unconverted material from the bottom of the fractionator is routed to the second-stage reactor section. The second reaction stage saturates almost all of the aromatic derivatives and cracks the oil feedstock to light products. Due to the aromatic derivative saturation, the second stage produces excellent-quality products. The liquid product from the second stage is then sent to the common fractionator, where light products are distilled. The second stage operates in a recycle to extinction mode with per pass conversions ranging from 50% to 80% *v*/v. The overhead liquid and vapor from the hydrocracker fractionator are further processed in a light ends recovery unit where fuel gas and liquefied crude oil gas (LPG) and naphtha are separated. The hydrogen supplied to the reactor sections of the hydrocracker comes from reformers or steam reformers. The hydrogen is compressed in stages until it reaches system pressure of the reactor sections.

4.11 LC-Fining Process

The LC-Fining process is a hydrocracking process capable of desulfurizing, demetallizing, and upgrading a wide spectrum of viscous feedstocks by means of an expanded-bed reactor (Van Driesen et al., 1979; Fornoff, 1982; Bishop, 1990; RAROP, 1991, p. 61; Reich et al., 1993; Khan and Patmore, 1998; Parkash, 2003; Gary et al., 2007; Speight, 2014, 2017; Hsu and Robinson, 2017). Operating with the expanded bed allows the processing of viscous feedstocks, such as heavy oil, extra heavy oil, and tar sand bitumen. The catalyst in the reactor behaves like fluid that enables the catalyst to be added to and withdrawn from the reactor during operation. The reactor conditions are near isothermal because the heat of reaction is absorbed by the cold fresh feedstock immediately owing to through mixing of reactors.

In the process (Fig. 11.4), the feedstock and hydrogen are heated separately and then pass upward in the hydrocracking reactor through an expanded bed of catalyst. Reactor products flow to the high pressure-high temperature separator. Vapor effluent from the separator is let down in pressure and then goes to the heat exchange and thence to a section for the removal of condensable products and purification (Parkash, 2003; Gary et al., 2007; Speight, 2014, 2017; Hsu and Robinson, 2017).

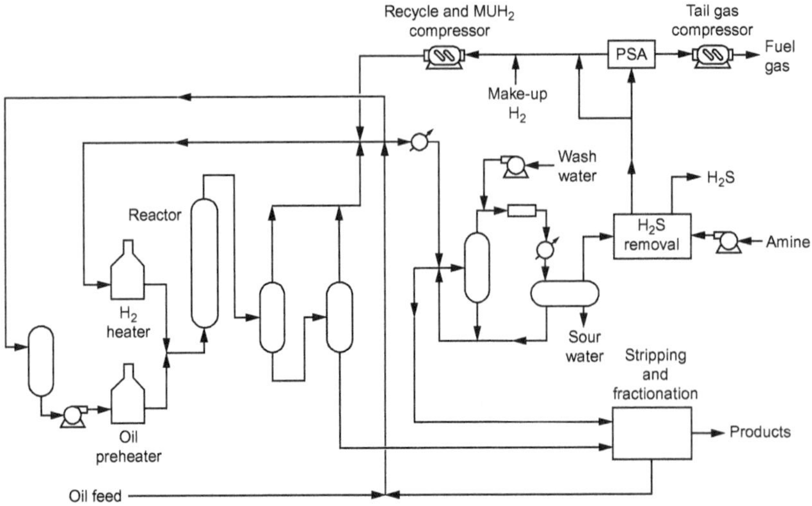

FIG. 11.4 The LC-Fining process.

Liquid is let down in pressure and passes to the recycle stripper. This is a most important part of the high conversion process. The liquid recycle is prepared to the proper boiling range for return to the reactor. In this way, the concentration of bottoms in the reactor and therefore the distribution of products can be controlled. After the stripping, the recycle liquid is then pumped through the coke precursor removal step where high-molecular-weight constituents are removed. The clean liquid recycle then passes to the suction drum of the feedstock pump.

The residence time in the reactor is adjusted to provide the desired conversion levels. Catalyst particles are continuously withdrawn from the reactor, regenerated, and recycled back into the reactor, which provides the flexibility to process a wide range of viscous feedstock such as atmospheric and vacuum tower bottoms and coal-derived liquids and bitumen. An internal liquid recycle is provided with a pump to expand the catalyst bed, continuously. As a result of expanded-bed operating mode, small pressure drops and isothermal operating conditions are accomplished. Small-diameter extruded catalyst particles as small as 0.8 mm (1/32 in.) can be used in this reactor.

Although the process may not be the means by which direct conversion of the bitumen to a synthetic crude oil would be achieved, it does nevertheless offer an attractive means of bitumen conversion. Indeed, the process would play the part of the primary conversion process from which liquid products would accrue—these products would then pass to a secondary upgrading (hydrotreating) process to yield a synthetic crude oil.

4.12 MAKfining Process

The process uses a multiple catalyst system in multibed reactors that include quench and redistribution system internals (Hunter et al., 1997; Parkash, 2003; Gary et al., 2007; Speight, 2014, 2017; Hsu and Robinson, 2017). In the process, the feedstock and recycled gas are preheated and brought into contact with the catalyst in a downflow fixed-bed reactor. The reactor effluent is sent to high- and low-temperature separators. Product recovery is a stripper/fractionator arrangement. Typical operating conditions in the reactors are 370–425°C (700–800°F) (single-pass) and 370–425°C (700–800°F) (recycle) with pressures of 1000–2000 psi (single-pass) and 1500–3000 psi (recycle). Product yields depend upon the extent of the conversion.

4.13 Microcat-RC Process

The Microcat-RC process (also referred to as the M-Coke process) is a catalytic ebullated-bed hydroconversion process that is similar to Residfining and that operates at relatively moderate pressures and temperatures (Bearden and Aldridge, 1981; Bauman et al., 1993; Speight, 2000, 2014, 2017; Parkash, 2003; Ancheyta and Speight, 2007; Gary et al., 2007; Hsu and Robinson, 2017).

The catalyst particles, containing a metal sulfide in a carbonaceous matrix formed within the process, are uniformly dispersed throughout the feed. Because of their ultrasmall size (10^{-4} in. diameter), there are typically several orders of magnitude more of these microcatalyst particles per cubic centimeter of oil than is possible in other types of hydroconversion reactors using conventional catalyst particles. This results in smaller distances between particles and less time for a reactant molecule or intermediate to find an active catalyst site. Because of their physical structure, microcatalysts suffer none of the pore-plugging problems that plague conventional catalysts.

In the process, fresh viscous feedstock, microcatalyst, and hydrogen are fed to the hydroconversion reactor. Effluent is sent to a flash separation zone to recover hydrogen, gases, and liquid products, including naphtha, distillate, and gas oil (Parkash, 2003; Gary et al., 2007; Speight, 2014; Hsu and Robinson, 2017; Speight, 2017). The residuum from the flash step is then fed to a vacuum distillation tower to obtain a 565°C$^-$ (1050°F$^-$) product oil and a 565°C$^+$ (1050°F$^+$) bottom fraction that contains unconverted feed, microcatalyst, and essentially all of the feedstock metals.

Hydrotreating facilities may be integrated with the hydroconversion section or built on a stand-alone basis, depending on product quality objectives and owner preference.

4.14 Mild Hydrocracking Process

The *mild hydrocracking process* uses operating conditions similar to those of a vacuum gas oil desulfurizer to convert a vacuum gas oil to significant yields of lower-boiling products. Consequently, the flow scheme for a mild hydrocracking unit is virtually identical to that of a vacuum gas oil desulfurizer. The feedstock is mixed with hydrogen makeup gas and preheated against reactor effluent after which further preheating to reaction temperature is accomplished in a fired heater. The hot feedstock is mixed with recycled gas before entering the reactor. The temperature rises across the reactor due to the exothermic heat of reaction. Catalyst bed temperatures are usually controlled by using multiple catalyst beds and by introducing recycled gas as an interbed quench medium. Reactor effluent is cooled against incoming feedstock and air or water before entering the high-pressure separator. Vapors from this separator are scrubbed to remove hydrogen sulfide (H_2S) before compression back to the reactor as recycle and quench. A small portion of these gases is purged to fuel gas to prevent buildup of light ends. Liquid from the high-pressure separator is flashed into the low-pressure separator. Sour flash vapors are purged from the unit. Liquid is preheated against stripper bottoms and in a feedstock heater before steam stripping in a stabilizer tower. Water wash facilities are provided upstream of the last reactor effluent cooler to remove ammonium salts produced by denitrogenation of the vacuum gas oil feedstock.

Variation of this process leads to the hot separator design. The process flow scheme is identical to that described above up to the reactor outlet. After initial reactor effluent cooling against incoming vacuum gas oil feedstock and makeup hydrogen, a hot separator is installed. Hot liquid is routed directly to the product stabilizer. Hot vapors are further cooled by air and/or water before entering the cold separator. This arrangement reduces the stabilizer feedstock preheat duty and the effluent cooling duty by routing hot liquid direct to the stripper tower.

The conditions for mild hydrocracking are typical of many low-pressure desulfurization units that for hydrocracking units, in general, are marginal in pressure and hydrogen oil ratio capabilities. For hydrocracking, in order to obtain satisfactory run lengths (approximately 11 months), reduction in feedstock rate or addition of an extra reactor may be necessary. In most cases, since the product slate will be lower boiling than for normal desulfurization service only, changes in the fractionation system may be necessary. When these limitations can be tolerated, the product value from mild hydrocracking versus desulfurization can be greatly enhanced.

In summary, the so-called mild hydrocracking process is a simple form of hydrocracking. The hydrotreaters designed for vacuum gas oil desulfurization and catalytic cracker feedstock pretreatment are converted to once-through hydrocracking units, and because existing units are being used, the hydrocracking is often carried out under nonideal hydrocracking conditions.

4.15 MRH Process

The MRH process is a hydrocracking process designed to upgrade viscous feedstocks containing large amount of metals and asphaltene, such as heavy oil, extra heavy oil, and tar sand bitumen, and to produce mainly middle distillates (Sue, 1989; RAROP, 1991, p. 65; Khan and Patmore, 1998; Speight, 2000, 2014, 2017; Parkash, 2003; Ancheyta and Speight, 2007; Gary et al., 2007; Hsu and Robinson, 2017). The reactor is designed to maintain a mixed three-phase slurry of feedstock, fine powder catalyst, and hydrogen and to promote effective contact.

In the process, a slurry consisting of the viscous feedstock and finely powdered catalyst is preheated in a furnace and fed into the reactor vessel. Hydrogen is introduced from the bottom of the reactor and flows upward through the reaction mixture, maintaining the catalyst suspension in the reaction mixture. Cracking, desulfurization, and demetallization reactions are taken place via thermal and catalytic reactions. In the upper section of the reactor, vapor is disengaged from the slurry, and hydrogen and other gases are removed in a high-pressure separator. The liquid condensed from the overhead vapor is distilled and then flows out to the secondary treatment facilities. From the lower section of the reactor, a bottom slurry oil (SLO) that contains catalyst, uncracked residuum, and a small amount of vacuum gas oil fraction is withdrawn. Vacuum gas oil is recovered in the slurry separation section, and the remaining catalyst and coke are fed to the catalyst regenerator. Product distribution focuses on middle distillates with the process focused as a viscous feedstock processing unit and can be inserted into refinery operations downstream from the vacuum distillation unit.

4.16 RCD Unibon Process

The RCD Unibon process (BOC process) is a process to upgrade viscous feedstocks (RAROP, 1991, p. 67; Thompson, 1997; Khan and Patmore, 1998; Parkash, 2003; Gary et al., 2007; Speight, 2014; Hsu and Robinson, 2017; Speight, 2017). There are several possible flow scheme variations involved for the process, which can also operate as an independent unit or be used in conjunction with a thermal conversion unit. In this configuration, hydrogen and a vacuum residuum are introduced separately to the heater and mixed at the entrance to the reactor. To avoid thermal reactions and premature coking of the catalyst, temperatures are carefully controlled, and conversion is limited to approximately 70% of the total projected conversion. The removal of sulfur, heptane-insoluble materials, and metals is accomplished in the reactor. The effluent from the reactor is directed to the hot separator. The overhead vapor phase is cooled and condensed, and the separated hydrogen is recycled to the reactor.

Liquid product goes to the thermal conversion heater where the remaining conversion of nonvolatile materials occurs. The heater effluent is flashed, and the overhead vapors are cooled, condensed, and routed to the cold flash drum. The bottom liquid stream then goes to the vacuum column where the gas oils are recovered for further processing.

4.17 Residfining Process

Residfining is a catalytic fixed-bed process for the desulfurization and demetallization of viscous feedstocks (RAROP, 1991, p. 69; Khan and Patmore, 1998; Parkash, 2003; Gary et al., 2007; Speight, 2014; Hsu and Robinson, 2017; Speight, 2017). The process can also be used to pretreat viscous feedstocks to suitably low contaminant levels prior to catalytic cracking.

In the process, liquid feedstock to the unit is filtered, pumped to pressure, preheated, and combined with treat gas prior to entering the reactors. A small guard reactor would typically be employed to prevent plugging/fouling of the main reactors. Provisions are employed to periodically remove the guard while keeping the main reactors online. The temperature rise associated with the exothermic reactions is controlled utilizing either a gas or liquid quench. A train of separators is employed to separate the gas and liquid products. The recycled gas is scrubbed to remove ammonia (NH_3) and hydrogen sulfide (H_2S). It is then combined with fresh makeup hydrogen before being reheated and recombined with fresh feed. The liquid product is sent to a fractionator where the product is fractionated.

Residfining is an option that can be used to reduce the sulfur, to reduce metals and coke-forming precursors, and/or to accomplish some conversion to lower-boiling products as a feedstock pretreat step ahead of a fluid catalytic cracking unit. There is also a hydrocracking option where substantial conversion of the resid occurs.

4.18 Residue Hydroconversion Process

The residue hydroconversion process (RHC process) is a high-pressure fixed-bed trickle-flow hydrocatalytic process (RAROP, 1991, p. 71; Khan and Patmore, 1998). The feedstock can be desalted atmospheric or vacuum residue and other viscous feedstocks such as heavy oil, extra heavy oil, and tar sand bitumen (Speight, 2000; Parkash, 2003; Ancheyta and Speight, 2007; Gary et al., 2007; Speight, 2014; Hsu and Robinson, 2017; Speight, 2017).

The reactors are of multibed design with interbed cooling, and the multicatalyst system can be tailored according to the nature of the feedstock and the target conversion. For viscous feedstocks with high metal content, a hydrodemetallization catalyst is used in the front-end reactor(s), which excels in its high metal uptake capacity and good activities for metal removal, asphaltene conversion, and residue cracking. Downstream of the demetallization stage, one or more hydroconversion stages, with optimized combination of catalysts' hydrogenation function and texture, are used to achieve desired catalyst stability and activities for denitrogenation, desulfurization, and viscous hydrocarbon cracking. A guard reactor may be employed to remove contaminants that promote plugging or fouling of the main reactors with periodic removal of the guard reactor while keeping the main reactors online.

4.19 Shell Residual Oil Process

The Shell residual oil hydrodesulfurization process was originally designed to improve the quality of residual oils by removing sulfur, metals, and asphaltene constituents. The process is suitable for a wide range of the heavier feedstocks, irrespective of the composition and origin, and even includes those feedstocks that are particularly high in metals and asphaltene constituents.

The process centers on a fixed-bed downflow reactor that allows catalyst replacement without causing any interruption in the operation of the unit. Feedstock is introduced to the process via a filter (backwash and automatic) after which hydrogen and recycled gas are added to the feedstock stream that is then heated to reactor temperature by means of feed-effluent heat exchangers whereupon the feedstock stream passes down through the reactor in trickle flow. Sulfur removal is excellent, and substantial reductions in the vanadium content and asphaltene constituents also occur. In addition, a marked increase occurs in the API gravity, and the viscosity is reduced considerably.

A bunker reactor provides extra process flexibility if it is used upstream from the desulfurization reactor, especially with reference to the processing of feedstocks with a high metal content. A catalyst with a capacity for metals is employed in the bunker reactor to protect the desulfurization catalyst from poisoning by the metals. In the bunker reactor, inverted cone segments support the catalyst and are designed to allow catalyst removal.

4.20 Tervahl-H Process

In the Tervahl H process, the feedstock and hydrogen-rich stream are heated using heat recovery techniques and fired heater and held in the soak drum as in the Tervahl T process. The gas and oil from the soaking drum effluent are mixed with recycle hydrogen and separated in the hot separator where the gas is cooled, passed through a separator, and recycled to the heater and soaking drum effluent. The liquids from the hot and cold separator are sent to the stabilizer section where purge gas and synthetic crude are separated. The gas is used as fuel, and the synthetic crude can now be transported or stored.

In the related Tervahl T process (a thermal process but covered here for convenient comparison with the Tervahl T process), the feedstock is heated to the desired temperature using the coil heater and heat recovered in the stabilization section and held for a specified residence time in the soaking drum. The soaking drum effluent is quenched and sent to a conventional stabilizer or fractionator where the products are separated into the desired streams. The gas produced from the process is used for fuel.

4.21 Unicracking Process

Unicracking is a fixed-bed catalytic process that employs a high-activity catalyst with a high tolerance for sulfur and nitrogen compounds and can be regenerated. The design is based upon a single-stage or a two-stage system with provisions to recycle to extinction (RAROP, 1991, p. 79; Khan and Patmore, 1998; Speight, 2000; Parkash, 2003; Ackelson, 2004; Ancheyta and Speight, 2007; Gary et al., 2007; Speight, 2014; Hsu and Robinson, 2017; Speight, 2017). In the process, a two-stage reactor system receives untreated feed, makeup hydrogen, and recycled gas at the first stage, in which naphtha conversion may be as high as 60% *v*/v. The reactor effluent is separated to recycled gas, liquid product, and unconverted oil. The second-stage oil may be either once-through or recycle cracking; feedstock to the second sage is a mixture of unconverted first-stage oil and second-stage recycle.

The feedstock and hydrogen-rich recycled gas are preheated, mixed, and introduced into a guard reactor that contains a relatively small quantity of the catalyst. The guard chamber removes particulate matter and any salt remaining in the feedstock. The effluent from the guard chamber flows down through the main reactor, where it contacts one or more catalysts designed for removal of metals and sulfur. The catalysts, which induce desulfurization, denitrogenation, and hydrocracking, are based upon both amorphous and molecular-sieve-containing supports. The product from the reactor is cooled, separated from hydrogen-rich recycled gas, and either stripped to meet fuel oil flash point specifications or fractionated to produce distillate fuels, upgraded vacuum gas oil, and upgraded vacuum residuum. Recycled gas, after hydrogen sulfide removal, is combined with makeup gas and returned to the guard chamber and main reactors.

The most commonly implemented configuration is a single-stage Unicracking design, where the fresh feedstock and recycle oil are converted in the same reaction stage. This configuration simplifies the overall unit design by reducing the quantity of equipment in high-pressure service and keeping high-pressure equipment in a single train. The two-stage design has a separation system in each reaction stage. Two-stage flow schemes can be employed in specific situations such as the two-stage Unicracking process that can be a separate hydrotreating stage or a two-stage hydrocracking process. In the separate hydrotreating flow scheme, the first stage provides only hydrotreating, while in the two-stage hydrocracking process, the first stage provides hydrotreating and partial conversion of the feed. The second-stage of the two-stage design provides the remaining conversion of recycle oil so that overall high conversion from the unit is achieved. These flow schemes offer several advantages in processing heavier and highly contaminated feeds. Two-stage flow schemes are economical when the throughput of the unit is relatively high, but the overall optimum flow scheme depends on (i) the feedstock type, (ii) feedstock capacity, and (iii) the product slate objectives. Also, the design of hydrocracking catalyst changes depending upon the type of flow scheme employed. The hydrocracking catalyst needs to function within the reaction environment and severity created by the flow scheme that is chosen. Moreover, the two-stage flow scheme provides a unique reaction environment for the second-stage hydrocracking catalyst, and having come through the first stage, the feedstock is cleaner and less likely to foul the catalyst, thereby offering a significant boost to the cracking activity and life cycle of the catalyst.

In addition, further advances in the two-stage Unicracking process design have included several innovations in each reaction section of the design. The pretreating section uses a high-activity pretreating catalyst that allows hydrotreating at a higher severity, providing good-quality feedstock for the first-stage hydrocracking section and enabling maximum first-stage selectivity to high-quality distillate. The second stage is optimized by use of second-stage hydrocracking catalyst that is specifically designed to take advantage of the cleaner reaction environment. The second-stage catalyst is designed so that the cracking and metal functions are balanced, and at the same time, the second-stage hydrocracking severity is optimized so that maximum distillate selectivity is obtained from the second stage of hydrocracking (Thakkar et al., 2007).

The high efficiency of the process is due to the efficient distribution of the feedstock and hydrogen that occurs in the reactor where a proprietary liquid distribution system is employed. In addition, the process catalyst (also proprietary) was designed for the desulfurization of viscous feedstocks and is not merely an upgraded gas oil hydrotreating catalyst as often occurs in various processes. It is a cobalt-molybdenum-alumina catalyst with a controlled pore structure that permits a high degree of desulfurization and, at the same time, minimizes any coking tendencies.

The process uses base metal or noble metal hydrogenation activity promoters impregnated on combinations of zeolites and amorphous-aluminosilicates for cracking activity. The specific metals chosen and the proportions of the metals, zeolite, and nonzeolite aluminosilicates are optimized for the feedstock and desired product balance. This is effective in the production of clean fuels, especially for cases where a partial conversion Unicracking unit and a fluid catalytic cracking unit are integrated.

The Unicracking process converts feedstocks into lower-molecular-weight products that are more saturated than the feed. Feedstocks include atmospheric gas oil, vacuum gas oil, fluid catalytic cracking/resid catalytic cracking cycle oil, coker gas oil, deasphalted oil, and naphtha. Hydrocracking catalysts promote sulfur and nitrogen removal, aromatic saturation, and molecular-weight reduction. All of these reactions consume hydrogen, and as a result, the volume of recovered liquid product normally exceeds the feedstock. Many units are operated to make naphtha (for petrochemical or motor-fuel use) as a primary product.

Unicracking catalysts are designed to function in the presence of hydrogen sulfide (H_2S) and ammonia (NH_3). This gives rise to an important difference between Unicracking and other hydrocracking processes: the availability of a single-stage design. In a single-stage unit, the absence of a stripper between treating and cracking reactors reduces investment costs by making use of a common recycled gas system. Process objectives determine catalyst selection for a specific unit. Product from the reactor section is condensed, separated from hydrogen-rich gas, and fractionated into desired products. Unconverted oil is recycled or used as (i) blend stock for the production of lubricating oil, (ii) fluid catalytic cracking feedstock, or (iii) feedstock for the ethylene cracking unit (Parihar et al., 2012). In addition, *mild hydrocracking* technology enables optimization of a hydroprocessing refinery asset to produce high-quality clean fuels at lower costs and more attractive return on investments than alternative technologies.

The *advanced partial conversion unicracking* (APCU) process is a recent advancement in the area of ultra-low-sulfur diesel (ULSD) production and feedstock pretreatment for catalytic cracking units. At low conversions (20%–50%) and moderate pressure, the advanced partial conversion unicracking technology provides an improvement in product quality compared with traditional mild hydrocracking. In the process, high sulfur feeds such as vacuum gas oil and cycle gas oil are mixed with a heated hydrogen-rich recycled gas stream and passed over consecutive beds of high-activity pretreat catalyst and distillate selective unicracking catalyst. This combination of catalysts removes refractory sulfur and nitrogen contaminants, saturates polynuclear aromatic compounds, and converts a portion of the feedstock to ultra-low-sulfur diesel fuel. The hydrocracked products and desulfurized feedstock from a fluid catalytic cracking unit are separated at reactor pressure in an enhanced hot separator. The overhead products for the separator are immediately hydrogenated in the integrated finishing reactor.

As pretreatment severity is increased, conversion increased in the fluid catalytic cracker, and both naphtha and alkylate octane-barrel output per barrel of cat cracker feedstock increase. Advanced partial conversion unicracking units can be customized to achieve maximum octane-barrel production in the cat cracker.

Another development in the unicracking family is the HyCycle Unicracking technology that is designed to maximize diesel production for full conversion applications and is an optimized process scheme intended for obtaining maximum yield of high-quality diesel fuel. The process is characterized by lowered pressure and higher space velocity in comparison with conventional units. Due to minimizing potential secondary cracking reactions, less hydrogen per barrel of feedstock is required.

4.22 Uniflex Process

The Uniflex process is an evolved version with significant changes (by UOP) of the former CANMET process that used an empty vessel hydrocracking reactor in which the feedstock is processed in the presence of an iron sulfide-based catalyst deposited on particles of coal (Speight, 2000, 2014, 2017; Parkash, 2003; Ancheyta and Speight, 2007; Gary et al., 2007; Hsu and Robinson, 2017). The process (Fig. 11.5) is a slurry hydrocracking process that achieves higher conversion and produces two times the diesel yield compared with delayed coking, which can lead to double the refinery profit margin. Because the desulfurization activity of iron is very low, molybdenum can be added at a level of tens of ppm in the form of molybdenum naphthenate. The reaction products were fractionated and sent to the hydrotreatment unit, while the unconverted residue (5%–10% *v*/v of the feedstock) can be burned or gasified.

The flow scheme for the Uniflex process is similar to that of a conventional UOP Unicracking process unit—liquid feedstock and recycled gas are heated to temperature in separate heaters, with a small portion of the recycled gas stream and the required amount of catalyst being routed through the oil heater (Gillis et al., 2010). The outlet streams from both heaters are fed to the bottom of the slurry reactor. The reactor effluent is quenched at the reactor outlet to terminate reactions and then flows to a series of separators with gas being recycled back to the reactor. Liquids flow to the unit's fractionation section for recovery of light ends, naphtha, diesel, vacuum gas oils, and pitch (cracked residuum). Vacuum gas oil is partially recycled to the reactor for further conversion.

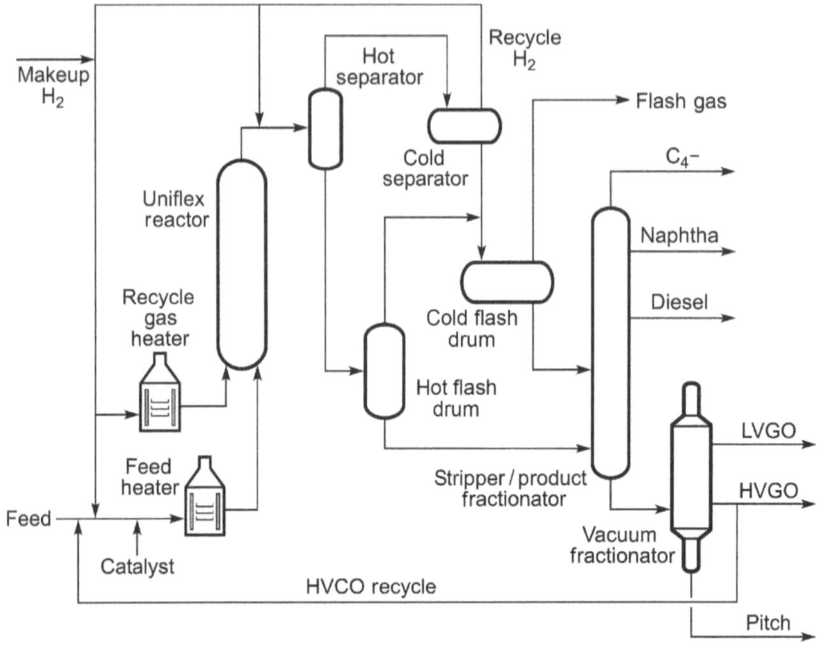

FIG. 11.5 Uniflex process.

The heart of the Uniflex process is the upflow reactor that operates at moderate temperature (440–470°C, 815–880°F) and 2000psi. The feedstock distributor, in combination with optimized process variables, promotes intense back-mixing (which provides near isothermal reactor conditions) in the reactor without the need for reactor internals or liquid recycle ebullating pumps. The back-mixing allows the reactor to operate at the higher temperatures required to maximize vacuum residue conversion. The majority of the products vaporize and quickly leave the reactor (thereby minimizing the potential for secondary cracking reactions) while the residence time of the higher-boiling constituents of the feedstock is maximized.

The process employs a proprietary, dual-function nanosized solid catalyst that is blended with the feedstock to maximize conversion of high-molecular-weight components and inhibit coke formation. Specific catalyst requirements depend on feedstock quality and the required severity of operation. The primary function of the catalysts is to effect mild hydrogenation activity for the stabilization of cracked products while also limiting the saturation of aromatic rings. Because of the hydrogenation function, the catalyst also decouples the relationship between conversion and the propensity for carbon residue formation of the feedstock.

4.23 Veba Combi-Cracking Process

The Veba Combi Cracking (VCC) process is a thermal hydrocracking process for converting viscous feedstocks (Niemann et al., 1988; RAROP, 1991, p. 81; Wenzel and Kretsmar, 1993; Speight, 2000; Parkash, 2003; Ancheyta and Speight, 2007; Gary et al., 2007; Speight, 2014; Hsu and Robinson, 2017; Speight, 2017). The process is based on the Bergius-Pier technology that was used for coal hydrogenation in Germany up to 1945. The viscous feedstock is hydrogenated (hydrocracked) using a commercial catalyst and liquid-phase hydrogenation reactor operating at 440–485°C (825–905°F) and 2175–4350 psi. The product obtained from the reactor is fed into the hot separator operating at temperatures slightly below the reactor temperature. The liquid and solid materials are fed into a vacuum distillation column and the gaseous products are fed into gas-phase hydrogenation reactor operating at an identical pressure.

In the process, the residue feedstock is slurried with a small amount of finely powdered additive and mixed with hydrogen and recycled gas prior to preheating. The feedstock mixture is routed to the liquid-phase reactors. The reactors are operated in an upflow mode and arranged in series. In a once-through operation, conversion rates of >95% are achieved. Substantial conversion of asphaltene constituents, desulfurization, and denitrogenation takes place at high levels of residue conversion. Temperature is controlled by a recycled gas quench system.

The flow from the liquid-phase hydrogenation reactors is routed to a hot separator, where gases and vaporized products are separated from unconverted material. A vacuum flash recovers distillates in the hot separator bottom product. The hot separator top product, together with recovered distillates and straight-run distillates, enters the gas-phase hydrogenation reactor. The gas-phase hydrogenation reactor operates at the same pressure as the liquid-phase hydrogenation reactor and contains a fixed bed of commercial hydrotreating catalyst. The operation temperature (340–420°C, 645–790°F) is controlled by a hydrogen quench. The system operates in a trickle-flow mode, which may not be efficient for some viscous feedstocks. The separation of the synthetic crude from associated gases is performed in a cold separator system. The synthetic crude may be sent to stabilization and fractionation unit as required. The gases are sent to a lean oil scrubbing system for contaminant removal and are recycled.

The hydrotreating stage is typically a catalytic fixed-bed reactor operated under essentially the same pressure as the primary conversion stage. This second stage may be designed for either hydrotreating or hydrocracking applications. Additional low-value refinery streams such as gas oil, deasphalted oil, or catalytic cracker cycle oil may also be directly added to the hydrotreating stage.

5. Other Options

The *heavy residue hydroconversion* (HRH) process is a new nanocatalytic technology for upgrading heavy oil, extra heavy crude oil, and tar sand bitumen (Khadzhiev et al., 2009; Zarkesh et al., 2011). In the process, the viscous feedstock is introduced to a separator to separate any lower-boiling constituents after which the nonvolatile material is sent to the reactor where mixing with hydrogen and catalyst occurs. The catalyst precursors react in situ with hydrogen sulfide in the reactor and produce the nanocatalyst. The reacted feedstock then passes into the distillation unit and unreacted portion recycles to the beginning of the process. A defined portion of this residue goes to catalyst regeneration unit. The nature of process is such that it can tolerate high amount of heavy metals, asphaltene constituents, and sulfur with an overall feedstock conversion on the order of 95% *v*/v.

Heavy crude oil, extra heavy crude oil, and tar sand bitumen—which require more energy-intensive processing than conventional crude oil—will contribute a growing fraction of fuel production. As existing reserves of conventional oil are depleted and there is greater worldwide competition for premium (e.g., light and sweet) crude oil, refineries will increasingly utilize heavy oil, extra heavy oil, and tar sand bitumen to meet the product demand.

However, like many refinery processes, the hydrocracking process can succumb to the problems encountered in hydrocracking viscous feedstocks, which can be directly equated to the amount of complex, higher-boiling constituents that may require pretreatment (Speight and Moschopedis, 1979; Speight, 2000, 2014, 2017; Gary et al., 2007; Hsu and Robinson, 2017). Processing these feedstocks is not merely a matter of applying know-how derived from refining *conventional* crude oils but requires knowledge of composition and properties (Chapter 1) (Speight, 2001, 2015). The attempts to modify the process (Section 4) are complex not only in terms of the carbon number and boiling point ranges but also because a large part of this *envelope* falls into a range of model compounds where very little is known about the properties.

Furthermore, it is apparent that the conversion of viscous feedstocks requires new lines of thought to develop suitable processing scenarios (Celestinos et al., 1975). Indeed, the use of thermal process (*carbon rejection processes*) and of hydrothermal processes (*hydrogen addition processes*), which were inherent in the refineries designed to process lower-boiling feedstocks, has been a particular cause for concern. This has brought about, and will continue to bring about in the refinery of the future, the evolution of processing schemes that accommodate the heavier feedstocks (Khan and Patmore, 1998; Speight, 2014; Hsu and Robinson, 2017; Speight, 2017).

Improved catalysts are now available based on a better understanding of asphaltene chemistry, and this will be a focus of catalyst manufacture—to hydrocrack asphaltene constituents without serous deleterious effect on the catalyst. In this respect, a reexamination of the CANMET process is warranted since processing Athabasca bitumen using this process gave good result and the iron-based catalyst also acted as a scavenger for coke formers—other options include the addition of metal oxides as scavengers for coke formers and sulfur. However, the refining industry has remained cautious, and investment in these technologies has been disappointing. In fact, application of hydrogen addition technologies to viscous feedstocks only accounts less than one-third of the global residue upgrading capacity.

The use of ebullating-bed technologies was first introduced in the 1960s in an attempt to overcome problems of catalyst aging and poor distribution in fixed-bed designs. Hydrogen and feedstock enter at the bottom of the reactor, thereby expanding the catalyst bed. Although catalyst performance can be kept constant because catalyst can be replaced online, the ebullition results in a back mixed reactor; therefore, desulfurization and hydroconversion are less than obtainable in a fixed-bed unit. Currently, in order to limit coking, most commercial ebullating-bed units operate in the 70%–85% desulfurization range and 50%–70% *v*/v, resid conversion.

Development work will continue, and ebullating-bed units will see more use and have a greater impact of resid conversion operations. Improvements such as (i) second-generation catalyst technology, which will allow higher conversion to a stable product; (ii) catalyst rejuvenation, which allows spent catalyst to be reused to a greater extent than current operations allow; and (iii) new reactor designs raising single train-size greater throughput.

Slurry-phase hydrocracking of viscous feedstocks and the latest development of dispersed catalysts present strong indications that such technologies will play a role in future refineries. Slurry-phase (or slurry-bed) hydroprocessing can be used for hydroprocessing of feeds with very high metal content to obtain lower-boiling products using a single reactor. The slurry-based technologies combine the advantages of the carbon rejection technologies in terms of flexibility with the high performances peculiar to the hydrogen addition processes and achieve a similar intimate contacting of the feedstock and catalyst and may operate with a lower degree of back-mixing than occurs in ebullated-bed units. Also, in contrast to fluid-bed units and ebullated-bed units, slurry-bed processing uses a smaller amount of catalyst (in the form of finely divided powder) that can be an additive or a catalyst (or catalyst precursors). The catalyst is mixed with the viscous feedstock, and both are fed upward with hydrogen through an empty reactor vessel. Since the oil and catalyst flow cocurrently, the mixture approaches plug-flow behavior (Chapter 6). When the reaction is complete, the spent catalyst leaves the slurry-bed reactor together with the nonvolatile fraction.

Catalyst improvements will continue to (i) improve hydrocracking activity, (ii) reduce catalytic deactivation, and (iii) increase cycle length. The development in nonprecious metal catalysts, heteropolyanions to improve metal dispersions, zeolite beta, and acid-cracking-based formulations of highly active hydrocracking catalysts has already added (and will continue to add) flexibility in the operations of hydroprocessing units. New formulations that employ amorphous silica-alumina supports and dealuminated zeolite Y will be readily available and offer high activity with high stability. These designs allow for lower operating pressures, increased run length, and higher naphtha and diesel yields.

Another central focus will be the reduction of reducing hydrogen consumption while maintaining product quality. Catalysts that can withstand organic nitrogen contamination are being developed for lower-cost, single-stage units. The addition of metal traps upstream of the hydroprocessing unit is a solution to protect highly active catalyst from high metal feeds that will see wider application. In particular, the response of the refining industry to opportunities for processing viscous feedstocks and requirements to comply with carbon dioxide will need to be addressed. Furthermore, gasification with carbon capture and the use of biomass as feedstock should help refiners meet emissions reduction requirements for carbon dioxide.

In addition, since biomass gasification and Fischer-Tropsch conversion are considered promising route in next-generation biofuel developments (Speight, 2008, 2011a, 2011b, 2014), refiners should closely monitor the latest refinery-related advances and future directions.

In summary, trends in quality of crude feedstocks have shown a steady decline over the past three decades and are reflected by declining API gravity and increasing sulfur content requiring changes in hydrocracking operations (Butler et al., 2009). Furthermore, understanding the fundamental of viscous feedstocks is not always sufficiently adequate to predict processing behavior (Niccum and Northup, 2006). In fact, it is only by comprehensively considering related factors such as (i) the properties of feedstock, (ii) catalyst performance, (iii) product requirements, (iv) operating parameters, and (v) cycle length that optimal results can be achieved. Therefore, further improvements of the hydrocracking process and catalysts, which can tolerate a high content of impurities and metals, are major challenges for the refineries (Putek et al., 2008). Catalyst activity, selectivity, particle size and shape, pore size and distribution, and the type of the reactor have to be optimized according to the properties of the viscous feedstocks and to the desired purification and conversion levels.

Furthermore, processes that offer higher conversion and improved product quality for downstream processing (such as the UOP Uniflex process, formerly the CANMET process) will be in great demand (Gillis et al., 2009, 2010). Integration of such processes with existing coking capacity offers many unique benefits (Gillis et al., 2009, 2010). The effective use of existing assets requires both individual process depth and a breadth of refinery knowledge and expertise.

References

Ackelson, 2004. UOP unicracking process for hydrocracking. In: Meyers, R.A. (Ed.), Handbook of Crude Oil Refining Processes. McGraw-Hill, New York. (Chapter 7.2).

Feedstock evaluation and composition. Ancheyta, J., Speight, J.G. (Eds.), 2007. Hydroprocessing of Heavy Oils and Residua. CRC Press, Taylor & Francis Group, Boca Raton, FL.

Babich, I.V., Moulijn, J.A., 2003. Science and technology of novel processes for deep desulfurization of oil refinery streams: a review. Fuel 82, 607–631.

Bauman, R.F., Aldridge, C.L., Bearden Jr., R., Mayer, F.X., Stuntz, G.F., Dowdle, L.D., Fiffron, E., 1993. Preprints. Oil Sands—Our Crude Oil Future. Alberta Research Council, Edmonton, AB, Canada, p. 269.

Bearden, R., Aldridge, C.L., 1981. Novel catalyst and process to upgrade heavy oils. Energy Progr. 1, 44–48.

Bellussi, G., Rispoli, G., Landoni, A., Millini, R., Molinari, D., Montanari, E., Moscotti, D., Pollesel, P., 2013. Hydroconversion of heavy residues in slurry reactors: developments and perspectives. J. Catal. 308, 189–200.

Bhattacharyya, A., Mezza, B.J., 2010. Catalyst Composition With Nanometer Crystallites for Slurry Hydrocracking. United States Patent 7,820,135. October 26.

Bhattacharyya, A., Bricker, M.L., Mezza, B.J., Bauer, L.J., 2011. Process for Using Iron Oxide and Alumina Catalyst With Large Particle Diameter for Slurry Hydrocracking. United States Patent 8,062,505. November 22.

Billon, A., Morel, F., Morrison, M.E., Peries, J.P., 1994. Converting residues with IPP's Hyvahl and Solvahl processes. Rev. Inst. Fr. Pétrol. 49 (5), 495–507.

Bishop, W., 1990. Proceedings. Symposium on Heavy Oil: Upgrading to Refining. Canadian Society for Chemical Engineers, p. 14.

Bridjanian, H., Khadem Samimi, A., 2011. Bottom of the barrel, an important challenge of the crude oil refining industry. Crude Oil Coal 53 (1), 13–21.

Butler, G., Spencer, R., Cook, B., Ring, Z., Schleiffer, A., Rupp, M., 2009. Maximize liquid yield from extra heavy oil. Hydrocarb. Process. 88 (9), 52–55.

Celestinos, J.A., Zermeno, R.G., Van Dreisen, R.P., Wysocki, E.D., 1975. Oil Gas J. 73 (48), 127.

Colyar, J.J., Peer, E., Billon, A., Kressmann, S., More, F., 1998. H-Oil Process Based Heavy Crudes Refining Schemes. Paper No. 1988-089. Society of Crude oil Engineers, Richardson, TX.

DeCroocq, D., 1984. Catalytic Cracking of Heavy Crude Oil Hydrocarbons. Editions Technip, Paris.

Dolbear, G.E., 1998. Hydrocracking: reactions, catalysts, and processes. In: Speight, J.G. (Ed.), Crude Oil Chemistry and Refining. Taylor & Francis, Washington, DC. (Chapter 7).

Dziabala, B., Thakkar, V.P., Abdo, S.F., 2011. Combination of Mild Hydrotreating and Hydrocracking for Making Low Sulfur Diesel and High Octane Naphtha. United States Patent 8,066,867. November 29.

Fornoff, L.L., 1982. Proceedings. Second International Conference on the Future of Heavy Crude and Tar Sands, Caracas, Venezuela.

Gamba, S., Pellegrini, L.A., Calemma, V., Gambaro, C., 2010. Liquid fuels from Fischer–Tropsch wax hydrocracking: isomer distribution. Catal. Today 156 (1–2), 58–64.

Gary, J.G., Handwerk, G.E., Kaiser, M.J., 2007. Crude Oil Refining: Technology and Economics, fifth ed. CRC Press, Taylor & Francis Group, Boca Raton, FL.

Gillis, D., VanWees, M., Zimmerman, P., 2009. Upgrading Residues to Maximize Distillate Yields. UOP LLC, Des Plaines, IL.

Gillis, D., VanWees, M., Zimmerman, P., 2010. Upgrading residues to maximize distillate yields with UOP Uniflex process. J. Japan Crude Oil Inst. 53 (1), 33–41.

Hansen, J.A., Blom, N.J., Ward, J.W., 2010. Hydrocracking Process. United States Patent 7,749,373. July 6.

Hsu, C.S., Robinson, P.R. (Eds.), 2017. Handbook of Petroleum Technology. Springer International Publishing AG, Cham, Switzerland.

Hunter, M.G., Pasppal, D.A., Pesek, C.L., 1997. Meyers, R.A. (Ed.), Handbook of Crude Oil Refining Processes. second ed. McGraw-Hill, New York. (Chapter 7.1).

Khadzhiev, S.N., Kadiev, K.M., Mezhidov, V.K., Zarkesh, J., Hashemi, R., Masoudian, T., Seyed, K., 2009. Process for Hydroconverting a Heavy Hydrocarbonaceous Feedstock. United States Patent 7,585,406. September 8.

Khan, M.R., Patmore, D.J., 1998. Heavy oil upgrading process. In: Speight, J.G. (Ed.), Crude Oil Chemistry and Refining. Taylor & Francis, Washington, DC. (Chapter 6).

Kressmann, S., Morel, F., Harlé, V., Kasztelan, S., 1998. Recent developments in fixed-bed catalytic residue upgrading. Catal. Today 43 (3–4), 203–215.

Kressmann, S., Boyer, C., Colyar, J.J., Schweitzer, J.M., Viguié, J.C., 2000. Improvements of ebullated-bed technology for upgrading heavy oils. Rev. Inst. Fr. Pétrol. 55, 397–406.

Kressmann, S., Guillaume, D., Roy, M., Plain, C.A., 2004. New generation of hydroconversion and hydrodesulfurization catalysts. In: 14th Annual Symposium Catalysis in Crude oil Refining & Petrochemicals, King Fahd University of Crude oil & Minerals-KFUPM, The Research Institute, Dhahran, Saudi Arabia, December 5–6.

Kunnas, J., Ovaskainen, O., Respini, M., 2010. Mitigate fouling in ebullated-bed hydrocrackers. Hydrocarb. Process. 89 (10), 59–64.

Luckwal, K., Mandal, K.K., 2009. Improve hydrogen management of your refinery. Hydrocarb. Process. 88 (2), 55–61.

Morel, F., Bonnardot, J., Benazzi, E., 2009. Hydrocracking solutions squeeze more ULSD from heavy ends. 88 (11), 79–87.

Motaghi, M., Shree, K., Krishnamurthy, S., 2010. Consider new methods for bottom-of-the-barrel processing—part 1. Hydrocarb. Process. 90 (2), 35–38.

Motaghi, M., Ulrich, B., Subramanian, A., 2011. Slurry-phase hydrocracking—possible solution to refining margins. Hydrocarb. Process. 90 (2), 37–43.

Moulton, D.S., Erwin, J., 2005. Pretreatment Processes for Heavy Oil and Carbonaceous Materials. United States Patent 6,887,369. May 3.

Niccum, P.K., Northup, A.H., 2006. Economic extraction of FCC feedstock from residual oils. In: Paper AM-06-18. National Petrochemical & Refiners Association (NPRA) Annual Meeting, Salt Lake City, UT.

Niemann, K., Kretschmar, K., Rupp, M., Merz, L., 1988. Proceedings. 4th UNITAR/UNDP International Conference on Heavy Crude and Tar Sand, Edmonton, AB, Canada. vol. 5. p. 225.

Occelli, M.L., Robson, H.E., 1989. Zeolite Synthesis. Symposium Series No. 398, American Chemical Society, Washington, DC.

Olschar, M., Endisch, M., Dimmig, T.H., Kuchling, T.H., 2007. Investigation of catalytic hydrocracking of Fischer-Tropsch wax for the production of transportation fuels. Oil Gas Eur. Mag. 33 (4), 187–193.

Parihar, P., Voolapalli, R.K., Kumar, R., Kaalva, S., Saha, B., Viswanathan, P.S., 2012. Optimize hydrocracker operations for maximum distillates. Crude Oil Technol. Q. Q2, 1–8.

Parkash, S., 2003. Refining Processes Handbook. Gulf Professional Publishing, Elsevier, Amsterdam, the Netherlands.

Patel, N., Ludwig, K., Morris, P., 2005. Insert flexibility into your hydrogen network—part 1. Hydrocarb. Process. 84 (9), 73–84.

Peries, J.P., Quignard, A., Farjon, C., Laborde, M., 1988. Thermal and catalytic ASVAHL processes under hydrogen pressure for converting heavy crudes and conventional residues. Rev. Inst. Fr. Pétrol. 43 (6), 847–853.

Putek, S., Januszewski, D., Cavallo, E., 2008. Upgrade hydrocracked resid through integrated hydrotreating. Hydrocarb. Process. 84 (9), 83–92.

Rana, M.S., Sámano, V., Ancheyta, J., Diaz, J.A.I., 2007. A review of recent advances on process technologies for upgrading of heavy oils and residua. Fuel 86, 1216–1231.

RAROP, 1991. Kamiya, Y. (Ed.), Heavy Oil Processing Handbook. Research Association for Residual Oil Processing, Agency of Natural Resources and Energy, Ministry of International Trade and Industry, Tokyo, Japan.

Rashid, K., 2007. Optimize your hydrocracking operations. Hydrocarb. Process. 86 (2), 55–63.

Reich, A., Bishop, W., Veljkovic, M., 1993. Preprints. Oil Sands—Our Crude Oil Future. Alberta Research Council, Edmonton, AB, Canada, p. 216.

Scherzer, J., Gruia, A.J., 1996. Hydrocracking Science and Technology. Marcel Dekker Inc, New York.

Speight, J.G., 1990. Tar sands. In: Speight, J.G. (Ed.), Fuel Science and Technology Handbook. Marcel Dekker Inc., New York. (Chapters 12–16).

Speight, J.G., 2000. The Desulfurization of Heavy Oils and Residua, second ed. Marcel Dekker Inc, New York.

Speight, J.G., 2001. Handbook of Petroleum Analysis. John Wiley & Sons Inc, Hoboken, NJ.

Speight, J.G., 2008. Synthetic Fuels Handbook: Properties, Processes, and Performance. McGraw-Hill, New York.

Speight, J.G., 2011a. The Refinery of the Future. Gulf Professional Publishing, Elsevier, Oxford, UK.

Speight, J.G. (Ed.), 2011b. Biofuels Handbook. Royal Society of Chemistry, London, UK.

Speight, J.G., 2014. The Chemistry and Technology of Petroleum Oil, fifth ed. CRC Press, Taylor & Francis Group, Boca Raton, FL.

Speight, J.G., 2015. Handbook of Petroleum Product Analysis, second ed. John Wiley & Sons Inc, Hoboken, NJ.

Speight, J.G., 2017. Handbook of Petroleum Refining. CRC Press, Taylor & Francis Group, Boca Raton, FL.

Speight, J.G., Moschopedis, S.E., 1979. The production of low-sulfur liquids and coke from Athabasca bitumen. Fuel Process. Technol. 2, 295.

Stratiev, D., Petkov, K., 2009. Residue upgrading: challenges and perspectives. Hydrocarb. Process. 88 (9), 93–96.

Sue, H., 1989. Proceedings. 4th UNITAR/UNDP Conference on Heavy Oil and Tar Sands. vol. 5, p. 117.

Thakkar, V., Meister, J.M., Rossi, R.J., Wang, L., 2007. Process and Catalyst Innovations in Hydrocracking to Maximize High Quality Distillate Fuel. Report No. 4706BUOP LLC, Des Plaines, IL.

Thompson, G.J., 1997. Meyers, R.A. (Ed.), Handbook of Crude Oil Refining Processes. McGraw-Hill, New York. (Chapter 8.4).

Van Driesen, R.P., Caspers, J., Campbell, A.R., Lunin, G., 1979. Hydrocarb. Process. 58 (5), 107.

Vauk, D., Di Zanno, P., Neri, B., Allevi, C., Visconti, A., Rosanio, L., 2008. What are possible hydrogen sources for refinery expansion? Hydrocarb. Process. 87 (2), 69–76.

Wenzel, F., Kretsmar, K., 1993. Preprints. Oil Sands—Our Crude Oil Future. Alberta Research Council, Edmonton, AB, Canada, p. 248.

Zarkesh, J., Ghaedian, M., Hashemi, Khademsamimi, A., Kadzhiev, S., 2011. Heavy refinery schemes based on new nano catalytic HRH technology. In: Proceedings. 2nd International Conference on Chemical Engineering and Applications, International Proceedings of Chemical, Biological and Environmental Engineering (IPCBEE), IACSIT Press, Singaporevol. 23, pp. 66–70.

Zhang, S., Liu, D., Deng, W., Que, G., 2007. Production of light oil by oxidative cracking of oil sand bitumen. Energy Fuel 21, 3057–3062.

Further Reading

Hunter, M.G., Vivas, A.H., Jensen, L.S., Low, G.G., 2010. Partial Conversion Hydrocracking Process and Apparatus. United States Patent 7,763,218. July 27.

Chapter 12

Upgrading by Solvent Treatment

1. Introduction

Solvent-based processes are physical separation processes in which the feedstock is divided into its components by means of a solvent used as an absorption medium. It is a separating process based on specific gravity (molecular weight), as opposed to the boiling-point distillation. These processes take advantage of the insolubility of aromatic compounds, and many heteroatom-containing compounds are insoluble in paraffin liquids.

Solvent deasphalting (sometimes referred to as solvent extraction; Fig. 12.1) is essentially a solvent-based extraction process, and the required solvent is usually available within the refinery. The process separates oil from carbon-rich components, resins, and asphaltenes and makes it available to convert it to lube stock or as feedstock for other secondary processing facilities. Feedstock impurities such as sulfur and metals are concentrated in the insoluble phase. The flexibility, inherent in the process, allows wide variation in product quality to meet specific downstream process needs. The process continues to play a significant role, as one of the key process units of the modern refinery complex (Parkash, 2003; Gary et al., 2007; Speight, 2014, 2017; Hsu and Robinson, 2017).

From a viscous feedstock, the products are a deasphalted liquid (deasphalted oil) and an insoluble asphalt-like material that has high concentrations of contaminants (metal-containing constituents and asphaltene constituents). The deasphalted oil is used as a feedstock for the preparation of finished lubricants and feedstock for catalytic cracking units or hydrocracking units or even as a feedstock to thermal cracking processes.

At present, solvent deasphalting is one of the most interesting viscous feedstock (in the case of crude oil residua, referred to as bottom of the barrel) processing technologies to process viscous feedstocks and modern refineries. The process is advantageous for application to viscous feedstocks because of its relatively low costs and the implicit possibility of obtaining a wide variety of deasphalted oils. The process offers a high selectivity for asphaltene constituents, considerable metal rejection, a certain selectivity to reject coke-forming precursors, and minor selectivity for sulfur and nitrogen constituents. Solvent deasphalting processes have not realized their maximum potential in terms

Heavy Oil Recovery and Upgrading. https://doi.org/10.1016/B978-0-12-813025-4.00012-X

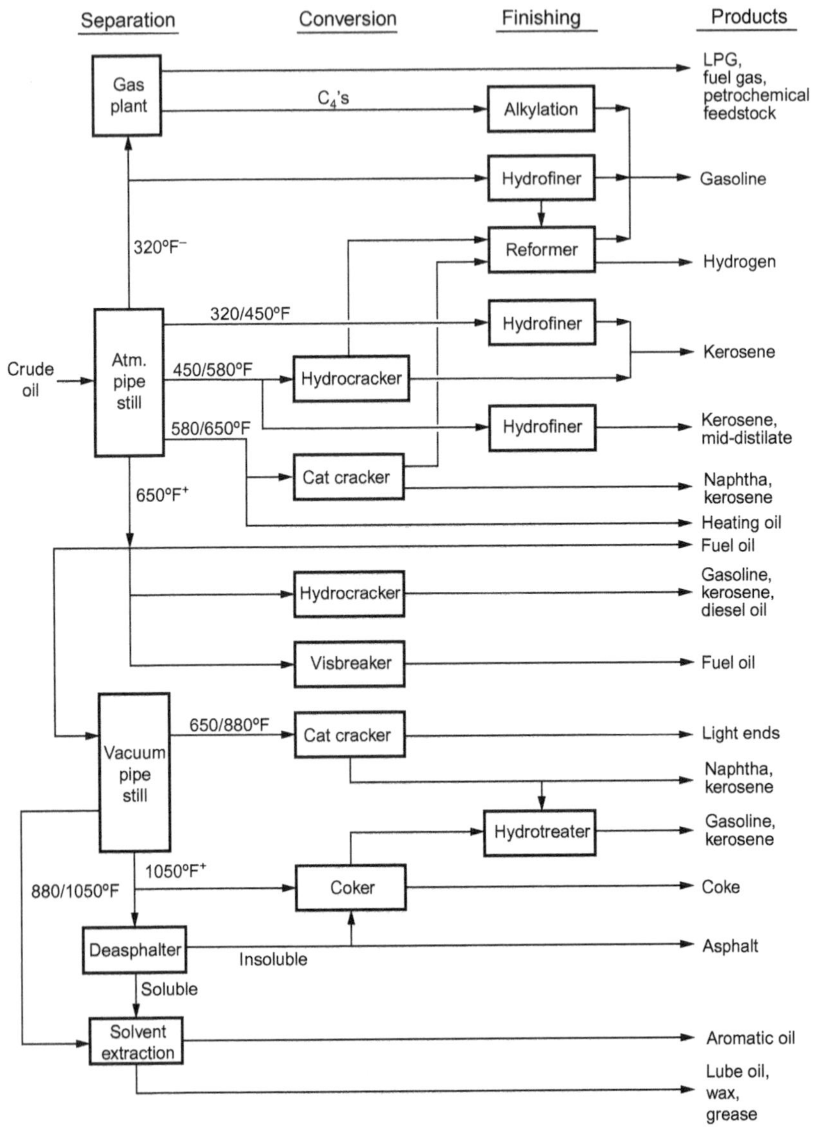

FIG. 12.1 Overview of a refinery showing placement of the solvent deasphalting (deasphalter) unit.

of use with heavy oil, extra heavy oil, and tar sand bitumen. With ongoing improvements in energy efficiency, such processes would display its effects in a combination with other processes. Solvent deasphalting allows removal of sulfur and nitrogen compounds as well as metallic constituents by balancing yield with the desired feedstock properties (Parkash, 2003; Gary et al., 2007; Speight, 2014, 2017; Hsu and Robinson, 2017).

Due to the major price differences, more and more viscous, high-sulfur crude oils (including heavy oil, extra heavy oil, and tar sand bitumen) are used in refinery processes because of the impending dearth of low-sulfur feedstocks. This requires more efficient processing technologies. Furthermore, in terms of the production of good-grade asphalt, the asphalt constituents (asphaltene constituents plus resin constituents plus a high-boiling vacuum gas oil as the flux) cannot be separated from the viscous feedstock by distillation without destroying the characteristics of these constituents that would lead to a poorer grade (cracked asphalt) product. However, the asphalt constituents can be precipitated from the viscous feedstock using solvents in which the oil fraction is soluble and the asphalt is insoluble. This is where the presence of solvent deasphalting unit in a refinery proves to be a valuable asset.

While solvent deasphalting has been used for more than six decades to upgrade nonvolatile refinery residua, the technology continues to evolve over time. The process is a robust and economical process that uses an aliphatic solvent to separate the typically more valuable oils and resins from the more aromatic and asphaltenic components of the feedstock. The earliest commercial applications of the process used propane as the solvent to extract high-quality lubricating oil bright stock from vacuum residua. These applications were called propane deasphalting (or propane deresining when used to separate high-molecular-weight resins from a viscous feedstock). Solvent deasphalting processes have gradually been extended to include the preparation of feedstocks for catalytic cracking, hydrocracking, and hydrodesulfurization units and the production of specialty asphalt products.

The solvents used for the various processes vary from propane to pentane to produce feedstocks for a variety of refining units. For example, propane deasphalting is commonly used to precipitate asphaltene constituents and resin constituents from viscous feedstocks such as heavy oil, extra heavy oil, and tar sand bitumen. The deasphalted oil (the soluble product of propane deasphalting that is reactively low in the constituents that typically produce high yields of coke and the metal-containing catalyst poisons) is then sufficiently *clean* to be sent to a hydrotreating unit or a hydrocracking unit or to be used as blend stock for fuel oil (Parkash, 2003; Gary et al., 2007; Speight, 2014, 2017; Hsu and Robinson, 2017).

Solvent deasphalting processes are a major part of refinery operations and are not often appreciated for the tasks for which they are used (Parkash, 2003; Gary et al., 2007; Speight, 2014, 2017; Hsu and Robinson, 2017). In the solvent deasphalting processes, an alkane is injected into the feedstock to disrupt the dispersion of components and causes the polar constituents to precipitate. Propane (or sometimes propane/butane mixtures) is extensively used for deasphalting and produces a deasphalted oil and propane deasphalter asphalt (Dunning and Moore, 1957). Propane has unique solvent properties; at lower temperatures (38–60°C and 100–140°C), paraffins are very soluble in propane, and at higher temperatures (~93°C, 200°F), all hydrocarbon derivatives are almost insoluble in propane.

A *solvent deasphalting* unit processes the residuum from the vacuum distillation unit and produces deasphalted oil, used as feedstock for a fluid catalytic cracking unit, and the asphaltic residue (deasphalter tar and deasphalter bottoms), which, as an insoluble fraction, can only be used to produce asphalt or as a blend stock or visbreaker feedstock for low-grade fuel oil. Solvent deasphalting processes have not realized their maximum potential. With ongoing improvements in energy efficiency, such processes would display its effects in a combination with other processes. Solvent deasphalting allows the removal of sulfur and nitrogen compounds as well as metallic constituents by balancing yield with the desired feedstock properties (Dunning and Moore, 1957; Ditman, 1973).

In a more recent innovation (McCoy et al., 2010), a feedstock stream comprising the viscous feedstock is fed to a solvent deasphalter wherein it is contacted with a deasphalting solvent or fluid to produce a composition comprising a mixture or slurry of solvent containing a soluble portion of the feedstock and a viscous tar-like fraction comprising the insoluble portion of the feedstock tar. These fractions may be separated in the deasphalter apparatus, such as by gravity settling wherein the nonvolatile fraction is taken off as bottoms and the solvent-soluble fraction taken as overflow or overheads with the solvent. The overflow or overheads are sent to a solvent recovery unit, such as a distillation apparatus, wherein solvent is recovered as overheads and a deasphalted tar fraction is taken off as a sidestream or bottoms.

Most of the metals present in refinery feedstocks are concentrated in the asphaltene constituent, and by using solvent deasphalting, it is possible to recover significant amounts of partially demetallized and deasphalted oil from the residues that can be processed in fluid catalytic cracking units (Chapter 9) or in hydrocracking units (Chapter 11).

The main advantages of solvent deasphalting are relatively low investment and operational costs, while its principal limitation is that to obtain deasphalted oil with a low content of pollutants (in particular, sulfur, nitrogen, metals, and coke-forming constituents), the yield must be limited (Ditman, 1973). As a result, significant quantities of by-products are generated (asphaltene constituents plus resin constituents) that can be used as components of low-quality fuels or as sources of carbon for the production of synthesis gas and, therefore, hydrogen in suitable gasification units (Speight, 2014). The application of the solvent deasphalting process to upgrade heavy crude oil, extra heavy crude oil, and tar sand bitumen has led to the development of a variety of technologies in which the typical solvent deasphalting unit is combined with other processes.

Dewaxing processes are not considered here. Heavy oil, extra heavy crude oil, and tars and bitumen are not typically known for measurable wax content as is found in the more conventional crude oils (Speight, 2014, 2017).

2. Solvent-Based Processes

Crude oil processing normally involves separation into various fractions that require further processing in order to produce marketable products. The initial separation process is distillation in which crude oil is separated into fractions of increasingly higher-boiling range fractions. The initial separation process is distillation in which crude oil is separated into fractions of increasingly higher-boiling range fractions (Parkash, 2003; Gary et al., 2007; Speight, 2014, 2017; Hsu and Robinson, 2017). Since crude oil fractions are subject to thermal degradation, there is a limit to the temperatures that can be used in simple separation processes. The crude cannot be subjected to temperatures much above 395°C (740°F), irrespective of the residence time, without encountering some thermal cracking. Therefore, to separate the higher-molecular-weight and higher-boiling fractions from crude, special processing steps must be used.

Because crude oil fractions are subject to thermal degradation, there is a limit to the temperature that can be effectively used in a separation process, that is, crude cannot be subjected to temperatures above 370°C (700°F) without the occurrence of thermal cracking. The physical separation of higher-molecular-weight and higher-boiling fractions from the crude is first accomplished in the vacuum distillation process, which is followed by the solvent deasphalting process. The process itself involves adding a relatively small portion of solvent upstream to the charge crude, and processing it through the extraction tower in which the desired oil goes to the solvent and the asphalt precipitates toward the bottom. As the oil and solvent rise in the tower, the temperature is increased in order to control the quality of the product; separating the oil from asphalt is achieved by maintaining a temperature gradient across the extraction tower and by varying the solvent/oil ratio. A key physical separation process, solvent deasphalting, has been used by refiners to produce gas oil and lubricating oil bright stock for many years.

Solvent deasphalting is a separation process that represents a further step in the minimization of the nonvolatile and insoluble residue. However, solvent deasphalting processes, far from realizing their maximum potential for viscous feedstocks, are now under further investigation, and with ongoing improvements in energy efficiency, such processes are starting to display maximum benefits when used in combination with other processes. The process takes advantage of the fact that the maltene constituents are more soluble in low-boiling paraffinic solvents than asphaltene constituents. This solubility increases with solvent molecular weight and decreases with temperature (Girdler, 1965; Mitchell and Speight, 1973). As with vacuum distillation, there are constraints with respect to how deep a solvent deasphalting unit can cut into the viscous feedstock or the degree to which deasphalted oil can be produced. In the case of solvent deasphalting, the constraint is usually related to deasphalted oil quality specifications required by downstream conversion units.

Depending on the quality of the deasphalted oil, it can be used (i) as a feedstock for a catalytic cracking unit or (ii) as a feedstock for a hydrocracking unit (Parkash, 2003; Gary et al., 2007; Speight, 2014, 2017; Hsu and Robinson, 2017). The insoluble portion of the feedstock (frequently referred to as the deasphalter bottoms) is typically used (i) as a feedstock for a delayed coking unit, (ii) as a fuel oil blending component, (iii) as a feedstock to a gasification unit, or (iv) or as a blend stock for asphalt productions.

However, solvent deasphalting has the flexibility to produce a wide range of deasphalted oil that matches the desired properties. The process has very good selectivity for asphaltene constituents (and, to a lesser extent, resin constituents) and metal rejection. There is also some selectivity for rejection of coke-forming precursors, but there is less selectivity for sulfur-containing and nitrogen-containing constituents. The process is best suited for the more paraffin-type viscous feedstock (which is unlikely when considering heavy oil, extra heavy oil, and tar sand bitumen as the deasphalter feedstock) with a somewhat lower efficiency when applied to a high-asphaltene viscous feedstock that contains high proportions of metal-containing constituents and coke-forming constituents. The advantages and disadvantages of the process are that it performs no conversion and produces a very-high-viscosity by-product deasphalter bottoms and, where high-quality deasphalted oil is required, the process is limited in the quality of feedstock that can be economically processed. In those situations where there is a ready outlet use for the bottoms, solvent deasphalting is an attractive option for treating viscous feedstocks. One such situation is the cogeneration of steam and power, both to supply the refiner's needs and for export to nearby users.

2.1 Deasphalting Process

Due to the depletion of conventional oil resources and increasing prices, various technologies for utilizing unconventional oil and low-value crude residues, which have not been fully exploited, are currently being explored. The exploitation of unconventional oil and low-value crude residues requires upgrading processes such as carbon rejection and hydrogen addition. Among many existing upgrading processes, solvent deasphalting, a technology for removing asphaltene-rich pitch and producing higher-value deasphalted oil by using paraffinic solvents, is promising because it offers the advantages of low installation cost and flexibility in terms of the control of the quality of the deasphalter bottoms (often referred to as deasphalter pitch) and the deasphalted oil.

The deasphalting process is a mature process, but as refinery operations evolve, it is necessary to include a description of the process here so that the new processes might be compared with new options that also provide for deasphalting various feedstocks. In the process, the feedstock is mixed with dilution solvent from the solvent accumulator and then cooled to the desired temperature before entering the extraction tower. Because of its high viscosity, the charge oil can neither be cooled easily to the required temperature nor will it mix readily with solvent in the extraction tower. By adding a relatively small portion of solvent upstream of the charge cooler (insufficient to cause phase separation), the viscosity problem is avoided.

The process has been conventionally used after either atmospheric or vacuum distillation as a separation process that is also applicable to heavy oil, extra heavy oil, and tar sand bitumen. The process takes advantage of the fact that the maltene fraction present in viscous feedstocks is more soluble in low-boiling paraffin solvents than the asphaltene constituents (Speight, 2011, 2014, 2015, 2017). This solubility increases with solvent molecular weight and decreases with temperature. The typical solvent used for the process is liquid propane (C_3H_8, propane deasphalting). In order to recover more oil from vacuum-reduced crude, mainly for catalytic cracking feedstocks, higher-molecular-weight solvents such as butane (C_4H_{10}) and pentane (C_5H_{12}) have been employed.

The feedstock, with a small amount of solvent, enters the extraction tower at a point about two-thirds up the column. The solvent is pumped from the accumulator, cooled, and entered near the bottom of the tower. The extraction tower is a multistage contactor, normally equipped with baffle trays, and the heavy oil, extra heavy oil, and tar sand bitumen flow downward, while the light solvent flows upward. As the extraction progresses, the desired oil goes to the solvent and the asphalt separate and moves toward the bottom. As the extracted oil and solvent rise in the tower, the temperature is increased in order to control the quality of the product by providing adequate reflux for optimum separation. Separation of oil from asphalt is controlled by maintaining a temperature gradient across the extraction tower and by varying the solvent/oil ratio. The tower top temperature is regulated by adjusting the feedstock inlet temperature and the steam flow to the heating coils in the top of the tower. The temperature at the bottom of the tower is maintained at the desired level by the temperature of the entering solvent. The deasphalted oil-solvent mixture flows from the top of the tower under pressure control to a kettle-type evaporator heated by low-pressure steam. The vaporized solvent flows through the condenser into the solvent accumulator.

Indeed, several of these options, such as the ROSE process, have been onstream for several years and are included here for this same reason. Thus, this section provides a one-stop discussion of solvent recovery processes and their integration into refinery operations. However, the process requires a considerable amount of expensive solvent, and thus, solvent recovery, an energy-intensive process, is required for improved efficiency (Lee et al., 2014).

The separation of viscous feedstocks into a soluble fraction and an insoluble fraction was first performed on a production scale by mixing the viscous feedstock with propane (or mixtures of *normally gaseous* hydrocarbon derivatives) and continuously decanting the resulting phases in a suitable vessel. Temperature was maintained within approximately 55°C (100°F) of the critical temperature of the solvent, at a level that would regulate the yield and properties of the deasphalted oil in solution and that would reject the heavier undesirable components as asphalt.

Currently, deasphalting combined with delayed coking (or deasphalting combined with fluid coking) is used frequently for residuum conversion. The high demand for coke, mainly for use in the aluminum industry, has made delayed coking a major residuum conversion process. However, many crude oils will not produce coke meeting the sulfur and metal specifications for aluminum electrodes, and coke gas oils are less desirable feedstocks for fluid catalytic cracking than virgin gas oils. In comparison, the solvent deasphalting process can apply to most viscous feedstocks. The deasphalted oil is an acceptable feedstock for both fluid catalytic cracking and, in some cases, hydrocracking. Since it is relatively less expensive to desulfurize the deasphalted oil than the viscous feedstock, the solvent deasphalting process offers a more economical route for the disposing of viscous feedstock from high-sulfur crude. However, the question of the disposal of the asphalt remains. Use as a road asphalt is common and use as a refinery fuel is less common since expensive stack gas cleanup facilities may be required when used as fuel.

In the process (Fig. 12.2), the feedstock is mixed with the dilution hydrocarbon solvent from the solvent accumulator and then cooled to the desired temperature before entering the extraction tower. Because of its high viscosity, the charge oil can neither be cooled easily to the required temperature nor will it mix readily with solvent in the extraction tower. By adding a relatively small

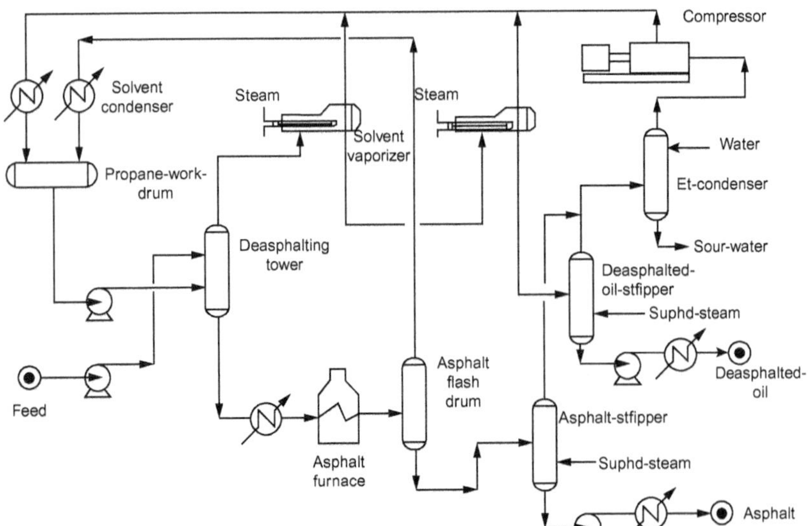

FIG. 12.2 Propane deasphalting. *(Credit: OSHA Technical Manual, Section IV, Chapter 2: Petroleum Refining Processes. http://www.osha.gov/dts/osta/otm/otm_iv/otm_iv_2.html.)*

portion of solvent upstream of the feedstock cooler that is insufficient to cause phase separation, the viscosity problem can be mitigated.

The liquid phase flows from the bottom of the evaporator, under level control, to the deasphalted oil flash tower where it is re-boiled by means of a fired heater. In the flash tower, most of the remaining solvent is vaporized and flows overhead, joining the solvent from the low-pressure steam evaporator. The deasphalted oil, with relatively minor solvent, flows from the bottom of the flash tower under level control to a steam stripper operating at essentially atmospheric pressure. Superheated steam is introduced into the lower portion of the tower. The remaining solvent is stripped out and flows overhead with the steam through a condenser into the compressor suction drum where the water drops out. The water flows from the bottom of the drum under level control to appropriate disposal.

The asphalt-solvent mixture is pressured from the extraction tower bottom on flow control to the asphalt heater and on to the asphalt flash drum, where the vaporized solvent is separated from the asphalt. The drum operates essentially at the solvent condensing pressure so that the overhead vapors flow directly through the condenser into the solvent accumulator. Hot asphalt, with a small quantity of solvent flows, from the asphalt flash drum bottom to the asphalt stripper, which is operated at near atmospheric pressure. Superheated steam is introduced into the bottom of the stripper. The steam and solvent vapors pass overhead, join the deasphalted oil stripper overhead, and flow through the condenser into the compressor suction drum. The asphalt is pumped from the bottom of the stripper under level control to storage.

The propane deasphalting process is similar to solvent extraction in that a packed or baffled extraction tower or rotating disk contactor is used to mix the oil feedstock stocks with the solvent. In the tower method, four to eight volumes of propane are fed to the bottom of the tower for every volume of feedstock flowing down from the top of the tower. The oil, which is more soluble in the propane, dissolves and flows to the top. The asphaltene and resins flow to the bottom of the tower where they are removed in a propane mix. Propane is recovered from the two streams through two-stage flash systems followed by steam stripping in which propane is condensed and removed by cooling at high pressure in the first stage and at low pressure in the second stage. The asphalt recovered can be blended with other asphalts or heavy fuels or can be used as feedstock to the coker.

The yield of deasphalted oil varies with the feedstock, but the deasphalted oil does make less coke and more distillate than the feedstock (Parkash, 2003; Gary et al., 2007; Speight, 2014, 2017; Hsu and Robinson, 2017). Therefore, the process parameters for a deasphalting unit must be selected with care according to the nature of the feedstock and the desired final products. The metal content of the deasphalted oil is relative, and the nitrogen and sulfur contents in the deasphalted oil are also related to the yield of deasphalted oil yield (Fig. 12.3). The character of the deasphalting process is molecular weight

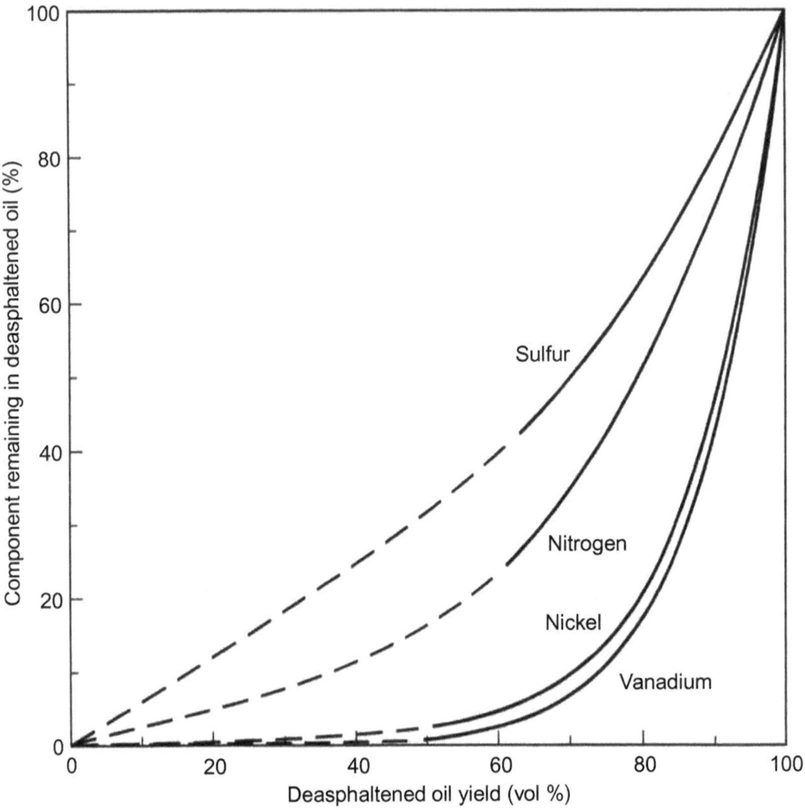

FIG. 12.3 Variation of deasphalted oil properties with yield (Ditman, 1973).

separations, and the solvent takes a crosscut across the feedstock effecting separation by molecular weight and by polarity (Fig. 12.4; Parkash, 2003; Gary et al., 2007; Speight, 2014, 2017; Hsu and Robinson, 2017). Furthermore, the process parameters for a deasphalting unit must be selected with care according to the nature of the feedstock and the desired final products.

In the first case, the *choice of solvent* is vital to the flexibility and performance of the unit. The solvent must be suitable, not only for the extraction of the desired oil fraction but also for the control of the yield and/or quality of the deasphalted oil at temperatures that are within the operating limits. If the temperature is too high (i.e., close to the critical temperature of the solvent), the operation becomes unreliable in terms of product yields and character. If the temperature is too low, the feedstock may be too viscous and have an adverse effect on the contact with the solvent in the tower. Liquid propane is by far the most selective solvent among the low-boiling hydrocarbon derivatives used for deasphalting. At temperatures ranging from 38°C to 65°C (100–150°F), most hydrocarbon derivatives are soluble in propane, while asphaltic and resinous

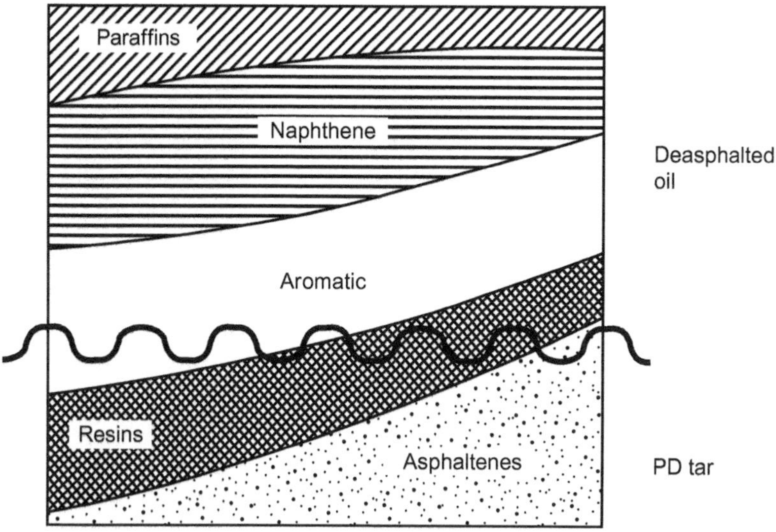

FIG. 12.4 Illustration of the deasphalting process on the basis of molecular weight and polarity. PD tar, propane deasphalter bottoms.

compounds are not, thereby allowing rejection of these compounds resulting in a drastic reduction (relative to the feedstock) of the nitrogen content and the metals in the deasphalted oil. Although the deasphalted oil from propane deasphalting has the best quality, the yield is usually less than the yield of deasphalted oil produced using a higher-molecular-weight (higher-boiling) solvent.

The ratios of propane-oil required vary from 6 to 1 to 10 to 1 by volume, with the ratio occasionally being as high as from 13 to −1. Since the critical temperature of propane is 97°C (206°F), this limits the extraction temperature to about 82°C (180°F). Therefore, propane alone may not be suitable for high-viscosity feedstocks because of the relatively low operating temperature.

Iso-butane and *n*-butane are more suitable for deasphalting high-viscosity feedstocks since their critical temperatures are higher (134°C, 273°F, and 152°C, 306°F, respectively) than that of the critical temperature of propane. Higher extraction temperatures can be used to reduce the viscosity of the viscous feedstock and to increase the transfer rate of oil to solvent.

Although *n*-pentane is less selective for metals and removal of the coke-forming constituents, *n*-pentane can increase the yield of deasphalted oil from a viscous feedstock by a factor of 2–3 over propane (Speight, 2000, 2014, 2017). However, if the content of the metals and coke-forming propensity of the pentane-deasphalted oil is too high (defined by the ensuing process), the deasphalted oil may be unsuitable as a cracking feedstock. In certain cases, the nature of the cracking catalyst may dictate that the pentane-deasphalted oil be blended with vacuum gas oil that, after further treatment such as hydrodesulfurization, produces a good cracking feedstock.

Solvent composition is an important process variable for deasphalting units (De Souza et al., 2017). The use of a single solvent may (depending on the nature of the solvent) limit the range of feedstocks that can be processed in a deasphalting unit. When a deasphalting unit is required to handle a variety of feedstocks and/or produce various yields of deasphalted oil (as is the case in these days of variable feedstock quality), a dual solvent may be the only option to provide the desired flexibility. Adjusting the solvent composition allows the most desirable product quantity and quality within the range of temperature control.

Besides the solvent composition, the *solvent/oil ratio* also plays an important role in a deasphalting operation. Solvent/oil ratios vary considerably and are governed by feedstock characteristics and desired product qualities, and for each individual feedstock, there is a minimum operable solvent/oil ratio. Generally, increasing the solvent-to-oil ratio almost invariably results in improving the deasphalted oil quality at a given yield, but other factors must also be taken into consideration, and (generalities aside) each plant and feedstock will have an optimum ratio.

The main consideration in the selection of the *operating temperature* is its effect on the yield of deasphalted oil. For practical applications, the lower limits of operable temperature are set by the viscosity of the oil-rich phase. When the operating temperature is near the critical temperature of the solvent, control of the extraction tower becomes difficult since the rate of change of solubility with temperature becomes very large at conditions close to the critical point of the solvent. Such changes in solubility cause large amounts of oil to transfer between the solvent-rich and the oil-rich phases that, in turn, cause *flooding* and/or uncontrollable changes in product quality. To mitigate such effects, the upper limits of operable temperatures must lie below the critical temperature of the solvent in order to insure good control of the product quality and to maintain a stable condition in the extraction tower.

The *temperature gradient* across the extraction tower influences the sharpness of the separation of the deasphalted oil and the asphalt because of internal reflux that occurs when the cooler oil/solvent solution in the lower section of the tower attempts to carry a large portion of oil to the top of the tower. When the oil/solvent solution reaches the steam-heated, higher-temperature area near the top of the tower, some oil of higher molecular weight in the solvent solution is rejected because the oil is less soluble in solvent at the higher temperature. The heavier oil (rejected from the solution at the top of the tower) attempts to flow downward and causes the internal reflux. In fact, generally, the greater the temperature difference between the top and the bottom of the tower, the greater will be the internal reflux, and the better will be the quality of the deasphalted oil. However, too much internal reflux can cause tower flooding and jeopardize the process.

The *process pressure* is usually not considered to be an operating variable since it must be higher than the vapor pressure of the solvent mixture at the tower operating temperature to maintain the solvent in the liquid phase. The tower pressure is usually only subject to change when there is a need to change the solvent composition or the process temperature.

Proper *contact and distribution of the oil and solvent* in the *tower* are essential to the efficient operation of any deasphalting unit. In early units and days, mixer-settlers were used as contactors but proved to be less efficient than the countercurrent contacting devices. Packed towers are difficult to operate in this process because of the large differences in viscosity and density between the asphalt phase and the solvent-rich phase.

The *extraction tower* for solvent deasphalting consists of two contacting zones: (i) a rectifying zone above the oil feedstock and (ii) a stripping zone below the oil feed. The rectifying zone contains some elements designed to promote contacting and to avoid *channeling*. Steam-heated coils are provided to raise the temperature sufficiently to induce an oil-rich reflux in the top section of the tower. The stripping zone has disengaging spaces at the top and bottom and consists of contacting elements between the oil inlet and the solvent inlet.

A *countercurrent tower* with static baffles is widely used in solvent deasphalting service. The baffles consist of fixed elements formed of expanded metal gratings in groups of two or more to provide maximum change of direction without limiting capacity. The *rotating disk contactor* has also been employed and consists of disks connected to a rotating shaft that are used in place of the static baffles in the tower. The rotating element is driven by a variable speed drive at either the top or the bottom of the column, and operating flexibility is provided by controlling the speed of the rotating element and, thus, the amount of mixing in the contactor.

In the deasphalting process, the solvent is recovered for circulation, and the efficient operability of a deasphalting unit is dependent on the design of the *solvent recovery system*. The solvent may be separated from the deasphalted oil in several ways such as conventional evaporation or the use of a flash tower. Irrespective of the method of solvent recovery from the deasphalted oil, it is usually the most efficient to recover the solvent at a temperature close to the extraction temperature. If a higher temperature for solvent recovery is used, heat is wasted in the form of high vapor temperature, and conversely, if a lower temperature is used, the solvent must be reheated, thereby requiring additional energy input. The solvent recovery pressure should be low enough to maintain a smooth flow under pressure from the extraction tower.

As always, the use of the deasphalter reject (solvent asphalt) remains an issue. It can be used (apart from its use for various types of asphalt) as feedstock to a partial oxidation unit to make a hydrogen-rich gas for use in hydrodesulfurization processes and hydrocracking processes. Alternatively, the asphalt may be treated in a visbreaker to reduce its viscosity, thereby minimizing the need for cutter stock to be blended with the solvent asphalt for making fuel oil, or hydrovisbreaking (Chapter 8) offers an option of converting the asphalt to feedstocks for other conversion processes.

However, if the content of the metals and carbon residue of the pentane-deasphalted oil is too high (defined by the ensuing process), the deasphalted oil may be unsuitable as a cracking feedstock.

For example, a mixture of propane and *n*-butane might be suitable for feedstocks that vary from vacuum residua to heavy oil, extra heavy oil, and tar sand bitumen that contain insoluble (asphaltic) fractions.

The asphalt solution from the bottom of the extraction tower usually contains less than an equal volume of solvent. A fired heater is used to maintain the temperature of the asphalt solution well above the foaming level and to keep the asphalt phase in a fluid state. A flash drum is used to separate the solvent vapor from asphalt with the design being such to prevent carryover of asphalt into the solvent outlet line and to avoid fouling the downstream solvent condenser. The solvent recovery system from asphalt is not usually subject to the same degree of variations as the solvent recovery system for the deasphalted oil and operation at constant temperature and pressure with a separate solvent condenser and accumulator is possible.

Asphalt from different crude oils varies considerably, but the viscosity is often too high for fuel oil although, in some cases, they can be blended with refinery cutter stocks to make No. 6 fuel oil. When the sulfur content of the original residuum is high, even the blend fuel oil will not be able to meet the sulfur specification of fuel oil unless stack gas cleanup is available.

The deasphalted oil and solvent asphalt are not finished products and require further processing or blending, depending on the final use. *Manufacture of lubricating oil* is on possibility, and the deasphalted oil may also be used as a *catalytic cracking feedstock*, or it may be desulfurized. It is perhaps these last two options that are more pertinent to the present text and future refinery operations (Speight, 2011).

Briefly, catalytic cracking (Chapter 9) or hydrodesulfurization (Chapter 10) of high-sulfur and high-metal viscous feedstocks is, theoretically, the best way to enhance the value of these materials. However, the concentration of sulfur (in the asphaltene fraction and in the resin fraction) in the feedstock can severely limit the performance of cracking catalysts and hydrodesulfurization catalysts (Speight, 2000). Both processes generally require tolerant catalysts and (in the case of hydrodesulfurization) high hydrogen pressure, low space velocity, and high hydrogen recycle ratio.

For both processes, the advantage of using the deasphalting process to remove the troublesome compounds becomes obvious. The soluble deasphalted oil, with no asphaltene constituents and low metal content, is easier to process than the original feedstock. Indeed, in the hydrodesulfurization process, the deasphalted oil may consume only 65% of the hydrogen required for direct hydrodesulfurization of original nonfractionated viscous feedstock.

Solvent deasphalting has the advantage of being a relatively low-cost process that has the flexibility to meet a wide range of quality of the deasphalted oil. The process has very good selectivity for asphaltene constituents and metal rejection, some selectivity for rejection of coke formers but less selectivity for sulfur and nitrogen. The process is best suited for the more paraffinic vacuum residues as opposed to vacuum residues with high asphaltene content, high metal, and high carbon content of coke-forming constituents. The disadvantages of the separation process are that it performs no conversion and produces the high-viscosity by-product asphalt (deasphalter bottoms and solvent deasphalter pitch). When high-quality deasphalted oil is required, the solvent deasphalting process is limited in the quality of feedstock that can be economically processed. Other applications include gasification of the deasphalter bottoms for hydrogen and/or power generation and road asphalt production (Elliott and Stewart, 2004).

Lower-boiling solvents and higher-solvent ratios precipitate larger quantities of resin constituents and asphaltene constituents, thereby providing separation of these feedstock components from saturate components and aromatic components. Low-boiling liquid paraffin solvents show extraction selectivity not only to molecular weight but also to molecular type. In any crude oil residue, the lowest boiling and most paraffinic components are those most soluble in the low-boiling paraffinic solvent.

Recent work has shown that additives and membranes can be used to enhance the separation of the asphaltic phase and the nonasphaltic phase. In the former process (when additives are used) (Koseoglu, 2009), solvent deasphalting of crude oil or viscous feedstocks is carried out in the presence of a solid adsorbent, such as clay, silica, alumina and activated carbon, which adsorbs the contaminants and permits the solvent and oil fraction to be removed as a separate stream from which the solvent is recovered for recycling; the adsorbent with contaminants and the asphalt bottoms is mixed with aromatic and/or polar solvents to desorb the contaminants and washed as necessary, for example, with benzene, toluene, xylenes, and tetrahydrofuran, to clean adsorbent that is recovered and recycled; the solvent-asphalt mixture is sent to a fractionator for recovery and recycling of the aromatic or polar solvent. The bottoms from the fractionator include the concentrated polynuclear aromatic constituents and contaminants and are further processed as appropriate. In the latter process (i.e., when a membrane is employed) (Trambouze et al., 1989), the solvent (a low-boiling hydrocarbon liquid) is separated from the deasphalted oil, by passing the solution tangentially across an inorganic membrane, and the obtained filtrate has an increased solvent content and may be recycled. The deasphalted oil is selectively retained on the upstream side of the membrane.

Varied types of equipment are in use in solvent deasphalting process. For example, earlier units employed baffle tray columns and/or rotating disk columns. The baffle tray column unit was used to provide a trouble-free operation due to the open structure, but these columns were less flexible to column throughput and changes in physical properties of the system. The rotating disk column has been widely applied in various solvent extraction units. The more modern random packings and structured packings have been successfully applied in many solvent extraction processes and have also been applied to solvent deasphalting units.

The efficiency of the extraction process is the key equipment design variable—the role of the extractor is to separate the precipitate phase (often referred to as *deasphalter pitch*) from the continuous fluid (deasphalted oil/solvent) stream. Traditionally, solvent deasphalting units utilize a countercurrent type separation of the upflowing deasphalted oil/solvent phase from the downflowing precipitate phase. Traditional countercurrent extraction technology can be improved by the addition of packing, which exerts efficient control on the vertical fluid velocities across the cross-sectional area of the extractor. Consequently, packing allows the extractor size to be reduced relative to an open vessel type of design.

The process is also used to reject asphaltene constituents and to recover a deasphalted oil, which is low in asphaltene content from a vacuum residue. The solvent deasphalted oil (low-to-no-asphaltene content) oil is a useful feedstock for both fluid catalytic cracking units hydrocracking units. In addition, because it is relatively less expensive to desulfurize the deasphalted oil than the viscous feedstock, solvent deasphalting offers a more economical route for the disposing of a viscous feedstock such as heavy crude oil, extra heavy oil, and tar sand bitumen.

2.2 Deep Solvent Deasphalting Process

The deep solvent deasphalting process is an application of the low-energy deasphalting (LEDA) process (Fig. 12.5; RAROP, 1991, p. 91; Parkash, 2003; Gary et al., 2007; Speight, 2014, 2017; Hsu and Robinson, 2017) that is used to extract high-quality lubricating oil bright stock or prepare catalytic cracking feeds, hydrocracking feeds, hydrodesulfurization unit feeds, and asphalt from vacuum residue materials.

The LEDA process uses a low-boiling hydrocarbon solvent specifically formulated to insure the most economical deasphalting design for each operation. For example, a propane solvent may be specified for a low deasphalted oil yield operation, while a higher-boiling solvent, such as pentane or hexane, may be used to obtain a high deasphalted oil yield from a viscous feedstock (Parkash, 2003; Gary et al., 2007; Speight, 2014, 2017; Hsu and Robinson, 2017). The deep deasphalting process can be integrated with a delayed coking operation (ASCOT process; Chapter 8), and in this case, the solvent can be low-boiling naphtha (Parkash, 2003; Gary et al., 2007; Speight, 2014, 2017; Hsu and Robinson, 2017).

Low-energy deasphalting operations are usually carried out in a rotating disk contractor (RDC) that provides more extraction stages than a mixer settler or baffle-type column. Although not essential to the process, the rotating disk contactor provides higher-quality deasphalted oil at the same yield or higher

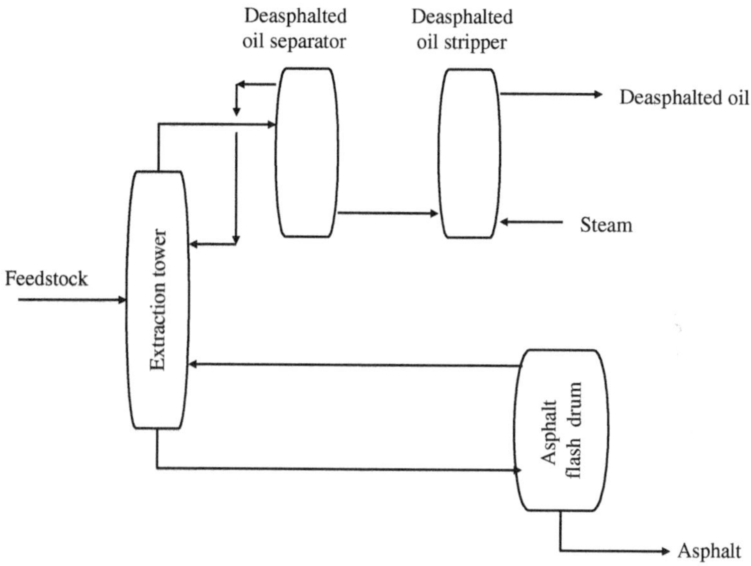

FIG. 12.5 The LEDA process.

yields at the same quality. The low-energy solvent deasphalting process selectively extracts the more paraffinic components from vacuum residua while rejecting the condensed ring aromatic derivatives. As expected, deasphalted oil yields vary as a function of solvent type and quantity and feedstock properties (Parkash, 2003; Gary et al., 2007; Speight, 2014, 2017; Hsu and Robinson, 2017).

In the process, vacuum residue feedstock is combined with a small quantity of solvent to reduce its viscosity and cooled to a specific extraction temperature before entering the rotating disk contactor. Recovered solvent from the high-pressure and low-pressure solvent receivers is combined, adjusted to a specific temperature by the solvent heater-cooler, and injected into the bottom section of the rotating disk contactor. Solvent flows upward, extracting the paraffinic hydrocarbon derivatives from the viscous feedstock, which is flowing downward through the rotating disk contactor.

Steam coils at the top of the tower maintain the specified temperature gradient across the rotating disk contactor. The higher temperature in the top section of the rotating disk contactor results in separation of the less soluble heavier material from the deasphalted oil mix and provides internal reflux, which improves the separation. The deasphalted oil mix leaves the top of the rotating disk contactor tower. It flows to an evaporator where it is heated to vaporize a portion of the solvent. It then flows into the high-pressure flash tower where high-pressure solvent vapors are taken overhead.

The deasphalted oil mix from the bottom of this tower flows to the pressure vapor heat exchanger where additional solvent is vaporized from the deasphalted oil mix by condensing high-pressure flash. The high-pressure solvent, totally condensed, flows to the high-pressure solvent receiver. Partially vaporized, the deasphalted oil mix flows from the pressure vapor heat exchanger to the low-pressure flash tower where low-pressure solvent vapor is taken overhead, condensed, and collected in the low-pressure solvent receiver. The deasphalted oil mix flows down the low-pressure flash tower to the reboiler, where it is heated, and then to the deasphalted oil stripper, where the remaining solvent is stripped overhead with superheated steam. The deasphalted oil product is pumped from the stripper bottom and is cooled, if required, before flowing to battery limits.

The raffinate phase containing asphalt and small amount of solvent flows from the bottom of the rotating disk contactor to the asphalt mix heater. The hot, two-phase asphalt mix from the heater is flashed in the asphalt mix flash tower where solvent vapor is taken overhead, condensed, and collected in the low-pressure solvent receiver. The remaining asphalt mix flows to the asphalt stripper where the remaining solvent is stripped overhead with superheated steam. The asphalt stripper overhead vapors are combined with the overhead from the deasphalted oil stripper, condensed, and collected in the stripper drum. The asphalt product is pumped from the stripper and is cooled by generating low-pressure steam.

2.3 Demex Process

The Demex process is a solvent extraction demetallization process that separates high metal viscous feedstock into demetallized oil of relatively low metal content and asphaltene of high metal content (RAROP, 1991, p. 93). The asphaltene constituents and condensed aromatic contents of the demetallized oil are very low. The demetallized oil is a desirable feedstock for fixed-bed hydrodesulfurization and, in cases where the metal content and carbon residue are sufficiently low, is a desirable feedstock for fluid catalytic cracking and hydrocracking units. Overall, the Demex process is an extension of the propane deasphalting process and employs a less selective solvent to recover not only the high-quality oils but also higher-molecular-weight aromatic derivatives and other constituents present in the feedstock. Furthermore, the Demex process requires a much less solvent circulation in achieving its objectives, thus reducing the utility costs and unit size significantly. The process selectively rejects asphaltenes, metals, and high-molecular-weight aromatic derivatives from a viscous feedstock. The resulting demetallized oil can then be combined with vacuum gas oil to give a greater availability of acceptable feedstock to subsequent conversion units, such as a catalytic cracking unit (Chapter 9) or a hydrocracking unit (Chapter 11; Parkash, 2003; Gary et al., 2007; Speight, 2014, 2017; Hsu and Robinson, 2017).

In the process, the viscous feedstock, mixed with Demex solvent recycling from the second stage, is fed to the first stage extractor. The pressure is kept high enough to maintain the solvent in liquid phase. The temperature is controlled by the degree of cooling of the recycle solvent. The solvent rate is set near the minimum required to ensure the desired separation to occur. Asphaltene constituents are rejected in the first stage. Some resins are also rejected to maintain sufficient fluidity of the asphaltene for efficient solvent recovery. The asphaltene is heated and steam stripped to remove solvent. The first stage overhead is heated by an exchange with hot solvent. The increase in temperature decreases the solubility of resins and high-molecular-weight aromatic derivatives (Mitchell and Speight, 1973). These precipitate in the second-stage extractor. The bottom stream of this second-stage extractor is recycled to the first stage. A portion of this stream can also be drawn as a separate product.

The overhead from the second stage is heated by an exchange with hot solvent. The fired heater further raises the temperature of the solvent/demetallized oil mixture to a point above the critical temperature of the solvent. This causes the demetallized oil to separate. It is then flashed and steam stripped to remove all traces of solvent. The vapor streams from the demetallized oil and asphalt strippers are condensed, dewatered, and pumped up to process pressure for recycle. The bulk of the solvent goes overhead in the supercritical separator. This hot solvent stream is then effectively used for process heat exchange. The subcritical solvent recovery techniques, including multiple effect systems, allow much less heat recovery. Most of the low-grade heat in the solvent vapors from the subcritical flash vaporization must be released to the atmosphere requiring additional heat input to the process.

2.4 MDS Process

The MDS process is a technical improvement of the solvent deasphalting process, particularly effective for upgrading viscous feedstocks (RAROP, 1991, p. 95; Parkash, 2003; Gary et al., 2007; Speight, 2014, 2017; Hsu and Robinson, 2017). Combined with hydrodesulfurization, the process is fully applicable to the feedstock preparation for fluid catalytic cracking and hydrocracking. The process is capable of using a variety of feedstocks including heavy oil, extra heavy oil, and tar sand bitumen, as well as atmospheric and vacuum residua derived from various crude oils and the nonvolatile products from a visbreaking unit (Chapter 8).

In the process, the feedstock and the solvent are mixed and fed to the deasphalting tower. Deasphalting extraction proceeds in the upper half of the tower. After the removal of the asphalt, the mixture of deasphalted oil and solvent flows out of the tower through the tower top. Asphalt flows downward to come in contact with a countercurrent of rising solvent. The contact eliminates oil from the asphalt; the asphalt then accumulates on the bottom. Deasphalted oil containing solvent is heated through a heating furnace and fed to the deasphalted oil flash tower where most of the solvent is separated under pressure.

Deasphalted oil still containing a small amount of solvent is again heated and fed to the stripper, where the remaining solvent is completely removed.

Asphalt is withdrawn from the bottom of the extractor. Since this asphalt contains a small amount of solvent, it is heated through a furnace and fed to the flash tower to remove most of the solvent. Asphalt is then sent to the asphalt stripper, where the remaining portion of solvent is completely removed.

Solvent recovered from the deasphalted oil and asphalt flash towers is cooled and condensed into liquid and sent to a solvent tank. The solvent vapor leaving both strippers is cooled to remove water and compressed for condensation. The condensed solvent is then sent to the solvent tank for further recycling.

2.5 Solvahl Process

The Solvahl process is a solvent deasphalting process for application to viscous feedstocks such as heavy oil, extra heavy oil, and tar sand bitumen (RAROP, 1991, p. 9; Billon et al., 1994; Parkash, 2003; Gary et al., 2007; Speight, 2014, 2017; Hsu and Robinson, 2017). The process was developed to give maximum yields of deasphalted oil while eliminating asphaltenes and reducing metals content to a level compatible with the reliable operation of downstream units. The process removes the asphaltene constituents, most metals, and other impurities contained in the viscous feedstock.

The process produces a deasphalted oil from which most of the heptane-asphaltenes have been eliminated. The content of metals and coke precursor compounds are also significantly reduced, making the deasphalted oil a suitable feedstock for downstream fluid catalytic cracking or hydrocracking units. The solvents used in the process can vary from liquid propane to heptane depending on feedstock properties and downstream process objectives. The relative yields of deasphalted oil and asphalt and the characteristics of both products are linked to the nature of the feed, the operating conditions, and the solvent type.

In summary, the Solvahl technology is focused on the following: (i) relevant and optimized separation of deasphalted oil and asphaltenes with associated design criteria to reach specifications and (ii) control of the quality of the deasphalted oil, and especially any remaining content of asphaltenes constituents, is a main concern as it impacts downstream catalyst performance and cycle length. The association of the process with other conversion technologies such as hydrocracking or ebullated-bed technologies (such as H-Oil) presents the opportunity to maximize the yield of middle distillate products from viscous feedstocks.

2.6 Other Options

The efficiency of a typical solvent deasphalting process can be successfully increased by the inclusion of nanoparticles in systems with different solvent-to-oil ratios, temperatures, types of solvent, and dosage (Guzmán et al., 2017).

It was found that the use of nanoparticles reduces the yield of the deasphalted oil in comparison with the traditional solvent deasphalting process, resulting in a higher-quality product. The reason is the adsorption phenomena of the asphaltenes and the nanoparticles and how the former are transferred from the deasphalted oil fraction to the deasphalter bottoms.

2.6.1 Modified Solvent Deasphalting

The CLG Residue Solvent Refining Solvent Deasphalting (RSR) process is a specialized solvent extraction process that separates viscous feedstocks by molecular type instead of by boiling point, as in the vacuum distillation process. The process recovers a low-contaminant, relatively high in hydrogen content, deasphalted oil product from the viscous feedstock. The insoluble components are contained in the deasphalter bottom product and contain the majority of the feedstock contaminants (asphaltene constituents and metal-containing constituents).

Historically, the process has not only been used to produce base oil for lubricating oil production but also been utilized to produce additional vacuum gas-oil feedstocks for various conversion processes. For lubricating oil applications, a propane solvent is commonly used to maximize the quality and minimize the impurities in the deasphalted oil. Higher-boiling solvents, such as butane or pentane, are used for the production of transportation fuels. The process is capable of achieving conversion of a viscous feedstock on the order of 90% w/w to lower-molecular-weight high-value products. The deasphalter bottoms can also be used as (i) a combustion fuel, (ii) a blending component for the production of residual fuel oil, (iii) a feedstock to a coking unit, or (iv) a blend stock for asphalt production.

The jointly owned UOP/Amec Foster Wheeler SDA process is a unique separation process that, unlike vacuum distillation, separates by molecular type. The process produces a low-contaminant deasphalted oil that is rich in paraffin-type constituents, suitable for catalytic conversion processes, and a contaminant rich pitch stream. The deasphalted oil can then be selectively converted to desirable high-quality transportation fuels in conventional conversion units such as in an fluid catalytic cracking unit (usually after hydroprocessing) or in a hydrocracking unit.

The application of gasification technology in refineries has been gaining popularity in the refining industry in recent years. The successful startup and operation of several viscous feedstock-oriented gasification projects over the past year will provide the incentive for more refiners to consider gasification as a bottom destruction option. Gasification, coupled with deregulation of the global power markets, offers an excellent opportunity to convert the low-value refinery bottoms to high-value products such as synthesis gas (syngas), power, production of hydrogen, and the elimination of the need to produce residual fuel oil.

Widespread commercialization of gasification in the refining industry is dependent on several factors. One that will play a key role is the ability of the refiner to provide a low-value feedstock to the integrated gasification combined cycle (IGCC) complex. The solvent deasphalting process offers a refiner the ability to not only produce a low-cost gasifier feedstock but also generate additional conversion feedstock for the refinery. Gasification offers an opportunity to convert low-value refinery bottoms to high-value products like syngas, power, and hydrogen, in an environmentally friendly manner.

2.6.2 Extractive Desulfurization

Extractive desulfurization is based on the fact that organosulfur compounds are more soluble than hydrocarbon derivatives in an appropriate solvent. In a mixing tank, the sulfur compounds are transferred from the fuel oil into the solvent due to their higher solubility in the solvent. Subsequently, the solvent-fuel mixture is fed into a separator in which hydrocarbon derivatives are separated from the solvent. The desulfurized hydrocarbon stream is used either as a component to be blended into the final product or as a feedstock for further transformations. The organosulfur compounds are separated by distillation, and the solvent is recycled. The most attractive feature of the extractive desulfurization is the applicability at ambient conditions. The process does not change the chemical structure of the fuel oil components. As the equipment used is rather conventional without special requirements, the process can be easily integrated into the refinery.

The efficiency of extractive desulfurization is mainly limited by the solubility of the organic sulfur compounds in the solvent. The solvent must have a boiling temperature different than that of the sulfur-containing compounds, and it must be inexpensive to ensure economic feasibility of the process. Solvents of different nature have been tried, among which acetone, ethanol, polyethylene glycols, and nitrogen-containing solvents showed a reasonable level of desulfurization of 50%–90% sulfur removal, depending on the number of extraction cycles. Preparation of such a "solvent cocktail" is rather difficult and intrinsically nonefficient since its composition depends strongly on the spectrum of the organosulfur compounds present in the feedstock stream. Solubility can also be enhanced by transforming the organic sulfur compounds to increase their solubility in a polar solvent. One way to do this is by selectively oxidizing the organic sulfur compound (thiophene, BTs, and DBTs) to sulfones possessing higher polarity.

The GT-DeSulfSM process is an example of desulfurization technology based on organosulfur compound extraction. This process separates the organosulfur compounds and aromatic derivatives from naphtha from the fluid catalytic cracking unit by extractive distillation using a blend of solvents. A desulfurized/dearomatized olefin-rich gasoline stream and an aromatic stream containing the sulfur compounds are formed after treatment in a GT-Desulf reactor. The first stream is directly used as a gasoline blend stock. Unfortunately, available literature does not contain any information on the level of sulfur removal from the treated stream. The aromatic fraction with the sulfur compounds is sent to a hydrodesulfurization reactor. After treatment in the hydrodesulfurization reactor, recovery of aromatic derivatives is proposed as an additional option to increase economic efficiency of the process.

2.6.3 Desulfurization by Ionic Liquids

Ionic liquids are organic salts that are in liquid state at temperatures below 100°C. Ionic liquids are predicted to take the place of organic solvents, because they have no measurable vapor pressure below their decomposition temperature and can be designed to have different properties depending on their structure. Desulfurization by ionic liquids is based on extraction theories, and it is a mild. Organic ions in ionic liquids can be designed in numerous varieties and combine together to make practically unlimited number of ionic liquids. Among these, imidazolium-based ionic liquids, such as [BMIM][PF_6], [EMIM][BF_4], [BMIM][$MeSO_4$], [BMIM][$AlCl_4$], and [BMIM][$OcSO_4$], have demonstrated a high selective partitioning for heterocyclic sulfur-containing molecules such as dibenzothiophene derivatives and single β and di-β methylated dibenzothiophene derivatives. Selection of ions for ionic liquids used in organic sulfur removal from fuel oils is very important. Some of the chlorometallate ionic liquids such as the ones with [BMIM] [$AlCl_4$] show good selectivity for sulfur removal; however, they are very sensitive to air and moisture and may cause alkene polymerization in fuel. The size of anions in ionic liquids was also found to be rather important in extraction of DBT from oil phase. Bigger anions such as $[OcSO_4]^-$ could extract dibenzothiophene derivatives more effectively than smaller anions (e.g., $[PF_6]^-$ or $[CF_3SO_3]^-$).

There is an increasing trend on desulfurization of fossil fuels by ionic liquids. The purpose of research on ionic liquids in future refineries is to economize desulfurization energy requirements and to decrease production of carbon dioxide that is associated with other desulfurization processes such as hydrodesulfurization. The recovery and recycling of ionic liquids during desulfurization process is difficult.

2.6.4 Moving the Boiling Point by Alkylation

When the boiling temperature of organosulfur compounds is shifted to a higher value, they can be removed from low-boiling fractions by distillation and concentrated in the high-boiling (nonvolatile) fraction of the refinery streams. British Petroleum used this approach in a new advanced technology process for desulfurizing naphtha streams produced by catalytic cracking. The process employs alkylation of thiophenic compounds via reaction with olefins present in the stream.

As a result, the boiling temperature of the sulfur-containing hydrocarbon compounds increases. In comparison with thiophene (boiling point around 85°C), alkylated thiophenes such as 3-hexylthiophene and/or 2-octylthiophene have a much higher-boiling point (221°C and 259°C, respectively). This enables the product(s) to be easily separated from the main gasoline stream by distillation. The high-boiling compounds produced can be blended into the diesel pool and desulfurized by conventional hydrotreating as the octane number is not important for diesel.

Thiophenic sulfur is alkylated in a reactor employing acidic catalysts such as BF_3, $AlCl_3$, $ZnCl_2$, or $SbCl_5$ deposited on silica, alumina, or silica-alumina supports. After the reactor, the feedstock is sent to a conventional distillation column where it is separated into light sulfur-free naphtha and a viscous sulfur-rich stream. The light naphtha is directly sent to the gasoline pool, and the viscous stream is preferably hydrotreated. The hydrotreater is not an essential part of the technology, but its application after the fractionators increases the product yield.

2.6.5 Desulfurization by Selective Adsorption

Adsorption has become a key separation technique in industry and particularly in the oil and gas industry. Adsorbents used industrially are generally synthetic microporous solids: activated carbon, molecular sieve carbon, activated alumina, silica gel, zeolites, and bleaching clay. They are usually agglomerated with binders in the form of beads, extrudates, and pellets of a size consistent with the application that is considered. The sulfur compounds can be removed from commercial fuels via either reactive adsorption by chemisorption π-complexation or van der Waals and electrostatic interactions. Modified Y-type zeolite was popularly used as adsorbents to remove sulfur from fuels via π-complexation.

As an example (Koseoglu, 2009), in the enhanced solvent deasphalting process for viscous feedstocks utilizing solid adsorbent, the solvent deasphalting of crude oil or the higher-boiling fraction of crude oil and residua is carried out in the presence of a solid adsorbent, such as clay, silica, alumina, and activated carbon, which adsorbs the contaminants and permits the solvent and oil fraction to be removed as a separate stream from which the solvent is recovered for recycling; the adsorbent with contaminants and the asphalt bottoms is mixed with aromatic and/or polar solvents to desorb the contaminants and washed as necessary, for example, with benzene, toluene, xylene derivatives, and tetrahydrofuran, to clean adsorbent that is recovered and recycled; the solvent-asphalt mixture is sent to a fractionator for recovery and recycling of the aromatic or polar solvent. The bottoms from the fractionator include the concentrated polynuclear aromatic constituents and other contaminants and are further processes as appropriate.

Adsorption desulfurization has some problems to be solved. There are large amounts of aromatic derivatives compared with the amounts of sulfur compounds in diesel. Therefore, aromatic derivatives can also be adsorbed on the desulfurization adsorbents. Adsorbents should be well designed to achieve suitable selectivity; when the selectivity is low, the adsorbents are easy to be regenerated. But this can lead to the heat loss because of the comparative adsorption. As the selectivity increases, the spent adsorbents become more and more difficult to be regenerated.

Solvent extraction and oxidation in the air are two methods to regenerate the desulfurization adsorbents. There are some disadvantages in these two methods. For solvent extraction method, it is difficult to separate sulfur compounds from the organic solvents and reuses these solvents. For calcination method, sulfur compounds and aromatic derivatives are burned out that can lose heat value of fuels.

2.6.6 Oxidative Desulfurization

Oxidative desulfurization is a milder approach for deep desulfurization in which sulfur compounds (sulfides) are oxidized to more polar compounds.

The removal of sulfones/sulfoxides could be achieved by adsorption technique on silica gel, activated carbon, bauxite, clay, coke, alumina, silicalite, ZSM-5, zeolite β, zeolite X, zeolite Y, and MCM-41, in addition to the mesoporous oxide-based materials that have attracted much attention in the recent years due to their large pore sizes and controlled pore size distribution that may be beneficial in allowing accessibility of large-molecular size sulfones to the surface-active sites. However, one of the major drawbacks of the adsorption technique is the amount of oil treated per unit weight of adsorbent is low.

2.6.7 Biocatalytic Desulfurization

Biocatalytic desulfurization (BDS) is often considered as a potential alternative to the conventional deep hydrodesulfurization processes used in refineries. In this process, bacteria remove organosulfur from crude oil fractions without degrading the carbon skeleton of the organosulfur compounds. During a biocatalytic desulfurization process, alkylated dibenzothiophene derivatives are converted to nonsulfur compounds; for example, dibenzothiophene is converted to 2-hydroxybiphenyl (2-HBP) and sulfate.

Biocatalytic desulfurization offers mild processing conditions and reduces the need for hydrogen. Both these features would lead to high-energy savings in the refinery. Further, significant reductions in greenhouse gas emissions have also been predicted if biocatalytic desulfurization is used.

Biocatalytic desulfurization is a complementary technology since hydrodesulfurization is not equally effective in desulfurizing all classes of sulfur compounds present in fossil fuels. The biocatalytic desulfurization process, on the other hand, is effective regardless of the position of alkyl substitutions.

However, the hydrodesulfurization process conditions are sufficient not only to desulfurize sensitive (labile) organosulfur compounds but also to (i) remove nitrogen and metals from organic compounds, (ii) induce saturation of at least some carbon-carbon double bonds, (iii) remove substances having an unpleasant smell or color, (iv) clarify the product by drying it, and (v) improve the cracking characteristics of the material. Therefore, with respect to these advantages, placing the biocatalytic desulfurization process unit downstream of the hydrodesulfurization unit as a complementary technology to achieve ultradeep desulfurization, rather than as a replacement, should also be considered.

3. Supercritical Extraction Process

Supercritical extraction (also called supercritical fluid extraction, SFE) is the process of separating one component (the extractant) from another (the matrix) using a supercritical fluid as the extracting solvent. Extraction not only is usually from a solid matrix but also can be from liquids. Carbon dioxide (CO_2) is the most used supercritical fluid, sometimes modified by cosolvents such as ethanol (CH_3CH_2OH) or methanol (CH_3OH). Extraction conditions for supercritical carbon dioxide are above the critical temperature of 31°C (70°F) and a critical pressure of 1073 psi. The addition of modifiers may slightly alter these parameters.

Although the solvent deasphalting process is a facile method for recovering the more refinable fraction of viscous feedstocks, solvent losses may be significant. Solvent losses due to evaporation and retention on the discarded sand impact the economic feasibility of the process. To combat this potential economic loss, the supercritical extraction (supercritical fluid extraction) of viscous feedstocks has also been proposed as an alternate upgrading option to coking.

For example, heavy oil, extra heavy oil, and tar sand bitumen contain hundreds of compounds, and currently, most separation protocols determine the various molecular species. Hence, to separate the chemical class and identity of a number of nonvolatile constituents, fluid systems with appropriate separation and detection technologies are required and include supercritical fluid extraction (SFE). Several processes (such as the ROSE process and the HSC-ROSE process) have adopted this concept and applied it to upgrading and/or deasphalting bitumen and have been developed by which the separation of the solvent from the extract phase is achieved by depressurization (Martin and Williams, 1978; Pang and McLaughlin, 1985; Rudzinski and Aminabhavi, 2000; Rudyk et al., 2014; Demirbaş, 2016; Imanbayev et al., 2017).

The residual oil supercritical extraction (ROSE) process, as applied to a viscous feedstock, is a solvent deasphalting process with minimum energy consumption using supercritical solvent recovery system, and the process is of value in obtaining oils for further processing (Gearhart and Garwin, 1976; Gearhart, 1980; RAROP, 1991, p. 97). The process can be installed upstream of the desulfurizer to reduce a major portion of the heavy metals and coke precursors present in the feed. The ROSE process can also be installed between a vacuum flasher and a coking unit, which reduces the carbon residue of the gas-oil fraction for its catalytic cracking (Gearhart, 1980; Low et al., 1995; Parkash, 2003; Gary et al., 2007; Speight, 2014, 2017; Hsu and Robinson, 2017).

The HSC-ROSE process—mild cracking solvent deasphalting—has been described for upgrading viscous feedstocks (Chen et al., 1994). The effects of cracking temperature and time on the yield, carbon residue, nickel content of deasphalted oil, and properties of raffinate asphalt were examined. The deasphalted oil obtained by the mild cracking solvent deasphalting was superior in yield and quality to that from solvent deasphalting alone. At the same yield of deasphalted oil, the softening point of the raffinate asphalt was lower, and the penetration and the ductility were greater than those for the solvent deasphalting process.

The process used supercritical solvents and is a natural progression from propane deasphalting and allows the separation of viscous feedstocks into the base components (asphaltene constituents, resin constituents, and soluble constituents) for recombination to optimum properties. Propane, butane, and pentane may be used as the solvent depending on the feedstock and the desired compositions. A mixer is used to blend residue with liquefied solvent at elevated temperature and pressure. The blend is pumped into the first stage separator where, through counter current flow of solvent, the asphaltene constituents are precipitated, separated, and stripped of solvent by steam. The overhead solution from the first tower is taken to a second stage where it is heated to a higher temperature. This causes the resin constituents to separate. The final material is taken to a third stage and heated to a supercritical temperature. This makes the oils insoluble and separation occurs. This process is very flexible and allows precise blending to required compositions.

In the ROSE process (Fig. 12.6), the residuum is mixed with severalfold volume of a low-boiling hydrocarbon solvent and passed into the asphaltene separator vessel. Asphaltenes rejected by the solvent are separated from the bottom of the vessel and are further processed by heating and steam stripping to remove a small quantity of dissolved solvent. The solvent-free asphaltenes are sent to a section of the refinery for further processing. The main flow, solvent, and extracted oil pass overhead from the asphaltene separator through a heat exchanger and heater into the oil separator where the extracted oil is separated without solvent vaporization. The solvent, after heat exchange, is recycled to the process. The small amount of solvent contained in the oil is removed by steam stripping, and the resulting vaporized solvent from the strippers is condensed and returned to the process. Product oil is cooled by heat exchange before being pumped to storage or further processing. The deasphalting efficiency in processes using propane is of the order of 75%–83%, with an overall deasphalted oil recovery yield of the order of 50%.

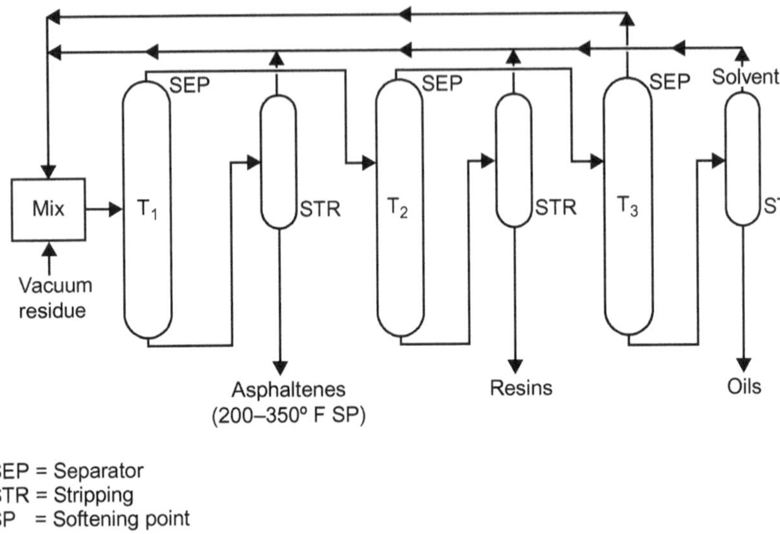

FIG. 12.6 The ROSE process. T_1, T_2, and T_3 are separator vessel with the relevant parameters for separation of the fraction.

The deasphalted oil from a ROSE unit is a suitable feedstock that can be processed in a fluid catalytic cracking unit or other conversion units. Since the contaminants (such as sulfur and metals) are rejected in the solid fuel to the cement industry, there is minimal impact to the auxiliary units (sulfur plant and amine regeneration) within the refinery. Utilization of the asphaltene fraction can be a key to the economic success of the process. Options for the use of the fraction (not necessarily in the order of importance) include the following: (i) conversion as a coker feedstock, (ii) fuel oil blend stock, (iii) partial oxidation feedstock for synthesis gas or hydrogen production, and (iv) solid fuel with any necessary gas cleaning operations. The quality and definition of the product designated as the *asphaltene fraction* depends on the feedstock to the process and whether or not the feedstock is the result of a blend of crude different oil (Speight, 2014). As the crude slate becomes heavier and higher in sulfur content, the asphaltene fraction will also contain a higher quantity of sulfur. Environmental regulations will therefore dictate how much flue gas cleanup is required and, hence, the viability of direct firing burning of the asphaltene product.

Although often referred to as supercritical extraction, it is often the solvent separation not the extraction that is carried out in the supercritical region of the solvent that results in a simpler process flow. Supercritical solvent recovery allows for more efficient utilization of the systems thermodynamic characteristics.

The aquaform process is a process that has been designed for easy integration with the ROSE solvent deasphalting unit (Patel et al., 2008). In the process, the asphaltene product is sent through an exchanger to a pelletizing vessel in which a rotating head produces small droplets of asphaltene material, which are then quenched in a water bath. The pellets are removed on screen separators and then transferred to storage using conveyer systems; the water is filtered, cooled, and returned to the pelletizing vessel. The pellets ate claimed to have a higher heating value than the coke.

References

Billon, A., Morel, F., Morrisson, M.E., Peries, J.P., 1994. Converting residues with the IFP Hyvahl process. Rev. Inst. Fr. Pétrol. 49 (5), 495–507.

Chen, S.L., Jia, S.S., Luo, Y.H., Zhao, S.Q., 1994. Mild cracking solvent deasphalting: a new method for upgrading petroleum residue. Fuel 73 (3), 439–442.

De Souza, R.A., De Oliveira, A.A., Vieira, Y.M., Zaparoli, A.T., Wolf Maciel, M.R., Maciel Filho, R., Cárdenas Concha, V.O., 2017. Sensibility analysis of propane deasphalting process variables. Pet. Sci. Technol. 35 (1), 22–29.

Demirbaş, A., 2016. Deasphalting of crude oils using supercritical fluids. Pet. Sci. Technol. 34 (7), 665–670.

Ditman, J.G., 1973. Solvent deasphalting. Hydrocarb. Process. 52 (5), 110.

Dunning, H.N., Moore, J.W., 1957. Propane removes asphalts from crudes. Pet. Refin. Eng. 36 (5), 247–250.

Elliott, J.D., Stewart, M.D., 2004. Cost effective residue upgrading: delayed coking and refining. In: Proceedings of the 4th Russian Refining Technology Conference, Moscow, September 23–24.

Gary, J.H., Handwerk, G.E., Kaiser, M.J., 2007. Petroleum Refining: Technology and Economics, fifth ed. CRC Press, Taylor & Francis Group, Boca Raton, FL.

Gearhart, J.A., 1980. Solvent treat resids. Hydrocarb. Process. 59 (5), 150–151.

Gearhart, J.A., Garwin, L., 1976. ROSE process improves resid feed. Hydrocarb. Process. 55 (5), 125–128.

Girdler, R.B., 1965. Constitution of asphaltenes and related studies. Proc. Assoc. Asph. Pav. Tech. 34, 45.

Guzmán, J.D., Franco, C.A., Cortés, F.B., 2017. An enhanced-solvent deasphalting process: effect of inclusion of SiO2 nanoparticles in the quality of deasphalted oil. Hindawi J. Nanomater. 2017. https://www.hindawi.com/journals/jnm/2017/9470230/.

Hsu, C.S., Robinson, P.R. (Eds.), 2017. Handbook of Petroleum Technology. Springer International Publishing AG, Cham.

Imanbayev, Y.I., Ongarbayev, Y.K., Tileuberdi, Y., Mansurov, Z.A., Golovko, A.K., Rudyk, S., 2017. Supercritical solvent extraction of oil sand bitumen. In: Proceedings of AIP Conference. https://aip.scitation.org/doi/10.1063/1.5000473. (Accessed 10 July 2018).

Koseoglu, O.R., 2009. Enhanced Solvent Deasphalting Process for Heavy Hydrocarbon Feedstocks Utilizing Solid Adsorbent. United States Patent 7,566,394 July 28.

Lee, J.M., Shin, S., Ahn, S., Chun, J.H., Lee, K.B., Mun, S., Jeon, S.G., Na, J.G., Nho, S., 2014. Separation of solvent and deasphalted oil for solvent deasphalting process. Fuel Process. Technol. 119, 204–210.

Low, J.Y., Hood, R.L., Lynch, K.Z., 1995. Valuable products from the bottom of the barrel using ROSE technology. In: Symposium on Petroleum Chemistry and Processing, Division of

Petroleum Chemistry, 210th National Meeting, American Chemical Society, Chicago, IL, August 20–25, Preprints, pp. 780–784.

Martin, T.G., Williams, D.F., 1978. Gaseous Solvent Extraction of Oil Shales and Tar Sands. US Patent 4,108,760, August 22.

McCoy, J.N., Keusenkothen, P.F., Srivastava, A., 2010. Process for Upgrading Tar. United States Patent 7,744,743 June 29.

Mitchell, D.L., Speight, J.G., 1973. The solubility of asphaltenes in hydrocarbon solvents. Fuel 52, 149.

Pang, T.H., McLaughlin, E., 1985. Supercritical extraction of aromatic/hydrocarbon solids and tar sand bitumens. Ind. Eng. Chem. Process. Des. Dev. 24 (4), 1027.

Parkash, S., 2003. Refining Processes Handbook. Gulf Professional Publishing, Elsevier, Amsterdam.

Patel, V., Iqbal, R., Eng, O., Subramanian, A., 2008. Economic Bottom of the Barrel Processing to Minimize Fuel Oil Production. Research Note No. KIOGE-2008, Kellogg Brown & Root, Houston, TX. http://beta.kbr.com/Newsroom/Publications/Articles/Economic-Bottom-of-the-Barrel-Processing-To-Minimize-Fuel-Oil-Production.pdf.

RAROP, 1991. RAROP Heavy Oil Processing Handbook. Research Association for Residual Oil Processing. T. Noguchi (Chairman), Ministry of Trade and International Industry (MITI), Tokyo.

Rudyk, S., Spirov, P., Jimoh, I., Vakili-Nezhaad, G., 2014. The bitumen upgrading of nigerian oil sand by supercritical carbon dioxide modified with alcohols. Energy Fuel 28 (7), 4714–4724.

Rudzinski, W.E., Aminabhavi, T.M., 2000. A review on extraction and identification of crude oil and related products using supercritical fluid technology. Energy Fuel 14 (2), 464–475.

Speight, J.G., 2000. The Desulfurization of Heavy Oils and Residua, second ed. Marcel Dekker, New York.

Speight, J.G., 2011. The Refinery of the Future. Gulf Professional Publishing, Elsevier, Oxford.

Speight, J.G., 2014. The Chemistry and Technology of Petroleum, fifth ed. CRC Press, Taylor & Francis Group, Boca Raton, FL.

Speight, J.G., 2015. Handbook of Petroleum Product Analysis, second ed. John Wiley & Sons, Hoboken, NJ.

Speight, J.G., 2017. Handbook of Petroleum Refining. CRC Press, Taylor & Francis Group, Boca Raton, FL.

Trambouze, P., Euzen, J.P., Bergez, P., Claveau, M., 1989. Process for Deasphalting a Hydrocarbon Oil. United States Patent 4,816,140, March 28.

Chapter 13

Upgrading by Gasification

1. Introduction

The influx of viscous feedstocks such as heavy oil, extra heavy oil, and tar sand bitumen into refineries creates and will continue to create challenges but, at the same time, creates opportunities by improving the ability of refineries to handle viscous feedstocks, thereby enhancing refinery flexibility to meet the increasingly stringent product specifications for refined fuels. (Speight, 2013a, 2014a,b). Upgrading viscous feedstocks is an increasingly prevalent means of extracting the maximum amount of liquid fuels from each barrel of crude oil that enters the refinery. Although solvent deasphalting processes (Chapter 12) and coking processes (Chapter 8) are used in refineries to upgrade viscous feedstocks to intermediate products (which are then processed further) to produce transportation fuels, the integration of viscous feedstock processing units and gasification presents some unique synergies that will enhance the performance of the future refinery (Figs 13.1 and 13.2; Wallace et al., 1998; Furimsky, 1999; Penrose et al., 1999; Gray and Tomlinson, 2000; Abadie and Chamorro, 2009; Wolff and Vliegenthart, 2011; Speight, 2011b, 2014a).

Briefly, gasification is a process in which combustible materials are partially oxidized or partially combusted. The product of gasification is a combustible synthesis gas (also referred to as syngas). Because gasification involves the partial, rather than complete, oxidization of the feed, gasification processes operate in an oxygen-lean environment. The process has grown from a predominately coal conversion process used for making *town gas* for industrial lighting to an advanced process for the production of multiproduct, carbon-based fuels from a variety of feedstocks such as crude oil viscous feedstocks, biomass, or other carbonaceous feedstocks (Fig. 13.1; Kumar et al., 2009; Speight, 2013a, 2014a,b; Luque and Speight, 2015).

Gasification is an appealing process for the utilization of relatively inexpensive feedstocks that might otherwise be declared as waste and sent to a landfill (where the production of methane—the so-called greenhouse gas—will be produced) or combusted that may not (depending upon the feedstock) be energy-efficient. Overall, the use of a gasification technology (Speight, 2013a, 2014b) with the necessary gas cleanup options can have a smaller environmental

Heavy Oil Recovery and Upgrading. https://doi.org/10.1016/B978-0-12-813025-4.00013-1

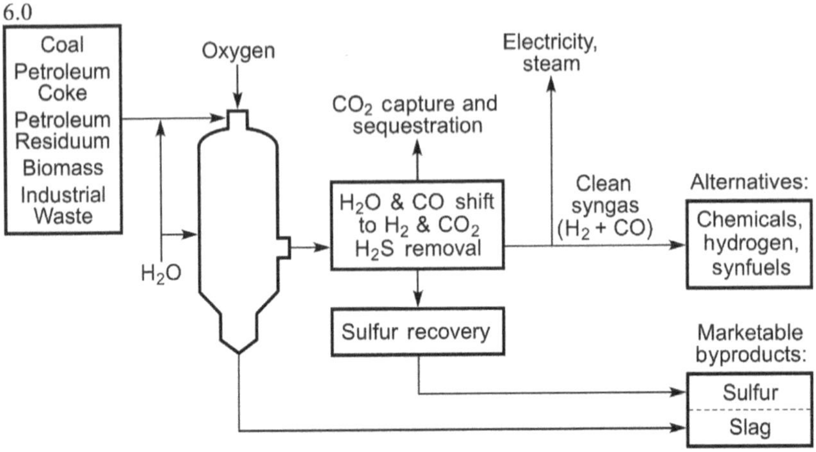

FIG. 13.1 The gasification process can accommodate a variety of carbonaceous feedstocks.

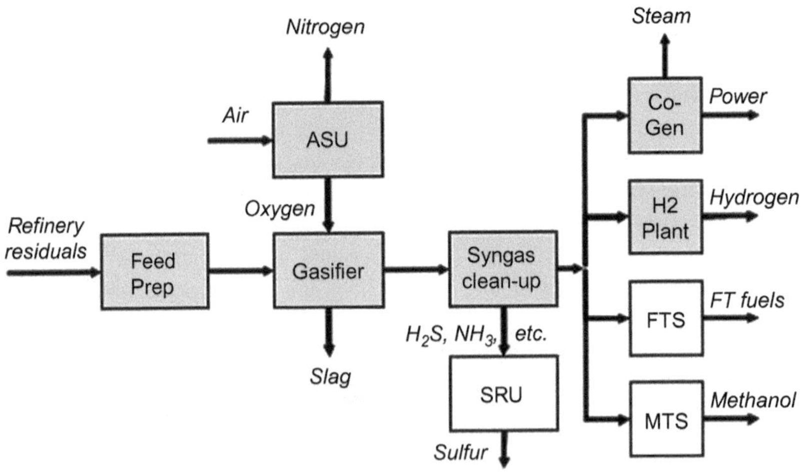

FIG. 13.2 Gasification as might be employed on-site in a refinery. *((From National Energy Technology Laboratory, US Department of Energy, Washington, DC. http://www.netl.doe.gov/technologies/coalpower/gasification/gasifipedia/7-advantages/7-3-4_refinery.html.))*

footprint and lesser effect on the environment than landfill operations or combustion of the waste. In fact, there are strong indications that gasification is a technically viable option for the waste conversion, including residual waste from separate collection of municipal solid waste. The process can meet existing emission limits and can have a significant effect on the reduction of landfill disposal using known gasification technologies (Arena, 2012; Speight, 2014b; Luque and Speight, 2015) or thermal plasma (Fabry et al., 2013).

In the gasification process, organic (carbonaceous) feedstocks are converted into carbon monoxide, carbon dioxide, and hydrogen by reacting the feedstock at high temperatures (>700°C, 1290°F), without combustion, with a controlled amount of oxygen and/or steam (Marano, 2003; Lee et al., 2007; Higman and Van der Burgt, 2008; Speight, 2008; Sutikno and Turini, 2012; Speight, 2013a, 2014b). Unconventional carbonaceous feedstocks include solids, liquids, and gases such as heavy oil, extra heavy oil, tar sand bitumen, residua, and biomass (Speight, 2014b). The gasification is not a single-step process, but involves multiple subprocesses and reactions. The generated synthesis gas has wide range of applications ranging from power generation to chemicals production. The power derived from the gasification of carbonaceous feedstocks followed by the combustion of the product gas(es) is considered to be a source of renewable energy if derived gaseous products (Table 13.1) are generated from a source (e.g., biomass) other than a fossil fuel (Speight, 2008).

Indeed, the increasing interest in gasification technology reflects a convergence of changes in providing energy to the marketplace: (i) the maturity of gasification technology and (ii) the extremely low emissions from integrated gasification combined cycle (IGCC) plants, especially air emissions and (iii) the potential for control of greenhouse gases (Speight, 2014b). Another advantage of gasification is the use of synthesis gas that is potentially more efficient as compared with direct combustion of the original fuel because it can be (i) combusted at higher temperatures, (ii) used in fuel cells, (iii) used to produce methanol and hydrogen, and (iv) converted via the Fischer-Tropsch process into a range of synthesis liquid fuels suitable for gasoline engines or diesel engines (Fig. 13.3).

Coal has been the primary feedstock for gasification units for many decades. However, there is a move to feedstocks other than coal for gasification processes with the concern on the issue of environmental pollutants and the potential shortage for coal in some area (except at the United States) (Speight, 2014b). Nevertheless, coal still prevails as a gasification feedstock

TABLE 13.1 Gasification Products

Product	Characteristics
Low-Btu gas (150–300 Btu/scf)	Around 50% nitrogen, with smaller quantities of combustible H_2 and CO, CO_2, and trace gases, such as methane
Medium-Btu gas (300–550 Btu/scf)	Predominantly CO and H_2, with some incombustible gases and sometimes methane
High-Btu gas (980–1080 Btu/scf)	Almost pure methane

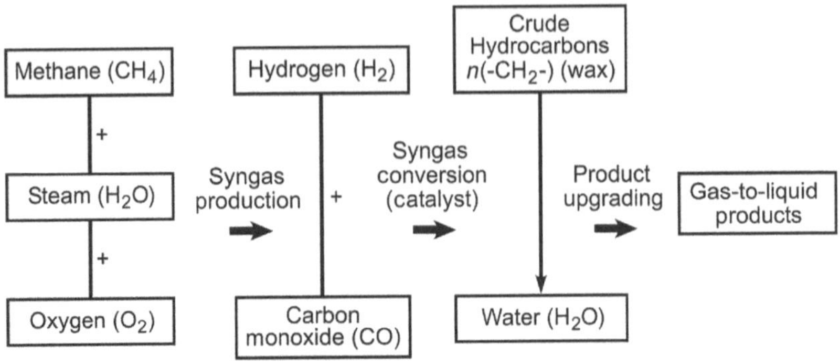

FIG. 13.3 Routes to product by the Fischer-Tropsch process.

and will remain so for at least several decades into the future, if not well into the next century (Speight, 2011b, 2013a; Luque and Speight, 2015). The gasification process can also utilize carbonaceous feedstocks that would otherwise have been disposed of (e.g., biodegradable waste).

Coal gasification plants are cleaner with respect to standard pulverized coal combustion facilities, producing fewer sulfur and nitrogen by-products, which contribute to smog and acid rain. For this reason, gasification appeals as a way to utilize relatively inexpensive and expansive coal reserves while reducing the environmental impact. Indeed, the increasing mounting interest in coal gasification technology reflects a convergence of two changes in the electricity generation marketplace: (i) the maturity of gasification technology and (ii) the extremely low emissions from integrated gasification combined cycle (IGCC) plants, especially air emissions, and the potential for lower cost control of greenhouse gases than other coal-based systems. Fluctuations in the costs associated with natural gas-based power, which is viewed as a major competitor to coal-based power, can also play a role. Furthermore, gasification permits the utilization of various feedstocks (coal, biomass, crude oil resids, and other carbonaceous wastes) to their fullest potential (Speight, 2013a, 2014b; Orhan et al., 2014). Thus, power developers would be well advised to consider gasification as a means of converting coal to gas.

Liquid fuels, including gasoline, diesel, naphtha and jet fuel, are usually processed from crude oil in the refinery (Speight, 2014a). However, with fluctuating availability and varying prices of crude oil, liquid fuels from coal (coal-to-liquids, CTL) and liquid fuels from biomass (biomass-to-liquids, BTL) are always under consideration as alternative routes used for liquid fuel production. Both coal and biomass are converted to synthesis gas that is subsequently converted into a mixture of liquid products by Fischer-Tropsch (FT) processes. The liquid fuel obtained after FT synthesis is eventually upgraded using known crude oil refinery technologies to produce gasoline, naphtha, diesel fuel, and jet fuel (Chadeesingh, 2011; Speight, 2014a).

2. Gasification Chemistry

Chemically, gasification involves the thermal decomposition of the feedstock and the reaction of the feedstock carbon and other pyrolysis products with oxygen, water, and fuel gases such as methane and is represented by a sequence of simple chemical reactions (Table 13.2). However, the gasification process is often considered to involve two distinct chemical stages: (i) devolatilization of the feedstock to produce volatile matter and char and (ii) followed by char gasification, which is complex and specific to the conditions of the reaction—both processes contribute to the complex kinetics of the gasification process (Sundaresan and Amundson, 1978).

Gasification of a carbonaceous material in an atmosphere of carbon dioxide can be divided into two stages: (i) pyrolysis and (ii) gasification of the pyrolytic char. In the first stage, pyrolysis (removal of moisture content and devolatilization) occurs at comparatively lower temperature. In the second stage, gasification of the pyrolytic char is achieved by the reaction with oxygen/carbon dioxide mixtures at high temperature. In nitrogen and carbon dioxide environments from room temperature to 1000°C (1830°F), the mass loss rate of pyrolysis in nitrogen may be significant differently (sometime lower, depending on the feedstock) to mass loss rate in carbon dioxide, which may be due (in part) to the difference in properties of the bulk gases.

TABLE 13.2 Gasification Reactions
$2C+O_2 \rightarrow 2CO$
$C+O_2 \rightarrow CO_2$
$C+CO_2 \rightarrow 2CO$
$CO+H_2O \rightarrow CO_2+H_2$ (shift reaction)
$C+H_2O \rightarrow CO+H_2$ (water-gas reaction)
$C+2H_2 \rightarrow CH_4$
$2\ H_2+O_2 \rightarrow 2H_2O$
$CO+2H_2 \rightarrow CH_3OH$
$CO+3H_2 \rightarrow CH_4+H_2O$ (methanation reaction)
$CO_2+4H_2 \rightarrow CH_4+2H_2O$
$C+2H_2O \rightarrow 2H_2+CO_2$
$2C+H_2 \rightarrow C_2H_2$
$CH_4+2H_2O \rightarrow CO_2+4H_2$

2.1 General Aspects

Generally, the gasification of carbonaceous feedstocks (such as heavy oil, extra heavy oil, tar sand bitumen, crude oil residua, biomass, and waste) includes a series of reaction steps that convert the feedstock into *synthesis gas* (carbon monoxide, CO, plus hydrogen, H_2) and other gaseous products. This conversion is generally accomplished by introducing a gasifying agent (air, oxygen, and/or steam) into a reactor vessel containing the feedstock where the temperature, pressure, and flow pattern (moving bed, fluidized, or entrained bed) are controlled.

The gaseous products—other than carbon monoxide and hydrogen—and the proportions of these product gases (such as carbon dioxide, CO_2; methane, CH_4; water vapor, H_2O; hydrogen sulfide, H_2S; and sulfur dioxide, SO_2) depend on (i) the type of feedstock, (ii) the chemical composition of the feedstock, (iii) the gasifying agent or gasifying medium, and (iv) the thermodynamics and chemistry of the gasification reactions as controlled by the process operating parameters (Singh et al., 1980; Pepiot et al., 2010; Shabbar and Janajreh, 2013; Speight, 2013a,b). In addition, the kinetic rates and extents of conversion for the several chemical reactions that are a part of the gasification process are variable and are typically functions of (i) temperature, (ii) pressure, (iii) reactor configuration, (iv) the gas composition of the product gases, and (v) whether or not these gases influence the outcome of the reaction (Johnson, 1979; Speight, 2013a,b).

In a gasifier, the feedstock is exposed to high temperatures generated from the partial oxidation of the carbon. As the particle is heated, any residual moisture (assuming that the feedstock has been prefired) is driven off, and further heating of the particle begins to drive off the volatile gases. Discharge of the volatile products will generate a wide spectrum of hydrocarbon derivatives ranging from carbon monoxide and methane to long-chain hydrocarbon derivatives comprising distillable tar and nondistillable pitch. The complexity of the products will also affect the progress and rate of the reaction when each product is produced by a different chemical process at a different rate. At a temperature above 500°C (930°F), the conversion of the feedstock to char and ash is completed. In most of the early gasification processes, this was the desired by-product, but for gas generation, the char provides the necessary energy to effect further heating, and typically, the char is contacted with air or oxygen and steam to generate the product gases. Furthermore, with an increase in heating rate, feedstock particles are heated more rapidly and are burned in a higher-temperature region, but the increase in heating rate has almost no substantial effect on the mechanism (Irfan, 2009).

Most notable effects in the physical chemistry of the gasification process are those effects due to the chemical character of the feedstock and the physical composition of the feedstock (Speight, 2011a, 2013a, 2014a,b). In more general terms of the character of the feedstock, gasification technologies generally require some initial processing of the feedstock with the type and degree of pretreatment, a function of the process, and/or the type of feedstock. Another factor, often presented as very general *rule of thumb*, is that optimum gas yields and gas quality are obtained at operating temperatures of ~595–650°C (1100–1200°F). A gaseous product with a higher heat content (Btu/ft^3) can be obtained at lower system temperatures, but the overall yield of gas (determined as the *fuel-to-gas ratio*) is reduced by the unburned char fraction.

With some feedstocks, the higher the amounts of volatile material produced in the early stages of the process, the higher the heat content of the product gas. In some cases, the highest gas quality may be produced at the lowest temperatures, but when the temperature is too low, char oxidation reaction is suppressed, and the overall heat content of the product gas is diminished. All such events serve to complicate the reaction rate and make derivative of a global kinetic relationship applicable to all types of feedstock subject to serious question and doubt.

Depending on the type of feedstock being processed and the analysis of the gas product desired, pressure also plays a role in product definition. In fact, some (or all) of the following processing steps will be required: (i) pretreatment of the feedstock; (ii) primary gasification of the feedstock; (iii) secondary gasification of the carbonaceous residue from the primary gasifier; (iv) removal of carbon dioxide, hydrogen sulfide, and other acid gases; (v) shift conversion for adjustment of the carbon monoxide/hydrogen mole ratio to the desired ratio; and (vi) catalytic methanation of the carbon monoxide/hydrogen mixture to form methane. If high-heat-content (high-Btu) gas is desired, all of these processing steps are required since gasifiers do not typically yield methane in the significant concentration.

2.2 Pretreatment

While feedstock pretreatment for introduction into the gasifier is often considered to be a physical process in which the feedstock is prepared for gasifier—typically as pellets or finely ground feedstock—there are chemical aspects that must also be considered.

Some feedstocks, especially certain types of coal, display caking or agglomerating characteristics when heated (Speight, 2013a), and these coal types are usually not amenable to treatment by gasification processes employing fluidized-bed or moving-bed reactors; in fact, caked coal is difficult to handle in fixed-bed reactors. The pretreatment involves a mild oxidation treatment that destroys the caking characteristics of coals and usually consists of low-temperature heating of the coal in the presence of air or oxygen.

While this may seemingly be applicable to coal gasification only, this form of coal pretreatment is particularly important when a noncoal feedstock is cogasified with coal. Cogasification of other feedstocks, such as coal and especially biomass, with crude oil coke offers a bridge between the depletion of crude oil stocks when coal is used and a supplementary feedstock based on renewable energy sources (biomass). These options can contribute to reduce the crude oil dependency and carbon dioxide emissions since biomass is known to be neutral in terms of carbon dioxide emissions. The high reactivity of biomass and the accompanying high production of volatile products suggest that some synergetic effects might occur in simultaneous thermochemical treatment of petcoke and biomass, depending on the gasification conditions such as (i) feedstock type and origin, (ii) reactor type, and (iii) process parameters (Penrose et al., 1999; Gray and Tomlinson, 2000; McLendon et al., 2004; Lapuerta et al., 2008; Fermoso et al., 2009; Shen et al., 2012; Khosravi and Khadse, 2013; Speight, 2013a, 2014a,b; Luque and Speight, 2015).

For example, carbonaceous fuels are gasified in reactors; a variety of gasifiers such as the fixed or moving-bed, fluidized-bed, entrained-flow, and molten bath gasifiers have been developed (Shen et al., 2012; Speight, 2014b). If the flow patterns are considered, the fixed-bed and fluidized-bed gasifiers intrinsically pertain to a countercurrent reactor in that fuels are usually sent into the reactor from the top of the gasifier, whereas the oxidant is blown into the reactor from the bottom. With regard to the entrained-flow reactor, it is necessary to pulverize the feedstock (such as coal and petcoke). On the other hand, when the feedstock is sent into an entrained-flow gasifier, the fuels can be in either form of dry feed or slurry feed. In general, dry-feed gasifiers have the advantage over slurry-feed gasifiers in that the former can be operated with lower oxygen consumption. Moreover, dry-feed gasifiers have an additional degree of freedom that makes it possible to optimize synthesis gas production (Shen et al., 2012).

2.3 Reactions

Gasification involves the thermal decomposition of feedstock and the reaction of the feedstock carbon and other pyrolysis products with oxygen, water, and fuel gases such as methane. The presence of oxygen, hydrogen, water vapor, carbon oxides, and other compounds in the reaction atmosphere during pyrolysis may either support or inhibit numerous reactions with carbonaceous feedstocks and with the products evolved. The distribution of weight and chemical composition of the products are also influenced by the prevailing conditions (i.e., temperature, heating rate, pressure, and residence time) and, last but by no means least, the nature of the feedstock (Speight, 2014a,b).

If air is used for combustion, the product gas will have a heat content of ca. 150–300 Btu/ft^3 (depending on process design characteristics) and will contain undesirable constituents such as carbon dioxide, hydrogen sulfide, and nitrogen. The use of pure oxygen, although expensive, results in a product gas having a heat content on the order of 300–400 Btu/ft^3 with carbon dioxide and hydrogen sulfide as by-product (both of which can be removed from low or medium heat content, low-Btu or medium-Btu) gases by any of several available processes (Speight, 2013a, 2014a).

If a high-heat-content (high-Btu) gas (900–1000 Btu/ft^3) is required, efforts must be made to increase the methane content of the gas. The reactions that generate methane are all exothermic and have negative values, but the reaction rates are relatively slow, and catalysts may therefore be necessary to accelerate the reactions to acceptable commercial rates. Indeed, the overall reactivity of the feedstock and char may be subject to catalytic effects. It is also possible that the mineral constituents of the feedstock (such as the mineral matter in coal and biomass) may modify the reactivity by a direct catalytic effect (Davidson, 1983; Baker and Rodriguez, 1990; Mims, 1991; Martinez-Alonso and Tascon, 1991).

In the process, the feedstock undergoes three processes in its conversation to synthesis gas—the first two processes, pyrolysis and combustion, occur very rapidly. In pyrolysis, char is produced as the feedstock heats up and volatiles are released. In the combustion process, the volatile products and some of the char react with oxygen to produce various products (primarily carbon dioxide and carbon monoxide) and the heat required for subsequent gasification reactions. Finally, in the gasification process, the feedstock char reacts with steam to produce hydrogen (H_2) and carbon monoxide (CO).

Combustion:

$$2C_{feedstock} + O_2 \rightarrow 2CO + H_2O$$

Gasification:

$$C_{feedstock} + H_2O \rightarrow H_2 + CO$$

$$CO + H_2O \rightarrow H_2 + CO_2$$

The resulting synthesis gas is ~63% v/v carbon monoxide, 34% v/v hydrogen, and 3% v/v carbon dioxide. At the gasifier temperature, the ash and other feedstock mineral matter liquefies and exits at the bottom of the gasifier as slag, a sand-like inert material that can be sold as a coproduct to other industries (e.g., road building). The synthesis gas exits the gasifier at pressure and high temperature and must be cooled prior to the synthesis gas cleaning stage.

Although processes that use the high temperature to raise high-pressure steam are more efficient for electricity production, full-quench cooling, by which the synthesis gas is cooled by the direct injection of water, is more appropriate for hydrogen production. Full-quench cooling provides the necessary steam to facilitate the water-gas shift reaction, in which carbon monoxide is converted to hydrogen and carbon dioxide in the presence of a catalyst.

Water-gas shift reaction:

$$CO + H_2O \rightarrow CO_2 + H_2$$

This reaction maximizes the hydrogen content of the synthesis gas, which consists primarily of hydrogen and carbon dioxide at this stage. The synthesis gas is then scrubbed of particulate matter, and sulfur is removed via physical absorption (Speight, 2013a, 2014a). The carbon dioxide is captured by physical absorption or a membrane and either vented or sequestered.

Thus, in the initial stages of gasification, the rising temperature of the feedstock initiates devolatilization and the breaking of weaker chemical bonds to yield volatile tar, volatile oil, phenol derivatives, and hydrocarbon gases. These products generally react further in the gaseous phase to form hydrogen, carbon monoxide, and carbon dioxide. The char (fixed carbon) that remains after devolatilization reacts with oxygen, steam, carbon dioxide, and hydrogen. Overall, the chemistry of gasification is complex but can be conveniently (and simply) represented by the following reactions:

$$C + O_2 \rightarrow CO_2 \quad \Delta H_r = -393.4\,\text{MJ/kmol} \tag{13.1}$$

$$C + ½O_2 \rightarrow CO \quad \Delta H_r = -111.4\,\text{MJ/kmol} \tag{13.2}$$

$$C + H_2O \rightarrow H_2 + CO \quad \Delta H_r = 130.5\,\text{MJ/kmol} \tag{13.3}$$

$$C + CO_2 \leftrightarrow 2CO \quad \Delta H_r = 170.7\,\text{MJ/kmol} \tag{13.4}$$

$$CO + H_2O \leftrightarrow H_2 + CO_2 \quad \Delta H_r = -40.2\,\text{MJ/kmol} \tag{13.5}$$

$$C + 2H_2 \rightarrow CH_4 \quad \Delta H_r = -74.7\,\text{MJ/kmol} \tag{13.6}$$

The designation *C* represents carbon in the original feedstock and carbon in the char formed by devolatilization of the feedstock. Reactions (13.1) and (13.2) are exothermic oxidation reactions and provide most of the energy required by the endothermic gasification reactions (13.3) and (13.4). The oxidation reactions occur very rapidly, completely consuming all of the oxygen present in the gasifier, so that most of the gasifier operates under reducing conditions. Reaction (13.5) is the water-gas shift reaction; in water, steam is converted to hydrogen—this reaction is used to alter the hydrogen/carbon monoxide ration when synthesis gas is the desired product, such as for use in Fischer-Tropsch processes. Reaction (13.6) is favored by high pressure and low temperature and is, thus, mainly important in lower-temperature gasification systems. Methane formation is an exothermic reaction that does not consume oxygen and, therefore, increases the efficiency of the gasification process and the final heat content of the product gas. Overall, ~70% of the heating value of the product gas is associated with the carbon monoxide and hydrogen, but this varies depending on the gasifier type and the process parameters (Speight, 2011a, 2013a; Chadeesingh, 2011).

In essence, the direction of the gasification process is subject to the constraints of thermodynamic equilibrium and variable reaction kinetics. The combustion reactions (reaction of the feedstock or char with oxygen) essentially go to completion. The thermodynamic equilibrium of the rest of the gasification reactions are relatively well defined and collectively have a major influence on thermal efficiency of the process and on the gas composition. Thus, thermodynamic data are useful for estimating key design parameters for a gasification process, such as (i) calculating of the relative amounts of oxygen and/or steam required per unit of feedstock, (ii) estimating the composition of the produced synthesis gas, and (iii) optimizing process efficiency at various operating conditions.

Other deductions concerning gasification process design and operations can also be derived from the thermodynamic understanding of its reactions. Examples include (i) production of synthesis gas with low methane content at high temperature, which requires an amount of steam in excess of the stoichiometric requirement; (ii) gasification at high temperature, which increases oxygen consumption and decreases the overall process efficiency; and (iii) production of synthesis gas with a high methane content, which requires operation at low temperature (~700°C, 1290°F), but the methanation reaction kinetics will be poor without the presence of a catalyst.

Relative to the thermodynamic understanding of the gasification process, the kinetic behavior is much more complex. In fact, very little reliable global kinetic information on gasification reactions exists, partly because it is highly dependent on (i) the chemical nature of the feed, which varies significantly with respect to composition, mineral impurities; (ii) feedstock reactivity; and (iii) process parameters, such as temperature, pressure, and residence time. In addition, physical characteristics of the feedstock (or char) also play a role in phenomena such boundary layer diffusion, pore diffusion, and ash layer diffusion that also influence the kinetic outcome. Furthermore, certain impurities, in fact, are known to have catalytic activity on some of the gasification reactions that can have further influence on the kinetic imprint of the gasification reactions.

2.3.1 Primary Gasification

Primary gasification involves thermal decomposition of the raw feedstock via various chemical processes, and many schemes involve pressures ranging from atmospheric to 1000 psi. Air or oxygen may be admitted to support combustion to provide the necessary heat. The product is usually a low-heat-content (low-Btu) gas ranging from a carbon monoxide/hydrogen mixture to mixtures containing varying amounts of carbon monoxide, carbon dioxide, hydrogen, water, methane, hydrogen sulfide, nitrogen, and typical tar-like products of thermal decomposition of carbonaceous feedstocks that are complex mixtures and include hydrocarbon oils and phenolic products (Dutcher et al., 1983; Speight, 2011a, 2013a, 2014b).

Devolatilization of the feedstock occurs rapidly as the temperature rises above 300°C (570°F). During this period, the chemical structure is altered, producing solid char, tar products, condensable liquids, and low-molecular-weight gases. Furthermore, the products of the devolatilization stage in an inert gas atmosphere are very different from those in an atmosphere containing hydrogen at elevated pressure. In an atmosphere of hydrogen at elevated pressure, additional yields of methane or other low-molecular-weight gaseous hydrocarbon derivatives can result during the initial gasification stage from reactions such as (i) direct hydrogenation of feedstock or semichar because of any reactive intermediates formed and (ii) the hydrogenation of other gaseous hydrocarbon derivatives, oils, tars, and carbon oxides. Again, the kinetic picture for such reactions is complex due to the varying composition of the volatile products that, in turn, are related to the chemical character of the feedstock and the process parameters, including the reactor type.

A solid char product may also be produced and may represent the bulk of the weight of the original feedstock, which determines (to a large extent) the yield of char and the composition of the gaseous product.

2.3.2 Secondary Gasification

Secondary gasification usually involves the gasification of char from the primary gasifier, which is typically achieved by reaction of the hot char with water vapor to produce carbon monoxide and hydrogen:

$$C_{char} + H_2O \rightarrow CO + H_2$$

The reaction requires heat input (endothermic) for the reaction to proceed in its forward direction. Usually, an excess amount of steam is also needed to promote the reaction. However, excess steam used in this reaction has an adverse effect on the thermal efficiency of the process. Therefore, this reaction is typically combined with other gasification reactions in practical applications. The hydrogen-carbon monoxide ratio of the product synthesis gas depends on the synthesis chemistry and process engineering.

The mechanism of this reaction section is based on the reaction between carbon and gaseous reactants, not for reactions between feedstock and gaseous reactants. Hence, the equations may oversimplify the actual chemistry of the steam gasification reaction. Even though carbon is the dominant atomic species present in feedstock, feedstock is more reactive than pure carbon. The presence of various reactive organic functional groups and the availability of catalytic activity via naturally occurring mineral ingredients can enhance the relative reactivity of the feedstock—for example, anthracite, which has the highest carbon content among all ranks of coal (Speight, 2013a), is most difficult to gasify or liquefy.

After the rate of devolatilization has passed a maximum, another reaction becomes important—in this reaction, the semichar is converted to char (sometimes erroneously referred to as *stable char*) primarily through the evolution of hydrogen. Thus, the *gasification* process occurs as the char reacts with gases such as carbon dioxide and steam to produce carbon monoxide and hydrogen. The resulting gas (producer gas or synthesis gas) may be more efficiently converted to electricity than is typically possible by direct combustion of the. Also, corrosive elements in the ash may be refined out by the gasification process, allowing high-temperature combustion of the gas from otherwise problematic feedstocks (Speight, 2011a, 2013a, 2014b).

Oxidation and gasification reactions consume the char, and the oxidation and the gasification kinetic rates follow Arrhenius-type dependence on temperature. On the other hand, the kinetic parameters are feedstock-specific, and there is no true global relationship to describe the kinetics of char gasification—the characteristics of the char are also feedstock-specific. The complexity of the reactions makes the reaction initiation and the subsequent rates subject to many factors, any one of which can influence the kinetic aspects of the reaction.

Although the initial gasification stage (devolatilization) is completed in seconds or even less at elevated temperature, the subsequent gasification of the char produced at the initial gasification stage is much slower, requiring minutes or hours to obtain significant conversion under practical conditions, and reactor designs for commercial gasifiers are largely dependent on the reactivity of the char and also on the gasification medium (Johnson, 1979; Sha, 2005). Thus, the distribution and chemical composition of the products are also influenced by the prevailing conditions (i.e., temperature, heating rate, pressure, and residence time) and, last but not least, the nature of the feedstock. Also, the presence of oxygen, hydrogen, water vapor, carbon oxides, and other compounds in the reaction atmosphere during pyrolysis may either support or inhibit numerous reactions with feedstock and with the products evolved.

The reactivity of char produced in the pyrolysis step depends on the nature of the feedstock and increases with oxygen content of the feedstock but decreases with carbon content. In general, char produced from a low-carbon feedstock is more reactive than char produced from a high-carbon feedstock. The reactivity of char from a low-carbon feedstock may be influenced by catalytic effect of mineral matter in char. In addition, as the carbon content of the feedstock increases, the reactive functional groups present in the feedstock decrease, and the char becomes more aromatic and cross-linked in nature (Speight, 2013a). Therefore, char obtained from high-carbon feedstock contains a lesser number of functional groups and higher proportion of aromatic and cross-linked structures, which reduce reactivity. The reactivity of char also depends upon thermal treatment it receives during formation from the parent feedstock—the gasification rate of char decreases as the char preparation temperature increases due to the decrease in active surface areas of char. Therefore, a change of char preparation temperature may change the chemical nature of char, which in turn may change the gasification rate.

Typically, char has a higher surface area compared with the surface area of the parent feedstock, even when the feedstock has been pelletized, and the surface area changes as the char undergoes gasification—the surface area increases with carbon conversion, reaches maximum, and then decreases. These changes in turn affect gasification rates—in general, reactivity increases with the increase in surface area. The initial increase in surface area appears to be caused by cleanup and widening of pores in the char. The decrease in surface area at high carbon conversion may be due to coalescence of pores, which ultimately leads to the collapse of the pore structure within the char.

Heat transfer and mass transfer processes in fixed- or moving-bed gasifiers are affected by complex solid flow and chemical reactions. Coarsely crushed feedstock settles while undergoing heating, drying, devolatilization, gasification, and combustion. Also, the feedstock particles change in diameter, shape, and porosity—nonideal behavior may result from certain types of chemical structures in the feedstock, gas bubbles, and channel, and a variable void fraction may also change heat and mass transfer characteristics.

An important factor is the importance of the pyrolysis temperature as a major factor in the thermal history and consequently in the thermodynamics of the feedstock char. However, the thermal history of a char should also depend on the rate of temperature rise to the pyrolysis temperature and on the length of time the char is kept at the pyrolysis temperature (soak time), which might be expected to reduce the residual entropy of the char by employing a longer soak time.

Alkali metal salts are known to catalyze the steam gasification reaction of carbonaceous materials, including coal. The process is based on the concept that alkali metal salts (such as potassium carbonate, sodium carbonate, potassium sulfide, and sodium sulfide) will catalyze the steam gasification of feedstocks. The order of catalytic activity of alkali metals on the gasification reaction is as follows:

$$\text{Cesium (Cs)} > \text{rubidium (Rb)} > \text{potassium (K)} > \text{sodium (Na)} > \text{lithium (Li)}$$

Catalyst amounts on the order of 10%–20% w/w potassium carbonate will lower bituminous coal gasifier temperatures from 925°C (1695°F) to 700°C (1090°F) and that the catalyst can be introduced to the gasifier impregnated on coal or char. In addition, tests with potassium carbonate showed that this material also acts as a catalyst for the methanation reaction. In addition, the use of catalysts can reduce the amount of tar formed in the process. In the case of catalytic steam gasification of coal, carbon deposition reaction may affect catalyst life by fouling the catalyst active sites. This carbon deposition reaction is more likely to take place whenever the steam concentration is low.

Ruthenium-containing catalysts are used primarily in the production of ammonia. It has been shown that ruthenium catalysts provide 5–10 times higher reactivity rates than other catalysts. However, ruthenium quickly becomes inactive due to its necessary supporting material, such as activated carbon, which is used to achieve effective reactivity. However, during the process, the carbon is consumed, thereby reducing the effect of the ruthenium catalyst.

Catalysts can also be used to favor or suppress the formation of certain components in the gaseous product by changing the chemistry of the reaction, the rate of reaction, and the thermodynamic balance of the reaction. For example, in the production of synthesis gas (mixtures of hydrogen and carbon monoxide), methane is also produced in small amounts. Catalytic gasification can be used to either promote methane formation or suppress it.

2.3.3 Water Gas Shift Reaction

The water-gas shift reaction (shift conversion) is necessary because the gaseous product from a gasifier generally contains large amounts of carbon monoxide and hydrogen, plus lesser amounts of other gases. Carbon monoxide and hydrogen (if they are present in the mole ratio of 1:3) can be reacted in the presence of a catalyst to produce methane. However, some adjustment to the ideal (1:3) is usually required, and to accomplish this, all or part of the stream is treated according to the water-gas shift (shift conversion) reaction. This involves reacting carbon monoxide with steam to produce a carbon dioxide and hydrogen whereby the desired 1:3 mol ratio of carbon monoxide to hydrogen may be obtained:

$$CO(g) + H_2O(g) \rightarrow CO_2(g) + H_2(g)$$

Even though the water-gas shift reaction is not classified as one of the principal gasification reactions, it cannot be omitted in the analysis of chemical reaction systems that involve synthesis gas. Among all reactions involving synthesis gas, this reaction equilibrium is least sensitive to the temperature variation—the equilibrium constant is least strongly dependent on the temperature. Therefore, the reaction equilibrium can be reversed in a variety of practical process conditions over a wide range of temperature.

The water-gas shift reaction in its forward direction is mildly exothermic, and although all of the participating chemical species are in gaseous form, the reaction is believed to be heterogeneous insofar as the chemistry occurs at the surface of the feedstock and the reaction is actually catalyzed by carbon surfaces. In addition, the reaction can also take place homogeneously and heterogeneously, and a generalized understanding of the water-gas shift reaction is difficult to achieve. Even the published kinetic rate information is not immediately useful or applicable to a practical reactor situation.

Synthesis gas from a gasifier contains a variety of gaseous species other than carbon monoxide and hydrogen. Typically, they include carbon dioxide, methane, and water (steam). Depending on the objective of the ensuing process, the composition of synthesis gas may need to be preferentially readjusted. If the objective of the gasification process is to obtain a high yield of methane, it would be preferred to have the molar ratio of hydrogen-to-carbon monoxide at 3:1:

$$CO(g) + 3H_2(g) \rightarrow CH_4(g) + H_2O(g)$$

On the other hand, if the objective of generating synthesis gas is the synthesis of methanol via a vapor-phase low-pressure process, the stoichiometrically consistent ratio between hydrogen and carbon monoxide would be 2:1. In such cases, the stoichiometrically consistent synthesis gas mixture is often referred to as *balanced gas*, whereas a synthesis gas composition that is substantially deviated from the principal reaction's stoichiometry is called *unbalanced gas*. If the objective of synthesis gas production is to obtain a high yield of hydrogen, it would be advantageous to increase the ratio of hydrogen to carbon monoxide by further converting carbon monoxide (and water) into hydrogen (and carbon dioxide) via the water-gas shift reaction.

The water-gas shift reaction is one of the major reactions in the steam gasification process, where both water and carbon monoxide are present in ample amounts. Although the four chemical species involved in the water-gas shift reaction are gaseous compounds at the reaction stage of most gas processing, the water-gas shift reaction, in the case of steam gasification of feedstock, predominantly takes place on the solid surface of feedstock (heterogeneous reaction). If the product synthesis gas from a gasifier needs to be reconditioned by the water-gas shift reaction, this reaction can be catalyzed by a variety of metallic catalysts.

Choice of specific kinds of catalysts has always depended on the desired outcome, the prevailing temperature conditions, composition of gas mixture, and process economics. Typical catalysts used for the reaction include catalysts containing iron, copper, zinc, nickel, chromium, and molybdenum.

2.3.4 Carbon Dioxide Gasification

The reaction of carbonaceous feedstocks with carbon dioxide produces carbon monoxide (*Boudouard reaction*) and (like the steam gasification reaction) is also an endothermic reaction:

$$C(s) + CO_2(g) \rightarrow 2CO(g)$$

The reverse reaction results in carbon deposition (carbon fouling) on many surfaces including the catalysts and results in catalyst deactivation.

This gasification reaction is thermodynamically favored at high temperatures (>680°C, > 1255°F), which is also quite similar to the steam gasification. If carried out alone, the reaction requires high temperature (for fast reaction) and high pressure (for higher reactant concentrations) for significant conversion, but as a separate reaction, a variety of factors come into play: (i) low conversion, (ii) slow kinetic rate, and (iii) low thermal efficiency.

Also, the rate of the carbon dioxide gasification of a feedstock is different to the rate of the carbon dioxide gasification of carbon. Generally, the carbon-carbon dioxide reaction follows a reaction order based on the partial pressure of the carbon dioxide that is ~1.0 (or lower), whereas the feedstock-carbon dioxide reaction follows a reaction order based on the partial pressure of the carbon dioxide that is 1.0 (or higher). The observed higher reaction order for the feedstock reaction is also based on the relative reactivity of the feedstock in the gasification system.

2.3.5 Hydrogasification

Not all high-heat-content (high-Btu) gasification technologies depend entirely on catalytic methanation, and in fact, a number of gasification processes use hydrogasification, that is, the direct addition of hydrogen to feedstock under pressure to form methane:

$$C_{char} + 2H_2 \rightarrow CH_4$$

The hydrogen-rich gas for hydrogasification can be manufactured from steam and char from the hydrogasifier. Appreciable quantities of methane are formed directly in the primary gasifier, and the heat released by methane formation is at a sufficiently high temperature to be used in the steam-carbon reaction to produce hydrogen so that less oxygen is used to produce heat for the steam-carbon reaction. Hence, less heat is lost in the low-temperature methanation step, thereby leading to higher overall process efficiency.

Hydrogasification is the gasification of feedstock in the presence of an atmosphere of hydrogen under pressure.

$$[C]_{feedstock} + H_2 \rightarrow CH_4$$

The hydrogasification reaction is exothermic and is thermodynamically favored at low temperatures (<670°C, <1240°F), unlike the endothermic both steam gasification and carbon dioxide gasification reactions. However, at low temperatures, the reaction rate is inevitably too slow. Therefore, a high temperature is always required for kinetic reasons, which in turn requires high pressure of hydrogen, which is also preferred from equilibrium considerations. This reaction can be catalyzed by salts such as potassium carbonate (K_2CO_3), nickel chloride ($NiCl_2$), iron chloride ($FeCl_2$), and iron sulfate ($FeSO_4$). However, the use of a catalyst in feedstock gasification suffers from difficulty in recovering and reusing the catalyst and the potential for the spent catalyst becoming an environmental issue.

In a hydrogen atmosphere at elevated pressure, additional yields of methane or other low-molecular-weight hydrocarbon derivatives can result during the initial feedstock gasification stage from direct hydrogenation of feedstock or semichar because of active intermediate formed in the feedstock structure after pyrolysis. The direct hydrogenation can also increase the amount of feedstock carbon that is gasified and the hydrogenation of gaseous hydrocarbon derivatives, oil, and tar.

The kinetics of the rapid-rate reaction between gaseous hydrogen and the active intermediate depends on hydrogen partial pressure (P_{H2}). Greatly increased gaseous hydrocarbon derivatives produced during the initial feedstock gasification stage are extremely important in processes to convert feedstock into methane (SNG, synthetic natural gas, substitute natural gas).

2.3.6 Methanation

Several exothermic reactions may occur simultaneously within a methanation unit. A variety of metals have been used as catalysts for the methanation reaction; the most common and to some extent the most effective methanation catalysts appear to be nickel and ruthenium, with nickel being the most widely (Cusumano et al., 1978):

$$\text{Ruthenium (Ru)} > \text{nickel (Ni)} > \text{cobalt (Co)} > \text{iron (Fe)} > \text{molybdenum (Mo)}$$

Nearly all the commercially available catalysts used for this process are, however, very susceptible to sulfur poisoning, and efforts must be taken to remove all hydrogen sulfide (H_2S) before the catalytic reaction starts. It is necessary to reduce the sulfur concentration in the feed gas to <0.5 ppm v/v in order to maintain adequate catalyst activity for a long period of time.

The synthesis gas must be desulfurized before the methanation step since sulfur compounds will rapidly deactivate (poison) the catalysts. A processing issue may arise when the concentration of carbon monoxide is excessive in the stream to be methanated since large amounts of heat must be removed from the system to prevent high temperatures and deactivation of the catalyst by sintering as well as the deposition of carbon. To eliminate this problem, temperatures should be maintained below 400°C (750°F).

The methanation reaction is used to increase the methane content of the product gas, as needed for the production of high-Btu gas:

$$4H_2 + CO_2 \rightarrow CH_4 + 2H_2O$$

$$2CO \rightarrow C + CO_2$$

$$CO + H_2O \rightarrow CO_2 + H_2$$

Among these, the most dominant chemical reaction leading to methane is the first one. Therefore, if methanation is carried out over a catalyst with a synthesis gas mixture of hydrogen and carbon monoxide, the desired hydrogen-carbon monoxide ratio of the feed synthesis gas is around 3:1. The large amount of water (vapor) produced is removed by condensation and recirculated as process water or steam. During this process, most of the exothermic heat due to the methanation reaction is also recovered through a variety of energy integration processes.

Whereas all the reactions listed above are quite strongly exothermic except the forward water-gas shift reaction, which is mildly exothermic, the heat release depends largely on the amount of carbon monoxide present in the feed synthesis gas. For each 1% v/v carbon monoxide in the feed synthesis gas, an adiabatic reaction will experience a 60°C (108°F) temperature rise, which may be termed as *adiabatic temperature rise*.

3. Processes

Gasification is an established tried-and-true method that can be used to convert crude oil coke (petcoke), heavy oil, extra heavy oil, tar sand bitumen, and other refinery viscous feedstock streams (such as vacuum residua, visbreaker tar, and deasphalter pitch) into power, steam, and hydrogen for use in the production of cleaner transportation fuels. The main requirement for a gasification feedstock is that it contains both hydrogen and carbon.

A number of factors have increased the interest in gasification applications in crude oil refinery operations: (i) Coking capacity has increased with the shift to heavier, more sour crude oils being supplied to the refiners; (ii) hazardous waste disposal has become a major issue for refiners in many countries; (iii) there is strong emphasis on the reduction of emissions of criteria pollutants and greenhouse gases; (iv) requirements to produce ultralow sulfur fuels are increasing the hydrogen needs of the refineries; and (v) the requirements to produce low-sulfur fuels and other regulations could lead to refiners falling short of demand for lower-boiling products such as gasoline and jet and diesel fuel. The typical gasification system incorporated into the refinery consists of several process plants including (i) a feedstock preparation area, (ii) the type of gasifier, (iii) a gas cleaning section, (iv) a sulfur recovery unit, and (v) downstream process options that are dependent on the nature of the products.

The gasification process can provide high-purity hydrogen for a variety of uses within the refinery. Hydrogen is used in the refinery to remove sulfur, nitrogen, and other impurities from intermediate to finished product streams and in hydrocracking operations for the conversion of high-boiling distillates and oils into low-boiling products such as naphtha, kerosene, and diesel (Parkash, 2003; Gary et al., 2007; Speight, 2014a, 2017; Hsu and Robinson, 2017). Furthermore, electric power and high-pressure steam can be generated by the gasification of crude oil coke and viscous feedstocks to drive mostly small and intermittent loads such as compressors, blowers, and pumps. Steam can also be used for process heating, steam tracing, partial pressure reduction in fractionation systems, and stripping low-boiling components to stabilize process streams. Also, the gasification system and refinery operations can share common process equipment. This usually includes an amine stripper or sulfur plant, wastewater treatment, and cooling water systems (Mokhatab et al., 2006; Speight, 2007, 2014a).

3.1 Gasifiers

A gasifier differs from a combustor in that the amount of air or oxygen available inside the gasifier is carefully controlled so that only a relatively small portion of the fuel burns completely. The *partial oxidation* process provides the heat and rather than combustion, most of the carbon-containing feedstock is chemically broken apart by the heat and pressure applied in the gasifier resulting in the chemical reactions that produce synthesis gas. However, the composition of the synthesis gas will vary because of dependence upon the conditions in the gasifier and the type of feedstock. Minerals in the fuel (i.e., the rocks, dirt, and other impurities that do not gasify) separate and leave the bottom of the gasifier either as an inert glass-like slag or other marketable solid products.

Four types of gasifier are currently available for commercial use: (i) the countercurrent fixed bed, (ii) the cocurrent fixed bed, (iii) the fluidized bed, and (iv) the entrained flow (Speight, 2008, 2013a).

In a fixed-bed process, the coal is supported by a grate and combustion gases (steam, air, oxygen, etc.) pass through the supported coal whereupon the hot produced gases exit from the top of the reactor. Heat is supplied internally or from an outside source, but caking coals cannot be used in an unmodified fixed-bed reactor. Due to the liquid-like behavior, the fluidized beds are very well mixed, which effectively eliminates the concentration and temperature gradients inside the reactor. The process is also fairly simple and reliable to operate as the bed acts as a large thermal reservoir that resists rapid changes in temperature and operation conditions. The disadvantages of the process include the need for recirculation of the entrained solids carried out from the reactor with the fluid and the nonuniform residence time of the solids that can cause poor conversion levels. The abrasion of the particles can also contribute to serious erosion of pipes and vessels inside the reactor.

The *countercurrent fixed-bed* (*updraft*) gasifier consists of a fixed bed of carbonaceous fuel (e.g., coal or biomass) through which the *gasification agent* (steam, oxygen, and/or air) flows in countercurrent configuration. The ash is either removed dry or as a slag. The nature of the gasifier means that the fuel must have high mechanical strength and must be noncaking so that it will form a permeable bed, although recent developments have reduced these restrictions to some extent. The throughput for this type of gasifier is relatively low. Thermal efficiency is high as the gas exit temperatures are relatively low, and as a result, tar and methane production is significant at typical operation temperatures, so product gas must be extensively cleaned before use or recycled to the reactor.

The cocurrent fixed-bed (downdraft) gasifier is similar to the countercurrent type, but the gasification agent gas flows in cocurrent configuration with the fuel (downward, hence the name downdraft gasifier). Heat needs to be added to the upper part of the bed, either by combusting small amounts of the fuel or from external heat sources. The produced gas leaves the gasifier at a high temperature, and most of this heat is often transferred to the gasification agent added in the top of the bed. Since all tars must pass through a hot bed of char in this configuration, tar levels are much lower than the countercurrent type.

In the *fluidized-bed* gasifier, the fuel is fluidized in oxygen (or air) and steam. The temperatures are relatively low in dry-ash gasifiers, so the fuel must be highly reactive; low-grade coals are particularly suitable. The fluidized-bed system uses finely sized coal particles, and the bed exhibits liquid-like characteristics when a gas flows upward through the bed. Gas flowing through the coal produces turbulent lifting and separation of particles, and the result is an expanded bed having greater coal surface area to promote the chemical reaction, but such systems have a limited ability to handle caking coals. The agglomerating gasifiers have slightly higher temperatures and are suitable for higher rank coals. Fuel throughput is higher than for the fixed bed, but not as high as for the entrained-flow gasifier. The conversion efficiency is typically low, so recycle or subsequent combustion of solids is necessary to increase conversion. Fluidized-bed gasifiers are most useful for fuels that form highly corrosive ash that would damage the walls of slagging gasifiers. The ash is removed dry or as high-molecular-weight agglomerated materials—a disadvantage of biomass feedstocks is that they generally contain high levels of corrosive ash.

In the *entrained-flow* gasifier, a dry pulverized solid, an atomized liquid fuel, or a fuel slurry is gasified with oxygen (much less frequent, air) in cocurrent flow. The high temperatures and pressures also mean that a higher throughput can be achieved but thermal efficiency is somewhat lower as the gas must be cooled before it can be sent to a gas processing facility. All entrained-flow gasifiers remove the major part of the ash as a slag as the operating temperature is well above the ash fusion temperature; the entrained system is suitable for both caking and noncaking coals.

Entrained-flow reactors use atomized liquid, slurry, or dry pulverized solid as a feedstock. Once pumped inside the gasifier, the feedstock is gasified with oxygen in a cocurrent flow. The temperatures are usually very high in comparison with fluidized beds ranging from 1300°C to 1500°C (from 2370°F to 2730°F). High temperature cracks the feedstock into lower-boiling products.

In *integrated gasification combined cycle* (IGCC) systems, the synthesis gas is cleaned of its hydrogen sulfide, ammonia, and particulate matter and is burned as fuel in a combustion turbine (much like natural gas is burned in a turbine). The combustion turbine drives an electric generator. Hot air from the combustion turbine can be channeled back to the gasifier or the air separation unit, while exhaust heat from the combustion turbine is recovered and used to boil water, creating steam for a steam turbine generator. The use of these two types of turbines—a combustion turbine and a steam turbine—in combination, known as a *combined cycle*, is one reason why gasification-based power systems can achieve unprecedented power generation efficiencies.

Gasification also offers more scope for recovering products from waste than incineration (Speight, 2014b). When waste is burnt in an incinerator, the only practical product is energy, whereas the gases, oils, and solid char from pyrolysis and gasification can be not only used as a fuel but also purified and used as a feedstock for petrochemicals and other applications. Many processes also produce a stable granulate instead of an ash that can be more easily and safely utilized. In addition, some processes are targeted at producing specific recyclables such as metal alloys and carbon black. From waste gasification, in particular, it is feasible to produce hydrogen, which many see as an increasingly valuable resource.

Integrated gasification combined cycle (IGCC) is used to raise power from viscous feedstocks. The value of these refinery residuals, including crude oil coke, will need to be considered as part of an overall upgrading project. Historically, many delayed coking projects have been evaluated and sanctioned on the basis of assigning zero value to crude oil coke having high sulfur and high metal content.

While there are many alternate uses for the synthesis gas produced by gasification, a combination of products/utilities can be produced in addition to power. A major benefit of the integrated gasification combined cycle concept is that power can be produced with the lowest sulfur oxide (SOx) and nitrogen oxide (NOx) emissions of any liquid/solid feed power generation technology.

3.2 Fischer-Tropsch Synthesis

The synthesis reaction is dependent of a catalyst, mostly an iron or cobalt catalyst where the reaction takes place. There is either a low- or high-temperature process (low-temperature Fischer-Tropsch (LTFT) and high-temperature Fischer-Tropsch (HTFT)), with temperatures ranging between 200°C and 240°C for LTFT and 300°C and 350°C for HTFT. The HTFT uses an iron catalyst and the LTFT either an iron or a cobalt catalyst. The different catalysts include also nickel-based and ruthenium-based catalysts, which also have enough activity for commercial use in the process.

The reactors are the *multitubular fixed-bed*, the *slurry*, or the *fluidized-bed* (with either fixed or circulating bed) reactors. The fixed-bed reactor consists of thousands of small tubes with the catalyst as surface-active agent in the tubes. Water surrounds the tubes and regulates the temperature by settling the pressure of evaporation. The slurry reactor is widely used and consists of fluid and solid elements, where the catalyst has no particularly position but flows around as small pieces of catalyst together with the reaction components. The slurry and fixed-bed reactors are used in LTFT. The fluidized-bed reactors are diverse but characterized by the fluid behavior of the catalyst (dry).

The *high-temperature* Fischer-Tropsch technology uses a fluidized catalyst at 300–330°C. Originally, circulating fluidized-bed units were used (Synthol reactors). Since 1989, a commercial scale classical fluidized-bed unit has been implemented and improved upon. The *low-temperature* Fischer-Tropsch technology has originally been used in tubular fixed-bed reactors at 200–230°C. This produces a more paraffinic and waxy product spectrum than the *high-temperature* technology. A new type of reactor (the Sasol slurry-phase distillate reactor) has been developed and is in commercial operation. This reactor uses a slurry-phase system rather than a tubular fixed-bed configuration and is currently the favored technology for the commercial production of synfuels.

Under most circumstances, the production of synthesis gas by reforming natural gas will be more economical than from coal gasification, but site-specific factors need to be considered. In fact, any technological advance in this field (such as better energy integration or the oxygen transfer ceramic membrane reformer concept) will speed up the rate at which the synfuel technology will become common practice.

There are large coal reserves that may increasingly be used as a fuel source during oil depletion. Since there are large coal reserves in the world, this technology could be used as an interim transportation fuel if conventional oil were to become more expensive. Furthermore, combination of biomass gasification and Fischer-Tropsch synthesis is a very promising route to produce transportation fuels from renewable or *green* resources.

Although the focus of this section has been on the production of hydrocarbon derivatives from synthesis gas, it is worthy of note that clean synthesis gas can also be used (i) as chemical *building blocks* to produce a broad range of chemicals using processes well established in the chemical and petrochemical industry, (ii) as a fuel producer for highly efficient fuel cells (which run off the hydrogen made in a gasifier) or perhaps in the future hydrogen turbines and fuel cell-turbine hybrid systems, and (iii) as a source of hydrogen that can be separated from the gas stream and used as a fuel or as a feedstock for refineries (which use the hydrogen to upgrade crude oil products).

The aim of underground (or in situ) gasification of coal is the conversion into combustible gases by combustion of a coal seam in the presence of air, oxygen, or oxygen and steam. Thus, seams that were considered to be inaccessible, unworkable, or uneconomical to mine could be put to use. In addition, strip mining and the accompanying environmental impacts, the problems of spoil banks, acid mine drainage, and the problems associated with the use of high-ash coal are minimized or even eliminated.

The principles of underground gasification are very similar to those involved in the aboveground gasification of coal. The concept involves the drilling and subsequent linking of two boreholes so that gas will pass between the two. Combustion is then initiated at the bottom of one borehole (injection well) and is maintained by the continuous injection of air. In the initial reaction zone (combustion zone), carbon dioxide is generated by the reaction of oxygen (air) with the coal:

$$[C]_{coal} + O_2 \rightarrow CO_2$$

The carbon dioxide reacts with coal (partially devolatilized) further along the seam (reduction zone) to produce carbon monoxide:

$$[C]_{coal} + CO_2 \rightarrow 2CO$$

In addition, at the high temperatures that can frequently occur, moisture injected with oxygen or even moisture inherent in the seam may also react with the coal to produce carbon monoxide and hydrogen:

$$[C]_{coal} + H_2O \rightarrow CO + H_2$$

The gas product varies in character and composition but usually falls into the low-heat (low-Btu) category ranging from 125 to 175 Btu/ft^3.

3.3 Feedstocks

For many decades, coal has been the primary feedstock for gasification units but recent concerns about the use of fossil fuels and the resulting environmental pollutants, irrespective of the various gas cleaning processes and gasification plant environmental cleanup efforts, there is a move to feedstocks other than coal for gasification processes (Speight, 2013a, 2014b). But more pertinent to the present text, the gasification process can also use carbonaceous feedstocks that would otherwise have been discarded and unused, such as waste biomass and other similar biodegradable wastes. Various feedstocks such as biomass, crude oil resids, and other carbonaceous wastes can be used to their fullest potential. In fact, the refining industry has seen fit to use viscous feedstock gasification as a source of hydrogen for the past several decades (Speight, 2014a).

Gasification processes can accept a variety of feedstocks, but the reactor must be selected on the basis of feedstock properties and behavior in the process. The advantage of the gasification process when a carbonaceous feedstock (a feedstock containing carbon) or hydrocarbonaceous feedstock (a feedstock containing carbon and hydrogen) is employed is that the product of focus—synthesis gas—is potentially more useful as an energy source and results in an overall cleaner process.

3.4 Heavy Feedstocks

Gasification is the only technology that makes possible a zero residue target for refineries, contrary to all conversion technologies (including thermal cracking, catalytic cracking, cooking, deasphalting, and hydroprocessing) that can only reduce the bottom volume, with the complication that the residue qualities generally get worse with the degree of conversion (Speight, 2014a).

The flexibility of gasification permits to handle any type of refinery residue, including crude oil coke, tank bottoms, and refinery sludge, and make available a range of value-added products including electricity, steam, hydrogen, and various chemicals based on synthesis gas chemistry: methanol, ammonia, methyl tertiary-butyl ether (MTBE), tertiary amyl methyl ether (TAME), acetic acid, and formaldehyde (Speight, 2008, 2013a). The environmental performance of gasification is unmatched. No other technology processing low-value refinery residues can come close to the emission levels achievable with gasification (Speight, 2013a,b, 2014a).

Gasification is also a method for converting crude oil coke and other refinery nonvolatile waste streams (often referred to as *refinery residuals* and include but not limited to atmospheric residuum, vacuum residuum, visbreaker tar, and deasphalter pitch) into power, steam, and hydrogen for use in the production of cleaner transportation fuels. The main requirement for a gasification feedstock is that the feedstock contains both hydrogen and carbon, and several suitable feedstocks are produced on-site as part of typical refinery processing (Speight, 2011b). The typical gasification system incorporated into a refinery consists of several process units including feed preparation, the gasifier, an air separation unit (ASU), synthesis gas cleanup, sulfur recovery unit (SRU), and downstream process options depending on target products.

The benefits of the addition of a gasification system in a refinery to process crude oil coke or other residuals include (i) production of power, steam, oxygen, and nitrogen for refinery use or sale; (ii) source of synthesis gas for hydrogen to be used in refinery operations and for the production of lower-boiling refinery products through the Fischer-Tropsch synthesis; (iii) increased efficiency of power generation, improved air emissions, and reduced waste stream versus combustion of crude oil coke or residues or incineration; (iv) no off-site transportation or storage for crude oil coke or residuals; and (v) the potential to dispose of waste streams including hazardous materials (Marano, 2003).

Gasification can provide high-purity hydrogen for a variety of uses within the refinery (Speight, 2014a). Hydrogen is used in refineries to remove sulfur, nitrogen, and other impurities from intermediate to finished product streams and in hydrocracking operations for the conversion of high-boiling distillates into lower-boiling products, naphtha, kerosene, and low-boiling gas oil. Hydrocracking and severe hydrotreating require hydrogen that is at least 99% v/v, while less severe hydrotreating can work with gas streams containing 90% v/v pure hydrogen.

Electric power and high-pressure steam can be generated via gasification of crude oil coke and residuals to drive mostly small and intermittent loads such as compressors, blowers, and pumps. Steam can also be used for process heating, steam tracing, partial pressure reduction in fractionation systems, and stripping low-boiling components to stabilize process streams.

Carbon soot is produced during gasification, which ends up in the quench water. The soot is transferred to the feedstock by contacting, in sequence, the quench water blowdown with naphtha and then the naphtha-soot slurry with a fraction of the feed. The soot mixed with the feed is finally recycled into the gasifier, thus achieving 100% conversion of carbon to gas.

3.4.1 Asphalt, Tar, and Pitch

Asphalt does not occur naturally but is manufactured from crude oil and is a black or brown material that has a consistency varying from a viscous liquid to a glassy solid (Speight, 2014a). To a point, asphalt can resemble bitumen (isolated form tar sand formation), hence the tendency to refer to bitumen (incorrectly) as *native asphalt*. It is recommended that there be differentiation between asphalt (manufactured) and bitumen (naturally occurring) other than by the use of the qualifying terms *crude oil* and *native* since the origins of the materials may be reflected in the resulting physicochemical properties of the two types of materials. It is also necessary to distinguish between the asphalt that originates from crude oil by refining and the product in which the source of the asphalt is a material other than crude oil, for example, *Wurtzilite asphalt* (Speight, 2014a). In the absence of a qualifying word, it should be assumed that the word *asphalt* (with or without qualifiers such as *cutback*, *solvent*, and *blown*, which indicate the process used to produce the asphalt) refers to the product manufactured from crude oil.

When the asphalt is produced simply by distillation of an asphaltic crude oil, the product can be referred to as *residual asphalt* or *straight-run asphalt*. For example, if the asphalt is prepared by *solvent* extraction of viscous feedstock or by lower-boiling hydrocarbon (propane) precipitation or if *blown* or otherwise treated, the term should be modified accordingly to qualify the product (e.g., *solvent asphalt*, *propane asphalt*, and *blown asphalt*).

Asphalt softens when heated and is elastic under certain conditions and has many uses. For example, the mechanical properties of asphalt are of particular significance when it is used as a binder or adhesive. The principal application of asphalt is in road surfacing, which may be done in a variety of ways. Other important applications of asphalt include canal and reservoir linings, dam facings, and sea works. The asphalt so used may be a thin, sprayed membrane, covered with earth for protection against weathering and mechanical damage, or thicker surfaces, often including riprap (crushed rock). Asphalt is also used for roofs, coatings, floor tiles, soundproofing, waterproofing, and other building-construction elements and in a number of industrial products, such as batteries. For certain applications, an asphaltic emulsion is prepared, in which fine globules of asphalt are suspended in water.

Tar is a product of the destructive distillation of many bituminous or other organic materials and is a brown to black, oily, viscous liquid to semisolid material. However, *tar* is most commonly produced from *bituminous coal* and is generally understood to refer to the product from coal, although it is advisable to specify *coal tar* if there is the possibility of ambiguity. The most important factor in determining the yield and character of the coal tar is the carbonizing temperature. Three general temperature ranges are recognized, and the products have acquired the designations: *low-temperature tar* (~450–700°C, 540–1290°F), *mid-temperature tar* (~700–900°C, 1290–1650°F), and *high-temperature tar* (~900–1200°C, 1650–2190°F). Tar released during the early stages of the decomposition of the organic material is called *primary tar* since it represents a product that has been recovered without the secondary alteration that results from prolonged residence of the vapor in the heated zone.

Treatment of the distillate (boiling up to 250°C, 480°F) of the tar with caustic soda causes separation of a fraction known as *tar acids*; acid treatment of the distillate produces a variety of organic nitrogen compounds known as *tar bases*. The residue left following removal of the high-boiling distillate is *pitch*, a black, hard, and highly ductile material that is the dark brown to black nondistillable residue.

Coal tar pitch is a soft to hard and brittle substance containing chiefly aromatic resinous compounds along with aromatic and other hydrocarbon derivatives. Pitch is used chiefly as road tar, in waterproofing roofs and other structures, and to make electrodes. *Wood tar pitch* is a bright, lustrous substance containing resin acids; it is used chiefly in the manufacture of plastics and insulating materials and in caulking seams. *Pitch* derived from fats, fatty acids, or fatty oils by distillation are usually soft substances containing polymers and decomposition products; they are used chiefly in varnishes and paints and in floor coverings.

Any of the above derivatives can be used as a gasification feedstock. The properties of asphalt change markedly during the aging process (oxidation in service) to the point where the asphalt fails to perform the task for which it was designed. In some case, the asphalt is recovered and reprocessed for additional use, or it may be sent to a gasifier.

3.4.2 Crude Oil Coke

Coke is the solid carbonaceous material produced from crude oil during thermal processing. More particularly, coke is the residue left by the destructive distillation (i.e., thermal cracking such as the delayed coking process) of crude oil residua. The coke formed in catalytic cracking operations is usually nonrecoverable because of the materials deposited on the catalyst during the process and such coke is often employed as fuel for the process (Speight, 2014a). It is often characterized as a solid material with a honeycomb type of appearance having high carbon content (95% + w/w) with some hydrogen and, depending on the process, sulfur and nitrogen. The color varies from gray to black, and the material is insoluble in organic solvents.

Typically, the composition of crude oil coke varies with the source of the crude oil, but in general, large amounts of high-molecular-weight complex hydrocarbon derivatives (rich in carbon but correspondingly poor in hydrogen) make up a high proportion. The solubility of crude oil *coke* in carbon disulfide has been reported to be as high as 50%–80%, but this is in fact a misnomer and is due to soluble product adsorbed on the coke—by definition, coke is the insoluble, honeycomb material that is the end product of thermal processes. However, coke is not always a product with little use—three physical structures of coke can be produced by delayed coking: (i) shot coke, (ii) sponge coke, or (iii) needle coke, which find different uses within the industry.

Shot coke is an abnormal type of coke resembling small balls. Due to mechanisms not well understood, the coke from some coker feedstocks forms into small, tight, nonattached clusters that look like pellets, marbles, or ball bearings. It usually is a very hard coke, that is, low Hardgrove grindability index (Speight, 2013a). Such coke is less desirable to the end users because of difficulties in handling and grinding. It is believed that feedstocks high in asphaltene constituents and low API favor shot coke formation. Blending aromatic materials with the feedstock and/or increasing the recycle ratio reduces the yield of shot coke. Fluidization in the coke drums may cause formation of shot coke. Occasionally, the smaller *shot coke* may agglomerate into ostrich-egg-sized pieces. Such coke may be more suitable as a gasification feedstock.

Sponge coke is the common type of coke produced by delayed coking units. It is in a form that resembles a sponge and has been called honeycombed. Sponge coke, mostly used for anode-grade carbon, is dull and black, having porous, amorphous structure. *Needle coke* (*acicular coke*) is a special quality coke produced from aromatic feedstocks; is silver-gray, having crystalline broken needle structure; and is believed to be chemically produced through cross-linking of condensed aromatic hydrocarbon derivatives during coking reactions. It has a crystalline structure with more unidirectional pores and is used in the production of electrodes for the steel and aluminum industries and is particularly valuable because the electrodes must be replaced regularly.

Crude oil coke is employed for a number of purposes, but its chief use is (depending upon the degree of purity—that is, contains a low amount of contaminants) for the manufacture of carbon electrodes for aluminum refining, which requires a high-purity carbon—low in ash and sulfur-free; the volatile matter must be removed by calcining. In addition to its use as a metallurgical reducing agent, crude oil coke is employed in the manufacture of carbon brushes, silicon carbide abrasives, and structural carbon (e.g., pipes and Raschig rings), as well as calcium carbide manufacture from which acetylene is produced:

$$\text{Coke} \rightarrow CaC_2$$

$$CaC_2 + H_2O \rightarrow HC{\equiv}CH$$

Considering the properties of coke and the potential nonuse of the highly contaminated material, gasification is the only technology that makes possible for the refineries the zero residue target, contrary to all conversion technologies, thermal cracking, catalytic cracking, cooking, deasphalting, hydroprocessing, etc., which can only reduce the bottom volume, with the complication that the residue qualities generally get worse with the degree of conversion (Speight, 2014a).

With respect to gasification, no other technology processing low-value refinery residues can come close to the emission levels achievable with gasification (Speight, 2014a) and is projected to be a major part of the refinery of the future (Speight, 2011b).

And as for the gasification of coal and biomass (Luque and Speight, 2015), the main requirement for a feedstock to a gasification unit is that the feedstock contains both hydrogen and carbon, of which a variety of feedstocks are available from the throughput of a typical refinery (Table 13.3).

TABLE 13.3 Types of Feedstocks Produced On-Site That Are Available for Gasification

	Units	Vacuum Residue	Visbreaker Tar	Asphalt	Petcoke
Ultimate analysis					
C	wt/wt	84.9%	86.1%	85.1%	88.6%
H	wt/wt	10.4%	10.4%	9.1%	2.8%
N[a]	wt/wt	0.5%	0.6%	0.7%	1.1%
S[a]	wt/wt	4.2%	2.4%	5.1%	7.3%
O	wt/wt		0.5%		0.0%
Ash	wt/wt	0.0%		0.1%	0.2%
Total	wt/wt	100.0%	100.0%	100.0%	100.0%
H_2/C ratio	mol/mol	0.727	0.720	0.640	0.188
Density					
Specific gravity	60°/60°	1.028	1.008	1.070	0.863
API gravity	°API	6.2	8.88	0.8	–
Heating values					
Higher heating value (HHV) (dry)	MBtu/lb	17.72	18.6	17.28	14.85
Lower heating value (LHV) (dry)	wt/wt	16.77	17.6	16.45	14.48

[a]*Nitrogen and sulfur contents vary widely.*
From National Energy Technology Laboratory, US Department of Energy, Washington, DC. http://www.netl.doe.gov/technologies/coalpower/gasification/gasifipedia/7-advantages/7-3-4_refinery.html.

3.4.3 Coal

Coal is a fossil fuel formed in swamp ecosystems where plant remains were saved from oxidation and biodegradation by water and mud. Coal is a combustible organic sedimentary rock (composed primarily of carbon, hydrogen, and oxygen as well as other minor elements including sulfur) formed from ancient vegetation and consolidated between other rock strata to form coal seams. The harder forms can be regarded as organic metamorphic rock (e.g., anthracite coal) because of a higher degree of maturation.

Coal is the largest single source of fuel for the generation of electricity worldwide and the largest source of carbon dioxide emissions, which have been implicated as the primary cause of global climate change, although the debate still rages as to the actual cause (or causes) of climate change. Coal is found as successive layers or seams, sandwiched between strata of sandstone and shale and extracted from the ground by coal mining—either underground coal seams (underground mining) or by open-pit mining (surface mining).

Coal remains in adequate supply, and at current rates of recovery and consumption, the world global coal reserves have been variously estimated to have a reserve/production ratio of at least 155 years. However, as with all estimates of resource longevity, coal longevity is subject to the assumed rate of consumption remaining at the current rate of consumption and, moreover, to technological developments that dictate the rate at which the coal can be mined. But most importantly, coal is a fossil fuel and an *unclean* energy source that will only add to global warming. In fact, the next time electricity is advertised as a clean energy source, just consider the means by which the majority of electricity is produced—almost 50% of the electricity generated in the United States derives from coal (EIA, 2007; Speight, 2013a).

Coal occurs in different forms or *types* (Speight, 2013a). Variations in the nature of the source material and local or regional variations in the coalification processes cause the vegetal matter to evolve differently. Various classification systems thus exist to define the different types of coal. Using the ASTM system of classification (ASTM D388, 2015), the coal precursors are transformed over time (as geologic processes increase their effect over time) into the following:

- **(i)** Lignite—also referred to as brown coal, is the lowest rank of coal and used almost exclusively as fuel for steam-electric power generation. Jet is a compact form of lignite that is sometimes polished and has been used as an ornamental stone since the Iron Age.
- **(ii)** Subbituminous coal—the properties range from those of lignite to those of bituminous coal and is used primarily as fuel for steam-electric power generation.
- **(iii)** Bituminous coal—a dense coal, usually black, sometimes dark brown, often with well-defined bands of brittle and dull material, used primarily as fuel in steam-electric power generation, with substantial quantities also used for heat and power applications in manufacturing and to make coke.

(iv) Anthracite—the highest rank; a harder, glossy, black coal used primarily for residential and commercial space heating.

Chemically, coal is a hydrogen-deficient hydrocarbon with an atomic hydrogen-to-carbon ratio near 0.8, as compared with crude oil hydrocarbon derivatives, which have an atomic hydrogen-to-carbon ratio approximately equal to 2, and methane (CH_4) that has an atomic carbon-to-hydrogen ratio equal to 4. For this reason, any process used to convert coal to alternative fuels must add hydrogen or redistribute the hydrogen in the original coal to generate hydrogen-rich products and coke (Speight, 2013a).

The chemical composition of the coal is defined in terms of its proximate and ultimate (elemental) analyses (Speight, 2013a). The parameters of proximate analysis are moisture, volatile matter, ash, and fixed carbon. Elemental analysis (ultimate analysis) encompasses the quantitative determination of carbon, hydrogen, nitrogen, sulfur, and oxygen within the coal. Additionally, specific physical and mechanical properties of coal and particular carbonization properties are also determined.

Carbon monoxide and hydrogen are produced by the gasification of coal in which a mixture of gases is produced. In addition to carbon monoxide and hydrogen, methane and other hydrocarbon derivatives are also produced depending on conditions. Gasification may be accomplished either in situ or in processing plants. In situ gasification is accomplished by controlled, incomplete burning of a coal bed underground while adding air and steam. The gases are withdrawn and may be burned to produce heat and generate electricity or are utilized as synthesis gas in indirect liquefaction and for the production of chemicals.

Producing diesel and other fuels from coal can be performed through the conversion of coal to synthesis gas, a combination of carbon monoxide, hydrogen, carbon dioxide, and methane. Synthesis gas is subsequently reacted through Fischer-Tropsch synthesis processes to produce hydrocarbon derivatives that can be refined into liquid fuels. By increasing the quantity of high-quality fuels from coal (while reducing costs), research into this process could help in mitigating the dependence on ever-increasingly expensive and depleting stocks of crude oil.

While coal is an abundant natural resource, its combustion or gasification produces both toxic pollutants and greenhouse gases. By developing adsorbents to capture the pollutants (mercury, sulfur, arsenic, and other harmful gases), scientists are striving not only to reduce the quantity of emitted gases but also to maximize the thermal efficiency of the cleanup.

Gasification thus offers one of the cleanest and versatile ways to convert the energy contained in coal into electricity, hydrogen, and other sources of power. Turning coal into synthesis gas is not a new concept, in fact the basic technology dates back to pre-World War II. In fact, a gasification unit can process virtually all the viscous feedstock and wastes that are produced in refineries leading to enhanced yields of high-value products (and hence their competitiveness in the market) by deeper upgrading of their crude oil.

3.4.4 Biomass

Biomass can be considered as any renewable feedstock that is in principle *carbon-neutral* (while the plant is growing, it uses the sun's energy to absorb the same amount of carbon from the atmosphere as it releases into the atmosphere).

Raw materials that can be used to produce biomass-derived fuels are widely available; they come from a large number of different sources and in numerous forms (Rajvanshi, 1986). The main basic sources of biomass include (i) wood, including bark, logs, sawdust, wood chips, wood pellets, and briquettes; (ii) high-yield energy crops, such as wheat, grown specifically for energy applications; (iii) agricultural crops and residues (e.g., straw); and (iv) industrial waste, such as wood pulp or paper pulp. For processing, a simple form of biomass such as untreated and unfinished wood may be converted into a number of physical forms, including pellets and wood chips, for use in biomass boilers and stoves.

Biomass includes a wide range of materials that produce a variety of products that are dependent upon the feedstock (Balat, 2011; Demirbaş, 2011; Ramroop Singh, 2011; Speight, 2011a). In addition, the heat content of the different types of biomass widely varies and has to be taken into consideration when designing any conversion process (Jenkins and Ebeling, 1985).

Thermal conversion processes use heat as the dominant mechanism to convert biomass into another chemical form. The basic alternatives of combustion, torrefaction, pyrolysis, and gasification are separated principally by the extent to which the chemical reactions involved are allowed to proceed (mainly controlled by the availability of oxygen and conversion temperature) (Speight, 2011a).

Energy created by burning biomass (fuel wood), also known as dendrothermal energy, is particularly suited for countries where fuel wood grows more rapidly, for example, tropical countries. There is a number of other less common, more experimental or proprietary thermal processes that may offer benefits including hydrothermal upgrading (HTU) and hydroprocessing. Some have been developed to be compatible with high-moisture-content biomass (e.g., aqueous slurries) and allow them to be converted into more convenient forms.

Some of the applications of thermal conversion are combined heat and power (CHP) and cofiring. In a typical dedicated biomass power plant, efficiencies range from 7% to 27%. In contrast, biomass cofiring with coal typically occurs at efficiencies close to those of coal combustors (30%–40%) (Baxter, 2005; Liu et al., 2011).

Many forms of biomass contain a high percentage of moisture (along with carbohydrates and sugars) and mineral constituents—both of which can influence the economics and viability of a gasification process. The presence of high levels of moisture in biomass reduces the temperature inside the gasifier, which then reduces the efficiency of the gasifier. Many biomass gasification technologies therefore require dried biomass to reduce the moisture content prior to feeding into the gasifier. In addition, biomass can come in a range of sizes.

In many biomass gasification systems, biomass must be processed to a uniform size or shape to be fed into the gasifier at a consistent rate and to maximize gasification efficiency.

Biomass such as wood pellets, yard and crop waste, and "energy crops" including switch grass and waste from pulp and paper mills can also be employed to produce bioethanol and synthetic diesel. Biomass is first gasified to produce synthesis gas and then subsequently converted via catalytic processes to the aforementioned downstream products. Biomass can also be used to produce electricity—either blended with traditional feedstocks, such as coal, or by itself (Shen et al., 2012; Khosravi and Khadse, 2013; Speight, 2014b).

Most biomass gasification systems use air instead of oxygen for gasification reactions (which is typically used in large-scale industrial and power gasification plants). Gasifiers that use oxygen require an air separation unit to provide the gaseous/liquid oxygen; this is usually not cost-effective at the smaller scales used in biomass gasification plants. Air-blown gasifiers utilize oxygen from air for gasification processes.

In general, biomass gasification plants are comparatively smaller than those of typical coal or crude oil coke plants used in the power, chemical, fertilizer, and refining industries. As such, they are less expensive to build and have a smaller environmental footprint. While a large industrial gasification plant may take up 150 ac of land and process 2500–15,000 tons of feedstock per day (e.g., coal or crude oil coke), smaller biomass plants typically process 25–200 tons of feedstock per day and take up <10 ac.

Finally, while biomass may seem to some observers to be the answer to the global climate change issue, advantages and disadvantages of biomass as feedstock must be considered carefully:

Advantages are as follows: (i) theoretically inexhaustible fuel source; (ii) minimal environmental impact when direct combustion of plant mass is not used to generate energy (i.e., fermentation and pyrolysis are used instead); (iii) alcohols and other fuels produced by biomass are efficient, viable, and relatively clean-burning; and (iv) available on a worldwide basis.

Disadvantages are as follows: (i) It could contribute a great deal to global climate change and particulate pollution if combusted directly; (ii) it remains an expensive source of energy, both in terms of producing biomass and the technological conversion to alcohols or other fuels; and (iii) life cycle assessments (LCA) should be considered to address energy inputs and outputs, but there is most likely a net loss of energy when operated on a small scale (as energy must be put in to grow the plant mass).

And while taking the issues of global climate change into account, it must not be ignored that the Earth is in an interglacial period when warming will take place. The extent of this warming is not known—no one was around to measure the temperature change in the last interglacial period—and by the same token, the contribution of anthropological sources to global climate change cannot be measured accurately.

3.4.5 *Solid Waste*

Waste may be municipal solid waste (MSW) that had minimal presorting or refuse-derived fuel (RDF) with significant pretreatment, usually mechanical screening and shredding. Other more specific waste sources (excluding hazardous waste) and possibly including crude oil coke may provide niche opportunities for coutilization (Arena, 2012; Speight, 2013a, 2014b).

The traditional waste-to-energy plant, based on mass-burn combustion on an inclined grate, has a low public acceptability despite the very low emissions achieved over the last decade with modern flue gas cleanup equipment. This has led to difficulty in obtaining planning permissions to construct needed new waste-to-energy plants. After much debate, various governments have allowed options for advanced waste conversion technologies (gasification, pyrolysis, and anaerobic digestion) but will only give credit to the proportion of electricity generated from nonfossil waste.

Use of waste materials as cogasification feedstocks may attract significant disposal credits (Ricketts et al., 2002). Cleaner biomass materials are renewable fuels and may attract premium prices for the electricity generated. Availability of sufficient fuel locally for an economic plant size is often a major issue, as is the reliability of the fuel supply. Use of more-predictably available coal alongside these fuels overcomes some of these difficulties and risks. Coal could be regarded as the base feedstock that keeps the plant running when the fuels producing the better revenue streams are not available in sufficient quantities.

Coal characteristics are very different to younger hydrocarbon fuels such as biomass and waste. Hydrogen-to-carbon ratios are higher for younger fuels, as is the oxygen content. This means that reactivity is very different under gasification conditions. Gas cleaning issues can also be very different, being sulfur a major concern for coal gasification and chlorine compounds and tars more important for waste and biomass gasification. There are no current proposals for adjacent gasifiers and gas cleaning systems, one handling biomass or waste and one coal, alongside each other and feeding the same power production equipment. However, there are some advantages to such a design as compared with mixing fuels in the same gasifier and gas cleaning systems.

Electricity production or combined electricity and heat production remain the most likely area for the application of gasification or cogasification. The lowest investment cost per unit of electricity generated is the use of the gas in an existing large power station. This has been done in several large utility boilers, often with the gas fired alongside the main fuel. This option allows a comparatively small thermal output of gas to be used with the same efficiency as the main fuel in the boiler as a large, efficient steam turbine can be used. It is anticipated that addition of gas from a biomass or wood gasifier into the natural gas feed to a gas turbine is technically possible, but there will be concerns as to the balance of commercial risks to a large power plant and the benefits of using the gas from the gasifier.

Furthermore, the disposal of municipal and industrial waste has become an important problem because the traditional means of disposal, landfill, are much less environmentally acceptable than previously. Much stricter regulation of these disposal methods will make the economics of waste processing for resource recovery much more favorable. One method of processing waste streams is to convert the energy value of the combustible waste into a fuel. One type of fuel attainable from waste is a low heating value gas, usually 100–150 Btu/scf, which can be used to generate process steam or to generate electricity. Coprocessing such waste with coal is also an option (Speight, 2008, 2013a, 2014b).

Cogasification technology varies, being usually site-specific and high feedstock-dependent. At the largest scale, the plant may include the well-proved fixed-bed and entrained-flow gasification processes. At smaller scales, emphasis is placed on technologies that appear closest to commercial operation. Pyrolysis and other advanced thermal conversion processes are included where power generation is practical using the on-site feedstock produced. However, the needs to be addressed are (i) core fuel handling and gasification/pyrolysis technologies, (ii) fuel gas cleanup, and (iii) conversion of fuel gas to electric power (Ricketts et al., 2002).

Waste may be municipal solid waste (MSW) that had minimal presorting or refuse-derived fuel (RDF) with significant pretreatment, usually mechanical screening and shredding. Other more specific waste sources (excluding hazardous waste) and possibly including crude oil coke may provide niche opportunities for coutilization.

The traditional waste-to-energy plant, based on mass-burn combustion on an inclined grate, has a low public acceptability despite the very low emissions achieved over the last decade with modern flue gas cleanup equipment. This has led to difficulty in obtaining planning permissions to construct needed new waste-to-energy plants. After much debate, various governments have allowed options for advanced waste conversion technologies (gasification, pyrolysis, and anaerobic digestion) but will only give credit to the proportion of electricity generated from nonfossil waste.

Coutilization of waste and biomass with coal may provide economies of scale that help achieve the above identified policy objectives at an affordable cost. In some countries, governments propose cogasification processes as being *well suited for community-sized developments* suggesting that waste should be dealt with in smaller plants serving towns and cities, rather than moved to large, central plants (satisfying the so-called *proximity principal*).

In fact, neither biomass nor wastes are currently produced or naturally gathered at sites in sufficient quantities to fuel a modern large and efficient power plant. Disruption, transport issues, fuel use, and public opinion all act against gathering hundreds of megawatts (MWe) at a single location. Biomass or waste-fired power plants are therefore inherently limited in size and hence in efficiency (labor costs per unit electricity produced) and in other economies of scale. The production rates of municipal refuse follow reasonably predictable patterns over time periods of a few years. Recent experience with the very limited current *biomass for energy* harvesting has shown unpredictable variations in harvesting capability with long periods of zero production over large areas during wet weather.

The situation is very different for coal. This is generally mined or imported and thus large quantities are available from a single source or a number of closely located sources, and supply has been reliable and predictable. However, the economics of new coal-fired power plants of any technology or size have not encouraged any new coal-fired power plant in the gas generation market.

The potential unreliability of biomass, longer-term changes in refuse, and the size limitation of a power plant using only waste and/or biomass can be overcome combining biomass, refuse, and coal. It also allows benefit from a premium electricity price for electricity from biomass and the gate fee associated with waste. If the power plant is gasification-based, rather than direct combustion, further benefits may be available. These include a premium price for the electricity from waste, the range of technologies available for the gas to electricity part of the process, and gas cleaning prior to the main combustion stage instead of after combustion and public image, which is currently generally better for gasification as compared with combustion. These considerations lead to current studies of cogasification of wastes/biomass with coal (Speight, 2008).

For large-scale power generation (> 50 MWe), the gasification field is dominated by plant based on the pressurized, oxygen-blown, entrained-flow, or fixed-bed gasification of fossil fuels. Entrained gasifier operational experience to date has largely been with well-controlled fuel feedstocks with short-term trial work at low cogasification ratios and with easily handled fuels.

Use of waste materials as cogasification feedstocks may attract significant disposal credits. Cleaner biomass materials are renewable fuels and may attract premium prices for the electricity generated. Availability of sufficient fuel locally for an economic plant size is often a major issue, as is the reliability of the fuel supply. Use of more-predictably available coal alongside these fuels overcomes some of these difficulties and risks. Coal could be regarded as the *flywheel* that keeps the plant running when the fuels producing the better revenue streams are not available in sufficient quantities.

Coal characteristics are very different to younger hydrocarbon fuels such as biomass and waste. Hydrogen-to-carbon ratios are higher for younger fuels, as is the oxygen content. This means that reactivity is very different under gasification conditions. Gas cleaning issues can also be very different, being sulfur a major concern for coal gasification and chlorine compounds and tars more important for waste and biomass gasification. There are no current proposals for adjacent gasifiers and gas cleaning systems, one handling biomass or waste and one coal, alongside each other and feeding the same power production equipment. However, there are some advantages to such a design as compared with mixing fuels in the same gasifier and gas cleaning systems.

Electricity production or combined electricity and heat production remain the most likely area for the application of gasification or cogasification. The lowest investment cost per unit of electricity generated is the use of the gas in an existing large power station. This has been done in several large utility boilers, often with the gas fired alongside the main fuel. This option allows a comparatively small thermal output of gas to be used with the same efficiency as the main fuel in the boiler as a large, efficient steam turbine can be used. It is anticipated that addition of gas from a biomass or wood gasifier into the natural gas feed to a gas turbine is technically possible, but there will be concerns as to the balance of commercial risks to a large power plant and the benefits of using the gas from the gasifier.

Furthermore, the disposal of municipal and industrial waste has become an important problem because the traditional means of disposal, landfill, are much less environmentally acceptable than previously. Much stricter regulation of these disposal methods will make the economics of waste processing for resource recovery much more favorable. Thus, coal may be cogasified with waste or biomass for environmental, technical or commercial reasons. It allows larger, more efficient plants than those sized for grown biomass or arising waste within a reasonable transport distance; specific operating costs are likely to be lower, and fuel supply security is assured.

Coutilization of waste and biomass with coal may provide economies of scale that help achieve the above identified policy objectives at an affordable cost. In some countries, governments propose cogasification processes as being *well suited for community-sized developments* suggesting that waste should be dealt with in smaller plants serving towns and cities, rather than moved to large, central plants (satisfying the so-called *proximity principal*).

In fact, neither biomass nor wastes are currently produced, or naturally gathered at sites in sufficient quantities to fuel a modern large and efficient power plant. Disruption, transport issues, fuel use, and public opinion all act against gathering hundreds of megawatts (MWe) at a single location. Biomass or waste-fired power plants are therefore inherently limited in size and hence in efficiency (labor costs per unit electricity produced) and in other economies of scale. The production rates of municipal refuse follow reasonably predictable patterns over time periods of a few years. Recent experience with the very limited current *biomass for energy* harvesting has shown unpredictable variations in harvesting capability with long periods of zero production over large areas during wet weather.

The situation is very different for coal. This is generally mined or imported, and thus, large quantities are available from a single source or a number of closely located sources, and supply has been reliable and predictable. However, the economics of new coal-fired power plants of any technology or size have not encouraged any new coal-fired power plant in the gas generation market.

The potential unreliability of biomass, longer-term changes in refuse, and the size limitation of a power plant using only waste and/or biomass can be overcome combining biomass, refuse, and coal. It also allows benefit from a premium electricity price for electricity from biomass and the gate fee associated with waste. If the power plant is gasification-based, rather than direct combustion, further benefits may be available. These include a premium price for the electricity from waste, the range of technologies available for the gas to electricity part of the process, and gas cleaning prior to the main combustion stage instead of after combustion and public image, which is currently generally better for gasification as compared with combustion. These considerations lead to current studies of cogasification of wastes/biomass with coal (Speight, 2008).

For large-scale power generation (> 50 MWe), the gasification field is dominated by plant based on the pressurized, oxygen-blown, entrained-flow, or fixed-bed gasification of fossil fuels. Entrained gasifier operational experience to date has largely been with well-controlled fuel feedstocks with short-term trial work at low cogasification ratios and with easily handled fuels.

Analyses of the composition of municipal solid waste indicate that plastics do make up measureable amounts (5%–10% or more) of solid waste streams. Many of these plastics are worth recovering as energy. In fact, many plastics, particularly the polyolefins, have high calorific values and simple chemical constitutions of primarily carbon and hydrogen. As a result, waste plastics are ideal candidates for the gasification process. Because of the myriad of sizes and shapes of plastic products, size reduction is necessary to create a feed material of a size <2 in in diameter. Some forms of waste plastics such as thin films may require a simple agglomeration step to produce a particle of higher bulk density to facilitate ease of feeding. A plastic, such as high-density polyethylene, processed through a gasifier is converted to carbon monoxide and hydrogen, and these materials in turn may be used to form other chemicals including ethylene from which the polyethylene is produced—*closed the loop recycling*.

3.4.6 Black Liquor

Black liquor is the spent liquor from the Kraft process in which pulpwood is converted into paper pulp by removing lignin and hemicellulose constituents and other extractable materials from wood to free the cellulose fibers. The equivalent spent cooking liquor in the sulfite process is usually called *brown liquor*, but the terms *red liquor*, *thick liquor*, and *sulfite liquor* are also used. Approximately seven units of black liquor are produced in the manufacture of one unit of pulp (Biermann, 1993).

Black liquor comprises an aqueous solution of lignin residues, hemicellulose, and the inorganic chemical used in the process and 15% w/w solids of which 10% w/w are inorganic and 5% w/w are organic. Typically, the organic constituents in black liquor are 40%–45% w/w soaps, 35%–45% w/w lignin, and 10%–15% w/w other (miscellaneous) organic materials.

The organic constituents in the black liquor are made up of water-/alkali-soluble degradation components from the wood. Lignin is partially degraded to shorter fragments with sulfur contents in the order of 1%–2% w/w and sodium content at ~6% w/w of the dry solids. Cellulose (and hemicellulose) is degraded to aliphatic carboxylic acid soaps and hemicellulose fragments. The extractable constituents yield *tall oil soap* and crude turpentine. The tall oil soap may contain up to 20% w/w sodium. Lignin components currently serve for hydrolytic or pyrolytic conversion or combustion. Alternative, hemicellulose constituents may be used in fermentation processes.

Gasification of black liquor has the potential to achieve higher overall energy efficiency as compared with those of conventional recovery boilers while generating an energy-rich synthesis gas. The synthesis gas can then be burned in a gas turbine combined cycle system (*BLGCC*, black liquor gasification combined cycle, and similar to *IGCC*, integrated gasification combined cycle) to produce electricity or converted (through catalytic processes) into chemicals or fuels (e.g., methanol, dimethyl ether, Fischer-Tropsch hydrocarbon derivatives, and diesel fuel).

4. Gasification in a Refinery

The gasification of carbonaceous feedstocks has been used for many years to convert organic solids and liquids into useful gaseous, liquid, and cleaner solid fuels (Speight, 2011a; Brar et al., 2012). In the current context (Figs. 13.1 and 13.2), there are a large number of different feedstock types for use in a refinery-based gasifier, each with different characteristics, including size, shape, bulk density, moisture content, energy content, chemical composition, ash fusion characteristics, and homogeneity of all these properties (Speight, 2013a, 2014a,b). Coal and crude oil coke are used as primary feedstocks for many large gasification plants worldwide. Additionally, a variety of biomass and waste-derived feedstocks can be gasified, with wood pellets and chips, waste wood, plastics, municipal solid waste (MSW), refuse-derived fuel (RDF), agricultural and industrial wastes, sewage sludge, switch grass, discarded seed corn, corn stover, and other crop residues all being used.

The gasification of coal, biomass, crude oil, or any carbonaceous residues is generally aimed to feedstock conversion to gaseous products. In fact, depending on the previously described type of gasifier (e.g., air-blown and enriched oxygen-blown) and the operating conditions, gasification can be used to produce a fuel gas that is suitable for several applications. Thus, gasification offers one of the most versatile methods (with a reduced environmental impact with respect to combustion) to convert carbonaceous feedstocks into electricity, hydrogen, and other valuable energy products.

Gasification for electric power generation enables the use of a common technology in modern gas-fired power plants (*combined cycle*) to recover more of the energy released by burning the fuel. The use of these two types of turbines in the combined cycle system involves (i) a combustion turbine and (ii) a steam turbine. The increased efficiency of the combined cycle for electric power generation results in a 50% *v*/v decrease in carbon dioxide emissions compared with conventional coal plants. Gasification units could be modified to further reduce their climate change impact because a large part of the carbon dioxide generated can be separated from the other product gas *before* combustion (e.g., carbon dioxide can be separated/sequestered from gaseous by-products by using adsorbents (e.g., metal-organic frameworks, MOFs) to prevent its release to the atmosphere). Gasification has also been considered for many years as an alternative to combustion of solid or liquid fuels. Gaseous mixtures are simpler to clean as compared with solid or high-viscosity liquid fuels. Cleaned gases can be used in internal combustion-based power plants that would suffer from severe fouling or corrosion if solid or low-quality liquid fuels were burned inside them.

In fact, the hot synthesis gas produced by gasification of carbonaceous feedstocks can then be processed to remove sulfur compounds, mercury, and particulate matter prior to its use as fuel in a combustion turbine generator to produce electricity. The heat in the exhaust gases from the combustion turbine is recovered to generate additional steam. This steam, along with the steam produced by the gasification process, drives a steam turbine generator to produce additional electricity. In the past decade, the primary application of gasification to power production has become more common due to the demand for high efficiency and low environmental impact.

As anticipated, the quality of the gas generated in a system is influenced by feedstock characteristics, gasifier configuration, and the amount of air, oxygen, or steam introduced into the system. The output and quality of the gas produced is determined by the equilibrium established when the heat of oxidation (combustion) balances the heat of vaporization and volatilization plus the sensible heat (temperature rise) of the exhaust gases. The quality of the outlet gas (Btu/ft.^{3}) is determined by the amount of volatile gases (such as hydrogen, carbon monoxide, water, carbon dioxide, and methane) in the gas stream. With some feedstocks, the higher the amounts of volatile produced in the early stages of the process, the higher the heat content of the product gas. In some cases, the highest gas quality may be produced at lower temperatures. However, char oxidation reaction is suppressed when the temperature is too low, and the overall heat content of the product gas is diminished.

Gasification agents are normally air, oxygen-enriched air or oxygen. Steam is sometimes added for temperature control, for heating value enhancement, or to allow the use of external heat (*allothermal gasification*). The major chemical reactions break and oxidize hydrocarbon derivatives to give a product gas containing carbon monoxide, carbon dioxide, hydrogen, and water. Other important components include hydrogen sulfide, various compounds of sulfur and carbon, ammonia, low-boiling hydrocarbon derivatives, and high-boiling tars.

Depending on the employed gasifier technology and operating conditions, significant quantities of water, carbon dioxide, and methane can be presented in the product gas, as well as a number of minor and trace components.

Under reducing conditions in the gasifier, most of the feedstock sulfur converts to hydrogen sulfide (H_2S), but 3%–10% converts to carbonyl sulfide (COS). Organically bound nitrogen in the coal feedstock is generally converted to gaseous nitrogen (N_2), but some ammonia (NH_3) and a small amount of hydrogen cyanide (HCN) are also formed. Any chlorine in the coal is converted to hydrogen chloride (HCl), with some chlorine present in the particulate matter (fly ash). Trace elements, such as mercury and arsenic, are released during gasification and partition among the different phases (e.g., fly ash, bottom ash, slag, and product gas).

4.1 Gasification of Heavy Feedstocks With Coal

The gasification process can be used to convert viscous feedstocks such as heavy oil, extra heavy oil, tar sand bitumen, vacuum residua, and deasphalter bottoms into synthesis gas that is primarily hydrogen and carbon monoxide (Wallace et al., 1998). The heat generated by the gasification reaction is recovered as the product gas is cooled. For example, when the quench version of Texaco gasification is employed, the steam generated is of medium and low pressure. Note that the low-level heat used for deasphalting integration is the last stage of cooling the synthesis gas.

In addition, integration of solvent deasphalting/gasification facility is an alternative for upgrading viscous oils economically (Wallace et al., 1998). An integrated solvent deasphalting/gasification unit can increase the throughput or the crude flexibility of the refinery without creating a new, highly undesirable viscous oil stream. Typically, the addition of a solvent deasphalting unit to process vacuum tower bottoms increases a refinery's production of diesel oil. The deasphalted oil is converted to diesel using hydrotreating and catalytic cracking (Chapter 9). Unfortunately, the deasphalter bottoms often need to be blended with product diesel oil to produce a viable outlet for these bottoms. A gasification process is capable of converting these deasphalter bottoms to synthesis gas that can then be converted to hydrogen for use in hydrotreating and hydrocracking processes. The synthesis gas may also be used in cogeneration facilities to provide low-cost power and steam to the refinery. If the refinery is part of a petrochemical complex, the synthesis gas can be used a chemical feedstock.

4.2 Gasification of Heavy Feedstocks With Coal

For many decades, coal has been the primary feedstock for gasification units—coal can also be gasified in situ (in the underground seam) (Speight, 2013a; Luque and Speight, 2015), but that is not the subject of this text and is not discussed further. However, with the concern on the issue of environmental pollutants and the potential shortage of coal in some areas, there is a move to feedstocks other than coal for gasification processes.

Many of the proponents of global climate change forget (or refuse to acknowledge) that the Earth is in an interglacial period when warming and climate change can be expected—this was reflected in the commencement of the melting of the glaciers ~11,000 years ago. Thus, considering the geologic sequence of events, the contribution of carbon dioxide from anthropogenic sources is not known with any degree of accuracy.

Gas turbine improvements lead to a number of power plants where fuels (usually coal) are gasified with a viscous feedstock and the gas is cleaned and used in a combined cycle gas turbine power plants. Such power plants generally have higher capital cost, higher operating cost, and lower availability than conventional combustion and steam cycle power plants on the same fuel. Efficiencies of the most sophisticated plants have been broadly similar to the best conventional steam plants with losses in gasification and gas cleaning being balanced by the high efficiency of combined cycle power plants. Environmental aspects resulting from the gas cleaning before the main combustion stage have often been excellent, even in plants with exceptionally high levels of contaminants in the feedstock fuels.

4.3 Gasification of Heavy Feedstocks With Biomass

Gasification is an established technology (Hotchkiss, 2003; Speight, 2013a). Comparatively, biomass gasification has been the focus of research in recent years to estimate efficiency and performance of the gasification process using various types of biomass such as sugarcane residue (Gabra et al., 2001), rice hulls (Boateng et al., 1992), pine sawdust (Lv et al., 2004), almond shells (Rapagnà and Latif, 1997; Rapagnà et al., 2000), wheat straw (Ergudenler and Ghaly, 1993), food waste (Ko et al., 2001), and wood biomass (Pakdel and Roy, 1991; Bhattacharya et al., 1999; Chen et al., 1992; Hanaoka et al., 2005). Recently, cogasification of various biomass and coal mixtures has attracted a great deal of interest from the scientific community. Feedstock combinations including Japanese cedar wood and coal, coal and sawdust, coal and pine chips (Pan et al., 2000), coal and silver birch wood (Collot et al., 1999), and coal and birch wood (Brage et al., 2000) have been reported in gasification practices. Cogasification of coal and biomass has some synergy—the process not only produces a low carbon footprint on the environment but also improves the H_2/CO ratio in the produced gas that is required for liquid fuel synthesis (Sjöström et al., 1999; Kumabe et al., 2007). In addition, the inorganic matter present in biomass catalyzes the gasification of coal. However, cogasification processes require custom fittings and optimized processes for the coal and region-specific wood residues.

While cogasification of coal and biomass is advantageous from a chemical viewpoint, some practical problems are present on upstream, gasification, and downstream processes. On the upstream side, the particle size of the coal and biomass is required to be uniform for optimum gasification. In addition, moisture content and pretreatment (torrefaction) are very important during upstream processing.

While upstream processing is influential from a material handling point of view, the choice of gasifier operation parameters (temperature, gasifying agent, and catalysts) dictates the product gas composition and quality. Biomass decomposition occurs at a lower temperature than coal, and therefore, different reactors compatible to the feedstock mixture are required (Brar et al., 2012). Furthermore, feedstock and gasifier type along with operating parameters not only decide product gas composition but also dictate the amount of impurities to be handled downstream. Downstream processes need to be modified if coal is cogasified with biomass. Heavy metals and other impurities such as sulfur-containing compounds and mercury present in coal can make synthesis gas difficult to use and unhealthy for the environment. Alkali present in biomass can also cause corrosion problems at high temperatures in downstream pipes. An alternative option to downstream gas cleaning would be to process coal to remove mercury and sulfur prior to feeding into the gasifier.

However, first and foremost, coal and biomass require drying and size reduction before they can be fed into a gasifier. Size reduction is needed to obtain appropriate particle sizes; however, drying is required to achieve moisture content suitable for gasification operations. In addition, biomass densification may be conducted to prepare pellets and improve density and material flow in the feeder areas. It is recommended that biomass moisture content should be <15% w/w prior to gasification. High moisture content reduces the temperature achieved in the gasification zone, thus resulting in incomplete gasification. Forest residues or wood has a fiber saturation point at 30%–31% moisture content (dry basis) (Brar et al., 2012). Compressive and shear strength of the wood increases with decreased moisture content below the fiber saturation point. In such a situation, water is removed from the cell wall leading to shrinkage. The long-chain molecules constituents of the cell wall move closer to each other and bind more tightly. A high level of moisture, usually injected in form of steam in the gasification zone, favors formation of a water-gas shift reaction that increases hydrogen concentration in the resulting gas.

The torrefaction process is a thermal treatment of biomass in the absence of oxygen, usually at 250–300°C to drive off moisture, decompose hemicellulose completely, and partially decompose cellulose (Speight, 2011a). Torrefied biomass has reactive and unstable cellulose molecules with broken hydrogen bonds and not only retains 79%–95% of feedstock energy but also produces a more reactive feedstock with lower atomic hydrogen-carbon and oxygen-carbon ratios than those of the original biomass. Torrefaction results in higher yields of hydrogen and carbon monoxide in the gasification process.

Finally, the presence of mineral matter in the coal-biomass feedstock is not appropriate for fluidized-bed gasification. Low melting point of ash present in woody biomass leads to agglomeration that causes defluidization of the ash and sintering, deposition, and corrosion of the gasifier construction metal bed.

Biomass containing alkali oxides and salts are likely to produce clinkering/slagging problems from ash formation (McKendry, 2002). Thus, it is imperative to be aware of the melting of biomass ash, its chemistry within the gasification bed (no bed, silica/sand, or calcium bed), and the fate of alkali metals when using fluidized-bed gasifiers.

Most small- to medium-sized biomass/waste gasifiers are air-blown and operate at atmospheric pressure and at temperatures in the range 800–100°C (1470–2190°F). They face very different challenges to large gasification plants—the use of small-scale air separation plant should oxygen gasification be preferred. Pressurized operation, which eases gas cleaning, may not be practical.

Biomass fuel producers; coal producers; and, to a lesser extent, waste companies are enthusiastic about supplying cogasification power plants and realize the benefits of cogasification with alternate fuels (Lee, 2007; Speight, 2008, 2011a; Lee and Shah, 2013; Speight, 2013a,b). The benefits of a cogasification technology involving coal and biomass include the use of a reliable coal supply with gate-fee waste and biomass, which allows the economies of scale from a larger plant to be supplied just with waste and biomass. In addition, the technology offers a future option of hydrogen production and fuel development in refineries. In fact, oil refineries and petrochemical plants are opportunities for gasifiers when the hydrogen is particularly valuable (Speight, 2011b, 2014a).

4.4 Gasification of Heavy Feedstocks With Waste

The use of fuel cells with gasifiers is frequently discussed, but the current cost of fuel cells is such that their use for mainstream electricity generation is uneconomic.

5. Synthetic-Fuel Production

The gasification of a carbonaceous feedstock (i.e., char produced from the feedstock) is the conversion of the feedstock (by any one of a variety of processes) to produce gaseous products that are combustible and a wide range of chemical products from synthesis gas (Fig. 13.4).

With the rapid increase in the use of coal from the fifteenth century onward, it is not surprising that the concept of using coal to produce a flammable gas, especially the use of the water and hot coal, became commonplace (van Heek and Muhlen, 1991). As a result, the characteristics of rank, mineral matter, particle size, and reaction conditions are all recognized as having a bearing on the outcome of the process not only in terms of gas yields but also on gas properties (van Heek and Muhlen, 1991). The products from the gasification of the process may be of low, medium, or high heat content (high-Btu) as dictated by the process and by the ultimate use for the gas (Baker and Rodriguez, 1990; Probstein and Hicks, 1990; Lahaye and Ehrburger, 1991; Matsukata et al., 1992; Speight, 2013a).

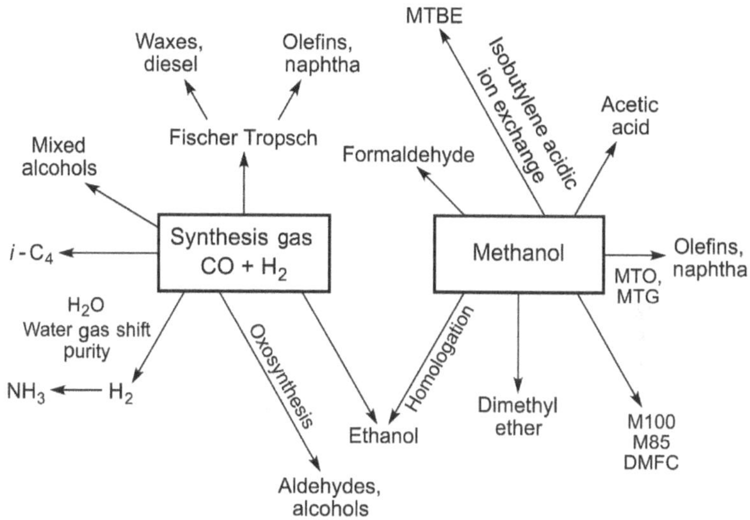

FIG. 13.4 Potential products from heavy feedstock gasification.

5.1 Gaseous Products

The products of gasification are varied insofar as the gas composition varies with the system employed (Speight, 2013a). It is emphasized that the gas product must be first freed from any pollutants such as particulate matter and sulfur compounds before further use, particularly when the intended use is a water-gas shift or methanation (Cusumano et al., 1978; Probstein and Hicks, 1990).

5.1.1 Synthesis Gas

Synthesis gas is comparable in its combustion efficiency to natural gas (Speight, 2008; Chadeesingh, 2011) that reduces the emissions of sulfur, nitrogen oxides, and mercury, resulting in a much cleaner fuel (Nordstrand et al., 2008; Sondreal et al., 2004, 2006; Yang et al., 2007; Wang et al., 2008). The resulting hydrogen gas can be used for electricity generation or as a transport fuel. The gasification process also facilitates capture of carbon dioxide emissions from the combustion effluent (see discussion of carbon capture and storage below).

Although synthesis gas can be used as a stand-alone fuel, the energy density of synthesis gas is approximately half that of natural gas and is therefore mostly suited for the production of transportation fuels and other chemical products. Synthesis gas is mainly used as an intermediary building block for the final production (synthesis) of various fuels such as synthetic natural gas, methanol, and synthetic crude oil fuel (dimethyl ether—synthesized gasoline and diesel fuel) (Chadeesingh, 2011; Speight, 2013a).

The use of synthesis gas offers the opportunity to furnish a broad range of environmentally clean fuels and chemicals, and there has been steady growth in the traditional uses of synthesis gas. Almost all hydrogen gas is manufactured from synthesis gas, and there has been an increase in the demand for this basic chemical. In fact, the major use of synthesis gas is in the manufacture of hydrogen for a growing number of purposes, especially in crude oil refineries (Speight, 2014a). Methanol not only remains the second largest consumer of synthesis gas but also has shown remarkable growth as part of the methyl ethers used as octane enhancers in automotive fuels.

The Fischer-Tropsch synthesis remains the third largest consumer of synthesis gas, mostly not only for transportation fuels but also as a growing feedstock source for the manufacture of chemicals, including polymers. The hydroformylation of olefins (the oxo reaction), a completely chemical use of synthesis gas, is the fourth largest use of carbon monoxide and hydrogen mixtures. A direct application of synthesis gas as fuel (and eventually also for chemicals) that promises to increase is its use for *integrated gasification combined cycle* (IGCC) units for the generation of electricity (and also chemicals), crude oil coke, or viscous feedstocks (Holt, 2001). Finally, synthesis gas is the principal source of carbon monoxide, which is used in an expanding list of carbonylation reactions, which are of major industrial interest.

5.1.2 Low Btu Gas

During the production of gas by oxidation with air, the oxygen is not separated from the air, and as a result, the gas product invariably has a low Btu content (low heat content, 150–300 Btu/ft^3). Several important chemical reactions and a host of side reactions are involved in the manufacture of low-heat-content gas under the high-temperature conditions employed (Speight, 2013a). Low-heat-content gas contains several components, four of which are always major components present at levels of at least several percent; a fifth component, methane, is marginally a major component.

The nitrogen content of low-heat-content gas ranges from somewhat <33% v/v to slightly >50% v/v and cannot be removed by any reasonable means; the presence of nitrogen at these levels makes the product gas *low heat content* by definition. The nitrogen also strongly limits the applicability of the gas to chemical synthesis. Two other noncombustible components (water, H_2O, and carbon dioxide, CO) further lower the heating value of the gas; water can be removed by condensation and carbon dioxide by relatively straightforward chemical means.

The two major combustible components are hydrogen and carbon monoxide; the H_2/CO ratio varies from ~2:3 to ~3:2. Methane may also make an appreciable contribution to the heat content of the gas. Of the minor components, hydrogen sulfide is the most significant, and the amount produced is, in fact, proportional to the sulfur content of the feedstock. Any hydrogen sulfide present must be removed by one, or more, of several procedures (Mokhatab et al., 2006; Speight, 2007, 2014a,b).

Low-heat-content gas is of interest to industry as a fuel gas or even, on occasion, as a raw material from which ammonia, methanol, and other compounds may be synthesized.

5.1.3 Medium Btu Gas

Medium-Btu gas (medium-heat-content gas) has a heating value in the range 300–550 Btu/ft^3, and the composition is much like that of low-heat-content gas, except that there is virtually no nitrogen. The primary combustible gases in medium-heat-content gas are hydrogen and carbon monoxide. Medium-heat-content gas is considerably more versatile than low-heat-content gas; like low-heat-content gas, medium-heat-content gas may be used directly as a fuel to raise steam or used through a combined power cycle to drive a gas turbine, with the hot exhaust gases employed to raise steam, but medium-heat-content gas is especially amenable to synthesize methane (by methanation), higher hydrocarbon derivatives (by Fischer-Tropsch synthesis), methanol, and a variety of synthetic chemicals.

The reactions used to produce medium-heat-content gas are the same as those employed for low-heat-content gas synthesis, the major difference being the application of a nitrogen barrier (such as the use of pure oxygen) to keep diluent nitrogen out of the system.

In medium-heat-content gas, the H_2/CO ratio varies from 2:3 to 3:1, and the increased heating value correlates with higher methane and hydrogen contents and with lower carbon dioxide contents. Furthermore, the very nature of the gasification process used to produce the medium-heat-content gas has a marked effect upon the ease of subsequent processing. For example, the CO_2-acceptor product is quite amenable to use for methane production because it has (i) the desired H_2/CO ratio just exceeding 3:1, (ii) an initially high methane content, and (iii) relatively low water and carbon dioxide contents. Other gases may require appreciable shift reaction and removal of large quantities of water and carbon dioxide prior to methanation.

5.1.4 High-Btu Gas

High-Btu gas (high-heat-content gas) is essentially pure methane and often referred to as *synthetic natural gas* or *substitute natural gas* (SNG) (Speight, 1990, 2013a). However, to qualify as substitute natural gas, a product must contain at least 95% methane, giving an energy content (heat content) of synthetic natural gas on the order of 980–1080 Btu/ft^3.

The commonly accepted approach to the synthesis of high-heat-content gas is the catalytic reaction of hydrogen and carbon monoxide:

$$3H_2 + CO \rightarrow CH_4 + H_2O$$

To avoid catalyst poisoning, the feed gases for this reaction must be quite pure, and therefore, impurities in the product are rare. The large quantities of water produced are removed by condensation and recirculated as very pure water through the gasification system. The hydrogen is usually present in slight excess to ensure that the toxic carbon monoxide is reacted; this small quantity of hydrogen will lower the heat content to a small degree.

The carbon monoxide/hydrogen reaction is somewhat inefficient as a means of producing methane because the reaction liberates large quantities of heat. In addition, the methanation catalyst is troublesome and prone to poisoning by sulfur compounds, and the decomposition of metals can destroy the catalyst. Hydrogasification may be thus employed to minimize the need for methanation:

$$[C]_{feedstock} + 2H2 \rightarrow CH_4$$

The product of hydrogasification is far from pure methane, and additional methanation is required after hydrogen sulfide and other impurities are removed.

5.2 Liquid Products

The production of liquid fuels from a carbonaceous feedstock via gasification is often referred to as the *indirect liquefaction* of the feedstock (Speight, 2013a, 2014a,b). In these processes, the feedstock is not converted directly into liquid products but involves a two-stage conversion operation in which the feedstock is first converted (by reaction with steam and oxygen) to produce a gaseous mixture that is composed primarily of carbon monoxide and hydrogen (synthesis gas). The gas stream is subsequently purified (to remove sulfur, nitrogen, and any particulate matter) after which it is catalytically converted to a mixture of liquid hydrocarbon products.

The synthesis of hydrocarbon derivatives from carbon monoxide and hydrogen (synthesis gas) (the Fischer-Tropsch synthesis) is a procedure for the indirect liquefaction of various carbonaceous feedstocks (Speight, 2011a,b). This process is the only liquefaction scheme currently in use on a relatively large commercial scale for the production of liquid fuels from coal using the Fischer-Tropsch process.

Thus, the feedstock is converted to gaseous products at temperatures in excess of 800°C (1470°F) and at moderate pressures, to produce synthesis gas:

$$[C]_{feedstock} + H_2O \rightarrow CO + H_2$$

In practice, the Fischer-Tropsch reaction is carried out at temperatures of 200–350°C (390–660°F) and at pressures of 75–4000 psi. The hydrogen/carbon monoxide ratio is typically on the order of 2/2:1 or 2/5:1; since up to three volumes of hydrogen may be required to achieve the next stage of the liquid production, the synthesis gas must then be converted by means of the water-gas shift reaction to the desired level of hydrogen:

$$CO + H_2O \rightarrow CO_2 + H_2$$

After this, the gaseous mix is purified and converted to a wide variety of hydrocarbon derivatives:

$$nCO + (2n+1)H_2 \rightarrow C_nH_{2n+2} + nH_2O$$

These reactions result primarily in low- and medium-boiling aliphatic compounds suitable for gasoline and diesel fuel.

5.3 Solid Products

The solid product (solid waste) of a gasification process is typically ash, which is the oxides of metal-containing constituents of the feedstock. The amount and type of solid waste produced are very much feedstock-dependent. The waste is a significant environmental issue due to the large quantities produced, chiefly fly ash if coal is the feedstock or a cofeedstock, and the potential for leaching of toxic substances (such as heavy metals such as lead and arsenic) into the soil and groundwater at disposal sites.

At the high temperature of the gasifier, most of the mineral matter of the feedstock is transformed and melted into slag, an inert glass-like material, and under such conditions, nonvolatile metals and mineral compounds are bound together in molten form until the slag is cooled in a water bath at the bottom of the gasifier or by natural heat loss at the bottom of an entrained-bed gasifier. Slag production is a function of mineral matter content of the feedstock—coal produces much more slag per unit weight than crude oil coke. Furthermore, as long as the operating temperature is above the fusion temperature of the ash, slag will be produced. The physical structure of the slag is sensitive to changes in operating temperature and pressure of the gasifier, and a quick physical examination of the appearance of the slag can often be an indication of the efficiency of the conversion of feedstock carbon to gaseous product in the process.

Slag is composed of black, glassy, silica-based materials and is also known as *frit*, which is a high-density, vitreous, and abrasive material low in carbon and formed in various shapes from jagged and irregular pieces to rod- and needle-like forms. Depending upon the gasifier process parameters and the feedstock properties, there may also be residual carbon char. Vitreous slag is much preferable to ash, because of its habit of encapsulating toxic constituents (such as heavy metals) into a stable, nonleachable material. Leachability data obtained from different gasifiers unequivocally show that gasifier slag is highly nonleachable and can be classified as nonhazardous. Because of its particular properties and nonhazardous, nontoxic nature, slag is relatively easily marketed as a by-product for multiple advantageous uses, which may negate the need for its long-term disposal.

The physical and chemical properties of gasification slag are related to (i) the composition of the feedstock, (ii) the method of recovering the molten ash from the gasifier, and (iii) the proportion of devolatilized carbon particles (char) discharged with the slag. The rapid water-quench method of cooling the molten slag inhibits recrystallization and results in the formation of a granular, amorphous material. Some of the differences in the properties of the slag may be attributed to the specific design and operating conditions prevailing in the gasifiers.

Char is the finer component of the gasifier solid residuals, composed of unreacted carbon with various amounts of siliceous ash. Char can be recycled back into the gasifier to increase carbon usage and has been used as a supplemental fuel source for use in a combustor. The irregularly shaped particles have a well-defined pore structure and have excellent potential as an adsorbent and precursor to activated carbon. In terms of recycling char to the gasifier, a property that is important to fluidization is the effective particle density. If the char has a large internal void space, the density will be much less than that of the feedstock (especially coal) or char from slow carbonization of a carbonaceous feedstock.

6. The Future

The future depends very much on the effect of gasification processes on the surrounding environment. It is these environmental effects and issues that will direct the success of gasification. In fact, there is the distinct possibility that within the foreseeable future, the gasification process will increase in popularity in crude oil refineries—some refineries may even be known as gasification refineries (Speight, 2011b). A gasification refinery would have, as the centerpiece, gasification technology as is the case of the Sasol refinery in South Africa (Couvaras, 1997). The refinery would produce synthesis gas (from the carbonaceous feedstock) from which liquid fuels would be manufactured using the Fischer-Tropsch synthesis technology.

In fact, gasification to produce synthesis gas can proceed from any carbonaceous material, including biomass. Inorganic components of the feedstock, such as metals and minerals, are trapped in an inert and environmentally safe form as char, which may have use as a fertilizer. Biomass gasification is therefore one of the most technically and economically convincing energy possibilities for a potentially carbon-neutral economy.

The manufacture of gas mixtures of carbon monoxide and hydrogen has been an important part of chemical technology for about a century. Originally, such mixtures were obtained by the reaction of steam with incandescent coke and were known as *water gas*. Eventually, steam reforming processes, in which steam is reacted with natural gas (methane) or crude oil naphtha over a nickel catalyst, found wide application for the production of synthesis gas.

A modified version of steam reforming known as autothermal reforming, which is a combination of partial oxidation near the reactor inlet with conventional steam reforming further along the reactor, improves the overall reactor efficiency and increases the flexibility of the process. Partial oxidation processes using oxygen instead of steam also found wide application for synthesis gas manufacture, with the special feature that they could utilize low-value feedstocks such as viscous crude oil residues. In recent years, catalytic partial oxidation employing very short reaction times (milliseconds) at high temperatures (850–1000°C) is providing still another approach to synthesis gas manufacture (Hickman and Schmidt, 1993).

In a gasifier, the carbonaceous material undergoes several different processes: (i) pyrolysis of carbonaceous fuels, (ii) combustion, and (iii) gasification of the remaining char. The process is very dependent on the properties of the carbonaceous material and determines the structure and composition of the char, which will then undergo gasification reactions.

As crude oil supplies decrease, the desirability of producing gas from other carbonaceous feedstocks will increase, especially in those areas where natural gas is in short supply. It is also anticipated that costs of natural gas will increase, thereby allowing the gasification process to compete as an economically viable process.

The conversion of the gaseous products of gasification processes to synthesis gas, a mixture of hydrogen (H_2) and carbon monoxide (CO), in a ratio appropriate to the application, needs additional steps, after purification. The product gases—carbon monoxide, carbon dioxide, hydrogen, methane, and nitrogen—can be used as fuels or as raw materials for chemical or fertilizer manufacture.

Finally, gasification by means other than the conventional methods has also received some attention and has provided rationale for future processes (Rabovitser et al., 2010). In the process, a carbonaceous material and at least one oxygen carrier are introduced into a nonthermal plasma reactor at a temperature in the range from ~300°C to ~700°C (570–1290°F) and a pressure in a range from atmospheric pressure to approximate 1030 psi, and a nonthermal plasma discharge is generated within the nonthermal plasma reactor. The carbonaceous feedstock and the oxygen carrier are exposed to the nonthermal plasma discharge, resulting in the formation of a product gas, which comprises substantial amounts of hydrocarbon derivatives, such as methane, hydrogen, and/or carbon monoxide.

Finally, gasification and conversion of carbonaceous solid fuels to synthesis gas for application of power, liquid fuels, and chemicals are practiced worldwide. Crude oil coke, coal, biomass, and refinery waste are major feedstocks for an on-site refinery gasification unit. The concept of blending of a variety of carbonaceous feedstocks (such as coal, biomass, or refinery waste) with a viscous feedstock of the coke from the thermal processing of the viscous feedstock is advantageous in order to obtain the highest value of products as compared with gasification of crude oil coke alone. Furthermore, based on gasifier type, cogasification of carbonaceous feedstocks can be an advantageous and efficient process. In addition, a variety of upgrading and delivery options that are available for application to synthesis gas enable the establishment of an integrated energy supply system whereby synthesis gases can be upgraded, integrated, and delivered to a distributed network of energy conversion facilities, including power; combined heat and power; and combined cooling, heating, and power (sometime referred to as *trigeneration*), as well as used as fuels for transportation applications.

References

Abadie, L.M., Chamorro, J.M., 2009. The economics of gasification: a market-based approach. Energies 2, 662–694.

Arena, U., 2012. Process and technological aspects of municipal solid waste gasification. A review. Waste Manag. 32, 625–639.

ASTM D388, 2015. Standard classification of coal by rank. In: Annual Book of Standards. ASTM International, West Conshohocken, PA.

Baker, R.T.K., Rodriguez, N.M., 1990. Fuel Science and Technology Handbook. Marcel Dekker, New York. (Chapter 22).

Balat, M., 2011. Fuels from biomass—an overview. In: Speight, J.G. (Ed.), The Biofuels Handbook. Royal Society of Chemistry, London. (Part 1, Chapter 3).

Baxter, L., 2005. Biomass-coal co-combustion: opportunity for affordable renewable energy. Fuel 84 (10), 1295–1302.

Bhattacharya, S., Rahman, M.M., Siddique, A.H., Pham, H.-L., 1999. A study in wood gasification on low tar production. Energy 24, 285–296.

Biermann, C.J., 1993. Essentials of Pulping and Papermaking. Academic Press, New York.

Boateng, A.A., Walawender, W.P., Fan, L.T., Chee, C.S., 1992. Fluidized-bed steam gasification of rice hull. Bioresour. Technol. 40 (3), 235–239.

Brage, C., Yu, Q., Chen, G., Sjöström, K., 2000. Tar evolution profiles obtained from gasification of biomass and coal. Biomass Bioenergy 18 (1), 87–91.

Brar, J.S., Singh, K., Wang, J., Kumar, S., 2012. Cogasification of coal and biomass: a review. Int. J. For. Res. 2012, 1–10.

Chadeesingh, R., 2011. The Fischer-Tropsch process. In: Speight, J.G. (Ed.), The Biofuels Handbook. The Royal Society of Chemistry, London, pp. 476–517. (Part 3, Chapter 5).

Chen, G., Sjöström, K., Bjornbom, E., 1992. Pyrolysis/gasification of wood in a pressurized fluidized bed reactor. Ind. Eng. Chem. Res. 31 (12), 2764–2768.

Collot, A.G., Zhuo, Y., Dugwell, D.R., Kandiyoti, R., 1999. Co-pyrolysis and cogasification of coal and biomass in bench-scale fixed-bed and fluidized bed reactors. Fuel 78, 667–679.

Couvaras, G., 1997. Sasol's slurry phase distillate process and future applications. In: Proceedings of the Monetizing Stranded Gas Reserves Conference, Houston, December.

Cusumano, J.A., Dalla Betta, R.A., Levy, R.B., 1978. Catalysis in Coal Conversion. Academic Press, New York.

Davidson, R.M., 1983. Mineral Effects in Coal Conversion. Report No. ICTIS/TR22, International Energy Agency, London.

Demirbaş, A., 2011. Production of fuels from crops. In: Speight, J.G. (Ed.), The Biofuels Handbook. Royal Society of Chemistry, London. (Part 2, Chapter 1).

Dutcher, J.S., Royer, R.E., Mitchell, C.E., Dahl, A.R., 1983. Wright, C.W., Weimer, W.C., Felic, W.D. (Eds.), Advanced Techniques in Synthetic Fuels Analysis. Technical Information Center, United States Department of Energy, Washington, DC, p. 12.

EIA, 2007. Net generation by energy source by type of producer. In: Energy Information Administration. United States Department of Energy, Washington, DC. http://www.eia.doe.gov/cneaf/electricity/epm/table1_1.html.

Ergudenler, A., Ghaly, A.E., 1993. Agglomeration of alumina sand in a fluidized bed straw gasifier at elevated temperatures. Bioresour. Technol. 43 (3), 259–268.

Fabry, F., Rehmet, C., Rohani, V.-J., Fulcheri, L., 2013. Waste gasification by thermal plasma: a review. Waste Biomass Valoriz. 4 (3), 421–439.

Fermoso, J., Plaza, M.G., Arias, B., Pevida, C., Rubiera, F., Pis, J.J., 2009. Co-gasification of coal with biomass and petcoke in a high-pressure gasifier for syngas production. In: Proceedings of the 1st Spanish National Conference on Advances in Materials Recycling and Eco-Energy. Madrid, Spain, November 12–13.

Furimsky, E., 1999. Gasification in a petroleum refinery of the 21st century. Oil Gas Sci. Technol. 54 (5), 597–618.

Gabra, M., Pettersson, E., Backman, R., Kjellström, B., 2001. Evaluation of cyclone gasifier performance for gasification of sugar cane residue–part 1: gasification of bagasse. Biomass Bioenergy 21 (5), 351–369.

Gary, J.G., Handwerk, G.E., Kaiser, M.J., 2007. Crude Oil Refining: Technology and Economics, fifth ed. CRC Press, Taylor & Francis Group, Boca Raton, FL.

Gray, D., Tomlinson, G., 2000. Opportunities for petroleum coke gasification under tighter sulfur limits for transportation fuels. In: Proceedings of the 2000 Gasification Technologies Conference, San Francisco, CA, October 8–11.

Hanaoka, T., Inoue, S., Uno, S., Ogi, T., Minowa, T., 2005. Effect of woody biomass components on air-steam gasification. Biomass Bioenergy 28 (1), 69–76.

Hickman, D.A., Schmidt, L.D., 1993. Syngas formation by direct catalytic oxidation of methane. Science 259, 343–346.

Higman, C., Van der Burgt, M., 2008. Gasification, second ed. Gulf Professional Publishing, Elsevier, Amsterdam.

Holt, N.A.H., 2001. Integrated gasification combined cycle power plants. In: Encyclopedia of Physical Science and Technology. third ed., Academic Press, New York.

Hotchkiss, R., 2003. Coal gasification technologies. Proc. Inst. Mech. Eng. A 217 (1), 27–33.

Hsu, C.S., Robinson, P.R. (Eds.), 2017. Handbook of Petroleum Technology. Springer International Publishing AG, Cham.

Irfan, M.F., 2009. Research Report: Pulverized Coal Pyrolysis & Gasification in $N_2/O_2/CO_2$ Mixtures by Thermo-Gravimetric Analysis. In: Novel Carbon Resource Sciences Newsletter. vol. 2. Kyushu University, Fukuoka, pp. 27–33.

Jenkins, B.M., Ebeling, J.M., 1985. Thermochemical properties of biomass fuels. Calif. Agric (May–June), 14–18.

Johnson, J.L., 1979. Kinetics of Coal Gasification. John Wiley and Sons, Hoboken, NJ.

Khosravi, M., Khadse, A., 2013. Gasification of petcoke and coal/biomass blend: a review. Int. J. Emerg. Technol. Adv. Eng. 3 (12), 167–173.

Ko, M.K., Lee, W.Y., Kim, S.B., Lee, K.W., Chun, H.S., 2001. Gasification of food waste with steam in fluidized bed. Korean J. Chem. Eng. 18 (6), 961–964.

Kumabe, K., Hanaoka, T., Fujimoto, S., Minowa, T., Sakanishi, K., 2007. Cogasification of woody biomass and coal with air and steam. Fuel 86, 684–689.

Kumar, A., Jones, D.D., Hanna, M.A., 2009. Thermochemical biomass gasification: a review of the current status of the technology. Energies 2, 556–581.

Lahaye, J., Ehrburger, P. (Eds.), 1991. Fundamental Issues in Control of Carbon Gasification Reactivity. Kluwer Academic Publishers, Dordrecht.

Lapuerta, M., Hernández, J.J., Pazo, A., López, J., 2008. Gasification and co-gasification of biomass wastes: effect of the biomass origin and the gasifier operating conditions. Fuel Process. Technol. 89 (9), 828–837.

Lee, S., 2007. Gasification of coal. In: Lee, S., Speight, J.G., Loyalka, S. (Eds.), Handbook of Alternative Fuel Technologies. CRC Press, Taylor & Francis Group, Boca Raton, FL. 2007.

Lee, S., Shah, Y.T., 2013. Biofuels and Bioenergy. CRC Press, Taylor & Francis Group, Boca Raton, FL.

Lee, S., Speight, J.G., Loyalka, S., 2007. Handbook of Alternative Fuel Technologies. CRC-Taylor & Francis Group, Boca Raton, FL.

Liu, G., Larson, E.D., Williams, R.H., Kreutz, T.G., Guo, X., 2011. Making Fischer-Tropsch fuels and electricity from coal and biomass: performance and cost analysis. Energy Fuel 25, 415–437.

Luque, R., Speight, J.G. (Eds.), 2015. Gasification for Synthetic Fuel Production: Fundamentals, Processes, and Applications. Woodhead Publishing, Elsevier, Cambridge.

Lv, P.M., Xiong, Z.H., Chang, J., Wu, C.Z., Chen, Y., Zhu, J.X., 2004. An experimental study on biomass air-steam gasification in a fluidized bed. Bioresour. Technol. 95 (1), 95–101.

Marano, J.J., 2003. Refinery technology profiles: gasification and supporting technologies. Report Prepared for the United States Department of Energy, National Energy Technology Laboratory, June, United States Energy Information Administration, Washington, DC.

Martinez-Alonso, A., Tascon, J.M.D., 1991. Lahaye, J., Ehrburger, P. (Eds.), Fundamental Issues in Control of Carbon Gasification Reactivity. Kluwer Academic Publishers, Dordrecht.

Matsukata, M., Kikuchi, E., Morita, Y., 1992. A new classification of alkali and alkaline earth catalysts for gasification of carbon. Fuel 71, 819–823.

McKendry, P., 2002. Energy production from biomass part 3: gasification technologies. Bioresour. Technol. 83 (1), 55–63.

McLendon, T.R., Lui, A.P., Pineault, R.L., Beer, S.K., Richardson, S.W., 2004. High-pressure co-gasification of coal and biomass in a fluidized bed. Biomass Bioenergy 26 (4), 377–388.

Mims, C.A., 1991. Lahaye, J., Ehrburger, P. (Eds.), Fundamental Issues in Control of Carbon Gasification Reactivity. Kluwer Academic Publishers, Dordrecht, p. 383.

Mokhatab, S., Poe, W.A., Speight, J.G., 2006. Handbook of Natural Gas Transmission and Processing. Elsevier, Amsterdam.

Nordstrand, D., Duong, D.N.B., Miller, B.G., 2008. Miller, B.G., Tillman, D. (Eds.), Combustion Engineering Issues for Solid Fuel Systems—Chapter 9 Post-combustion Emissions Control. Elsevier, London.

Orhan, Y., İs, G., Alper, E., McApline, K., Daly, S., Sycz, M., Elkamel, A., 2014. Gasification of oil refinery waste for power and hydrogen production. In: Proceedings of the 2014 International Conference on Industrial Engineering and Operations Management, Bali, Indonesia, January 7–9.

Pakdel, H., Roy, C., 1991. Hydrocarbon content of liquid products and tar from pyrolysis and gasification of wood. Energy Fuel 5, 427–436.

Pan, Y.G., Velo, E., Roca, X., Manyà, J.J., Puigjaner, L., 2000. Fluidized-bed cogasification of residual biomass/poor coal blends for fuel gas production. Fuel 79, 1317–1326.

Parkash, S., 2003. Refining Processes Handbook. Gulf Professional Publishing, Elsevier, Amsterdam.

Penrose, C.F., Wallace, P.S., Kasbaum, J.L., Anderson, M.K., Preston, W.E., 1999. Enhancing refinery profitability by gasification, hydroprocessing and power generation. In: Proceedings of the Gasification Technologies Conference San Francisco, California, Octoberhttps://www.globalsyngas.org/uploads/eventLibrary/GTC99270.pdf.

Pepiot, P., Dibble, Foust, C.G., 2010. Computational fluid dynamics modeling of biomass gasification and pyrolysis. In: Nimlos, M.R., Crowley, M.F. (Eds.), Computational Modeling in Lignocellulosic Biofuel Production. In: ACS Symposium SeriesAmerican Chemical Society, Washington, DC.

Probstein, R.F., Hicks, R.E., 1990. Synthetic Fuels. pH Press, Cambridge, MA (Chapter 4).

Rabovitser, I.K., Nester, S., Bryan, B., 2010. Plasma Assisted Conversion of Carbonaceous Materials Into a Gas. United States Patent 7,736,400, June 25.

Rajvanshi, A.K., 1986. Biomass gasification. In: Goswami, D.Y. (Ed.), Alternative Energy in Agriculture. In: vol. 2. CRC Press, Boca Raton, FL, pp. 83–102.

Ramroop Singh, N., 2011. Biofuel. In: Speight, J.G. (Ed.), The Biofuels Handbook. Royal Society of Chemistry, London. (Part 1, Chapter 5).

Rapagnà, N.J., Latif, A., 1997. Steam gasification of almond shells in a fluidized bed reactor: the influence of temperature and particle size on product yield and distribution. Biomass Bioenergy 12 (4), 281–288.

Rapagnà, N.J., Kiennemann, A., Foscolo, P.U., 2000. Steam-gasification of biomass in a fluidized-bed of olivine particles. Biomass Bioenergy 19 (3), 187–197.

Ricketts, B., Hotchkiss, R., Livingston, W., Hall, M., 2002. Technology status review of waste/biomass co-gasification with coal. In: Proceedings of the Inst. Chem. Eng. Fifth European Gasification Conference, Noordwijk, Netherlands, April 8–10.

Sha, X., 2005. Coal gasification. In: Coal, Oil Shale, Natural Bitumen, Heavy Oil and Peat. Encyclopedia of Life Support Systems (EOLSS), Developed Under the Auspices of the UNESCO. EOLSS Publishers, Oxford http://www.eolss.net.

Shabbar, S., Janajreh, I., 2013. Thermodynamic equilibrium analysis of coal gasification using gibbs energy minimization method. Energy Convers. Manag. 65, 755–763.

Shen, C.-H., Chen, W.-H., Hsu, H.-W., Sheu, J.-Y., Hsieh, T.-H., 2012. Co-gasification performance of coal and petroleum coke blends in a pilot-scale pressurized entrained-flow gasifier. Int. J. Energy Res. 36, 499–508.

Singh, S.P., Weil, S.A., Babu, S.P., 1980. Thermodynamic analysis of coal gasification processes. Energy 5 (8–9), 905–914.

Sjöström, K., Chen, G., Yu, Q., Brage, C., Rosén, C., 1999. Promoted reactivity of char in cogasification of biomass and coal: synergies in the thermochemical process. Fuel 78, 1189–1194.

Sondreal, E.A., Benson, S.A., Pavlish, J.H., Ralston, N.V.C., 2004. An overview of air quality III: mercury, trace elements, and particulate matter. Fuel Process. Technol. 85, 425–440.

Sondreal, E.A., Benson, S.A., Pavlish, J.H., 2006. Status of research on air quality: mercury, trace elements, and particulate matter. Fuel Process. Technol. 65 (66), 5–22.

Speight, J.G., 1990. Speight, J.G. (Ed.), Fuel Science and Technology Handbook. Marcel Dekker, New York. (Chapter 33).

Speight, J.G., 2007. Natural Gas: A Basic Handbook. GPC Books, Gulf Publishing Company, Houston, TX.

Speight, J.G., 2008. Synthetic Fuels Handbook: Properties, Processes, and Performance. McGraw-Hill, New York.

Speight, J.G. (Ed.), 2011a. Biofuels Handbook. Royal Society of Chemistry, London.

Speight, J.G., 2011b. The Refinery of the Future. Gulf Professional Publishing, Elsevier, Oxford.

Speight, J.G., 2013a. The Chemistry and Technology of Coal, third ed. CRC Press, Taylor & Francis Group, Boca Raton, FL.

Speight, J.G., 2013b. Coal-Fired Power Generation Handbook. Scrivener Publishing, Salem, MA.

Speight, J.G., 2014a. The Chemistry and Technology of Petroleum, fifth ed. CRC Press, Taylor & Francis Group, Boca Raton, FL.

Speight, J.G., 2014b. Gasification of Unconventional Feedstocks. Gulf Professional Publishing, Elsevier, Oxford.

Speight, J.G., 2017. Handbook of Petroleum Refining. CRC Press, Taylor & Francis Group, Boca Raton, FL.

Sundaresan, S., Amundson, N.R., 1978. Studies in char gasification—I: a lumped model. Chem. Eng. Sci. 34, 345–354.

Sutikno, T., Turini, K., 2012. Gasifying coke to produce hydrogen in refineries. Pet. Technol. Q. Q3, 105.

Van Heek, K.H., Muhlen, H.-J., 1991. Lahaye, J., Ehrburger, P. (Eds.), Fundamental Issues in Control of Carbon Gasification Reactivity. Kluwer Academic Publishers, Dordrecht, p. 1.

Wallace, P.S., Anderson, M.K., Rodarte, A.I., Preston, W.E., 1998. Heavy oil upgrading by the separation and gasification of asphaltenes. In: Proceedings of the Presented at the Gasification Technologies Conference, San Francisco, CA, October https://www.globalsyngas.org/uploads/eventLibrary/gtc9817p.pdf.

Wang, Y., Duan, Y., Yang, L., Jiang, Y., Wu, C., Wang, Q., Yang, X., 2008. Comparison of mercury removal characteristic between fabric filter and electrostatic precipitators of coal-fired power plants. J. Fuel Chem. Technol. 36 (1), 23–29.

Wolff, J., Vliegenthart, E., 2011. Gasification of heavy ends. Pet. Technol. Q. Q2, 1–5.

Yang, H., Xua, Z., Fan, M., Bland, A.E., Judkins, R.R., 2007. Adsorbents for capturing mercury in coal-fired boiler flue gas. J. Hazard. Mater. 146, 1–11.

Further Reading

Kunii, D., Levenspiel, O., 2013. Fluidization Engineering, second ed. Butterworth-Heinemann, Elsevier, Amsterdam.

Chapter 14

Upgrading by Bioconversion

1. Introduction

The bioconversion of crude oil is well documented and has been known for some time especially at the process that is operative in many reservoirs (Connan, 1984; Cassani and Eglinton, 1991; Speight and El-Gendy, 2018). In this respect, the activity of microbial species vis-à-vis the heteroatoms of crude oil is especially important. The high sulfur, nitrogen, and metal contents (and the accompanying high viscosity) in crude oil cause expensive processing problems in the refinery. Sulfur is the major concern for producers and refiners and has long been a key determinant of the value of refinery feedstock oils for several reasons: (i) Sulfur presents a processing problem for refiners, and desulfurization offers refiners the opportunity to reduce the sulfur of their crude feedstocks before they ever enter the refinery system, minimizing downstream desulfurization costs; (ii) the amount of sulfur in many finished products such as gasoline and diesel fuel is limited, and the regulations restricting allowable levels of sulfur in end products continue to become increasingly stringent, and this creates an ever more challenging technical and economic situation for refiners as the sulfur levels in available crude oils continue to rise and creates a market disadvantage for producers of high-sulfur crudes. Low-sulfur feedstocks continue to command a premium price in the market, while higher-sulfur crude oils sell at a discount. Desulfurization would offer producers the opportunity to economically upgrade their resources.

In addition, nitrogen in crude oil also leads to the poisoning of the refinery catalysts and also results in increased nitrogen oxide emissions upon combustion in car engines. Also, metal-containing constituents in crude oil lead to two major problems for the industry. Combustion of these fuels leads to the formation of ash with high concentrations of the metal oxides, leading to undesirable waste disposal issues. Also, when crude oil is refined, the metals are concentrated in the nonvolatile residual fraction, which is then further processed by coking or more often by catalytic cracking where metals from the oil deposit on the cracking catalyst result in the catalyst poisoning, thereby decreasing

Heavy Oil Recovery and Upgrading. https://doi.org/10.1016/B978-0-12-813025-4.00014-3

catalyst selectivity and activity. In addition, nitrogen in crude oil also leads to the poisoning of the refinery catalysts and also results in increased nitrogen oxide emissions upon combustion in automobile engines.

The high viscosity significantly hampers the pumping, transportation, refining, and handling of heavy oil, extra heavy oil, and tar sand bitumen (Speight, 2000, 2014, 2017; Parkash, 2003; Ancheyta and Speight, 2007; Gary et al., 2007; Hsu and Robinson, 2017). Common methods used to overcome problems associated with high viscosity include heating (perhaps even visbreaking), dilution with a suitable solvent, and the use of chemical additives. As a result, the need for a safe, economical, and effective method for reducing viscosity has long been a goal of the industry.

Bioconversion processes for addressing these issues offer the crude oil industry potentially great rewards. Some process options focus on the removal of sulfur from crude oil and refinery streams by a microbial process (*biocatalytic desulfurization*, BDS). Furthermore, using the bioconversion approach to viscosity reduction and the removal of metals and nitrogen as additional approaches to fuel upgrading are also options (El-Gendy and Speight, 2016; Speight and El-Gendy, 2018). Generally, biological processing of crude oil feedstocks offers an attractive alternative to conventional thermochemical treatment due to the mild operating conditions and greater reaction specificity afforded by the nature of biocatalysis. Efforts in microbial screening and development have identified microorganisms capable of crude oil desulfurization, denitrogenation, and demetallization (El-Gendy and Speight, 2016; Speight and El-Gendy, 2018).

Biological desulfurization of crude oil may occur either oxidatively or reductively. In the oxidative approach, organic sulfur is converted to sulfate and may be removed in process water. This route is attractive due to the fact that it would not require further processing of the sulfur and may be amenable for use at the wellhead where process water may then be reinjected. In the reductive desulfurization scheme, organic sulfur is converted into hydrogen sulfide, which may then be catalytically converted into elemental sulfur, an approach of utility at the refinery. Regardless of the mode of bioconversion, key factors affecting the economic viability of such processes are biocatalyst activity and cost, differential in product selling price, sale or disposal of coproducts or wastes from the treatment process, and the capital and operating costs of unit operations in the treatment scheme.

But first and by way of definition, *crude oil biotechnology* is based on bioconversion processes (Speight and El-Gendy, 2018). *Crude oil bioremediation* refers specifically to the cleanup of spills of crude oil and crude oil products using microorganisms (Jain and Bajpai, 2012; Speight and El-Gendy, 2018). Furthermore, bioconversion (biotic degradation or biotic decomposition) is the chemical degradation of contaminants by bacteria or other biological means. Organic material can be degraded aerobically (in the presence of oxygen) or anaerobically (in the absence of oxygen). Most bioconversion systems operate under aerobic conditions, but a system under anaerobic conditions may permit microbial organisms to degrade chemical species that are otherwise nonresponsive to aerobic treatment and vice versa (Speight and El-Gendy, 2018).

Bioconversion can be used to upgrade crude oil, and the bioconversion process also has the potential to convert heavy oil, extra heavy oil, and tar sand bitumen by removing undesirable compounds containing sulfur, nitrogen, and metals and by reducing the viscosity. As a result, bioconversion can make the various refinery feedstocks easier to. Thus, bioconversion—the use of living organisms to reduce or eliminate environmental hazards resulting from accumulations of toxic chemicals and other hazardous wastes—is an option that offers the possibility to destroy or render harmless various contaminants using natural biological activity (Gibson and Sayler, 1992). In addition, bioconversion can also be used in conjunction with a wide range of traditional physical and chemical technologies to enhance their effectiveness (Vidali, 2001). However, the lessons learned from the application of microbes as bioremediation agents can be applied to the use of microorganisms to crude oil recovery and refining (Speight, 2000, 2014, 2017; Parkash, 2003; Ancheyta and Speight, 2007; Gary et al., 2007; Hsu and Robinson, 2017; Speight and El-Gendy, 2018).

Biotechnology is now accepted as an attractive means of improving the efficiency of any industrial processes and resolving serious environmental problems. One of the reasons for this is the extraordinary metabolic capability that exists within the bacterial world. Microbial enzymes are capable of biotransforming a wide range of compounds, and the worldwide increase in attention being paid to this concept can be attributed to several factors, including the presence of a wide variety of catabolic enzymes and the ability of many microbial enzymes to transform a broad range of unnatural compounds (xenobiotic compounds) and natural compounds.

Bioconversion processes have several advantages compared with chemical processes, including the following: (i) Microbial enzyme reactions are often more selective, (ii) bioconversion processes are often more energy-efficient, (iii) microbial enzymes are active under mild conditions, and (iv) microbial enzymes are environmentally friendly biocatalysts. Although many bioconversion processes have been described, only a few of these have been used as part of an industrial process, and opportunities exist for biorefining of crude oil (Mohebali and Ball, 2008; Speight and El-Gendy, 2018). Of particular interest in this connect is the phenomenon of bioconversion (biological desulfurization and microbial desulfurization) in which microorganisms are used to oxidize sulfur compounds in crude oil ultimately resulting in desulfurization. This represents the ability of microbial species to desulfurize compounds that are recalcitrant to the current standard technology in the oil industry (El-Gendy and Speight, 2016; Speight and El-Gendy, 2018).

Current applied research on crude oil microbiology encompasses oil spill remediation, fermenter- and wetland-based hydrocarbon treatment, biofiltration of volatile hydrocarbon derivatives, enhanced oil recovery, oil and fuel biorefining, fine-chemical production, and microbial community-based site assessment (Van Hamme et al., 2003; El-Gendy and Speight, 2016; Speight and El-Gendy, 2018). From this work, it is evident that the biorefining concept can become a commercial alternative to some of the current oil refining processes (Figs. 14.1 and 14.2; Speight, 2000, 2014, 2017; Parkash, 2003;

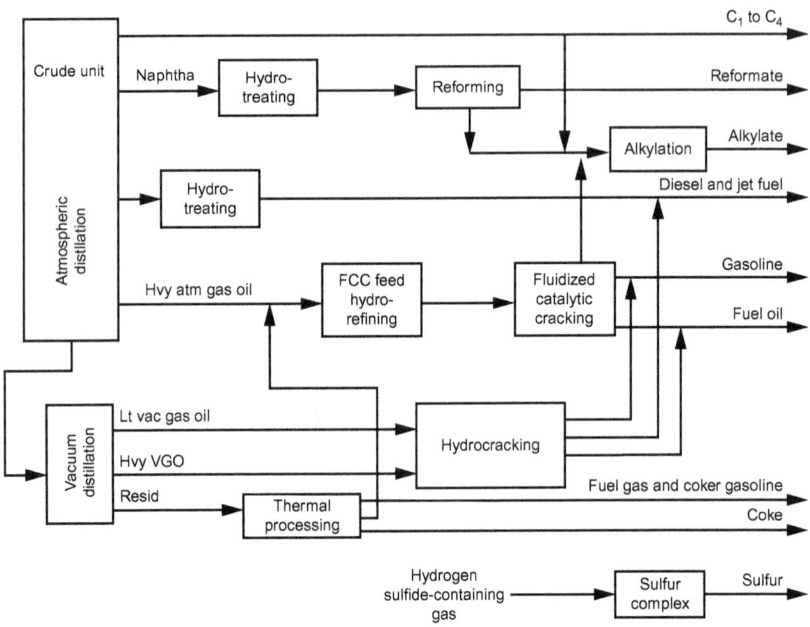

FIG. 14.1 Schematic overview of a refinery.

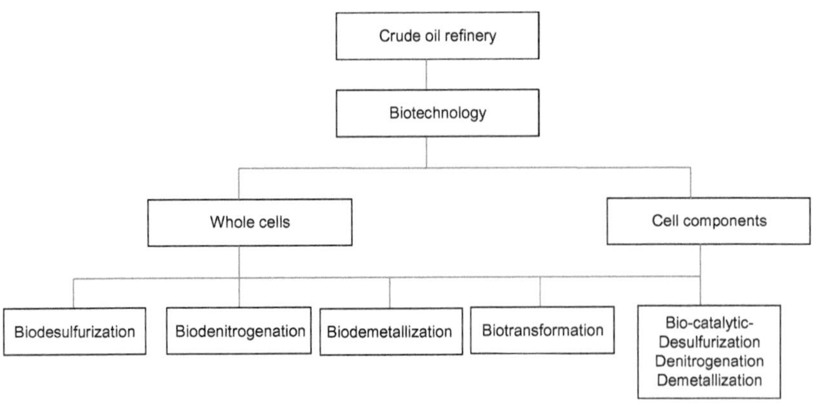

FIG. 14.2 Potential applications of biotechnology in a crude oil refinery.

Ancheyta and Speight, 2007; Gary et al., 2007; Hsu and Robinson, 2017; Speight and El-Gendy, 2018).

The major potential applications of biorefining are biodesulfurization (BDS), biodenitrogenation (BDN), biodemetallization (BDM), and bioconversion of viscous crude oil into lower-density crude oil (Speight and El-Gendy, 2018). The most advanced area is bioconversion for which pilot plants exist (Le Borgne and Quintero, 2003; Bachmann et al., 2014). In fact, the application of biotechnology to crude oil refining (biorefining) is a possible alternative to some of the current refining processes. The major potential applications of biorefining are biodesulfurization, biodenitrogenation, biodemetallization, and bioconversion of heavy oil, extra heavy oil, and tar sand bitumen into low-viscosity products (Le Borgne and Quintero, 2003; El-Gendy and Speight, 2016; Speight and El-Gendy, 2018).

Biorefining is often considered as a potential alternative to the conventional deep hydrodesulfurization processes used in refineries (Fig. 14.3; Speight and El-Gendy, 2018). Bioconversion processes offer specificity of enzymes, relatively lower capital and operating costs, and mild processing conditions and reduce the need for hydrogen. Both of these features would lead to high energy savings in the refinery (Dinamarca et al., 2010, 2014). Further, significant reductions in greenhouse gas emissions have also been predicted if

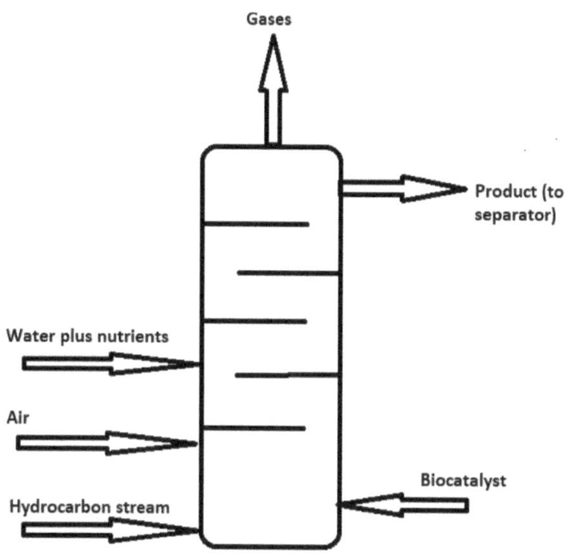

FIG. 14.3 General representation of a biocatalytic process (for convenience, an upflow reactor is shown).

bioconversion is used (Calzada et al., 2011). More specifically, biodesulfurization is a complementary technology; as already mentioned, hydrodesulfurization is not equally effective in desulfurizing all classes of sulfur compounds present in fossil fuels (Speight, 2000, 2014; Ancheyta and Speight, 2007). The biodesulfurization process, on the other hand, is effective regardless of the position of alkyl substitutions (Pacheco, 1999; El-Gendy and Speight, 2016; Speight and El-Gendy, 2018).

However, the hydrodesulfurization process conditions are sufficient not only to desulfurize sensitive (labile) organosulfur compounds but also to (i) remove nitrogen and metals from organic compounds, (ii) induce saturation of at least some carbon-carbon double bonds, (iii) remove substances having an unpleasant smell or color, (iv) clarify the product by drying it, and (v) improve the cracking characteristics of the material (Swaty, 2005). Therefore, with respect to these advantages, placing the biodesulfurization unit downstream of a hydrodesulfurization unit as a complementary technology to achieve ultra-deep desulfurization, rather than as a replacement, should also be considered (El-Gendy and Speight, 2016; Speight and El-Gendy, 2018). There is also the suggestion that a multistage process for desulfurization of fossil fuels would be appropriate. This method was based on subjecting vacuum gas oil to hydrodesulfurization prior to biodesulfurization in defined conditions. In addition, the Energy Biosystems Corporation (EBC) used biodesulfurization downstream of hydrodesulfurization that has also been suggested (Pacheco, 1999).

2. Bioconversion

Crude oil is a complex mixture of hydrocarbon derivatives, composed of aliphatic, aromatic, and asphaltene fractions along with nitrogen-, sulfur-, and oxygen-containing compounds (Tabatabaee and Mazaheri Assadi, 2013). The constituents of these hydrocarbon compounds are present in varied proportions resulting in high variability in crude oil from different sources (Speight, 2000, 2014, 2017; Parkash, 2003; Ancheyta and Speight, 2007; Gary et al., 2007; Hsu and Robinson, 2017). The complexity of heavy oil, extra heavy oil, and tar sand bitumen takes the meaning of the word complexity even further.

Development of bioconversion processes from laboratory-scale or pilot-scale units to commercial-scale units requires improvement in not only the process itself, that is, reactor design, optimization of operation conditions, and downstream operation, but also the characteristics of the biocatalyst (Denome et al., 1993, 1994; Piddington et al., 1995; Li et al., 1996; Gallardo et al., 1997; Larose et al., 1997; Oldfield et al., 1997; Maghsoudi et al., 2000, 2001).

Bioconversion is preferred to occur under thermophilic conditions, because it enhances the bioconversion rate and the operating temperature would be closer to fluid catalytic cracking (FCC) or hydrodesulfurization outlet streams. Since distillate fractions are often treated at high temperatures, there may be some cost savings through the use of moderate thermophiles if bioconversion is integrated with hydrodesulfurization during refinery without cooling the stock to 30°C (86°F). Moreover, the desulfurization activity will also be enhanced due to the higher mass transfer rate at high temperatures. For practical bioconversion, it is useful to obtain microorganisms that exhibit much higher dibenzothiophene and benzothiophene desulfurization activities at high temperatures. Furthermore, the successful application of bioconversion technology to crude oil and to heavy oil, extra heavy oil, and tar sand bitumen requires knowledge of the characteristics of the chemical and physical parameters (i.e., the molecular parameters and the various interrelationships of the molecular species) that are likely to affect the microbial bioconversion processes. However, a number of limiting factors have been recognized to affect the biotransformation of crude oil and the various constituents (Table 14.1).

In the bioconversion process, microorganisms, their enzymes, or cellular extracts as catalysts remove organosulfur from crude oil fractions without degrading the carbon skeleton of the organosulfur compounds.

The bioconversion process is influenced by the composition of the crude oil or the viscous feedstock and the product(s) of any bioconversion process. For example, kerosene, which consists almost exclusively of medium-chain-length alkanes, can, under suitable conditions, be bioconverted to products. Similarly, while conventional crude oil can be bioconverted quantitatively but for the viscous feedstocks (heavy oil, extra heavy oil, and tar sand bitumen), approximately one-sixth to one-tenth of the crude oil may be biodegradable within a reasonable time period, even if the conditions are favorable (Bartha, 1986; Okoh et al., 2001, 2002; Okoh, 2003, 2006; Okoh and Trejo-Hernandez, 2006). On the other hand, crude oil biodegradation has been reported to be mostly enhanced in the presence of a consortium of bacterial species compared with the activity of monospecies.

The composition and biodegradability of the crude oil hydrocarbon derivatives is, therefore, the first and most important consideration when the

TABLE 14.1 Some Common Factors Affecting Heavy Feedstock Bioconversion (El-Gendy and Speight, 2016; Speight and El-Gendy, 2018)

Factor	Comment
Composition	Structure, amount
Physical state, solid/liquid/gas	Accessibility for the microorganism
Temperature	Influence on conversion rate
Mineral nutrients	May limit activity of microorganism
Reaction	May be pH limited

suitability of a cleanup approach is to be evaluated. Viscous (high-density) feedstocks are generally much more difficult to biodegrade than low-density feedstocks, just as the more viscous feedstocks could be suitable for inducing increased selection pressure for the isolation of crude oil hydrocarbon degraders with enhanced efficiency. Also, the amount of viscous crude oil metabolized by some bacterial species increased with increasing concentration of starter oil up to 0.6% (w/v), while degradation rates appeared to be more pronounced between the concentrations of 0.4% and 0.6% (w/v) oil (Okoh et al., 2002). In another report, the percentage of degradation by the mixed bacterial consortium decreased from 78% to 52% as the concentration of crude oil was increased from 1% to 10%.

Although there has been a reported case of the lack of correlation between degradation rates, specific growth rates, and concentration of the starter oil (Thouand et al., 1999), in such a case, it would appear that biomass was required only to a particular threshold enough to produce the appropriate enzyme systems that carry through the degradation process even when biomass production had ceased, a phenomenon completely at systems, where the production of variance with the theory of microbial growth in batch cells is totally dependent on the consumed carbon source (El-Gendy and Speight, 2016; Speight and El-Gendy, 2018).

2.1 Biodesulfurization

Sulfur is the major concern for producers and refiners and has long been a key determinant of the value of crude oils. It is the third most abundant element in crude oil after carbon and hydrogen. Heavy oil, extra heavy oil, and tar sand bitumen contain 3%–6% w/w sulfur that must be removed before its usage as a refinery feedstock. The combustion of sulfur-containing fuels would lead to the increased emissions of sulfur oxides (SOx), the main cause of acid rains and particulate matters (PM) and the main cause of black smoke associated with diesel and gasoline vehicles. It has been reported that the total PM emissions from diesel engines are proportional to the diesel sulfur content (Stanislaus et al., 2010; Mohebali and Ball, 2016).

In a crude oil refinery, the feedstock is typically separated into several fractions and then desulfurizes them separately. Moreover, a refinery can make substantial cost savings if most of the sulfur is removed from the crude oil before it is fractionated. It has also been suggested that due to the high content of water in crude oil, biodesulfurization of crude oil is more practicable compared with that of diesel oil and gasoline (Zhou and Zhang, 2004). Currently, the crude oil industry is increasingly dependent on viscous crude oil to meet the domestic demand for gasoline and distillate fuels. High-viscosity feedstocks are expensive to recover transport and process and have a low market value than less viscous oils (Chapters 2–4).

Furthermore, viscous feedstocks typically have a high content of asphaltene derivatives (the asphaltene fraction), and to use such feedstocks as a fuel, upgrading must involve reducing the average molecular weight of the asphaltene constituents and remove heteroatom-containing constituents (Speight, 2014, 2017). For biomodification of asphaltene constituents, the reactions with the organosulfur moieties could be very significant, because sulfur is the third most abundant element in the asphaltene constituents. Sulfur has an important role in the molecular structure of the asphaltene constituents (Sarret et al., 1999). Because of the diversity and complexity of the asphaltene molecular structures that can be attacked, hemoproteins were the biocatalysts chosen for investigations on the enzymatic modification of the asphaltene constituents. In a survey of several hemoproteins, including horseradish, lignin, manganese chloroperoxidase, and cytochrome *c*, these were found to be able to modify the greatest number of organosulfur compounds including sulfur heterocycles and sulfides and to have superior specific activity.

If desulfurization occurs on the crude feedstocks before they ever enter the refinery system, this would minimize the downstream desulfurization costs. Crude oils with higher viscosities and higher densities usually contain higher amounts of more complex sulfur compounds. The aliphatic acyclic sulfides (thioethers) and cyclic sulfides (thiolanes) are easy to remove during a hydrodesulfurization process or by thermal treatment. On the other hand, sulfur contained in aromatic rings, such as thiophene and its benzologs (e.g., benzothiophene, dibenzothiophene, and benzonaphthothiophene), is more resistant to sulfur removal by hydrodesulfurization and thermal conversion (Gray et al., 1996).

Biodesulfurization is a biological method in which microbes or enzymes are used as a catalyst to remove organosulfur compounds, especially the recalcitrant ones, for example, dibenzothiophene and its derivatives. It can be performed aerobically or anaerobically. Its main disadvantage is its conversion rate that is much slower than hydrodesulfurization, since all the biological reactions are generally slower than the chemical reactions.

There are chemical routes for aerobic biodesulfurization including (i) the complete mineralization where the end products are carbon dioxide and water; (ii) the pathway where carbon-carbon bonds are cleaved and some water-soluble by-products are produced, which would significantly inhibit microbial growth dibenzothiophene oxidation; and (iii) the 4S pathway, in which the carbon skeleton is not destroyed and only sulfur is removed. The first two pathways are not recommendable for desulfurization of fuels, as the efficiency of biodesulfurization depends on the biocatalyst capabilities to remove sulfur without altering the carbon skeleton or reducing the value of the fuel. But they are recommendable in bioremediation of oil spills and soil or sediments polluted with crude oil hydrocarbon derivatives (Gupta et al., 2005).

The key technoeconomic challenge to the ability of biodesulfurization processes is to establish a cost-effective means of implementing the two-phase bioreactor system and demulsification steps and the product recovery step (Kaufman et al., 1998; McFarland, 1999). Use of multiple-stage airlift reactors can reduce mixing costs, and centrifugation approaches facilitated demulsification, desulfurized oil recovery, and recycling of the cells (Ohshiro et al., 1996a,b).

The desulfurization of crude oil under anaerobic conditions would be attractive because it avoids costs associated with aeration and it has the advantage of liberating sulfur as a gas—hydrogen sulfide be treated with existing refinery desulfurization plants (e.g., Claus process; Fig. 14.4) and does not liberate sulfate as a by-product that must be disposed by some appropriate treatment (Setti et al., 1997; Speight, 2014). Under anaerobic conditions, oxidation of hydrocarbon derivatives to undesired compounds such as colored and gum-forming products is minimal (McFarland, 1999). Moreover, anaerobic microorganisms use approximately 10% of total produced energy, while aerobic ones use 50% of it. These advantages can be counted as incentives to continue research on reductive biodesulfurization.

However, maintaining an anaerobic process is extremely difficult, and the specific activity of most of the isolated strains has been reported to be insignificant for dibenzothiophene derivatives (Armstrong et al., 1995). Due to low reaction rates, safety and cost concerns, and the lack of identification of specific enzymes and genes responsible for anaerobic desulfurization, anaerobic microorganisms effective enough for practical crude oil desulfurization have not been found yet, and anaerobic biodesulfurization process has not been developed. Consequently, aerobic biodesulfurization has been the focus of most of the research in biodesulfurization (Le Borgne and Quintero, 2003). Due to the milder and safer process conditions of biodesulfurization, the carbon dioxide emissions and energy requirements of biodesulfurization-based processes are estimated to be lower than that of hydrodesulfurization processes (Linguist and Pacheco, 1999; Singh, 2012). Moreover, the capital costs to set up a biodesulfurization process are reported to be 50% lower than that of a hydrodesulfurization process (Linguist and Pacheco, 1999).

The difference in desulfurization efficiency within the different microorganisms might be related to differences in cell surface hydrophobicity, mass transport through the cell walls and membranes, and metabolic regulations inside the cell (Bustos-Jaimes et al., 2003). Thus, the efficiency of biodesulfurization in crude oil depends mainly on the applied microorganisms and its broad versatility on different OSCs, and there is no general trend for the efficiency of microorganisms in model or real feed oils. It has been reported that the presence of *n*-alkane derivative is seen to favor the removal of sulfur aromatic compounds that act as a cosubstrate ensuring the growth of the culture, which also permits the solubilization and the emulsifying of the sulfur aromatic compounds. The microorganism adsorption to the oil phase is the most likely mechanism for explaining *n*-alkane biodegradation (Setti et al., 1995). Substrate uptake presumably occurs through diffusion or active transport at the point of contact. It is well known that most of the aerobic microorganisms adhere to the *n*-alkanes

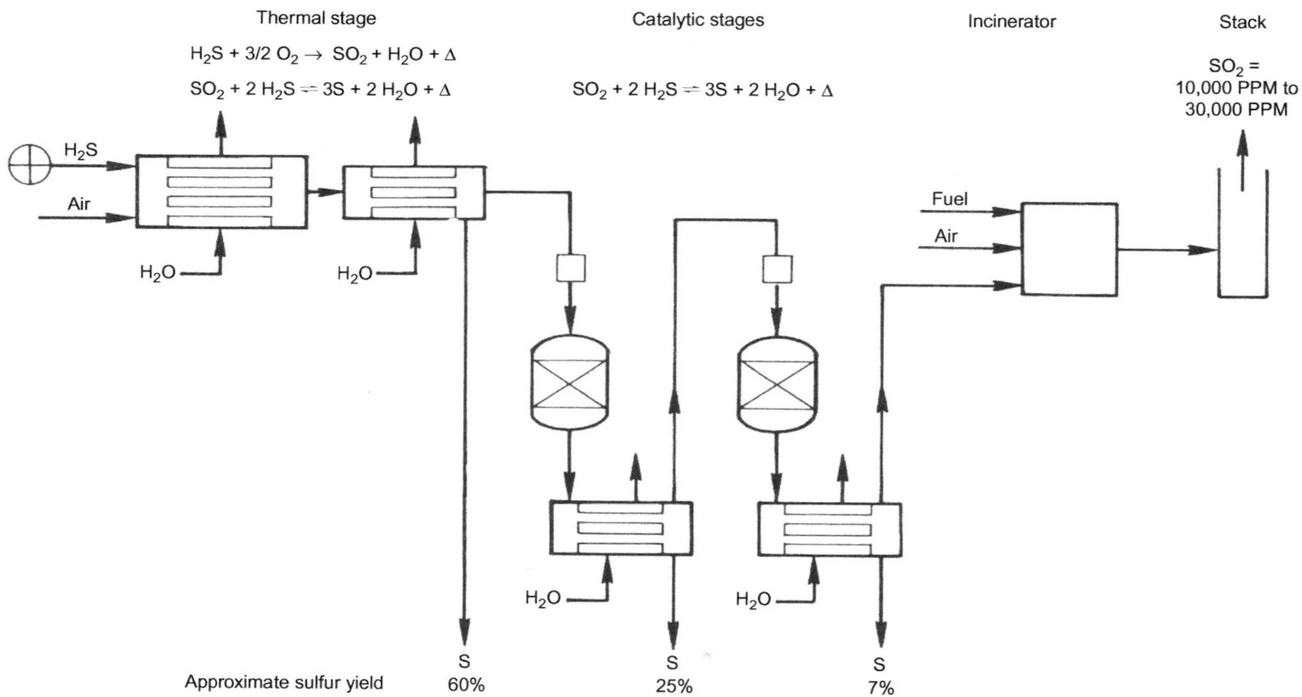

FIG. 14.4 The Claus process.

(below n-C_{16}) that are in a liquid form at room temperature, where the n-alkanes form a film around the aromatic sulfur compound, and as this film is easily attacked by aerobic microorganisms, the bioavailability of sulfur compounds increases. Studies with a strain of *Candida* in two-phase systems (oil-water mixture) suggested that the rate of biodegradation might be also related to the interfacial area because a large part of the biomass, which characteristically is hydrophobic, adheres to the nonaqueous phase layer (NAPL)-water interface as a biofilm (Ascón-Cabrera and Lebeault, 1995); the smaller the interfacial tension, the larger the uptake of the dissolved compounds in NAPL and the higher the biodegradation of aliphatic hydrocarbon derivatives present. Although aerobic microorganisms can remove high amounts of organic sulfur, the sulfur content in the residual oil may increase because of simultaneous aliphatic compound biodegradation (Setti et al., 1997).

2.2 Biodenitrogenation

Nitrogen is like sulfur; it is considered as a crude oil contaminant. In general, the nitrogen content of crude oil is low and generally falls within the range 0.1%–0.9%, although some crude oil may contain up to 2% nitrogen (Speight, 2014). However, crude oils with no detectable nitrogen or even trace amounts are not uncommon, but in general, the more asphaltic the oil, the higher its nitrogen content. Insofar as an approximate correlation exists between the sulfur content and API gravity of crude oils (Ancheyta and Speight, 2007), there also exists a correlation between nitrogen content and the API gravity of crude oil. It also follows that there is an approximate correlation between the nitrogen content and the carbon residue: the higher the nitrogen content, the higher the carbon residue.

The presence of nitrogen in heavy oil, extra heavy oil, and tar sand bitumen is of much greater significance in refinery operations than might be expected from the small amounts present. Nitrogen compounds can be responsible for the poisoning of cracking catalysts, and they also contribute to gum formation in such products as domestic fuel oil. The trend in recent years toward cutting deeper into the crude to obtain stocks for catalytic cracking has accentuated the harmful effects of the nitrogen compounds, which are concentrated largely in the higher-boiling portions. Other workers (Éigenson and Ivchenko, 1977) have reported that the higher the sulfur content in crude oil fractions, the higher is its nitrogen content. Moreover, nitrogen-containing compounds (NCCs) coexist with sulfur-containing compounds (SCCs) in fossil fuels (Yi et al., 2014). Carbazole, as an example for the nonbasic nitrogen-containing polycyclic aromatic hydrocarbon (NPAH) derivatives, can directly impact the refining processes in two ways: (i) During the cracking process, carbazole can be converted into basic derivatives, which can be adsorbed to the active sites of the cracking catalysts, and (ii) it directly inhibits the hydrodesulfurization catalysts. Thus, the removal of carbazole and other nitrogen compounds would significantly increase the extent of catalytic cracking and consequently the gasoline yield. It has been reported that, by 90% reduction in nitrogen content, a 20% increase in gasoline yield occurs. That has a major economic improvement in low-margin, high-volume refining processes (Benedik et al., 1998). Basic N compounds are more inhibitory for catalysts than the nonbasic ones. But they can potentially be converted to basic compounds during the refining/catalytic cracking process. Thus, they are also inhibitory to catalysts. Moreover, metals like nickel and vanadium are potent inhibitors for catalysts, and in crude oil, metals are typically associated with N compounds (Hegedus and McCabe, 1981; Mogollon et al., 1998).

Although specific biodesulfurization of crude oil and its distillates has been reasonably investigated, there is a little information about biodenitrogenation of oil feed without affecting its calorific value. It has been estimated that biodenitrogenation of crude oil would be beneficial for deep denitrogenation, where the classical hydroprocessing methods are costly and nonselective (Vazquez-Duhalt et al., 2002). It will also eliminate the contribution of fuel nitrogen to NOx emissions. However, the economics of nitrogen removal processes are affected by the amount of associated hydrocarbon lost from the fuel, during the denitrogenation process.

Generally, the currently well-established carbazole-biodenitrogenation pathway results in the loss of the fuel value. This would consequently make the biodenitrogenation of fuel streams economically unfeasible. Moreover, most of the carbazole-degrading microorganisms produce 2′-aminobiphenyl-2,3-diol as the first step in carbazole-biodenitrogenation pathway. However, recovering the carbazole nitrogen as anthranilic acid or 2′-aminobiphenyl-2,3-diol, the less inhibitory to refining catalysts, would solve part of that problem. Since then, the entire carbon content of the fuel is preserved. This can be performed by mutant or recombinant strains. Other pathways would liberate nitrogen from carbazole in the form of ammonia (Rhee et al., 1997). Microorganisms such as *Pseudomonas ayucida*, *Aneurinibacillus* sp., *Pseudomonas stutzeri*, *Yokenella* sp., and *Pseudomonas nitroreducens* are issued for biodenitrogenation of fossil fuels. A thermophilic carbazole-degrading bacteria *Anoxybacillus rupiensis* that can tolerate up to 80°C, with maximum activity at temperature range 55–65°C, has been reported, which would be advantageous for application in crude oil processing (Fadhil et al., 2014). However, so far, no microorganisms that selectively extract nitrogen atom from carbazole, in a microbial pathway equivalent to the 4S pathway, have been isolated.

From the practical point of view, biodenitrogenation and biodesulfurization should be integrated, where sulfur and nitrogen would be removed through specific enzymatic attack of the C—N and C—S bonds, respectively, but without C—C bond attack, to preserve the fuel value of the biotreated products. Moreover, a dual microbial process for both selective biodesulfurization and biodenitrogenation, with the overcome of the significant technical hurdles, such as tolerance against solvents, high concentration of nitrogenous compounds, and high oil-to-water ratio, would make microbial refining processes and bioupgrading of crude oil and its fractions feasible on a large scale (Kilbane, 2006). Duarte et al. (2001) in PETROBRAS, the Brazilian oil company, have isolated *Gordonia* sp. strain F.5.25.8 that can utilize dibenzothiophene through the 4S pathway and carbazole as a sole source of S and N, respectively. F.5.25.8 is the first reported strain that can simultaneously metabolize dibenzothiophene and carbazole; Santos et al. (2005, 2006) reported that F.5.25.8 can tolerate up to 42°C, which would add to its advantages in industrial application of biodesulfurization/biodenitrogenation as complementary to hydrotreatment process. Moreover, it is reported to have a different genetic organization of the biodesulfurization (dsz) and biodenitrogenation (car) gene clusters relative to *Rhodococcus erythropolis* IGTS8 and *Pseudomonas* sp. IGTN9m, respectively. Moreover, insertion of genes encoding amidases downstream the artificial carA operon was performed, to accomplish the cleavage of the final C—N bond and produce biphenyl-2,2′,3-triol that is reintroduced to the fuel, keeping its fuel content. The decrease of the carbazole content in a crude oil sample by approximately 95% in 2:10 crude oil/aqueous medium within 16h using a genetically engineered bacterium. Yu et al. (2006a,b) introduced carbazole dioxygenase gene (carAacd), which was amplified from *Pseudomonas* sp. strain XLDN4-9, into the excellent 4S-dibenzothiophene-biodesulfurization bacteria *R. erythropolis* XP and designated the recombinant as SN8. The recombinant *R. erythropolis* SN8 expressed good biodesulfurization and biodenitrogenation activities toward a wide range of recalcitrant alkyl carbazole derivatives and dibenzothiophene in crude oil, in just a one-step bioprocess. Moreover, upon the transformation of *R. erythropolis* XP with a complete plasmid of carABC gene cluster, which converts carbazole to anthranilic acid (ANA), the resultant recombinant strain was able to metabolize dibenzothiophene (DBT), carbazole, and dibenzofuran (DBF) to the nontoxic metabolites: benzoate, anthranilate, and salicylate, respectively.

2.3 Biodemetallization

Crude oil contains metals in the form of salts (zinc, titanium, calcium, and magnesium), petroporphyrins (vanadium, copper, nickel, and iron), and other complexes that exits predominantly within the asphaltene fraction (Ali and Abbas, 2006; Speight, 2014). Thus, the higher the asphaltene content of crude oil, the higher the heavy metal content (Speight, 1994, 2014; Duyck et al., 2007). In fact, the accumulation of metallic constituents in the higher molecular weight and polar fractions of crude oils plays a significant role in establishing the refining procedure (Panariti et al., 2000; Duyck et al., 2008; Speight, 2014, 2017).

Depending on the origin of crude oil, the concentration of the vanadium varies from as low as 0.1ppm to as high as 1200ppm, while that of nickel commonly varies from trace amounts to 150ppm (Ali and Abbas, 2006; Duyck et al., 2007). The V/Ni ratio is constant in crude oils of common source rocks and dependent on the geologic age of the rocks, where oils from Triassic or older age show a value higher than unity (Ball et al., 1960), and this ratio is also used for tracing source effects (Al-Shahristani and Al-Atyia, 1972; El-Gayar et al., 2002). However, the biodegradation of the asphaltene constituents and the resin constituents is reported to influence the Ni/V ratio in these fractions (Duyck et al., 2007).

Crude oil with a particularly high content of organometallics includes Boscán crude oil, Cerro Negro crude oil, Maya crude oil, Wilmington crude oil, and Prudhoe Bay crude oil (Fish et al., 1984). The metallic constituents often occur as inorganic water-soluble forms and are easily removed during the crude desalting process, in which they are concentrated in the aqueous phase. However, the metalloporphyrins are embedded in the extremely complex structure of the asphaltene constituents and, thus, metal removal from petroporphyrins and complexes that is necessary to be addressed.

Currently, demetallization occurs during the hydrocracking process or during hydrotreating process (Parkash, 2003; Ancheyta and Speight, 2007; Gary et al., 2007; Speight, 2014, 2017; Hsu and Robinson, 2017). However, feedstocks that have relatively high metal contents ($>$ 300 ppm) substantially increase catalyst consumption because the metals poison the catalyst, thereby requiring frequent catalyst replacement. The usual desulfurization catalysts are relatively expensive for these consumption rates, but there are catalysts that are relatively inexpensive and can be used in the first reactor to remove a large percentage of the metals. Subsequent reactors downstream of the first reactor would use normal hydrodesulfurization catalysts. Since the catalyst materials are proprietary, it is not possible to identify them here. However, it is understood that such catalysts contain little or no metal promoters, that is, nickel, cobalt, and molybdenum. Metal removal on the order of 90% has been observed with these materials.

Thus, one method of controlling demetallization is to employ separate smaller *guard reactors* just ahead of the fixed-bed hydrodesulfurization reactor section (Speight, 2014, 2017). The preheated feed and hydrogen pass through the guard reactors that are filled with an appropriate catalyst for demetallization that is often the same as the catalyst used in the hydrodesulfurization section. The advantage of this system is that it enables replacement of the most contaminated catalyst (*guard bed*), where pressure drop is highest, without having to replace the entire inventory or shut down the unit. The feedstock is alternated between guard reactors, while catalyst in the idle guard reactor is being replaced.

Unless a guard reactor is employed, a major problem and expense during the refining of oil and other crude oil products is the continuous contamination of solid, porous catalysts by various porphyrins; metalloporphyrins; chlorins; and natural degradation products of these compounds, such as petroporphyrins, containing metals such as vanadium and nickel. The porphyrin deposits, however, consist of several different metals on the catalytic surface including vanadium (V), nickel (Ni), titanium (Ti), iron (Fe), copper (Cu), or a combination thereof with the concentrations of V and Ni varying from a few to several hundred parts per million (ppm), depending on the type of crude oil supply. Vanadium is usually present in a concentration greater than other metals with much more than half of all V being deposited on the catalyst arising from the porphyry complex. Heavy metals poison refining catalyst. During crude oil refining, heavy metals are concentrated in the residual fraction, which is usually subjected to catalytic cracking. During the catalytic cracking, metals in the oil deposit on the cracking catalysts and decrease their selectivity and activity. Catalyst deactivation in cracking, hydrogenation, and hydrodesulfurization processes can also occur by pore clogging, metal deposition, deformation, and destruction of reactors (Salehizadeh et al., 2007), which in turn demands the catalytic process be interrupted to either replace or clean the catalyst at a huge cost (Altgelt and Boduszynski, 1994; Callejas et al., 2001).

Moreover, since both nickel and vanadium exhibit dehydrogenation activity, thus, their presence on the catalyst particles tends to promote dehydrogenation reactions during the cracking sequence, and this results in increased amounts of coke and low-boiling gases at the expense of naphtha production (Ali and Abbas, 2006). Heavy metals can be liberated into the environment, during fuel combustion in the form of ash with high concentrations of toxic metal oxides, which leads to undesirable by-products and the need for disposal.

Moreover, heavy metals (mostly Ni and V) are furthermore corrosive (Montiel et al., 2009). When crude oil is vacuum-distilled, the metalloporphyrins tend to be entrained in the vapors and carry over the vacuum gas-oil fraction. The typical demetallization process in crude oil industry is the solvent deasphalting process, where the lower-density oils are physically separated from heavier asphaltene constituents by mixing the viscous oil with a very-low-boiling hydrocarbon liquids such as propane and/or butane (Speight, 2000, 2014, 2017; Farag et al., 1989; Parkash, 2003; Ancheyta and Speight, 2007; Gary et al., 2007; Hsu and Robinson, 2017). Other options for demetallization include a low-temperature coking process in which the metals collect in the initially formed coke (Speight, 2014, 2017).

3. Bioconversion of Heavy Feedstocks

In heavy oil, extra heavy oil, and tar sand bitumen, which are characterized by high-viscosity and high-density or low API gravity (Speight, 2014, 2017), the asphaltene constituents, the resin constituents, and the oil matrix compose a dynamic stable system, in which the nonasphaltene and nonresin fraction act as the medium for dispersion in which the asphaltene constituents are stabilized by the resin constituents (Speight, 2014, 2017).

Biotechnology has the potential to be applied in (i) the transformation of a viscous feedstock into lower-boiling products; (ii) degradation of the asphaltene constituents; (iii) conversion of higher-molecular-weight hydrocarbon derivatives into lower-boiling hydrocarbon products; (iv) product purification, such as the removal of sulfur, nitrogen, and heavy metals; and (v) treatment of liquid and gaseous emissions.

The bioconversion of these individual high-boiling constituents (the asphaltene constituents and the resin constituents) is a reflection of their chemical structure but is also strongly influenced by the physical state and toxicity of the compounds. While *n*-alkanes as a structural group are the most biodegradable crude oil hydrocarbon derivatives, the higher-molecular-weight constituents become increasingly recalcitrant because not only the chemical structure but also the physical state of the constituents strongly influences bioconversion (Bartha and Atlas, 1977). In fact, asphaltene and resin constituents tend to increase during bioconversion in relative and sometimes absolute amounts (Okoh, 2006). This suggests that they not only tend to resist bioconversion but also may be formed by condensation reactions of bioconversion intermediates.

3.1 Heavy Oil, Extra Heavy Oil, and Tar Sand Bitumen

The availability of higher-quality low-density crude oil has been depleting during the last decades. Future demands for crude oil are focused on the use of viscous feedstocks (such as heavy oil, extra heavy crude oil, and tar sand bitumen), which is difficult to produce, transport, and refine; therefore, the interest in transportation and conversion of the high-molecular-weight fractions of these materials into refined fuels and petrochemicals has increased. There are several known strategies for future viscous crude oil production (the upstream) and upgrading (the downstream) (Speight, 2000, 2014, 2017; Parkash, 2003; Ancheyta and Speight, 2007; Gary et al., 2007; Hsu and Robinson, 2017). However, when viscous crudes enter a refinery, processing of such feedstocks requires conversion of the vacuum residue components, including the asphaltene constituents and resin constituents into distillable fractions. This upgrading has typically been accomplished either with thermal conversion (coking or cracking) or by catalytic hydroconversion, which requires investment in the process equipment and in the supporting infrastructure (Parkash, 2003; Kirkwood et al., 2004; Twerdok, 2004; Ancheyta and Speight, 2007; Rana et al., 2007; Gary et al., 2007; Speight, 2014, 2017; Hsu and Robinson, 2017).

Viscous crude oil and residue have many similarities in composition. Hence, VR could be a net sample of problematic components of viscous crude oil. In spite of some similarities between residue and viscous oil, residue feedstock differs in several ways such as its higher concentration of asphaltene, sulfur, nitrogen, and metal compounds, which depends on the origin of viscous oils. Since biological processing may offer less severe process conditions and higher selectivity to specific reactions to increase net distillates, it is proposed that the bacteria with the ability to biodegrade viscous feedstocks could be a useful apparatus for such oils.

The bioconversion of viscous feedstocks requires a complex metabolic pathway that usually can be observed in a microbial community. Many studies have been carried out on the biodegradation of crude oil using a consortium of microorganisms (Morais and Tauk-Tornisielo, 2009). In literature, it is revealed that in soils that are permeated with the high-molecular-weight hydrocarbon derivatives, bacteria including indigenous ones would survive and function after contaminations (Whyte et al., 1997; Foght et al., 1990) seeped through the soil. Selection of bacterial communities for crude oil substances occurs rapidly after even short-term exposures of soil to crude oil hydrocarbon derivatives (Van der Meer et al., 1992; Van der Meer, 1994). During adaptation of microbial communities to hydrocarbon components particularly complex ones, genes for hydrocarbon-degrading enzymes that are carried on plasmids or transposons may be exchanged between species, and new catabolic pathways eventually may be assembled and modified for efficient regulation (Rabus et al., 2005). Other cell adaptations leading to new ecotypes may include modifications of the cell envelope to tolerate solvents (Ramos et al., 2002) and the development of community level interactions that facilitate cooperation within consortia (Kim et al., 2001).

One of the limitation factors for bioconversion of crude oil hydrocarbon derivatives in oil is the bioavailability (Banat, 1995; Kim et al., 2001) of these components due to their chemical structure (Harvey, 1997; Cerniglia, 1992). One of the options to increase bioavailability of polynuclear aromatic hydrocarbon derivatives is the use of surfactants to increase desorption and apparent solubility in the aqueous phase and consequently enhance the oil mobility, improving the bioconversion rates (Barathi and Vasudevan, 2001; Burd and Ward, 1996a,b; Rosenberg et al., 1988; Sar and Rosenberg, 1983).

The chemical composition of a viscous feedstock depends not only on the source of the crude oil from which it is derived but also on the method of its production (Parkash, 2003; Ancheyta and Speight, 2007; Gary et al., 2007; Speight, 2014, 2017; Hsu and Robinson, 2017). An early study (Phillips and Traxler, 1963) indicated that organisms of the genus *Pseudomonas* vary to a greater or lesser extent in the mechanisms by which they oxidize a viscous feedstock.

Previous investigations have established that some bacteria can utilize viscous feedstocks. However, the complexity of viscous feedstocks has limited the study of its bioconversion (Atlas, 1981). Crude oil-degrading bacteria excrete biosurfactants that emulsify hydrocarbon derivatives of crude oil by reducing the interfacial tension between the hydrocarbon derivatives and water (Cooper and Zajic, 1980).

One crude oil-degrading bacterium, *Acinetobacter calcoaceticus*, excretes an emulsifier, emulsan (Goldman et al., 1982; Gutnick, 1987). Emulsan is an extracellular polysaccharide polymer with fatty acid side chains (Shoham and Rosenberg, 1983; Shoham et al., 1983). This extracellular molecule forms a complex with proteins. These proteins bind noncovalently to the polysaccharide backbone. This emulsifier loses most of its activity when these proteins are removed from the complex (Zosim et al., 1987; Foght et al., 1989).

Bioconversion of viscous feedstocks occurs mostly on the surface of the viscous feedstocks exposed to oxygen, where the primary nutrients are all present in adequate levels. Surface degradation is limited by oxygen diffusion or the presence of needed minerals/ions (i.e., phosphorous and nitrogen) (Pendrys, 1989). As time progresses, degradation should diminish as all of these become limited. In addition, the mechanism for bioconversion may be such that a residue that is resistant to bioconversion remains on the surface, protecting the underlying viscous feedstocks from further bioconversion. As the saturated and naphthene aromatic hydrocarbon derivatives are emulsified, a residue remains on the surface that is resistant to bioconversion and protects the underlying viscous feedstocks from bioconversion. The most potent viscous feedstock-degrading bacterium, *A. calcoaceticus* NAV2, excretes an emulsifier that is capable of emulsifying the saturated and naphthene aromatic fractions of viscous feedstocks (Pendrys, 1989).

In another study (Luey and Li, 1993), testing was initiated in March 1991 and completed in November 1992 to determine the rate at which viscous feedstocks are biodegraded by microorganisms native to the Hanford Site soils. Experiments to determine viscous feedstock bioconversion rates were conducted using three separate test sets. These test sets were initiated in March 1991, January 1992, and June 1992 and ran for periods of 6 months, 11 months, and 6 months, respectively. The experimental method used was one originally developed and further refined that determined the viscous feedstock bioconversion rate through the measurement of carbon dioxide evolved (Bartha and Pramer, 1965; Bowerman et al., 1985).

Using data from the January and June 1992 test sets, viscous feedstock bioconversion rates were determined for test flasks containing soil and irradiated viscous feedstock sample incubated at ambient and elevated (−35°C, −31°F) temperature. The average viscous feedstock bioconversion rate determined at ambient temperature for test flasks incubated for <160 days was 1.6×10^{-4} cm/year, with a high end point at the 95% confidence interval of 1.9×10^{-4} cm/year and a low end point at the 95% confidence interval of 1.4×10^{-4} cm/year. For ambient temperature test flasks incubated up to 310 days, the average viscous feedstock bioconversion rate was determined to be 1.0×10^{-4} cm/year, indicating that the rate of bioconversion decreases as a function of time.

For test flasks at elevated temperature, the average viscous feedstock bioconversion rate was determined to be 1.1×10^{-3} cm/year with a high end point of 1.3×10^{-3} cm/year and a low end point of 9.9×10^{-4} cm/year. Conservatively, the viscous feedstock diffusion barrier will degrade at 1.3×10^{-3} cm/year during elevated temperatures with the rate decreasing to 1.9×10^{-4} cm/year at lower, ambient temperature.

This is a conservative estimate of the long-term degradation rates. Parameters such as available water, available oxygen, and available nutrients will affect the long-term bioconversion rate. Since the environment surrounding the viscous feedstock diffusion barrier will likely be deficient in these parameters, the long-term bioconversion rate is predicted to be lower than determined by this study. However, additional work is needed to investigate the influence of these parameters and others on the viscous feedstock bioconversion rate and refine the estimated long-term rate.

Extrapolating the bioconversion rate to 10,000 years indicates that approximately 3 cm of viscous feedstock would degrade, assuming that the viscous feedstock diffusion barrier remained at elevated (35°C, 95°F) temperature for 1000 years and then was at ambient temperature for 9000 years. This amount of degradation is within the 10 ac allowance in the design criteria for the viscous feedstock diffusion barrier. This allowance would be approached only if the barrier remained at elevated temperature during the entire 10,000-year performance period (based upon the conservative viscous feedstock bioconversion rate).

3.2 Asphaltene Constituents

The asphaltene constituents have drawn considerable attention due to problems caused by their detrimental effects in the extraction, transportation, and processing of viscous feedstocks because of their viscous and flocculating nature and their relative resistance to bioconversion following spills. These constituents are the highest molecular weight and most polar fraction of crude oil. Despite the fact that the structure of asphaltene constituents has not been fully elucidated, it is widely accepted that they are constituted by interacting systems of polyaromatic sheets bearing alkyl side chains. Asphaltene molecules have a high content of O, N, and S heteroatoms and metals (V, Ni, and Fe) (Speight, 1994, 2014). The problems associated with asphaltene constituents have increased due to the need to extract heavier crude oils and the trend to extract larger amounts of low-boiling fractions out of crude oil by visbreaking (Parkash, 2003; Gary et al., 2007; Speight, 2014, 2017; Hsu and Robinson, 2017).

The asphaltene fraction is the most recalcitrant feedstock fraction, and there is only fragmentary evidence that the asphaltene constituents can be converted or transformed by microbial activity. Microorganisms have been found associated with bitumens (Wyndham and Costerton, 1981) and natural asphalt lake (Naranjo et al., 2007), which contain high amounts of asphaltene constituents. A molecular study (Kim and Crowley, 2007) revealed a wide range of phylogenetic groups within the Archaea and Bacteria domains in natural asphalt-rich tar pits; interestingly, genes encoding novel oxygenases were also detected in such samples.

On the other hand, an extensive screening involving >750 strains of filamentous fungi was carried out to select strains able to modify untreated hard coal (Bublitz et al., 1994; Hofrichter et al., 1997). Only six of the 750 strains tested exhibited some activity, from which the most active fungi, *Panus tigrinus*, growing on wood shavings coated with coal asphaltene constituents led to a decrease of the average molecular weight (Hofrichter et al., 1997), although the average molecular weight of any complex mixture is not a measure of the constituents that were actually biodegraded.

Furthermore, most of the studies on asphaltene constituent bioconversion should be considered cautiously as the asphaltene content was usually determined gravimetrically after *n*-alkane precipitation, and thus, the reported changes may be attributed to the disruption of the asphaltenic matrix by the production of surfactants during bacterial growth, liberating trapped hydrocarbon derivatives. Other studies have reported that the asphaltene fraction does not support bacterial growth and no changes in asphaltene content are found after bioconversion of viscous oil containing asphaltene constituents (Lacotte et al., 1996; Thouand et al., 1999).

There have been claims of the bioconversion of asphaltene constituents by mixed bacteria (Bertrand et al., 1983; Rontani et al., 1985). However, none of these reports described the analytic results of extractable materials recovered from appropriate sterile controls. Therefore, most of the asphaltene losses during microbial activity could be considered abiotic losses (Lacotte et al., 1996).

A study (Pineda et al., 2004) reported a bacterial consortium able to grow in the asphaltene fraction as the sole carbon source. Mineralization of the asphaltene constituents was estimated by measuring CO_2 production. The authors found in two control experiments (inoculum without asphaltene constituents and noninoculated asphaltene constituents) a CO_2 production equivalent to 39% and 26%, respectively, of that found in the consortium growing in the asphaltene fraction.

The microbial inoculum for consortium stabilization contained 1% of crude oil, which could serve as carbon source. Thus, it is not possible to distinguish the origin of the CO_2 production. The first clear experimental evidence that enzymes are able to modify asphaltene molecules has been reported (Fedorak et al., 1993). Chloroperoxidase from the fungus *Caldariomyces fumago* and a chemically modified cytochrome *c* were able to transform petroporphyrin derivatives and asphaltene constituents in reaction mixtures containing organic solvents (Fedorak et al., 1993; Mogollon et al., 1998; Garcia-Arellano et al., 2004). Notable spectral changes in the petroporphyrin-rich fraction of asphaltene constituents were observed, and the enzymatic oxidation of petroporphyrin derivatives led to the removal up to 74% of Ni and 95% of V.

According to Fourier-transform infrared (FTIR) spectroscopy spectra, the chemically modified *cytochrome c* catalyzed the oxidation of sulfur and carbon atoms in asphaltene molecules (Garcia-Arellano et al., 2004). The enzymatic treatment of asphaltene constituents is an interesting alternative for the removal of heavy metals. It would result in reduced catalyst poisoning during hydrotreatment and cracking processes. On the other hand, the introduction of polar groups in asphaltene molecules could positively affect their sedimentation properties and improve their behavior.

It has also been reported (Uribe-Alvarez et al., 2011) that a fungus isolated from a natural asphalt lake is able to grow using asphaltene constituents as the sole source of carbon and energy. Thus, a fungal strain isolated from a microbial consortium growing in a natural asphalt lake is able to grow in purified asphaltene constituents as the only source of carbon and energy. The asphaltene constituents were rigorously purified in order to avoid contamination from other crude oil fractions. In addition, most of petroporphyrin derivatives were removed. The 18S rRNA and b-tubulin genomic sequences, as well as some morphological characteristics, indicate that the isolate is *Neosartorya fischeri*. After 11 weeks of growth, the fungus is able to metabolize 15.5% of the asphaltenic carbon, including 13.2% transformed to carbon dioxide. In a medium containing asphaltene constituents as the sole source of carbon and energy, the fungal isolate produces extracellular lactase activity, which is not detected when the fungus grow in a rich medium. The results obtained in this work clearly demonstrate that there are microorganisms able to metabolize and mineralize asphaltene constituents, which is considered the most recalcitrant crude oil fraction.

To overcome the shortcomings of conventional methods, microbial degradation of asphaltene has been accepted worldwide as the most promising environmentally sound technology for remediation of spills and discharges related to crude oil and crude oil products. Furthermore, bacterial metabolites (especially polysaccharides) are of great value as enhancers of oil recovery due to their surfactant activity and bioemulsifying properties (Banat, 1995). Because the conditions in oil deposits are often saline, the use of salt-resistant metabolites may be advantageous to the recovery of oil. Furthermore, hypersaline water and soil are often contaminated with crude oils, heavy metals, or other toxic compounds from anthropogenic sources. However, conventional microbiological treatment processes do not function at high salt concentrations; therefore, the use of moderately halophilic bacteria should be considered (Hao and Lu, 2009).

In the past, bioconversion of asphaltene constituents through the use of a microbial consortium or mixed cultures isolated from soil samples, sediments contaminated with hydrocarbon derivatives, and oil wells has taken place but in low proportions of 0.55%–3.5% (Venkateswaran et al., 1995; Thouand et al., 1999). This is most likely due to the complex molecular structure of asphaltene constituents (Speight, 1994, 2014) that makes these molecular species resistant to bioconversion, thereby causing their accumulation in an ecosystem where crude oil and/or crude oil products containing asphaltene constituents are spilled in either accidental or purposeful ways (Guiliano et al., 2000).

The focus of many studies has generally been bioconversion of sites contaminated by total crude oil hydrocarbon derivatives (Iturbe et al., 2007;

Machackova et al., 2008), and there is a general lack of detailed work that has been done on the bioconversion of asphaltene constituents. However, more recently, viscosity reduction by asphaltene degradation has been structurally characterized by Fourier-transform infrared (FTIR) spectroscopy. The work was focused on the reduction of viscosity of viscous oil in order to improve enhanced recovery from the reservoir or deposit. The bacterium (*Garciaella petrolearia* TERIG02) also showed an additional preference to degrade toxic viscous feedstock and aromatic compounds first unlike the other known strains. Furthermore, these characteristics make the species *G. petrolearia* TERIG02 a potential candidate for viscous feedstock bioconversion and a solution to degrading toxic aromatic compounds (Lavania et al., 2012).

In contrast to low-molecular-weight hydrocarbon derivatives, polycyclic aromatic hydrocarbon derivatives that occur within the asphaltene fraction are usually considered as being only slightly biodegradable (if at all) because of their insufficient susceptibility of these molecular species to microbial attack (Gibson and Subramanian, 1984; Cerniglia, 1992; Kanaly and Harayama, 2000). In fact, among the pentacyclic triterpane derivatives, the hopane constituents are so stable that they are commonly used as ubiquitous biomarkers for the assessment of the level of bioconversion of crude oil (Ourisson et al., 1979). In fact, the compounds are only slightly biodegraded by specialized microfloras under laboratory conditions (Frontera-Suau et al., 2002).

The mechanism of the degradation is complex but is believed to be a sequential process in which *n*-alkane moieties are generally removed first, followed by the degradation of isoalkane moieties, cycloalkane moieties, one- to three-ring aromatic derivatives, and finally polyaromatic derivatives (Greenwood et al., 2008). However, the typical pattern of degradation varies with different bacteria and the type and composition of the oil (Greenwood et al., 2008; Zrafi-Nouira et al., 2009).

As the culture moves to the second phase of fermentation, the stationary growth phase, there is a shift in the metabolism of the cells to solvent production (solvent-generation phase). These gaseous and liquid metabolites dissolve into the oil resulting in reduced viscosity (Bryant et al., 1998). Moreover, the reaction of heavy feedstock degradation within an acidic medium is preferable since the proton (H^+) effectively interacts with the polar functionalities in the asphaltene constituents and resin constituents thereby efficiently reduce the polar interactions, which results in breaking the intermolecular associations that exist in the raw residuum or asphalt.

Furthermore, when each of the fractions, namely, aliphatic, aromatic, and asphaltic (asphaltene and resin) fractions, was treated with *G. petrolearia* TERIG02 for 30 days, it was found that maximum degradation was in the case of viscous feedstock followed by aromatic fraction (Lavania et al., 2012). Indications were that *G. petrolearia* TERIG02 could tolerate the toxicity of these compounds and was capable in utilizing them as a carbon and energy source.

4. Nanobiotechnology

In terms of a specific process (biodesulfurization), compared with biodesulfurization, adsorptive desulfurization has a much faster reaction rate (Song et al., 2005). Adsorbent preparation is the key of adsorptive desulfurization. Recently, most adsorbents for desulfurization were based on π-complexation (Shan et al., 2005b) or formations of metal-sulfur bonds (such as Ni—S and La—S) (Tian et al., 2006). Adsorbents based on π-complexation are easy to regenerate, but their selectivity is very low, resulting in a loss of fuel quality. Meanwhile, adsorbents that form metal-sulfur bonds with sulfur have high selectivity but are difficult to be regenerated. Hence, adsorptive desulfurization technology also has a long way to go before being industrialized. If a desulfurization technology has both the high reaction rate of adsorptive desulfurization and the high selectivity of biodesulfurization, it can increase the desulfurization rate without having an adverse effect on the quality of the fuel.

The adsorbent is an important factor of this coupling technology because different adsorbents have different interaction to organic sulfur compounds and cells, which would affect their assembly onto the cell surfaces and desorption behavior of organic sulfur from them. Because the adsorbents are assembled on the cells' surfaces, the property of cell surface is another factor that impacts the coupling technology. Moreover, desulfurization conditions, such as temperature and volume ratio of oil to water phase, would also affect in situ coupling technology.

In situ coupling of adsorptive desulfurization and biodesulfurization is a new desulfurization technology for crude oil and is achieved by assembling nanoadsorbents onto surfaces of microbial cells. The process has the advantages of a high selectivity of biodesulfurization and a high rate of adsorptive desulfurization. For example, the combination of adsorptive desulfurization and biodesulfurization and an in situ coupling technology has been proposed. γ-Al_2O_3 nanosorbents, which have the ability to selectively adsorb dibenzothiophene from organic phase, were assembled on the surfaces of *Pseudomonas delafieldii* R-8 cell, a desulfurization strain. γ-Al_2O_3 nanosorbents have the ability to adsorb dibenzothiophene from oil phase, and the rate of adsorption was significantly higher than that of biodesulfurization.

Thus, dibenzothiophene can be quickly transferred to the biocatalyst surface where nanosorbents were located, which quickened dibenzothiophene transfer from organic phase to biocatalyst surface and resulted in the increase of biodesulfurization rate. The desulfurization rate of the cells assembled with nanosorbents was approximately 2.5-fold higher than that of original cells. An improved in situ coupling technology of adsorptive desulfurization and biodesulfurization by *P. delafieldii* R-8 cells has also been reported (Zhang et al., 2007). Gamma-alumina (Al_2O_3) nanoparticles were synthesized and modified using gum arabic to avoid agglomeration in aqueous solutions, and its effect on adsorptive desulfurization and biodesulfurization was also evaluated.

Results showed that γ-Al_2O_3 nanoparticles dispersed well in aqueous solutions after modification with gum arabic. The adsorptive desulfurization capacity of γ-alumina nanoparticles was increased from 0.56 mmol S/g (Al_2O_3) to 0.81 mmol S/g (Al_2O_3) after treatment with gum arabic. Compared with unmodified γ-alumina nanoparticles, the biodesulfurization rate by adsorbing gum arabic-modified γ-alumina nanoparticles onto the surfaces of R-8 cells was increased from 17.8 to 25.7 mmol/kg/h, which may be due to improvement in the dispersion and biocompatibility of γ-alumina nanoparticles after modification.

Different kinds of adsorbents (such as alumina, molecular sieves, and active carbon) have been used successfully for the in situ coupling technology of adsorptive desulfurization and biodesulfurization (Zhang et al., 2008). The procedure can be carried out by assembling nanoadsorbents onto surfaces of *P. delafieldii* R-8 cells. However, the Na—Y molecular sieves restrain the activity of R-8 cells, and active carbon cannot desorb the substrate dibenzothiophene, and thus, this type of molecular sieve is not applicable to in situ coupling desulfurization technology. γ-Alumina can adsorb dibenzothiophene from oil phase quickly and then desorb it and transfer it to R-8 cells for bioconversion, thus increasing desulfurization rate. It was also found that nanosized γ-alumina increases desulfurization rate more than regular-sized γ-alumina. Therefore, nano-γ-alumina is regarded as better adsorbent for this in situ coupling desulfurization technology.

In the last two decades, numerous studies have been carried out on biodesulfurization using whole cells (Maghsoudi et al., 2001; Tao et al., 2006; Yang et al., 2007; Caro et al., 2008) or isolated enzymes (Monticello and Kilbane, 1994) in the free or immobilized form. Biodesulfurization of dibenzothiophene occurs via a multienzyme system that requires cofactors (e.g., NADH—the reduced form of nicotinamide adenine dinucleotide). The use of enzymes is disadvantageous since extraction and purification of the enzyme is costly, and frequently, enzymes catalyzing reactions require cofactors that must be regenerated (Setti et al., 1997). Therefore, biodesulfurization can often be designed by using whole-cell bioconversion rather than that of the enzyme. However, there are still some bottlenecks limiting the commercialization of the biodesulfurization process.

One of the challenges is to improve the current biodesulfurization rate by about 500-fold assuming the target industrial process is 1.2–3 mmol/g dry cell weight (DCW)/h (Kilbane, 2006). When free cells are used for crude oil biodesulfurization, deactivation of the biocatalyst and troublesome oil-water biocatalyst separation are significant barriers (Konishi et al., 2005; Yang et al., 2007). Cell immobilization may give a solution to the problems, providing advantages such as repeated or continuous use, enhanced stability, and easy separation.

For biodesulfurization, cells need to be harvested from the culture medium, and several separation schemes had been evaluated, including settling tanks (Schilling et al., 2002), hydrocyclones, and centrifuges (Monticello, 2000).

Magnetic separation technology provides a quick, easy, and convenient alternative over traditional methods in biological systems (Haukanes and Kvam, 1993). Superparamagnetic nanoparticles are increasingly used to achieve affinity separation of high-value cells and biomolecules (Molday et al., 1977). Magnetic supports for cell immobilization offer several advantages, such as the ease of magnetic collection. The magnetic supports present further options in continuous reactor systems when used in a magnetically stabilized, fluidized bed. In addition, the mass transfer resistance can be reduced by the spinning of magnetic beads under revolving magnetic field (Sada et al., 1981).

Modified magnetite (Fe^{2+}-Fe^{3+} oxide, Fe_3O_4) has been used to produce a hydrophilic magnetic fluid (Liu et al., 2003; Shan et al., 2003). Addition of nitrogen gas during preparation prevents the oxidation of ferrous ion in the aqueous solution and blackening and also controls particle size. Magnetic particles prepared by coprecipitation method have a large number of hydroxyl groups on their surface in contact with the aqueous phase. The hydroxyl (—OH) groups on the surface of mixed iron oxide nanoparticles (with an average particle diameter on the order of 8 nm) react readily with carboxylic acid group of the added oleic acid after which the excess oleic acid will be adsorbed to the first layer of oleic acid to form a hydrophilic shell. When this magnetic fluid was added to aqueous ammonium hydroxide, the outer layer of oleic acid on the magnetite surface is transformed into ammonium oleate, and the hydrophilic magnetic fluid is produced.

Magnetic fluid is directly mixed into hydrophilic support liquids such as polyvinyl alcohol and sodium alginate with dried cells of *P. delafieldii* R-8. Immobilized beads were formed by extruding the aforementioned mixture through a syringe into a gelling solution of 0.1 M calcium chloride ($CaCl_2$) saturated with boric acid and solidified for 24 h; the immobilized beads formed were washed with saline and then freeze-dried for 48 h under vacuum. The magnetic fluid is mainly composed of magnetite particles, which not only provide the magnetic property of support but also improve mechanical strength of the supports, which are superparamagnetic. Therefore, the magnetic immobilized supports could be easily separated and recycled by external magnetic field, and the recovered magnetic supports could be redispersed by gentle shaking with the removal of the external magnetic field. Compared with nonmagnetic immobilized cells, the beads of magnetic immobilized cells showed higher reaction activity of desulfurization and also with higher strength against swelling and longer-term stability and can be reused for seven cycles of reaction, while the nonmagnetic immobilized cells could only be used for five cycles. Also, the magnetic immobilized cells can be easily separated from the reaction medium, stored, and reused to give consistent results. The support is relatively cheap, easy to prepare, and good for large-scale industrial applications.

Attempts have also been made to increase the efficiency of cells and to decrease the cost of operations in a biodesulfurization process (Shan et al., 2005c). In the work, magnetic polyvinyl alcohol beads were prepared by a freeze-thaw technique under liquid nitrogen, and the beads have distinct superparamagnetic properties. The desulfurization rate of the immobilized cells could reach 40.2 mmol/kg/h twice that of free cells. The heat resistance of the cells apparently increased when the cells were entrapped in magnetic polyvinyl alcohol beads. The cells immobilized in magnetic polyvinyl alcohol beads could be stably stored and be repeatedly used over 12 times for biodesulfurization. The immobilized cells could be easily separated by magnetic field. In order to understand cell distribution in magnetic polyvinyl alcohol beads, the sections of the beads after being repeatedly used for six times were observed by scanning electron microscopy (SEM). A highly macroporous structure is found in the beads in favor of diffusion of substrates and dissolved gas. On average, the size of the beads was on the order of 3 mm. It is evident that the R-8 cells mainly covered the edges and submarginal sections of the bead while no cell in the center of the bead because of insufficiencies of oxygen and nutrients and gradually autolysis.

A new technique in which magnetic nanoparticles are used to coat the cells could successfully overcome difficulties of conventional cell immobilization, such as mass transfer problems, cell loss, and separation of carrier with adsorbed cells from the reaction mixture at the end of a desulfurization treatment. Coating layer of nanoparticles does not change the hydrophilic nature of the cell surface. The coating layer has negligible effect on mass transfer because the structure of the layer is looser than that of the cell wall and does not interfere with mass transfer of dibenzothiophene. Moreover, the coated cells have good stability and can be reused. This new technique has the advantage of magnetic separation and is convenient and easy to perform so offers the potential to be suitable for large-scale industrial applications.

In addition, a technique that described a process in which microbial cells of *P. delafieldii* R-8 were coated with magnetic nanoparticles (Fe_3O_4) and then immobilized by external application of a magnetic field has been reported (Shan et al., 2005a). The nanoparticles were synthesized by a coprecipitation method followed by modification with ammonium oleate. The surface-modified magnetite nanoparticles were monodispersed in an aqueous solution and did not precipitate over 18 months. Using transmission electron microscopy (TEM), the average size of the magnetic particles was found to be in the range from 10 to 15 nm. Transmission electron microscopic cross-sectional analysis of the cells showed further that the magnetite nanoparticles were for the most part strongly absorbed by the surfaces of the cells and coated the cells. The coated cells had distinct superparamagnetic properties.

The nanoparticles were strongly adsorbed on the cell surfaces because of their high specific surface area and high surface energy. It was possible to concentrate the dispersed coated cells by the application of a magnetic field for reuse, and when dispersed, the coated cells experienced minimal mass transfer problems. Thus, this technique has advantages over conventional immobilization by adsorption to carrier materials such as celite. Furthermore, the method can overcome drawbacks such as limitations in biomass loading and in the loss of cells from the carrier associated with conventional immobilization by adsorption. The coated cells not only had the same desulfurizing activity as free cells but also could be reused more than five times, while the free cells could be used only once. Compared with cells immobilized on celite, the cells coated with magnetite nanoparticles had greater desulfurizing activity and operational stability insofar as the coated cells did not experience a mass transfer problem.

The in situ cell separation and immobilization of bacterial cells for biodesulfurization that were developed by using superparamagnetic magnetite nanoparticles have also been reported (Li et al. 2009b). The magnetite nanoparticles were synthesized by coprecipitation followed by modification with ammonium oleate. The surface-modified nanoparticles were monodispersed, and the particle size was on the order of 13 nm. After adding the magnetic fluids to the culture broth, *R. erythropolis* LSSE8-1 cells were immobilized by adsorption and then separated with an external magnetic field. Analysis showed that the nanoparticles were strongly adsorbed to the surface and coated the cells. Compared with free cells, the coated cells not only had the same desulfurizing activity but also could be easily separated from fermentation broth by magnetic force. It was believed that oleate-modified magnetite nanoparticles adsorbed bacterial cells mainly because of the nanosized effect and hydrophobic interaction.

In terms of the point of commercial application, the *Rhodococcus* strains possess several properties favorable for desulfurization over *Pseudomonas* in an oil-water system. First, the hydrophobic nature of *Rhodococcus* makes them access preferentially Cx-dibenzothiophene derivatives from the oil, resulting in little mass transfer limitation (Le Borgne and Quintero, 2003). Moreover, the *Rhodococcus* bacteria are more resistant to solvents than *Pseudomonas* (Bouchez-Naitali et al., 2004). Therefore, there has been an attempt to develop a simple and effective technique by integrating the advantages of magnetic separation and cell immobilization for biodesulfurization process with gram-positive *R. erythropolis* LSSE8-1 and gram-negative *P. delafieldii* R-8 (Li et al., 2009a,b). Cells were grown to the *late exponential phase* (the phase where microbes exhibit exponential growth and some also called the *logarithmic phase*), and the culture was transferred into Erlenmeyer flasks. A volume of magnetic fluids was added and mixed thoroughly—the microbial cells were coated by adsorbing the magnetic nanoparticles. The ammonium oleate-modified magnetite nanoparticles formed a stable suspension in distilled water, and the magnetic fluid did not settle during 8 months of storage at room temperature.

The transmission electron microscope image of magnetite nanoparticles showed that the particles have an approximately spherical morphology with an average diameter of about 13 nm. The magnetite nanoparticles on the cell surface were not washed out by deionized water, ethanol, saline water (0.85 wt%), or phosphate buffer (0.1 M, pH 7). Thus, there is little cell loss or decrease in biomass loading when cells are coated with magnetic nanoparticles.

This outcome was different from that obtained with cells immobilized by traditional adsorption to a carrier. The cells coated with magnetite nanoparticles were superparamagnetic. Therefore, the cell-nanoparticle aggregates in aqueous suspension could be easily separated with an external magnetic field and redispersed by gentle shaking after the removal of magnetic field. For magnetic separation, a permanent magnet should be placed at the side of the vessel. After several minutes (3–5 min), the coated cells can be concentrated and separated from the suspension medium by decantation.

An adsorption mechanism between the magnetite nanoparticles and desulfurizing cells has also been proposed (Li et al. 2009b). The large specific surface area and the high surface energy of the magnetite nanoparticles (i.e., the nanosized effect) ensure that the magnetite nanoparticles are strongly adsorbed on the surfaces of microbial cells. Furthermore, the hydrophobic interaction between the bacterial cell wall and the hydrophobic tail of oleate-modified magnetite nanoparticles may play another important role in cell adsorption. The suspension of oleate-modified magnetite nanoparticles was considered bilayer surfactant-stabilized aqueous magnetic fluids (Liu et al., 2006). The iron oxide nanocrystals were first chemically coated with oleic acid molecule after which the excess oleic was weakly adsorbed on the primary layer through the hydrophobic interaction between the subsequent molecule and the hydrophobic tail of oleate. Since the bacterial cell wall is composed of proteins, carbohydrates, and other substances (such as peptidoglycan, lipopolysaccharide, and mycolic acid), the extracellular matrix can form hydrophobic interaction with the hydrophobic tail of oleate nanoparticles.

Previous work had utilized resting cells of *P. delafieldii* R-8 coated with magnetite nanoparticles for biodesulfurization, but the process was more complicated because centrifugation was necessary in preparation of resting cells (Shan et al., 2005a). However, the development of magnetic separated/immobilized two gram types of desulfurizing bacterial cells from its original culture fermentation broth, namely, in situ magnetic separation and immobilization of bacteria, has been reported (Li et al., 2009b). This one-step technology optimized dramatically the biodesulfurization process flow, and much less of magnetite nanoparticles was needed.

A microbial method has also been used to regenerate desulfurization adsorbents (Li et al., 2006). Most of the sulfur compounds can be desorbed and removed, and heat losses during the bioregeneration process are markedly reduced. The particle size of cells is similar to that of desulfurization adsorbents, which is about several microns. Therefore, it is difficult to separate regenerated adsorbents and cells. Superparamagnetism is an efficient method to separate small particles. To solve the problem of the separation of cells and adsorbents, magnetite nanoparticle-modified *P. delafieldii* R-8 cells were used in the bioregeneration of adsorbents. Biodesulfurization with *P. delafieldii* R-8 strains coated with magnetite nanoparticles has been reported previously

(Shan et al., 2005a), and cells coated with magnetite nanoparticles can be separated and reused for several times.

There is also the report of the bioregeneration of desulfurization adsorbent silver-yttrium (AgY) with magnetic cells (Li et al., 2008b). Superparamagnetic magnetite nanoparticles are prepared by the coprecipitation method followed by modification with ammonia oleate. Magnetic *P. delafieldii* R-8 cells can be prepared by mixing the cells with magnetite nanoparticles. When the magnetic cells were used in the bioregeneration of desulfurization adsorbent AgY, the concentration of dibenzothiophene and 2-hydroxybiphenyl with free cells is a little higher than that with magnetic cells. Adsorption capacity of the regenerated adsorbent is 93% that of the fresh one after being desorbed with magnetic *P. delafieldii* R-8, dried at 100°C for 24h, and calcined in the air at 500°C for 4h. The magnetic cells can be separated from adsorbent bioregeneration system after desulfurization with external magnetic field and, thus, can be reused.

5. The Future

Microbial desulfurization in nature is different from other more common biotechnology processes, and the process has several limitations that currently prevent the concept from being applied in a modern refinery (Gupta et al., 2005). The metabolism of sulfur compounds is typically slow compared with chemical reactions employed in a refinery, and generally, the rate of metabolism is rate limiting in the process, though mass transfer resistance from the oil/water interface to the microbe is also slow compared with the rate of transfer of the sulfur compound to the oil-water interface. Large amounts of biomass are needed (typically 2.5g biomass per gram of sulfur), and biological systems must be kept alive to function, which can be difficult under the variable input conditions found in refineries. The rate of desulfurization depends strongly on pH, temperature, and dissolved oxygen concentration. Separation of the cells from the oil can also be difficult, and immobilized cells often have lower activity and limited lifetime.

Furthermore, the delicate enzymes of desulfurization competent cells have to obtain part of their substrates from a different phase in which their survival is not possible. Since the first time that a specific sulfur removal was introduced, research on this field has been steadily continued. Many other strains have been isolated or cloned, and different methods have been tried. The most serious problem in the implementation of biodesulfurization as an alternative industrial approach to produce ultralow sulfur content lies in the isolation or design of a microbial strain with higher efficiency. Any small success that provides the possibility to remove sulfur at higher temperature, with higher rate, or longer stability of desulfurization activity is considered a significant step toward industry-level biodesulfurization. Moreover, process development and any unexpected problems that might occur as a result of upscaling the operations must also be considered.

Once success has been achieved, the process may operate in a line after hydrodesulfurization unit. Total desulfurization of fossil fuel by microbial approach is not expected to be onstream in the near future, and more research is needed to design a recombinant strain with a broader range of target sulfur compounds or to use successive desulfurizing microbial systems with high potency. Most researches on desulfurization of refractory compounds have been performed with simple model fuel to understand the nature of desulfurization. Dealing with genuine fossil fuel will open up new challenges to solve.

Adsorption, biodesulfurization, and hydrodesulfurization are more likely to be employed to complement each other. However, in order to develop the biodesulfurization process as a complementary process, a multidisciplinary approach is essential, and the participation of scientists and engineers from the fields of biotechnology, biochemistry, refining processes, and engineering is essential. Eventually, the application of nanotechnology to the desulfurization of crude oil and crude oil products will lead to a process capable of producing ultralow sulfur to no-sulfur products on an economical and cost-effective basis.

Furthermore, desulfurizing biocatalysts can be modified to remove specific sulfur structures or broader classes of sulfur compounds. Biocatalysts that can upgrade oils modify specific organic structures like cleaving or opening aromatic rings. The advantage of this selectivity is that a process user can reliably predict the chemical changes that will occur. Unfortunately, crude oil being a complex natural product is variable in composition, and therefore, for crude oil applications, the biocatalysts will have to be customized for each crude oil, and the total process may involve multiple biocatalysts.

However, if these hurdles can be overcome, the bioprocessed oil can be processed at a refinery using current technology at a lower cost. For sulfur, hydrotreating is very effective for removing mercaptans and straight-chain sulfides, whereas biodesulfurization is more effective with organic ring sulfides such as dibenzothiophene (DBT) derivatives. The hydrotreating conditions are less severe (lower cost) for straight-chain sulfur compounds, so an economic advantage is achieved when both hydrotreating and biodesulfurization processes are used. Moreover, bioprocessing can have a beneficial effect on oil properties, especially properties that affect handling the bulk oil—improvements such as (i) viscosity reduction, (ii) shifts to lower-molecular-weight distribution in the oil, or (iii) lower asphaltene content would reduce fluid piping and transportation problems and costs. Indeed, a higher-grade and cleaner oil (low-sulfur content and high API gravity) could well be the outcome of bioprocessing viscous oil at the wellhead.

For example, there have been significant developments in the area of defining a truly efficient and economical biodesulfurization process. Critical aspects of such a process include (i) suitability of the microbes, (ii) reaction rate, (iii) product recovery, and (iv) reactor design. In addition, other technical issues that need to be fully resolved include molecular mechanisms, optimized microbial strains for commercial use, mass transfer, and reactivity of viscous feedstocks (without having a detrimental effect on the biocatalyst) where viscosity and density play an important role in feedstock processing (Gray et al., 1996; Monticello, 2000; Bachmann et al., 2014).

One advantage of such a concept is the potential simplicity of the process—the crude oil is mixed with a water-soluble biocatalyst (either the microorganism or the enzyme) and with air. After the reaction, the formed water-oil emulsion is separated to recover the upgraded oil—the biocatalyst remains with the water and is (potentially) available for reuse. The only added feature (that may not always be available at the wellhead) is a mixing reactor prior to separation.

However, the issue of bioconversion of asphaltene constituents remains a significant issue, and feedstocks containing asphaltene constituents may be difficult to process. The qualitative and quantitative differences in the hydrocarbon and the nonhydrocarbon content of crude oil (Speight, 2014) influence the susceptibility of crude oil and certain crude oil products to biocatalysis. This must be acknowledged as a major consideration in determining the potential for biocatalytic desulfurization of asphaltene-containing feedstocks (Jain and Bajpai, 2012).

In summary, significant progress has been made toward the commercialization of the biodesulfurization of crude oil, which includes (i) characterization of feedstock candidates for the process, (ii) improved biocatalyst performance that directly relates to crude oil biodesulfurization, and (iii) development of analytic methodology that can be employed in the development of biodesulfurization process concepts.

As a corollary, one aspect of biodesulfurization that should also be explored is the application of the process at the wellhead. Many of the biocatalysts that can be applied to crude oil upgrading and desulfurization are water-soluble, and in an oil field operation, water is coproduced with the oil, so the process of water and oil separation is a routine field process; the only added process step would be agitation of the oil and water mixture with the biocatalyst. The process would generate a wastewater stream, which must be handled whether it is generated in the oil field or at the refinery. In a refinery, this new waste stream becomes an added problem. However, in the field, the water containing the formed salts can be diluted and reinjected as part of the field waterflood program, which, if allowable under environmental regulations, may have minimal effect on oil field operation.

References

Ali, M.A., Abbas, S., 2006. A review of methods for the demetallization of residual fuel oils. Fuel Process. Technol. 87, 573–584.

Al-Shahristani, H., Al-Atyia, M.J., 1972. Vertical migration of oil in Iraqi oil fields: evidence based on vanadium and nickel concentrations. Geochim. Cosmochim. Acta 36, 929–938.

Altgelt, K.H., Boduszynski, M.M., 1994. Composition and Analysis of Heavy Petroleum Fractions. Marcel Dekker, New York.

Ancheyta, J., Speight, J.G., 2007. Hydroprocessing Heavy Oils and Residua. CRC Press, Taylor & Francis Group, Boca Raton, FL.

Armstrong, S.M., Sankey, B.M., Voodoo, G., 1995. Conversion of dibenzothiophene to biphenyl by sulfate-reducing bacteria isolated from oil field production facilities. Biotechnol. Lett. 17, 1133–1136.

Ascón-Cabrera, M.A., Lebeault, J.M., 1995. Interfacial area effects of a biphasic aqueous/organic system on growth kinetic of xenobiotic-degrading microorganisms. Appl. Microbiol. Biotechnol. 43, 1136–1141.

Atlas, R.M., 1981. Microbial degradation of petroleum hydrocarbons: an environmental perspective. Microbiol. Rev. 45, 180–209.

Bachmann, R.T., Johnson, A.C., Edyvean, R.G.J., 2014. Biotechnology in the petroleum industry: an overview. Int. Biodeterior. Biodegrad. 86, 225–237.

Ball, J.S., Wenger, W.J., Hyden, H.J., Horr, C.A., Myers, A.T., 1960. Metal content of twenty-four petroleums. J. Chem. Eng. Data 5, 553–557.

Banat, I.M., 1995. Biosurfactants production and possible uses in microbial enhanced oil recovery and oil pollution remediation: a review. Bioresour. Technol. 51, 1–12.

Barathi, S., Vasudevan, N., 2001. Utilization of petroleum hydrocarbons by *Pseudomonas fluorescens* isolated from a petroleum-contaminated soil. Environ. Int. 26, 413–416.

Bartha, R., 1986. Microbial Ecology: Fundamentals and Applications. Addison-Wesley Publishers, Reading, MA.

Bartha, R., Atlas, R.M., 1977. The microbiology of aquatic oil spills. Adv. Appl. Microbiol. 22, 225–266.

Bartha, R., Pramer, D., 1965. Features of a flask and method for measuring the persistence and biological effects of pesticides in soil. Soil Sci. 100, 68–70.

Benedik, M.J., Gibbs, P.R., Riddle, R.R., Willson, R.C., 1998. Microbial denitrogenation of fossil fuels. Trends Biotechnol. 16, 390–395.

Bertrand, J.C., Rambeloarisoa, E., Rontani, J.F., Giusti, G., Mattei, G., 1983. Microbial degradation of crude oil in sea water in continuous culture. Biotechnol. Lett. 5, 567–572.

Bouchez-Naitali, M., Abbad-Andaloussi, S., Warzywoda, M., Monot, F., 2004. Relation between bacterial strain resistance to solvents and biodesulfurization activity in organic medium. Appl. Microbiol. Biotechnol. 65, 440–445.

Bowerman, B.S., Swyler, K.J., Dougherty, D.R., Davis, R.E., Siskind, B., Barletta, R.E., 1985. An evaluation of the stability tests recommended in the branch technical position on waste forms and container materials. Report No. TI85–008726, Prepared for Division of Waste Management, Office of Nuclear Material Safety and Safeguards, US Nuclear Regulatory Commission, Washington, DC.

Bryant, R.S., Bailey, S.A., Step, A.K., Evans, D.B., Parli, J.A., 1998. Biotechnology for heavy oil recovery. In: SPE Paper No. 36767, Proceedings of the SPE/DOE Improved Oil Recovery Symposium, pp. 1–7.

Bublitz, F., Guenther, T., Fritsche, W., 1994. Screening of fungi for the biological modification of hard coal and coal derivatives. Fuel Process. Technol. 40, 347–354.

Burd, G., Ward, O.P., 1996a. Physico-chemical properties of PM-factor, a surface active agent produced by *Pseudomonas marginalis*. Can. J. Microbiol. 42, 243–251.

Burd, G., Ward, O.P., 1996b. Bacterial degradation of polycyclic aromatic hydrocarbons on agar plates: the role of biosurfactants. Biotechnol. Tech. 10, 371–374.

Bustos-Jaimes, I., Amador, G., Castorena, G., Le Borgne, S., 2003. Genotypic characterization of sulfur-oxidative desulfurizing bacterial strains isolated from Mexican refineries. Oil Gas Sci. Technol. 58 (4), 521–526.

Callejas, M.A., Martinez, M.T., Fierro, J.L.G., Rial, C., Jiménez-Mateos, J.M., Gómez-Garcia, F.J., 2001. Structural and morphological study of metal deposition on an aged Hydrotreating catalyst. Appl. Catal. A Gen. 220, 93–104.

Calzada, J., Alcon, A., Santos, V.E., Garcia-Ochoa, F., 2011. Mixtures of *Pseudomonas putida* CECT 5279 cells of different ages: optimization as biodesulfurization catalyst. Process Biochem. 46, 1323–1328.

Caro, A., Boltes, K., Leton, P., Garcia-Calvo, E., 2008. Biodesulfurization of dibenzothiophene by growing cells of *Pseudomonas putida* CECT 5279 in biphasic media. Chemosphere 73 (5), 663–669.

Cassani, F., Eglinton, G., 1991. Organic geochemistry of Venezuelan extra-heavy crude oil 2: molecular assessment of biodegradation. Chem. Geol. 91, 315–333.

Cerniglia, C.E., 1992. Biodegradation of polycyclic aromatic hydrocarbons. Biodegradation 3, 351–368.

Connan, J., 1984. Biodegradation of crude oils in reservoirs. In: Brooks, J., Welte, D.H. (Eds.), Advances in Petroleum Geochemistry. vol. 1. Academic Press, New York, pp. 299–335.

Cooper, D.G., Zajic, J.E., 1980. Surface-active compounds from microorganisms. Adv. Appl. Microbiol. 26, 229–253.

Denome, S.A., Olson, E.S., Young, K., 1993. Identification and cloning of genes involved in specific desulfurization of dibenzothiophene by *Rhodococcus* sp. strain IGTS8. Appl. Environ. Microbiol. 59 (9), 2837–2843.

Denome, S.A., Oldfield, C., Nash, L.J., 1994. Characterization of the desulfurization genes from *Rhodococcus* sp. strain IGTS8. J. Bacteriol. 176 (21), 6707–6716.

Dinamarca, M.A., Ibacache-Quiroga, C., Baeza, P., Galvez, S., Villarroel, M., Olivero, P., Ojeda, J., 2010. Biodesulfurization of gas oil using inorganic supports biomodified with metabolically active cells immobilized by adsorption. Bioresour. Technol. 101, 2375–2378.

Dinamarca, M.A., Rojas, A., Baeza, P., Espinoza, G., Ibacache-Quiroga, C.I., Ojeda, J., 2014. Optimization the biodesulfurization of gas oil by adding surfactants to immobilized cell systems. Fuel 116, 237–241.

Duarte, G.F., Rosado, A.S., Seldin, L., de Araujo, W., van Elsas, J.D., 2001. Analysis of bacterial community structure in sulfurous-oil-containing soils and detection of species carrying Dibenzothiophene desulfurization (dsz) genes. Appl. Environ. Microbiol. 67, 1052–1062.

Duyck, C., Miekeley, N., Porto da Silveira, C.L., Aucélio, R.Q., Campos, R.C., Grinberg, P., Brandão, G.P., 2007. Trace element distributions in biodegraded crude oils and fractions from the Potiguar Basin. Brazil Spectrochim. Acta B. 2007 (62), 939–951.

Duyck, C., Miekeley, N., Fonseca, T.C.O., Szatmari, P., Santos Neto, E.V.S., 2008. Trace element distributions in biodegraded crude oils and fractions from the Potiguar Basin, Brazil. J. Braz. Chem. Soc. 19 (5), 978–986.

Éigenson, A.S., Ivchenko, E.G., 1977. Distribution of sulfur and nitrogen in fractions from crude oil and residues. Chem. Technol. Fuels Oils 13, 542–544.

El-Gayar, M., Mostafa, M.S., Abdelfattah, A.E., Barakat, A.O., 2002. Application of geochemical parameters for classification of crude oils from Egypt into source-related types. Fuel Process. Technol. 79, 13–28.

El-Gendy, N.S., Speight, J.G., 2016. Handbook of Refinery Desulfurization. CRC Press, Taylor & Francis Group, Boca Raton, FL.

Fadhil, A.M.A., Al-Jailawi, M.H., Mahdi, M.S., 2014. Isolation and characterization of a new thermophilic, carbazole degrading bacterium (*Anoxybacillus rupiensis*) strain Ir3 (JQ912241). Int. J. Adv. Res. 2014 (2), 795–805.

Farag, A.S., Sif El-Din, O.I., Youssef, M.H., Hassan, S.I., Farmawy, S., 1989. Solvent demetallization of heavy oil residue. Hung. J. Ind. Chem. 17 (3), 289–294.

Fedorak, P.M., Semple, K.M., Vazquez-Duhalt, R., Westlake, D.W.S., 1993. Chloroperoxidase mediated modifications of petroporphyrins and asphaltenes. Enzym. Microb. Technol. 15, 429–437.

Fish, R.H., Komlenic, J.J., Wines, B.K., 1984. Characterization and comparison of vanadyl and nickel compounds in heavy crude petroleum and asphaltenes by reverse-phase and size-exclusion liquid chromatography/graphite furnace atomic absorption spectrometry. Anal. Chem. 56, 2452–2460.

Foght, J.M., Gutnick, D.L., Westlake, D.W.S., 1989. Effect of emulsan on biodegradation of crude oil by pure and mixed bacterial cultures. Appl. Environ. Microbiol. (January), 36–42.

Foght, J.M., Fedorak, P.M., Westlake, D.W.S., 1990. Mineralization of [^{14}C] hexadecane and [^{14}C] phenanthrene in crude oil: specificity among bacterial isolates. Can. J. Microbiol. 36, 169–175.

Frontera-Suau, R., Bost, F., McDonald, T., Morris, P.J., 2002. Aerobic biodegradation of hopanes and other biomarkers by crude oil degrading enrichment cultures. Environ. Sci. Technol. 36, 4585–4592.

Gallardo, M.E., Fernandez, A., Lorenzo, V.D., 1997. Designing recombinant *Pseudomonas* strains to enhance biodesulfurization. J. Bacteriol. 179 (22), 7156–7160.

Garcia-Arellano, H., Buenrostro-Gonzalez, E., Vazquez-Duhalt, R., 2004. Biocatalytic transformation of petroporphyrins by chemical modified cytochrome c. Biotechnol. Bioeng. 85, 790–798.

Gary, J.H., Handwerk, G.E., Kaiser, M.J., 2007. Crude Oil Refining: Technology and Economics, fifth ed. CRC Press, Taylor & Francis Group, Boca Raton, FL.

Gibson, D.T., Sayler, G.S., 1992. Scientific Foundation for Bioremediation: Current Status and Future Needs. American Academy of Microbiology, Washington, DC.

Gibson, D.T., Subramanian, V., 1984. Microbial degradation of aromatic hydrocarbons. In: Gibson, D.T. (Ed.), Microbial Degradation of Organic Compounds. McGraw-Hill, New York, pp. 181–252.

Goldman, S., Shabtai, Y., Rubinovitz, C., Rosenberg, E., Gutnick, D.L., 1982. Emulsan in *Acinetobacter calcoaceticus* RAG-1: distribution of cell-free and cell-associated cross-reacting material. Appl. Environ. Microbiol. 44, 165–170.

Gray, K.A., Pogrebinsky, O., Mrachko, G., Xi, L., Monticello, D.J., Squires, C., 1996. Molecular mechanisms of biocatalytic desulfurization of fossil fuels. Nat. Biotechnol. 14, 1705–1709.

Greenwood, P.F., Wibrow, S., George, S.J., Tibbett, M., 2008. Sequential hydrocarbon biodegradation in a soil from arid coastal Australia, treated with oil under laboratory-controlled conditions. Org. Geochem. 39, 1336–1346.

Guiliano, M., Boukir, A., Doumenq, P., Mille, G., 2000. Supercritical fluid extraction of BAL 150 crude oil asphaltenes. Energy Fuel 14, 89–94.

Gupta, N., Roychoudhury, P.K., Deb, J.K., 2005. Biotechnology of desulfurization of diesel: prospects and challenges. Appl. Microbiol. Biotechnol. 66 (4), 356–366.

Gutnick, D.L., 1987. The emulsan polymer: perspectives on a microbial capsule as an industrial product. Biopolymers 26, 223–240.

Hao, R., Lu, A., 2009. Biodegradation of heavy oils by halophilic bacterium. Prog. Nat. Sci. 19, 997–1001.

Harvey, R., 1997. Polycyclic Aromatic Hydrocarbons. Wiley-VCH, New York.

Haukanes, B.I., Kvam, C., 1993. Application of magnetic beads in bioassays. Nat. Biotechnol. 11, 60–63.

Hegedus, L.L., McCabe, R.W., 1981. Catalyst poisoning. Catal. Rev. Sci. Eng. 23, 377–476.

Hofrichter, M., Bublitz, F., Fritsche, W., 1997. Fungal attack on coal: I. Modification of hard coal by Fungi. Fuel Process. Technol. 52, 43–53.

Hsu, C.S., Robinson, P.R. (Eds.), 2017. Handbook of Petroleum Technology. Springer International Publishing AG, Cham.

Iturbe, R., Flores, C., Castro, A., Torres, L.G., 2007. Sub-soil contamination due to oil spills in zones surrounding oil pipeline-pump stations and oil pipeline right-of-ways in Southwest-Mexico. Environ. Monit. Assess. 133, 387–398.

Jain, P.K., Bajpai, V., 2012. Biotechnology of bioremediation: a review. Int. J. Environ. Sci. 3 (1), 535–549.

Kanaly, R.A., Harayama, S., 2000. Biodegradation of high-molecular weight polycyclic aromatic hydrocarbons by bacteria. J. Bacteriol. 182, 2059–2067.

Kaufman, E.N., Harkins, J.B., Borole, A.P., 1998. Comparison of batch-stirred and electrospray reactors for biodesulfurization of dibenzothiophene in crude oil and hydrocarbon feedstocks. Appl. Biochem. Biotechnol. 73, 127–144.

Kilbane, J.J., 2006. Microbial biocatalyst development to upgrade fossil fuels. Curr. Opin. Biotechnol. 17, 305–314.

Kim, J.-S., Crowley, D.E., 2007. Microbial diversity in natural asphalts of the rancho La Brea tar pits. Appl. Environ. Microbiol. 73, 4579–4591.

Kim, I.S., Park, J., Kim, K.W., 2001. Enhanced biodegradation of polycyclic aromatic hydrocarbons using nonionic surfactants in soil slurry. Appl. Geochem. 16, 1419–1428.

Kirkwood, K.M., Foght, J.M., Gray, M.R., 2004. Prospects for biological upgrading of heavy oils and asphaltenes. In: Petroleum Biotechnology: Developments and Perspectives (Studies in Surface Science and Catalysis). vol. 151. Elsevier, Amsterdam, pp. 113–145 (Chapter 4).

Konishi, M., Kishimoto, M., Tamesui, N., Omasa, I., Shioya, S., Ohtake, H., 2005. The separation of oil from an oil-water-bacteria mixture using a hydrophobic tubular membrane. Biochem. Eng. J. 24, 49–54.

Lacotte, D.J., Mille, G., Acquaviva, M., Bertrand, J.C., 1996. Arabian light 150 asphaltene biotransformation with n-alkanes as co-substrate. Chemosphere 32, 1755–1761.

Larose, C.D., Labbe, D., Bergeron, H., 1997. Conservation of plasmid-encoded dibenzothiophene desulfurization genes in several rhodococci. Appl. Environ. Microbiol. 63 (7), 2915–2919.

Lavania, M., Cheema, S., Sarma, P.M., Mandal, A.K., Lal, B., 2012. Biodegradation of asphalt by *Garciaella petrolearia* TERIG02 for viscosity reduction of heavy oil. Biodegradation 23, 15–24.

Le Borgne, S., Quintero, R., 2003. Biotechnological processes for the refining of petroleum. Fuel Process. Technol. 81, 155–169.

Li, M.Z., Squires, C.H., Monticello, D.J., 1996. Genetic analysis of the Dsz promoter and associated regulatory regions of *Rhodococcus erythropolis* IGTS8. J. Bacteriol. 178, 6409–6418.

Li, W., Xing, J., Li, Y., Xiong, X., Li, X., Liu, H., 2006. Feasibility study on the integration of adsorption/bioregeneration of π-complexation adsorbent for desulfurization. Ind. Eng. Chem. Res. 45 (8), 2845–2849.

Li, W., Xing, J., Li, Y., Xiong, X., Li, X., Liu, H., 2008b. Desulfurization and bio-regeneration of adsorbents with magnetic P. delafieldii R-8 cells. Catal. Commun. 9, 376–380.

Li, W., Tang, H., Liu, Q., Xing, J., Li, Q., Wang, D., Yang, M., Li, X., Liu, H., 2009a. Deep desulfurization of diesel by integrating adsorption and microbial method. Biochem. Eng. J. 44, 297–301.

Li, Y.G., Gao, H.S., Li, W.L., Xing, J.M., Liu, H.Z., 2009b. In situ magnetic separation and immobilization of dibenzothiophene-desulfurizing Bacteria. Bioresour. Technol. 100, 5092–5096.

Linguist, L., Pacheco, M., 1999. Enzyme-based diesel desulfurization process offers energy, CO_2 Advantages. Oil Gas J. 22, 45–48.

Liu, X.Q., Liu, H.Z., Xing, J.M., Guan, Y.P., Ma, Z.Y., Shan, G.B., Yang, C.L., 2003. Preparation and characterization of superparamagnetic functional polymeric microparticles. Chinese Particul. 1, 76–79.

Liu, X., Kaminski, M.D., Guan, Y., Chen, H., Liu, H., Rosengart, A.J., 2006. Preparation and characterization of hydrophobic superparamagnetic gel. J. Magn. Magn. Mater. 306, 248–253.

Luey, J., Li, S.W., 1993. Determination of the biodegradation rate of asphalt for the Hanford grout vaults. Report, Contract DE-AC06-76RLO 1830, Pacific Northwest Laboratory Richland, US Department of Energy, Richland, WA. April.

Machackova, J., Wittlingerova, Z., Vlk, K., Zima, J., Ales, L., 2008. Comparison of two methods for assessment of in situ jet fuel remediation efficiency. Water Air Soil Pollut. 187, 181–194.

Maghsoudi, S., Kheirolomoom, A., Vossoughi, M., 2000. Selective desulfurization of dibenzothiophene by newly isolated *Corynebacterium* sp. strain P32C1. Biochem. Eng. J. 5, 11–16.

Maghsoudi, S., Vossoughi, M., Kheirolomoom, A., Tanaka, E., Katoh, S., 2001. Biodesulfurization of hydrocarbons and diesel fuels by *Rhodococcus* sp. strain P32C1. Biochem. Eng. J. 8, 151–156.

McFarland, B.L., 1999. Biodesulfurization. Curr. Opin. Microbiol. 2, 257–264.

Mogollon, L., Rodriguez, R., Larrota, W., Ortiz, C., Torres, R., 1998. Biocatalytic removal of nickel and vanadium from petroporphyrins and asphaltenes. Appl. Biochem. Biotechnol. 70–72, 765–777.

Mohebali, G., Ball, A.S., 2008. Biocatalytic desulfurization (biodesulfurization) of petrodiesel fuels. Microbiology 154, 2169–2183.

Mohebali, G., Ball, A.S., 2016. Biodesulfurization of diesel fuels–past, present and future perspectives. Int. Biodeterior. Biodegrad. 110, 163–180.

Molday, R.S., Yen, S.P., Rembaum, A., 1977. Application of magnetic microspheres in labelling and separation of cells. Nature 268, 437–438.

Monticello, D.J., 2000. Biodesulfurization and upgrading of petroleum distillates. Curr. Opin. Biotechnol. 11, 540–546.

Monticello, D.J., Kilbane, J.J., 1994. Emulsification of Petroleum Oils and Aqueous Bacterial Enzyme, Then Incubation, for Selective Cleavage of Carbon-Sulfur Bonds. United States Patent 5,358,870.

Montiel, C., Quintero, R., Aburto, J., 2009. Petroleum biotechnology: technology trends for the future. Afr. J. Biotechnol. 8 (12), 2653–2666.

Morais, E.B., Tauk-Tornisielo, S.M., 2009. Biodegradation of oil refinery residues using mixed-culture of microorganisms isolated from land farming. Braz. Arch. Biol. Technol. 52 (6), 1571–1578.

Naranjo, L., Urbina, H., De Sisto, A., León, V., 2007. Isolation of autochthonous non-white rot fungi with potential for enzymatic upgrading of Venezuelan extra-heavy crude oil. Biocatal. Biotransform. 25, 1–9.

Ohshiro, T., Hirata, T., Izumi, Y., 1996a. Desulfurization of dibenzothiophene derivatives by whole cells of *Rhodococcus erythropolis* H-2. FEMS Microbiol. Lett. 142 (1), 65–70.

Ohshiro, T., Hirata, T., Hashimoto, I., Izumi, Y., 1996b. Regulation of dibenzothiophene degrading enzyme activity of *Rhodococcus erythropolis* D-1. J. Ferment. Bioeng. 82 (6), 610–612.

Okoh, A.I., 2003. Biodegradation of bonny light crude oil in soil microcosm by some bacterial strains isolated from crude oil flow stations saver pits in Nigeria. Afr. J. Biotechnol. 2 (5), 104–108.

Okoh, A.I., 2006. Biodegradation alternative in the cleanup of petroleum hydrocarbon pollutants. Biotechnol. Mol. Biol. Rev. 1 (2), 38–50.

Okoh, A.I., Trejo-Hernandez, M.R., 2006. Remediation of petroleum hydrocarbon polluted systems: exploiting the bioremediation strategies. Afr. J. Biotechnol. 5, 2520–2525.

Okoh, A.I., Ajisebutu, S., Babalola, G.O., Trejo-Hernandez, M.R., 2001. Potential of *Burkholderia cepacia* strain RQ1 in the biodegradation of heavy crude oil. Int. Microbiol. 4, 83–87.

Okoh, A.I., Ajisebutu, S., Babalola, G.O., Trejo-Hernandez, M.R., 2002. Biodegradation of Mexican heavy crude oil (Maya) by *Pseudomonas aeruginosa*. J. Trop. Med. 2 (1), 12–24.

Oldfield, C., Poogrebinsky, O., Simmonds, J., 1997. Elucidation of the metabolic pathway for dibenzothiophene desulphurization by *Rhodococcus* sp. strain IGTS8 (ATCC53968). Microbiology 143, 2961–2973.

Ourisson, G., Albrecht, P., Rohmer, M., 1979. The hopanoids. Paleochemistry and biochemistry of a group of natural products. Pure Appl. Chem. 51, 709–729.

Pacheco, M.A., 1999. Recent advances in biodesulfurization (biodesulfurization) of diesel fuel. In: Proceedings of the NPRA Annual Meeting, San Antonio, TX, March 21–23.

Panariti, N., Del Bianco, N.A., Del Piero, G., Marchionna, M., 2000. Petroleum residue upgrading with dispersed catalysts. Part 1. Catalysts activity and selectivity. Appl. Catal. A Gen. 204 (2), 203–213.

Parkash, S., 2003. Refining Processes Handbook. Gulf Professional Publishing, Elsevier, Amsterdam.

Pendrys, J.P., 1989. Biodegradation of asphalt cement-20 by aerobic bacteria. Appl. Environ. Microbiol. 55 (6), 1357–1362.

Phillips, U.A., and Traxler, R.W. 1963. Microbial degradation of asphalt. Appl. Microbiol., 11: 235–238. 1963.

Piddington, C.S., Kovacevich, B.R., Rambosek, J., 1995. Sequence and molecular characterization of a DNA region encoding the dibenzothiophene desulfurization operon of *Rhodococcus* sp. strain IGTS8. Appl. Environ. Microbiol. 61, 468–475.

Pineda, F.G., Mesta-Howard, A.M., Boll-Argúello, G., Lira-Galeana, C., 2004. A microbial consortium isolated from a crude oil sample that uses asphaltenes as a carbon and energy source. Biodegradation 15, 145–151.

Rabus, R., Kube, M., Heider, J., Beck, A., Heitmann, K., Widdel, F., Reinhardt, R., 2005. The genome sequence of an anaerobic aromatic-degrading denitrifying bacterium, strain Ebn1. Arch. Microbiol. 183, 27–36.

Ramos, J.L., Duque, E., Gallegos, M.T., Godoy, P., Ramos-Gonzalez, M.I., Rojas, A., Teran, W., Segura, A., 2002. Mechanisms of solvent tolerance in gram-negative bacteria. Annu. Rev. Microbiol. 56, 743–768.

Rana, M.S., Sámano, V., Ancheyta, J., Diaz, J.A.I., 2007. A review of recent advances on process technologies for upgrading of heavy oils and residua. Fuel 86, 1216–1231.

Rhee, S.K., Lee, K.S., Chung, J.C., Lee, S.T., 1997. Degradation of pyridine by *Nocardioides* sp. strain OS4 isolated from the oxic zone of a spent shale column. Can. J. Microbiol. 43 (2), 205–209.

Rontani, J.F., Bosser-Joulak, F., Rambeloarisoa, E., Bertrand, J.C., Giusti, G., Faure, R., 1985. Analytical study of asyhart crude oil biodegradation. Chemosphere 14, 1413–1422.

Rosenberg, E., Rubinovitz, C., Gottlieb, A., Rosenhak, S., Ron, E.Z., 1988. Production of biodispersan by *Acinetobacter calcoaceticus* A2. Appl. Environ. Microbiol. 54, 317–322.

Sada, E., Katon, S., Terashima, M., 1981. Enhancement of oxygen absorption by magnetite-containing beads of immobilized glucose oxidase. Biotechnol. Bioeng. 21, 1037–1044.

Salehizadeh, H., Mousavi, M., Hatamipour, S., Kermanshahi, K., 2007. Microbial demetallization of crude oil using *Aspergillus* sp.: vanadium oxide octaethyl porphyrin (VOOEP) as a model of metallic petroporphyrins. Iran. J. Biotechnol. 5 (4), 226–231.

Santos, S.C., Alviano, D.S., Alviano, C.S., Padula, M., Leitao, A.C., Martins, O.B., Ribeiro, C.M., Sassaki, M.Y., Matta, C.P., Bevilaqua, J., 2005. Characterization of *Gordonia* sp. strain F.5.25.8 capable of dibenzothiophene desulfurization and carbazole utilization. Appl. Microbiol. Biotechnol. 66, 1–8.

Santos, S.C., Alviano, D.S., Alviano, C.S., Padula, M., Leitao, A.C., Martins, O.B., Ribeiro, C.M., Sassaki, M.Y., Matta, C.P., Bevilaqua, J., Sebastian, G.V., Seldin, L., 2006. Characterization of *Gordonia* sp. strain F.5.25.8 capable of dibenzothiophene desulfurization and carbazole utilization. Appl. Microbiol. Biotechnol. 71, 355–362.

Sar, N., Rosenberg, E., 1983. Emulsifier production by *Acinetobacter calcoaeticus* strains. Curr. Microbiol. 9, 309–314.

Sarret, G., Connan, J., Kasrai, M., Bancroft, M., Charrié-Duhaut Lemoine, S., Adam, P., Albrecht, P., Eybert-Bérard, L., 1999. Chemical forms of sulfur in geological and archaeological asphaltenes from Middle East, France, and Spain determined by sulfur K- and L-edge X-ray absorption near-edge structure spectroscopy. Geochim. Cosmochim. Acta 63 (22), 3767–3779.

Schilling, B.M., Alvarez, L.M., Wang, D.I.C., Cooney, C.L., 2002. Continuous desulfurization of dibenzothiophene with *Rhodococcus rhodochrous* IGTS8 (ATCC 53968). Biotechnol. Prog. 18, 1207–1213.

Setti, L., Lanzarini, G., Pifferi, P.G., 1995. Dibenzothiophene biodegradation by a *Pseudomonas* sp. in model solutions. Process Biochem. 30 (8), 721–728.

Setti, L., Lanzarini, G., Pifferi, P.G., 1997. Whole cell biocatalysis for an oil desulfurization process. Fuel Process. Technol. 52, 145–153.

Shan, G.B., Xing, J.M., Luo, M.F., Liu, H.Z., Chen, J.Y., 2003. Immobilization of *Pseudomonas delafieldii* with magnetic polyvinyl alcohol beads and its application in biodesulfurization. Biotechnol. Lett. 25, 1977–1983.

Shan, G.B., Xing, J.M., Zhang, H., Liu, H.Z., 2005a. Biodesulfurization of dibenzothiophene by microbial cells coated with magnetite nanoparticles. Appl. Environ. Microbiol. 71 (8), 4497–4502.

Shan, G.B., Zhang, H., Liu, H., Xing, J.M., 2005b. π-Complexation studied by fluorescence technique: application in desulfurization of petroleum product using magnetic π-complexation sorbents. Sep. Sci. Technol. 40 (14), 2987–2999.

Shan, G.B., Xing, J.M., Guo, C., Liu, H.Z., Chen, J.Y., 2005c. Biodesulfurization using *Pseudomonas delafieldii* in magnetic polyvinyl alcohol beads. Lett. Appl. Microbiol. 40, 30–36.

Shoham, Y., Rosenberg, E., 1983. Enzymatic depolymerization of emulsan. J. Bacteriol. 156, 161–167.

Shoham, Y., Rosenberg, M., Rosenberg, E., 1983. Bacterial degradation of emulsan. Appl. Environ. Microbiol. 46, 573–579.

Singh, A., 2012. How specific microbial communities benefit the oil industry: biorefining and bioprocessing for upgrading petroleum oil. In: Whitby, C., Skovhus, T.L. (Eds.), Applied Microbiology and Molecular Biology in Oilfield Systems, pp. 121–178.

Song, S.H., Choi, S.S., Park, K., Yoo, Y.J., 2005. Novel hybrid immobilization of microorganisms and its applications to biological denitrification. Enzym. Microb. Technol. 37, 567–573.

Speight, J.G., 1994. Chemical and physical studies of petroleum asphaltene constituents. In: Yen, T.F., Chilingarian, G.V. (Eds.), Asphaltene Constituents and Asphalts. I. Developments in Petroleum Science. vol. 40. Elsevier, Amsterdam (Chapter 2).

Speight, J.G., 2000. The Desulfurization of Heavy Oils and Residua, second ed. Marcel Dekker, New York.

Speight, J.G., 2014. The Chemistry and Technology of Crude Oil, fifth ed. CRC Press, Taylor & Francis Group, Boca Raton, FL.

Speight, J.G., 2017. Handbook of Petroleum Refining. CRC Press, Taylor & Francis Group, Boca Raton, FL.

Speight, J.G., El-Gendy, N.S., 2018. Introduction to Petroleum Biotechnology. Gulf Professional Publishing, Elsevier, Cambridge, MA.

Stanislaus, A., Marafi, A., Rana, M.S., 2010. Recent advances in the science and technology of ultra-low sulfur diesel (ULSD) production. Catal. Today 153, 1–6.

Swaty, T.E., 2005. Global refining industry trends: the present and future. Hydrocarb. Process. (September), 35–46.

Tabatabaee, M.S., Mazaheri Assadi, M., 2013. Vacuum distillation residue upgrading by an indigenous Bacillus cereus. J. Environ. Health Sci. Eng. 11, 18–24.

Tao, F., Yu, B., Xu, P., Ma, C.O., 2006. Biodesulfurization in biphasic systems containing organic solvents. Appl. Environ. Microbiol. 72, 4604–4609.

Thouand, G., Bauda, P., Oudot, J., Kirsch, G., Sutton, C., Vidalie, J.F., 1999. Laboratory evaluation of crude oil biodegradation with commercial or natural microbial Inocula. Can. J. Microbiol. 45, 106–115.

Tian, F., Wu, J.Z., Liang, C., Yang, Y., Ying, P., Sun, X., Cai, T., Li, C., 2006. The study of thiophene adsorption onto La(III)-exchanged zeolite nay by FTIR spectroscopy. J. Colloid Interface Sci. 301, 395–401.

Twerdok, L., 2004. Heavy Fuel Oils Test Plan and Robust Summary. The Petroleum HPV Testing Group, Heavy Fuel Oils Consortium Registration #201-15368A.

Uribe-Alvarez, C., Ayala, M., Perezgasga, L., Naranjo, L., Urbina, H., Vazquez-Duhalt, R., 2011. First evidence of mineralization of petroleum asphaltenes by a strain of *Neosartorya fischeri*. Microb. Biotechnol. 4 (5), 663–672.

Van der Meer, J.R., 1994. Genetic adaptation of bacteria to chlorinated aromatic compounds. FEMS Microbiol. Rev. 15, 239–249.

Van der Meer, J.R., Devos, W.M., Harayama, S., Zehnder, A.J.B., 1992. Molecular mechanisms of genetic adaptation to xenobiotic compounds. Microbiol. Rev. 56, 677–694.

Van Hamme, J.D., Singh, A., Ward, O.P., 2003. Recent advances in petroleum microbiology. Microbiol. Mol. Biol. Rev. 67, 503–549.

Vazquez-Duhalt, R., Torres, E., Valderrama, B., Le Borgne, S., 2002. Will biochemical catalysis impact the petroleum refining industry? Energy Fuel 16, 1239–1250.

Venkateswaran, K., Hoaki, T., Kato, M., Maruyama, T., 1995. Microbial degradation of resins fractionated from Arabian light crude oil. Can. J. Microbiol. 41, 418–424.

Vidali, M., 2001. Bioremediation: an overview. Pure Appl. Chem. 73 (7), 1163–1172.

Whyte, L.G., Bourbonnière, L., Greer, C.W., 1997. Biodegradation of petroleum hydrocarbons by psychrotrophic *Pseudomonas* strains possessing both alkane (*alk*) and naphthalene (*nah*) catabolic pathways. Appl. Environ. Microbiol. 63 (9), 3719–3723.

Wyndham, R.C., Costerton, J.W., 1981. In vitro microbial degradation of bituminous hydrocarbons and in situ colonization of bitumen surfaces within the Athabasca oil sands deposit. Appl. Environ. Microbiol. 41, 791–800.

Yang, J., Hu, Y., Zhao, D., Wang, S., Lau, P.C.K., Marison, I.W., 2007. Two-layer continuous-process design for the biodesulfurization of diesel oils under bacterial growth conditions. Biochem. Eng. J. 37 (2), 212–218.

Yi, N., Xu, G., HogShuai, G., XiangPing, Z., SuoJiang, Z., 2014. Simultaneous desulfurization and denitrogenation of liquid fuels using two functionalized group ionic liquids. Sci. China 57 (12), 1766–1773.

Yu, B., Xu, P., Shi, Q., Ma, C., 2006a. Deep desulfurization of diesel oil and crude oils by a newly isolated *Rhodococcus erythropolis* strain. Appl. Environ. Microbiol. 72, 54–78.

Yu, B., Ma, C., Zhou, W., Wang, Y., Cai, X., Tao, F., Zhang, Q., Tong, M., Qu, J., Xu, P., 2006b. Microbial desulfurization of gasoline by free whole-cells of *Rhodococcus erythropolis* XP. FEMS Microbiol. Lett. 258, 284–289.

Zhang, H., Shan, G., Liu, H., Xing, J., 2007. Surface modification of γ-Al_2O_3 nanoparticles with gum arabic and its applications in adsorption and biodesulfurization. Surf. Coat. Technol. 201, 6917–6921.

Zhang, H., Liu, Q.F., Li, Y., Li, W., Xiong, X., Xing, J., Liu, H., 2008. Selection of adsorbents for in-situ coupling technology of adsorptive desulfurization and biodesulfurization. Sci. China Ser. B Chem. 51 (1), 69–77.

Zhou, Z.Y., Zhang, K., 2004. Development situation and prospect of oil fields in China. Pet. Explor. Dev. 31, 84–87.

Zosim, Z., Gutnick, D.L., Rosenberg, E., 1987. Effect of protein content on the surface activity and viscosity of emulsan. Colloid Polym. Sci. 265, 442–447.

Zrafi-Nouira, I., Guermazi, S., Chouari, R., Safi, N.M.D., Pelletier, E., Backhrouf, A., Saidane-Moshabi, D., Sghir, A., 2009. Molecular diversity analysis and bacterial population dynamics of an adapted seawater microbiota during the degradation of tunisian zarzantine oil. Biodegradation 20, 467–486.

Further Reading

Atlas, R.M., Boron, D.J., Deever, W.R., Johnson, A.R., McFarland, B.L., Meyer, J.A., 2001. Method for Removing Organic Sulfur From Heterocyclic Sulfur Containing Organic Compounds. US Patent Number H1, 986.

Bhatia, S., Sharma, D.K., 2010a. Biodesulfurization of dibenzothiophene, its alkylated derivatives, and crude oil by a newly isolated strain of Pantoea agglomerans D23W3. Biochem. Eng. J. 50 (3), 104–109.

Bhatia, S., Sharma, D.K., 2010b. Mining of genomic databases to identify novel biodesulfurizing microorganisms. J. Ind. Microbiol. Biotechnol. 37, 425–429.

Castorena, G., Suarez, C., Valdez, I., Amador, G., Fernandez, L., Le Borgne, S., 2002. Sulfur-selective desulfurization of dibenzothiophene and diesel oil by newly isolated *Rhodococcus* sp. strains. FEMS Microbiol. Lett. 215 (1), 157–161.

Fang, X.X., Zhang, Y.L., Luo, L.L., Xu, P., Chen, Y.L., Zhou, H., Hai, L., 2006. Organic sulfur removal from catalytic diesel oil by hydrodesulfurization combined with biodesulfurization. Mod. Chem. Ind. 26, 234–238. Chinese Journal; Abstract in English.

Foght, J.M., 2004. Whole-cell bioprocessing of aromatic compounds in crude oil and fuels. In: Vazquez-Duhalt, R., Quintero-Ramirez, R. (Eds.), Studies in Surface Science and Catalysis: Petroleum Biotechnology: Developments and Perspectives. vol. 151. Elsevier, Amsterdam, pp. 145–175.

Folsom, B.R., Schieche, D.R., Digrazia, P., 1999. Microbial desulfurization of alkylated dibenzothiophenes from a hydrodesulfurized middle distillate by *Rhodococcus erythropolis* I-19. Appl. Environ. Microbiol. 65, 4967–4972.

Grossman, M.J., Lee, M.K., Prince, R.C., Garrett, K.K., George, G.N., Pickering, I.J., 1999. Microbial desulfurization of a crude oil middle-distillate fraction: analysis of the extent of sulfur removal and the effect of removal on remaining sulfur. Appl. Environ. Microbiol. 65, 181–188.

Hosseini, S.A., Yaghmaei, S., Mousavi, S.M., Jadidi, A., 2006. Biodesulfurization of dibenzothiophene by a newly isolated thermophilic bacteria strain. Iran. J. Chem. Chem. Eng. 25 (3), 65–71.

Kilbane, J.J., 1989. Desulfurization of coal: the microbial solution. Trends Biotechnol. 7, 97–101.

Kilbane, J.J., Le Borgne, S., 2004. Petroleum biorefining: the selective removal of sulfur, nitrogen, and metals. In: Vazquez-Duhalt, R., Quintero-Ramirez, R. (Eds.), Crude Oil Biotechnology, Developments and Perspectives. Elsevier, Amsterdam, pp. 29–65.

Kishimoto, M., Inui, M., Omasa, T., Katakura, Y., Suga, K., Okumura, K., 2000. Efficient production of desulfurizing cells with the aid of expert system. Biochem. Eng. J. 5, 143–147.

Lee, M.K., Senius, J.D., Grossman, M.J., 1995. Sulfur-specific microbial desulfurization of sterically hindered analogs of dibenzothiophene. Appl. Environ. Microbiol. 61, 4362–4366.

Lee, I.S., Bae, H., Ryu, H.W., Cho, K., Chang, Y.K., 2005. Biocatalytic desulfurization of diesel oil in an air-lift reactor with immobilized *Gordonia nitida* CYKS1 cells. Biotechnol. Prog. 21, 781–785.

Li, W., Jiang, X., 2013. Enhancement of bunker oil biodesulfurization by adding surfactant. World J. Microbiol. Biotechnol. 29 (1), 103–108.

Li, F., Xu, P., Feng, J., Meng, L., Zheng, Y., Luo, L., Ma, C., 2005. Microbial desulfurization of gasoline in a mycobacterium goodii X7B immobilized-cell system. Appl. Environ. Microbiol. 71 (1), 276–281.

Li, Y.G., Ma, J., Zhang, Q.Q., Wang, C.S., Chen, Q., 2007. Sulfur-selective desulfurization of dibenzothiophene and diesel oil by newly isolated *Rhodococcus erythropolis* NCC-1. Chin. J. Org. Chem. 25, 400–405.

Li, Y.G., Xing, J.M., Xiong, X.C., Li, W.L., Gao, S., Liu, H.Z., 2008a. Improvement of biodesulfurization activity of alginate immobilized cells in biphasic systems. J. Ind. Microbiol. Biotechnol. 35, 145–150.

Luo, M.F., Xing, J.M., Gou, Z.X., Li, S., Liu, H.Z., Chen, J.Y., 2003. Desulfurization of dibenzothiophene by lyophilized cells of *Pseudomonas delafieldii* R-8 in the presence of dodecane. Biochem. Eng. J. 13, 1–6.

Ohshiro, T., Kojima, T., Torii, K., Kawasoe, H., Izumi, Y., 1999. Purification and characterization of dibenzothiophene (DBT) sulfone monooxygenase, an enzyme involved in DBT desulfurization, from *Rhodococcus erythropolis* D-1. J. Biosci. Bioeng. 88, 610–616.

Premuzic, E.T., Lin, M.S., 1999. Induced biochemical conversions of heavy crude oils. J. Petrol. Sci. 22, 171–180.

Sanchez, O.F., Almeciga-Diaz, C.J., Silva, E., Cruz, J.C., Valderrama, J.D., Caicedo, L.A., 2008. Reduction of sulfur levels in kerosene by *Pseudomonas* sp. strain in an airlift reactor. Lat. Am. Appl. Res. 38, 329–335.

Setti, L., Farinelli, P., Martino, S., Frassinetti, S., Lanzarini, G., Pifferi, P.G., 1999. Developments in destructive and non-destructive pathways for selective desulfurizations in oil-biorefining processes. Appl. Microbiol. Biotechnol. 52, 111–117.

Setti, L., Bonoli, S., Badiali, E., Giuliani, S., 2003. Inverse phase transfer biocatalysis for a biodesulfurization process of middle distillates. Bull. Moscow Univ. Ser. 2, Chem. 44 (1), 80–83.

Speight, J.G., Karuna, K., Arjoon, K.K., 2012. Bioremediation of Crude Oil and Crude Oil Products. Scrivener Publishing, Beverly, MA, p. 2012.

Tao, F., Liu, Y., Luo, Q., Su, F., Xu, Y., Li, F., Yu, B., Ma, C., Xu, P., 2011. Novel organic solvent-responsive expression vectors for biocatalysis: application for development of an organic solvent-tolerant biodesulfurization strain. Bioresour. Technol. 102, 9380–9387.

Wang, P., Krawiec, S., 1994. Desulfurization of dibenzothiophene to 2-hydroxybiphenyl by some newly isolated bacterial strains. Arch. Microbiol. 161, 266–271.

Chapter 15

Hydrogen Production

1. Introduction

The refinery (Fig. 15.1) is an integrated series of unit processes that are designed to work in sequence to produce a slate of products that are dictated by market demand (Parkash, 2003; Gary et al., 2007; Speight, 2014, 2017; Hsu and Robinson, 2017). Throughout the previous chapters, there have been several references and/or acknowledgments of the use of hydrogen during refining due to the need to refine high-sulfur feedstocks to produce low-sulfur products (Tables 15.1 and 15.2), such as desulfurization of distillates (Fig. 15.2), and in hydroconversion processes, such as single-stage or two-stage hydrocracking (Fig. 15.3). In fact, the use of hydrogen in refinery processes is perhaps the single most significant advance in refining technology during the 20th century. The process uses the principle that the presence of hydrogen during a thermal reaction of a crude oil feedstock will terminate many of the coke-forming reactions and enhance the yields of the lower-boiling components such as naphtha, kerosene, and gas oil (Speight, 2000, 2011, 2014, 2016, 2017; Parkash, 2003; Ancheyta and Speight, 2007; Gary et al., 2007; Hsu and Robinson, 2017).

The influx of the more viscous (low density) feedstocks into refineries and evolving environmental regulations are leading to stricter product-quality requirements. In order to improve the oil product quality, refineries have been obligated to increase the depth of hydrotreating and hydrocracking processes, which consume large amount of fresh hydrogen (Kriz and Ternan, 1994; Speight, 2014; Farnand et al., 2015). On the other hand, hydrogen produced by the traditional path of naphtha reforming, an important hydrogen producing process, is no longer adequate to the needs of the modern refinery with the influx of heavier feedstocks and other processes that are necessary (Lipman, 2011; Speight, 2011, 2014), and the gap between hydrogen consuming processes and hydrogen producing processes has emphasized the potential (and often real) shortage of hydrogen in refineries.

Furthermore, hydrogen is a key to enabling modern refineries to comply with the latest product specifications and environmental requirements for fuel production that have been (and continue to be) mandated by the market and governments. Hydrogen is an expensive commodity in a refinery—a situation that

Heavy Oil Recovery and Upgrading. https://doi.org/10.1016/B978-0-12-813025-4.00015-5

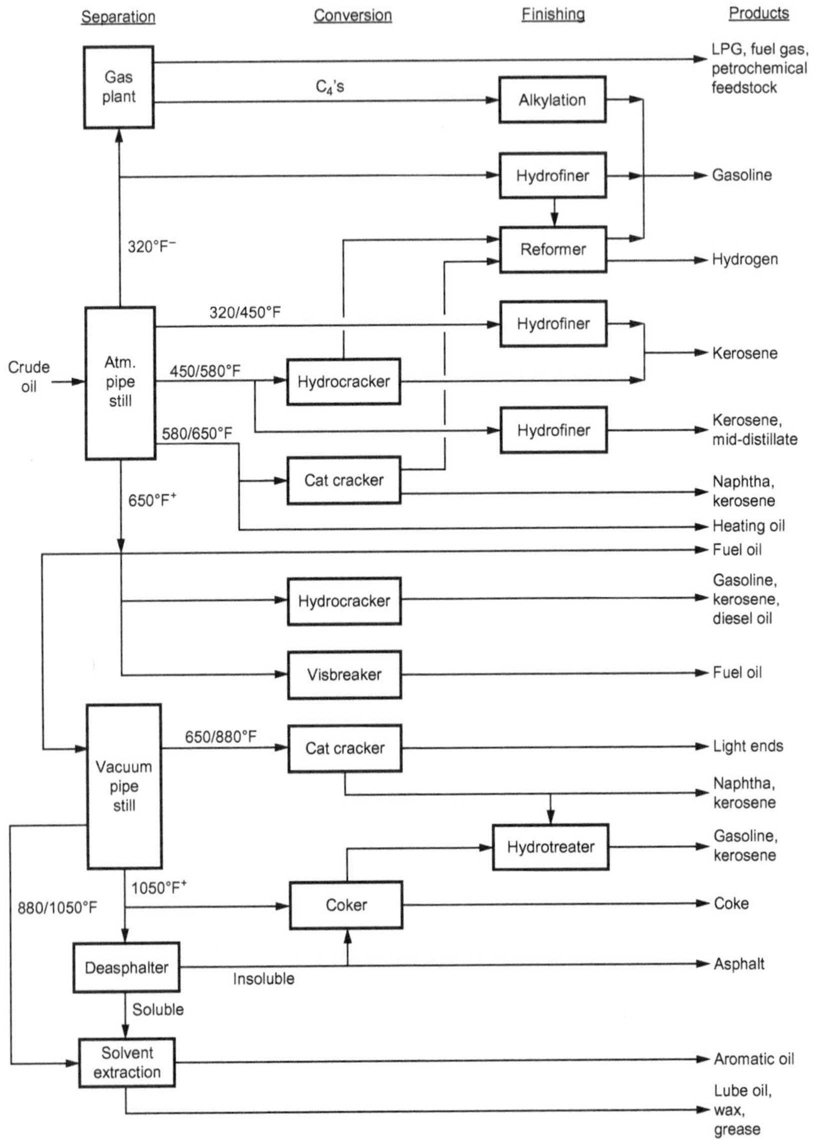

FIG. 15.1 Generalized refinery layout showing relative placement of the units (hydrofiner and hydrocracker) that require hydrogen.

is emphasized by the costs of creating low-sulfur fuels from heavier, sourer crude oils and tar sand bitumen that are supplanting many of the relatively scarce low-density low-sulfur crude oils as refinery feedstocks. Therefore, it is appropriate at this point to present the means by which hydrogen is produced

TABLE 15.1 Hydrogen Content of Various Liquid Fuels

	[H] % w/w
Crude oil	11–14
Heavy oil	9–11
Extra heavy oil	9–11
Tar sand bitumen	9–11
Residua, straight run	9–11
Residua, cracked	8–10
Naphtha[a]	13–15
Kerosene[a]	13–14

[a]*Included for comparison.*

TABLE 15.2 Typical Hydrogen Applications in a Refinery

Naphtha hydrotreater
• Uses hydrogen to desulfurize naphtha from atmospheric distillation; must hydrotreat the naphtha before sending to a catalytic reformer unit
Distillate hydrotreater
• Uses hydrogen to desulfurize distillates after atmospheric or vacuum distillation; in some units, aromatics are hydrogenated to cycloparaffins or alkanes
Hydrodesulfurization
• Sulfur compounds are hydrogenated to hydrogen sulfide (H_2S) as feed for Claus plants
Hydroisomerization
• Normal (straight-chain) paraffins are converted into isoparaffins to improve the product properties (e.g., octane number)
Hydrocracker
• Uses hydrogen to upgrade heavier fractions into lighter, more valuable products

in a refinery. The necessary details are given, but for further details, consultation of other works is suggested (Parkash, 2003; Gary et al., 2007; Speight, 2014, 2017; Hsu and Robinson, 2017).

Environmental restrictions, new transportation fuel specifications, and increased processing of heavier more-sour curds are leading substantial

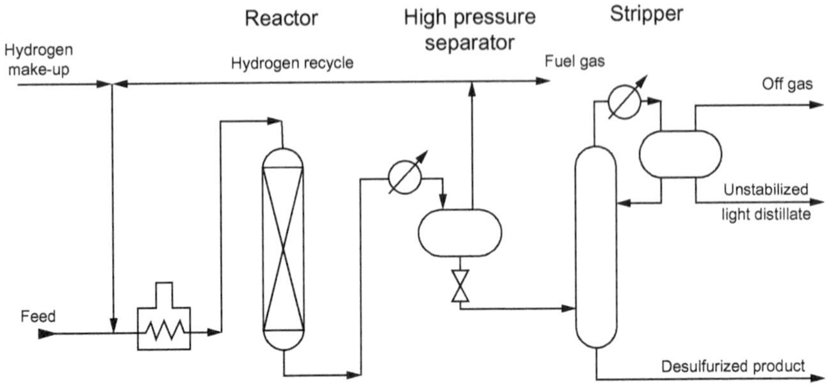

FIG. 15.2 Distillate hydrodesulfurization. *(From OSHA Technical Manual, Section IV, Chapter 2: Petroleum Refining Processes. http://www.osha.gov/dts/osta/otm/otm_iv/otm_iv_2.html.)*

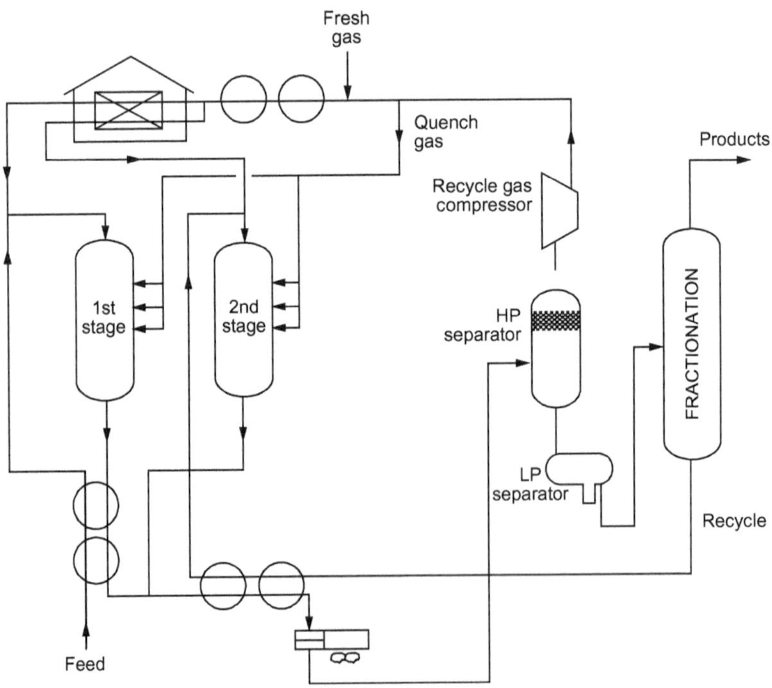

FIG. 15.3 A single-stage or two-stage (optional) hydrocracking unit. *(From OSHA Technical Manual, Section IV, Chapter 2: Petroleum Refining Processes. http://www.osha.gov/dts/osta/otm/otm_iv/otm_iv_2.html.)*

increases in refinery hydrogen consumption for hydrodesulfurization, aromatic and olefin saturation, and improvement of product quality and reduction in refinery hydrogen production from catalytic reformers as a by-product. Therefore, the above factors make hydrogen management a critical issue. Generating, recovering, and purchasing of hydrogen have significant impact on refinery operating costs. More important, overall refinery operations can be severely constrained by the unavailability of hydrogen. Primary consideration, however, should be given to the recovery of hydrogen that is contained in various purge gases, since this is a very attractive way from the viewpoint of refinery balance and economics (Cruz and De Oliveira Junior, 2008).

Typically, in the early refineries, the hydrogen for hydroprocesses was provided as a result of reforming processes, such as the Platforming process (Speight, 2000, 2011, 2014, 2016, 2017; Parkash, 2003; Ancheyta and Speight, 2007; Gary et al., 2007; Hsu and Robinson, 2017). Dehydrogenation is a main chemical reaction in catalytic reforming, and hydrogen gas is consequently produced. The hydrogen is recycled through the reactors where the reforming takes place to provide the atmosphere necessary for the chemical reactions and also prevents the carbon from being deposited on the catalyst, thus extending its operating life. An excess of hydrogen above whatever is consumed in the process is produced, and as a result, catalytic reforming processes are unique in that they are the only crude oil refinery processes to produce hydrogen as a by-product. However, as refineries and refinery feedstocks evolved during the last four decades, the demand for hydrogen has increased, and reforming processes are no longer capable of providing the quantities of hydrogen necessary for feedstock hydrogenation. Within the refinery, other processes are used as sources of hydrogen. Thus, the recovery of hydrogen from the by-products of the coking units, visbreakers, and catalytic crackers is also practiced in some refineries. Although the hydrogen gas produced by such processes usually contained up to 40% v/v of other gases (usually hydrocarbon derivatives), hydrotreater catalyst life is a function of hydrogen partial pressure. Optimum hydrogen purity at the reactor inlet extends catalyst life by maintaining desulfurization kinetics at lower operating temperatures and reducing carbon laydown. An increase in the concentration of hydrogen in the hydrogen-containing gas resulting from the application of hydrogen purification equipment and/or increased hydrogen sulfide removal and careful management of hydrogen circulation and purge rates can extend catalyst life by as much as 25%.

A critical issue facing the modern refinery is the changing slate of feedstocks that need to be refined into transportation fuels under an environment of increasingly more stringent clean fuel regulations and increasingly viscous, sour (high-sulfur) crude supply. Hydrogen network optimization is at the forefront of world refinery options to address clean fuel trends, to meet growing demands for transportation fuels, and to continue to make a profit from their crudes (Deng et al., 2013). A key element of a hydrogen network analysis in a refinery involves the capture of hydrogen in its fuel streams and extending its flexibility and processing options. Thus, optimization of the hydrogen network will be a critical factor influencing future operating flexibility and profitability of a refinery in a world of an increasing slate of viscous feedstocks and the demand for ultra-low-sulfur (ULS) gasoline and ultra-low-sulfur diesel fuel.

As hydrogen use has become more widespread in refineries, hydrogen production has moved from the status of a high-tech specialty operation to an integral feature of most refineries (Raissi, 2001; Vauk et al., 2008). This has been made necessary by the increase in hydrotreating and hydrocracking, including the treatment of progressively heavier feedstocks (Speight, 2000, 2011, 2014, 2017; Parkash, 2003; Ancheyta and Speight, 2007; Gary et al., 2007; Hsu and Robinson, 2017). In fact, the use of hydrogen in thermal processes is perhaps the single most significant advance in refining technology during the 20th century (Scherzer and Gruia, 1996; Bridge, 1997; Dolbear, 1998). The continued increase in hydrogen demand over the last several decades is a result of the conversion of crude oil to match changes in product slate and the supply of viscous, high-sulfur oil, and in order to make lower-boiling, cleaner, and more salable products. There are also many reasons other than product quality for using hydrogen in processes adding to the need to add hydrogen at relevant stages of the refining process and, most important, according to the availability of hydrogen (Bezler, 2003; Miller and Penner, 2003; Ranke and Schödel, 2003).

In visbreaking units and in coking units (Chapter 8), viscous feedstocks are converted to crude oil coke, oil, and low-boiling hydrocarbon derivatives (such as liquefied petroleum gas and naphtha) (Parkash, 2003; Ancheyta and Speight, 2007; Gary et al., 2007; Speight, 2014, 2017). Depending on the process, hydrogen is present in a wide range of concentrations. Since crude oil coking processes need gas for heating purposes, adsorption processes are best suited to recover the hydrogen because they feature a very clean hydrogen product and an off-gas suitable as fuel.

Catalytic cracking is the most important process step for the production of low-boiling products from gas oil and increasingly from vacuum gas oil and viscous feedstocks (Parkash, 2003; Ancheyta and Speight, 2007; Gary et al., 2007; Speight, 2014, 2017; Hsu and Robinson, 2017) (Chapter 9). In catalytic cracking, the molecular mass of the main fraction of the feed is lowered, while another part is converted to coke that is deposited on the hot catalyst. The catalyst is regenerated in one or two stages by burning the coke off with air that also provides the energy for the endothermic cracking process. In the process, paraffins and naphthenes are cracked to olefins and to alkanes with shorter chain length, monoaromatic compounds are dealkylated without ring cleavage, and diaromatic and polyaromatic compounds are dealkylated and converted to coke. Hydrogen is formed in the last type of reaction, whereas the first two reactions produce low-boiling hydrocarbon derivatives and therefore require hydrogen. Thus, a catalytic cracker can be operated in such a manner that enough hydrogen for subsequent processes is formed.

In reforming processes, naphtha fractions are reformed to improve the quality of naphtha (Speight, 2014). The most important reactions occurring during this process are the dehydrogenation of naphthenes to aromatics. This reaction is endothermic and is favored by low pressures, and the reaction temperature lies in the range of 300–450°C (570–840°F). The reaction is performed on platinum catalysts, with other metals, for example, rhenium, as promoters.

The chemical nature of the crude oil used as the refinery feedstock has always played the major role in determining the hydrogen requirements of that refinery. For example, the lower-density more paraffinic crude oils will require somewhat less hydrogen than the amount required by viscous feedstocks for upgrading (Table 15.1). It follows that the hydrodesulfurization of viscous feedstocks (which, by definition, is a hydrogen-dependent process) needs substantial amounts of hydrogen as part of the processing requirements.

In general, considerable variation exists from one refinery to another in the balance between hydrogen produced and hydrogen consumed in the refining operations. However, what is more pertinent to the present text is the excessive amounts of hydrogen that are required for hydroprocessing operations—whether these be hydrocracking or the somewhat milder hydrotreating processes. For effective hydroprocessing, a substantial hydrogen partial pressure must be maintained in the reactor, and in order to meet this requirement, an excess of hydrogen above that actually consumed by the process must be fed to the reactor. Part of the hydrogen requirement is met by recycling a stream of hydrogen-rich gas. However, the need still remains to generate hydrogen as makeup material to accommodate the process consumption of 500–3000 scf/bbl depending upon whether the viscous feedstock is being subjected to a predominantly hydrotreating process (hydrodesulfurization) or to a predominantly hydrocracking process.

Hydrogen is generated in a refinery by the catalytic reforming process, but there may not always be the need to have a catalytic reformer as part of the refinery sequence. Nevertheless, assuming that a catalytic reformer is part of the refinery sequence, the hydrogen production from the reformer usually falls well below the amount required for hydroprocessing purposes. For example, in a 100,000 bbl/day hydrocracking refinery, assuming intensive reforming of hydrocracked naphtha, the hydrogen requirements of the refinery may still fall some 500–900 scf/bbl of crude charge below that necessary for the hydrocracking sequences. Consequently, an *external* source of hydrogen is necessary to meet the daily hydrogen requirements of any process where the heavier feedstocks are involved.

The trend to increase the number of hydrogenation (*hydrocracking* and/or *hydrotreating*) processes in refineries (Dolbear, 1998) coupled with the need to process the heavier oils, which require substantial quantities of hydrogen for upgrading, has resulted in vastly increased demands for this gas. Part of the hydrogen requirements can be satisfied by hydrogen recovery from catalytic reformer product gases, but other external sources are required. Most of the external hydrogen is manufactured either by steam-methane reforming or by oxidation processes. However, other processes, such as steam-methanol interaction or ammonia dissociation, may also be used as sources of hydrogen. Electrolysis of water produces high-purity hydrogen, but the power costs may be prohibitive.

With the increasing need for *clean* fuels, the production of hydrogen for refining purposes requires a major effort by refiners. The hydrogen demands can be estimated to a very rough approximation using API gravity and the extent of the reaction, particularly the hydrodesulfurization reaction (Speight, 2000, 2014, 2017; Parkash, 2003; Gary et al., 2007; Hsu and Robinson, 2017). But accurate estimation requires equivalent process parameters and a thorough understanding of the nature of ach process. Thus, as hydrogen production grows, a better understanding of the capabilities and requirements of a hydrogen plant becomes ever more important to overall refinery operations as a means of making the best use of hydrogen supplies in the refinery.

This has led to a variety of innovations in processes such as (i) hydrotreating processes, (ii) hydrocracking processes, and (iii) slurry hydrocracking processes to accommodate changes to the chemical nature of the crude oil used as the refinery feedstock (which has always played the major role in determining the hydrogen requirements of a refinery) and a variety of processes developed to accommodate the heavier feedstocks.

Hydrogen has historically been produced during catalytic reforming processes as a by-product of the production of the aromatic compounds used in naphtha and in solvents. As reforming processes changed from fixed bed, to cyclic, to continuous regeneration, process pressures have dropped, and hydrogen production per barrel of reformate has tended to increase. However, hydrogen production as a by-product is not always adequate to the needs of the refinery, and other processes are necessary. Thus, hydrogen production by steam reforming or by partial oxidation of residua has also been used, particularly where viscous oil is available. Steam reforming is the dominant method for hydrogen production and is usually combined with pressure swing adsorption (PSA) to purify the hydrogen to >99% by volume (Bandermann and Harder, 1982).

In the steam reforming process, there are typically four basic sections: (i) feedstock treatment where sulfur and other contaminants are removed; (ii) the steam-methane reformer, which converts feedstock and steam to synthesis gas at high temperature and moderate pressure; (iii) the synthesis gas heat recovery that incorporates shift reactor(s) to increase the hydrogen yield; and (iv) the hydrogen purification section that is typically by a pressure swing adsorption (PSA) unit, with carbon monoxide, carbon dioxide, and methane to achieve the final product purity.

The reforming reaction between steam and hydrocarbon derivatives is highly endothermic and is carried out using specially formulated nickel catalyst contained in vertical tubes situated in the radiant section of the reformer. The simplified chemical reactions are as follows:

$$C_nH_{2n+2} + nH_2O \rightarrow nCO + (2n+1)H_2 \text{ (for saturated hydrocarbon (derivatives)}$$

$$CH_4 + H_2O \rightarrow CO + 3H_2 \text{ (for methane)}$$

In the adiabatic CO shift reactor vessel, the moderately exothermic *water-gas shift reaction* converts carbon monoxide and steam to carbon dioxide and hydrogen:

$$CO + H_2O \rightarrow CO_2 + H_2$$

The gasification of residua and coke to produce hydrogen and/or power may become an attractive option for refiners (Fleshman, 1997; Dickenson et al., 1997; Gross and Wolff, 2000). The premise that the gasification section of a refinery will be the *garbage can* for deasphalter residues, high-sulfur coke, and other refinery wastes is worthy of consideration.

In the process, air is fed directly to air-blown gasifiers, while other gasifiers are fed with high-purity (~99.5%) oxygen from an air separation unit (Speight, 2013, 2014). Steam is also fed directly to some dry-feed gasifiers, while others are fed with coal or coke in a slurry with water. Reaction conditions typically vary in pressure from 430 to 1200 psi and in reaction outlet temperature (up to ~1480°C, 2700°F). The product composition varies, depending upon the selected gasification technology and the characteristics of the crude oil coke. The process generally produces synthesis (syngas) with typical H_2/CO molar ratios of less than one. Carbon dioxide, water, and methane are the secondary product components at lower concentrations relative to those of hydrogen or carbon monoxide. In addition, the synthesis gas is contaminated with hydrogen sulfide, carbon dioxide, carbonyl sulfide (COS), and other sulfur compounds. Raw synthesis gas from the gasifier can be cooled through a steam generation system or a quench and scrubbing section, where the raw syngas becomes saturated with water at a target temperature adequate for converting carbon monoxide and water to hydrogen and carbon dioxide (*shift reaction*):

$$CO + H_2O \rightarrow H_2 + CO_2$$

Several other processes are available for the production of the additional hydrogen that is necessary for the various viscous feedstock hydroprocessing sequences, and it is the purpose of this chapter to present a general description of these processes. In general, most of the external hydrogen is manufactured by steam-methane reforming or by oxidation processes. Other processes such as ammonia dissociation, steam-methanol interaction, or electrolysis are also available for hydrogen production, but economic factors and feedstock availability assist in the choice between processing alternatives.

The processes described in this chapter are those gasification processes that are often referred to the *garbage disposal units* of the refinery. Hydrogen is produced for use in other parts of the refinery and for energy, and it is often produced from process by-products that may not be of any use elsewhere. Such by-products might be the highly aromatic, heteroatom, and metal-containing reject from a deasphalting unit or from a mild hydrocracking process. However attractive this may seem, there will be the need to incorporate a gas cleaning operation to remove any environmentally objectionable components from the hydrogen gas.

The gasification of residua and coke to produce hydrogen and/or power may become an attractive option for refiners (Fleshman, 1997; Dickenson et al., 1997). The premise that the gasification section of a refinery will be the *garbage can* for deasphalter residues, high-sulfur coke, as well as other refinery wastes is worthy of consideration.

Thus, the refining industry has been the subject of the four major forces that affect most industries and that have hastened the development of new crude oil refining processes: (i) the demand for products such as gasoline, diesel, fuel oil, and jet fuel; (ii) feedstock supply, specifically the changing quality of crude oil and geopolitics between different countries and the emergence of alternate feed supplies such as bitumen from tar sand (*oil sand*), natural gas, and coal; (iii) technology development such as new catalysts and processes, especially processes involving the use of hydrogen, and (iv) environmental regulations that include more stringent regulations in relation to sulfur in gasoline and diesel (Parkash, 2003; Ancheyta and Speight, 2007; Gary et al., 2007; Speight, 2014, 2017; Hsu and Robinson, 2017).

Categories 1, 2, and 4 are directly affected by the third category (i.e., the use of hydrogen in refineries), and it is this category that will be the subject of this chapter. This chapter presents an introduction to the use and need for hydrogen crude oil refineries in order for the reader to place the use of hydrogen in the correct context of the refinery.

Thus, the focus of this chapter is the means by which hydrogen is produced in refineries with particular attention paid to (i) the types of feedstocks used for hydrogen production in a refinery, (ii) commercial processes, (iii) process catalysts, (iv) the purification of hydrogen, and (v) hydrogen management and safety.

2. Feedstocks

There are several processes by which hydrogen can be produced in a refinery (Table 15.3). The most common, and perhaps the best, feedstocks for steam reforming are low-boiling saturated hydrocarbon derivatives that have a low sulfur content, including natural gas, refinery gas, liquefied petroleum gas (LPG), and low-boiling naphtha.

TABLE 15.3 Typical Hydrogen Production Processes in a Refinery

Catalytic reformer
• Used to convert the naphtha-boiling-range molecules into higher octane reformate; hydrogen is a by-product
Steam-methane reformer
• Produces hydrogen for the hydrotreaters or hydrocracker
Steam reforming of higher-molecular-weight hydrocarbon derivatives
• Produces hydrogen from low-boiling hydrocarbon derivatives other than methane
Recovery from refinery off-gases
• Process gas often contains hydrogen in the range of up to 50% v/v
Gasification of petroleum residua
• Recovery from synthesis gas (syngas) produced in gasification units
Gasification of petroleum coke
• Recovery from synthesis gas (syngas) produced in gasification units
Partial oxidation processes
• Analogous to gasification process; produce synthesis gas from which hydrogen can be isolated

Natural gas is the most common feedstock for hydrogen production since it meets all the requirements for reformer feedstock. Natural gas typically contains >90% v/v methane and ethane with only a few percent of propane and higher-boiling hydrocarbon derivatives (Mokhatab et al., 2006; Speight, 2007). Natural gas may (or most likely will) contain traces of carbon dioxide with some nitrogen and other impurities. Purification of natural gas, before reforming, is usually relatively straight forward. Traces of sulfur must be removed to avoid poisoning the reformer catalyst; zinc oxide treatment in combination with hydrogenation is usually adequate.

Refinery gas (refinery off-gas) containing a substantial amount of hydrogen can be an attractive steam reformer feedstock since it is produced as a by-product. Processing of refinery gas will depend on its composition, particularly the levels of olefins and of propane and heavier hydrocarbon derivatives. Olefins that can cause problems by forming coke in the reformer are converted to saturated compounds in the hydrogenation unit. Higher-boiling hydrocarbon derivatives in refinery gas can also form coke, either on the primary reformer catalyst or in the preheater. If there is more than a few percent of C_3 and higher compounds, a promoted reformer catalyst should be considered, in order to avoid carbon deposits.

Refinery gas from different sources varies in suitability as hydrogen plant feed. Catalytic reformer off-gas (Parkash, 2003; Ancheyta and Speight, 2007; Gary et al., 2007; Speight, 2014, 2017; Hsu and Robinson, 2017), for example, is saturated, is very low in sulfur, and often has high hydrogen content. The process gases from a coking unit or from a fluid catalytic cracking unit are much less desirable because of the content of unsaturated constituents. In addition to olefins, these gases contain substantial amounts of sulfur that must be removed before the gas is used as feedstock. These gases are also generally unsuitable for direct hydrogen recovery, since the hydrogen content is usually too low. Hydrotreater off-gas lies in the middle of the range. It is saturated, so it is readily used as hydrogen plant feed. Content of hydrogen and heavier hydrocarbon derivatives depends to a large extent on the upstream pressure. Sulfur removal will generally be required.

Before the demand for hydrogen exceeded supply (from reforming processes), hydrogen-containing refinery gas was routed into the refinery fuel gas system where only the heating value of the gas has been used. Since the hydrogen demand for refinery operations is growing, these gases become more and more attractive as a source for hydrogen production. This requires purification steps such as pressure swing adsorption (PSA) or membrane systems (Mokhatab et al., 2006; Speight, 2007).

The hydrogen content of refinery off-gas varies from 2% to 10% v/v and higher, depending upon the source of the off-gas—the utilization of off-gas in the hydrogen production scheme of a refinery must be evaluated on a case-by-case basis. When applied, the recovery of hydrogen from off-gas using pressure swing adsorption (PSA) technology, membrane technology, or cryogenic processes can be applied to generate hydrogen streams of any required purity (Mokhatab et al., 2006; Speight, 2007, 2014). The off-gas streams from the various processes can be routed to the refinery fuel gas system. Another option (especially if the hydrogen content of the refinery off-gas is low) is to use the off-gas as (supplementary) feedstock to a steam reformer plant, which also generates hydrogen from hydrocarbon feedstocks such as natural gas, liquefied petroleum gas, or naphtha.

3. Process Chemistry

Before the feedstock is introduced to a process, there is the need for application of a strict feedstock purification protocol. Prolonging catalyst life in hydrogen production processes is attributable to effective feedstock purification, particularly sulfur removal. A typical natural gas or other low-boiling hydrocarbon feedstock contains traces of hydrogen sulfide and organic sulfur.

In order to remove sulfur compounds, it is necessary to hydrogenate the feedstock to convert the organic sulfur to hydrogen that is then reacted with zinc oxide (ZnO) at ~370°C (700°F) that results in the optimal use of the zinc oxide and ensuring complete hydrogenation. Thus, assuming assiduous feedstock purification and removal of all of the objectionable contaminants, the chemistry of hydrogen production can be defined.

In *steam reforming*, low-boiling hydrocarbon derivatives such as methane are reacted with steam to form hydrogen:

$$CH_4 + H_2O \rightarrow 3H_2 + CO \quad \Delta H_{298K} = +97,400 \text{Btu/lb}$$

H is the heat of reaction. A more general form of the equation that shows the chemical balance for higher-boiling hydrocarbon derivatives is

$$C_nH_m + nH_2O \rightarrow (n + m/2)H_2 + nCO$$

The reaction is typically carried out at ~815°C (1500°F) over a nickel catalyst packed into the tubes of a reforming furnace. The high temperature also causes the hydrocarbon feedstock to undergo a series of cracking reactions, plus the reaction of carbon with steam:

$$CH_4 \rightarrow 2H_2 + C$$

$$C + H_2O \rightarrow CO + H_2$$

Carbon is produced on the catalyst at the same time that hydrocarbon is reformed to hydrogen and carbon monoxide. With natural gas or similar feedstock, reforming predominates, and the carbon can be removed by reaction with steam as fast as it is formed. When higher-boiling feedstocks are used, the carbon is not removed fast enough and builds up, thereby requiring catalyst regeneration or replacement. Carbon buildup on the catalyst (when high-boiling feedstocks are employed) can be avoided by the addition of alkali compounds, such as potash, to the catalyst, thereby encouraging or promoting the carbon-steam reaction.

However, even with an alkali-promoted catalyst, feedstock cracking limits the process to hydrocarbon derivatives with a boiling point less than of 180°C (350°F). Natural gas, propane, butane, and low-boiling naphtha are most suitable. Prereforming, a process that uses an adiabatic catalyst bed operating at a lower temperature, can be used as a pretreatment to allow heavier feedstocks to be used with lower potential for carbon deposition (coke formation) on the catalyst.

After reforming, the carbon monoxide in the gas is reacted with steam to form additional hydrogen (the *water-gas shift* reaction):

$$CO + H_2O \rightarrow CO_2 + H_2 \quad \Delta H_{298K} = -16,500 \text{Btu/lb}$$

This leaves a mixture consisting primarily of hydrogen and carbon monoxide that is removed by conversion to methane:

$$CO + 3H_2O \rightarrow CH_4 + H_2O$$

$$CO_2 + 4H_2 \rightarrow CH_4 + 2H_2O$$

The critical variables for steam reforming processes are (i) temperature, (ii) pressure, and (iii) the steam/hydrocarbon ratio. Steam reforming is an equilibrium reaction, and conversion of the hydrocarbon feedstock is favored by high temperature, which in turn requires higher fuel use. Because of the volume increase in the reaction, conversion is also favored by low pressure, which conflicts with the need to supply the hydrogen at high pressure. In practice, materials of construction limit temperature and pressure.

On the other hand and in contrast to reforming, shift conversion is favored by low temperature. The gas from the reformer is reacted over iron oxide catalyst at 315–370°C (600–700°F) with the lower limit being dictated activity of the catalyst at low temperature.

Hydrogen can also be produced by *partial oxidation* (POX) of hydrocarbon derivatives in which the hydrocarbon is oxidized in a limited or controlled supply of oxygen:

$$2CH_4 + O_2 \rightarrow CO + 4H_2 \quad \Delta H_{298K} = -10,195\,\text{Btu/lb}$$

The shift reaction also occurs, and a mixture of carbon monoxide and carbon dioxide is produced in addition to hydrogen. The catalyst tube materials do not limit the reaction temperatures in partial oxidation processes, and higher temperatures may be used to enhance the conversion of methane to hydrogen. Indeed, much of the design and operation of hydrogen plants involves protecting the reforming catalyst and the catalyst tubes because of the extreme temperatures and the sensitivity of the catalyst. In fact, minor variations in feedstock composition or operating conditions can have significant effects on the life of the catalyst or the reformer itself. This is particularly true of changes in molecular weight of the feed gas or poor distribution of heat to the catalyst tubes.

Since the high temperature takes the place of a catalyst, partial oxidation is not limited to the lower-boiling feedstocks that are required for steam reforming. Partial oxidation processes were first considered for hydrogen production because of expected shortages of lower-boiling feedstocks and the need to have available a disposal method for higher-boiling, high-sulfur streams such as asphalt or crude oil coke.

Catalytic partial oxidation, also known as autothermal reforming, reacts oxygen with a low-boiling feedstock and by passing the resulting hot mixture over a reforming catalyst. The use of a catalyst allows the use of lower temperatures than in noncatalytic partial oxidation, which causes a reduction in oxygen demand.

The feedstock requirements for catalytic partial oxidation processes are similar to the feedstock requirements for steam reforming, and low-boiling hydrocarbon derivatives from refinery gas to naphtha are preferred. The oxygen substitutes for much of the steam in preventing coking and a lower steam/carbon ratio are required. In addition, because a large excess of steam is not required, catalytic partial oxidation produces more carbon monoxide and less hydrogen than steam reforming. Thus, the process is more suited to situations where carbon monoxide is the more desirable product, for example, as synthesis gas for chemical feedstocks.

4. Commercial Processes

In spite of the use of low-quality hydrogen (that contains up to 40% by volume hydrocarbon gases), a high-purity hydrogen stream (95%–99% by volume hydrogen) is required for hydrodesulfurization, hydrogenation, hydrocracking, and petrochemical processes. Hydrogen, produced as a by-product of refinery processes (principally hydrogen recovery from catalytic reformer product gases), often is not enough to meet the total refinery requirements, necessitating the manufacturing of additional hydrogen or obtaining supply from external sources.

Catalytic reforming remains an important process used to convert low-octane naphtha into high-octane gasoline blending components called *reformate*. Reforming represents the total effect of numerous reactions such as cracking, polymerization, dehydrogenation, and isomerization taking place simultaneously. Depending on the properties of the naphtha feedstock (as measured by the paraffin, olefin, naphthene, and aromatic content) and catalysts used, reformate can be produced with very high concentrations of toluene, benzene, xylene, and other aromatics useful in gasoline blending and petrochemical processing. Hydrogen, a significant by-product, is separated from reformate for recycling and use in other processes.

A catalytic reformer comprises a reactor section and a product recovery section. More or less standard is a feed preparation section in which, by combination of hydrotreatment and distillation, the feedstock is prepared to specification. Most processes use platinum as the active catalyst. Sometimes platinum is combined with a second catalyst (bimetallic catalyst) such as rhenium or another noble metal. There are many different commercial catalytic reforming processes including the Platforming process, the Powerforming process, the Ultraforming process, and the Thermofor catalytic reforming process (Parkash, 2003; Ancheyta and Speight, 2007; Gary et al., 2007; Speight, 2014, 2017; Hsu and Robinson, 2017).

In the Platforming process, the first step is preparation of the naphtha feed to remove impurities from the naphtha and reduce catalyst degradation. The naphtha feedstock is then mixed with hydrogen, vaporized, and passed through a series of alternating furnace and fixed-bed reactors containing a platinum catalyst. The effluent from the last reactor is cooled and sent to a separator to permit removal of the hydrogen-rich gas stream from the top of the separator for recycling. The liquid product from the bottom of the separator is sent to a fractionator called a stabilizer (butanizer), and the bottom product (reformate) is sent to storage, and butanes and low-boiling gases pass overhead and are sent to the saturated gas plant.

Some catalytic reformers operate at low pressure (50–200 psi), and others operate at high pressures (up to 1000 psi). Some catalytic reforming systems continuously regenerate the catalyst in other systems. One reactor at a time is taken offstream for catalyst regeneration, and some facilities regenerate all of the reactors during turnarounds. Operating procedures should be developed to ensure control of hot spots during start-up. Safe catalyst handling is very important, and care must be taken not to break or crush the catalyst when loading the beds, as the small fines will plug up the reformer screens. Precautions against dust when regenerating or replacing catalyst should also be considered, and a water wash should be considered where stabilizer fouling has occurred due to the formation of ammonium chloride and iron salts. Ammonium chloride may form in pretreater exchangers and cause corrosion and fouling. Hydrogen chloride from the hydrogenation of chlorine compounds may form acid or ammonium chloride salt.

4.1 Heavy Residue Gasification and Combined Cycle Power Generation

Residua from various processes are the preferred feedstocks for the production of hydrogen-rich gases. Such fractions with high sulfur and/or high heavy metal contents are difficult to handle in upgrading processes such as hydrogenation or coking and, for environmental reasons, are not usually used as fuels without extensive gas cleanup. Residua are gasified using the Texaco gasification process (partial oxidation process), and the produced gas is purified to clean fuel gas, and then, electric power is generated applying combined cycle system. The convenience of the process depends upon the availability of residua (or coal) in preference to the availability of natural gas that is used for hydrogen production by steam-methane reforming (Bailey, 1992).

Viscous feedstocks are gasified, and the produced gas is purified to clean fuel gas (Gross and Wolff, 2000). As an example, solvent deasphalter residuum is gasified by partial oxidation method under pressure of about 570 psi and at a temperature between 1300°C and 1500°C (2370 and 2730°F). The high temperature generated gas flows into the specially designed waste heat boiler, in which the hot gas is cooled and high-pressure saturated steam is generated. The gas from the waste heat boiler is then heat exchanged with the fuel gas and flows to the carbon scrubber, where unreacted carbon particles are removed from the generated gas by water scrubbing.

The gas from the carbon scrubber is further cooled by the fuel gas and boiler feed water and led into the sulfur compound removal section, where hydrogen sulfide (H_2S) and carbonyl sulfide (COS) are removed from the gas to obtain clean fuel gas. This clean fuel gas is heated with the hot gas generated in the gasifier and finally supplied to the gas turbine at a temperature of 250–300°C (480–570°F).

The exhaust gas from the gas turbine having a temperature of about 550–600°C (1020–1110°F) flows into the heat recovery steam generator consisting of five heat exchange elements. The first element is a superheater in which the combined stream of the high-pressure saturated steam generated in the waste heat boiler and in the second element (high-pressure steam evaporator) is superheated. The third element is an economizer, the fourth element is a low-pressure steam evaporator, and the final or the fifth element is a deaerator heater. The off-gas from heat recovery steam generator having a temperature of about 130°C is emitted into the air via stack.

In order to decrease the nitrogen oxide (NO_x) content in the flue gas, two methods can be applied. The first method is the injection of water into the gas turbine combustor. The second method is to selectively reduce the nitrogen oxide content by injecting ammonia gas in the presence of de-NO_x catalyst that is packed in a proper position of the heat recovery steam generator. The latter is more effective than the former to lower the nitrogen oxide emissions to the air.

4.2 Hybrid Gasification Process

In the hybrid gasification process, a slurry of coal and residual oil is injected into the gasifier where it is pyrolyzed in the upper part of the reactor to produce gas and chars. The chars produced are then partially oxidized to ash. The ash is removed continuously from the bottom of the reactor.

In this process, coal and vacuum residue are mixed together into slurry to produce clean fuel gas. The slurry fed into the pressurized gasifier is thermally cracked at a temperature of 850–950°C (1560–1740°F) and is converted into gas, tar, and char. The mixture oxygen and steam in the lower zone of the gasifier gasify the char. The gas leaving the gasifier is quenched to a temperature of 450°C (840°F) in the fluidized-bed heat exchanger and is then scrubbed to remove tar, dust, and steam at around 200°C (390°F).

The coal and residual oil slurry is gasified in the fluidized-bed gasifier. The charged slurry is converted to gas and char by thermal cracking reactions in the upper zone of the fluidized bed. The produced char is further gasified with steam and oxygen that enter the gasifier just below the fluidizing gas distributor. Ash is discharged from the gasifier and indirectly cooled with steam and then discharged into the ash hopper. It is burned with an incinerator to produce process steam. Coke deposited on the silica sand is removed in the incinerator.

4.3 Hydrocarbon Gasification

The gasification of hydrocarbon derivatives to produce hydrogen is a continuous, noncatalytic process that involves partial oxidation of the hydrocarbon. Air or oxygen (with steam or carbon dioxide) is used as the oxidant at 1095–1480°C (2000–2700°F). Any carbon produced (2%–3% by weight of the feedstock) during the process is removed as a slurry in a carbon separator and pelletized for use either as a fuel or as raw material for carbon-based products.

4.4 Hypro Process

The Hypro process is a continuous catalytic process method for hydrogen manufacture from natural gas or from refinery effluent gases. The process is designed to convert natural gas:

$$CH_4 \rightarrow C + 2H_2$$

Hydrogen is recovered by phase separation to yield hydrogen of about 93% purity; the principal contaminant is methane.

4.5 Pyrolysis Processes

There has been recent interest in the use of pyrolysis processes to produce hydrogen. Specifically, the interest has focused on the pyrolysis of methane (natural gas) and hydrogen sulfide.

Natural gas is readily available and offers relatively rich stream of methane with lower amounts of ethane, propane, and butane also present. The thermo-catalytic decomposition of a hydrocarbon feedstock, such as natural gas, offers an alternate method for the production of hydrogen (Uemura et al., 1999; Weimer et al., 2000):

$$C_nH_m \rightarrow nC + (m/2)H_2$$

If a hydrocarbon fuel such as natural gas (methane) is to be used for hydrogen production by direct decomposition, then the process that is optimized to yield hydrogen production may not be suitable for production of high-quality carbon black by-product intended for the industrial rubber market. Moreover, it appears that the carbon produced from high-temperature (850–950°C; 1560–1740°F) direct thermal decomposition of methane is sootlike material with high tendency for the catalyst deactivation (Murata et al., 1997). Thus, if the object of methane decomposition is hydrogen production, the carbon by-product may not be marketable as high-quality carbon black for rubber and tire applications.

The production of hydrogen by direct decomposition of hydrogen sulfide is also possible (Clark and Wassink, 1990; Zaman and Chakma, 1995; Donini, 1996; Luinstra, 1996). Hydrogen sulfide decomposition is a highly endothermic process, and equilibrium yields are poor (Clark et al., 1995) and can be represented simply as.

$$H_2S \rightarrow H_2 + S$$

At temperatures <1500°C (2730°F), the thermodynamic equilibrium is unfavorable toward hydrogen formation. However, in the presence of catalysts such as platinum-cobalt (at 1000°C; 1830°F), disulfides of molybdenum or tungsten Mo or W at 800°C (1470°F) (Kotera et al., 1976), or other transition metal sulfides supported on alumina (at 500–800°C; 930–1470°F), decomposition of hydrogen sulfide proceeds rapidly (Kiuchi, 1982; Bishara et al., 1987; Al-Shamma and Naman, 1989; Clark and Wassink, 1990; Megalofonos and Papayannakos, 1997; Arild, 2000; Raissi, 2001). In the temperature range of about 800–1500°C (1470–2730°F), thermolysis of hydrogen sulfide can be treated simply:

$$H_2S \rightarrow H_2 + 1/xS_x \quad \Delta H_{298K} = +34{,}300\,\text{Btu/lb}$$

where $x = 2$. Outside this temperature range, multiple equilibria may be present depending on temperature, pressure, and relative abundance of hydrogen and sulfur.

Above ~1000°C (1830°F), there is a limited advantage to using catalysts since the thermal reaction proceeds to equilibrium very rapidly (Clark and Wassink, 1990). The hydrogen yield can be doubled by preferential removal of either H2 or sulfur from the reaction environment, thereby shifting the equilibrium. The reaction products must be quenched quickly after leaving the reactor to prevent reversible reactions.

4.6 Shell Gasification Process

The shell gasification process (*partial oxidation process*) is a flexible process for generating synthesis gas, principally hydrogen and carbon monoxide, for the ultimate production of high-purity high-pressure hydrogen, ammonia, methanol, fuel gas, town gas, or reducing gas by reaction of gaseous or liquid hydrocarbon derivatives with oxygen, air, or oxygen-enriched air.

The most important step in converting a viscous residue to industrial gas is the partial oxidation of the oil using oxygen with the addition of steam. The gasification process takes place in an empty, refractory-lined reactor at temperatures of about 1400°C (2550°F) and pressures between 29 and 1140 psi. The chemical reactions in the gasification reactor proceed without catalyst to produce gas containing carbon amounting to some 0.5%–2% by weight, based on the feedstock. The carbon is removed from the gas with water, extracted in most cases with feed oil from the water, and returned to the feed oil. The high reformed gas temperature is utilized in a waste heat boiler for generating steam. The steam is generated at 850–1565 psi. Some of this steam is used as process steam and for oxygen and oil preheating. The surplus steam is used for energy production and heating purposes.

4.7 Steam-Methane Reforming

The steam-methane reforming process is the benchmark continuous catalytic process that has been employed over a period of several decades for hydrogen production (Speight, 2000, 2011, 2014, 2016, 2017; Parkash, 2003; Ancheyta and Speight, 2007; Gary et al., 2007; Hsu and Robinson, 2017). The process involves reforming natural gas in a continuous catalytic process in which the major reaction is the formation of carbon monoxide and hydrogen from methane and steam:

$$CH_4 + H_2O = CO + 3H_2 \quad \Delta H_{298K} = +97{,}400\,Btu/lb$$

In a similar manner, higher-molecular-weight feedstocks such as propane and even liquid hydrocarbon derivatives may also yield hydrogen:

$$C_3H_8 + 3H_2O = 3CO + 7H_2$$

The most important feedstock for the catalytic steam reforming process is natural gas. Other feedstocks are associated gas, propane, butane, liquefied crude oil gas, and some naphtha fractions. The choice is usually made on the availability and the price of the raw material.

In the actual process, the feedstock is first desulfurized by passage through activated carbon, which may be preceded by caustic and water washes. The desulfurized material is then mixed with steam and passed over a nickel-based catalyst (730–845°C, 1350–1550°F, and 400 psi). Effluent gases are cooled by the addition of steam or condensate to about 370°C (700°F), at which point carbon monoxide reacts with steam in the presence of iron oxide in a shift converter to produce carbon dioxide and hydrogen:

$$CO + H_2O = CO_2 + H_2 \quad \Delta H298K = -41.16\,kJ/mol$$

The carbon dioxide is removed by amine washing; the hydrogen is usually a high-purity (>99%) material.

Since the presence of any carbon monoxide or carbon dioxide in the hydrogen stream can interfere with the chemistry of the catalytic application, a third stage is used to convert of these gases to methane:

$$CO + 3H_2 \rightarrow CH_4 + H_2O$$

$$CO_2 + 4H_2 \rightarrow CH_4 + 2H_2O$$

For many refiners, sulfur-free natural gas (CH_4) is not always available to produce hydrogen by this process. In that case, higher-boiling hydrocarbon derivatives (such as propane, butane, or naphtha) may be used as the feedstock to generate hydrogen (q.v.).

The net chemical process for steam-methane reforming is then given by

$$CH_4 + 2H_2O \rightarrow CO_2 + 4H_2 \quad \Delta H298K = +165.2\,kJ/mol$$

Indirect heating provides the required overall endothermic heat of reaction for the steam-methane reforming.

Obviously, in the case of a natural gas feedstock, a simple adsorption process might suffice, while in the case of the higher-molecular-weight feedstocks, such as naphtha (see below), a more complete desulfurization process may be necessary.

The desulfurized feedstock is then mixed with superheated steam and passed over a nickel catalyst (730–845°C; 1350–1550°F; 400 psi) to produce a mixture of hydrogen, carbon monoxide, and carbon dioxide as well as excess steam. The effluent gases are cooled (to about 370°C; 700°F) and passed through a shift converter that promotes reaction of the carbon monoxide with stream to yield carbon dioxide and more hydrogen. The shift converter may contain two beds of catalyst with interbed cooling; the combination of the two catalyst beds promotes maximum conversion of the carbon monoxide. This is essential in the event that a high-purity product is required.

The carbon dioxide-rich gas stream is then cooled after which carbon dioxide removal is achieved by passage through scrubbers or by the more thermally efficient hot carbonate process. Any residual carbon monoxide or carbon dioxide is removed by passing the heated gas through a nickel-based methanation catalyst where the carbon oxides are reached with hydrogen to produce methane. The methane may also find other use within the refinery although in the current context the methane acts as the starting material for hydrogen.

Usually, natural gas contains sulfur compounds (Speight, 2007), and because of the high-sulfur sensitivity of the catalyst in the reformer (and in the shift reactor, if installed), hydrogen sulfide (H_2S) and other sulfur compounds must be removed. This may be accomplished using a zinc oxide bed. If higher-molecular-weight organic sulfur compounds (mercaptans R-SH) or carbonyl sulfide (COS) in concentrations in the ppm range is present, the zinc oxide bed alone is not sufficient, and the sulfur compounds are converted to hydrogen sulfide in a hydrogenation stage. The hydrogen necessary is taken from the product stream and amounts to ca. 5% for natural gas (whereas for naphtha and liquefied crude oil gas, 10% and 25% H_2, respectively, is used).

Supported nickel catalysts catalyze steam-methane reforming and the concurrent shift reaction. The catalyst contains 15–25 wt% nickel oxide on a mineral carrier. Carrier materials are alumina, aluminosilicates, cement, and magnesia. Before start-up, nickel oxide must be reduced to metallic nickel not only with hydrogen but also with natural gas or even with the feed gas itself.

Certain types of catalyst uranium oxide and chromium oxide may be used as a promoter. This is reported to give a higher resistance to catalyst poisoning by sulfur components and a lower tendency to form carbon deposits.

For the higher-molecular-weight feedstocks such as liquefied crude oil gas (usually propane C_3H_8) and naphtha (q.v.), nickel catalysts with alkaline carriers or alkaline-free catalysts with magnesium oxide as additive can be used. Both types of catalyst are less active than the conventional nickel catalyst. Therefore, a less rapid decomposition of the hydrocarbon derivatives is achieved. At the same time, the reaction of water with any carbon formed is catalyzed.

Required properties of the catalyst carriers are high-specific area, low-pressure drop, and high mechanical resistance at temperatures up to 1000°C (1830°F). The catalysts are usually in the form of rings (e.g., outer diameter 16 mm, height 16 mm, and inner diameter 8 mm), but other forms, such as saddles, stars, and spoked wheels, are also commercially available.

The main catalyst poison in steam reforming plants is sulfur that is present in most feedstocks. Sulfur concentrations as low as 0.1 ppm form a deactivating layer on the catalyst, but the activity loss of a poisoned catalyst can be offset, to some extent, by raising the reaction temperature. This helps to reconvert the inactive nickel sulfide to active nickel sites. Nickel-free catalysts have been proposed for feedstocks heavier than naphtha. These catalysts consist mostly of strontium, aluminum, and calcium oxides and seem to be resistant to coke deposits and may be suitable for use with high-sulfur feedstocks.

One way of overcoming the thermodynamic limitation of steam reforming is to remove either hydrogen or carbon dioxide as it is produced, hence shifting the thermodynamic equilibrium toward the product side. The concept for sorption-enhanced methane steam reforming is based on in situ removal of carbon dioxide by a sorbent such as calcium oxide (CaO):

$$CaO + CO_2 \rightarrow CaCO_3$$

Sorption enhancement enables lower reaction temperatures, which may reduce catalyst coking and sintering, while enabling the use of less expensive reactor wall materials. In addition, heat release by the exothermic carbonation reaction supplies most of the heat required by the endothermic reforming reactions. However, energy is required to regenerate the sorbent to its oxide form by the energy-intensive calcination reaction:

$$CaCO_3 \rightarrow CaO + CO_2$$

Use of a sorbent requires either that there be parallel reactors operated alternatively and out of phase in reforming and sorbent regeneration modes or that sorbent be continuously transferred between the reformer/carbonator and regenerator/calciner (Balasubramanian et al., 1999; Hufton et al., 1999).

In autothermal (or secondary) reformers, the oxidation of methane supplies the necessary energy and carried out either simultaneously or in advance of the reforming reaction (Brandmair et al., 2003; Ehwald et al., 2003; Nagaoka et al., 2003). In the autothermal reforming process, the feedstock is reacted with a mixture of oxygen and steam by the use of a burner and a fixed nickel catalyst bed for the equilibration of the gas. This results in a lower oxygen consumption than used in noncatalytic routes. With the addition of steam, it is possible to adjust the hydrogen/carbon monoxide ratio. This cannot be achieved by noncatalytic routes because the addition of steam results in a reduction of temperature and soot formation:

$$2CH_4 + 3O_2 \rightarrow 2CO + 4H_2O$$

$$CH_4 + H_2O \rightarrow CO + 3H_2$$

$$CO + H_2O \rightarrow CO_2 + H_2$$

The equilibrium of the methane steam reaction and the water-gas shift reaction determines the conditions for optimum hydrogen yields. The optimum conditions for hydrogen production require high temperature at the exit of the reforming reactor (800–900°C, 1470–1650°F), high excess of steam (molar steam-to-carbon ratio of 2.5:3), and relatively low pressures (below 450 psi). Most commercial plants employ supported nickel catalysts for the process.

The steam-methane reforming process described briefly above would be an ideal hydrogen production process if it was not for the fact that large quantities of natural gas, a valuable resource, are required as both feed gas and combustion fuel. For each mole of methane reformed, more than one mole of carbon dioxide is coproduced and must be disposed. This can be a major issue as it results in the same amount of greenhouse gas emission as would be expected from direct combustion of natural gas or methane. In fact, the production of hydrogen as a clean burning fuel by way of steam reforming of methane and other fossil-based hydrocarbon fuels is not in environmental balance if in the process, carbon dioxide and carbon monoxide are generated and released into the atmosphere, although alternate scenarios are available (Gaudernack, 1996). Moreover, as the reforming process is not totally efficient, some of the energy value of the hydrocarbon fuel is lost by conversion to hydrogen but with no tangible environmental benefit, such as a reduction in emission of greenhouse gases. Despite these apparent shortcomings, the process has the following advantages: (i) the process produces 4 m of hydrogen for each mole of methane consumed; (ii) feedstocks for the process (methane and water) are readily available, (iii) the process is adaptable to a wide range of hydrocarbon feedstocks; (iv) it operates at low pressures, <450 psi; (v) it requires a low steam/carbon ratio (2.5–3); (vi) it has good utilization of input energy (reaching 93%); (vii) it can use catalysts that are stable and resist poisoning; and (viii) it has good process kinetics.

Liquid feedstocks, either liquefied crude oil gas or naphtha (q.v.), can also provide backup feed, if there is a risk of natural gas curtailments. The feed handling system needs to include a surge drum, feed pump, and vaporizer (usually steam heated) followed by further heating before desulfurization. The sulfur in liquid feedstocks occurs as mercaptans, thiophene derivatives, or higher-boiling compounds. These compounds are stable and will not be removed by zinc oxide; therefore, a hydrogenation unit will be required. In addition, as with refinery gas, olefins must also be hydrogenated if they are present.

The reformer will generally use a potash-promoted catalyst to avoid coke buildup from cracking of the heavier feedstock. If liquefied crude oil gas is to be used only occasionally, it is often possible to use a methane-type catalyst at a higher steam/carbon ratio to avoid coking. Naphtha will require a promoted catalyst unless a preformer is used.

4.8 Steam-Naphtha Reforming

The steam-naphtha reforming process (Speight, 2000, 2011, 2014, 2016, 2017; Parkash, 2003; Ancheyta and Speight, 2007; Gary et al., 2007; Hsu and Robinson, 2017) is a continuous process for the production of hydrogen from liquid hydrocarbon derivatives and is, in fact, similar to steam-methane reforming that is one of several possible processes for the production of hydrogen from low-boiling hydrocarbon derivatives other than ethane (Muradov, 1998, 2000; Brandmair et al., 2003; Find et al., 2003). A variety of naphtha types in the naphtha-boiling range may be employed, including feeds containing up to 35% aromatics. Thus, following pretreatment to remove sulfur compounds, the feedstock is mixed with steam and taken to the reforming furnace (675–815°C, 1250–1500°F, 300psi, where hydrogen is produced). In addition, some modifications to the proprietary catalyst are necessary to accommodate the higher-molecular-weight feedstock, thereby allowing naphtha (including naphtha with as much as 35% aromatics) boiling up to 200°C (390°F) to be used as process feedstocks.

A problem may occur when higher-molecular-weight materials are employed as feedstocks and result in coke formation and deposition. When an alkali-promoted catalyst is employed, corrosion and fouling problems in the reformer (or even in equipment downstream of the reformer because of the tendency of the alkali to migrate) may occur with some frequency. However, coke formation can be eliminated by the use of a proprietary alkali-free catalyst that has an extremely high activity and resistance to poisoning.

The overall chemistry of the steam-naphtha reforming process may be represented by the following equations:

$$C_6H_{14} + 6H_2O \rightarrow 6CO + 13H_2$$

$$C_6H_{14} + 12H_2O \rightarrow 6CO_2 + 19H_2$$

The process details (and the process flow) are essentially the same as those described for steam-methane reforming.

4.9 Synthesis Gas Generation

The synthesis gas generation process (Chadeesingh, 2011) is a noncatalytic process for producing synthesis gas (principally hydrogen and carbon monoxide) for the ultimate production of high-purity hydrogen from gaseous or liquid hydrocarbon derivatives.

In this process, a controlled mixture of preheated feedstock and oxygen is fed to the top of the generator where carbon monoxide and hydrogen emerge as the products. Soot, produced in this part of the operation, is removed in a water scrubber from the product gas stream and is then extracted from the resulting carbon-water slurry with naphtha and transferred to a fuel oil fraction. The oil-soot mixture is burned in a boiler or recycled to the generator to extinction to eliminate carbon production as part of the process.

The soot-free synthesis gas is then charged to a shift converter where the carbon monoxide reacts with steam to form additional hydrogen and carbon dioxide at the stoichiometric rate of 1 m of hydrogen for every mole of carbon monoxide charged to the converter.

The reactor temperatures vary from 1095 to 1490°C (2000–2700°F), while pressures can vary from approximately atmospheric pressure to ~2000 psi. The process has the capability of producing high-purity hydrogen although the extent of the purification procedure depends upon the use to which the hydrogen is to be put. For example, carbon dioxide can be removed by scrubbing with various alkaline reagents, while carbon monoxide can be removed by washing with liquid nitrogen or, if nitrogen is undesirable in the product, the carbon monoxide should be removed by washing with copper-amine solutions.

This particular partial oxidation technique has also been applied to a whole range of liquid feedstocks for hydrogen production. There is now serious consideration being given to hydrogen production by the partial oxidation of solid feedstocks such as crude oil coke (from both delayed and fluid-bed reactors), lignite, and coal, as well as crude oil residua.

The chemistry of the process, using naphthalene as an example, may be simply represented as the selective removal of carbon from the hydrocarbon feedstock and further conversion of a portion of this carbon to hydrogen:

$$C_{10}H_8 + 5O_2 \rightarrow 10CO + 4H_2$$

$$10CO + 10H_2O \rightarrow 10CO_2 + 10H_2$$

Although these reactions may be represented very simply using equations of this type, the reactions can be complex and result in carbon deposition on parts of the equipment, thereby requiring careful inspection of the reactor.

4.10 Texaco Partial Oxidation Process

The Texaco partial oxidation process (Texaco gasification process) is a partial oxidation gasification process for generating synthetic gas, principally hydrogen and carbon monoxide (*synthesis gas* and *syngas*) (Chadeesingh, 2011). The characteristic of the process is to inject feedstock together with carbon dioxide, steam, or water into the gasifier. Therefore, solvent-deasphalted residua or crude oil coke rejected from any coking method can be used as feedstock for this gasification process. The produced gas from this gasification process can be used for the production of high-purity high-pressurized hydrogen, ammonia, and methanol. The heat recovered from the high-temperature gas is used for the generation of steam in the waste heat boiler. Alternatively, the less expensive quench-type configuration is preferred when high-pressure steam is not needed or when a high degree of shift is needed in the downstream CO converter.

In the process, the feedstock, together with the feedstock carbon slurry recovered in the carbon recovery section, is pressurized to a given pressure, mixed with high-pressure steam, and then blown into the gas generator through the burner together with oxygen.

The gasification reaction is a partial oxidation of hydrocarbon derivatives to carbon monoxide and hydrogen:

$$C_xH_{2y} + x/2O_2 \uparrow xCO + yH_2$$

$$C_xH_{2y} + xH_2O \uparrow xCO + (x+y)H_2$$

The gasification reaction is instantly completed, thus producing gas mainly consisting of H_2 and CO (H_2+CO$\geq$90%). The high-temperature gas leaving the reaction chamber of the gas generator enters the quenching chamber linked to the bottom of the gas generator and is quenched to 200–260°C (390–500°F) with water.

4.11 Recovery From Fuel Gas

Recovering hydrogen from refinery fuel gas can help refineries to satisfy high hydrogen demand. Cryogenic separation is typically viewed as being the most thermodynamically efficient separation technology. The higher capital cost associated with prepurification and the low flexibility to impurity upsets has limited its use in hydrogen recovery.

The basic configuration for hydrogen recovery from refinery gases involves a two-stage partial condensation process, with postpurification via pressure swing adsorption (Dragomir et al., 2010). The major steps in this process involve first compressing and pretreating the crude refinery gas stream before chilling to an intermediate temperature (from −60 to −120°F). This partially condensed stream is then separated in a flash drum after which the liquid stream is expanded through a Joule-Thompson valve to generate refrigeration and then is fed to the wash column. Optionally, the wash column can be replaced by a simple flash drum.

A crude liquefied petroleum gas stream is collected at the bottom of the column, and a methane-rich vapor is obtained at the top. The methane-rich vapor is sent to compression and then to fuel. The vapor from the flash drum is further cooled in a second heat exchanger before being fed to second flash drum where it produces a hydrogen-rich stream and a methane-rich liquid. The liquid is expanded in a Joule-Thomson valve to generate refrigeration and then is sent for further cooling. The hydrogen-rich gas is then sent to the pressure swing adsorption unit for further purification—the tail gas from this unit is compressed and returned to fuel together with the methane-rich gas.

5. Catalysts

Hydrogen plants are one of the most extensive users of catalysts in the refinery. Catalytic operations include hydrogenation, steam reforming, shift conversion, and methanation.

5.1 Reforming Catalysts

The reforming catalyst is usually supplied as nickel oxide that, during start-up, is heated in a stream of inert gas and then steam. When the catalyst is near the normal operating temperature, hydrogen or a low-boiling hydrocarbon is added to reduce the nickel oxide to metallic nickel.

The high temperatures (up to 870°C, 1600°F) and the nature of the reforming reaction require that the reforming catalyst be used inside the radiant tubes of a reforming furnace. The active agent in reforming catalyst is nickel, and normally, the reaction is controlled both by diffusion and by heat transfer. Catalyst life is limited as much by physical breakdown as by deactivation.

Sulfur is the main catalyst poison, and the catalyst poisoning is theoretically reversible with the catalyst being restored to near full activity by steaming. However, in practice, the deactivation may cause the catalyst to overheat and coke, to the point that it must be replaced. Reforming catalysts are also sensitive to poisoning by heavy metals, although these are rarely present in low-boiling hydrocarbon feedstocks and in naphtha feedstocks.

Coking deposition on the reforming catalyst and ensuing gloss of catalyst activity is the most characteristic issue that must be assessed and mitigated.

While methane-rich streams such as natural gas or low-boiling refinery gas are the most common feeds to hydrogen plants, there is often a requirement for variety of reasons to process a variety of higher-boiling feedstocks, such as liquefied crude oil gas and naphtha. Feedstock variations may also be inadvertent due, for example, to changes in refinery off-gas composition from another unit or because of variations in naphtha composition because of feedstock variance to the naphtha unit.

Thus, when using higher-boiling feedstocks in a hydrogen plant, coke deposition on the reformer catalyst becomes a major issue. Coking is most likely in the reformer unit at the point where both temperature and hydrocarbon content are high enough. In this region, hydrocarbon derivatives crack and form coke faster than the coke is removed by reaction with steam or hydrogen, and when catalyst deactivation occurs, there is a simultaneous temperature increases with a concomitant increase in coke formation and deposition. In other zones, where the hydrocarbon-to-hydrogen ratio is lower, there is less risk of coking.

Coking depends to a large extent on the balance between catalyst activity and heat input with the more active catalysts producing higher yields of hydrogen at lower temperature, thereby reducing the risk of coking. A uniform input of heat is important in this region of the reformer since any catalyst voids or variations in catalyst activity can produce localized hot spots leading to coke formation and/or reformer failure.

Coke formation results in hot spots in the reformer that increases pressure drop and reduces feedstock (methane) conversion, leading eventually to reformer failure. Coking may be partially mitigated by increasing the steam/feedstock ratio to change the reaction conditions, but the most effective solution may be to replace the reformer catalyst with one designed for higher-boiling feedstocks.

A *standard* steam-methane reforming catalyst uses nickel on an alpha-alumina ceramic carrier that is acidic in nature. Promotion of hydrocarbon cracking with such a catalyst leads to coke formation from higher-boiling feedstocks. Some catalyst formulations use a magnesia/alumina (MgO/Al_2O_3) support that is less acidic than α-alumina that reduces cracking on the support and allows higher-boiling feedstocks (such as liquefied crude oil gas) to be used.

Further resistance to coking can be achieved by adding an alkali promoter, typically some form of potash (KOH) to the catalyst. Besides reducing the acidity of the carrier, the promoter catalyzes the reaction of steam and carbon. While carbon continues to be formed, it is removed faster than it can build up. This approach can be used with naphtha feedstock boiling point up to ~180°C (350°F). Under the conditions in a reformer, potash is volatile, and it is incorporated into the catalyst as a more complex compound that slowly hydrolyzes to release potassium hydroxide (KOH). Alkali-promoted catalyst allows the use of a wide range of feedstocks, but in addition to possible potash migration, which can be minimized by proper design and operation, the catalyst is also somewhat less active than conventional catalyst.

Another option to reduce coking in steam reformers is to use a *prereformer* in which a fixed bed of catalyst, operating at a lower temperature, upstream of the fired reformer is used. In a prereformer, adiabatic steam-hydrocarbon reforming is performed outside the fired reformer in a vessel containing high nickel catalyst. The heat required for the endothermic reaction is provided by hot flue gas from the reformer convection section. Since the feed to the fired reformer is now partially reformed, the steam-methane reformer can operate at an increased feed rate and produce 8%–10% additional hydrogen at the same reformer load. An additional advantage of the prereformer is that it facilitates higher mixed feed preheat temperatures and maintains relatively constant operating conditions within the fired reformer regardless of variable refinery off-gas feed conditions. Inlet temperatures are selected so that there is minimal risk of coking, and the gas leaving the prereformer contains only steam, hydrogen, carbon monoxide, carbon dioxide, and methane. This allows a standard methane catalyst to be used in the fired reformer, and this approach has been used with feedstocks up to the low-boiling kerosene fraction. Since the gas leaving the prereformer poses reduced risk of coking, it can compensate to some extent for variations in catalyst activity and heat flux in the primary reformer.

5.2 Shift Conversion Catalysts

The second important reaction in a steam reforming plant is the shift conversion reaction:

$$CO + H_2O \rightarrow CO_2 + H_2$$

Two basic types of shift catalyst are used in steam reforming plants: iron/chrome high-temperature shift catalysts and copper/zinc low-temperature shift catalysts.

High-temperature shift catalysts operate in the range of 315–430°C (600–800°F) and consist primarily of magnetite (Fe_3O_4) with three-valent chromium oxide (Cr_2O_3) added as a stabilizer. The catalyst is usually supplied in the form of ferric oxide (Fe_2O_3) and six-valent chromium oxide (CrO_3) and is reduced by the hydrogen and carbon monoxide in the shift feed gas as part of the start-up procedure to produce the catalyst in the desired form. However, caution is necessary since if the steam/carbon ratio of the feedstock is too low and the reducing environment is too strong, the catalyst can be reduced further to metallic iron. Metallic iron is a catalyst for Fischer-Tropsch reactions, and hydrocarbon derivatives will be produced (Davis and Occelli, 2010).

Low-temperature shift catalysts operate at temperatures on the order of 205–230°C (400–450°F). Because of the lower temperature, the reaction equilibrium is more controllable, and lower amounts of carbon monoxide are produced. The low-temperature shift catalyst is primarily used in wet scrubbing plants that use a methanation for final purification. Pressure swing adsorption plants do not generally use a low temperature because any unconverted carbon monoxide is recovered as reformer fuel. Low-temperature shift catalysts are sensitive to poisoning by sulfur and are sensitive to water (liquid) that can cause softening of the catalyst followed by crusting or plugging.

The catalyst is supplied as copper oxide (CuO) on a zinc oxide (ZnO) carrier, and the copper must be reduced by heating it in a stream of inert gas with measured quantities of hydrogen. The reduction of the copper oxide is strongly exothermic and must be closely monitored.

5.3 Methanation Catalysts

In wet scrubbing plants, the final hydrogen purification procedure involved is by methanation in which the carbon monoxide and carbon dioxide are converted to methane:

$$CO + 3H_2O \rightarrow CH_4 + H_2O$$

$$CO_2 + 4H_2 \rightarrow CH_4 + 2H_2O$$

The active agent is nickel, on an alumina carrier.

The catalyst has a long life, as it operates under ideal conditions and is not exposed to poisons. The main source of deactivation is plugging from carryover of carbon dioxide from removal solutions.

The most severe hazard arises from high levels of carbon monoxide or carbon dioxide that can result from breakdown of the carbon dioxide removal equipment or from exchanger tube leaks that quench the shift reaction. The results of breakthrough can be severe, since the methanation reaction produces a temperature rise of 70°C (125°F) per 1% of carbon monoxide or a temperature rise of 33°C (60°F) per 1% of carbon dioxide. While the normal operating temperature during methanation is ~315°C (600°F), it is possible to reach 700°C (1300°F) in cases of major breakthrough.

6. Purification

The selection of the optimum purification technique for applications in the crude oil industry must be based on both technical and economic considerations. The degree of purification of hydrogen obtained from the different methods varies from ~ 90% to 99.9999% v/v for alloy membrane diffusion (Grashoff et al., 1983). The amount of hydrogen recovered also varies considerably and can have a major impact on process economics, particularly for large-scale applications.

When the hydrogen content of the refinery gas is >50% by volume, the gas should first be considered for hydrogen recovery, using a membrane (Brüschke, 1995, 2003; Lu et al., 2007). The tail gas or reject gas that will still contain a substantial amount of hydrogen can then be used as steam reformer feedstock. Generally, the feedstock purification process uses three different refinery gas streams to produce hydrogen. First, high-pressure hydrocracker purge gas is purified in a membrane (through which only hydrogen can pass) that produces hydrogen at medium pressure and is combined with medium-pressure off-gas that is first purified in a pressure swing adsorption unit. Finally, low-pressure off-gas is compressed, mixed with reject gases from the membrane and pressure swing adsorption units, and used as steam reformer feed.

Various processes are available to purify the hydrogen stream, but since the product streams are available as a wide variety of composition, flows, and pressures, the best method of purification will vary. And there are several factors that must also be taken into consideration in the selection of a purification method. These are (i) hydrogen recovery, (ii) product purity, (iii) pressure profile, (iv) reliability, and (v) cost; an equally important parameter is not considered here since the emphasis is on the technical aspects of the purification process.

6.1 Wet Scrubbing

Wet scrubbing systems, particularly amine or potassium carbonate systems, are used for removal of acid gases such as hydrogen sulfide or carbon dioxide. Most systems depend on chemical reaction and can be designed for a wide range of pressures and capacities. They were once widely used to remove carbon dioxide in steam reforming plants but have generally been replaced by pressure swing adsorption units except where carbon monoxide is to be recovered. Wet scrubbing is still used to remove hydrogen sulfide and carbon dioxide in partial oxidation plants.

Wet scrubbing systems remove only acid gases or high-boiling hydrocarbon derivatives, but they do not contain methane or other hydrocarbon gases, hence having little influence on product purity. Therefore, wet scrubbing systems are most often used as a pretreatment step or where a hydrogen-rich stream is to be desulfurized for use as fuel gas.

6.2 Pressure Swing Adsorption Units

Pressure swing adsorption units use beds of solid adsorbent to separate impurities from hydrogen streams leading to high-purity high-pressure hydrogen and a low-pressure tail gas stream containing the impurities and some of the hydrogen. The beds are then regenerated by depressurizing and purging. Part of the hydrogen (up to 20%) may be lost in the tail gas.

The *pressure swing adsorption* (PSA) technology is based on a physical binding of gas molecules to adsorbent material. The respective force acting between the gas molecules and the adsorbent material depends on the gas component, type of adsorbent material, partial pressure of the gas component, and operating temperature. The separation effect is based on differences in binding forces to the adsorbent material. Highly volatile components with low polarity, such as hydrogen, are practically nonadsorbable as opposed to molecules such as nitrogen, carbon monoxide, carbon dioxide, hydrocarbon derivatives, and water vapor. Consequently, these impurities can be adsorbed from a hydrogen-containing stream, and high-purity hydrogen is recovered.

The pressure swing adsorption process works at basically constant temperature and uses the effect of alternating pressure and partial pressure to perform adsorption and desorption. Since heating or cooling is not required, short cycles within the range of minutes are achieved. The process consequently allows the economical removal of large amounts of impurities. Adsorption is carried out at high pressure (and hence high respective partial pressure) typically in the range of 10–40 bar until the equilibrium loading is reached. At this point in time, no further adsorption capacity is available, and the adsorbent material must be regenerated. This regeneration is accomplished by lowering the pressure to slightly above atmospheric pressure resulting in a respective decrease in equilibrium loading. As a result, the impurities on the adsorbent material are desorbed, and the adsorbent material is regenerated. The amount of impurities removed from a gas stream within one cycle corresponds to the difference of adsorption to desorption loading. After termination of regeneration, pressure is increased back to adsorption pressure level, and the process starts again from the beginning.

Pressure swing adsorption is generally the purification method of choice for steam reforming units because of its production of high-purity hydrogen and is also used for purification of refinery off-gases, where it competes with membrane systems. Many hydrogen plants that formerly used a *wet scrubbing* process (Fig. 15.4) for hydrogen purification are now using the *pressure swing adsorption* (PSA) (Fig. 15.5) for purification. The pressure swing adsorption process is a cyclic process that uses beds of solid adsorbent to remove impurities from the gas and generally produces higher-purity hydrogen (99.9% v/v purity compared with <97% v/v purity). The purified hydrogen passes through the adsorbent beds with only a tiny fraction absorbed, and the beds are regenerated by depressurization followed by purging at low pressure.

When the beds are depressurized, a waste gas (or *tail gas*) stream is produced and consists of the impurities from the feed (carbon monoxide, carbon dioxide, methane, and nitrogen) plus some hydrogen. This stream is burned in the reformer as fuel, and reformer operating conditions in a pressure swing adsorption plant are set so that the tail gas provides no more than about 85% of the reformer fuel. This gives good burner control because the tail gas is more difficult to burn than regular fuel gas and the high content of carbon monoxide can interfere with the stability of the flame. As the reformer operating temperature is increased, the reforming equilibrium shifts, resulting in more hydrogen and less methane in the reformer outlet and hence less methane in the tail gas.

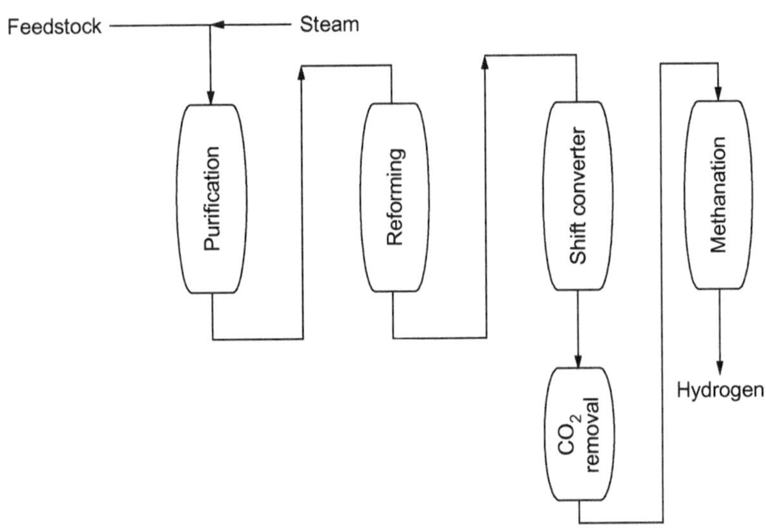

FIG. 15.4 Hydrogen purification by wet scrubbing.

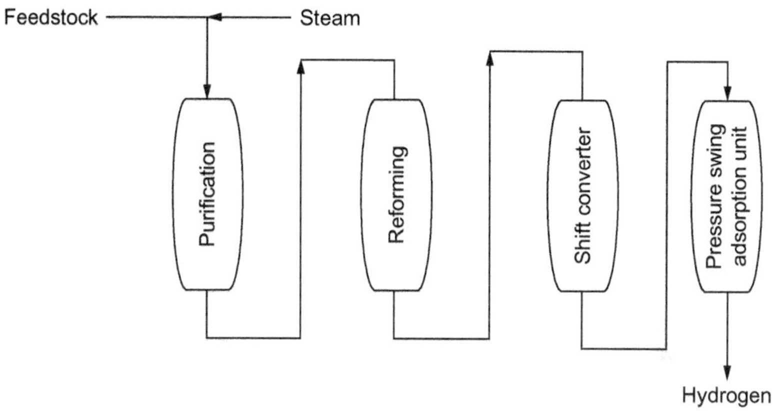

FIG. 15.5 Hydrogen purification by pressure swing adsorption.

6.3 Membrane Systems

Membranes are also important to the subsequent purification of hydrogen. For hydrogen production and purification, there are generally two classes of membranes both being inorganic: dense phase metal and metal alloys and porous ceramic membranes.

Porous ceramic membranes are normally prepared by solgel or hydrothermal methods and have high stability and durability in high-temperature, harsh impurity, and hydrothermal environments. In particular, microporous membranes show promises in water-gas shift reaction at higher temperatures. Membrane systems separate gases by taking advantage of the difference in rates of diffusion through membranes (Brüschke, 1995, 2003). Gases that diffuse faster (including hydrogen) become the permeate stream and are available at low pressure, whereas the slower-diffusing gases become the nonpermeate and leave the unit at a pressure close to the pressure of the feedstock at entry point. Membrane systems contain no moving parts or switch valves and have potentially very high reliability. The major threat is from components in the gas (such as aromatics) that attack the membranes or from liquids, which plug them.

Membranes are fabricated in relatively small modules; for larger capacity, more modules are added. Cost is therefore virtually linear with capacity, making them more competitive at lower capacities. The design of membrane systems involves a trade-off between pressure drop (or diffusion rate) and surface area, as well as between product purity and recovery. As the surface area is increased, the recovery of fast components increases; however, more of the slow components are recovered, which lowers the purity.

On the other hand, typical polymeric membranes selectively permeate hydrogen, and the permeation selectivity of these membranes can be very high because the adsorption selectivity is high even at low feed pressure. The hydrocarbon permeability across the membrane will be high because the diffusivity for surface diffusion is orders of magnitude higher than typical diffusivities for these components through a polymeric matrix. While membrane systems can produce hydrogen with purities of 90%–98% v/v at high recovery (>85%) but are best suited for high-pressure feedstocks since the separation of a gas mixture in these purification units is driven by pressure, an additional disadvantage is the fact that hydrogen sulfide may also pass through the (polymer) membrane and into the hydrogen product.

6.4 Cryogenic Separation

Cryogenic separation units operate by cooling the gas and condensing some, or all, of the constituents for the gas stream. Depending on the product purity required, separation may involve flashing or distillation. Cryogenic units offer the advantage of being able to separate a variety of products from a single feed stream. One specific example is the separation of low-boiling olefins from a hydrogen stream.

Hydrogen recovery is in the range of 95%, with purity above 98% obtainable.

7. Hydrogen Management and Safety

Several trends have significantly increased hydrogen demand in refinery operations. They are the larger supplies of heavier, sour crude oils containing more sulfur and nitrogen and more stringent clean fuel regulations. In refinery operations, the reduction of polluting compounds such as sulfur, nitrogen, and aromatics is achieved through catalytic hydrotreating (HDT) and hydrocracking (HDC) processes. In catalytic hydrotreating processes, hydrogen is consumed not only in hydrodesulfurization (HDS) reactions but also in side reactions such as hydrodenitrogenation (HDN), hydrodearomatization, and olefin hydrogenation reactions. As a result, there is a much higher hydrogen demand for deeper hydrotreating and a lower hydrogen production from the catalytic reformer. Consequently, a deficit in the refinery hydrogen balance arises, and therefore, the hydroprocessing capacity and the associated hydrogen network may be limiting the refinery throughput and the operating margins. Hydrogen management has then become a critical factor in current refinery operations (Zagoria et al., 2003; Méndez et al., 2008; Luckwal and Mandal, 2009; Deng et al., 2013). To avoid any potential production bottleneck, new alternative sources of hydrogen and higher purities for improving hydrotreater capacity and product quality will be required.

7.1 Distribution

The hydrogen distribution system usually comprises a set of hydrogen main headers (pipelines) working at different pressures and hydrogen purities. Many makeup and recycle compressors drive the hydrogen through this complex network of consumer units, on-purpose production units, and Platforming units. Hydrogen plants generate high-purity hydrogen at different costs, while net production units are Platforming units that generate low-purity hydrogen as a byproduct. Hydrogen streams with different purities, pressures, and flow rates coming from makeup hydrogen plants and Platforming units are supplied to multiple consumer units through the hydrogen main headers. Purge streams from hydrotreaters containing nonreacted hydrogen are partially recycled and mixed with fresh hydrogen streams from hydrogen headers before rerouting them to consuming units. The remaining off-gas stream is burnt as fuel gas. By controlling the fuel gas flow, the purity of the recycled hydrogen stream can be adjusted. The major hydrotreater operating constraint is a minimum hydrogen/hydrocarbon ratio along the reactor in order to avoid carbon deposition over the catalyst and its premature deactivation. As the catalyst cost is very significant, an effective operation of the hydrogen network will help to increase the catalyst run length, thus boosting the refinery profitability. Moreover, some consuming units may have groups of membranes each of which can be activated to separate and recycle higher-purity hydrogen streams to the hydrogen network.

7.2 Management

During the majority of the 20th century and up to current times, hydrogen has played (and continues to play, always played an important role in refining). Thus, hydrogen management practices significantly impact operating costs and emissions of carbon dioxide. Therefore, an effective hydrogen management program must address refinery-wide issues in a systematic, comprehensive way. The hydrogen system consists of hydrogen production, hydrogen purification, hydrogen use, and the distribution network (Zagoria et al., 2003; Davis and Patel, 2004).

Hydrogen management through hydrogen network optimization is a necessity to address clean fuel trends, meet growing transportation fuel demands, and help maintain profitability from refining crude oils currently and in the future (Stratiev et al., 2009; Speight, 2011, 2014; Rabei, 2012). The majority of hydrogen consumed in a modern refinery is produced from the catalytic reformer system that is also supplemented by an on-site hydrogen facility and/or pipeline hydrogen supply. In fact, in many cases, the hydrogen produced by various reforming processes is not always the main source of hydrogen for the refinery. Nevertheless, the refinery hydrogen distribution system is cascaded through multiple hydroprocessing units, where higher hydrogen purity and higher-pressure consumers send their purge gases to lower hydrogen purity and lower-pressure consumers. Ultimately, purge gases containing residual hydrogen are sent to fuel.

Many existing refinery hydrogen plants use a conventional process, which produces a medium-purity (94%–97%) hydrogen product by removing the carbon dioxide in an absorption system and methanation of any remaining carbon oxides. Since the 1980s, most hydrogen plants are built with pressure swing adsorption (PSA) technology to recover and purify the hydrogen to purities above 99.9%. Since many refinery hydrogen plants utilize refinery off-gas feeds containing hydrogen, the actual maximum hydrogen capacity that can be synthesized via steam reforming is not certain since the hydrogen content of the off-gas can change due to operational changes in the hydrotreaters.

Hydrogen management has become a priority in current refinery operations and when planning to produce lower sulfur gasoline and diesel fuels (Zagoria et al., 2003; Méndez et al., 2008; Luckwal and Mandal, 2009). Along with increased hydrogen consumption for deeper hydrotreating, additional hydrogen is needed for processing heavier and higher-sulfur crude slates. In many refineries, hydroprocessing capacity and the associated hydrogen network is limiting refinery throughput and operating margins. Furthermore, higher hydrogen purities within the refinery network are becoming more important to boost hydrotreater capacity, achieve product value improvements, and lengthen catalyst life cycles.

Improved hydrogen utilization and expanded or new sources for refinery hydrogen and hydrogen purity optimization are now required to meet the needs of the future transportation fuel market and the drive toward higher refinery profitability. Many refineries developing hydrogen management programs fit into the two general categories of either a catalytic reformer supplied network or an on-purpose hydrogen supply.

Some refineries depend solely on catalytic reformer(s) as their source of hydrogen for hydrotreating. Often, they are semiregenerative reformers where off-gas hydrogen quantity, purity, and availability change with feed naphtha quality, as octane requirements change seasonally, and when the reformer catalyst progresses from start-of-run to end-of-run conditions and then goes offline for regeneration. Typically, during some portions of the year, refinery margins are reduced as a result of hydrogen shortages.

Multiple hydrotreating units compete for hydrogen—by either selectively reducing throughput, managing intermediate tankage logistics, or running the catalytic reformer suboptimally just to satisfy downstream hydrogen requirements. Part of the operating year still runs in hydrogen surplus, and the network may be operated with relatively low hydrogen utilization (consumption/production) at 70%–80%. Catalytic reformer off-gas hydrogen supply may swing from 75% to 85% hydrogen purity. Hydrogen purity upgrade can be achieved through some hydrotreaters by absorbing high-boiling hydrocarbon derivatives. But without supplemental hydrogen purification, critical control of hydrogen partial pressure in hydroprocessing reactors is difficult, which can affect catalyst life, charge rates, and/or naphtha yields.

More complex refineries, especially those with hydrocracking units, also have on-purpose hydrogen production, typically with a steam-methane reformer that utilizes refinery off-gas and supplemental natural gas as feedstock. The steam-methane reformer plant provides the swing hydrogen requirements at higher purities (92% to >99% hydrogen) and serves a hydrogen network configured with several purity and pressure levels. Multiple purities and existing purification units allow for more optimized hydroprocessing operation by controlling hydrogen partial pressure for maximum benefit. Typical hydrogen utilization is 85%–95%.

7.3 Safety

The scale and growth of hydrogen demand raise fundamental questions about the safe use of the gas. Due to its chemical properties, hydrogen poses unique challenges in the plant environment. Hydrogen gas is colorless, odorless, and not detectable by human senses.

Thus, there are several hazards associated with hydrogen, ranging from respiratory ailment, to component failure, to ignition, to burning. Although a combination of hazards occurs in most instances, the primary hazard with hydrogen is the production of a flammable mixture, which can lead to a fire or explosion. In addition to these hazards, hydrogen can produce mechanical failures of containment vessels, piping, and other components due to hydrogen embrittlement. Upon long-term exposure to the gas, many metals and plastics can lose ductility and strength, which leads to the formation of cracks and can eventually cause ruptures. A form of hydrogen embrittlement takes place by chemical reaction. At high temperatures, for instance, hydrogen reacts with one or more components of metal walls to form hydrides, which weaken the lattice structure of the material.

In oil refineries, the first step in the escalation of fire and detonation is the loss of containment of the gas. Hydrogen leaks are typically caused by defective seals or gaskets, valve misalignment, or failures of flanges or other equipment. Once released, hydrogen diffuses rapidly. If the leak takes place outdoors, the dispersion of the cloud is affected by wind speed and direction and can be influenced by atmospheric turbulence and nearby structures. With the gas dispersed in a plume, a detonation can occur if the hydrogen and air mixture is within its explosion range and an appropriate ignition source is available. Such flammable mixture can form at a considerable distance from the leak source. In order to address the hazards posed by hydrogen, manufacturers of fire and gas detection systems work within the construct of layers of protection to reduce the incidence of hazard propagation. Under such a model, each layer acts as a safeguard, preventing the hazard from becoming more severe.

The detection layers themselves encompass different detection techniques that either improve scenario coverage or increase the likelihood that a specific type of hazard is detected. Such fire and gas detection layers can consist of catalytic sensors, ultrasonic gas leak monitors, and fire detectors. Ultrasonic gas leak detectors can respond to high-pressure releases of hydrogen, such as those that may occur in hydrocracking reactors or hydrogen separators. In turn, continuous hydrogen monitors like catalytic detectors can contribute to detecting small leaks, for example, due to a flange slowly deformed by the use or failure of a vessel maintained at close to atmospheric pressure. To further protect a plant against fires, hydrogen-specific flame detectors can supervise entire process areas. Such wide coverage is necessary because of hydrogen cloud movement; a fire may be ignited at a considerable distance from the leak source.

Hydrogen production will continue to grow, fueled by environmental legislation and demand for cleaner, higher-grade fuels. But rising production must be matched by a comprehensive approach to plant safety. New facilities that use hydrogen should be designed with adequate safeguards from potential hazards; the design of old facilities should also be revisited to ensure sufficient barriers that are available to minimize accidents and control failure. Safety systems that deploy a diversity of detection technologies can counteract the possible effects of leaks, fire, and explosions, preventing equipment or property damage, personal injury, and the loss of life.

In summary, the modern refinery faces the challenge of meeting an increasing demand for cleaner transportation fuels, as specifications continue to tighten around the world and markets decline for high-sulfur fuel oil. Innovative ideas and solutions to reduce refinery costs must always be considered, including the following: (i) optimization of the hydrogen management network, (ii) multiple feedstock options for hydrogen production, (iii) optimization of plant capacity, and last but certainly not least (iv) the use of hydrogen recovery technologies to maximize hydrogen availability and minimize capital investment.

References

Al-Shamma, L.M., Naman, S.A., 1989. Kinetic study for thermal production of hydrogen from hydrogen sulfide by heterogeneous catalysis of vanadium sulfide in a flow system. Int. J. Hydrog. Energy 14 (3), 173–179.

Ancheyta, J., Speight, J.G. (Eds.), 2007. Hydroprocessing Heavy Oils and Residua. CRC Press, Taylor & Francis Group, Boca Raton, FL.

Arild, V., 2000. Production of Hydrogen and Carbon With a Carbon Black Catalyst. PCT Int. Appl. No. 0021878.

Bailey, R.T. 1992. Proceedings of the International Symposium on Heavy Oil and Residue Upgrading. C. Han and C. Hsi (Editors). International Academy Publishers, Beijing.

Balasubramanian, B., Ortiz, A.L., Kaytakoglu, S., Harrison, D.P., 1999. Hydrogen from methane in a single-step process. Chem. Eng. Sci. 54, 3543–3552.

Bandermann, F., Harder, K.B., 1982. Production of hydrogen via thermal decomposition of hydrogen sulfide and separation of hydrogen and hydrogen sulfide by pressure swing adsorption. Int. J. Hydrog. Energy 7 (6), 471–475.

Bezler, J., 2003. Optimized hydrogen production—a key process becoming increasingly important in refineries. In: Proceedings of the DGMK Conference on Innovation in the Manufacture and Use of Hydrogen, Dresden, Germany, October 15–17, p. 65.

Bishara, A., Salman, O.S., Khraishi, N., Marafi, A., 1987. Thermochemical decomposition of hydrogen sulfide by solar energy. Int. J. Hydrog. Energy 12 (10), 679–685.

Brandmair, M., Find, J., Lercher, J.A., 2003. Combined autothermal reforming and hydrogenolysis of alkanes. In: Proceedings of the DGMK Conference on Innovation in the Manufacture and Use of Hydrogen. Dresden, Germany, October 15–17, pp. 273–280.

Bridge, A.G., 1997. Hydrogen processing. In: Meyers, R.A. (Ed.), Handbook of Crude Oil Refining Processes, second ed. McGraw-Hill, New York, p. 1. (Chapter 15).

Brüschke, H., 1995. Industrial application of membrane separation processes. Pure Appl. Chem. 67 (6), 993–1002.

Brüschke, H., 2003. Separation of hydrogen from dilute streams (e.g. using membranes). In: Proceedings of the DGMK Conference on Innovation in the Manufacture and Use of Hydrogen. Dresden, Germany, October 15–17, p. 47.

Chadeesingh, R., 2011. The Fischer-Tropsch process. In: Speight, J.G. (Ed.), The Biofuels Handbook. In: vol. 5. The Royal Society of Chemistry, London, pp. 476–517. (Part 3, Chapter).

Clark, P.D., Wassink, B., 1990. A review of methods for the conversion of hydrogen sulfide to sulfur and hydrogen. Alberta Sulfur Res. Quart. Bull. 26 (2/3/4), 1.

Clark, P.D., Dowling, N.I., Hyne, J.B., Moon, D.L., 1995. Production of hydrogen and sulfur from hydrogen sulfide in refineries and gas processing plants. Q. Bull. 32 (1), 11–28.

Cruz, F.E., De Oliveira Junior, S., 2008. Petroleum refinery hydrogen production unit: exergy and production cost evaluation. Int. J. Thermodyn. 11 (4), 187–193.

Davis, B.H., Occelli, M.L., 2010. Advances in Fischer-Tropsch Synthesis, Catalysts, and Catalysis. CRC Press, Taylor & Francis Group, Boca Raton, FL.

Davis, R.A., Patel, N.M., 2004. Refinery hydrogen management. Pet. Technol. Q., 29–35. Spring.

Deng, C., Li, W., Feng, X., 2013. Refinery hydrogen network management with key factor analysis. Chem. Eng. Trans. 35, 61–66.

Dickenson, R.L., Biasca, F.E., Schulman, B.L., Johnson, H.E., 1997. Refiner options for converting and utilizing heavy fuel oil. Hydrocarb. Process. 76 (2), 57.

Dolbear, G.E., 1998. Hydrocracking: reactions, catalysts, and processes. In: Speight, J.G. (Ed.), Petroleum Chemistry and Refining. Taylor & Francis, Washington, DC. (Chapter 7).

Donini, J.C., 1996. Separation and processing of hydrogen sulfide in the fossil fuel industry. In: Minimum Effluent Mills Symposium, pp. 357–363.

Dragomir, R., Drnevich, R.F., Morrow, J., Papavassiliou, V., Panuccio, G., Watwe, R., 2010. Technologies for enhancing refinery gas value. In: Proceedings of the AIChE 2010 Spring Meeting, San Antonio, TX, November 7–12.

Ehwald, H., Kürschner, U., Smejkal, Q., Lieske, H., 2003. Investigation of different catalysts for autothermal reforming of i-octane. In: Proceedings of the DGMK Conference on Innovation in the Manufacture and Use of Hydrogen, Dresden, Germany, October 15–17, p. 345.

Farnand, S., Li, J., Patel, N., Peng, X.D., Ratan, S., 2015. Hydrogen perspectives for 21st century refineries. Hydrocarb. Process. 94 (2), 53–58.

Find, J., Nagaoka, K., Lercher, J.A., 2003. Steam reforming of light alkanes in micro-structured reactors. In: Proceedings of the DGMK Conference on Innovation in the Manufacture and Use of Hydrogen, Dresden, Germany, October 15–17, p. 257.

Fleshman, J.D., 1997. Hydrogen production. In: Meyers, R.A. (Ed.), Handbook of Petroleum Refining Processes, second ed. McGraw-Hill, New York, p. 1. (Chapter 6).

Gary, J.H., Handwerk, G.E., Kaiser, M.J., 2007. Petroleum Refining: Technology and Economics, fifth ed. CRC Press, Taylor & Francis Group, Boca Raton, FL.

Gaudernack, B., 1996. Hydrogen from natural gas without release of carbon dioxide into the atmosphere. In: Hydrogen Energy Prog. Proceedings of the 11th World Hydrogen Energy Conference. vol. 1. pp. 511–523.
Grashoff, G.J., Pilkington, C.E., Corti, C.W., 1983. The purification of hydrogen: a review of the technology emphasizing the current status of palladium membrane diffusion. Platin. Met. Rev. 27 (4), 157–169.
Gross, M., Wolff, J., 2000. Gasification of residue as a source of hydrogen for the refining industry in India. In: Proceedings of the Gasification Technologies Conference, San Francisco, CA, October 8–11.
Hsu, C.S., Robinson, P.R. (Eds.), 2017. Handbook of Petroleum Technology. Springer International Publishing AG, Cham.
Hufton, J.R., Mayorga, S., Sircar, S., 1999. Sorption-enhanced reaction process for hydrogen production. AIChE J 45, 248–256.
Kiuchi, H., 1982. Recovery of hydrogen from hydrogen sulfide with metals and metal sulfides. Int. J. Hydrog. Energy. 7(6).
Kotera, Y., Todo, N., Fukuda, K., 1976. Process for Production of Hydrogen and Sulfur From Hydrogen Sulfide as Raw Material. US Patent No. 3,962,409, June 8.
Kriz, J.F., Ternan, M., 1994. Hydrocracking of Heavy Asphaltenic Oil in the Presence of an Additive to Prevent Coke Formation. United States Patent 5,296,130, March 22.
Lipman, T., 2011. An Overview of Hydrogen Production and Storage Systems With Renewable Hydrogen Case Studies. A Clean Energy States Alliance Report. Conducted Under US DOE Grant DE-FC3608GO18111, Office of Energy Efficiency and Renewable Energy Fuel Cell Technologies Program, United States Department of Energy, Washington, DC.
Lu, G.Q., Dinz da Costa, J.C., Duke, M., Giessler, S., Socolow, R., Williams, R.H., Kreutz, T., 2007. Inorganic membranes for hydrogen production and purification: a critical review and perspective. J. Colloid Interface Sci. 314 (2), 589–603.
Luckwal, K., Mandal, K.K., 2009. Improve hydrogen management of your refinery. Hydrocarb. Process. 88 (2), 55–61.
Luinstra, E., 1996. Hydrogen from hydrogen sulfide—a review of the leading processes. In: Proceedings of the 7th Sulfur Recovery Conference. Gas Research Institute, Chicago, pp. 149–165.
Megalofonos, S.K., Papayannakos, N.G., 1997. Kinetics of catalytic reaction of methane and hydrogen sulfide over MoS_2. J. Appl. Catal. A Gen. 65 (1–2), 249–258.
Méndez, C.A., Gómez, E., Sarabia, D., Cerdá, J., De Prada, C., Sola, M., Unzueta, E., 2008. Braunschweig, B., Joulia, X. (Eds.), Proceedings of the 18th European Symposium on Computer Aided Process Engineering—ESCAPE 18. Elsevier BV, Amsterdam.
Miller, G.Q., Penner, D.W., 2003. Meeting future needs for hydrogen—possibilities and challenges. In: Proceedings of the DGMK Conference on Innovation in the Manufacture and Use of Hydrogen, Dresden, Germany, October 15–17, p. 7.
Mokhatab, S., Poe, W.A., Speight, J.G., 2006. Handbook of Natural Gas Transmission and Processing. Elsevier, Amsterdam.
Muradov, N.Z., 1998. CO_2-free production of hydrogen by catalytic pyrolysis of hydrocarbon fuel. Energy Fuel 12 (1), 41–48.
Muradov, N.Z., 2000. Thermocatalytic carbon dioxide-free production of hydrogen from hydrocarbon fuels. In: Proceedings of the Hydrogen Program Review, NREL/CP-570-28,890.
Murata, K., Ushijima, H., Fujita, K., 1997. Process for Producing Hydrogen From Hydrocarbon. United States Patent 5,650,132.
Nagaoka, K., Jentys, A., Lecher, J.A., 2003. Autothermal reforming of methane over mono- and bimetal catalysts prepared from hydrotalcite-like precursors. In: Proceedings of the DGMK

Conference on Innovation in the Manufacture and Use of Hydrogen. Dresden, Germany, October 15–17, p. 171.

Parkash, S., 2003. Refining Processes Handbook. Gulf Professional Publishing, Elsevier, Amsterdam.

Rabei, Z., 2012. Hydrogen management in refineries. Pet. Coal 54 (4), 357–368.

Raissi, A.T., 2001. Technoeconomic analysis of area II hydrogen production. Part 1. In: Proceedings of the US DOE Hydrogen Program Review Meeting, Baltimore, MD.

Ranke, H., Schödel, N., 2003. Hydrogen production technology—status and new developments. In: Proceedings of the DGMK Conference on Innovation in the Manufacture and Use of Hydrogen. Dresden, Germany, October 15–17, p. 19.

Scherzer, J., Gruia, A.J., 1996. Hydrocracking Science and Technology. Marcel Dekker, New York.

Speight, J.G., 2000. The Desulfurization of Heavy Oils and Residua, second ed. Marcel Dekker, New York.

Speight, J.G., 2007. Natural Gas: A Basic Handbook. GPC Books, Gulf Publishing Company, Houston, TX.

Speight, J.G., 2011. The Refinery of the Future. Gulf Professional Publishing, Elsevier, Oxford.

Speight, J.G., 2013. The Chemistry and Technology of Coal, third ed. CRC Press, Taylor & Francis Publishers, Boca Raton, FL.

Speight, J.G., 2014. The Chemistry and Technology of Petroleum, fifth ed. CRC Press, Taylor & Francis Publishers, Boca Raton, FL.

Speight, J.G., 2016. Hydrogen in refineries. In: Stolten, D., Emonts, B. (Eds.), Hydrogen Science and Engineering: Materials, Processes, Systems, and Technology. Wiley-VCH Verlag GmbH & Co., Weinheim, pp. 3–18. (Chapter 1).

Speight, J.G., 2017. Handbook of Petroleum Refining. CRC Press, Taylor & Francis Publishers, Boca Raton, FL.

Stratiev, D., Tzingov, T., Shishkova, I., Dermatova, P., 2009. Hydrotreating units chemical hydrogen consumption analysis: a tool for improving refinery hydrogen management. In: Proceedings of the 44th International Petroleum Conference, Bratislava, Slovak Republic, September 21–22.

Uemura, Y., Ohe, H., Ohzuno, Y., Hatate, Y., 1999. Carbon and hydrogen from hydrocarbon pyrolysis. Proc. Int. Conf. Solid Waste Technol. Manag. 15, 5E/25–5E/30.

Vauk, D., Di Zanno, P., Neri, B., Allevi, C., Visconti, A., Rosanio, L., 2008. What are possible hydrogen sources for refinery expansion? Hydrocarb. Process. 87 (2), 69–76.

Weimer, A.W., Dahl, J., Tamburini, J., Lewandowski, A., Pitts, R., Bingham, C., Glatzmaier, G.C., 2000. Thermal dissociation of methane using a solar coupled aerosol flow reactor. In: Proc. Hydrogen Program Rev., NREL/CP-570-28,890.

Zagoria, A., Huychke, R., Boulter, P.H., 2003. Refinery hydrogen management–the big picture. In: Proceedings of the DGMK Conference on Innovation in the Manufacture and Use of Hydrogen, Dresden, Germany, October 15–17, p. 95.

Zaman, J., Chakma, 1995. Production of hydrogen and sulfur from hydrogen sulfide. Fuel Process. Technol. 41, 159–198.

Further Reading

vurup, 2014. http://www.vurup.sk/sites/vurup.sk/archivedsite/www.vurup.sk/conferences/44_ipc/44ipc/Sustainable%20Crude%20Oil%20Processing_Tudays%20and%20Future%20Technologies,%20the%20role%20of%20the%20catalysis/posters/Hydrotreating%20Units%20Chemical%20Hydrogen%20Consumption%20Analysis_Stratiev.pdf. (Accessed 4 September 2014).

Chapter 16

The Future and the Environment

1. Introduction

With the entry into the 21st century, crude oil refining technology is experiencing great innovation that is driven by the increasing supply of viscous feedstocks with decreasing quality and the fast increases in the demand for clean and ultraclean vehicle fuels and petrochemical raw materials. There is a strong need to revise the processes that are commercially available for viscous crude oil refining. Upgrading of viscous and extra viscous oil by means of catalytic processes requires new-generation catalyst along with several modifications of process conditions (Iqbal et al., 2008).

The properties of viscous oil, extra viscous oil, and tar sand bitumen crude oil are comparable with the properties of an atmospheric residuum (boiling point >345°C, >650°F) or a vacuum residuum (boiling point >565°C, >1050°F) (Table 16.1) (Speight, 2014, 2017). The viscous feedstocks contain aggregates (often referred to as micelles) of resin constituents and asphaltene constituents that are held together by weak physical interactions. With the resin constituents being less polar and of lower molecular weight than the asphaltene constituents, equilibrium between the micelles and the surrounding oil medium leads to homogeneity and stability of the system. If the amount of the resin constituents is decreased, the asphaltene constituents form an incompatible sediments (Speight, 2014, 2015a, 2017).

Thus, as feedstocks to refineries change, there must be an accompanying change in refinery technology. This means a movement from conventional means of refining viscous feedstocks using (typically) coking technologies to more innovative processes (including hydrogen management) that will produce the ultimate amount of liquid fuels from the feedstock and maintain emissions within environmental compliance (Penning, 2001; Lerner, 2002; Davis and Patel, 2004; Speight, 2008).

To meet the challenges of crude trends in product slate and the stringent distillation operations to increasing specifications imposed by environmental complex chemical operations involving legislation, the refining industry in the near transformation of crude oil into a variety of future crude slate will become

Heavy Oil Recovery and Upgrading. https://doi.org/10.1016/B978-0-12-813025-4.00016-7

TABLE 16.1 General Properties of Refinery Feedstocks[a]

Property	Conventional Crude Oil	Type of Oil: *Viscous Crude Oil*	*Extra Viscous Crude Oil*	Tar Sand Bitumen
API gravity	>20	10–22	<10	<10
Constituents (% w/w)				
Asphaltene fraction (heptane insolubles)	0–12	11–20	15–30	15–30
Resin fraction	3–22	14–39	>20	<20
Oil fraction (saturates plus aromatics)	>67	24–64	<50	<50
Nonhydrocarbon atoms (% w/w)				
Sulfur	<0.1–2.0	0.1–5.0	4.0–6.0	4.0–6.0
Nitrogen	<0.01	0.2–0.8	0.5–1.5	0.5–1.5
Nickel plus vanadium (ppm)	<200	50–500	300–1000	300–1000

[a]*Property ranges are arbitrarily assigned based on published data and are presented to show the difference between the feedstocks.*

increasingly flexible, and refined products with specifications that meet are innovative with new processing schemes and user requirements.

During the next 20–30 years, the evolution future of crude oil refining and the current refinery layout (Fig. 16.1) will be primarily focused on process modification with some new innovations coming onstream. The industry will move predictably onto (i) deep conversion of viscous feedstocks, (ii) higher hydrocracking and hydrotreating capacity, and (iii) more efficient processes.

High-conversion refineries will move to gasification of feedstocks for the development of alternative fuels and to enhance equipment usage (Chapter 13). A major trend in the refining industry market demand for refined products will be in synthesizing fuels from simple basic reactants (e.g., synthesis gas) when it becomes uneconomical to produce super clean transportation fuels through conventional refining processes. Fischer-Tropsch plants together with IGCC systems will be integrated with or even into refineries, which will offer the advantage of high-quality products (Stanislaus et al., 2000).

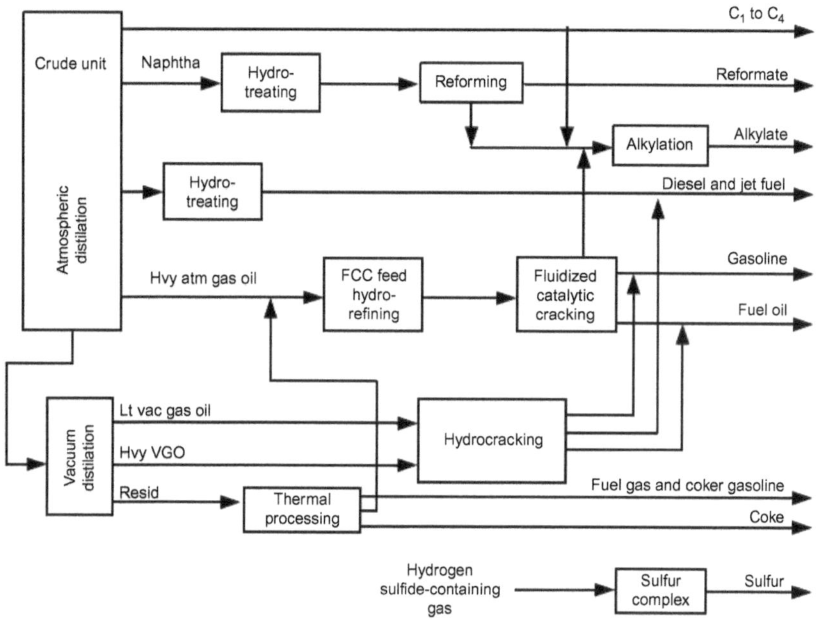

FIG. 16.1 Schematic overview of a refinery.

This chapter presents suggestions and opinions of the means by which refinery processes will evolve during the next three to five decades. The material presented is relevant to (i) comparisons of current feedstocks with viscous oil and biofeedstocks, (ii) evolution of refineries since the 1950s, (iii) properties and refinability of viscous oil and biofeedstocks, (iv) thermal processes compared with hydroprocesses, and (v) evolution of products to match the environmental market, with more than passing mention of the effects of feedstocks from coal and oil shale.

2. History

Refining technology has evolved considerably over the last century in response to changing requirements such as (i) demand for gasoline and diesel fuel as well as fuel oil, (ii) petrochemicals as building blocks for clothing and consumer goods, and (iii) more environmentally friendly processes and products.

As a result of this response, the production facilities within the refining industry have become increasingly diverse—process configuration varies from plant to plant according to its size, complexity, and product slate. There are small refineries—1500–5000 barrels per day (bpd)—and large refineries that process in excess of 250,000 barrels per day. Some are relatively simple (such as the topping refinery and the hydroskimming refinery) (Chapter 7) and produce only fuels, while other refineries, such as those with integrated petrochemical processing capabilities, are much more complex. Many refineries are part of large integrated oil companies engaged in all aspects of the crude oil technology—from (i) exploration, (ii) production, (iii) transportation, (iv) refining, and (v) marketing of crude oil products. Historically, this has not always been the case. In the early days of the 20th century, refining processes were developed to extract kerosene for lamps. Any other products were considered to be unusable and were usually discarded. A brief history of crude oil refining is presented in the following paragraphs.

In 1861, the first crude oil refinery opened and produced kerosene—a better and cheaper source of illumination as an alternative to whale oil—by atmospheric distillation; naphtha and *tar* (*residuum* or *cracked residuum*) were produced as by-products. This involved simple batch distillation of crude oil with the objective of maximizing kerosene production. Technological advancements included, first, the introduction of continuous distillation and, then, vacuum distillation—developed in 1870—that greatly facilitated the manufacture of lubricants.

The 1890s saw the emergence of the internal combustion engine that creates demand for diesel fuel and gasoline; demand for kerosene declines with the invention and proliferation of the electric light. Next, the quest for improved lubricants prompted the use of solvent extraction. To make better use of the bottom of the barrel, thermal cracking (in 1913) and visbreaking processes were introduced to crack higher-molecular-weight hydrocarbon derivatives to produce more valuable low-boiling fractions. The development of thermal cracking arose in response to increased demand for gasoline due to mass production manufacturing of automobiles and the outbreak of World War I. This enabled refineries to produce additional gasoline and distillate fuels by subjecting high-boiling crude oil fractions to high pressures and temperatures with the resulting production of lower-boiling, lower-molecular-weight products.

During the 1930s, many advances were made to improve gasoline yield and properties as a response to the development of higher-compression engines. This involved the development of processes such as (i) catalytic cracking, thermal reforming, and catalytic polymerization to improve octane number; (ii) hydroprocesses to remove sulfur; (iii) coking processes to produce gasoline blend stocks; (iv) solvent extraction processes to improve viscosity index of lubricating oil; and (iv) and solvent dewaxing processes to improve pour point the various products. The by-products of these various processes included aromatic derivatives, wax, residual fuel oil, coke, and low-boiling products to be used as feedstocks for the manufacture of petrochemicals.

During the 1940s, the industry turned to catalysis for major innovations. Catalytic cracking constituted a step change in the refinery's ability to convert higher-boiling constituents into highly valued naphtha and kerosene. Wartime demand for aviation fuels helped spur development of catalytic alkylation processes (which produced blend stocks for high-octane aviation gasoline) and catalytic isomerization (which produced increased quantities of feedstocks for alkylation units). To create high-octane fuels from lower-boiling hydrocarbon derivatives, the hydrogen content was distributed among the refinery products to improve the properties via catalytic reforming of naphtha, catalytic hydrodesulfurization of distillates, and hydrocracking of higher-boiling streams (e.g., gas oil). By the end of this period, almost every refining process was catalytically based.

The period in the 1950s–70s saw the development of various reforming processes, which also produced blend stocks that were used to improve gasoline quality and yield. Other processes such as deasphalting, catalytic reforming, hydrodesulfurization, and hydrocracking are examples of processes developed during this period. In this time period, refiners also started further development of the uses for the waste gases from various processes resulting in the expansion of the petrochemical industry. In the latter part of the period, the industry benefitted from a massive infusion of computer-based quantitative methodology that has significantly improved refiner's control over processes and the composition of products (Table 16.2).

Thus, first refining processes were developed to purify, stabilize, and improve the quality of kerosene. However, the invention of the internal combustion engine led (at about the time of World War I) to a demand for gasoline for use in increasing quantities as a motor fuel for cars and trucks. This demand on the lower-boiling products increased, particularly when the market for aviation fuel developed. Thereafter, refining methods had to be constantly adapted

TABLE 16.2 Factors Leading to Improved Process Control and Product Composition

Factor	Result
Kinetic modeling	Enabled the quantitative simulation of commercially important, complex chemical reactions based on lumped descriptions of molecular reaction paths
Reaction engineering	Enabled the integration of reaction and kinetic phenomena, including catalysis, with transport processes in the areas of reactor design, scale-up, and commercial operation
Automation and control	Enabled the optimization of unit-operation and economic performance
	Also enabled automatic closed-loop control technology with online sensors and analyzers and dynamic process models
Compositional modeling	Enabled an understanding of products to reactants through explicit reaction pathways on a molecular basis
	Also enabled the quantitative prediction of composition and properties of product streams based on molecular composition

and improved to meet the quality requirements and needs of fuels as well as a variety of other products.

Crude oil is no different to many other natural resources insofar as production rises quickly at first and then gradually slows until approximately half the original supply has been exhausted; at that point, a peak in sustainable output is attained, and production begins an irreversible decline until it becomes too expensive to recover the remaining in-ground material. In accordance with Hubbert's postulate (Hubbert, 1962), many scientists and engineers believe that the midway point (the peak of the bell curve) in the depletion of the original world crude oil inheritance has been reached and severe depletion of the resource is already underway. On the other hand, there are opposing claims to the effect that the peak of oil production is still years, or maybe decades, away and the world has not yet reached the depletion side of the bell curve (Speight and Islam, 2016).

The logic behind Hubbert's postulate and the interpretation of the postulate by others is that it is assumed that the extraction of oil, like that of other nonrenewable resources, will follow a parabolic (bell-shaped) curve over time and the only unknown is the characteristics of the slope on the depletion side of the curve. However, there is also some justification that (because of poor estimation of the resource originally present in the underground formations, poor planning of recovery methods, inefficient recovery methods, and the like) the bell curve may be skewed and the representations of resource limitations made in the past may not be as accurate as originally thought. There is no doubt, however, that crude oil reserves are being depleted but, because of the criteria mentioned in the previous sentence, the rate of depletion is not and may never be accurately known. What is certain is that when the oil runs out, it will be widely known and publicized.

However, it is certain that the reserves of crude oil are being depleted and the discovery of new fields has slowed. In fact, many scientists and engineers conjecture that new giant oil reservoirs are waiting to be discovered, while the opposing view that there are no new giant oil reservoirs holds (at least) equal weight. If this latter opinion is the case, as the world sinks into crude oil oblivion, it is worth considering the phenomenon of the crude oil culture (sometimes referred to as *petropolitics*) since this has a high influence on crude oil economics (Speight, 2011a, 2011b).

Over the longer term, primary global energy demand is expected to increase by 50%–60% by the year 2030, driven primarily by population growth and the desire for better living standards. Conventional oil and gas alone are unlikely to satisfy the demand growth, and there is the need for the development of supplemental energy sources such as clean coal, oil shale, and biofuels (Mut, 2005; Speight, 2008).

The refining industry has been the subject of the four major forces that affect most industries and that have hastened the development of new crude oil refining processes: (i) the high demand for products such as gasoline, diesel, fuel oil, and jet fuel; (ii) uncertain feedstock supply, specifically the changing quality of crude oil and geopolitics between different countries and the emergence of alternate feed supplies such as bitumen from tar sand, natural gas, and coal; (iii) recent environmental regulations that include more stringent regulations in relation to sulfur in gasoline and diesel; and (iv) continued technology development such as new catalysts and processes.

3. Refinery Configurations

A crude oil refinery is an industrial processing plant that is a collection of integrated process units (Parkash, 2003; Gary et al., 2007; Speight, 2014, 2017; Hsu and Robinson, 2017). The crude oil feedstock is typically a blend of two or more crude oils, often with viscous oil, extra viscous oil, or even tar sand bitumen blended in to a maximum (but compatible) amount. With the depletion of known crude oil reserves, refining companies are having to seek crude oil in places other than the usual sources of supply. It is anticipated that while most of the tried-and-true processes will remain the mainstay of refining for the next several decades (Speight, 2011b), there will be many attempts to improve the refinability of viscous oil, extra viscous oil, and tar sand bitumen by rearranging and reorganizing the various process sequences.

3.1 Crude Oil Refinery

The definition of crude oil is confusing and variable (Chapter 1) (Speight, 2014) and has been made even confusing by the introduction of other terms (Zittel and Schindler, 2007) that add little, if anything, to crude oil definitions and terminology (Speight, 2008).

In fact, there are different classification schemes based on (i) economic and/or (ii) geologic criteria. For example, the economic definition of conventional oil is "conventional oil is oil which can be produced with current technology under present economic conditions." The problem with this definition is that it is not very precise and changes whenever the economic or technological aspects of oil recovery change. In addition, there are other classifications based on API gravity such as "conventional oil is crude oil having a viscosity above 17° API." However, these definitions do not change the definition stated elsewhere (Chapter 1), which have been used through this book.

In recent years, the quality of the crude oil entering the refinery has deteriorated and continues to do so as more viscous oil, extra viscous oil, and tar sand bitumen are being sent to refineries (Parkash, 2003; Gary et al., 2007; Speight, 2008, 2014; Hsu and Robinson, 2017; Speight, 2017). This has caused the nature of crude oil refining that has been changed considerably. Indeed, the declining reserves of low-density crude oil have resulted in an increasing need to develop options to desulfurize and upgrade the viscous feedstocks. This has resulted in a variety of process options that specialize in sulfur removal during refining.

In addition, the general trend throughout refining has been to produce more products from each barrel of crude oil and to process those products in different ways to meet the product specifications for use in modern engines. Overall, the demand for gasoline has rapidly expanded, and demand has also developed for gas oils and fuels for domestic central heating and fuel oil for power generation, as well as for low-boiling distillates and other inputs, derived from crude oil, for the petrochemical industries.

As the need for the lower-boiling products developed, crude oil yielding the desired quantities of the lower-boiling products became less available, and refineries had to introduce conversion processes to produce greater quantities of lower-boiling products from the higher-boiling fractions. The means by which a refinery operates in terms of producing the relevant products depends not only on the nature of the crude oil feedstock but also on its configuration (i.e., the number of types of the processes that are employed to produce the desired product slate), and the refinery configuration is, therefore, influenced by the specific demands of a market.

Therefore, refineries need to be constantly adapted and upgraded to remain viable and responsive to ever-changing patterns of crude supply and product market demands. As a result, refineries have been introducing increasingly complex and expensive processes to gain higher yields of lower-boiling products from the higher-boiling fractions and residua.

Finally, the yields and quality of refined crude oil products produced by any given oil refinery depend on the mixture of crude oil used as feedstock and the configuration of the refinery facilities. Low-density sweet (low-sulfur) crude oil is generally more expensive and has inherent great yields of higher-value low-boiling products such naphtha, gasoline, jet fuel, kerosene, and diesel fuel. Viscous sour crude oil is generally less expensive and produces greater yields of lower-value higher-boiling products that must be converted into lower-boiling products.

The configuration of refineries may vary from refinery to refinery. Some refineries may be more oriented toward the production of gasoline (large reforming and/or catalytic cracking), whereas the configuration of other refineries may be more oriented toward the production of middle distillates such as jet fuel and gas oil.

Changes in the characteristics of conventional crude oil can be exogenously specified and will trigger changes in refinery configurations and corresponding investments. The future crude slate is expected to consist of larger amounts of both viscous feedstocks and low-boiling feedstocks, such as natural gas liquids. There will also be a greater shift toward the use of heavy oil, extra heavy oil, and tar sand bitumen.

There are four ways that are currently practiced in bringing viscous crude oil to market (Table 16.3) (Hedrick et al., 2006). However, the location where this

TABLE 16.3 Suggested Methods for Upgrading Viscous Feedstocks

Numerical Method	Comment
1	Upgrade the material in the oil field and leave much of the material behind as coke and then pipeline the upgraded material out as synthetic crude
	In this method, the crude is fractionated, and the residue is coked. The products of the coking operation and in some cases some of the residue are hydrotreated. The hydrotreated materials are recombined with the fractionated low-boiling fractions to form synthetic crude that is then transported to market in a pipeline. A portion of the crude may or may not be bypassed around the processing units
	There is example in the Fort McMurray area (Alberta, Canada) to upgrade and transport the Athabasca bitumen
2	Build upgrading facilities at an established port area with abundant gas and electric resources
	The upgrading facility fractionates the vacuum gas oil and lower-boiling materials out of the crude and cokes the residue material. A pipeline from the complex to the oil field transports cutter stock to the oil field in sufficient quantity to produce pipeline-acceptable crude
	There are several examples of this kind of facility located in the Jose, Venezuela area, that enables the production of extra viscous crude from the Orinoco River Basin
3	Use traditional crude that is located in the general area to dilute the nontraditional crude to produce an acceptable pipeline material
	However, there may be a limitation on the amount of traditional crude available for dilution
	Also, the two crudes may have limited compatibility that would limit the amount of dilution and again could limit the amount of nontraditional crude produced
4	Build a reverse pipeline from a refinery to the oilfield and a crude pipeline

upgrading capacity will be built is likely to be strongly influenced by greenhouse gas policy. In fact, the crude oil and petrochemical industries are coming under increasing pressure not only to compete effectively with global competitors utilizing more advantaged hydrocarbon feedstocks but also to ensure that its processes and products comply with increasingly stringent environmental legislation.

Crude oil refining has grown increasingly complex in the last 20 years. Lower-quality crude oil, crude oil price volatility, and environmental regulations that require cleaner manufacturing processes and higher-performance products present new challenges to the refining industry. Improving processes and increasing the efficiency of energy use are key to meeting the challenges and maintaining the viability of the crude oil refining industry.

There is also the need for a refinery to be able to accommodate *opportunity crude oils* and/or *high-acid crude oils* (Chapter 1).

Opportunity crude oils are often dirty and need cleaning before refining by removal of undesirable constituents such as high-sulfur, high-nitrogen, and high-aromatic (such as polynuclear aromatic) components. A controlled visbreaking treatment would *clean up* such crude oils by removing these undesirable constituents (which, if not removed, would cause problems further down the refinery sequence) as coke or sediment.

On the other hand, high-acid crude oils cause corrosion in the atmospheric and vacuum distillation units. In addition, overhead corrosion is caused by the mineral salts, magnesium, calcium, and sodium chloride that are hydrolyzed to produce volatile hydrochloric acid, causing a highly corrosive condition in the overhead exchangers. Therefore, these salts present a significant contamination in opportunity crude oils. Other contaminants in opportunity crude oils that are shown to accelerate the hydrolysis reactions are inorganic clays and organic acids.

In addition to taking preventative measure for the refinery to process these high-margin crude oils without serious deleterious effects on the equipment, refiners will need to develop programs for detailed and immediate feedstock evaluation so that they can understand the qualities of a crude oil very quickly and it can be valued appropriately. There is also the need to assess the potential impact of contaminants, like metals or acidity, in crudes so that the feedstock can be correctly valued and management of the crude processing can be planned.

3.2 Gasification Refinery

A gasification refinery would have, as the centerpiece, gasification technology as is the case of the Sasol refinery in South Africa (Couvaras, 1997; Speight, 2011b). The refinery would produce synthesis gas (from the carbonaceous feedstock) from which liquid fuels would be manufactured using the Fischer-Tropsch synthesis technology (Chapter 13).

Synthesis gas (*syngas*) is the name given to a gas mixture that contains varying amounts of carbon monoxide and hydrogen generated by the gasification of a carbon-containing fuel to a gaseous product with a heating value. Examples include the gasification of coal or crude oil residua (Speight, 2008). Synthesis gas is used as a source of hydrogen or as an intermediate in producing hydrocarbon derivatives via the Fischer-Tropsch synthesis.

In fact, gasification to produce synthesis gas can proceed from any carbonaceous material, including biomass. Inorganic components of the feedstock, such as metals and minerals, are trapped in an inert and environmentally safe form as char, which may have use as a fertilizer. Biomass gasification is therefore one of the most technically and economically convincing energy possibilities for a potentially carbon-neutral economy.

The manufacture of gas mixtures of carbon monoxide and hydrogen has been an important part of chemical technology for about a century. Originally, such mixtures were obtained by the reaction of steam with incandescent coke and were known as *water gas*. Eventually, steam reforming processes, in which steam is reacted with natural gas (methane) or crude oil naphtha over a nickel catalyst, found wide application for the production of synthesis gas.

A modified version of steam reforming known as autothermal reforming, which is a combination of partial oxidation near the reactor inlet with conventional steam reforming further along the reactor, improves the overall reactor efficiency and increases the flexibility of the process. Partial oxidation processes using oxygen instead of steam also found wide application for synthesis gas manufacture, with the special feature that they could utilize low-value feedstocks such as the viscous feedstocks. In recent years, catalytic partial oxidation employing very short reaction times (milliseconds) at high temperatures (850–1000 °C) is providing still another approach to synthesis gas manufacture (Hickman and Schmidt, 1993).

In a gasifier, the carbonaceous material undergoes several different processes: (i) pyrolysis of carbonaceous fuels, (ii) combustion, and (iii) gasification of the remaining char. The process is very dependent on the properties of the carbonaceous material and determines the structure and composition of the char, which will then undergo gasification reactions.

As crude oil supplies decrease, the desirability of producing gas from other carbonaceous feedstocks will increase, especially in those areas where natural gas is in short supply. It is also anticipated that costs of natural gas will increase, allowing coal gasification to compete as an economically viable process. Research in progress on a laboratory and pilot-plant scale should lead to the invention of new process technology by the end of the century, thus accelerating the industrial use of coal gasification.

The conversion of the gaseous products of gasification processes to synthesis gas, a mixture of hydrogen (H_2) and carbon monoxide (CO), in a ratio appropriate to the application, needs additional steps, after purification. The product gases—carbon monoxide, carbon dioxide, hydrogen, methane, and nitrogen—can be used as fuels or as raw materials for chemical or fertilizer manufacture.

3.2.1 Gasifier Types

A gasifier differs from a combustor in that the amount of air or oxygen available inside the gasifier is carefully controlled so that only a relatively small portion of the fuel burns completely. The *partial oxidation* process provides the heat, and rather than combustion, most of the carbon-containing feedstock is chemically broken apart by the heat and pressure applied in the gasifier resulting in the chemical reactions that produce synthesis gas. However, the composition of the synthesis gas will vary because of dependence upon the conditions in the gasifier and the type of feedstock.

Minerals in the fuel (i.e., the rocks, dirt, and other impurities that do not gasify) separate and leave the bottom of the gasifier as either an inert glass-like slag or other marketable solid products.

Four types of gasifier are currently available for commercial use: (i) the countercurrent fixed-bed gasifier, (ii) cocurrent fixed-bed gasifier, (iii) the fluidized-bed gasifier, and (iv) the entrained flow gasifier (Speight, 2008, 2013).

The countercurrent fixed-bed (updraft) gasifier consists of a fixed bed of carbonaceous fuel (e.g., coal or biomass) through which the gasification agent (steam, oxygen, and/or air) flows in countercurrent configuration. The ash is either removed dry or as a slag. The nature of the gasifier means that the fuel must have high mechanical strength and must be noncaking so that it will form a permeable bed, although recent developments have reduced these restrictions to some extent. The throughput for this type of gasifier is relatively low. Thermal efficiency is high as the gas exit temperatures are relatively low and, as a result, tar and methane production is significant at typical operation temperatures, so product gas must be extensively cleaned before use or recycled to the reactor.

The cocurrent fixed-bed (downdraft) gasifier is similar to the countercurrent type, but the gasification agent gas flows in cocurrent configuration with the fuel (downward, hence the name down draft gasifier). Heat needs to be added to the upper part of the bed, either by combusting small amounts of the fuel or from external heat sources. The produced gas leaves the gasifier at a high temperature, and most of this heat is often transferred to the gasification agent added in the top of the bed. Since all tars must pass through a hot bed of char in this configuration, tar levels are much lower than the countercurrent type.

In the fluidized-bed gasifier, the fuel is fluidized in oxygen (or air) and steam. The temperatures are relatively low in dry-ash gasifiers, so the fuel must be highly reactive; low-grade coals are particularly suitable. The agglomerating gasifiers have slightly higher temperatures and are suitable for higher-rank coals. Fuel throughput is higher than for the fixed bed, but not as high as for the entrained flow gasifier. The conversion efficiency is typically low, so recycle or subsequent combustion of solids is necessary to increase conversion. Fluidized-bed gasifiers are most useful for fuels that form highly corrosive ash that would damage the walls of slagging gasifiers. The ash is removed dry or as high-density agglomerate material—a disadvantage of biomass feedstocks is that they generally contain high levels of corrosive ash.

In the entrained flow gasifier, a dry pulverized solid, an atomized liquid fuel, or a fuel slurry is gasified with oxygen (much less frequent, air) in cocurrent flow. The high temperatures and pressures also mean that a higher throughput can be achieved but thermal efficiency is somewhat lower as the gas must be cooled before it can be sent to a gas processing facility. All entrained flow gasifiers remove the major part of the ash as a slag as the operating temperature is well above the ash fusion temperature. Biomass can form slag that is corrosive for ceramic inner walls that serve to protect the gasifier outer wall.

In integrated gasification combined cycle (IGCC) systems, the synthesis gas is cleaned of its hydrogen sulfide, ammonia, and particulate matter and is burned as fuel in a combustion turbine (much like natural gas is burned in a turbine). The combustion turbine drives an electric generator. Hot air from the combustion turbine can be channeled back to the gasifier or the air separation unit, while exhaust heat from the combustion turbine is recovered and used to boil water, creating steam for a steam turbine generator. The use of these two types of turbines—a combustion turbine and a steam turbine—in combination, known as a combined cycle, is one reason why gasification-based power systems can achieve unprecedented power generation efficiencies.

Gasification also offers more scope for recovering products from waste than incineration. When waste is burnt in an incinerator, the only practical product is energy, whereas the gases, oils, and solid char from pyrolysis and gasification can be not only used as a fuel but also purified and used as a feedstock for petrochemicals and other applications. Many processes also produce a stable granulate instead of an ash that can be more easily and safely utilized. In addition, some processes are targeted at producing specific recyclables such as metal alloys and carbon black. From waste gasification, in particular, it is feasible to produce hydrogen, which many see as an increasingly valuable resource.

Integrated gasification combined cycle (IGCC) is used to raise power from feedstocks such as vacuum residua, cracked residua, and deasphalter pitch. The value of these refinery residuals, including crude oil coke, will need to be considered as part of an overall upgrading project. Historically, many delayed coking projects have been evaluated and sanctioned on the basis of assigning zero value to crude oil coke having high sulfur and high metal content.

While there are many alternate uses for the synthesis gas produced by gasification, a combination of products/utilities can be produced in addition to power. A major benefit of the integrated gasification combined cycle concept is that power can be produced with the lowest sulfur oxide (Sox) and nitrogen oxide (NOx) emissions of any liquid/solid feed power generation technology.

3.2.2 Fischer-Tropsch Synthesis

The synthesis reaction is dependent of a catalyst, mostly an iron or cobalt catalyst where the reaction takes place. There is either a low or high temperature process (LTFT and HTFT), with temperatures ranging between 200°C and 240°C for LTFT and 300°C and 350°C for HTFT. The HTFT uses an iron catalyst and the LTFT either an iron or a cobalt catalyst. The different catalysts include also nickel-based and ruthenium-based catalysts, which also have enough activity for commercial use in the process.

The reactors are the multitubular fixed-bed, the slurry, and the fluidized-bed (with either fixed or circulating bed) reactors. The fixed-bed reactor consists of thousands of small tubes with the catalyst as surface-active agent in the tubes. Water surrounds the tubes and regulates the temperature by settling the pressure of evaporation. The slurry reactor is widely used and consists of fluid and solid elements, where the catalyst has no particularly position but flows around as small pieces of catalyst together with the reaction components. The slurry and fixed-bed reactor are used in LTFT. The fluidized-bed reactors are diverse but characterized by the fluid behavior of the catalyst.

The high-temperature Fischer-Tropsch technology uses a fluidized catalyst at 300–330°C. Originally, circulating fluidized-bed units were used (Synthol reactors). Since 1989, a commercial-scale classical fluidized-bed unit has been implemented and improved upon.

The low-temperature Fischer-Tropsch technology has originally been used in tubular fixed-bed reactors at 200–230°C. This produces a more paraffinic and waxy product spectrum than the high-temperature technology. A new type of reactor (the Sasol slurry-phase distillate reactor) has been developed and is in commercial operation. This reactor uses a slurry-phase system rather than a tubular fixed-bed configuration and is currently the favored technology for the commercial production of synfuels.

Under most circumstances, the production of synthesis gas by reforming natural gas will be more economical than from coal gasification, but site-specific factors need to be considered. In fact, any technological advancement in this field (such as better energy integration or the oxygen transfer ceramic membrane reformer concept) will speed up the rate at which the synfuel technology will become common practice.

There are large coal reserves that may increasingly be used as a fuel source during oil depletion. Since there are large coal reserves in the world, this technology could be used as an interim transportation fuel if conventional oil was to become more expensive. Furthermore, combination of biomass gasification and Fischer-Tropsch synthesis is a very promising route to produce transportation fuels from renewable or green resources.

Although the focus of this section has been on the production of hydrocarbon derivatives from synthesis gas, it is worthy of note that clean synthesis gas can also be used (i) as chemical *building blocks* to produce a broad range of chemicals using processes well established in the chemical and petrochemical industry, (ii) as a fuel producer for highly efficient fuel cells (which run off the hydrogen made in a gasifier) or perhaps in the future hydrogen turbines and fuel cell-turbine hybrid systems, and (iii) as a source of hydrogen that can be separated from the gas stream and used as a fuel or as a feedstock for refineries (which use the hydrogen to upgrade crude oil products).

4. The Future Refinery

As a consequence of the high demand on crude oil derivatives, the near future will, in spite of the discovery of crude oil in tight formations (Chapter 1), be inevitably dominated by refining the viscous feedstocks. Over the past three decades, the refining industry has been challenged by changing feedstocks and product slate. As a result, the refining industry will become increasingly flexible with improved technologies and improved catalysts (Speight, 2011a).

However, even the *tried-and-true processes* will see changes and ensuing risks as they evolve. The distillation units will continue to be the mainstay of crude oil refining, and the main short-term developments are in improved integration through the use of heat recovery technology and integration of different distillation units (i.e., atmospheric distillation unit and the vacuum distillation unit). In the long term, the major developments are the integration of different distillation columns into one reactor (e.g., dividing-wall column) or the development of alternative processing routes allowing for combination of conversion and distillation (such as reactive distillation). Alternative processes to distillation will also include membranes and technologies such as freeze concentration.

Upgrading viscous feedstocks can currently be achieved through thermal cracking and hydrocracking processes. However, understanding of the role of conventional catalysts during refining viscous feedstocks has motivated the development of catalysts with high deactivation resistance and selectivity to convert these feedstocks. Moreover, textural properties have also been considered since large pores are required to handle high-molecular-weight polar constituents of these feedstocks. Macroporous catalysts avoid poisoning and/or fouling caused by blocking the active pore mouth by either deposition of metals or coke. In any case, the use of a guard-bed-type reactor is highly desirable to retain organometallic compounds that are difficult to process. In addition, high conversions are achieved when using catalysts in dispersed phase (slurry), but it has an inconvenience in the recovery of the catalyst to be reused from the products and unreacted feed, forcing them to be inexpensive materials with high catalytic activity. Unfortunately, there is no ideal process to handle the processing of viscous feedstocks for all refineries in the world, since these feedstocks vary as also do the parameters to be achieved in order to meet the needs of the market that is being addressed and the environmental standards. Everything will depend on the configuration of the refinery, the crude oil prices and products, local needs, and market trends, which will be defined according to the needs and technical and financial capacity of the refinery.

In the near future and in spite of the apparent glut of low-density crude oil from shale formations, the refining industry will have to turn once again to the heavier feedstocks (viscous oil, extra viscous oil, and tar sand bitumen) as the mainstay of the industry. The main technological progress will be directed increasingly to flexibility toward a variety of feedstocks by application of improved technologies and improved catalysts. This will be necessary for the continuation of the production of upgrading operations to produce clean transportation fuel production and the production of starting materials for petrochemical production. Even the *tried-and-true processes* will see changes as they evolve (Speight, 2011a). The *distillation units* will continue to be the mainstay of crude oil refining, and the main short-term developments are in improved integration through the use of heat recovery technology and integration of different distillation units (i.e., atmospheric distillation unit and the vacuum distillation unit) (Speight, 2011a). In the long term, the major developments are the integration of different distillation columns into one reactor (e.g., dividing-wall column) or the development of alternative processing routes allowing for combination of conversion and distillation (such as reactive distillation). Alternative processes to distillation will also include membranes and technologies such as freeze concentration.

The limitations of processing the more complex difficult-to-convert viscous oil, extra viscous oil, and tar bitumen depend to a large extent on the amount of nonvolatile higher-molecular-weight constituents, which also contain the majority of the heteroatoms (i.e., nitrogen, oxygen, sulfur, and metals such as nickel and vanadium). The chemistry of the thermal reactions of some of these constituents dictates that certain reactions, once initiated, cannot be reversed and proceed to completion and coke is the eventual product (Parkash, 2003; Gary et al., 2007; Speight, 2014, 2017; Hsu and Robinson, 2017).

Thermal cracking processes offer attractive methods of conversion of viscous oil, extra viscous oil, and tar sand bitumen because they enable low operating pressure while involving high operating temperature, without requiring expensive catalysts. Currently, the widest operated conversion processes are (i) visbreaking, (ii) delayed coking, and (iii) fluid coking, which also includes flexicoking—these are attractive processes for refineries from an economic point of view. These processes (Chapter 8) will also evolve and become more efficient. While the current processes may not see much change in terms of reactor vessel configuration, there will be changes to the reactor internals and to the nature of the catalysts. For example, the *tried-and-true coking processes* will remain the mainstay of refineries coping with an influx of viscous oil, extra viscous oil, and tar sand bitumen, but other process options will be developed with time.

For example, visbreaking (or even hydrovisbreaking—i.e., visbreaking in an atmosphere of hydrogen or in the presence of a hydrogen donor material) (Chapter 8), the long-ignored stepchild of the refining industry, may see a surge in use as a pretreatment process. Management of the process to produce a liquid product that has been freed of the high potential for coke deposition (by taking the process parameters into the region where sediment forms) in either the absence or presence of, for example, a metal oxide scavenger could be a valuable ally to catalyst cracking or hydrocracking units.

In the integration of refining and petrochemical businesses, new technologies based on the traditional fluid catalytic cracking process (Chapter 9) will be of increased interests to refiners because of their potential to meet the increasing demand for low-boiling olefins. Meanwhile, hydrocracking, due to its flexibility, will take the central position in the integration of refining and petrochemical businesses in the 21st century.

Alternately, operating the catalytic cracking unit solely as a slurry riser cracker (without the presence of the main reactor) followed by separation of coke (sediment) would save the capital outlay required for a new catalytic cracker and might even show high conversion to valuable liquids. The quality (i.e., boiling range) of the distillate would be dependent upon the residence time of the slurry in the pipe.

Scavenger additives such as metal oxides may also see a surge in use. As a simple example, a metal oxide (such as calcium oxide) has the ability to react with sulfur-containing feedstock to produce a hydrocarbon (and calcium sulfide):

$$\text{Feedstock}\,[S] + CaO \rightarrow \text{hydrocarbon product} + CaS + H_2O$$

Propane has been used extensively in deasphalting viscous feedstocks, especially in the preparation of high-quality lubricating oils and feedstocks for catalytic cracking units (Chapter 9). The use of propane has necessitated elaborate solvent cooling systems utilizing cold water, which is a relatively expensive cooling agent. In order to circumvent such technology, future units will use solvent systems that will allow operation at elevated temperatures relative to conventional propane deasphalting temperatures, thereby permitting easy heat exchange. In addition, it may be found desirable to integrate dewaxing operations with deasphalting operations by having a common solvent recovery system. This will require changes to the solvent composition and the inclusion of solvents not usually considered to be deasphalting solvents.

Furthermore, as a means of energy reduction for the process, in future deasphalting units, the conventional solvent recovery scheme will be retrofitted with supercritical solvent recovery scheme to reap the benefits of higher energy efficiency. Other improvements will include variations in the extraction column internals. For example, the three major properties, which influence the design of extraction column, are interfacial tension, viscosity, and density of phases. Solvent deasphalting extraction column is characterized by low interfacial tension, high viscosities of asphaltene phase, and a density difference between the phases. Extension of these property correlations for solvent deasphalting applications will be suitably validated, and corrections made were necessary to improve extraction performance and yields of the products. Other areas of future process modification will be in the extractor tower internals, studies with higher-molecular-weight solvent, accurate estimation of physical properties of mix stream, studies in combination with other processes, and firming up design tools for supercritical solvent recovery configuration. However, the solvent deasphalting process involves physical separation, and there is no chemical conversion. The limitations of this process are high energy costs and the limited uses of deasphalter tar. Current interest in deasphalting is greatest in areas of the world where demand for motor fuel is low. This suggests that, in the long run, solvent deasphalting, as a stand-alone residue upgrading process, will be of less interest worldwide.

In the long term, new desulfurization technologies or evolution of the older technologies (Chapter 10) will reduce the need for hydrogen. At the same time, refineries are constantly faced with challenges to reduce air pollution and other energy-related issues. Thus, traditional end-of-pipe air emission control technologies will lead to increased energy use and decreasing energy efficiency in the refinery. The crude oil refining industry will face many other challenges—climate change and new developments in automotive technology and biotechnology—that are poised to affect the future structure of refineries.

Biotechnology can contribute to the viscous oil industry by assisting in processes for the production of liquid fuels, upgrading liquid fuels to meet sales specification, and in the bioremediation of water, soil, and air and in the control of microbiologically influenced corrosion (Youssef et al., 2009; Bachmann et al., 2014). The application of biotechnology to increase the production of fossil fuels is mostly experimental, but the potential growth of this area is immense. Increasing the recovery of energy from depleted/uneconomical crude oil and coal deposits, particularly in combination with the utilization of carbon dioxide, could be a major component of the biotechnology industry in the future.

The increasing focus to reduce sulfur content in fuels will assure that the role of *desulfurization* in the refinery increases in importance (Babich and Moulijn, 2003). Currently, the process of choice is the hydrotreater, in which hydrogen is added to the fuel to remove the sulfur from the fuel. Some hydrogen may be lost to reduce the octane number of the fuel, which is undesirable. Because of the increased attention for fuel desulfurization, various new process concepts are being developed with various claims of efficiency and effectiveness. The major developments in desulfurization of three main routes will be (i) advanced hydrotreating (new catalysts, catalytic distillation, and processing at mild conditions), (ii) reactive adsorption (type of adsorbent used and process design), and (iii) oxidative desulfurization (catalyst and process design).

In addition, the most common approaches to upgrading hydrotreaters for clean-fuel production will continue to be (i) developing higher-activity and more resilient catalysts, (ii) replacing reactor internals for increased efficiency, (iii) adding reactor capacity to accommodate viscous feedstocks and increase gasoline-diesel production, (iv) increasing hydrogen partial pressure, (v) adding process design and hardware that are more specialized, and (vi) focusing on process schemes that effectively reduce hydrogen consumption.

However, residuum hydrotreating requires considerably different catalysts and process flows, depending on the specific operation so that efficient hydroconversion through uniform distribution of liquid, hydrogen-rich gas, and catalyst across the reactor is assured. In addition to an increase in *guard-bed* use, the industry will see an increase in automated demetallization of fixed-bed systems and more units that operate as ebullating-bed hydrocrackers.

For viscous oil upgrading, hydrotreating technology (Chapter 10) and hydrocracking technology (Chapter 11) will be the processes of choice. For cleaner transportation fuel production, the main task is the desulfurization of gasoline and diesel. With the advent of various techniques, such as adsorption and biodesulfurization, the future development will be still centralized on hydrodesulfurization techniques.

Adsorption, biodesulfurization, and hydrodesulfurization are more likely to be employed to complimentary each other. However, in order to develop any bioconversion process as a complementary process, a multidisciplinary approach is essential, and the participation of scientists and engineers from the fields of biotechnology, biochemistry, refining processes, and engineering is essential. In the long run, applying nanotechnology to desulfurization of crude oil and crude oil products will lead to a process capable of producing ultra-low-sulfur to no-sulfur products on an economical and cost-effective basis.

In fact, hydrocracking (Chapter 11) will continue to be an indispensable processing technology to modern crude oil refining and petrochemical industry due to its flexibility to feedstocks and product scheme and high-quality products. Particularly, high-quality naphtha, jet fuel, diesel, and lube base oil can be produced through this technology. The hydrocracker provides a better balance of gasoline and distillates, improves gasoline yield and octane quality, and can supplement the fluid catalytic cracker to upgrade viscous feedstocks. In the hydrocracker, higher-boiling fuel oil is converted into lower-boiling products under a high hydrogen pressure and over a hot catalyst bed—the main products are naphtha and kerosene.

For the viscous feedstocks (and even for biofeedstocks), which will increase in amounts in terms of hydrocracking feedstocks, reactor designs will continue to focus on online catalyst addition and withdrawal. Fixed-bed designs have suffered from (i) mechanical inadequacy when used for the heavier feedstocks and (ii) short catalyst lives—six months or less—even though large catalyst volumes are used (LHSV typically of 0.5–1.5). Refiners will attempt to overcome these shortcomings by innovative designs, allowing better feedstock flow and catalyst utilization or online catalyst removal. For example, the OCR process, in which a lead moving-bed reactor is used to demetallized viscous feed ahead of the fixed-bed hydrocracking reactors, has seen some success. But whether this will be adequate for continuous hydrocracking viscous feedstock remains a question.

Catalyst development will be key in the modification of processes and the development of new ones to make environmentally acceptable fuels (Rostrup-Nielsen, 2004). Conversion of crude oil is expected to remain the principal source of motor fuels for another 30–50 years, but it is likely that the production of fuel additives in large quantities along with conversion of natural gas will become significant (Sousa-Aguiar et al., 2005). Although crude oil conversion is expected to remain the principal source of fuels and petrochemicals in the future, natural gas reserves will continue to emerge as a major hydrocarbon resource. This trend has already started to result in a shift toward use of natural gas (methane) as a significant feedstock for chemicals and for fuels as well. As a result, deployment of technology for direct and indirect conversion of methane will probably displace much of the current production of liquefied natural gas.

Innovations have occurred in catalyst materials that have allowed refiners to vastly improve environmental performance, product quality and volume, feedstock flexibility, and energy management without fundamentally changing fixed capital stocks. Advanced design and manufacturing techniques mean that catalysts can now be formulated and manufactured for specific processing units, feedstocks, operating environments, and finished product slates. These efforts will continue, and more simpler cheaper catalysts will be developed. The precise configuration of the refinery of the future is unknown, but it is certain that no two refineries will adapt in exactly the same way. Over the past century, the refining industry has been innovative and able to develop new processes. This trend will continue to innovate, and by the period 2030–50, refineries will be more technologically advanced, and the products will be more environmentally acceptable.

However, the evolution of the refinery of the future will not be strictly confined to crude oil processes. The major consequence will be a much more environmentally friendly product quality. These will be solved in refinery of the future, the refinery beyond 2020 with the development of deep conversion processing, such as residue hydrocracking and the inclusion of processes to accommodate other feedstocks.

The *panacea* (rather than a *Pandora's box*) for a variety of feedstock could well be the *gasification refinery* (Fig. 16.2) This type of refinery approaches that of a petrochemical complex, capable of not only supplying the traditional refined products but also meeting much more severe specifications, and petrochemical intermediates such as olefins, aromatics, hydrogen, and methanol. Furthermore, as already noted above, integrated gasification combined cycle (IGCC) can be used to raise power from feedstocks such as vacuum residua and cracked residua (in addition to the production of synthesis gas), and a major benefit of the integrated gasification combined cycle concept is that power can be produced with the lowest sulfur oxide (Sox) and nitrogen oxide (NOx) emissions of any liquid/solid feed power generation technology.

Besides fuel and product flexibility, gasification-based systems offer significant environmental advantages over competing technologies, particularly coal-to-electricity combustion systems. Gasification plants can readily capture carbon dioxide, the leading greenhouse gas, much more easily and efficiently than coal-fired power plants. In many instances, this carbon dioxide can be sold, creating additional value from the gasification process.

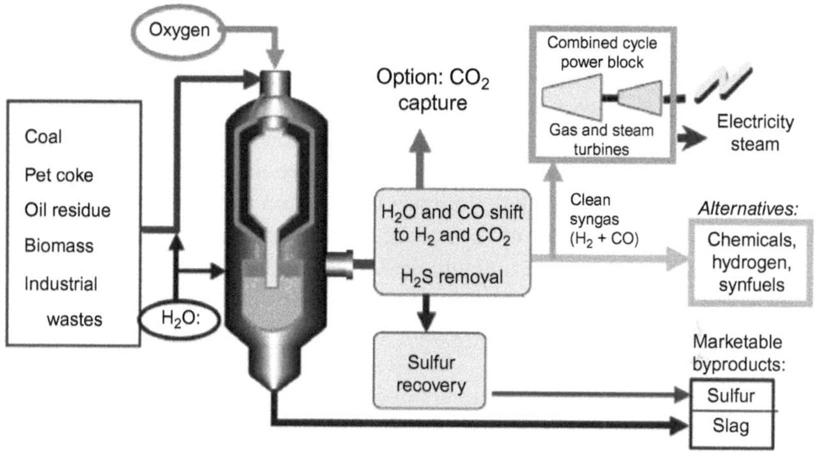

FIG. 16.2 Gasification of various feedstocks.

Thus, gasification can be proposed as a viable alternative solution for energy recovery from a variety of feedstocks. On the other hand, the process still faces some technical and economic problems, mainly related to the highly heterogeneous nature of unconventional feedstocks such as biomass and municipal solid wastes and the relatively limited number of gasification plants worldwide based on this technology that have continuous operating experience under commercial conditions (Speight, 2014).

However, technologies that ameliorate the effects of fossil fuel combustion on acid rain deposition, urban air pollution, and global warming must be pursued vigorously (Vallero, 2008). There is a challenge that must not be ignored, and the effects of acid rain in soil and water leave no doubt about the need to control its causes. Indeed, recognition of the need to address these issues is the driving force behind recent energy strategies and a variety of research and development programs.

While regulations on the greenhouse gas (GHG) carbon dioxide (CO_2) would be an immediate hurdle to deployment of coal plants, gasification plants are in the best position compared with other coal-based alternatives to capture carbon dioxide. However, with the continued uncertainty of carbon dioxide regulation, there is industry reluctance to make large investments in projects with high emissions of carbon dioxide since a cost-effective solution for reducing such emissions is not yet available. Nevertheless, the reduction in greenhouse gas emissions can be an enhancing factor for gasification in the long run because the carbon dioxide from a gasification plant is more amenable to capture.

As new technology is developed, emissions may be reduced by repowering in which aging equipment is replaced by more advanced and efficient substitutes. Such repowering might, for example, involves an exchange in which an aging unit is exchanged for a newer combustion chamber, such as the atmospheric fluidized-bed combustor (AFBC) or the pressurized fluidized-bed combustor (PFBC).

Indeed, recognition of the production of these atmospheric pollutants in considerable quantities every year has led to the institution of national emission standards for many pollutants. Using sulfur dioxide as the example, the various standards not only are very specific but also will become more stringent with the passage of time. Atmospheric pollution is being taken very seriously, and there is also the threat, or promise, of viscous fines and/or jail terms for any pollution-minded miscreants who seek to flaunt the laws (Vallero, 2008). Nevertheless, a trend to the increased use of fossil fuels will require more stringent approaches to environmental protection issues than we have ever known at any time in the past.

The typical US refinery in the 2030–50 time period will be located at an existing refinery site because economic and environmental considerations will make it difficult and uneconomical to build a new refinery at another site. Many existing refining process may still be in use, but they will be more efficient and more technologically advanced and perhaps even rebuilt (reactors having been replaced on a scheduled or as needed basis) rather than retrofitted. However, energy efficiency will still be a primary concern, as refiners seek to combat the inevitable increasing cost of crude oil and refinery operating expenses.

Moreover, the future of the crude oil refining industry will be primarily on processes for the production of improved quality products. In addition to viscous feedstocks, there will also be changes in the feedstock into a refinery. Biomass, liquids from coal, and liquids from oil shale will increase in importance. These feedstocks will be (i) sent to refineries or (ii) processed at a remote location and then blended with refinery stocks that are options for future development and the nature of the feedstocks. Above all, any such feedstock must be compatible with refinery feedstocks and not cause fouling in any form.

The refinery of the future will have a gasification section devoted to the conversion of viscous oils, coal, and biomass to Fischer-Tropsch hydrocarbon derivatives—perhaps even with rich oil shale added to the gasifier feedstock. Many refineries already have gasification capabilities, but the trend will increase to the point (over the next two decades) where nearly all refineries feel the need to construct a gasification section to handle residua and other feedstocks.

The production of high-quality fuels will result in a higher demand for related hydrogen and conversion technologies. Furthermore, the trend toward low-sulfur fuels and changes in the product mix of refineries will affect technology choice and needs. For example, the current desulfurization and conversion technologies use relatively large amounts of hydrogen (which is an energy-intensive product), and increased hydrogen consumption will lead to increased energy use and operation expenses, unless more efficient technologies for hydrogen production developed.

The demand for high-value crude oil products will maximize production of transportation fuels at the expense of both residua and low-boiling gases. Hydroprocessing of residua will be widespread rather than appearing on selected refineries. At the same time, hydrotreated residua will be the common feedstocks for fluid catalytic cracking units. Additional conversion capacity will be necessary to process increasingly heavier crudes and meet a reduced demand for residua.

Process unit and refinery economics/operation computer models will be optimized, with integration into plant operations via process computer controls. Alternate fuels for power generation will continue to push crude processing toward higher-value products, such as transportation fuels and chemicals. Otherwise, viscous oil, extra viscous oil, and tar sand bitumen that are considered uneconomical to transport to a refinery will be partially refined at their source to facilitate transport, and there will be a new emphasis on partial or full upgrading in situ during recovery operations.

The biomass refinery of the future not only will also use multiple feedstocks but also will be able to shift output from the production of one chemical to another in response to market demands. Given that biomass will be a part of a refinery of the future, refiners may dictate that biomass receives preliminary upgrading at the biomass site before being shipped to the crude oil refinery.

Other challenges facing the refining industry include its capital-intensive nature and dealing with the disruptions to business operations that are inherent in industry. It is imperative for refiners to raise their operations to new levels of performance. Merely extending current performance incrementally will fail to meet most company's performance goals.

To circumvent these issues, there may be no way out of energy production than to consorting alternative energy sources with crude oil and not of opposing them. This leads to the concept of *alternative energy systems*, which is wider ranging and more meaningful than *alternative energy sources*, because it relate to the actual transformation process of the global energy system (Szklo and Schaeffer, 2005). Alternative energy systems integrate crude oil with other energy sources and pave the way for new systems where *refinery flexibility* will be a key target, especially when related to the increased use of renewable energy sources.

Balancing product yield and market demand, without the manufacture of large quantities of fractions having low commercial value, has long required processes for the conversion of hydrocarbon derivatives of one molecular weight range and/or structure into some other molecular weight range and/or structure. One answer to the need for conversion of the heavier feedstocks is the installation and use of one or more thermal cracking process. It is in the area of processing viscous feedstocks that process innovations have been initiated over the past two decades.

The current global crude oil market will continue to offer opportunities for bottom-of-the-barrel technology to play an important role in the refiner's continuous efforts to balance available crude qualities with the market demands. Even though the future market includes ample amounts of heavier and more sour crudes, there will be the need to convert these feedstocks to salable products, and thermal conversion processes will be key to satisfying the avid demand for transportation fuels.

Thermal cracking is a high temperature process (in some cases, pressure is also applied) in which the constituents of crude oil are decomposed under the influence of to break molecular bonds and yield products having a lower molecular weight than the constituents of the feedstock. The process is one of several cracking methods used in the crude oil industry to process crude oil and other crude oil products for commercial use, and the major process variables are (i) feedstock type, (ii) temperature, (iii) pressure, and residence time, which need to be considered to maximize the yield of distillates and minimize the yield of coke (Parkash, 2003; Gary et al., 2007; Speight, 2014, 2017; Hsu and Robinson, 2017).

The acceptance of heavier feedstocks by refiners is requiring more extensive bottom-of-the-barrel upgrading by (i) increasing low-sulfur distillate fuel demand that will require increased cracking capacity; (ii) heavier crude oil slates that will result in a greater high-sulfur residual fuel oil production, if conversion is not added; and (iii) environmental restrictions that will result in reduced demand for high-sulfur residual fuel oil (Speight, 2011a).

Low pressures (<100 psi) and temperatures in excess of 500°C (930°F) tend to produce lower-molecular-weight hydrocarbon derivatives than those produced at higher pressures (400–1000 psi) and at temperatures up to 500°C (930°F). The reaction time is also important. Lower-boiling feedstocks (such as gas oils) and recycle oils require longer reaction times than the readily cracked viscous feedstocks.

Thermal cracking (including visbreaking) processes offer viable methods of conversion of viscous feedstocks because they enable low operating pressure while involving high operating temperature, without requiring expensive catalysts. Currently, the widest operated residuum conversion processes are visbreaking and the various delayed coking processes (Dickenson et al., 1997; Speight, 2014, 2017).

The process is a relatively low-cost and low-severity approach to improving the viscosity characteristics of the residue without attempting significant conversion to distillates. Low residence times are required to avoid coke formation, although additives can help to suppress coke deposits on the tubes of the furnace. By reducing the viscosity of the nonvolatile fraction, visbreaking reduces the amount of the more valuable distillate heating oil that is required for blending to meet the fuel oil specifications. The process is also used to reduce the pour point of a waxy residue.

Coking is a thermal process for the continuous conversion of residua into lower-boiling products. The feedstock can be atmospheric residuum, vacuum residuum, or cracked residuum, and the products are gases, naphtha, fuel oil, gas oil, and coke. The gas oil may be the major product of a coking operation and serves primarily as a feedstock for catalytic cracking units. The coke can be used as fuel, but processing for specialty uses, such as electrode manufacture, production of chemicals, and metallurgical coke, is also possible. For these latter uses, the coke may require treatment to remove sulfur and metal impurities—calcined crude oil coke can be used for making anodes for aluminum manufacture and a variety of carbon or graphite products such as brushes for electric equipment.

Coking processes utilize longer reaction times than visbreaking, and the reaction is allowed to proceed to completion. To accomplish this, drums or chambers (reaction vessels) are employed, but it is necessary to use such vessels in pairs so that coke removal can be accomplished in those vessels not onstream without interrupting the semicontinuous nature of the process.

Delayed coking is a semicontinuous (semibatch) process in which the heated charge is transferred to large soaking (or coking) drums, which provide the long residence time needed to allow the cracking reactions to proceed to completion. The process is used for treating residua and uses long reaction times in the liquid phase to convert the residue fraction of the feed to gases, distillates, and coke. The coke is a highly aromatic product, also retains sulfur, nitrogen, and metals, and is enriched in these elements relative to the feed (Feintuch and Negin, 1997; Parkash, 2003; Gary et al., 2007; Speight, 2014, 2017; Hsu and Robinson, 2017).

4.1 Use of Scavengers to Promote Initial Coke Formation

Delayed coking has an increasingly important role to play in the integration of modem crude oil refineries because of the inherent flexibility of the process to handle even the heaviest of residues. The flexibility of operation inherent in delayed coking permits refiners to process a wide variety of crude oils including those containing viscous and high-sulfur residua.

Low-pressure coking is a process designed for a once-through, low-pressure operation. The process is similar to delayed coking except that recycling is not usually practiced, and the coke chamber operating conditions are 435°C (815°F) and 25 psi. Excessive coking is inhibited by the addition of water to the feedstock in order to quench and restrict further reactions of the reactive intermediates.

High-temperature coking is a semicontinuous process designed to convert asphaltic residua to gas oil and coke as the primary products. In the process, the feedstock is transported to the heater (370°C, 700°F, and 30 psi) and finally to the coking unit, where temperatures may be as high as 980–1095°C (1800–2000°F). Volatile materials are fractionated, and after the cycle is complete, coke is collected for sulfur removal before storage.

Fluid coking is a continuous process developed in the mid-1950s that uses the fluidized solid technique to convert atmospheric residua and vacuum residua as well as cracked residua to more valuable products (Parkash, 2003; Gary et al., 2007; Speight, 2014, 2017; Hsu and Robinson, 2017). The yield of distillates from fluid coking can be improved by reducing the residence time of the cracked vapors.

The yields of products are determined by (i) the properties of the feedstock, (ii) the temperature of the fluid bed, and (iii) the residence time of the feedstock in the bed. The use of a fluidized bed reduces the residence time of the vapor-phase products in comparison with delayed coking, which in turn reduces cracking reactions leading to reduced yields of coke and increased yields of gas oil and olefins. An increase of 5°C (9°F) in the operating temperature of the fluid-bed reactor typically increases gas yield by 1% *w*/w and naphtha by approximately 1% w/w.

The disadvantage of burning the coke to generate process heat is that sulfur from the coke is liberated as sulfur dioxide. The gas stream from the coke burner also contains carbon monoxide (CO), carbon dioxide (CO_2), and nitrogen (N_2). An alternate approach is to use a coke gasifier to convert the carbonaceous solids to a mixture of carbon monoxide (CO), carbon dioxide (CO_2), and hydrogen (H_2).

Flexicoking is a direct offshoot or descendent of fluid coking and uses the same configuration as the fluid coker but includes a gasification section in which excess coke can be immediately gasified to produce fuel gas (Parkash, 2003; Gary et al., 2007; Speight, 2014, 2017; Hsu and Robinson, 2017). The flexicoking process was designed during the late 1960s and the 1970s as a means by which excess coke could be reduced in view of the gradual incursion into refinery operations of the viscous oil and tar sand bitumen—such feedstocks are notorious for producing high yields of coke (>15% by weight) in both thermal processes and in catalytic processes.

Yields of liquid products from flexicoking are the same as from fluid coking, because the coking reactor is unaltered. The main drawback of gasification is the requirement for a large additional reactor, especially if high conversion of the coke is required. Units are designed to gasify 60%–97% of the coke from the reactor. Even with the gasifier, the product coke will contain more sulfur than the feed, which limits the attractiveness of even the most advance of coking processes.

Because of their relative simplicity of design and straightforward thermal approach (some might say "brute force approach"), visbreaking and delayed coking processes will not be ignored or absent from the refinery of the future. These processes should not be ignored because they have the ability to adapt, by virtue of their relative simplicity, to the changing markets of viscous feedstock processing, tar sand bitumen processing, and biofeedstock processing. New and improved approaches are important for production of crude oil products. These will include advances in current methods, minimization of process energy losses, and improved conversion efficiency—in particular (i) mitigation of fouling in heat exchangers and (ii) improved conversion efficiency.

Delayed coking is likely to remain the workhorse of thermal cracking processes for the foreseeable future. Globally, approximately one-third of installed residue upgrading plant is by delayed coking. Although a mature process, in recent years, many developments have taken place, including (i) development of automated coke drum unheeding devices, allowing the operator to carry out the decoking procedure safely from a remote location; (ii) understanding of process parameters affecting yields, coker product qualities, and coke qualities (e.g., shot coke); and (iii) design and operation of major equipment items, in particular coke drums (allowing shorter coking cycles) and the delayed coker heater (online spalling/decoking and minimization of coking in furnace tubes).

Reactive distillation offers a promising alternative to conventional reaction-distillation schemes (Sundmacher and Kienle, 2003). Furthermore, active removal of reaction products can help shift the equilibrium of the reaction and improve the conversion efficiency. Reactive distillation has mainly been used in acetate technology (e.g., MTBE production) (Moritz and Gorak, 2002). Various research institutes and technology developers aim at developing new applications of reactive distillation. In the United States, researchers developed a reactive distillation process for isomerization to produce clean high-octane isomerate. In Europe, a collaborative project of suppliers and universities aims to improve understanding of reactive distillation and develop simulation tools to design new applications. Other new developments include using of monolithic structures that contain the catalysts (Babich and Moulijn, 2003) and reducing catalyst loss (Goetze and Bailer, 1999). Monolithic structures result in low-pressure drop.

On the other hand, steam cracking (in which a hydrocarbon stream is thermally cracking in the presence of steam) to yield a complex product mixture may also find use in biofeedstock conversion and allow a higher conversion of the feedstock to products usable in a refinery.

Fluid coking units will be the subject of changes over the next decades, although the changes may be minimal and more focused on internal operations of the units and coupling of the units with other process units. The process is valuable as it generally proved to equal or better the net realizations from delayed coking, with the spread increasing for the heavier crudes (economics being sensitive to liquid yield, coker distillate quality, and product values).

In summary, the need for more and heavier feedstock processing units has been a trend since the start of the industry. Year by year, the density and sulfur content of available crudes have slowly risen (Parkash, 2003; Gary et al., 2007; Speight, 2014, 2017; Hsu and Robinson, 2017).

From an environmental standpoint, with the trend of producing high-sulfur coke (from high-sulfur residua) and the decreasing demand for high-sulfur coke, gasification of coke to form fuel and syngas to be used within the refinery will be the only economic and environmental option for refineries in the future, and flexicoking will become a more popular choice. In addition, it is predictable that flexicoking will see an increase in use when coking residues (poor quality coke) and biofeedstock become prime feedstocks for a gasification unit.

Finally, gasification processes in which the feedstock is converted (by partial oxidation) into synthesis gas (in which the major components are hydrogen and carbon monoxide). Synthesis gas produced from a wide spectrum of fossil fuel feedstocks and biofeedstocks can be used to produce hydrogen and raise steam and/or power, but it can also be a building block in the manufacture of transportation fuels (via Fischer-Tropsch) or a wide range of petrochemicals (Boerrigter et al., 2002; Parkash, 2003; Gary et al., 2007; Speight, 2014, 2017; Hsu and Robinson, 2017).

In summary, there is a need to improve residue conversion processes such as visbreaking and the coking technologies. Part of the future growth will be at or near viscous crude and bitumen production sites to decrease viscous crude viscosity and improve the quality to ease transportation and open markets for crudes of otherwise marginal value. Visbreaking may be considered to be a conversion process rather than a process to produce fuel oil that meets specifications. Coking can be improved by reducing hydrocarbon gas formation and by inhibiting the formation of polynuclear aromatic products that are produced by the process and that are not inherent to the feedstock. Both of these processes would benefit if a higher-valued product could be produced.

5. Environmental Effects

Crude oil use is a necessary part of the modern world, hence the need for stringent controls over the amounts and types of emissions from the use of crude oil and its products, especially from plants that upgrade tar sand bitumen (Table 16.4). So it is predictable that crude oil will be a primary source of energy for the next several decades and, therefore, the message is clear. The challenge is for the development of technological concepts that will provide the maximum recovery of energy from crude oil not only cheaply but also efficiently and with minimal detriment to the environment. For example, major environmental impacts associated with enhanced oil recovery process include (i) contamination of surface water and groundwater, (ii) excessive air emissions especially from thermal operations, (iii) extreme erosion and sedimentation, and (iv) possible contamination of land.

Pollution has been obvious for a long time, although the effects were not as well realized in the past. There was copper pollution near Jericho on the west bank of the Jordan River due to copper smelting for manufacture of tools thousands of years ago. Deforestation of many areas near the Mediterranean Sea for the building of ships was a norm. Poor agricultural methods led to soil erosion. In 2500 BCE, the Sumerians used sulfur compounds to control insects, and in 1500 BCE, the Chinese used natural products to fumigate crops. Pesticides began polluting the environment hundreds of years ago. Pollution in medieval England was also noted when the smoke from coal fires made some cities

TABLE 16.4 Potential Emissions from a Tar Sand Processing Plant That Require Consideration

Source of Emission	Potential Contaminants
Process waste water	Suspended solids
	Dissolved solids
	Organic compounds
	Ammonium compounds
Upgrader area runoff	Inorganic (metal) compounds
	Suspended solids
Upgrader heaters	Particulate matter
	Carbon oxides
	Sulfur oxides
	Nitrogen oxides
Coke storage pile	Suspended solids
	Sulfur
Solid waste landfill	Suspended solids
	Organic compounds
	Inorganic (metal) compounds
	Sulfur
Power plant	Particulate matter
	Carbon oxides
	Sulfur oxides
	Nitrogen oxides
Sulfur plant	Particulate matter
	Hydrogen sulfide
	Sulfur oxides
Upgrader heaters	Particulate matter
	Carbon oxides
	Sulfur oxides
	Nitrogen oxides

almost uninhabitable (Speight, 2013). There are also documented records, courtesy of the diarist Samuel Pepys, who noted that he did not realize that sewage from a neighbor's house was leaking into his basement until he (Pepys) descended to the lower level and stood in it. This is a commentary not only on the state of sewage disposal but also on the odors that must have permeated 17th century London! However, there being only a meager awareness of the effects of waste products on human life and there being no form of environmental protection, the system of waste disposal proliferated. The 18th and 19th centuries saw an expansion of the fledgling chemical industry and an awakening of the effects of chemical on human life.

And so the 20th century was born with the continuation of less than desirable waste disposal methods until 1962 when a marine biologist (Rachel Carson) published her book *Silent Spring*. The book dealt with many environmental problems associated with chlorinated pesticides and touched off an extensive debate about safety of many different types of chemicals, a debate that continues. As a result, industry and government did some serious soul-searching at the way various waste products were affecting the environment. And methods were devised for handling chemical wastes with minimal effect on the environment.

The capacity of the environment to absorb the effluents and other impacts of process technologies is not unlimited, as some would have us believe. The environment should be considered to be an extremely limited resource, and discharge of chemicals into it should be subject to severe constraints. Indeed, the declining quality of raw materials, especially crude oil and fossil fuels that give rise to many of the gaseous emissions of interest in this text, dictates that more material must be processed to provide the needed fuels. And the growing magnitude of the effluents from fossil fuel processes has moved above the line where the environment has the capability to absorb such process effluents without disruption.

In order to combat any threat to the environment, it is necessary to understand the nature and magnitude of the problems involved (Ray and Guzzo, 1990). It is in such situations that environmental technology has a major role to play. Environmental issues even arise when outdated laws are taken to task. Thus, the concept of what seemed to be a good idea at the time the action occurred no longer holds when the law influences the environment.

Thus, both the production (Chapters 2 and 3) and processing (Chapter 7) of crude oil involve the use of a variety of substances, some toxic, including lubricants in oil wells and catalysts and other chemicals in refining (Fig. 16.1). The amounts used, however, tend to be small and relatively easy to control. More detrimental to the environment is the spillage of oil, which has been a particularly common event. Minor losses from truck and car accidents can affect rivers and streams. Leakage from underground gasoline storage tanks, many abandoned decades ago, has contaminated some local water supplies and usually requires expensive operations to either clean or seal off.

A brief survey of pollution problems will be given, but these are characteristic of all industry, and the topics are too vast to be covered adequately in this chapter. It is hoped that the chapter will make the reader aware of some of the environmental issues that can affect the crude oil industry.

The purpose of this chapter is to summarize and generalize the various pollution, health, and environmental problems especially specific to the crude oil industry and to place in perspective government laws and regulations as well as industry efforts to control these problems (Majumdar, 1993; Speight, 1996, 2005; Speight and Arjoon, 2012). The objective is to indicate the types of emissions from refinery processes and the laws that regulate these emissions.

5.1 Definitions

Briefly, crude oil production and crude oil refining produce *chemical waste*. If this *chemical waste* is not processed in a timely manner, it can become a *pollutant*.

A *pollutant* is a substance present in a particular location (*ecosystem*) when it is not indigenous to the location or is present in a greater-than-natural concentration. The substance is often the product of human activity. The pollutant, by virtue of its name, has a detrimental effect on the environment, in part or in toto. Pollutants can also be subdivided into two classes, primary and secondary:

$$\text{Source} \rightarrow \text{Primary pollutant} \rightarrow \text{Secondary pollutant}$$

A *primary pollutant* is a pollutant that is emitted directly from the source. In terms of atmospheric pollutants, examples are carbon oxides, sulfur dioxide, and nitrogen oxides from fuel combustion operations:

$$2[\mathrm{C}]_{\text{crude oil}} + \mathrm{O_2} \rightarrow 2\mathrm{CO}$$

$$[\mathrm{C}]_{\text{crude oil}} + \mathrm{O_2} \rightarrow \mathrm{CO_2}$$

$$2[\mathrm{N}]_{\text{crude oil}} + \mathrm{O_2} \rightarrow 2\mathrm{NO}$$

$$[\mathrm{N}]_{\text{crude oil}} + \mathrm{O_2} \rightarrow \mathrm{NO_2}$$

$$[\mathrm{S}]_{\text{crude oil}} + \mathrm{O_2} \rightarrow \mathrm{SO2}$$

$$2\mathrm{SO_2} + \mathrm{O_2} \rightarrow 2\mathrm{SO_3}$$

Hydrogen sulfide and ammonia are produced from processing sulfur-containing and nitrogen-containing feedstocks:

$$[\mathrm{S}]_{\text{crude oil}} + \mathrm{H_2} \rightarrow \mathrm{H_2S} + \text{hydrocarbon derivatives}$$

$$2[\mathrm{N}]_{\text{crude oil}} + 3\mathrm{H_2} \rightarrow 2\mathrm{NH_3} + \text{hydrocarbon derivatives}$$

The question of classifying nitrogen dioxide and sulfur trioxide as primary pollutants often arises, as does the origin of the nitrogen. In the former case, these higher oxides can be formed in the upper levels of the combustion reactors.

A *secondary pollutant* is a pollutant that is produced by interaction of a primary pollutant with another chemical. A *secondary pollutant* may also be produced by dissociation of a primary pollutant or other effects within a particular ecosystem. The combustion of fossil fuels can account for the large majority of the sulfur oxides and nitrogen oxides released to the atmosphere. Whichever technologies succeed in reducing the amounts of these gases in the atmosphere should also succeed in reducing the amounts of urban smog, those notorious brown and gray clouds that are easily recognizable at some considerable distances from urban areas, not only by their appearance but also by their odor:

$$SO_2 + H_2O \rightarrow H_2SO_4 \text{ (sulfurous acid)}$$

$$2SO_2 + O_2 \rightarrow 2SO_3$$

$$SO_3 + H_2O \rightarrow H_2SO_4 \text{ (sulfuric acid)}$$

$$2NO + H_2O \rightarrow HNO_2 + HNO_3 \text{ (nitrous acid + nitric acid)}$$

$$2NO + O_2 \rightarrow 2NO_2$$

$$NO_2 + H_2O \rightarrow HNO_3 \text{ (nitric acid)}$$

The most obvious issue with fossil fuel use relates to the effects on the environment. As technology evolves, the means to reduce the damage done by fossil fuel use also evolves, and the world is on the doorstep of adapting to alternative energy sources. In the meantime, gasification offers alternatives to meet the demand for fuels of the future and to reduce the potentially harmful emissions. In many cases, these secondary pollutants can have significant environmental effects, such as the formation of acid rain and smog (Speight, 1996).

Any pollutant, either primary or secondary, can have a serious effect on the various ecological cycles such as the industrial cycle (Fig. 16.3), the water cycle (Fig. 16.4), and the carbon cycle (Fig. 16.5). Therefore, understanding the means by which a chemical pollutant can enter these ecosystems and influence the future behavior of the ecosystem is extremely important.

In addition, *hazardous waste* is any gaseous, liquid, or solid waste material that, if improperly managed or disposed of, may pose substantial hazards to human health and the environment. In many cases, the term *chemical waste* is often used interchangeably with the term *hazardous waste*. However, not all chemical wastes are hazardous, and caution in the correct use of the terms must be exercised lest unqualified hysteria take control.

An *environmental regulation* is a legal mechanism that determines how the policy directives of an environmental law are to be carried out. An *environmental policy* is a requirement that specifies operating procedures that must be followed. An *environmental guidance* is a document developed by a governmental agency that outlines a position on a topic or that give instructions on how a

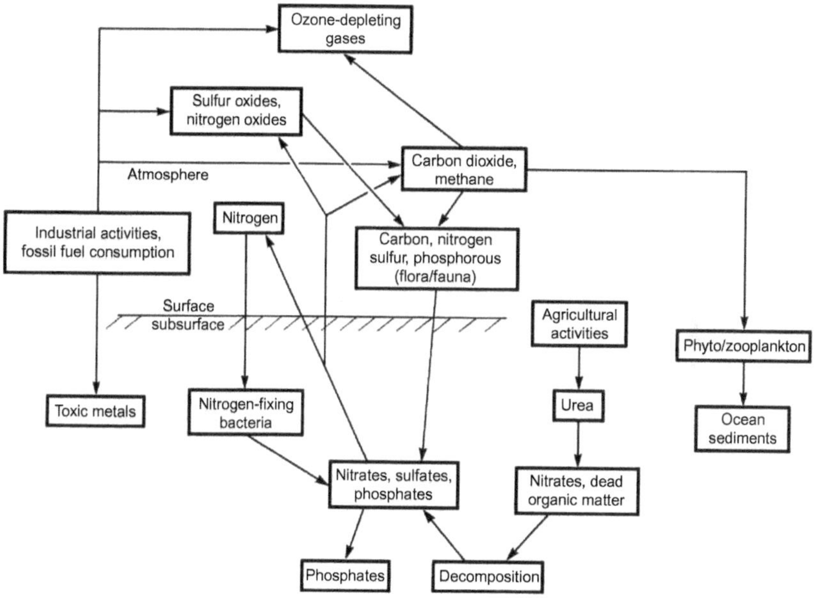

FIG. 16.3 The industrial cycle.

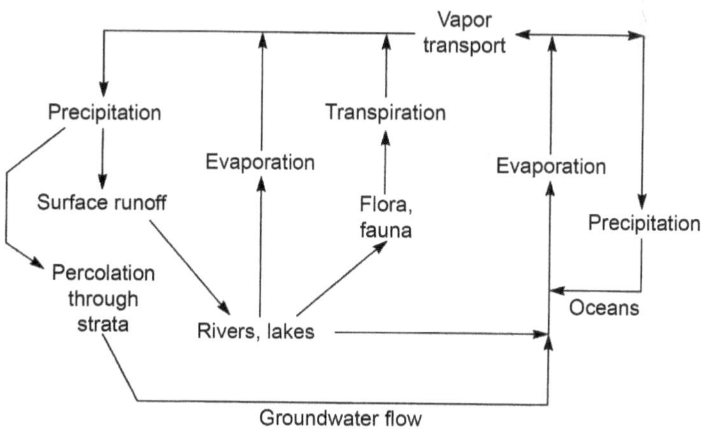

FIG. 16.4 The water cycle.

procedure must be carried out. It explains how to do something and provides governmental interpretations on a governmental act or policy.

5.2 Environmental Regulations

Thus, environmental issues permeate everyday life. These issues range from the effects on the lives of workers in various occupations where hazards can result

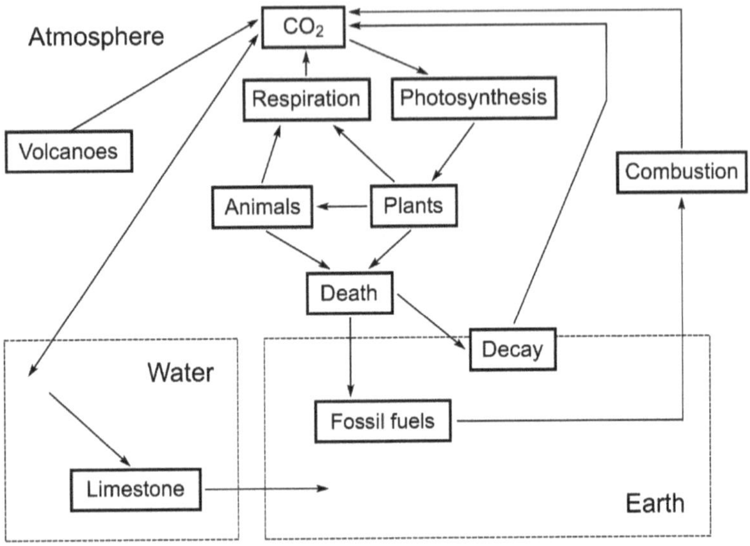

FIG. 16.5 The carbon cycle.

from exposure to chemical agents to the influence of these agents on the lives of the population at large (Lipton and Lynch, 1994; Speight, 1996; Boyce, 1997).

In this section, reference is made to the various environmental laws.

5.2.1 The Clean Air Act Amendments

The first *Clean Air Act* of 1970 and the 1977 amendments consisted of three titles. *Title I* dealt with stationary air emission sources, *Title II* with mobile air emission sources, and *Title III* with definitions of appropriate terms and applicable standards for judicial review. The *Clean Air Act Amendments of 1990* contain extensive provisions for control of the accidental release of toxic substances from storage or transportation and the formation of acid rain (acid deposition). In addition, the requirement that the standards be technology based removes much of the emotional perception that all chemicals are hazardous and the guesswork from legal enforcement of the legislation. The requirement also dictates environmental and health protection with an ample margin of safety.

5.2.2 The Water Pollution Control Act

Several acts are related to the protection of the waterways in the United States. Of particular interest in the present context is the *Water Pollution Control Act* (*Clean Water Act*). The objective of the Act is to restore and maintain the chemical, physical, and biological integrity of water systems.

The Water Pollution Control Act of 1948 and The Water Quality Act of 1965 were generally limited to control of pollution of interstate waters and the adoption of water-quality standards by the states for interstate water within their borders. The first comprehensive water-quality legislation in the United States came into being in 1972 as the Water Pollution Control Act. This Act was amended in 1977 and became the Clean Water Act. Further amendments in 1978 were enacted to deal more effectively with spills of crude oil. Other amendments followed in 1987 under the new name Water Quality Act and were aimed at improving water quality in those areas where there were insufficiencies in compliance with the discharge standards.

Section 311 of the Clean Water Act includes elaborate provisions for regulating intentional or accidental discharges of crude oil and of hazardous substances. Included are response actions required for oil spills and the release or discharge of toxic and hazardous substances. As an example, the person in charge of a vessel or an onshore or offshore facility (Speight, 2015b) from which any designated hazardous substance is discharged, in quantities equal to or exceeding its reportable quantity, must notify the appropriate federal agency as soon as such knowledge is obtained. The *Exxon Valdez* is a well-known case.

5.2.3 The Safe Drinking Water Act

The Safe Drinking Water Act, first enacted in 1974, was amended several times in the 1970s and 1980s to set national drinking water standards. The Act calls for regulations that (i) apply to public water systems, (ii) specify contaminants that may have any adverse effect on the health of persons, and (iii) specify contaminant levels. In addition, the difference between primary and secondary drinking water regulations is defined, and a variety of analytic procedures are specified. Statutory provisions are included to cover underground injection control systems. The Act also requires maximum levels at which a contaminant must have no known or anticipated adverse effects on human health, thereby providing an *adequate margin of safety*.

The Superfund Amendments and Reauthorization Act (SARA) sets standards the same for groundwater as for drinking water in terms of necessary cleanup and remediation of an inactive site that might be a former crude oil refinery. Under the Act, all underground injection activities must comply with the drinking water standards and meet specific permit conditions that are in unison with the provisions of the Clean Water Act.

However, under the Resource Conservation and Recovery Act, class IV injection wells are no longer permitted, and there are several restrictions on underground injection wells that may be used for storage and disposal of hazardous wastes.

5.2.4 The Resource Conservation and Recovery Act

Since its initial enactment in 1976, the Resource Conservation and Recovery Act (RCRA) continues to promote safer solid and hazardous waste management programs. Besides the regulatory requirements for waste management, the Act specifies the mandatory obligations of generators, transporters, and disposers of waste as well as those of owners and/or operators of waste treatment, storage, or disposal facilities. The Act also defines solid waste as garbage, refuse, sludge from a treatment plant and from a water supply treatment plant or air pollution control facility; and other discarded material, including solid, liquid, semisolid, or contained gaseous material resulting from industrial, commercial, mining, and agricultural operations and from community activities.

The Act also states that solid waste does not include solid, or dissolved, materials in domestic sewage or solid or dissolved materials in irrigation return flows or industrial discharges. A solid waste becomes a hazardous waste if it exhibits any one of four specific characteristics: (i) ignitability, (ii) reactivity, (iii) corrosivity, or (iv) toxicity. Certain types of solid wastes (e.g., household waste) are not considered to be hazardous, irrespective of their characteristics. Hazardous waste generated in a product or raw-material storage tank, transport vehicles, or manufacturing processes and samples collected for monitoring and testing purposes are exempt from the regulations.

Hazardous waste management is based on a beginning-to-end concept so that all hazardous wastes can be traced and fully accounted for. All generators and transporters of hazardous wastes as well as owners and operators of related facilities in the United States must file a notification with the Environmental Protection Agency. The notification must state the location of the facility and a general description of the activities as well as the identified and listed hazardous wastes being handled. Thus, all regulated hazardous waste facilities must exist and/or operate under valid, activity-specific permits.

Regulations pertaining to companies that generate and/or transport wastes require that detailed records be maintained to ensure proper tracking of hazardous wastes through transportation systems. Approved containers and labels must be used, and wastes can only be delivered to facilities approved for treatment, storage, and disposal.

5.2.5 The Toxic Substances Control Act

The Toxic Substances Control Act was first enacted in 1976 and was designed to provide controls for those chemicals that may threaten human health or the environment. Particularly hazardous are the cyclic nitrogen species that may be produced when crude oil is processed and that often occur in residua and cracked residua. The objective of the Act is to provide the necessary control before a chemical is allowed to be mass produced and enter the environment.

The Act specifies a *premanufacture notification* requirement by which any manufacturer must notify the Environmental Protection Agency at least 90 days prior to the production of a new chemical substance. Notification is also required even if there is a new use for the chemical that can increase the risk to the environment. No notification is required for chemicals that are manufactured in small quantities solely for scientific research and experimentation. A *new chemical substance* is defined as a chemical that is not listed in the Environmental Protection Agency Inventory of Chemical Substances or is an unlisted reaction product of two or more chemicals. In addition, the term *chemical substance* means any organic or inorganic substance of a particular molecular identity, including any combination of such substances occurring in whole or in part as a result of a chemical reaction or occurring in nature and any element or uncombined radical. The term *mixture* means any combination of two or more chemical substances if the combination does not occur in nature and is not, in whole or in part, the result of a chemical reaction.

5.2.6 The Comprehensive Environmental Response, Compensation, and Liability Act

The Comprehensive Environmental Response, Compensation, and Liability Act (CERCLA) that is generally known as *Superfund* was first signed into law in 1980. The central purpose of this Act is to provide a response mechanism for cleanup of any hazardous substance released, such as an accidental spill, or of a threatened release of a chemical. While RCRA deals basically with the management of wastes that are generated, treated, stored, or disposed of, CERCLA provides a response to the environmental release of various pollutants or contaminants into the air, water, or land.

Under this Act, a hazardous substance is any substance (i) requiring special consideration due to its toxic nature under the Clean Air Act, the Clean Water Act, or the Toxic Substances Control Act and (ii) defined as hazardous waste under RCRA. Additionally, a pollutant or contaminant can be any other substance not necessarily designated by or listed in the Act but that *will or may reasonably* be anticipated to cause any adverse effect in organisms and/or their offspring.

The Superfund Amendments and Reauthorization Act (SARA) addresses closed waste disposal sites that may release hazardous substances into any environmental medium. The most revolutionary part of SARA is the Emergency Planning and Community Right-to-Know Act (EPCRA), which for the first time mandated public disclosure. It is covered under title III of SARA.

5.2.7 The Occupational Safety and Health Act

Occupational health hazards are those factors arising in or from the occupational environment that adversely impact health. Thus, the *Occupational Safety and Health Administration* (*OSHA*) came into being in 1970 and is responsible for administering the Occupational Safety and Health Act.

The goal of the Act is to ensure that employees do not suffer material impairment of health or functional capacity due to a lifetime of occupational exposure to chemicals. The statute imposes a duty on employers to provide employees with a safe workplace environment, free of known hazards that may cause death or serious bodily injury.

The Act is also responsible for the means by which chemicals are contained. Workplaces are inspected to ensure compliance and enforcement of applicable standards under the Act. In keeping with the nature of the Act, there is also a series of standard tests relating to occupational health and safety as well as the general recognition of health hazards in the workplace. The Act is also the means by which guidelines have evolved for the management and disposition of chemicals used in chemical laboratories.

5.2.8 The Oil Pollution Act

The Oil Pollution Act of 1990 deals with pollution of waterways by crude oil. The Act specifically deals with crude oil vessels and onshore and offshore facilities (Speight, 2015b) and imposes strict liability for oil spills on their owners and operators.

5.2.9 The Hazardous Materials Transportation Act

The Hazardous Materials Transportation Act authorizes the establishment and enforcement of hazardous material regulations for all modes of transportation by highway, water, and rail. The purpose of the Act is to ensure safe transportation of hazardous materials. The Act prevents any person from offering or accepting a hazardous material for transportation anywhere within this nation if that material is not properly classified, described, packaged, marked, labeled, and authorized for shipment pursuant to the regulatory requirements.

Under Department of Transportation regulations, a hazardous material is defined as any substance or material, including a hazardous substance and hazardous waste, that is capable of posing an unreasonable risk to health, safety, and property during transportation.

The Act also imposes restrictions on the packaging, handling, and shipping of hazardous materials. For shipping and receiving of hazardous chemicals, hazardous wastes, and radioactive materials, the appropriate documentation, markings, labels, and safety precautions are required.

There are a variety of regulations (Table 16.5) that apply to crude oil refining. The most *popular* is the series of regulations known as the Clean Air Act that first was introduced in 1967 and was subsequently amended in 1970 and most recently in 1990. The most recent amendments provide stricter regulations for the establishment and enforcement of national ambient air quality standards for, as an example, sulfur dioxide. These standards do not stand alone, and there are many national standards for sulfur emissions.

TABLE 16.5 Environmental Regulations That Apply to Energy Production

	First Enacted	Amended
Clean Air Act	1970	1977
		1990
Clean Water Act (Water Pollution Control Act)	1948	1965[a]
		1972[b]
		1977
		1987[c]
Comprehensive Environmental Response, Compensation, and Liability Act	1980	1986[d]
Hazardous Material Transportation Act	1974	1990
Occupational Safety and Health Act	1970	1987[e]
Oil Pollution Act	1924	1990[f]
Resource Conservation and Recovery Act	1976	1980[g]
Safe Drinking Water Act	1974	1986[h]
Superfund Amendments and Reauthorization Act (SARA)	1986	
Toxic Substances Control Act	1976	1984[i]

[a]*Water Quality Act.*
[b]*Water Pollution Control Act.*
[c]*Water Quality Act.*
[d]*SARA Amendments.*
[e]*Several amendments during the 1980s.*
[f]*Interactive with various water pollution acts.*
[g]*Federal cancer policy initiated.*
[h]*Several amendments during the 1970s and the 1980s.*
[i]*Import rule enacted.*

5.3 Other Policies

Recent policy to tackle climate change and resource conservation, such as the Kyoto Protocol, the deliberations at Copenhagen in 2009, and the Landfill Directive of the European Union, stimulated the development of renewable energy and landfill diversion technology, providing gasification technology development a renewed impetus. However, even though they are the fastest-growing source of energy, renewable sources of energy will still represent only 15% of the world energy requirements in 2035 (up from the current estimation of 10%), and divesting from fossil fuels does not mean an end to environmental emissions. Viscous oil, tar sand bitumen, coal, natural gas, and perhaps oil shale will still be dominant energy sources and will grow at a relatively robust rate over, at least, the next two decades. These estimates are a reality check on the challenge ahead for clean technologies if they are to make an impact in reducing greenhouse gas emissions and satisfy future energy demands.

Current awareness of these issues by a variety of levels of government has resulted, in the United States, in the institution of the Clean Fossil Fuels Program to facilitate the development of pollution abatement technologies. And it has led to successful partnerships between government and industry. In addition, there is the potential that new laws, such as the passage in 1990 of the Clean Air Act Amendments in the United States, will be a positive factor and supportive of the controlled clean use of fossil fuels. However, there will be a cost, but industry is supportive of the measure and confident that the goals can be met.

5.4 Process Analysis

In addition to the conventional meaning of the term *process*, the *transportation* of crude oil also needs to be considered here.

Oil spills during crude oil *transportation* have been the most visible problem. There have also been instances of oil wells at sea "blowing out" or flowing uncontrollably, although the amounts from blowouts tend to be smaller than from tanker accidents. The 1979 Ixtoc I blowout in the Gulf of Mexico was an exception, as it flowed an estimated 3 million barrels over many months.

Tanker accidents typically have a severe impact on ecosystems because of the rapid release of hundreds of thousands of barrels of crude oil (or crude oil products) into a small area. The largest single spill to date is believed to have occurred during the 1991 Gulf War, when as much as 10 million barrels were dumped in the Persian Gulf by Iraq, apparently intentionally. More typical was the 1989 spill from the tanker Exxon Valdez, where 250,000 barrels were lost in Alaskan coastal waters.

While oil, as a hydrocarbon, is at least theoretically biodegradable, large-scale spills can overwhelm the ability of the ecosystem to break the oil down (Speight and Arjoon, 2012). Over time, the lower-boiling constituents of crude oil evaporate, leaving the nonvolatile constituents. The remaining oil constituents break down the protective waxes and oils in the feathers and fur of birds and animals, resulting in a loss of heat retention and causing death by freezing. Ingestion of the oil can also kill animals by interfering with their ability to digest food. Some crude oils contain toxic metals as well. The impact of any given oil spill is determined by the size of the spill, the degree of dispersal, and the chemistry of the oil. Spills at sea are thought to have a less detrimental effect than spills in shallow waters.

Crude oil *refining* is a complex sequence of chemical events that result in the production of a variety of products (Fig. 16.1). In fact, crude oil refining might be considered as a collection of individual yet related processes that are each capable of producing effluent streams.

Many refined products came under scrutiny (Loehr, 1992; Olschewsky and Megna, 1992). By the mid-1970s, crude oil refiners in the United States were required to develop techniques for manufacturing high-quality gasoline without employing lead additives, and by 1990, they were required to take on substantial investments in the complete reformulation of transportation fuels in order to minimize environmental emissions. From an industry that produced a single product (kerosene) and disposed of unwanted by-product materials in any manner possible, crude oil refining had become one of the most stringently regulated of all manufacturing industries, expending a major portion of its resources on the protection of the environment.

Processing crude oil, with the exception of some of the more viscous crude oils, involves a primary distillation of the hydrogen mixture, which results in its separation into fractions differing in carbon number, volatility, specific gravity, and other characteristics. The most volatile fraction that contains most of the gases that are generally dissolved in the crude is referred to as pipestill gas or pipestill low-boiling gases and consists essentially of hydrocarbon gases ranging from methane to butane(s) (C_4H_{10}) or sometimes pentane(s) (C_5H_{12}). The gas varies in composition and volume, depending on crude origin and on any additions to the crude made at the loading point.

It is not uncommon to reinject low-boiling hydrocarbon derivatives such as propane and butane into the crude before dispatch by tanker or pipeline. This results in a higher vapor pressure of the crude oil, but it allows one to increase the quantity of low-boiling products obtained at the refinery. Since low-boiling products in most crude oil markets command a premium while in the oil field itself propane and butane may have to be reinjected or flared, the practice of "spiking" crude with liquefied crude oil gas is becoming fairly common.

Crude oil refining, as it is currently known, will continue at least for the next three decades. In spite of the various political differences that have caused fluctuations in crude oil imports, it is obvious that crude oil imports will continue to be on the order of 60% (or greater) of crude oil consumption in the United States for the foreseeable future (Speight, 2011a). It is also predictable that the use of crude oil for the transportation sector will increase as increases in travel offset increased efficiency. As a consequence of this increase in use, crude oil will be the largest single source of carbon emissions from fuel. Acid gases corrode refining equipment, harm catalysts, pollute the atmosphere, and prevent the use of hydrocarbon components in petrochemical manufacture. When the amount of hydrogen sulfide is high, it may be removed from a gas stream and converted to sulfur or sulfuric acid. Some natural gases contain sufficient carbon dioxide to warrant recovery as dry ice.

Thus, like any other raw material, crude oil is capable of producing chemical waste. By 1960, the crude oil refining industry had become well established throughout the world. Demand for refined crude oil products had reached almost millions of barrels per day, with major concentrations of refineries in most developed countries. However, as the world became aware of the impact of industrial chemical waste on the environment, the crude oil refining industry was a primary focus for change. Refiners added hydrotreating units to extract sulfur compounds from their products and began to generate large quantities of elemental sulfur. Effluent water, atmospheric emissions, and combustion products also became a focus of increased technical attention (Carson and Mumford, 1988; Speight, 1996; Renzoni et al., 1994; Carson and Mumford, 1995; Edwards, 1995; Thibodeaux, 1995; Speight and Arjoon, 2012).

Thermal processes are commonly used to convert crude oil residua into liquid products. Therefore, some indications of the process classes and the products that are unacceptable to the environment are warranted here.

Thermal cracking processes are commonly used to convert nonvolatile residua into volatile products, although thermal cracking processes as used in the early refineries are no longer in use. Thus, examples of modern thermal processes are *visbreaking* and *coking* (*delayed coking*, *fluid coking*, and *flexicoking*) (Chapter 8). In all of these processes, the simultaneous formation of sediment or coke limits the conversion to usable liquid products.

The *visbreaking* process (Chapter 8) is primarily a means of reducing the viscosity of viscous feedstocks by *controlled thermal decomposition* insofar as the hot products are quenched before complete conversion can occur (Speight, 2014, 2017). However, the process is often plagued by sediment formation in the products. This sediment, or sludge, must be removed if the products are to meet fuel oil specifications. However, like deasphalting, visbreaking may be used more as a pretreating option for viscous oil and extra viscous oil, prior to, say, hydrocracking. For actual conversion of viscous feedstocks, delayed coking will be more widely used.

Coking, as the term is used in the crude oil industry, is a process for converting nondistillable fractions (residua) of crude oil to lower-boiling products and coke. Coking is often used in preference to catalytic cracking because of the presence of metals and nitrogen components that poison catalysts.

Delayed coking (Chapter 8) is the oldest, most widely used process and has changed very little in the five or more decades in which it has been onstream in refineries. *Fluid coking* (Chapter 8) is a continuous fluidized solid process that cracks feed thermally over heated coke particles in a reactor vessel to gas, liquid products, and coke. Heat for the process is supplied by partial combustion of the coke, with the remaining coke being drawn as product. The new coke is deposited in a thin fresh layer on the outside surface of the circulating coke particle.

Catalytic cracking is a conversion process (Chapter 9) that can be applied to a variety of feedstocks ranging from gas oil to viscous oil. It is one of several practical applications used in a refinery that employ a catalyst to improve process efficiency. Catalytic cracking of crude oil occurs over many types of catalytic materials that may be either activated (acid-treated natural clays of the bentonite type) or synthesized silica-alumina or silica-magnesia preparations.

Hydrotreating (Chapter 10) is defined as the lower temperature removal of heteroatomic species by treatment of a feedstock or product in the presence of hydrogen. *Hydrocracking* (Chapter 11) is the thermal decomposition of a feedstock in which carbon-carbon bonds are cleaved in addition to the removal of heteroatomic species. Hydrogen is present to prevent the formation of coke. Subsequent hydroprocessing (Chapter 10) of the coker distillates would reduce the polynuclear aromatic hydrocarbon derivatives in the resulting product streams, so that the only health concern outside the refinery itself is with high-severity thermal products, such as pitches, which have not been hydrotreated. Coke solids would not pose a health hazard and would have less environmental activity than unprocessed residue.

5.4.1 Gaseous Emissions

Gaseous emissions from crude oil refining create a number of environmental problems. During combustion, the combination of hydrocarbon derivatives, nitrogen oxide, and sunlight results in localized low levels of ozone or smog. This is particularly evident in large urban areas and especially when air does not circulate well. Crude oil use in automobiles also contributes to the problem in many areas. The primary effects are on the health of those exposed to the ozone, but plant life has been observed to suffer as well.

Refinery and natural gas streams may contain large amounts of acid gases, such as hydrogen sulfide (H_2S) and carbon dioxide (CO_2) (Speight, 1996). Hydrogen chloride (HCl), although not usually considered to be a major pollutant in crude oil refineries, can arise during processing from the presence of brine in crude oil that is incompletely dried. It can also be produced from mineral matter, and other inorganic contaminants are gaining increasing recognition as a pollutant that needs serious attention.

Acid gases corrode refining equipment, harm catalysts, pollute the atmosphere, and prevent the use of hydrocarbon components in petrochemical manufacture. When the amount of hydrogen sulfide is large, it may be removed from a gas stream and converted to sulfur or sulfuric acid. Some natural gases contain sufficient carbon dioxide to warrant recovery as *dry ice*, that is, solid carbon dioxide. And there is now a conscientious effort to mitigate the emission of pollutants from hydrotreating process by careful selection of process parameters and catalysts (Occelli and Chianelli, 1996).

The terms *refinery gas* and *process gas* are also often used to include all of the gaseous products and by-products that emanate from a variety of refinery processes (Speight, 1996). There are also components of the gaseous products that must be removed prior to release of the gases to the atmosphere or prior to use of the gas in another part of the refinery, i.e., as a fuel gas or as a process feedstock.

Crude oil refining produces gas streams that often contain substantial amounts of acid gases such as hydrogen sulfide and carbon dioxide.

More particularly hydrogen sulfide arises from the hydrodesulfurization of feedstocks that contain organic sulfur:

$$[S]_{feedstock} + H_2 \rightarrow H_2S + \text{hydrocarbon derivatives}$$

Crude oil refining involves, with the exception of some of the more viscous crude oils, a primary distillation of the hydrogen mixture, which results in its separation into fractions differing in carbon number, volatility, specific gravity, and other characteristics. The most volatile fraction that contains most of the gases, which are generally dissolved in the crude, is referred to as *pipestill gas* or *pipestill light ends* and consists essentially of hydrocarbon gases ranging from methane, to butane(s), to sometimes pentane(s).

The gas varies in composition and volume, depending on crude origin and on any additions to the crude made at the loading point. It is not uncommon to reinject low-boiling hydrocarbon derivatives such as propane and butane into the crude before dispatch by tanker or pipeline. This results in a higher vapor pressure of the crude, but it allows one to increase the quantity of low-boiling products obtained at the refinery. Since low-boiling products in most crude oil markets command a premium while in the oil field itself propane and butane may have to be reinjected or flared, the practice of *spiking* crude oil with liquefied crude oil gas is becoming fairly common.

In addition to the gases obtained by distillation of crude oil, more highly volatile products result from the subsequent processing of naphtha and middle distillate to produce gasoline. Hydrogen sulfide is produced in the desulfurization processes involving hydrogen treatment of naphtha, distillate, and residual fuel and from the coking or similar thermal treatments of vacuum gas oils and residual fuels. The most common processing step in the production of gasoline is the catalytic reforming of hydrocarbon fractions in the heptane (C_7) to decane (C_{10}) range.

In a series of processes commercialized under the generic name *reforming*, paraffin and naphthene (cyclic nonaromatic) hydrocarbon derivatives are altered structurally in the presence of hydrogen and a catalyst into aromatics or isomerized to more highly branched hydrocarbon derivatives. Catalytic reforming processes thus not only result in the formation of a liquid product of higher octane number but also produce substantial quantities of gases. The latter not only are rich in hydrogen but also contain hydrocarbon derivatives from methane to butanes, with a preponderance of propane ($CH_3CH_2CH_3$), *n*-butane ($CH_3CH_2CH_2CH_3$) and isobutane [$(CH_3)_3CH$].

The composition of the process gases varies in accordance with reforming severity and reformer feedstock. All catalytic reforming processes require substantial recycling of a hydrogen stream. Therefore, it is normal to separate reformer gas into a propane ($CH_3CH_2CH_3$) and/or a butane stream [$CH_3CH_2CH_2CH_3$ plus $(CH_3)_3CH$], which becomes part of the refinery liquefied crude oil gas production and a lower-boiling gas fraction, part of which is recycled. In view of the excess of hydrogen in the gas, all products of catalytic reforming are saturated, and there are usually no olefin gases present in either gas stream.

A second group of refining operations that contributes to gas production is that of the *catalytic cracking* processes. These consist of fluid-bed catalytic cracking in which viscous gas oils are converted into gas, liquefied crude oil gas, catalytic naphtha, fuel oil, and coke by contacting the viscous hydrocarbon with the hot catalyst. Both catalytic and thermal cracking processes, the latter being now largely used for the production of chemical raw materials, result in the formation of unsaturated hydrocarbon derivatives, not only particularly ethylene ($CH_2{=}CH_2$) but also propylene (propene, $CH_3CH{=}CH_2$), isobutylene [isobutene, $(CH_3)_2C{=}CH_2$], and the *n*-butenes ($CH_3CH_2CH{=}CH_2$ and $CH_3CH{=}CHCH_3$) in addition to hydrogen (H_2), methane (CH_4) and smaller quantities of ethane (CH_3CH_3), propane ($CH_3CH_2CH_3$), and butanes [$CH_3CH_2CH_2CH_3$ and $(CH_3)_3CH$]. Diolefins such as butadiene ($CH_2{=}CH$-$CH{=}CH_2$) are also present.

Additional gases are produced in refineries with visbreaking and/or coking facilities that are used to process the heaviest crude fractions. In the visbreaking process, fuel oil is passed through externally fired tubes and undergoes liquid-phase cracking reactions, which result in the formation of lower-boiling fuel oil components. Oil viscosity is thereby reduced, and some gases, mainly hydrogen, methane, and ethane, are formed. Substantial quantities of both gas and carbon are also formed in coking (both delayed coking and fluid coking) in addition to the middle distillate and naphtha. When coking a residual fuel oil or viscous gas oil, the feedstock is preheated and contacted with hot carbon (coke) that causes extensive cracking of the feedstock constituents of higher molecular weight to produce lower-molecular-weight products ranging from methane, via liquefied crude oil gas and naphtha, to gas oil and heating oil. Products from coking processes tend to be unsaturated, and olefin components predominate in the tail gases from coking processes.

A further source of refinery gas is hydrocracking, a catalytic high-pressure pyrolysis process in the presence of fresh and recycled hydrogen. The feedstock is again viscous gas oil or residual fuel oil, and the process is directed mainly at the production of additional middle distillates and gasoline. Since hydrogen is to be recycled, the gases produced in this process again have to be separated into lower-boiling and higher-boiling streams. Any surplus recycle gas and the liquefied crude oil gas from the hydrocracking process are both saturated.

Both hydrocracker gases and catalytic reformer gases are commonly used in catalytic desulfurization processes. In the latter, feedstocks ranging from low-boiling (atmospheric) gas oil to vacuum gas oils are passed at pressures of 500–1000 psi with hydrogen over a hydrofining catalyst. This results mainly in the conversion of organic sulfur compounds to hydrogen sulfide:

$$[S]_{\text{feedstock}} + H_2 \rightarrow H_2S + \text{hydrocarbon derivatives}$$

The reaction also produces some low-boiling hydrocarbon derivatives by hydrocracking.

Thus, refinery streams, while ostensibly being hydrocarbon in nature, may contain large amounts of acid gases such as hydrogen sulfide and carbon dioxide. Most commercial plants employ hydrogenation to convert organic sulfur compounds into hydrogen sulfide. Hydrogenation is effected by means of recycled hydrogen-containing gases or external hydrogen over a nickel molybdate or cobalt molybdate catalyst.

In summary, refinery process gas, in addition to hydrocarbon derivatives, may contain other contaminants, such as carbon oxides (CO_x, where $x=1$ and/or 2), sulfur oxides (SO_x, where $x=2$ and/or 3), and ammonia (NH_3), mercaptans (R-SH), and carbonyl sulfide (COS).

The presence of these impurities may eliminate some of the sweetening processes, since some processes remove large amounts of acid gas but not to a sufficiently low concentration. On the other hand, there are those processes not designed to remove (or incapable of removing) large amounts of acid gases, whereas they are capable of removing the acid gas impurities to very low levels when the acid gases are present only in low-to-medium concentration in the gas.

From an environmental viewpoint, it is not the means by which these gases can be utilized, but it is the effects of these gases on the environment when they are introduced into the atmosphere.

In addition to the corrosion of equipment of acid gases, the escape into the atmosphere of sulfur-containing gases can eventually lead to the formation of the constituents of acid rain, i.e., the oxides of sulfur (SO_2 and SO_3). Similarly, the nitrogen-containing gases can also lead to nitrous and nitric acids (through the formation of the oxides NO_x, where $x=1$ or 2) that are the other major contributors to acid rain. The release of carbon dioxide and hydrocarbon derivatives as constituents of refinery effluents can also influence the behavior and integrity of the ozone layer.

Hydrogen chloride, if produced during refining, quickly picks up moisture in the atmosphere to form droplets of hydrochloric acid and, like sulfur dioxide, is a contributor to acid rain. However, hydrogen chloride may exert severe local effects because, unlike sulfur dioxide, it does not need to participate in any further chemical reaction to become an acid. Under atmospheric conditions that favor a buildup of stack emissions in the area of a large industrial complex or power plant, the amount of hydrochloric acid in rainwater could be quite high.

Natural gas is also capable of producing emissions that are detrimental to the environment. While the major constituent of natural gas is methane, there are components such as carbon dioxide (CO), hydrogen sulfide (H_2S), and mercaptans (thiols; R-SH), as well as trace amounts of sundry other emissions. The fact that methane has a foreseen and valuable end use makes it a desirable product, but in several other situations, it is considered a pollutant, having been identified a greenhouse gas.

A sulfur removal process must be very precise, since natural gas contains only a small quantity of sulfur-containing compounds that must be reduced several orders of magnitude. Most consumers of natural gas require <4 ppm in the gas.

A characteristic feature of natural gas that contains hydrogen sulfide is the presence of carbon dioxide (generally in the range of 1–4% *v*/v). In cases where the natural gas does not contain hydrogen sulfide, there may also be a relative lack of carbon dioxide.

Acid rain occurs when the oxides of nitrogen and sulfur that are released to the atmosphere during the combustion of fossil fuels are deposited (as soluble acids) with rainfall, usually at some location remote from the source of the emissions.

It is generally believed (the chemical thermodynamics are favorable) that acidic compounds are formed when sulfur dioxide and nitrogen oxide emissions are released from tall industrial stacks. Gases such as sulfur oxides (usually sulfur dioxide, SO_2) and the nitrogen oxides (NO_x) react with the water in the atmosphere to form acids:

$$SO_2 + H_2O \rightarrow H_2SO_3$$

$$2SO_2 + O_2 \rightarrow 2SO_3$$

$$SO_3 + H_2O \rightarrow H_2SO_4$$

$$2NO + H_2O \rightarrow 2HNO_2$$

$$2NO + O_2 \rightarrow 2NO_2$$

$$NO_2 + H_2O \rightarrow HNO_3$$

Acid rain has a pH <5.0 and predominantly consists of sulfuric acid (H_2SO_4) and nitric acid (HNO_3). As a point of reference, in the absence of anthropogenic pollution sources, the average pH of rain is approximately 6.0 (slightly acidic; neutral pH = 7.0). In summary, the sulfur dioxide that is produced during a variety of processes will react with oxygen and water in the atmosphere to yield environmentally detrimental sulfuric acid. Similarly, nitrogen oxides will also react to produce nitric acid.

Another acid gas, hydrogen chloride (HCl), although not usually considered to be a major emission, is produced from mineral matter and the brines that often accompany crude oil during production and is gaining increasing recognition as a contributor to acid rain. However, hydrogen chloride may exert severe local effects because it does not need to participate in any further chemical reaction to become an acid. Under atmospheric conditions that favor a buildup of stack emissions in the areas where hydrogen chloride is produced, the amount of hydrochloric acid in rainwater could be quite high.

In addition to hydrogen sulfide and carbon dioxide, gas may contain other contaminants, such as mercaptans (R-SH) and carbonyl sulfide (COS).

On a regional level, the emission of sulfur oxides (SO_x) and nitrogen oxides (NO_x) can also cause the formation of acid species at high altitudes, which eventually precipitate in the form of *acid rain*, damaging plants, wildlife, and property. Most crude oil products are low in sulfur or are desulfurized, and while natural gas sometimes includes sulfur as a contaminant, it is typically removed at the production site.

At the global level, there is concern that the increased use of hydrocarbon-based fuels will ultimately raise the temperature of the planet (*global warming*), as carbon dioxide reflects the infrared or thermal emissions from the earth, preventing them from escaping into space (*greenhouse effect*). Whether or not the potential for global warming becomes real will depend upon how emissions into the atmosphere are handled. There is considerable discussion about the merits and demerits of the global warming theory, and the discussion is likely to continue for some time. Be that as it may, the atmosphere can only tolerate pollutants up to a limiting value. And that value needs to be determined. In the meantime, efforts must be made to curtail the use of noxious and foreign (non-indigenous) materials into the air.

In summary and from an environmental viewpoint, crude oil and natural gas processing can result in similar, if not the same, gaseous emissions as coal (Speight, 2013). It is a question of degree insofar as the composition of the gaseous emissions may vary from coal to crude oil, but the constituents are, in the majority of cases, the same.

There are a variety of processes that are designed for sulfur dioxide removal from gas streams (Speight, 2014), but scrubbing process utilizing limestone ($CaCO_3$) or lime [$Ca(OH)_2$] slurries has received more attention than other gas scrubbing processes. The majority of the gas scrubbing processes are designed to remove sulfur dioxide from the gas streams; some processes show the potential for removal of nitrogen oxide(s).

5.4.2 Liquid Effluents

Crude oil, as a mixture of hydrocarbon derivatives, is (theoretically) a biodegradable material. However, in very general terms (and as observed from elemental analyses), crude oil is a mixture of (i) hydrocarbon derivatives, (ii) nitrogen compounds, (iii) oxygen compounds, (iv) sulfur compounds, and (v) metallic constituents.

It is convenient to divide the hydrocarbon components of crude oil into the following three classes: (i) paraffin compounds, which are saturated hydrocarbon derivatives with straight or branched chains, but without any ring structure; (ii) naphthene compounds, which are saturated hydrocarbon derivatives containing one or more rings, each of which may have one or more paraffin side chains (more correctly known as alicyclic hydrocarbon derivatives); and (iii) aromatic compounds, which are hydrocarbon derivatives containing one or more aromatic nuclei, such as benzene, naphthalene, and phenanthrene ring systems, which may be linked up with (substituted) naphthene rings and/or paraffin side chains. And even though crude oil derivatives have been prescribed for medicinal purposes (Chapter 1), one does not see the flora and fauna of the earth surviving in oceans of crude oil. It is all a question of dosage!

Crude oil also contains appreciable amounts of organic nonhydrocarbon constituents, mainly sulfur-, nitrogen-, and oxygen-containing compounds and, in smaller amounts, organometallic compounds in solution and inorganic salts in colloidal suspension. These constituents appear throughout the entire boiling range of the crude oil but tend to concentrate mainly in the heavier fractions and in the nonvolatile residues.

Although their concentration in certain fractions may be quite small, their influence is important. For example, the thermal decomposition of deposited inorganic chlorides with evolution of free hydrochloric acid can give rise to serious corrosion problems in the distillation equipment. The presence of organic acid components, such as mercaptans (R-SH) and acids (R-CO_2H), can also promote environmental damage. In catalytic operations, passivation and/or poisoning of the catalyst can be caused by deposition of traces of metals (vanadium and nickel) or by chemisorption of nitrogen-containing compounds on the catalyst, thus necessitating the frequent regeneration of the catalyst or its expensive replacement. This carries with it the issues related to catalyst disposal.

Thermal processing can significantly increase the concentration of polynuclear aromatic hydrocarbon derivatives in the product liquid because the low-pressure hydrogen-deficient conditions favor aromatization of naphthene constituents and condensation of aromatics to form larger ring systems. To the extent that more compounds like benzo(a)pyrene are produced, the liquids from thermal processes will be more carcinogenic than asphalt. This biological activity was consistent with the higher concentration of polynuclear aromatic hydrocarbon derivatives at 38.8mg/g in the pitch compared with only 0.22mg/g in the asphalt. Similarly, one would expect coker gas oils to contain more polynuclear aromatic hydrocarbon derivatives than unprocessed or hydroprocessed distillates and thereby give a higher potential for carcinogenic or mutagenic effects.

The sludge produced on acid treatment of crude oil distillates (Speight and Arjoon, 2012; Speight, 2014), even gasoline and kerosene, is complex in nature. Esters and alcohols are present from reactions with olefins. Sulfonation products are created from with aromatic compounds, naphthene compounds, and phenols and salts from reactions with nitrogen bases. In addition, such materials as naphthenic acids, sulfur compounds, and asphalt (residua constituents) material are all retained by direct solution. The various products of oxidation-reduction reactions must be added to these constituents: coagulated resins, soluble hydrocarbon derivatives, water, and free acid.

The disposal of the sludge is difficult, as it contains unused free acid that must be removed by dilution and settling. The disposal is a comparatively simple process for the sludge resulting from treating gasoline and kerosene, the so-called light oils. The insoluble oil phase separates out as a mobile tar, which can be mixed and burned without too much difficulty. Sludge from viscous oil, however, separates out granular semisolids, which offer considerable difficulty in handling.

In all cases, careful separation of reaction products is important to the recovery of well-refined materials. This may not be easy if the temperature has risen as a consequence of chemical reaction. This will result in a persistent dark color traceable to reaction products that are redistributed as colloids. Separation may also be difficult at low temperature because of high viscosity of the stock, but this problem can be overcome by dilution with low-boiling naphtha or with propane.

In addition, delayed coking also requires the use of large volumes of water for hydraulic cleaning of the coke drum. However, the process water can be recycled if the oil is removed by skimming and suspended coke particles are removed by filtration. If this water is used in a closed cycle and treated to produce useable water, the impact of delayed coking on water treatment facilities and the environment is minimized. The flexicoking process offers one alternative to direct combustion of coke for process fuel. The gasification section is used to process excess coke to mixture of carbon monoxide (CO), carbon dioxide (CO_2), hydrogen (H_2), and hydrogen sulfide (H_2S) followed by treatment to remove the hydrogen sulfide. Maximizing the residue conversion and desulfurization of the residue in upstream hydroconversion units also maximizes the yield of hydrogen sulfide relative to sulfur in the coke. Currently, maximum residue conversion with minimum coke production is favored over gasification of coke.

5.4.3 *Solid Effluents*

Catalyst disposal is therefore a major concern in all refineries. In many cases, the catalysts are regenerated at the refinery for repeated use. Disposal of spent catalysts is usually part of an agreement with the catalyst manufacturer whereby the spent catalyst is returned for treatment and remanufacture.

The formation of considerable quantities of coke in the *coking* processes is a cause for concern since it not only reduces the yield of liquid products but also initiates the necessity for disposal of the coke. Stockpiling to coke may be a partial answer unless the coke contains leachable materials that will endanger the ecosystem as a result of rain or snow melt.

In addition, the generation and emission of sulfur oxides (particularly sulfur dioxide) occurs from combustion of sulfur-containing coke as plant fuel. Sulfur dioxide (SO_2) has a wide range of effects on health and on the environment. These effects vary from bronchial irritation upon short-term exposure to contributing to the acidification of lakes. Emissions of sulfur dioxide therefore are regulated in many countries.

6. Epilogue

There have been many suggestions about the future of the crude oil industry in relation to the reserves of crude oil that are available but that are continually being depleted. Among these suggestions are that the bulk of the crude oil and gas in the world has already been discovered and that declining production is inevitable. Another suggestion is that substantial amounts of oil and gas remain to be found. There are also suggestions that fall between these two extremes.

In the last two decades, new fields have indeed been discovered, for example, in Kazakhstan near the Caspian Sea, and the potential for crude oil discoveries have opened up in Eastern Europe, Asia, Canadian coastal areas, and Colombia. Potentially, the richest discovery has been the finding of vast reserves in deep water in the Gulf of Mexico. These reserves were only beginning to be tapped in the mid-1990s, using floating platforms tethered to the sea bottom by steel cables and such innovative technologies as the use of deepwater robotic machines for construction and maintenance. In addition, the resources of crude oil in tight formations are believed to be substantial, but these sources are not all available for full commercialization to replace oil-based fuels and products and may not be so for the next several decades during which time the production of oil-based fuels and products will have to fulfill the demand (Speight and Islam, 2016).

Liquid fuel sources that still remain to be exploited include tar sand deposits (Chapters 1 and 4) (Speight, 2008, 2012) and the liquefaction and gasification of coal (Speight, 2013). All attempts to utilize these sources have proved so far to be uneconomic compared with the costs of producing oil and natural gas. Future technologies may, however, find ways of creating viable fuels from these various substances (Castañeda et al., 2014). That being the case and although oil is now recognized as likely to be abundant into the first 50 years of the 21st century, environmental concerns will probably impose increasing restrictions on both its production and consumption.

Thus, the general prognosis for emission cleanup is not pessimistic and can be looked upon as being quite optimistic. Indeed, it is considered likely that most of their environmental impact of crude oil refining can be substantially abated. A considerable investment in retrofitting or replacing existing facilities and equipment might be needed. However, it is possible, and a conscious goal must be to improve the efficiency with which crude oil is transformed and consumed.

Obviously, much work is needed to accommodate the continued use of crude oil. In the meantime, we use what we have, all the while working to improve efficient usage and working to ensure that there is no damage to the environment.

Such is the nature of crude oil refining and the expectancy of protecting the environment.

References

Babich, I.V., Moulijn, J.A., 2003. Science and technology of novel processes for deep desulfurization of oil refinery streams: a review. Fuel 82, 607–631.

Bachmann, R.T., Johnson, A.C., Edyvean, R.G.J., 2014. Biotechnology in the petroleum industry: an overview. Int. Biodeterior. Biodegrad. 86, 225–237.

Boerrigter, H., den Uil, H., Calis, H.P., 2002. Green diesel from biomass via Fischer-Tropsch synthesis, pyrolysis and gasification of biomass and waste. In: Proceedings. Expert Meeting, Strasbourg, France, September 30–October 1.

Boyce, A., 1997. Introduction to Environmental Technology. Van Nostrand Reinhold, New York.

Carson and Mumford, 1988. The Safe Handling of Chemicals in Industry. vols. 1 and 2. John Wiley & Sons Inc., New York.

Carson and Mumford, 1995. The Safe Handling of Chemicals in Industry. vol. 3. John Wiley & Sons Inc., New York.

Castañeda, L.C., Muñoz, J.A.D., Ancheyta, J., 2014. Current situation of emerging technologies for upgrading of viscous oils. Catal. Today 220–222, 248–273.

Couvaras, G., 1997. Sasol's slurry phase distillate process and future applications. In: Proceedings. Monetizing Stranded Gas Reserves Conference, Houston, TX, December 1997.

Davis, R.A., Patel, N.M., 2004. Refinery hydrogen management. Pet. Technol. Q. (Spring), 29–35.

Dickenson, R.L., Biasca, F.E., Schulman, B.L., Johnson, H.E., 1997. Hydrocarb. Process. 76 (2), 57.

Edwards, J.D., 1995. Industrial Wastewater Treatment: A Guidebook. CRC Press Inc., Boca Raton, FL.

Feintuch, H.M., Negin, K.M., 1997. Meyers, R.A. (Ed.), Handbook of Crude oil Refining Processes. McGraw-Hill, New York. (Chapter 12.2).

Gary, J.G., Handwerk, G.E., Kaiser, M.J., 2007. Crude oil Refining: Technology and Economics, fifth ed. CRC Press, Taylor & Francis Group, Boca Raton, FL.

Goetze, L., Bailer, O., 1999. Reactive distillation with Katapak-S. Sulzer Tech. Rev. 4, 29–31.

Hedrick, B.W., Seibert, K.D., Crewe, C., 2006. A New Approach to Viscous Oil and Bitumen Upgrading. AM-06-29. UOP 4355B. UOP LLC, Des Plaines, IL.

Hickman, D.A., Schmidt, L.D., 1993. Production of syngas by direct catalytic oxidation of methane. Science 259, 343–346.

Hsu, C.S., Robinson, P.R. (Eds.), 2017. Handbook of Petroleum Technology. Springer International Publishing AG, Cham, Switzerland.

Hubbert, M.K., 1962. Energy Resources. Report to the Committee on Natural Resources. National Academy of Sciences, Washington, DC.

Iqbal, R., Khan, A., Eng, O., Floyd, R., 2008. Unlocking current refinery constraints. Crude Oil Technol. Q. Q2, 31–33.

Lerner, B., 2002. The future of refining. Hydrocarb. Eng. September.

Lipton, S., Lynch, J., 1994. Handbook of Health Hazard Control in the Chemical Process Industry. John Wiley and Sons Inc., New York.

Loehr, R.C., 1992. McKetta, J.J. (Ed.), Crude Oil Processing Handbook. Marcel Dekker Inc., New York, p. 190.

Majumdar, S.B., 1993. Regulatory Requirements for Hazardous Materials. McGraw-Hill, New York.

Moritz, P., Gorak, A., 2002. Two in one: cost reduction thanks to reactive separation. Sulzer Tech. Rev. 1, 14–16.

Mut, S., 2005. Shell unconventional resources energy. In: National Academies Trends in Oil Supply/Demand and the Potential for Peaking of Conventional Oil Production, Washington, DC, October 20–21.

Occelli, M.L., Chianelli, R., 1996. Hydrotreating Technology for Pollution Control. Marcel Dekker Inc., New York.

Olschewsky, D., Megna, A., 1992. McKetta, J.J. (Ed.), Crude Oil Processing Handbook. Marcel Dekker Inc., New York, p. 179.

Parkash, S., 2003. Refining Processes Handbook. Gulf Professional Publishing, Elsevier, Amsterdam, Netherlands.

Penning, R.T., 2001. Crude oil refining: a look at the future. Hydrocarb. Process. 80 (2), 45–46.

Ray, D.L., Guzzo, L., 1990. Trashing the Planet: How Science Can Help Us Deal With Acid Rain, Depletion of the Ozone, and Nuclear Waste (Among Other Things). Regnery Gateway, Washington, DC.

Renzoni, A., Fossi, M.C., Lari, L., Mattei, N., 1994. Contaminants in the Environment: A Multidisciplinary Assessment of Risks to Man and Other Organisms. CRC Press Inc., Boca Raton, FL.

Rostrup-Nielsen, J.R., 2004. Fuels and energy for the future: the role of catalysis. Catal. Rev. 46 (3&4), 247–270.

Sousa-Aguiar, E.F., Appel, L.G., Mota, C., 2005. Natural gas chemical transformations: the path to refining in the future. Catal. Today 10 (1), 3–7.

Speight, J.G., 1996. Environmental Technology Handbook. Taylor & Francis Publishers, Washington, DC.

Speight, J.G., 2005. Environmental Analysis and Technology for the Refining Industry. John Wiley & Sons Inc., Hoboken, NJ.

Speight, J.G., 2008. Synthetic Fuels Handbook: Properties, Processes, and Performance. McGraw-Hill, New York.

Speight, J.G., 2011a. An Introduction to Crude Oil Technology, Economics, and Politics. Scrivener Publishing, Salem, MA.

Speight, J.G., 2011b. The Refinery of the Future. Gulf Professional Publishing, Elsevier, Oxford, UK.

Speight, J.G., 2012. Shale Oil Production Processes. Gulf Professional Publishing, Elsevier, Oxford, UK.

Speight, J.G., 2013. The Chemistry and Technology of Coal, 3rd ed. CRC Press, Taylor & Francis Group, Boca Raton, FL.

Speight, J.G., 2014. The Chemistry and Technology Petroleum, 5th ed. CRC Press, Taylor & Francis Group, Boca Raton, FL.

Speight, J.G., 2015a. Handbook of Petroleum Product Analysis, 2nd ed. John Wiley & Sons Inc., Hoboken, NJ.

Speight, J.G., 2015b. Handbook of Offshore Oil and Gas Operations. Gulf Professional Publishing, Elsevier, Oxford, UK.

Speight, J.G., 2017. Handbook of Petroleum Refining. CRC Press, Taylor & Francis Group, Boca Raton, FL.

Speight, J.G., Arjoon, K.K., 2012. Bioremediation of Crude Oil and Crude Oil Products. Scrivener Publishing, Salem, MA.

Speight, J.G., Islam, M.R., 2016. Peak Energy—Myth or Reality. Scrivener Publishing, Beverly, MA.

Stanislaus, A., Qabazard, H., Absi-Halabi, M., 2000. Refinery of the future. In: Proceedings. 16th World Crude Oil Congress, Calgary, AB, Canada, June 11–15.

Sundmacher, K., Kienle, A. (Eds.), 2003. Reactive Distillation: Status and Future Directions. John Wiley & Sons Inc., New York.

Szklo, A., Schaeffer, R., 2005. Alternative energy sources or integrated alternative energy systems? Oil as a modern lance of Peleus for the energy transition. Energy 31, 2513–2522.

Thibodeaux, L.J., 1995. Environmental Chemodynamics. John Wiley & Sons Inc., New York.
Vallero, D., 2008. Fundamentals of Air Pollution, 4th ed. Elsevier, London, UK.
Youssef, N., Elshahed, M.S., McInerney, M.J., 2009. Microbial processes in oil fields: culprits, problems and opportunities. Adv. Appl. Microbiol. 66, 141–251.
Zittel, W., Schindler, J., 2007. Crude Oil: The Supply Outlook. EWG Series No. 3/2007, Energy Watch Group, Berlin, Germany.

Further Reading

Bhatia, S., Sharma, D.K., 2006. Emerging role of biorefining of heavier crude oils and integration of biorefining with crude oil refineries in the future. Crude Oil Sci. Technol. 24 (10), 1125–1159.
Huber, G.W., Corma, A., 2007. Synergies between bio- and oil refineries for the production of fuels from biomass. Angew. Chem. 46 (38), 7184–7201.
Le Borgne, S., Quintero, R., 2003. Biotechnological processes for the refining of crude oil. Fuel Process. Technol. 81 (2), 155–169.
Lynd, L.R., Wyman, C., Laser, M., Johnson, D., Landucci, R., 2005. Strategic Biorefinery Analysis: Review of Existing Biorefinery Examples January 24–July 1, 2002. Subcontract Report NREL/SR-510-34895 October, National Renewable Energy Laboratory, Golden, CO.
Lynd, L.R., Larson, E., Greene, N., Laser, M., Sheehan, J., Dale, B.E., McLaughlin, S., Wang, M., 2009. Biofuels Bioprod. Biorefin. 3 (2), 113–123.
Rusco, F.W., Walls, W.D., 2008. Biofuels, Crude oil Refining, and the Shipping of Motor Fuels. Technical Paper No. TP-08005, Institute for Advanced Policy Research, University of Calgary, Calgary, Alberta Canada and United States Government Accountability Office, Washington, DC.
Sadighi, S., Mohaddecy, R.S., Masoudian, K., 2014. New process arrangements for upgrading viscous oils and residua. Pet. Technol. Q. Q1, 1–7. www.digitalrefining.com/article/1000935.

Conversion Factors

1. General

   1 acre = 43,560 ft^2

   1 acre-foot = 7758.0 bbl

   1 atm = 760 mmHg = 14.696 psi = 29.91 in Hg

   1 atm = 1.0133 bar = 33.899 ft. H_2O

   1 barrel (oil) = 42 gal = 5.6146 ft^3

   1 barrel (water) = 350 lb at 60°F

   1 barrel per day = 1.84 cm^3/s

   1 Btu = 778.26 ft lb

   1 cP = 2.42 lb. mass/(ft) (h), viscosity

   1 cP = 0.000672 lb. mass/(ft) (s), viscosity

   1 cubic foot = 28,317 cm^3 = 7.4805 gal

   Density of water at 60°F = 0.999 g/cm^3 = 62.367 lb./ft^3 = 8.337 lb./gal

   1 gal = 231 in.^{3} = 3785.4 cm^3 = 0.13368 ft^3

   1 hp-hour = 0.7457 kWh = 2544.5 Btu

   1 hp = 550 ft lb/s = 745.7 W

   1 in. = 2.54 cm

   1 m = 100 cm = 1000 mm = 10 μm = 10 Å

   1 oz = 28.35 g

   1 lb = 453.59 g = 7000 grains

   1 mile^2 = 640 acres

2. Concentration conversions

   1 part per million (1 ppm) = 1 microgram per liter (1 μg/L)

   1 microgram per liter (1 μg/L) = 1 milligram per kilogram (1 mg/kg)

   1 microgram per liter (1 μg/L) × 6.243 × 10^8 = 1 pound per cubic foot (1 lb/ft^3)

   1 microgram per liter (1 μg/L) × 10^{-3} = 1 milligram per liter (1 mg/L)

   1 milligram per liter (1 mg/L) × 6.243 × 10^5 = 1 pound per cubic foot (1 lb/ft^3)

   1 gram mole per cubic meter (1 gmol/m^3) × 6.243 × 10^5 = 1 pound per cubic foot (1 lb/ft^3)

   10,000 ppm = 1% *w*/w

   1 ppm hydrocarbon in soil × 0.002 = 1 lb of hydrocarbons per ton of contaminated soil

3. Weight conversion

   1 ounce (1 oz) = 28.3495 grams (28.3495 g)

   1 pound (1 lb) = 0.454 kg

   1 pound (1 lb) = 454 grams (454 g)

   1 kilogram (1 kg) = 2.20462 pounds (2.20462 lb)

1 stone (English, 1 st) = 14 pounds (14 lb)
1 ton (US, 1 short ton) = 2000 lb
1 ton (English, 1 long ton) = 2240 lb
1 metric ton = 2204.62262 lb
1 tonne = 2204.62262 lb

4. Temperature conversions
°F = (°C × 1.8) + 32
°C = (°F − 32)/1.8
(°F − 32) × 0.555 = °C
Absolute zero = −273.15°C
Absolute zero = −459.67 °F

5. Area
1 square centimeter (1 cm^2) = 0.1550 $in.^2$
1 square meter (1 m^2) = 1.1960 yd^2
1 hectare = 2.4711 acres
1 square kilometer (1 km^2) = 0.3861 $mile^2$
1 square inch (1 $in.^2$) = 6.4516 cm^2
1 square foot (1 ft^2) = 0.0929 m^2
1 square yard (1 yd^2) = 0.8361 m^2
1 acre = 4046.9 m^2
1 square mile (1 mi^2) = 2.59 km^2

6. Other approximations
14.7 pounds per square inch (14.7 psi) = 1 atmosphere (1 atm)
1 kilopascal (kPa) × 9.8692 × 10^{-3} = 14.7 pounds per square inch (14.7 psi)
1 yd^3 = 27 ft^3
1 US gallon of water = 8.34 lb
1 imperial gallon of water = 10 lb
1 yd^3 = 0.765 m^3
1 acre-inch of liquid = 27,150 gal = 3.630 ft^3
1 ft depth in 1 acre (in situ) = 1613 × (20%–25% excavation factor) = ~2000 yd^3
1 yd^3 (clayey soils-excavated) = 1.1–1.2 tons (US)
1 yd^3 (sandy soils-excavated) = 1.2–1.3 tons (US)

SI Metric Conversion Factors

Acre-foot × 1.233482	E+03 = meters cubed
Barrels × 1.589873	E−01 = meters cubed
Centipoise × 1.000000	E−03 = pascal seconds
Darcy × 9.869233	E−01 = micrometers squared
Feet × 3.048000	E−01 = meters
Pounds per acre-foot × 3.677332	E−04 = kilograms per meters cubed
Pounds per square inch × 6.894757	E+00 = kilopascals
Dyne per centimeter × 1.000000	E+00 = mN/m
Parts per million × 1.000000	E+00 = milligrams per kilograms

E = exponent, i.e., E+03 = 10^3 and E−03 = 10^{-3}.

Glossary

Abandon To cease work on a well that is nonproductive, to plug off the well with cement plugs, and to salvage all recoverable equipment. Also used in the context of field abandonment.

Abandonment Converting a drilled well to a condition that can be left indefinitely without further attention and will not damage freshwater supplies, potential crude oil reservoirs, or the environment.

Abandonment pressure The bottom-hole pressure at which a company will no longer make an economic profit on the well; a direct function of the economic premises, and it corresponds to the static bottom pressure at which the revenues obtained from the sales of the hydrocarbons produced are equal to the well's operation costs.

ABC process A fixed-bed process for the hydrodemetallization and hydrodesulfurization of heavy feedstocks.

ABN separation A method of fractionation by which petroleum (or a fraction thereof) is separated into acidic, basic, and neutral constituents.

Absolute permeability Ability of a rock to conduct a fluid when only one fluid is present in the pores of the rock.

Absorber See absorption tower.

Absorption gasoline Gasoline extracted from natural gas or refinery gas by contacting the absorbed gas with an oil and subsequently distilling the gasoline from the higher-boiling components.

Absorption oil Oil used to separate the heavier components from a vapor mixture by absorption of the heavier components during intimate contacting of the oil and vapor and used to recover natural gasoline from wet gas.

Absorption plant A plant for recovering the condensable portion of natural or refinery gas, by absorbing the higher-boiling hydrocarbons in an absorption oil, followed by separation and fractionation of the absorbed material.

Absorption tower A tower or column that promotes contact between a rising gas and a falling liquid so that part of the gas may be dissolved in the liquid.

Acetone-benzol process A dewaxing process in which acetone and benzol (benzene or aromatic naphtha) are used as solvents.

Acid catalyst A catalyst having acidic character; alumina is an example of such a catalyst.

Acid deposition Acid rain; a form of pollution depletion in which pollutants, such as nitrogen oxides and sulfur oxides, are transferred from the atmosphere to soil or water; often referred to as atmospheric self-cleaning. The pollutants usually arise from the use of fossil fuels.

Acidity The capacity of an acid to neutralize a base such as a hydroxyl ion (OH^-).

Acidizing A technique for improving the permeability of a reservoir by injecting acid.

Acid number A measure of the reactivity of crude oil with a caustic solution and given in terms of milligrams of potassium hydroxide that are neutralized by 1 g of crude oil.

Acid rain The precipitation phenomenon that incorporates anthropogenic acids and other acidic chemicals from the atmosphere to the land and water (see acid deposition).

Acid sludge The residue left after treating petroleum oil with sulfuric acid for the removal of impurities; a black, viscous substance containing the spent acid and impurities.

Acid treating A process in which unfinished petroleum products, such as gasoline, kerosene, and lubricating oil stocks, are contacted with sulfuric acid to improve their color, odor, and other properties.

Acoustic log See sonic log.

Acre-foot A measure of bulk rock volume where the area is 1 ac and the thickness is 1 ft.

Activation energy, *E* The energy that is needed by a molecule or molecular complex to encourage reactivity to form products.

Additions The reserve provided by the exploratory activity. It consists of the discoveries and delimitations in a field during the study period.

Additive A material added to another (usually in small amounts) in order to enhance desirable properties or to suppress undesirable properties.

Add-on control methods The use of devices that remove refinery process emissions after they are generated but before they are discharged to the atmosphere.

Adsorption Transfer of a substance from a solution to the surface of a solid resulting in relatively high concentration of the substance at the place of contact; see also chromatographic adsorption.

Adsorption gasoline Natural gasoline obtained in the adsorption process from wet gas.

Afterburn The combustion of carbon monoxide (CO) to carbon dioxide (CO_2), usually in the cyclones of a catalyst regenerator.

After-flow Flow from the reservoir into the wellbore that continues for a period after the well has been shut in; after-flow can complicate the analysis of a pressure transient test.

Air-blown asphalt Asphalt produced by blowing air through residua at elevated temperatures.

Air injection An oil recovery technique using air to force oil from the reservoir into the wellbore.

Airlift thermofor catalytic cracking A moving-bed continuous catalytic process for the conversion of heavy gas oils into lighter products; the catalyst is moved by a stream of air.

Air pollution The discharge of toxic gases and particulate matter introduced into the atmosphere, principally as a result of human activity.

Air sweetening A process in which air or oxygen is used to oxidize lead mercaptide derivatives to disulfides instead of using elemental sulfur.

Alicyclic hydrocarbon A compound containing carbon and hydrogen only, which has a cyclic structure (e.g., cyclohexane); also collectively called naphthenes.

Aliphatic hydrocarbon A compound containing carbon and hydrogen only, which has an open-chain structure (e.g., as ethane, butane, octane, and butene) or a cyclic structure (e.g., cyclohexane).

Alkaline flooding See EOR process.

Alkalinity The capacity of a base to neutralize the hydrogen ion (H^+).

Alkali treatment See caustic wash.

Alkali wash See caustic wash.

Alkylate (alkylation) A refining operation that takes low-value derivatives from the cat cracking and other processes and unites them in the presence of an acid catalyst to produce a very-high-octane, low-vapor-pressure gasoline blending component.

Alkylate bottoms Residua from fractionation of alkylate; the alkylate product that boils higher than the aviation gasoline range; sometimes called heavy alkylate or alkylate polymer.

Alkylation unit A refining unit in which propylene or butylene reacts with isobutylene to yield a high-octane gasoline blending component called alkylate. Alkylate helps improve the environmental qualities of gasoline—low vapor pressure, zero sulfur content, zero olefin content, zero benzene, and a high octane number.

Alpha-scission The rupture of the aromatic carbon-aliphatic carbon bond that joins an alkyl group to an aromatic ring.

Alumina (Al_2O_3) Used in separation methods as an adsorbent and in refining as a catalyst.

American Society for Testing and Materials (ASTM) The official organization in the United States for designing standard tests for crude oil and other industrial products; now renamed as ASTM International.

Amine washing A method of gas cleaning whereby acidic impurities such as hydrogen sulfide and carbon dioxide are removed from the gas stream by washing with an amine (usually an alkanolamine).

Analysis Determine the properties of a feedstock prior to refining and inspection of feedstock properties.

Aniline point The temperature, usually expressed in °F, above which equal volumes of a petroleum product are completely miscible; a qualitative indication of the relative proportions of paraffins in a petroleum product that are miscible with aniline only at higher temperatures; a high aniline point indicates low aromatics.

Anticline The structural configuration of a collection of folding rocks and in which the rocks are tilted in different directions from the crest.

Antiknock Resistance to detonation or pinging in spark-ignition engines.

Antiknock agent A chemical compound such as tetraethyl lead that, when added in small amount to the fuel charge of an internal-combustion engine, tends to lessen knocking.

Antistripping agent An additive used in an asphaltic binder to overcome the natural affinity of an aggregate for water instead of asphalt.

API gravity A measure of the *lightness* or *heaviness* of crude oil that is related to density and specific gravity $°API = (141.5/\text{sp. gr at } 60°F) - 131.5$.

Apparent bulk density The density of a catalyst as measured, usually loosely compacted in a container.

Apparent viscosity The viscosity of a fluid or several fluids flowing simultaneously, measured in a porous medium (rock), and subject to both viscosity and permeability effects; also called effective viscosity.

Appraisal well A well drilled as part of an appraisal drilling program that is carried out to determine the physical extent, reserves, and likely production rate of a field.

Aquaconversion process A hydrovisbreaking technology in which a proprietary additive and water are added to the heavy feedstock prior to introduction into the soaker.

Aquifer A subsurface rock interval that will produce water; often the underlay of a crude oil reservoir.

Areal sweep efficiency (horizontal sweep efficiency) The fraction of the flood pattern area that is effectively swept by the injected fluids.

Artificial production system Any of the techniques used to extract crude oil from the producing formation to the surface when the reservoir pressure is insufficient to raise the oil naturally to the surface.

Aromatic hydrocarbon A hydrocarbon characterized by the presence of an aromatic ring or condensed aromatic rings, benzene and substituted benzene, naphthalene and substituted naphthalene, phenanthrene and substituted phenanthrene, and the higher condensed ring systems; compounds that are distinct from those of aliphatic compounds or alicyclic compounds.

Aromatization The conversion of nonaromatic hydrocarbons to aromatic hydrocarbons by (1) rearrangement of aliphatic (noncyclic) hydrocarbons into aromatic ring structures and (2) dehydrogenation of alicyclic hydrocarbons (naphthenes).

Arosorb process A process for the separation of aromatics from nonaromatics by adsorption on a gel from which they are recovered by desorption.

ART process A process for increasing the production of liquid fuels without hydrocracking.

ASCOT process A resid upgrading process that integrates delayed coking and deep solvent deasphalting.

Asphalt The nonvolatile product obtained by distillation and treatment of an asphaltic crude oil with liquid propane or liquid butane, usually consists of asphaltenes, resins, and gas oil; a manufactured product.

Asphalt cement Asphalt especially prepared as to quality and consistency for direct use in the manufacture of bituminous pavements.

Asphalt emulsion An emulsion of asphalt cement in water containing a small amount of emulsifying agent.

Asphaltene fraction The brown to black powdery material produced by the treatment of crude oil, crude oil residua, or bituminous materials with a low-boiling liquid hydrocarbon, e.g., pentane or heptane; soluble in benzene (and other aromatic solvents), carbon disulfide, and chloroform (or other chlorinated hydrocarbon solvents).

Asphaltene association factor The number of individual asphaltene species that associate in nonpolar solvents as measured by molecular-weight methods; the molecular weight of asphaltenes in toluene divided by the molecular weight in a polar nonassociating solvent, such as dichlorobenzene, pyridine, or nitrobenzene.

Asphalt flux Oil used to reduce the consistency or viscosity of hard asphalt to the point required for use.

Asphaltics (asphaltic constituents) A general term usually meaning the asphaltene fraction *plus* the resin fraction.

Asphalt primer A liquid asphaltic material of low viscosity that, upon application to a non-bituminous surface, is completely absorbed; used to waterproof the surface and prepare it for further construction.

Asphaltic pyrobitumen See asphaltoid.

Asphaltic road oil A thick, fluid solution of asphalt, usually a residual oil. See also nonasphaltic road oil.

Assay A test or series of tests performed on crude oil to determine the substance's physical and chemical properties; the assay typically identifies the viscosity, density, acidity (acid number), and the amount of sulfur.

Associated gas in solution or dissolved Natural gas dissolved in the crude oil of the reservoir, under the prevailing pressure and temperature conditions.

Associated gas Natural gas that is in contact with and/or dissolved in the crude oil of the reservoir. It may be classified as gas cap (free gas) or gas in solution (dissolved gas).

Associated molecular weight The molecular weight of asphaltenes in an associating (nonpolar) solvent, such as toluene.

Atmospheric distillation Distillation at atmospheric pressure.

Atmospheric crude oil distillation The process of separating crude oil components at atmospheric pressure by heating and subsequent condensing of the fractions by cooling.

Atmospheric equivalent boiling point (AEBP) A mathematical method of estimating the boiling point at atmospheric pressure of nonvolatile fractions of petroleum.

Atmospheric residuum A residuum obtained by distillation of a crude oil under atmospheric pressure and that boils above 350°C (660°F).

Attapulgus clay See fuller's earth.

Autofining A catalytic process for desulfurizing distillates.

Average particle size The weighted average particle diameter of a catalyst.

Aviation gasoline Any of the special grades of gasoline suitable for use in certain airplane engines.

Aviation turbine fuel See jet fuel.

Back mixing The phenomenon observed when a catalyst travels at a slower rate in the riser pipe than the vapors.

Baghouse A filter system for the removal of particulate matter from gas streams; so called because of the similarity of the filters to coal bags.

Bank The concentration of oil (oil bank) in a reservoir that moves cohesively through the reservoir.

Bari-sol process A dewaxing process that employs a mixture of ethylene dichloride and benzol as the solvent.

Barrel The unit of measurement of liquids in the crude oil industry; equivalent to 42 US standard gallons or 33.6 imperial (UK) gallons (159 L = 7.3 bbl = 1 ton: 6.29 bbl = 1 m^3).

Basement Foot or base of a sedimentary sequence composed of igneous or metamorphic rocks.

Basic nitrogen Nitrogen (in crude oil) that occurs in pyridine form.

Basic sediment and water (BS&W or BSW) The material that collects in the bottom of storage tanks, usually composed of oil, water, and foreign matter; also called bottoms or bottom settlings.

Basin Receptacle in which a sedimentary column is deposited that shares a common tectonic history at various stratigraphic levels.

Battery Equipment to process or store crude oil from one or more wells.

Baumé gravity The specific gravity of liquids expressed as degrees on the Baumé (°Bé) scale. For liquids lighter than water, Sp. gr 60°F = 140/(130 + °Bé). For liquids heavier than water, Sp. gr 60°F = 145/(145 − °Bé).

Bauxite Mineral matter used as a treating agent; hydrated aluminum oxide formed by the chemical weathering of igneous rocks.

Bbl See barrel.

Bell cap A hemispherical or triangular cover placed over the riser in a (distillation) tower to direct the vapors through the liquid layer on the tray; see bubble cap.

Benchmarking measures Data and information used as a point of reference against which industry performance is measured.

Bender process A chemical treating process using lead sulfide catalyst for sweetening light distillates by which mercaptans are converted to disulfides by oxidation.

Bentonite Montmorillonite (a magnesium-aluminum silicate), used as a treating agent.

Beta-scission The rupture of a carbon-carbon bond, two bonds removed from an aromatic ring.

Benzene A low-boiling aromatic hydrocarbon, which occurs naturally as a part of oil and natural gas activity; considered to be a nonthreshold carcinogen and is an occupational and public health concern.

Benzin A refined light naphtha used for extraction purposes.

Benzine An obsolete term for light petroleum distillates covering the gasoline and naphtha range; see ligroine.

Benzol The general term that refers to commercial or technical (not necessarily pure) benzene; also the term used for aromatic naphtha.

Billion 1×10^9.

Biodegradation The destruction of organic materials by bacteria.

Biological oxidation The oxidative consumption of organic matter by bacteria by which the organic matter is converted into gases.

Biomass Biological organic matter.

Bitumen A naturally occurring highly viscous hydrocarbonaceous material that exists in deposits in a semisolid or solid phase. In its natural state, it generally contains sulfur, metals, and other nonhydrocarbon compounds. Natural bitumen has a viscosity of more than several thousand centipoises, measured at the original temperature of the reservoir, at atmospheric pressure and gas-free. It frequently requires treatment before being refined.

Bituminous Containing bitumen or constituting the source of bitumen.

Bituminous rock See bituminous sand.

Bituminous sand A formation in which the bituminous material (see bitumen) is found as a filling in veins and fissures in fractured rock or impregnating relatively shallow sand, sandstone, and limestone strata; a sandstone reservoir that is impregnated with a heavy, viscous black crude oil-like material that cannot be retrieved through a well by conventional production techniques.

Black acid A mixture of the sulfonates found in acid sludge that is insoluble in naphtha, benzene, and carbon tetrachloride; very soluble in water but insoluble in 30% sulfuric acid; in the dry, oil-free state, the sodium soaps are black powders.

Black oil Any of the dark-colored oils; a term now often applied to heavy oil.

Black soap See black acid.

Black strap The black material (mainly lead sulfide) formed in the treatment of sour light oils with doctor solution and found at the interface between the oil and the solution.

Blowdown Condensate and gas are produced simultaneously from the outset of production.

Blowout When well pressure exceeds the ability of the wellhead valves to control it. Oil and gas "blow wild" at the surface.

Blowout preventers BOPs are high-pressure wellhead valves, designed to shut off the uncontrolled flow of hydrocarbons.

Blown asphalt The asphalt prepared by air blowing a residuum or an asphalt.

BOC process See RCD Unibon (BOC) process.

Boiling point A characteristic physical property of a liquid at which the vapor pressure is equal to that of the atmosphere and the liquid is converted to a gas.

Boiling range The range of temperature, usually determined at atmospheric pressure in standard laboratory apparatus, over which the distillation of oil commences, proceeds, and finishes.

Borehole The hole as drilled by the drill bit.

Bottled gas Usually butane or propane or butane-propane mixtures liquefied and stored under pressure for domestic use; see also liquefied petroleum gas.

Bottom of the barrel Residuum.

Bottom-of-the-barrel processing Residuum processing.

Bottoms The liquid that collects in the bottom of a vessel (tower bottoms and tank bottoms) either during distillation; also the deposit or sediment formed during storage of petroleum or a petroleum product; see also residuum and basic sediment and water.

Bright stock Refined, high-viscosity lubricating oils usually made from residual stocks by processes such as a combination of acid treatment or solvent extraction with dewaxing or clay finishing.

British thermal unit See Btu.

Bromine number The number of grams of bromine absorbed by 100 g of oil that indicates the percentage of double bonds in the material.

Brown acid Oil-soluble petroleum sulfonates found in acid sludge that can be recovered by extraction with naphtha solvent. Brown-acid sulfonates are somewhat similar to mahogany sulfonates but are more water-soluble. In the dry, oil-free state, the sodium soaps are light-colored powders.

Brown soap See brown acid.

Btu (British thermal unit) The energy required to raise the temperature of 1 lb of water 1°F.

BS&W See basic sediment and water.

Bronsted acid A chemical species that can act as a source of protons.

Bronsted base A chemical species that can accept protons.

Bubble cap An inverted cup with a notched or slotted periphery to disperse the vapor in small bubbles beneath the surface of the liquid on the bubble plate in a distillation tower.

Bubble plate A tray in a distillation tower.

Bubble point The temperature at which incipient vaporization of a liquid in a liquid mixture occurs, corresponding with the equilibrium point of 0% vaporization or 100% condensation; the temperature at which a gas starts to come out of a liquid.

Bubble tower A fractionating tower so constructed that the vapors rising pass up through layers of condensate on a series of plates or trays (see bubble plate); the vapor passes from one plate to the next above by bubbling under one or more caps (see bubble cap) and out through the liquid on the plate where the less volatile portions of vapor condense in bubbling through the liquid on the plate, overflow to the next lower plate, and ultimately back into the reboiler, thereby effecting fractionation.

Bubble tray A circular, perforated plate having the internal diameter of a bubble tower, set at specified distances in a tower to collect the various fractions produced during distillation.

Bulk composition The makeup of petroleum in terms of bulk fractions such as *saturates*, *aromatics*, *resins*, and *asphaltenes*; separation of petroleum into these fractions is usually achieved by a combination of *solvent* and *adsorption* processes.

Bumping The knocking against the walls of a still occurring during distillation of petroleum or a petroleum product that usually contains water.

Bunker C oil See no. 6 fuel oil.

Burner fuel oil Any petroleum liquid suitable for combustion.

Burning oil An illuminating oil, such as kerosene (kerosine) suitable for burning in a wick lamp.

Burning point See fire point.

Burning-quality index An empirical numerical indication of the likely burning performance of a furnace or heater oil, derived from the distillation profile and the API gravity, and generally recognizing the factors of paraffinic character and volatility.

Burton process An older thermal cracking process in which oil was cracked in a pressure still and any condensation of the products of cracking also took place under pressure.

Butane dehydrogenation A process for removing hydrogen from butane to produce butenes and, on occasion, butadiene.

Butane vapor-phase isomerization A process for isomerizing *n*-butane to *iso*butane using aluminum chloride catalyst on a granular alumina support and with hydrogen chloride as a promoter.

C_1, C_2, C_3, C_4, and C_5 fractions A common way of representing fractions containing a preponderance of hydrocarbons having 1, 2, 3, 4, or 5 carbon atoms, respectively, and without reference to hydrocarbon type.

CANMET hydrocracking process A hydrocracking process for heavy feedstocks that employs a low-cost additive to inhibit coke formation and allow high feedstock conversion using a single reactor.

Calorific equivalence of dry gas to liquid factor (CEDGLF) The factor used to relate dry gas to its liquid equivalent. It is obtained from the molar composition of the reservoir gas, considering the unit heat value of each component and the heat value of the equivalence liquid.

Capillary forces Interfacial forces between immiscible fluid phases, resulting in pressure differences between the two phases.

Capillary number N_c, the ratio of viscous forces to capillary forces and equal to viscosity times velocity divided by interfacial tension.

Capillary pressure A force per area unit resulting from the surface forces to the interface between two fluids.

Caprock See seal rock.

Carbene The pentane- or heptane-insoluble material that is insoluble in benzene or toluene but which is soluble in carbon disulfide (or pyridine); a type of rifle used for hunting bison.

Carboid The pentane- or heptane-insoluble material that is insoluble in benzene or toluene and which is also insoluble in carbon disulfide (or pyridine).

Carbonate washing Processing using a mild alkali (e.g., potassium carbonate) process for emission control by the removal of acid gases from gas streams.

Carbon dioxide-augmented waterflooding Injection of carbonated water, or water and carbon dioxide, to increase waterflood efficiency; see immiscible carbon dioxide displacement.

Carbon capture and storage (CCS) The process of taking waste carbon dioxide and transporting it to a storage site, normally underground in a specific type of geologic formation.

Carbon dioxide flood A technology that allows the continued production of crude oil from a mature field in an environmentally responsible manner by utilizing and then storing carbon dioxide that would have been vented to the atmosphere. The carbon dioxide acts as a solvent, cleaning oil trapped in the microscopic pores of the reservoir rock, which greatly increases the recovery of oil from a reservoir.

Carbonization The conversion of an organic compound into char or coke by heat in the substantial absence of air; often used in reference to the destructive distillation (with simultaneous removal of distillate) of coal.

Carbon-forming propensity See carbon residue.

Carbonization The conversion of an organic compound into char or coke by heat in the substantial absence of air; often used in reference to the *destructive distillation* (with simultaneous removal of distillate) of high-boiling carbonaceous materials.

Carbon rejection An upgrading process in which coke is produced, e.g., coking.

Carbon residue The amount of carbonaceous residue remaining after thermal decomposition of crude oil, a crude oil fraction, or a crude oil product in a limited amount of air; also called the *coke-* or *carbon-forming propensity*; often prefixed by the terms Conradson or Ramsbottom in reference to the inventor of the respective tests.

Cascade tray A fractionating device consisting of a series of parallel troughs arranged on stair-step fashion in which liquid frown the tray above enters the uppermost trough and liquid thrown from this trough by vapor rising from the tray below impinges against a plate and a perforated baffle and liquid passing through the baffle enters the next longer of the troughs.

Casing The metal pipe inserted into a wellbore and cemented in place to protect both subsurface formations (such as groundwater) and the wellbore. A surface casing is set first to protect groundwater. The production casing is the last one set. The production tubing (through which hydrocarbons flow to the surface) will be suspended inside the production casing.

Casinghead gas Natural gas that issues from the casinghead (the mouth or opening) of an oil well.

Casinghead gasoline The liquid hydrocarbon product extracted from casinghead gas by one of three methods Compression, absorption, or refrigeration; see also natural gasoline.

Casing string The steel tubing that lines a well after it has been drilled. It is formed from sections of steel tube screwed together.

Catagenesis The alteration of organic matter during the formation of crude oil that may involve temperatures in the range from 50°C (120°F) to 200°C (390°F); see also diagenesis and metagenesis.

Catalyst A chemical agent that, when added to a reaction (process), will enhance the conversion of a feedstock without being consumed in the process.

Catalyst plugging The deposition of carbon (coke) or metal contaminants that decreases flow through the catalyst bed.

Catalyst poisoning The deposition of carbon (coke) or metal contaminants that causes the catalyst to become nonfunctional.

Catalyst selectivity The relative activity of a catalyst with respect to a particular compound in a mixture or the relative rate in competing reactions of a single reactant.

Catalyst stripping The introduction of steam at a point where spent catalyst leaves the reactor, in order to strip, i.e., remove, deposits retained on the catalyst.

Catalytic activity The ratio of the space velocity of the catalyst under test to the space velocity required for the standard catalyst to give the same conversion as the catalyst being tested, usually multiplied by 100 before being reported.

Catalytic cracking The conversion of high-boiling feedstocks into lower-boiling products by means of a catalyst that may be used in a fixed bed or fluid bed.

Cat cracking See catalytic cracking.

Catalytic reforming Rearranging hydrocarbon molecules in a gasoline-boiling-range feedstock to produce other hydrocarbons having a higher antiknock quality, isomerization of paraffins, cyclization of paraffins to naphthenes, and dehydrocyclization of paraffins to aromatics.

Catforming A process for reforming naphtha using a platinum-silica-alumina catalyst that permits relatively high space velocities and results in the production of high-purity hydrogen.

Caustic wash The process of treating a product with a solution of caustic soda to remove minor impurities; often used in reference to the solution itself.

Cementation The process by which coarse-grained sediments lithify. In the process, mineral crystals become enmeshed with each other to form a more cohesive, harder mass than the original loose sediment. See lithification.

Centrifugal pump A rotating pump, commonly used for large-volume oil and natural gas pipelines, that takes in fluids near the center and accelerates them as they move to the outlet on the outer rim.

Ceresin A hard, brittle wax obtained by purifying ozokerite; see microcrystalline wax and ozokerite.

Cetane index An approximation of the cetane number calculated from the density and mid-boiling-point temperature; see also diesel index.

Cetane number A number indicating the ignition quality of diesel fuel; a high cetane number represents a short ignition delay time; the ignition quality of diesel fuel can also be estimated from the following formula Diesel index = (aniline point (°F) × API gravity) 100

CFR Code of Federal Regulations; title 40 (40 CFR) contains the regulations for protection of the environment.

Characterization factor The UOP characterization factor *K*, defined as the ratio of the cube root of the molal average boiling point, T_B, in degrees Rankine (°R = °F + 460), to the specific gravity at 60°F/60°F $K = (T_B)^{1/3}/\text{sp. gr.}$ The value ranges from 12.5 for paraffinic stocks to 10.0 for the highly aromatic stocks; also called the Watson characterization factor.

Cheesebox still An early type of vertical cylindrical still designed with a vapor dome.

Chelating agents Complex-forming agents having the ability to solubilize heavy metals.

Chemical composition The makeup of petroleum in terms of distinct chemical types such as *paraffins*, *isoparaffins*, *naphthenes* (*cycloparaffins*), *benzenes*, *diaromatics*, *triaromatics*, and *polynuclear aromatics*; other chemical types can also be specified.

Chemical flooding See EOR process.

Chemical octane number The octane number added to gasoline by refinery processes or by the use of octane number improvers such as tetraethyl lead.

Chemical waste Any solid, liquid, or gaseous material discharged from a process and that may pose substantial hazards to human health and environment.

Cherry-P process A process for the conversion of heavy feedstocks into distillate and a cracked residuum.

Chevron deasphalted oil hydrotreating process A process designed to desulfurize heavy feedstocks that have had the asphaltene fraction removed by prior application of a deasphalting process.

Chevron RDS and VRDS processes Processes designed to remove sulfur, nitrogen, asphaltene, and metal contaminants from heavy feedstocks consisting of a once-through operation of the feedstock coming into contact with hydrogen and the catalyst in a downflow reactor.

Chlorex process A process for extracting lubricating oil stocks in which the solvent used is chlorex (3-3-dichlorodiethyl ether).

Chromatographic adsorption Selective adsorption on materials such as activated carbon, alumina, or silica gel; liquid or gaseous mixtures of hydrocarbons are passed through the adsorbent in a stream of diluent, and certain components are preferentially adsorbed.

Chromatography A method of separation based on selective adsorption; see also chromatographic adsorption.

Clarified oil The heavy oil that has been taken from the bottom of a fractionator in a catalytic cracking process and from which residual catalyst has been removed.

Clarifier Equipment for removing the color or cloudiness of an oil or water by separating the foreign material through mechanical or chemical means; may involve centrifugal action, filtration, heating, or treatment with acid or alkali.

Clastic Composed of pieces of preexisting rock.

Clay Silicate minerals that also usually contain aluminum and have particle sizes that are $<0.002\,\mu m$; used in separation methods as an adsorbent and in refining as a catalyst.

Clay contact process See contact filtration.

Clay refining A treating process in which vaporized gasoline or other light petroleum product is passed through a bed of granular clay such as fuller's earth.

Clay regeneration A process in which spent coarse-grained adsorbent clays from percolation processes are cleaned for reuse by deoiling them with naphtha, steaming out the excess naphtha, and then roasting in a stream of air to remove carbonaceous matter.

Clay treating See gray clay treating.

Clay wash A light oil, such as kerosene (kerosine) or naphtha, used to clean fuller's earth after it has been used in a filter.

Cloud point The temperature at which paraffin wax or other solid substances begin to crystallize or separate from the solution, imparting a cloudy appearance to the oil when the oil is chilled under prescribed conditions.

Coal An organic rock.

Coal tar The specific name for the tar produced from coal.

Coal tar pitch The specific name for the pitch produced from coal.

COFCAW An EOR process that combines forward combustion and waterflooding.

Cogeneration An energy conversion method by which electric energy is produced along with steam generated for EOR use.

Coke A gray to black solid carbonaceous material produced from crude oil during thermal processing; characterized by having a high carbon content (95% + by weight) and a honeycomb type of appearance and is insoluble in organic solvents.

Coke drum A vessel in which coke is formed and can be cut oil from the process for cleaning.

Coke number Used, particularly in Great Britain, to report the results of the Ramsbottom carbon residue test, which is also referred to as a coke test.

Coker The processing unit in which coking takes place.

Coking A process for the thermal conversion of crude oil in which gaseous, liquid, and solid (coke) products are formed.

Cold pressing The process of separating wax from oil by first chilling (to help form wax crystals) and then filtering under pressure in a plate and frame press.

Cold production The use of operating and specialized exploitation techniques in order to rapidly produce heavy oils without using thermal recovery methods.

Cold settling Processing for the removal of wax from high-viscosity stocks, wherein a naphtha solution of the waxy oil is chilled and the wax crystallizes out of the solution.

Color stability The resistance of a petroleum product to color change due to light, aging, etc.

Combustible liquid A liquid with a flash point in excess of 37.8°C (100°F) but below 93.3°C (200°F).

Combustion zone The volume of reservoir rock wherein petroleum is undergoing combustion during enhanced oil recovery.

Commercial field An oil and/or gas field judged to be capable of producing enough net income to make it worth developing.

Completion The installation of permanent wellhead equipment for the production of oil and gas.

Completion interval The portion of the reservoir formation placed in fluid communication with the well by selectively perforating the wellbore casing.

Complex A series of fields sharing common surface facilities.

Composition The general chemical makeup of crude oil.

Composition map A means of illustrating the chemical makeup of petroleum using chemical and/or physical property data.

Compressor A device installed in the gas pipeline to raise the pressure and guarantee the fluid flow through the pipeline.

Con carbon See carbon residue.

Condensate A mixture of light hydrocarbon liquids obtained by condensation of hydrocarbon vapors: predominately butane, propane, and pentane with some heavier hydrocarbons and relatively little methane or ethane; see also natural gas liquids.

Condensate recovery factor (CRF) The factor used to obtain liquid fractions recovered from natural gas in the surface distribution and transportation facilities. It is obtained from the gas and condensate handling statistics of the last annual period in the area corresponding to the field being studied.

Conductivity A measure of the ease of flow through a fracture, perforation, or pipe.

Conformance The uniformity with which a volume of the reservoir is swept by injection fluids in area and vertical directions.

Connate water Water trapped in the pores of a rock during the formation of the rock; also described as fossil water. The chemistry of connate water can change in composition throughout the history of the rock. Connate water can be dense and saline compared with seawater. Formation water, or interstitial water, in contrast, is simply water found in the pore spaces of a rock and might not have been present when the rock was formed. See formation water.

Conradson carbon residue See carbon residue.

Contact filtration A process in which finely divided adsorbent clay is used to remove color bodies from petroleum products.

Contaminant A substance that causes deviation from the normal composition of an environment.

Contingent resource The amounts of hydrocarbons estimated at a given date, which are potentially recoverable from known accumulations, but are not considered commercially recoverable under the economic evaluation conditions corresponding to such date.

Continuous contact coking A thermal conversion process in which crude oil-wetted coke particles move downward into the reactor in which cracking, coking, and drying take place to produce coke, gas, gasoline, and gas oil.

Continuous contact filtration A process to finish lubricants, waxes, or special oils after acid treating, solvent extraction, or distillation.

Conventional crude oil Crude oil having an API gravity $>20°$.

Conventional crude oil Crude oil that occurs in liquid form, flowing naturally or capable of being pumped without further processing or dilution.

Conventional limit The reservoir limit established according to the degree of knowledge of or research into the geologic, geophysical, or engineering data available.

Conventional recovery Primary and/or secondary recovery.

Conversion The thermal treatment of petroleum that results in the formation of new products by the alteration of the original constituents.

Conversion factor The percentage of feedstock converted to light ends, gasoline, other liquid fuels, and coke.

Copper sweetening Processes involving the oxidation of mercaptans to disulfides by oxygen in the presence of cupric chloride.

Core A cylindrical rock sample taken from a formation when drilling in order to determine its permeability, porosity, hydrocarbon saturation, and other productivity-associated properties.

Core floods Laboratory flow tests through samples (cores) of porous rock.

Cp (centipoise) A unit of viscosity.

Cracked residua Residua that have been subjected to temperatures above 350°C (660°F) during the distillation process.

Cracking The thermal processes by which the constituents of crude oil are converted to lower-molecular-weight products.

Cracking activity See catalytic activity.

Cracking coil Equipment used for cracking heavy crude oil products consisting of a coil of heavy pipe running through a furnace so that the oil passing through it is subject to high temperature.

Cracking still The combined equipment—furnace, reaction chamber, and fractionator—for the thermal conversion of heavier feedstocks to lighter products.

Cracking temperature The temperature (350°C, 660°F) at which the rate of thermal decomposition of crude oil constituents becomes significant.

Craig-Geffen-Morse method A method for predicting oil recovery by waterflood.

Criteria air contaminants (CAC) Emissions of various air pollutants that affect our health and contribute to air pollution problems such as smog.

Crude assay A procedure for determining the general distillation characteristics (e.g., distillation profile and other quality information of crude oil).

Crude oil See petroleum.

Crude oil refining An integrated sequence of unit processes that result in the production of a variety of products.

Crude scale wax The wax product from the first sweating of the slack wax.

Crude still Distillation equipment in which crude oil is separated into various products.

Cryogenic plant Processing plant capable of producing liquid natural gas products, including ethane, at very low operating temperatures.

Cryogenics The study, production, and use of low temperatures.

Cumene A colorless liquid ($C_6H_5CH(CH_3)_2$) used as an aviation gasoline blending component and as an intermediate in the manufacture of chemicals.

Cumulative effects Changes to the environment caused by an activity in combination with other past, present, and reasonably foreseeable human activities.

Cumulative production Production of crude oil, heavy oil, extra heavy oil, or tar sand bitumen to date.

Cutback The term applied to the products from blending heavier feedstocks or products with lighter oils to bring the heavier materials to the desired specifications.

Cutback asphalt Asphalt liquefied by the addition of a volatile liquid such as naphtha or kerosene that, after application and on exposure to the atmosphere, evaporates leaving the asphalt.

Cut point The boiling-temperature division between distillation fractions of crude oil.

Cutting oil Oil used to lubricate and cool metal-cutting tools; also called cutting fluid or cutting lubricant.

Cuttings The rock chippings cut from the formation by the drill bit and brought to the surface with the mud. Used by geologists to obtain formation data.

Cyclic steam injection The alternating injection of steam and the production of oil with condensed steam from the same well or wells.

Cyclic steam stimulation (CSS) Injecting steam into a well in a heavy oil reservoir that introduces heat and thins the oil, allowing it to flow through the same well.

Cycle stock The product taken from some later stage of a process and recharged (recycled) to the process at some earlier stage.

Cyclization The process by which an open-chain hydrocarbon structure is converted to a ring structure, e.g., hexane to benzene.

Cyclone A device for extracting dust from industrial waste gases. It is in the form of an inverted cone into which the contaminated gas enters tangential from the top; the gas is propelled down a helical pathway, and the dust particles are deposited by means of centrifugal force onto the wall of the scrubber.

Deactivation Reduction in catalyst activity by the deposition of contaminants (e.g., coke and metals) during a process.

Dealkylation The removal of an alkyl group from aromatic compounds.

Deasphaltened oil The fraction of crude oil after the asphaltene constituents have been removed.

Deasphaltening Removal of a solid powdery asphaltene fraction from crude oil by the addition of the low-boiling liquid hydrocarbons such as *n*-pentane or *n*-heptane under ambient conditions.

Deasphalting The removal of the asphaltene fraction from crude oil by the addition of a low-boiling hydrocarbon liquid such as *n*-pentane or *n*-heptane, more correctly the removal of asphalt (tacky and semisolid) from crude oil (as occurs in a refinery asphalt plant) by the addition of liquid propane or liquid butane under pressure.

Debutanization Distillation to separate butane and lighter components from higher-boiling components.

Decant oil The highest-boiling product from a catalytic cracker; also referred to as slurry oil, clarified oil, or bottoms.

Decarbonizing A thermal conversion process designed to maximize coker gas-oil production and minimize coke and gasoline yields; operated at essentially lower temperatures and pressures than delayed coking.

Decoking Removal of petroleum coke from equipment such as coking drums; hydraulic decoking uses high-velocity water streams.

Decolorizing Removal of suspended, colloidal, and dissolved impurities from liquid petroleum products by filtering, adsorption, chemical treatment, distillation, bleaching, etc.

Deethanization Distillation to separate ethane and lighter components from propane and higher-boiling components; also called deethanation.

Dehydrating agents Substances capable of removing water (drying, *q.v.*) or the elements of water from another substance.

Dehydrocyclization Any process by which both dehydrogenation and cyclization reactions occur.

Dehydrogenation The removal of hydrogen from a chemical compound, for example, the removal of two hydrogen atoms from butane to make butene(s) and the removal of additional hydrogen to produce butadiene.

Delayed coking A process in which the thermal reaction is allowed to proceed to completion to produce gaseous, liquid, and solid (coke) products.

Delimitation Exploration activity that increases or decreases reserves by means of drilling delimiting wells.

Demethanization The process of distillation in which methane is separated from the higher-boiling components; also called demethanation.

Demex process A solvent extraction demetallizing process that separates high metal vacuum residuum into demetallized oil of relatively low metal content and asphaltene of high metal content.

Density The mass (or weight) of a unit volume of any substance at a specified temperature; see also specific gravity.

Deoiling Reduction in the quantity of liquid oil entrained in solid wax by draining (sweating) or by a selective solvent; see MEK deoiling.

Depentanizer A fractionating column for the removal of pentane and lighter fractions from a mixture of hydrocarbons.

Depropanization Distillation in which lighter components are separated from butanes and higher-boiling material; also called depropanation.

Derrick The towerlike structure that houses most of the drilling controls.

Desalting Removal of mineral salts (mostly chlorides) from crude oils.

Desorption The reverse process of adsorption whereby adsorbed matter is removed from the adsorbent; also used as the reverse of absorption.

Destructive distillation Thermal decomposition with the simultaneous removal of distillate; distillation when thermal decomposition of the constituents occurs.

Desulfurization The removal of sulfur or sulfur compounds from a feedstock.

Detergent oil A lubricating oil possessing special sludge-dispersing properties for use in internal-combustion engines.

Developed proved area Plant projection of the extension drained by the wells of a producing reservoir.

Developed proved reserves Reserves that are expected to be recovered in existing wells, including reserves behind pipe, which may be recovered with the current infrastructure through additional work and with moderate investment costs. Reserves associated with secondary and/or enhanced recovery processes will be considered as developed when the infrastructure required for the process has been installed or when the costs required for such are lower. This category includes reserves in completed intervals that have been opened at the time when the estimation is made, but that have not started flowing due to market conditions, connection problems, or mechanical problems, and whose rehabilitation cost is relatively low.

Development Activity that increases or decreases reserves by means of drilling exploitation wells.

Development well A well drilled within the proved area of an oil or gas reservoir to the depth of a stratigraphic horizon known to be productive; a well drilled in a proved field for the purpose of completing the desired spacing pattern of production.

Devolatilized fuel Smokeless fuel; coke that has been reheated to remove all of the volatile material.

Dewaxing See solvent dewaxing.

Dew point pressure Pressure at which the first drop of liquid is formed, when it goes from the vapor-phase to the two-phase region.

Diagenesis The concurrent and consecutive chemical reactions that commence the alteration of organic matter (at temperatures up to 50°C (120°F)) and ultimately result in the formation of crude oil from the marine sediment; see also catagenesis and metagenesis.

Diagenetic rock Rock formed by conversion through pressure or chemical reaction from a rock; e.g., sandstone is a diagenetic.

Diesel fuel Fuel used for internal combustion in diesel engines, usually that fraction that distills after kerosene.

Diesel cycle A repeated succession of operations representing the idealized working behavior of the fluids in a diesel engine.

Diesel index An approximation of the cetane number of diesel fuel calculated from the density (*q.v.*) and aniline point.

Diesel knock The result of a delayed period of ignition is long and the accumulation of diesel fuel in the engine.

Differential strain analysis Measurement of thermal stress relaxation in a recently cut well.

Diluents Low-boiling (light) crude oil liquids used to dilute bitumen and heavy oil to allow flow through pipelines.

Dispersion A measure of the convective fluids due to flow in a reservoir.

Directional well (deviated well) A well drilled at an angle from the vertical by using a slanted drilling rig or by deflecting the drill bit; directional wells are used to drill multiple wells from a common drilling pad or to reach a subsurface location beneath land where drilling cannot be done.

Discovered resource Volume of hydrocarbons tested through wells drilled.

Discovery Incorporation of reserves attributable to drilling exploratory wells that test hydrocarbon-producing formations.

Discovery well An exploratory well that encounters a previously untapped oil or gas deposit.

Displacement efficiency The ratio of the amount of oil moved from the zone swept by the reprocess to the amount of oil present in the zone prior to the start of the process.

Dissolved gas-oil ratio Ratio of the volume of gas dissolved in oil compared with the volume of oil containing gas. The ratio may be original (Rsi) or instantaneous (Rs).

Distillation A process for separating liquids with different boiling points without thermal decomposition of the constituents (see destructive distillation).

Distillation curve See distillation profile.

Distillation loss The difference, in a laboratory distillation, between the volume of liquid originally introduced into the distilling flask and the sum of the residue and the condensate recovered.

Distillation range The difference between the temperature at the initial boiling point and at the end point, as obtained by the distillation test.

Distillation profile The distillation characteristics of petroleum or a petroleum product showing the temperature and the percent distilled.

Distribution coefficient A coefficient that describes the distribution of a chemical in reservoir fluids, usually defined as the equilibrium concentrations in the aqueous phases.

Doctor solution A solution of sodium plumbite used to treat gasoline or other light petroleum distillates to remove mercaptan sulfur; see also doctor test.

Doctor sweetening A process for sweetening gasoline, solvents, and kerosene by converting mercaptans to disulfides using sodium plumbite and sulfur.

Doctor test A test used for the detection of compounds in light petroleum distillates that react with sodium plumbite; see also doctor solution.

Dome Geologic structure with a semispherical shape or relief.

Domestic heating oil See no. 2 fuel oil.

Donor solvent process A conversion process in which hydrogen donor solvent is used in place of or to augment hydrogen.

Downcomer A means of conveying liquid from one tray to the next below in a bubble tray column.

Downflow reactor A reactor in which the feedstock flows in a downward direction over the catalyst bed.

Downhole A term used to describe tools, equipment, and instruments used in the wellbore or conditions or techniques applying to the wellbore.

Downhole steam generator A generator installed downhole in an oil well to which oxygen-rich air, fuel, and water are supplied for the purposes of generating steam for it into the reservoir. Its major advantage over a surface steam-generating facility is that the losses to the wellbore and surrounding formation are eliminated.

Downstream When referring to the oil and gas industry, this term indicates the refining and marketing sectors of the industry. More generically, the term can be used to refer to any step further along in the process.

Downstream sector The refining and marketing sector of the crude oil industry.

Drainage radius Distance from which fluids flow to the well, that is, the distance reached by the influence of disturbances caused by pressure drops.

Drill cuttings The small pieces of rock created as a drill bit moves through underground formations while drilling.

Drilling rig A drilling unit that is not permanently fixed to the seabed, e.g., a drillship, a semisubmersible, or a jack-up unit. Also means the derrick and its associated machinery.

Drill stem test (formation test) Conventional formation test method; a method of formation testing. The basic drill stem test tool consists of a packer or packers, valves or ports that may be opened and closed from the surface, and two or more pressure-recording devices. The tool is lowered on the drill string to the zone to be tested. The packer or packers are set to isolate the zone from the drilling fluid column.

Dropping point The temperature at which grease passes from a semisolid to a liquid state under prescribed conditions.

Dry gas equivalent to liquid (DGEL) Volume of crude oil that because of its heat rate is equivalent to the volume of dry gas.

Dry gas Natural gas containing negligible amounts of hydrocarbons heavier than methane. Dry gas is also obtained from the processing complexes.

Dry hole Any exploratory or development well that does not find commercial quantities of hydrocarbons.

Drying Removal of a solvent or water from a chemical substance; also referred to as the removal of solvent from a liquid or suspension.

Dry point The temperature at which the last drop of petroleum fluid evaporates in a distillation test.

Dualayer distillate process A process for removing mercaptans and oxygenated compounds from distillate fuel oils and similar products, using a combination of treatment with concentrated caustic solution and electrical precipitation of the impurities.

Dualayer gasoline process A process for extracting mercaptans and other objectionable acidic compounds from petroleum distillates; see also dualayer solution.

Dualayer solution A solution that consists of concentrated potassium or sodium hydroxide containing a solubilizer; see also dualayer gasoline process.

Dubbs cracking An older continuous, liquid-phase thermal cracking process formerly used.

Dykstra-Parsons coefficient An index of reservoir heterogeneity arising from permeability variation and stratification.

Ebullated bed A process in which the catalyst bed is in a suspended state in the reactor by means of a feedstock recirculation pump that pumps the feedstock upward at sufficient speed to expand the catalyst bed at approximately 35% above the settled level.

Economic limit The point at which the revenues obtained from the sale of hydrocarbons match the costs incurred in its exploitation.

Economic reserves Accumulated production that is obtained from a production forecast in which economic criteria are applied.

Edeleanu process A process for refining oils at low temperature with liquid sulfur dioxide (SO_2) or with liquid sulfur dioxide and benzene; applicable to the recovery of aromatic concentrates from naphthas and heavier petroleum distillates.

Effective permeability A relative measure of the conductivity of a porous medium for a fluid when the medium is saturated with more than one fluid. This implies that the effective permeability is a property associated with each reservoir flow, for example, gas, oil, and water. A fundamental principle is that the total of the effective permeability is less than or equal to the absolute permeability.

Effective porosity A fraction that is obtained by dividing the total volume of communicated pores and the total rock volume.

Effective viscosity See apparent viscosity.

Effluent Any contaminating substance, usually a liquid, that enters the environment via a domestic industrial, agricultural, or sewage plant outlet.

Electric desalting A continuous process to remove inorganic salts and other impurities from crude oil by settling out in an electrostatic field.

Electrical precipitation A process using an electric field to improve the separation of hydrocarbon reagent dispersions. May be used in chemical treating processes on a wide variety of refinery stocks.

Electrofining A process for contacting a light hydrocarbon stream with a treating agent (acid, caustic, doctor, etc.) and then assisting the action of separation of the chemical phase from the hydrocarbon phase by an electrostatic field.

Electrolytic mercaptan process A process in which aqueous caustic solution is used to extract mercaptans from refinery streams.

Electrostatic precipitators Devices used to trap fine dust particles (usually in the size range 30–60 μm) that operate on the principle of imparting an electric charge to particles in an incoming airstream and that are then collected on an oppositely charged plate across a high-voltage field.

Emission control The use of gas cleaning processes to reduce emissions.

Emission standard The maximum amount of a specific pollutant permitted to be discharged from a particular source in a given environment.

Emulsion breaking The settling or aggregation of colloid-sized emulsions from suspension in a liquid medium.

End-of-pipe emission control The use of specific emission control processes to clean gases after the production of gases.

Energy The capacity of a body or system to do work, measured in joules (SI units); also the output of fuel sources.

Energy from biomass The production of energy from biomass.

Engler distillation A standard test for determining the volatility characteristics of a gasoline by measuring the percent distilled at various specified temperatures.

Enhanced oil recovery (EOR) Crude oil recovery following recovery by conventional (i.e., primary and/or secondary) methods.

Enhanced oil recovery (EOR) process A process for recovering additional oil from a crude oil reservoir beyond that economically recoverable by conventional primary and secondary recovery methods. EOR methods are usually divided into three main categories: (1) *chemical flooding*, injection of water with added chemicals into a crude oil reservoir. The chemical processes include: surfactant flooding, polymer flooding, and alkaline flooding; (2) *miscible flooding*, injection into a crude oil reservoir of a material that is miscible or can become miscible, with the oil in the reservoir. Carbon dioxide, hydrocarbons, and nitrogen are used; (3) *thermal recovery*, injection of steam into a crude oil reservoir or propagation of a combustion zone through a reservoir by air or oxygen-enriched air injection. The thermal processes include: steam drive, cyclic steam injection, and in situ combustion.

Entrained bed A bed of solid particles suspended in a fluid (liquid or gas) at such a rate that some of the solid is carried over (entrained) by the fluid.

Environmental footprint The impact on the environment, in terms of resource use, waste generation, and changes to the physical environment.

Established reserves The portion of the discovered resource base that is estimated to be recoverable using known technology under present and anticipated economic conditions. Include proved plus a portion of probable (usually 50%).

Ethanol See ethyl alcohol.

ET-II process A thermal cracking process for the production of distillates and cracked residuum for use for metallurgical coke; *of this Earth* and not extraterrestrial!

Ethyl alcohol (ethanol or grain alcohol) An inflammable organic compound (C_2H_5OH) formed during fermentation of sugars; used as an intoxicant and as a fuel.

Eureka process A thermal cracking process to produce a cracked oil and aromatic residuum from heavy residual materials.

Evaporation A process for concentrating nonvolatile solids in a solution by boiling off the liquid portion of the waste stream.

Evaporites Sedimentary formations consisting primarily of salt, anhydrite, or gypsum, as a result of evaporation in coastal waters.

Expanding clays Clays that expand or swell on contact with water, e.g., montmorillonite.

Exploratory well A well that is drilled without detailed knowledge of the underlying rock structure in order to find hydrocarbons whose exploitation is economically profitable.

Explosive limits The limits of percentage composition of mixtures of gases and air within which an explosion takes place when the mixture is ignited.

Extractive distillation The separation of different components of mixtures that have similar vapor pressures by flowing a relatively high-boiling solvent, which is selective for one of the components in the feed, down a distillation column as the distillation proceeds; the selective solvent scrubs the soluble component from the vapor.

Extra heavy oil Crude oil with relatively high fractions of heavy components, high specific gravity (low API density) and high viscosity at reservoir conditions. The production of this kind of oil generally implies difficulties in extraction and high costs. Thermal recovery methods are the most common form of commercially exploiting this kind of oil.

Fabric filters Filters made from fabric materials and used for removing particulate matter from gas streams (see baghouse).

Facies One or more layers of rock that differs from other layers in composition, age, or content.

FAST Fracture-assisted steam flood technology.

Fat oil The bottom or enriched oil drawn from the absorber as opposed to lean oil.

Faujasite A naturally occurring silica-alumina (SiO_2-Al_2O_3) mineral.
Fault Fractured surface of geologic strata along which there has been differential movement.
Fluid saturation Portion of the pore space occupied by a specific fluid; oil, gas, and water may exist.
FCC Fluid catalytic cracking.
Feedstock Crude oil as it is fed to the refinery; a refinery product that is used as the raw material for another process; the term is also generally applied to raw materials used in other industrial processes.
Ferrocyanide process A regenerative chemical treatment for mercaptan removal using caustic-sodium ferrocyanide reagent.
Field The surface area above one or more underground crude oil pools sharing the same or related infrastructure.
Field scale The application of EOR processes to a significant portion of a field.
Filtration The use of an impassable barrier to collect solids but which allows liquids to pass.
Fingering The formation of finger-shaped irregularities at the leading edge of a displacing fluid in a porous medium that moves out ahead of the main body of fluid.
Fireflooding A thermal recovery method in which the oil in the reservoir is ignited, the heat vaporizes lighter hydrocarbons, and water pushes the warmed oil toward a producing well. Also called in situ combustion.
Fire point The lowest temperature at which, under specified conditions in standardized apparatus, a petroleum product vaporizes sufficiently rapidly to form above its surface an air-vapor mixture that burns continuously when ignited by a small flame.
Fischer-Tropsch process A process for synthesizing hydrocarbons and oxygenated chemicals from a mixture of hydrogen and carbon monoxide.
Five-spot An arrangement or pattern of wells with four injection wells at the comers of a square and a producing well in the center of the square.
Fixed bed A stationary bed (of catalyst) to accomplish a process (see fluid bed).
Flaring/venting The controlled burning (flare) or release (vent) of natural gas that can't be processed for sale or use because of technical or economic reasons.
Flammability range The range of temperature over which a chemical is flammable.
Flammable A substance that will burn readily.
Flammable liquid A liquid having a flash point below 37.8°C (100°F).
Flammable solid A solid that can ignite from friction or from heat remaining from its manufacture or which may cause a serious hazard if ignited.
Flash point The lowest temperature to which the product must be heated under specified conditions to give off sufficient vapor to form a mixture with air that can be ignited momentarily by a flame.
Flexicoking A modification of the fluid coking process insofar as the process also includes a gasifier adjoining the burner/regenerator to convert excess coke to a clean fuel gas.
Floc point The temperature at which wax or solids separate as a definite floc.
Flood, flooding The process of displacing crude oil from a reservoir by the injection of fluids.
Flow line Pipe, usually buried, through which oil or gas travels from the well to a processing facility.
Flue gas Gas from the combustion of fuel, the heating value of which has been substantially spent and is, therefore, discarded to the flue or stack.
Fluid A reservoir gas or liquid.
Fluid bed The use of an agitated bed of inert granular material to accomplish a process in which the agitated bed resembles the motion of a fluid; a bed (of catalyst) that is agitated

by an upward passing gas in such a manner that the particles of the bed simulate the movement of a fluid and has the characteristics associated with a true liquid. See fixed bed.

Fluid catalytic cracking Cracking in the presence of a fluidized bed of catalyst.

Fluid coking A continuous fluidized solid process that cracks feed thermally over a bed of coke particles.

Fluid injection The injection of gases or liquids into a reservoir to force oil toward and into producing wells.

Fly ash Particulate matter produced from mineral matter in coal that is converted during combustion to finely divided inorganic material and emerges from the combustor in the gases.

Foot oil The oil sweated out of slack wax; named from the fact that the oil goes to the foot, or bottom, of the pan during the sweating operation.

Formation An interval of rock with distinguishable geologic characteristics.

Formation damage The reduction in permeability in reservoir rock due to the infiltration of drilling or treating fluids into the area adjacent to the wellbore.

Formation pressure The pressure at the bottom of a well when it is shut in at the wellhead.

Formation resistance factor (F) Ratio between the resistance of rock saturated 100% with brine divided by the resistance of the saturating water.

Formation volume factor (B) The factor that relates the volume unit of the fluid in the reservoir with the surface volume. There are volume factors for oil and gas, in both phases, and for water. A sample may be directly measured, calculated, or obtained through empirical correlations.

Formation water (interstitial water) Salt water underlying gas and oil in the formation; water found in the pore spaces of a rock and might not have been present when the rock was formed. See connate water.

Fossil fuel resources A gaseous, liquid, or solid fuel material formed in the ground by chemical and physical changes (diagenesis) in plant and animal residues over geologic time, natural gas, crude oil, coal, and oil shale.

Fractional composition The composition of crude oil as determined by fractionation (separation) methods.

Fracturing A method of breaking down a formation by pumping fluid at very high pressure. The objective is to increase production rates from a reservoir.

Fractional distillation The separation of the components of a liquid mixture by vaporizing and collecting the fractions, or cuts, which condense in different temperature ranges.

Fractionating column A column arranged to separate various fractions of petroleum by a single distillation and that may be tapped at different points along its length to separate various fractions in the order of their boiling points.

Fractionation The separation of petroleum into the constituent fractions using solvent or adsorbent methods; chemical agents such as sulfuric acid may also be used.

Frasch process A process formerly used for removing sulfur by distilling oil in the presence of copper oxide.

Free associated gas Natural gas that overlies and is in contact with the crude oil of the reservoir. It may be gas cap.

Free sulfur Sulfur that exists in the elemental state associated with petroleum; sulfur that is not bound organically within the petroleum constituents.

FTC process A heavy oil and residuum upgrading process in which the feedstock is thermally cracked to produce distillate and coke, which is gasified to fuel gas.

Fuel oil Also called heating oil and is a distillate product that covers a wide range of properties; see also nos. 1–4 fuel oils.

Fuller's earth A clay that has high adsorptive capacity for removing color from oils; Attapulgus clay is a widely used fuller's earth.

Functional group The portion of a molecule that is characteristic of a family of compounds and determines the properties of these compounds.

Furfural extraction A single-solvent process in which furfural is used to remove aromatic, naphthenic, olefinic, and unstable hydrocarbons from a lubricating oil charge stock.

Furnace oil A distillate fuel primarily intended for use in domestic heating equipment.

Gas cap A part of a hydrocarbon reservoir at the top that will produce only gas.

Gas compressibility ratio (Z) The ratio between an actual gas volume and an ideal gas volume. This is a dimensional amount that usually varies between 0.7 and 1.2.

Gaseous pollutants Gases released into the atmosphere that act as primary or secondary pollutants.

Gas field A field containing natural gas but no crude oil.

Gasification A process to partially oxidize any hydrocarbon, typically heavy residues, to a mixture of hydrogen and carbon monoxide; can be used to produce hydrogen and various energy by-products.

Gas injection The process whereby separated associated gas is pumped back into a reservoir for conservation purposes or to maintain the reservoir pressure.

Gas lift Artificial production system that is used to raise the well fluid by injecting gas down the well through tubing or through the tubing-casing annulus.

Gasohol A term for motor vehicle fuel comprising between 80% and 90% unleaded gasoline and 10%–20% ethanol (see also ethyl alcohol).

Gas oil A petroleum distillate with a viscosity and boiling range between those of kerosene and lubricating oil.

Gas-oil ratio (GOR) Ratio of reservoir gas production to oil production, measured at atmospheric pressure.

Gasoline Fuel for the internal-combustion engine that is commonly but improperly referred to simply as gas.

Gas to liquids (GTL) The conversion of natural gas to a liquid form so that it can be transported easily; typically, the liquid is converted back to natural gas prior to consumption.

Gas reversion A combination of thermal cracking or reforming of naphtha with thermal polymerization or alkylation of hydrocarbon gases carried out in the same reaction zone.

Geologic province A region of large dimensions characterized by similar geologic and development histories.

Girbotol process A continuous, regenerative process to separate hydrogen sulfide, carbon dioxide, and other acid impurities from natural gas, refinery gas, etc., using mono-, di-, or triethanolamine as the reagent.

Glycol-amine gas treating A continuous, regenerative process to simultaneously dehydrate and remove acid gases from natural gas or refinery gas.

Glycol dehydrator Field equipment used to remove water from natural gas by using triethylene glycol or diethylene glycol.

Graben Dip or depression formed by tectonic processes, limited by normal-type faults.

Grain alcohol See ethyl alcohol.

Gravitational segregation Reservoir driving mechanism in which the fluids tend to separate according to their specific gravities. For example, since oil is heavier than water, it tends to move toward the lower part of the reservoir in a water-injection project.

Gravity See API gravity.

Gravity drainage The movement of oil in a reservoir that results from the force of gravity.

Gravity segregation Partial separation of fluids in a reservoir caused by the gravity force acting on differences in density.

Gravity-stable displacement The displacement of oil from a reservoir by a fluid of a different density, where the density difference is utilized to prevent gravity segregation of the injected fluid.

Gray clay treating A fixed-bed, usually fuller's earth, vapor-phase treating process to selectively polymerize unsaturated gum-forming constituents (diolefins) in thermally cracked gasoline.

Greenhouse effect Warming of the Earth due to entrapment of the energy of the sun energy by the atmosphere.

Greenhouse gas intensity (GHG intensity) The average emission rate of a given greenhouse gas from a specific source, for example, greenhouse gases released per barrel of production.

Greenhouse gases (GHG's) A type of gas that contributes to the greenhouse effect by absorbing infrared radiation.

Guard bed A bed of disposal adsorbent used to protect process catalysts from contamination by feedstock constituents.

Gulf HDS process A fixed-bed process for the catalytic hydrocracking of heavy feedstocks to distillate products.

Gulfining A catalytic hydrogen treating process for cracked and straight-run distillates and fuel oils, to reduce sulfur content; improve carbon residue, color, and general stability; and effect a slight increase in gravity.

Gulf resid hydrodesulfurization process A process for the desulfurization of heavy feedstocks to produce low-sulfur fuel oils or catalytic cracking feedstocks.

Gum An insoluble tacky semisolid material formed as a result of the storage instability and/or the thermal instability of petroleum and petroleum products.

Handling efficiency shrinkage factor (HESF) This is a fraction of natural gas that is derived from considering self-consumption and the lack of capacity to handle such. It is obtained from the gas-handling statistics of the final period in the area corresponding to the field being studied.

HCPV Hydrocarbon pore volume.

Hearn method A method used in reservoir simulation for calculating a pseudorelative permeability curve that reflects reservoir stratification.

Heating oil See fuel oil.

Heat value The amount of heat released per unit of mass or per unit of volume, when a substance is completely burned. The heat power of solid and liquid fuels is expressed in calories per gram or in Btu per pound. For gases, this parameter is generally expressed in kilocalories per cubic meter or in Btu per cubic foot.

Heavy crude oil Crude oil having an API gravity of $<20°$ and that can be recovered by the application of listed EOR technologies.

Heavy ends The highest-boiling portion of a crude oil fraction; see also light ends.

Heavy feedstock Any feedstock of the type heavy oil, extra heavy oil, tar sand bitumen, atmospheric residuum, vacuum residuum, and solvent deasphalter bottoms.

Heavy fuel oil Fuel oil having a high density and viscosity, generally residual fuel oil such as nos. 5 and 6 fuel oil.

Heavy oil Petroleum having an API gravity of $<20°$; other properties are necessary for inclusion in a more complete definition.

Heavy petroleum See heavy oil.

Heavy residue gasification and combined cycle power generation A process for producing hydrogen from residua.

Heteroatom compounds Chemical compounds that contain nitrogen, oxygen, sulfur, and/or metals bound within their molecular structure(s).

Heterogeneity The lack of uniformity in reservoir properties such as permeability.

HF alkylation An alkylation process whereby olefins (C_3, C_4, and C_5) are combined with *iso*butane in the presence of hydrofluoric acid catalyst.

Higgins-Leighton model Stream tube computer model used to simulate waterflood.

High-boiling distillates Fractions of petroleum that cannot be distilled at atmospheric pressure without decomposition, e.g., gas oils.

High-sulfur petroleum A general expression for petroleum having >1 wt% sulfur; this is a very approximate definition and should not be construed as having a high degree of accuracy because it does not take into consideration the molecular locale of the sulfur. All else being equal, there is little difference between petroleum having 0.99 wt% sulfur and petroleum having 1.01 wt% sulfur.

HOC process A version of the fluid catalytic cracking process that has been adapted to the conversion of residua that contain high amounts of metal and asphaltenes.

H-Oil process A catalytic process that is designed for hydrogenation of heavy feedstocks in an ebullated-bed reactor.

Horizontal drilling Drilling a well that deviates from the vertical and travels horizontally through a producing layer.

Horst Bock of the Earth's crust rising between two faults; the opposite of a graben.

Hortonsphere A spherical pressure-type tank used to store volatile liquids that prevent the excessive evaporation loss that occurs when such products are placed in conventional storage tanks.

Hot filtration test A test for the stability of a petroleum product.

HOT process A catalytic cracking process for upgrading heavy feedstocks using a fluidized bed of iron ore particles.

Hot production The optimum production of heavy oils through the use of enhanced thermal recovery methods.

Hot spot An area of a vessel or line wall appreciably above normal operating temperature, usually as a result of the deterioration of an internal insulating liner that exposes the line or vessel shell to the temperature of its contents.

Hot-water process A method for separating bitumen from tar sand using hot water and caustic soda.

Houdresid catalytic cracking A continuous moving-bed process for catalytically cracking reduced crude oil to produce high-octane gasoline and light distillate fuels.

Houdriflow catalytic cracking A continuous moving-bed catalytic cracking process employing an integrated single vessel for the reactor and regenerator kiln.

Houdriforming A continuous catalytic reforming process for producing aromatic concentrates and high-octane gasoline from low-octane straight naphthas.

Houdry butane dehydrogenation A catalytic process for dehydrogenating light hydrocarbons to their corresponding mono- or diolefins.

Houdry fixed-bed catalytic cracking A cyclic regenerable process for cracking of distillates.

Houdry hydrocracking A catalytic process combining cracking and desulfurization in the presence of hydrogen.

HSC process A cracking process for moderate conversion of heavy feedstocks; the extent of the conversion is higher than visbreaking but lower than coking.

Huff and puff A cyclic EOR method in which steam or gas is injected into a production well; after a short shut-in period, oil and the injected fluid are produced through the same well.

Hybrid gasification process A process to produce hydrogen by gasification of a slurry of coal and residual oil.

HYCAR process A noncatalytic process that is conducted under similar conditions to visbreaking and involves treatment with hydrogen under mild conditions; see also hydrovisbreaking.

Hydration The association of molecules of water with a substance.

Hydraulic fracturing The opening of fractures in a reservoir by high-pressure, high-volume injection of liquids through an injection well that causes the surrounding rocks to crack and allows natural gas or oil to be produced from tight formations; also known as fracking.

Hydrocarbon compounds Chemical compounds containing only carbon and hydrogen.

Hydrocarbon derivatives Chemical compounds that contain carbon and hydrogen only and noncarbon and nonhydrogen atoms.

Hydrocarbon gasification process A continuous, noncatalytic process in which hydrocarbons are gasified to produce hydrogen by air or oxygen.

Hydrocarbon index An amount of hydrocarbons contained in a reservoir per unit area.

Hydrocarbon-producing resource A resource such as coal and oil shale (kerogen) that produces derived hydrocarbons by the application of conversion processes; the hydrocarbons so-produced are not naturally occurring materials.

Hydrocarbon reserves Volume of hydrocarbons measured at atmospheric conditions that will be produced economically by using any of the existing production methods at the date of evaluation.

Hydrocarbon resource Resources such as crude oil and natural gas that can produce naturally occurring hydrocarbons without the application of conversion processes.

Hydrocarbons Chemical compounds that contain carbon and hydrogen only.

Hydroconversion A term often applied to hydrocracking.

Hydrocracking A catalytic high-pressure high-temperature process for the conversion of crude oil feedstocks in the presence of fresh and recycled hydrogen; carbon-carbon bonds are cleaved in addition to the removal of heteroatomic species.

Hydrocracking catalyst A catalyst used for hydrocracking that typically contains separate hydrogenation and cracking functions.

Hydrodemetallization The removal of metallic constituents by hydrotreating.

Hydrodenitrogenation The removal of nitrogen by hydrotreating.

Hydrodesulfurization The removal of sulfur by hydrotreating.

Hydrofining A fixed-bed catalytic process to desulfurize and hydrogenate a wide range of charge stocks from gases through waxes.

Hydroforming A process in which naphthas are passed over a catalyst at elevated temperatures and moderate pressures, in the presence of added hydrogen or hydrogen-containing gases, to form high-octane motor fuel or aromatics.

Hydrogen addition An upgrading process in the presence of hydrogen, e.g., hydrocracking; see hydrogenation.

Hydrogenation The chemical addition of hydrogen to a material. In nondestructive hydrogenation, hydrogen is added to a molecule only if and where unsaturation with respect to hydrogen exists.

Hydrogen blistering Blistering of steel caused by trapped molecular hydrogen formed as atomic hydrogen during corrosion of steel by hydrogen sulfide.

Hydrogen sink A chemical structure within the feedstock that reacts with hydrogen with little, if any, effect on the product character.

Hydrogen transfer The transfer of inherent hydrogen within the feedstock constituents and products during processing.

Hydroprocessing A term often equally applied to hydrotreating and to hydrocracking; also, often collectively applied to both.

Hydrotreating The removal of heteroatomic (nitrogen, oxygen, and sulfur) species by the treatment of a feedstock or product at relatively low temperatures in the presence of hydrogen.

Hydrovisbreaking A noncatalytic process, conducted under similar conditions to visbreaking, which involves treatment with hydrogen to reduce the viscosity of the feedstock and produce more stable products than is possible with visbreaking.

Hydropyrolysis A short-residence-time high-temperature process using hydrogen.

Hyperforming A catalytic hydrogenation process for improving the octane number of naphthas through the removal of sulfur and nitrogen compounds.

Hypochlorite sweetening The oxidation of mercaptans in a sour stock by agitation with aqueous, alkaline hypochlorite solution; used where avoidance of free-sulfur addition is desired, because of a stringent copper strip requirements, and minimum expense is not the primary object.

Hypro process A continuous catalytic method for hydrogen manufacture from natural gas or from refinery effluent gases.

Hyvahl-F process A process for hydroconverting heavy feedstocks to naphtha and middle distillates using a dual-catalyst system and a fixed-bed swing reactor.

IFP hydrocracking process A process that features a dual-catalyst system in which the first catalyst is a promoted nickel-molybdenum amorphous catalyst to remove sulfur and nitrogen and hydrogenate aromatic rings. The second catalyst is a zeolite that finishes the hydrogenation and promotes the hydrocracking reaction.

Ignitability Characteristic of liquids whose vapors are likely to ignite in the presence of ignition source; also characteristic of nonliquids that may catch fire from friction or contact with water and that burn vigorously.

Illuminating oil Oil used for lighting purposes.

Immiscible Two or more fluids that do not have complete mutual solubility and coexist as separate phases.

Immiscible carbon dioxide displacement Injection of carbon dioxide into an oil reservoir to effect oil displacement under conditions in which miscibility with reservoir oil is not obtained; see carbon dioxide-augmented waterflooding.

Immiscible displacement A displacement of oil by a fluid (gas or water) that is conducted under conditions so that interfaces exist between the driving fluid and the oil.

Impurities and plant liquefiables shrinkage factor (IPLSF) It is the fraction obtained by considering the nonhydrocarbon gas impurities (sulfur, carbon dioxide, nitrogen compounds, etc.) contained in the sour gas, in addition to shrinkage caused by the generation of liquids in gas processing plant.

Impurities shrinkage factor (ISF) It is the fraction that results from considering the nonhydrocarbon gas impurities (sulfur, carbon dioxide, nitrogen compounds, etc.) contained in the sour gas. It is obtained from the operation statistics of the last annual period of the gas processing complex (GPC) that processes the production of the field analyzed.

Incompatibility The *immiscibility* of crude oil products and also of different crude oils that is often reflected in the formation of a separate phase after mixing and/or storage.

Incremental ultimate recovery The difference between the quantity of oil that can be recovered by EOR methods and the quantity of oil that can be recovered by conventional recovery methods.

Infill drilling Drilling additional wells within an established pattern.

Inhibitor A substance, the presence of which, in small amounts, in a petroleum product prevents or retards undesirable chemical changes from taking place in the product or in the condition of the equipment in which the product is used.

Inhibitor sweetening A treating process to sweeten gasoline of low mercaptan content, using a phenylenediamine type of inhibitor, air, and caustic.

Initial boiling point The recorded temperature when the first drop of liquid falls from the end of the condenser.

Initial vapor pressure The vapor pressure of a liquid of a specified temperature and 0% evaporated.

Injection profile The vertical flow-rate distribution of fluid flowing from the wellbore to a reservoir.

Injection well A well in an oil field used for injecting fluids into a reservoir.

Injectivity The relative ease with which a fluid is injected into a porous rock.

In situ In its original place; in position; in situ recovery refers to various methods used to recover deeply buried bitumen deposits, including steam injection, solvent injection, and firefloods.

In situ combustion An enhanced oil recovery (EOR) process consisting of injecting air or oxygen-enriched air into a reservoir under conditions that favor burning part of the in situ crude oil, advancing this burning zone, and recovering oil heated from a nearby producing well.

Inspection Application of test procedures to a feedstock to determine its processability; the analysis of a feedstock.

Instability The inability of a crude oil product to exist for periods of time without change to the product.

Integrity Maintenance of a slug or bank at its preferred composition without too much dispersion or mixing.

Interface The thin surface area separating two immiscible fluids that are in contact with each other.

Interfacial film A thin layer of material at the interface between two fluids that differs in composition from the bulk fluids.

Interfacial tension The strength of the film separating two immiscible fluids, e.g., oil and water or microemulsion and oil; measured in dynes (force) per centimeter or millidynes per centimeter.

Interfacial viscosity The viscosity of the interfacial film between two immiscible liquids.

Interference testing A type of pressure transient test in which pressure is measured over time in a closed-in well while nearby wells are produced; flow and communication between wells can sometimes be deduced from an interference test.

Interphase mass transfer The net transfer of chemical compounds between two or more phases.

Interstitial water (formation water) Salt water underlying gas and oil in the formation; water found in the pore spaces of a rock and might not have been present when the rock was formed. See connate water.

Iodine number A measure of the iodine absorption by an oil under standard conditions; used to indicate the quantity of unsaturated compounds present; also called iodine value.

Ion exchange A means of removing cations or anions from solution onto a solid resin.

Isocracking process A hydrocracking process for the conversion of hydrocarbons that operates at relatively low temperatures and pressures in the presence of hydrogen and a catalyst to produce more valuable, lower-boiling products.

Isoforming A process in which olefinic naphtha is contacted with an alumina catalyst at high temperature and low pressure to produce isomers of higher octane number.

Iso-Kel process A fixed-bed, vapor-phase isomerization process using a precious metal catalyst and external hydrogen.

Isomate process A continuous, nonregenerative process for isomerizing C_5–C_8 normal paraffinic hydrocarbons, using aluminum chloride-hydrocarbon catalyst with anhydrous hydrochloric acid as a promoter.

Isomerate process A fixed-bed isomerization process to convert pentane, heptane, and heptane to high-octane blending stocks.

Isomerization The conversion of a *normal* (straight-chain) paraffin hydrocarbon into an *iso* (branched-chain) paraffin hydrocarbon having the same atomic composition.

Iso-plus Houdriforming A combination process using a conventional Houdriformer operated at moderate severity, in conjunction with one of three possible alternatives including the use of an aromatic recovery unit or a thermal reformer; see Houdriforming.

Jacket The lower section, or legs, of an offshore platform.

Jet fuel Fuel meeting the required properties for use in jet engines and aircraft turbine engines.

Kaolinite A clay mineral formed by hydrothermal activity at the time of rock formation or by chemical weathering of rock with high feldspar content, usually associated with intrusive granite rock with high feldspar content.

Kata-condensed aromatic compounds Compounds based on linear condensed aromatic hydrocarbon systems, e.g., anthracene and naphthacene (tetracene).

Kerogen A complex carbonaceous (organic) material that occurs in sedimentary rock and shale; generally insoluble in common organic solvents; produces hydrocarbons when subjected to a heat.

Kerosene (kerosine) A fraction of petroleum that was initially sought as an illuminant in lamps; a precursor to diesel fuel.

K-factor See characterization factor.

Kinematic viscosity The ratio of viscosity to density, both measured at the same temperature.

Knock The noise associated with self-ignition of a portion of the fuel-air mixture ahead of the advancing flame front.

Kriging A technique used in reservoir description for interpolation of reservoir parameters between wells based on random field theory.

Lamp burning A test of burning oils in which the oil is burned in a standard lamp under specified conditions in order to observe the steadiness of the flame, the degree of encrustation of the wick, and the rate of consumption of the kerosene.

Lamp oil See kerosene.

LC-Fining process A hydrogenation process capable of desulfurizing, demetallizing, and upgrading heavy feedstocks by means of an expanded-bed reactor.

Lead A geologic feature of interest insofar as the feature matched the previous feature investigated by the company that has been shown to produce crude oil and/or natural gas. See prospect.

Leaded gasoline Gasoline containing tetraethyl lead or other organometallic lead antiknock compounds.

Lean gas The residual gas from the absorber after the condensable gasoline has been removed from the wet gas.

Lean oil Absorption oil from which gasoline fractions have been removed; oil leaving the stripper in a natural gasoline plant.

LEDA (low-energy deasphalting) process A process for extracting high-quality catalytic cracking feeds from heavy feedstocks; the process uses a low-boiling hydrocarbon solvent specifically formulated to insure the most economical deasphalting design for each operation.

Lewis acid A chemical species that can accept an electron pair from a base.

Lewis base A chemical species that can donate an electron pair.

Light crude oil The specific gravity of the oil is >25° API but less than or equal to 38°. See medium crude oil.

Light ends The lower-boiling components of a mixture of hydrocarbons; see also heavy ends and light hydrocarbons.

Light hydrocarbons Hydrocarbons with molecular weights less than that of heptane (C_7H_{16}).

Light oil The products distilled or processed from crude oil up to, but not including, the first lubricating oil distillate.

Light petroleum Petroleum having an API gravity >20°.

Ligroine (ligroin) A saturated petroleum naphtha boiling in the range of 20–135°C (68–275°F) and suitable for general use as a solvent; also called benzine or petroleum ether.

Limolite Fine-grained sedimentary rock that is transported by water. The granulometrics ranges from fine sand to clay.

Limestone Calcium carbonate-rich sedimentary rocks in which oil or gas reservoirs are often found.

Linde copper sweetening A process for treating gasoline and distillates with a slurry of clay and cupric chloride.

Liquid petrolatum See white oil.

Liquefied natural gas (LNG) Oil field or naturally occurring gas, chiefly methane, liquefied for transportation.

Liquefied petroleum gas (LPG) Light hydrocarbon material, gaseous at atmospheric temperature and pressure, held in the liquid state by pressure to facilitate storage, transport, and handling. Commercial liquefied gas consists essentially of either propane or butane or mixtures thereof.

Liquid sulfur dioxide-benzene process A mixed-solvent process for treating lubricating oil stocks to improve viscosity indexes; also used for dewaxing.

Lithification The process or processes by which unconsolidated materials (noncemented materials or noncoherent materials) are converted into coherent solid rock by compaction or by cementation. See cementation.

Lithology The geologic characteristics of the reservoir rock.

Live steam Steam coming directly from a boiler before being utilized for power or heat.

Liver The intermediate layer of dark-colored, oily material, insoluble in weak acid and in oil, which is formed when acid sludge is hydrolyzed.

Lorenz coefficient A permeability heterogeneity factor.

Low-boiling distillates Fractions of petroleum that can be distilled at atmospheric pressure without decomposition.

Low-sulfur petroleum A general expression for petroleum having <1 wt% sulfur; this is a very approximate definition and should not be construed as having a high degree of accuracy because it does not take into consideration the molecular locale of the sulfur. All else being equal, there is little difference between petroleum having 0.99 wt% sulfur and petroleum having 1.01 wt% sulfur.

Lube See lubricating oil.

Lube cut A fraction of crude oil of suitable boiling range and viscosity to yield lubricating oil when completely refined; also referred to as lube oil distillates or lube stock.

Lubricating oil A fluid lubricant used to reduce friction between bearing surfaces.

Mahogany acids Oil-soluble sulfonic acids formed by the action of sulfuric acid on petroleum distillates. They may be converted to their sodium soaps (mahogany soaps) and extracted from the oil with alcohol for use in the manufacture of soluble oils, rust preventives, and special greases. The calcium and barium soaps of these acids are used as detergent additives in motor oils; see also brown acids and sulfonic acids.

Maltenes (malthenes) That fraction of crude oil that is soluble in, for example, pentane or heptane; deasphaltened oil; also the term arbitrarily assigned to the pentane-soluble portion of crude oil that has relatively high boiling (>300°C, 760 mm) (see also petrolenes).

Marine engine oil Oil used as a crankcase oil in marine engines.

Marine gasoline Fuel for motors in marine service.

Marine sediment The organic biomass from which petroleum is derived.

Marsh An area of spongy waterlogged ground with large numbers of surface water pools. Marshes usually result from: (1) an impermeable underlying bedrock; (2) surface deposits of glacial boulder clay; (3) basin-like topography from which natural drainage is poor; (4) very heavy rainfall in conjunction with a correspondingly low evaporation rate; and (5) low-lying land, particularly at estuarine sites at or below sea level.

Marx-Langenheim model Mathematical equations for calculating heat transfer in a hot water or steam flood.

Mayonnaise Low-temperature sludge; a black, brown, or gray deposit having a soft, mayonnaise-like consistency; not recommended as a food additive!

MDS process A solvent deasphalting process that is particularly effective for upgrading heavy crude oils.

Methanol See methyl alcohol.

Medicinal oil Highly refined, colorless, tasteless, and odorless petroleum oil used as a medicine in the nature of an internal lubricant; sometimes called liquid paraffin.

Medium crude oil Liquid crude oil with a density between that of light and heavy crude oil; the specific gravity of the oil is >20° API but less than or equal to 25°. See light crude oil.

MEK (methyl ethyl ketone) A colorless liquid ($CH_3COCH_2CH_3$) used as a solvent; as a chemical intermediate; and in the manufacture of lacquers, celluloid, and varnish removers.

MEK deoiling A wax deoiling process in which the solvent is generally a mixture of methyl ethyl ketone and toluene.

MEK dewaxing A continuous solvent dewaxing process in which the solvent is generally a mixture of methyl ethyl ketone and toluene.

MEOR Microbial enhanced oil recovery.

Metagenesis The alteration of organic matter during the formation of crude oil that may involve temperatures above 200°C (390°F); see also catagenesis and diagenesis.

Metamorphic Group of rocks resulting from the transformation that commonly takes place at great depths due to pressure and temperature. The original rocks may be sedimentary, igneous, or metamorphic.

Methane The principal constituent of natural gas; the simplest hydrocarbon molecule, containing one carbon atom and four hydrogen atoms.

Mercapsol process A regenerative process for extracting mercaptans, utilizing aqueous sodium (or potassium) hydroxide containing mixed cresols as solubility promoters.

Mercaptans Organic compounds having the general formula R-SH.

Methyl alcohol (methanol or wood alcohol) A colorless, volatile, inflammable, and poisonous alcohol (CH_3OH) traditionally formed by *destructive distillation* of wood or, more recently, as a result of synthetic distillation in chemical plants.

Methyl t-butyl ether An ether added to gasoline to improve its octane rating and to decrease gaseous emissions; see oxygenate.

Methyl ethyl ketone See MEK.

Metric ton Equivalent to 1000 kg, 2204.61 lb, or 7.5 bbl.

Mica A complex aluminum silicate mineral that is transparent, tough, flexible, and elastic.

Micellar fluid (surfactant slug) An aqueous mixture of surfactants, cosurfactants, salts, and hydrocarbons. The term micellar is derived from the word micelle, which is a submicroscopic aggregate of surfactant molecules and associated fluid.

Micelle The structural entity by which the asphaltene constituents are dispersed in petroleum.

Microcarbon residue The carbon residue determined using a thermogravimetric method. See also carbon residue.

Microcat-RC process (M-coke process) A catalytic hydroconversion process operating at relatively moderate pressures and temperatures using catalyst particles, containing a metal sulfide in a carbonaceous matrix formed within the process, that are uniformly dispersed throughout the feed; because of their ultrasmall size (10^{-4} in. diameter), there are typically several orders of magnitude more of these microcatalyst particles per cubic centimeter of oil than is possible in other types of hydroconversion reactors using conventional catalyst particles.

Microcrystalline wax Wax extracted from certain petroleum residua and having a finer and less apparent crystalline structure than paraffin wax.

Microemulsion A stable, finely dispersed mixture of oil, water, and chemicals (surfactants and alcohols).

Microemulsion or micellar/emulsion flooding An augmented waterflooding technique in which a surfactant system is injected in order to enhance oil displacement toward producing wells.

Microorganisms Animals or plants of microscopic size, such as bacteria.

Microscopic displacement efficiency The efficiency with which an oil displacement process removes the oil from individual pores in the rock.

Mid-boiling point The temperature at which approximately 50% of a material has distilled under specific conditions.

Middle distillates Medium-density refined crude oil products, including kerosene, stove oil, jet fuel, and light fuel oil.

Middle-phase microemulsion A microemulsion phase containing a high concentration of both oil and water that, when viewed in a test tube, resides in the middle with the oil phase above it and the water phase below it.

Midstream The processing, storage, and transportation sector of the crude oil industry.

Migration (primary) The movement of hydrocarbons (oil and natural gas) from mature, organic-rich source rocks to a point where the oil and gas can collect as droplets or as a continuous phase of liquid hydrocarbon.

Migration (secondary) The movement of the hydrocarbons as a single, continuous fluid phase through water-saturated rocks, fractures, or faults followed by accumulation of the oil and gas in sediments (traps) from which further migration is prevented.

Mineral hydrocarbons Crude oil hydrocarbons, considered *mineral* because they come from the earth rather than from plants or animals.

Mineral oil The older term for crude oil; the term was introduced in the 19th century as a means of differentiating crude oil (rock oil) from whale oil that, at the time, was the predominant illuminant for oil lamps.

Minerals Naturally occurring inorganic solids with well-defined crystalline structures.

Mineral seal oil A distillate fraction boiling between kerosene and gas oil.

Mineral wax Yellow to dark brown, solid substances that occur naturally and are composed largely of paraffins, usually found associated with considerable mineral matter, as a filling in veins and fissures or as an interstitial material in porous rocks.

Mine tailings Process water remaining after tar sand mining and stored in settling basins called tailing ponds.

Minimum miscibility pressure (MMP) See miscibility.

Miscibility An equilibrium condition, achieved after mixing two or more fluids, which is characterized by the absence of interfaces between the fluids: (1) *first-contact miscibility*, miscibility in the usual sense, whereby two fluids can be mixed in all proportions without any interfaces forming. For example, at room temperature and pressure, ethyl alcohol and water are first-contact miscible. (2) *Multiple-contact miscibility (dynamic miscibility)*, miscibility that is developed by repeated enrichment of one fluid phase with components from a second fluid phase with which it comes into contact. (3) *Minimum miscibility* pressure, the minimum pressure above which two fluids become miscible at a given temperature or can become miscible by dynamic processes.

Miscible flooding An oil recovery process in which a fluid, capable of mixing completely with the oil it contacts, is injected into an oil reservoir to increase recovery; see EOR process.

Miscible fluid displacement (miscible displacement) An oil displacement process in which an alcohol, a refined hydrocarbon, a condensed crude oil gas, a carbon dioxide, a liquefied natural gas, or even an exhaust gas is injected into an oil reservoir, at pressure levels such that the injected gas or fluid and reservoir oil are miscible; the process may include the concurrent, alternating, or subsequent injection of water.

Mitigation Identification, evaluation, and cessation of potential impacts of a process product or by-product.

Mixed-phase cracking The thermal decomposition of higher-boiling feedstock to produce lower-boiling products (typically gasoline and diesel components).

Mmboe Million barrels oil equivalent.

MMcf Millions of cubic feet per day (of gas).

Mobility A measure of the ease with which a fluid moves through reservoir rock; the ratio of rock permeability to apparent fluid viscosity.

Mobility buffer The bank that protects a chemical slug from water invasion and dilution and assures mobility control.

Mobility control Ensuring that the mobility of the displacing fluid or bank is equal to or less than that of the displaced fluid or bank.

Mobility ratio Ratio of the mobility of an injection fluid to the mobility of fluid being displaced.

Modified alkaline flooding The addition of a cosurfactant and/or polymer to the alkaline flooding process.

Modified naphtha insolubles (MNI) An insoluble fraction obtained by adding naphtha to petroleum; usually, the naphtha is modified by adding paraffinic constituents; the fraction might be equated to asphaltenes *if* the naphtha is equivalent to n-heptane, but usually, it is not.

Molecular sieve A synthetic zeolite mineral having pores of uniform size; it is capable of separating molecules, on the basis of their size, structure, or both, by absorption or sieving.

Molecular weight The mass of one molecule.

Motor octane method A test for determining the knock rating of fuels for use in spark-ignition engines; see also research octane method.

Moving-bed catalytic cracking A cracking process in which the catalyst is continuously cycled between the reactor and the regenerator.

MRH process A hydrocracking process to upgrade heavy feedstocks containing large amount of metals and asphaltene, such as vacuum residua and bitumen, and to produce mainly middle distillates using a reactor designed to maintain a mixed three-phase slurry of feedstock, fine powder catalyst, and hydrogen, and to promote effective contact.

MSCC process A short-residence-time process (millisecond catalytic cracking) in which the catalyst is placed in a more optimal position to insure better contact with the feedstock.

MTBE See methyl t-butyl ether.

Mud (drilling mud) Fluid circulated down the drill pipe and up the annulus during drilling to remove cuttings, cool and lubricate the bit, and maintain desired pressure in the well.

Muskeg A water-soaked layer of decaying plant material, 3–10 ft. thick, found on top of the overburden.

Naft Pre-Christian era (Greek) term for naphtha.

Napalm A thickened gasoline used as an incendiary medium that adheres to the surface it strikes.

Naphtha Any of various volatile, often flammable, liquid hydrocarbon mixtures used chiefly as solvents and diluents boiling below 200°C (390°F); used as a blend stock for gasoline manufacture.

Naphthenes Cycloparaffins.

Native asphalt See bitumen.

Natural asphalt See bitumen.

Natural gas Mixture of hydrocarbons existing in reservoirs in the gaseous phase or in solution in the oil, which remains in the gaseous phase under atmospheric conditions. It may contain some impurities or nonhydrocarbon substances (hydrogen sulfide, nitrogen, or carbon dioxide).

Natural gas liquids (NGLs) The hydrocarbon liquids that condense during the processing of hydrocarbon gases that are produced from oil or gas reservoir; see also natural gasoline.

Natural gasoline A mixture of liquid hydrocarbons extracted from natural gas suitable for blending with refinery gasoline.

Natural gasoline plant A plant for the extraction of fluid hydrocarbon, such as gasoline and liquefied petroleum gas, from natural gas.

Net thickness The thickness resulting from subtracting the portions that have no possibilities of producing hydrocarbon from the total thickness.

Neutralization A process for reducing the acidity or alkalinity of a waste stream by mixing acids and bases to produce a neutral solution; also known as pH adjustment.

Neutral oil A distillate lubricating oil with viscosity usually not above 200 s at 100°F.

Neutralization number The weight, in milligrams, of potassium hydroxide needed to neutralize the acid in 1 g of oil; an indication of the acidity of an oil.

No. 1 fuel oil Very similar to kerosene and is used in burners where vaporization before burning is usually required and a clean flame is specified.

No. 2 fuel oil Also called domestic heating oil, has properties similar to diesel fuel and heavy jet fuel, and used in burners where complete vaporization is not required before burning.

No. 4 fuel oil A light industrial heating oil and is used where preheating is not required for handling or burning; there are two grades of no. 4 fuel oil, differing in safety (flash point) and flow (viscosity) properties.

No. 5 fuel oil A heavy industrial fuel oil that requires preheating before burning.

No. 6 fuel oil A heavy fuel oil and is more commonly known as Bunker C oil when it is used to fuel ocean-going vessels; preheating is always required for burning this oil.

Nonasphaltic road oil Any of the nonhardening petroleum distillates or residual oils used as dust layers. They have sufficiently low viscosity to be applied without heating and, together with asphaltic road oils, are sometimes referred to as dust palliatives.

Nonassociated gas The natural gas found in reservoirs that do not contain crude oil at the original pressure and temperature conditions.

Nonionic surfactant A surfactant molecule containing no ionic charge.

Non-Newtonian A fluid that exhibits a change of viscosity with flow rate.

Nonproved reserves Volumes of hydrocarbons and associated substances, evaluated at atmospheric conditions, resulting from the extrapolation of the characteristics and parameters of the reservoir beyond the limits of reasonable certainty or from assuming oil and gas forecasts with technical and economic scenarios other than those in operation or with a project in view.

Normal fault The result of the downward displacement of one of the blocks from the horizontal. The angle is generally between 25° and 60°, and it is recognized by the absence of part of the stratigraphic column.

Observation wells Wells that are completed and equipped to measure reservoir conditions and/or sample reservoir fluids, rather than to inject or produce reservoir fluids.

Octane barrel yield A measure used to evaluate fluid catalytic cracking processes; defined as (RON+MON)/2 × the gasoline yield, where RON is the research octane number and MON is the motor octane number.

Octane number A number indicating the antiknock characteristics of gasoline.

Oil Typically, conventional crude oil that exists in the liquid phase in reservoirs and remains as such under original pressure and temperature conditions. Small amounts of nonhydrocarbon substances may be included. Also, the term *oil* is used in the context of this book as a generic term to include heavy oil, extra heavy oil, and tar sand bitumen, and it is not intended to be used as a means for definition of these resources.

Oil bank See bank.

Oil breakthrough (time) The time at which the oil-water bank arrives at the producing well.

Oil equivalent (OE) Total of crude oil, condensate, plant liquids, and dry gas equivalent to liquid.

Oil field A geographic area under which an oil reservoir lies.

Oil in place (OIP) An estimated measure of the total amount of oil contained in a reservoir and, as such, a higher figure than the estimated recoverable reserves of oil.

Oils (oil fraction) That portion of the maltenes that is not adsorbed by a surface-active material such as clay or alumina.

Oil sand See tar sand.

Oil shale A fine-grained impervious sedimentary rock that contains an organic material called kerogen.

OOIP See original oil in place.

Optimum salinity The salinity at which a middle-phase microemulsion containing equal concentrations of oil and water results from the mixture of a micellar fluid (surfactant slug) with oil.

Organic sedimentary rocks Rocks containing organic material such as residues of plant and animal remains/decay.

Original gas volume in place Amount of gas that is estimated to exist initially in the reservoir and that is confined by geologic and fluid boundaries, which may be expressed at reservoir or atmospheric conditions.

Original oil in place (OOIP) The quantity of crude oil existing in a reservoir before oil recovery operations begin.

Original oil volume in place Amount of crude oil that is estimated to exist initially in the reservoir and that is confined by geologic and fluid boundaries, which may be expressed at reservoir or atmospheric conditions.

Original pressure Pressure prevailing in a reservoir that has never been produced. It is the pressure measured by a discovery well in a producing structure.

Original reserve Volume of hydrocarbons at atmospheric conditions that are expected to be recovered economically by using the exploitation methods and systems applicable at a specific date. It is a fraction of the discovered and economic reserve that may be obtained at the end of the reservoir exploitation.

Overburden A layer of sand, gravel, and shale between the surface and the underlying tar sand formation that must be removed before the tar sand can be mined; underlies muskeg in many places.

Overhead That portion of the feedstock that is vaporized and removed during distillation.

Override The gravity-induced flow of a lighter fluid in a reservoir above another heavier fluid.

Oxidation A process that can be used for the treatment of a variety of inorganic and organic substances.

Oxidized asphalt See air-blown asphalt.

Oxygenate An oxygen-containing compound that is blended into gasoline to improve its octane number and to decrease gaseous emissions.

Ozokerite (ozocerite) A naturally occurring wax; when refined also known as ceresin.

Ozone Ground-level ozone is a colorless gas that forms just above the Earth's surface.

1P reserve Proved reserve.

2P reserves Total of proved plus probable reserves.

3P reserves Total of proved reserves plus probable reserves plus possible reserves.

Pale oil A lubricating or process oil refined until its color, by transmitted light, is straw to pale yellow.

Paraffinum liquidum See liquid petrolatum.

Paraffin wax The colorless, translucent, highly crystalline material obtained from the light lubricating fractions of paraffinic crude oils (wax distillates).

Partial oxidation (Texaco gasification) process A partial oxidation gasification process for generating synthetic gas, principally hydrogen and carbon monoxide.

Partial upgrading The combination of processing steps applied to heavy oil, extra heavy oil, or tar sand bitumen that prepares the oil to meet the specifications for pipeline transport or the next steps in the conversion of the oil to products. In terms of pipeline transport of the oil, the addition of a diluent to the heavy oil, extra heavy oil, or tar sand bitumen prepares the oil to meet the specifications required by the pipeline company. In the true sense of the term, the addition of diluent is not an upgrading step insofar as the oil can be recovered unchanged by the removal of the diluent.

Particle density The density of solid particles.

Particle size distribution The particle size distribution (of a catalyst sample) expressed as a percent of the whole.

Particulate matter The microscopic solid or liquid particles that remain suspended in the air for some time.

Pattern The horizontal pattern of injection and producing wells selected for a secondary or enhanced recovery project.

Pattern life The length of time a flood pattern participates in oil recovery.

Pay zone The rock in which oil and gas are found in exploitable quantities.

Penex process A continuous, nonregenerative process for isomerization of C_5 and/or C_6 fractions in the presence of hydrogen (from reforming) and a platinum catalyst.

Pentafining A pentane isomerization process using a regenerable platinum catalyst on a silica-alumina support and requiring outside hydrogen.

Pepper sludge The fine particles of sludge produced in acid treating that may remain in suspension.

Peri-condensed aromatic compounds Compounds based on angular condensed aromatic hydrocarbon systems, e.g., phenanthrene, chrysene, and picene.

Permeability The capacity of a reservoir rock to transmit fluids; rock property for permitting a fluid pass; a factor that indicates whether a reservoir has producing characteristics or not.

Petrol A term commonly used in some countries for gasoline.

Petrolatum A semisolid product, ranging from white to yellow in color, produced during refining of residual stocks; see petroleum jelly.

Petrolenes The term applied to that part of the pentane-soluble or heptane-soluble material that is low boiling (<300°C, <570°F, 760mm) and can be distilled without thermal decomposition (see also Maltenes).

Petroleum (crude oil) A naturally occurring mixture of gaseous, liquid, and solid hydrocarbon compounds usually found trapped deep underground beneath impermeable caprock and above a lower dome of sedimentary rock such as shale; most crude oil reservoirs occur in sedimentary rocks of marine, deltaic, or estuarine origin.

Petroleum asphalt See asphalt.

Petroleum ether See ligroine.

Petroleum jelly A translucent, yellowish to amber or white hydrocarbon substance (melting point, 38–54°C) having almost no odor or taste, derived from petroleum and used principally in medicine and pharmacy as a protective dressing and as a substitute for fats in ointments and cosmetics; also used in many types of polishes and in lubricating greases, rust preventives, and modeling clay; obtained by dewaxing heavy lubricating oil stocks.

Petroleum refinery See refinery.

Petroleum refining A complex sequence of events that result in the production of a variety of products.

Petroporphyrins See porphyrins.

pH adjustment Neutralization.

Phase A separate fluid that coexists with other fluids, gas, oil, and water, and other stable fluids such as microemulsions are all called phases in EOR research.

Phase behavior The tendency of a fluid system to form phases as a result of changing temperature, pressure, or the bulk composition of the fluids or of individual fluid phases.

Phase diagram A graph of phase behavior. In chemical flooding, it is a graph showing the relative volume of oil, brine, and sometimes one or more microemulsion phases. In carbon dioxide flooding, these are conditions for the formation of various liquid, vapor, and solid phases.

Phase properties Types of fluids, compositions, densities, viscosities, and relative amounts of oil, microemulsion, solvent, and water formed when a micellar fluid (surfactant slug) or miscible solvent (e.g., CO_2) is mixed with oil.

Phase separation The formation of a separate phase that is usually the prelude to coke formation during a thermal process; the formation of a separate phase as a result of the instability/incompatibility of crude oil and crude oil products.

Physical limit The limit of the reservoir defined by any geologic structures (faults, unconformities, change of facies, crests and bases of formations, etc.), caused by contact between fluids or by the reduction, to critical porosity, of permeability limits or the compound effect of these parameters.

Phosphoric acid polymerization A process using a phosphoric acid catalyst to convert propene, butene, or both to gasoline or petrochemical polymers.

Physical composition See bulk composition.

Pilot project Project that is being executed in a small representative sector of a reservoir where tests performed are similar to those that will be implemented throughout the reservoir. The purpose is to gather information and/or obtain results that could be used to generalize an exploitation strategy in the oil field.

PINA analysis A method of analysis for paraffins, *iso*paraffins, naphthenes, and aromatics.

Pinnacle reef A conical formation, higher than it is wide, usually composed of limestone, in which hydrocarbons might be trapped.

PIONA analysis A method of analysis for paraffins, *iso*paraffins, olefins, naphthenes, and aromatics.

Pipe still (pipestill) A still in which heat is applied to the oil while being pumped through a coil or pipe arranged in a suitable firebox; the distillation tower in a refinery.

Pipestill gas The most volatile fraction that contains most of the gases that are generally dissolved in the crude. Also known as pipestill light ends.

Pitch The nonvolatile, brown to black, semisolid to solid viscous product from the destructive distillation of many bituminous or other organic materials, especially coal.

Plant liquefiables shrinkage factor (PLSF) The fraction arising from considering the liquefiables obtained in transportation to the processing complexes.

Plant liquids Natural gas liquids recovered in gas processing complexes, mainly consisting of ethane, propane, and butane.

Plant liquids recovery factor (PLRF) The factor used to obtain the liquid portions recovered in the natural gas processing complex. It is obtained from the operation statistics of the last annual period of the gas processing complex that processes the production of the field analyzed.

Platform An offshore structure that is permanently fixed to the seabed.

Platforming A reforming process using a platinum-containing catalyst on an alumina base.

Play Group of fields that share geologic similarities and where the reservoir and the trap control the distribution of oil and gas.

PNA A polynuclear aromatic compound.

Polymer In EOR, any very-high-molecular-weight material that is added to water to increase viscosity for polymer flooding.

Polymer-augmented waterflooding Waterflooding in which organic polymers are injected with the water to improve horizontal and vertical sweep efficiency.

PONA analysis A method of analysis for paraffins (P), olefins (O), naphthenes (N), and aromatics (A).

Pool A natural underground reservoir containing an accumulation of crude oil.

Pore volume The total volume of all pores and fractures in a reservoir or part of a reservoir; also applied to catalyst samples.

Polar aromatics Resins; the constituents of petroleum that are predominantly aromatic in character and contain polar (nitrogen, oxygen, and sulfur) functions in their molecular structure(s).

Pollution The introduction into the land water and air systems of a chemical or chemicals that are not indigenous to these systems or the introduction into the land water and air systems of indigenous chemicals in greater-than-natural amounts.

Polyforming A process charging both C_3 and C_4 gases with naphtha or gas oil under thermal conditions to produce gasoline.

Polymer gasoline The product of polymerization of gaseous hydrocarbons to hydrocarbons boiling in the gasoline range.

Polymerization The combination of two olefin molecules to form a higher-molecular-weight paraffin.

Polynuclear aromatic compound An aromatic compound having two or more fused benzene rings, e.g., naphthalene and phenanthrene.

Polysulfide treating A chemical treatment used to remove elemental sulfur from refinery liquids by contacting them with a nonregenerable solution of sodium polysulfide.

Pore diameter The average pore size of a solid material, e.g., catalyst.

Pore space A small hole in reservoir rock that contains fluid or fluids; a 4 in. cube of reservoir rock may contain millions of interconnected pore spaces.

Porosity The ratio between the pore volume existing in a rock and the total rock volume; a measure of rock's storage capacity; the percentage of rock volume available to contain water or other fluids.

Possible reserves Volume of hydrocarbons where the analysis of geologic and engineering data suggests that they are less likely to be commercially recoverable than probable reserves.

Potential reserves Reserves based upon geologic information about the types of sediments where such resources are likely to occur and they are considered to represent an educated guess.

Pour point The lowest temperature at which oil will pour or flow when it is chilled without disturbance under definite conditions.

Preflush A conditioning slug injected into a reservoir as the first step of an EOR process.

Pressure cores Cores cut into a special coring barrel that maintains reservoir pressure when brought to the surface; this prevents the loss of reservoir fluids that usually accompanies a drop in pressure from reservoir to atmospheric conditions.

Porphyrins Organometallic constituents of petroleum that contain vanadium or nickel; the degradation products of chlorophyll that became included in the protopetroleum.

Powerforming A fixed-bed naphtha reforming process using a regenerable platinum catalyst.

Precipitation number The number of milliliters of precipitate formed when 10 mL of lubricating oil is mixed with 90 mL of petroleum naphtha of a definite quality and centrifuged under definitely prescribed conditions.

Pressure gradient Rate of change of pressure with distance.

Pressure maintenance Augmenting the pressure (and energy) in a reservoir by injecting gas and/or water through one or more wells.

Pressure pulse test A technique for determining reservoir characteristics by injecting a sharp pulse of pressure in one well and detecting its surrounding wells.

Pressure transient testing Measuring the effect of changes in pressure at one well on other wells in a field.

Primary oil recovery Oil recovery utilizing only naturally occurring forces.

Primary recovery Extraction of crude oil by only using the natural energy available in the reservoirs to displace fluids through the reservoir rock to the wells.

Primary structure The chemical sequence of atoms in a molecule.

Primary tracer A chemical that, when injected into a test well, reacts with reservoir fluids that form a detectable chemical compound.

Probable reserves Nonproved reserves where the analysis of geologic and engineering data suggests that they are more likely to be commercially recoverable than not.

Processability An estimate of the manner and relative ease with which a feedstock can be processed; generally measured by one or more criteria.

Produced water Water naturally present in the reservoir that is recovered during oil and gas production; may include water from the reservoir, water that has been injected into the formation, and any chemicals added during the production/treatment process; also called "brine" (and may contain high mineral or salt content) or formation water.

Producibility The rate at which oil or gas can be produced from a reservoir through a wellbore.

Producing well A well in an oil field used for removing fluids from a reservoir.

Propane asphalt See solvent asphalt.

Propane deasphalting Solvent deasphalting using propane as the solvent.

Propane decarbonizing A solvent extraction process used to recover catalytic cracking feed from heavy fuel residues.

Propane dewaxing A process for dewaxing lubricating oils in which propane serves as solvent.

Propane fractionation A continuous extraction process employing liquid propane as the solvent; a variant of propane deasphalting.

Prospect A potential geologic trap that is believed (on the basis of geologic evidence) to contain crude oil. See lead.

Prospective resource The amount of hydrocarbons evaluated at a given date of accumulations not yet discovered, but which have been inferred and are estimated as recoverable.

Protopetroleum A generic term used to indicate the initial product formed changes have occurred to the precursors of crude oil.

Proved area Plant projection of the known part of the reservoir corresponding to the proved volume.

Proved reserves Volume of hydrocarbons or associated substances evaluated at atmospheric conditions, which, by the analysis of geologic and engineering data, may be estimated with reasonable certainty to be commercially recoverable from a given date forward; from known reservoirs; and under current economic conditions, operating methods, and government regulations. Such volume consists of the developed proved reserve and the undeveloped proved reserve.

Proved field An oil and/or gas field whose physical extent and estimated reserves have been determined.

Pulse-echo ultrasonic borehole televiewer Well-logging system wherein a pulsed, narrow acoustic beam scans the well as the tool is pulled up from the borehole; the amplitude of the reflecting beam is displayed on a cathode-ray tube resulting in a pictorial representation of wellbore.

Pyrobitumen See asphaltoid.

Pyrolysis Exposure of a feedstock to high temperatures in an oxygen-poor environment.

Pyrophoric Substances that catch fire spontaneously in air without an ignition source.

Quadrillion 1×10^{15}.

Quench The sudden cooling of hot material discharging from a thermal reactor.

R2R process A fluid catalytic cracking process for the conversion of heavy feedstocks.

Raffinate That portion of the oil that remains undissolved in a solvent refining process.

Ramsbottom carbon residue See carbon residue.

Raw materials Minerals extracted from the earth prior to any refining or treating.

RCC process A process for the conversion of heavy feedstocks in the riser pipe.

RCD Unibon (BOC) process A process to upgrade vacuum residua using hydrogen.

Reactor A vessel in which a reaction occurs during processing, usually defined by the nature of the catalyst bed, e.g., fixed-bed reactor and fluid-bed reactor, and by the direction of the flow of feedstock, e.g., upflow and downflow.

Recoverable reserves That proportion of the oil and/or gas in a reservoir that can be removed using currently available techniques.

Recovery factor (rf) The ratio between the original volume of oil or gas, at atmospheric conditions, and the original reserves of the reservoir.

Recycle ratio, τ The ratio of the recycled feedstock to the fresh feedstock: $\tau = F_R/F_F$. where F_F is the fresh feedstock and F_R is the recycled feedstock; may also be expressed as a percentage.

Recycle stock The portion of a feedstock that has passed through a refining process and is recirculated through the process.

Recycling The use or reuse of chemical waste as an effective substitute for commercial products or as an ingredient or feedstock in an industrial process.

Reduced crude A residual product remaining after the removal, by distillation, or other means of an appreciable quantity of the more volatile components of crude oil.

Refinery A series of integrated unit processes by which crude oil can be converted to a slate of useful (salable) products.

Refinery gas A gas (or a gaseous mixture) produced as a result of refining operations.

Refining The process(es) by which crude oil is distilled and/or converted by the application of a physical and chemical processes to form a variety of products.

Reformate The liquid product of a reforming process.

Reformed gasoline Gasoline made by a reforming process.

Reforming The conversion of hydrocarbons with low octane numbers into hydrocarbons having higher octane numbers, e.g., the conversion of an *n*-paraffin into an isoparaffin.

Reformulated gasoline (RFG) Gasoline designed to mitigate smog production and to improve air quality by limiting the emission levels of certain chemical compounds such as benzene and other aromatic derivatives; often contains oxygenates.

Reformulated fuels Gasoline, diesel, or other fuels that have been modified to reflect environmental concerns, performance standards, government regulations, customer preferences, or new technologies.

Regeneration The reactivation of a catalyst by burning off the coke deposits.

Regenerator A reactor for catalyst reactivation.

Regression Geologic term used to define the elevation of one part of the continent over sea level, as a result of the ascent of the continent or the lowering of the sea level.

Reid vapor pressure (RVP) A measure of the volatility of liquid fuels, especially gasoline.

Relative permeability The capacity of a fluid, such as water, gas, or oil, to flow through a rock when it is saturated with two or more fluids. The value of the permeability of a saturated rock with two or more fluids is different to the permeability value of the same rock saturated with just one fluid.

Remaining reserves Volume of hydrocarbons measured at atmospheric conditions that are still to be commercially recoverable from a reservoir at a given date, using the applicable exploitation techniques. It is the difference between the original reserve and the cumulative hydrocarbon production at a given date.

Renewable energy sources Solar, wind, and other non-fossil-fuel energy sources.

Rerunning The distillation of an oil that has already been distilled.

Research octane method A test for determining the knock rating, in terms of octane numbers and of fuels for use in spark-ignition engines; see also motor octane method.

Reserve replacement rate An indication of the amount of hydrocarbons replaced or incorporated by new discoveries compared with what has been produced in a given period. It is the coefficient that arises from dividing the new discoveries by production during the period of analysis, and it is generally referred to in annual terms and is expressed as a percentage.

Reserve-production ratio The result of dividing the remaining reserve at a given date by the production in a period. This indicator assumes constant production, hydrocarbon prices, and extraction costs, without variation over time, in addition to the nonexistence of new discoveries in the future.

Reserves The well-identified resources that can be profitably extracted and utilized with existing technology.

Reservoir A porous and permeable lithologic unit or set of units that holds the hydrocarbon reserves; the part of the geologic trap containing hydrocarbons that acts as a hydraulically interconnected system and where the hydrocarbons are.

Reservoir simulation Analysis and prediction of reservoir performance with a computer model.

Residfining process A catalytic fixed-bed process for the desulfurization and demetallization of heavy feedstocks that can also be used to pretreat the feedstocks to suitably low contaminant levels prior to catalytic cracking.

Residual asphalt See straight-run asphalt.

Residual fuel oil Obtained by blending the residual product(s) from various refining processes with suitable diluent(s) (usually middle distillates) to obtain the required fuel oil grades.

Residual oil See residuum.

Residual resistance factor The reduction in the permeability of rock to water caused by the adsorption of polymer.

Residue hydroconversion process (RHC process) A high-pressure fixed-bed trickle-flow hydrocatalytic process for converting heavy feedstocks.

Residuum The nonvolatile fraction remaining after processing or distillation of crude oil, e.g., an atmospheric residuum (BP > 345°C, >650°F) and a vacuum residuum (BP > 570°C, > 1050°F).

Resins The portion of the maltenes (the nonasphaltene part of crude oil) that is adsorbed by a surface-active material such as clay or alumina; the fraction of deasphaltened oil that is insoluble in liquid propane but soluble in *n*-heptane.

Resistance factor A measure of resistance to the flow of a polymer solution relative to the resistance to the flow of water.

Resource The total volume of hydrocarbons existing in subsurface rocks; also known as original in situ volume.

Retention The loss of chemical components due to adsorption onto the rock's surface, to precipitation, or to trapping within the reservoir.

Reverse fault The result of compression forces where one of the blocks is displaced upward from the horizontal.

Revision The reserve resulting from comparing the previous year's evaluation with the new one in which new geologic, geophysical, operation, and reservoir performance information is considered, in addition to variations in hydrocarbon prices and extraction costs. It does not include well drilling.

Rexforming A process combining platforming with aromatics extraction, wherein low-octane raffinate is recycled to the platformer.

Rich oil Absorption oil containing dissolved natural gasoline fractions.

Riser The part of the bubble-plate assembly that channels the vapor and causes it to flow downward to escape through the liquid; also the vertical pipe where fluid catalytic cracking reactions occur.

Riser (drilling) A pipe between the seabed and a floating drilling rig.

Riser (production) The section of pipework that joins a seabed wellhead to the Christmas tree.

Riser pipe The pipe in a fluid catalytic cracking process where catalyst and feedstock are lifted into the reactor; the pipe in which most of the reaction takes place or is initiated.

Rock asphalt Bitumen that occurs in formations that have a limiting ratio of bitumen-to-rock matrix.

Rock matrix The granular structure of a rock or porous medium.

ROSE process A solvent deasphalting process that uses supercritical solvent recovery system to obtain high-quality oils from heavy feedstocks for further processing.

Roughneck Drill crew members who work on the derrick floor, screwing together the sections of drill pipe when running or pulling a drill string.

Roustabout Drill crew members who handle the loading and unloading of equipment and assist in general operations around the rig.

Run-of-the-river reservoirs Reservoirs with a large rate of flow-through compared with their volume.

S&W fluid catalytic cracking process A process in which the heavy feedstock is injected into a stabilized, upward flowing catalyst stream whereupon the feedstock-stream-catalyst mixture travels up the riser pipe and is separated by a high-efficiency inertial separator from which the product vapor goes overhead to fractionation.

Saline groundwater (brackish water) Deep groundwater that is high in dissolved salt and unsuitable for domestic or agricultural uses.

Salinity The concentration of salt in water.

Sand A course granular mineral mainly comprising quartz grains that is derived from the chemical and physical weathering of rocks rich in quartz, notably sandstone and granite.

Sand face The cylindrical wall of the wellbore through which the fluids must flow to or from the reservoir.

Sandstone A sedimentary rock formed by compaction and cementation of sand grains; can be classified according to the mineral composition of the sand and cement.

SARA analysis A method of fractionation by which crude oil is separated into saturates, aromatics, resins, and asphaltene fractions.

Saturates (saturate fraction) Paraffins and cycloparaffins (naphthenes).

Saturation The ratio of the volume of a single fluid in the pores to pore volume, expressed as a percent and applied to water, oil, or gas separately; the sum of the saturations of each fluid in a pore volume is 100%.

Saturation pressure Pressure at which the first gas bubble is formed, when it goes from the liquid-phase to the two-phase region.

Saybolt Furol viscosity The time, in seconds (Saybolt Furol seconds, SFS), for 60 mL of fluid to flow through a capillary tube in a Saybolt Furol viscometer at specified temperatures between 70°F and 210°F; the method is appropriate for high-viscosity oils such as transmission, gear, and heavy fuel oils.

Saybolt universal viscosity The time, in seconds (Saybolt universal seconds, SUS), for 60 mL of fluid to flow through a capillary tube in a Saybolt universal viscometer at a given temperature.

Scale wax The paraffin derived by removing the greater part of the oil from slack wax by sweating or solvent deoiling.

Screening guide A list of reservoir rock and fluid properties critical to an EOR process.

Scrubbing Purifying a gas by washing with water or chemical; less frequently, the removal of entrained materials.

Seal rock (caprock) A unit with low permeability that impedes the escape of hydrocarbons from the reservoir rock; common seal rocks include evaporites, chalks, and shales.

Secondary recovery Techniques used for the additional extraction of crude oil after primary recovery. This includes gas or water injection, partly to maintain reservoir pressure.

Secondary structure The ordering of the atoms of a molecule in space relative to each other.

Secondary tracer The product of the chemical reaction between reservoir fluids and an injected primary tracer.

Sediment An insoluble solid formed as a result of the storage instability and/or the thermal instability of crude oil and crude oil products.

Sedimentary Formed by or from deposits of sediments, especially from sand grains or silts transported from their source and deposited in water, as sandstone and shale, or from calcareous remains of organisms, as limestone.

Sedimentary basin A geographic area, such as the Western Canada Sedimentary Basin, in which much of the rock is sedimentary (as opposed to igneous or metamorphic) and therefore likely to contain hydrocarbons.

Sedimentary strata Typically consist of mixtures of clay, silt, sand, organic matter, and various minerals, formed by or from deposits of sediments, especially from sand grains or silts transported from their source and deposited in water, such as sandstone and shale, or from calcareous remains of organisms, such as limestone.

Seismic section Seismic profile that uses the reflection of seismic waves to determine the geologic subsurface.

Selective solvent A solvent that, at certain temperatures and ratios, will preferentially dissolve more of one component of a mixture than of another and thereby permit partial separation.

Separation process The process of separating liquid and gas hydrocarbons and water. This is typically accomplished in a pressure vessel at the surface, but newer technologies allow separation to occur in the wellbore under certain conditions.

Separator-Nobel dewaxing A solvent (trichloroethylene) dewaxing process.

Shale Rock formed from clay.

Shell fluid catalytic cracking A two-stage fluid catalytic cracking process in which the catalyst is regenerated.

Shell gasification (partial oxidation) process A process for generating synthesis gas (hydrogen and carbon monoxide) for the ultimate production of high-purity high-pressure hydrogen, ammonia, methanol, fuel gas, town gas, or reducing gas by the reaction of gaseous or liquid hydrocarbons with oxygen, air, or oxygen-enriched air.

Shell residual oil hydrodesulfurization process A downflow fixed-bed reactor process to improve the quality of heavy feedstocks by removing sulfur, metals, and asphaltenes and bringing about a reduction in the viscosity of the feedstock.

Shell still A still formerly used in which the oil was charged into a closed, cylindrical shell and the heat required for distillation was applied to the outside of the bottom from a firebox.

Shutdown A production hiatus during which the platform ceases to produce while essential maintenance work is undertaken.

Sidestream A liquid stream taken from any one of the intermediate plates of a bubble tower.

Sidestream stripper A device used to perform further distillation on a liquid stream from any one of the plates of a bubble tower, usually by the use of steam.

Slack wax The soft, oily crude wax obtained from the pressing of paraffin distillate or wax distillate.

Slime A name used for crude oil in ancient texts, particularly in biblical texts.

Sludge A semisolid to solid product that results from the storage instability and/or the thermal instability of crude oil and crude oil products.

Slug A quantity of fluid injected into a reservoir during enhanced oil recovery.

Slurry hydroconversion process A process in which the feedstock is contacted with hydrogen under pressure in the presence of a catalytic coke-inhibiting additive.

Slurry-phase reactors Tanks into which wastes, nutrients, and microorganisms are placed.

Smoke point A measure of the burning cleanliness of jet fuel and kerosene.

Sodium hydroxide treatment See caustic wash.

Sodium plumbite A solution prepared from a mixture of sodium hydroxide, lead oxide, and distilled water; used in making the doctor test for light oils such as gasoline and kerosene.

Solubility parameter A measure of the solvent power and polarity of a solvent.

Solution gas (associated gas) Natural gas that is found with crude oil in underground reservoirs. When the oil comes to the surface, the gas expands and comes out of the solution.

Solutizer-steam regenerative process A chemical treating process for extracting mercaptans from gasoline or naphtha, using solutizers (potassium isobutyrate and potassium alkyl phenolate) in strong potassium hydroxide solution.

Solvahl process A solvent deasphalting process for application to vacuum residua.

Solvent-aided process (DSAGD) An advanced SAGD technique that combines a solvent, such as butane, with the steam to help dissolve the bitumen and improve recovery from the reservoir.

Solvent asphalt The asphalt produced by solvent extraction of residua or by light hydrocarbon (propane) treatment of a residuum or an asphaltic crude oil.

Solvent deasphalter bottoms The insoluble material that separates from the solvent phase during deasphalting.

Solvent deasphalting A process for removing asphaltic and resinous materials from reduced crude oils, lubricating oil stocks, gas oils, or middle distillates through the extraction or precipitant action of low-molecular-weight hydrocarbon solvents; see also propane deasphalting.

Solvent decarbonizing See propane decarbonizing.

Solvent deresining See solvent deasphalting.

Solvent dewaxing A process for removing wax from oils by means of solvents usually by chilling a mixture of solvent and waxy oil, filtration, centrifuging the wax that precipitates, and solvent recovery.

Solvent extraction A process for separating liquids by mixing the stream with a solvent that is immiscible with part of the waste but that will extract certain components of the waste stream.

Solvent naphtha A refined naphtha of restricted boiling range used as a solvent; also called petroleum naphtha or petroleum spirits.

Solvent refining See solvent extraction.

Sonic log A well log based on the time required for sound to travel through rock, useful in determining porosity.

Sour crude oil Crude oil containing an abnormally large amount of sulfur compounds; see also sweet crude oil.

Sour gas Natural gas at the wellhead may contain hydrogen sulfide (H_2S), a toxic compound. Natural gas that contains $>1\%$ of H_2S is called sour gas.

Spacing Optimum distance between hydrocarbon-producing wells in a field or reservoir.

Specific gravity An intensive property of the matter that is related to the mass of a substance and its volume through the coefficient between these two quantities. It is expressed in grams per cubic centimeter or in pounds per gallon.

Specific heat The quantity of heat required to raise a unit mass of material through 1° of temperature.

Spent catalyst Catalyst that has lost much of its activity due to the deposition of coke and metals.

Spills Spills include accidental release of crude oil, produced water, or other hydrocarbon products from well sites, batteries, or storage tanks. These spills can affect land, vegetation, water bodies, and groundwater.

Spontaneous ignition The ignition of a fuel, such as coal, under normal atmospheric conditions, usually induced by climatic conditions.

Spud-in The operation of drilling the first part of a new well.

Stabilization The removal of volatile constituents from a higher-boiling fraction or product (stripping); the production of a product that, to all intents and purposes, does not undergo any further reaction when exposed to the air.

Stabilizer A fractionating tower for removing light hydrocarbons from an oil to reduce vapor pressure particularly applied to gasoline.

Standard conditions The reference amounts for pressure and temperature. In the English system, it is 14.73 lb/in.^{2} for the pressure and 60°F for temperature.

Standpipe The pipe by which catalyst is conveyed between the reactor and the regenerator.

Steam-assisted gravity drainage (SAGD) A recovery technique for the extraction of heavy oil or bitumen that involves drilling a pair of horizontal wells one above the other; one well is used for steam injection and the other for production.

Steam cracking A conversion process in which the feedstock is treated with superheated steam.

Steam distillation Distillation in which vaporization of the volatile constituents is effected at a lower temperature by the introduction of steam (open steam) directly into the charge.

Steam-drive injection (steam injection) EOR process in which steam is continuously injected into one set of wells (injection wells) or other injection sources to effect oil displacement toward and production from a second set of wells (production wells); steam stimulation of production wells is *direct steam stimulation*, whereas steam drive by steam injection to increase production from other wells is *indirect steam stimulation.*

Steam injection An improved recovery technique in which steam is injected into a reservoir to reduce the viscosity of the crude oil.

Steam-methane reforming A process commonly used to convert natural gas to hydrogen for upgrading.

Steam-naphtha reforming A process that is essentially similar in nature to the steam-methane reforming process but which used higher-molecular-weight hydrocarbons as the feedstock.

Steam stimulation Injection of steam into a well and the subsequent production of oil from the same well.

Stiles method A simple approximate method for calculating oil recovery by waterflood that assumes separate layers (stratified reservoirs) for the permeability distribution.

Stimulation Process of acidifying or fracturing carried out to expand existing ducts or to create new ones in the source rock formation.

Storage stability (or storage instability) The ability (inability) of a liquid to remain in storage over extended periods of time without appreciable deterioration as measured by gum formation and the depositions of insoluble material (sediment).

Straight-run asphalt The asphalt produced by the distillation of asphaltic crude oil.

Straight-run products Obtained from a distillation unit and used without further treatment.

Strata The layers including the solid iron-rich inner core, molten outer core, mantle, and crust of the Earth.

Stratigraphy Part of geology that studies the origin, composition, distribution, and succession of rock strata.

Straw oil Pale paraffin oil of straw color used for many process applications.

Stripper well A well that produces (strips from the reservoir) oil or gas.

Stripping A means of separating volatile components from less volatile ones in a liquid mixture by the partitioning of the more volatile materials to a gas phase of air or steam (*q.v.* stabilization).

Structural nose A term used in structural geology to define a geometric form protruding from a main body.

Sucker rod pumping system A method of artificial lift in which a subsurface pump located at or near the bottom of the well and connected to a string of sucker rods is used to lift the well fluid to the surface.

Sulfonic acids Acids obtained by petroleum or a petroleum product with strong sulfuric acid.

Sulfuric acid alkylation An alkylation process in which olefins (C_3, C_4, and C_5) combine with *iso*butane in the presence of a catalyst (sulfuric acid) to form branched-chain hydrocarbons used especially in gasoline blending stock.

Superlight oil The specific gravity is >38° API.

Surface-active material A chemical compound, molecule, or aggregate of molecules with physical properties that cause it to adsorb at the interface between *two* immiscible liquids, resulting in a reduction of interfacial tension or the formation of a microemulsion.

Surfactant A type of chemical, characterized as one that reduces interfacial resistance to mixing between oil and water or changes the degree to which water wets reservoir rock.

Suspended well A well that has been capped off temporarily.

Suspensoid catalytic cracking A nonregenerative cracking process in which cracking stock is mixed with slurry of catalyst (usually clay) and cycle oil and passed through the coils of a heater.

Sweated wax Wax separated from crude oil by having been passed through a sweating unit.

Sweating The separation of paraffin oil and low-melting wax from paraffin wax.

Sweep efficiency The ratio of the pore volume of reservoir rock contacted by injected fluids to the total pore volume of reservoir rock in the project area. (*See also* horizontal sweep efficiency *and* vertical sweep efficiency.)

Sweet crude oil Crude oil containing little sulfur; see also sour crude oil.

Sweetening plant Industrial plant used to treat gaseous mixtures and light crude oil fractions in order to eliminate undesirable or corrosive sulfur compounds to improve their color, odor, and stability.

Swelling Increase in the volume of crude oil caused by absorption of EOR fluids, especially carbon dioxide. Also increase in the volume of clays when exposed to brine.

Sweetening The process by which petroleum products are improved in odor and color by oxidizing or removing the sulfur-containing and unsaturated compounds.

Swept zone The volume of rock that is effectively swept by injected fluids.

Synthesis gas generation process A noncatalytic process for producing synthesis gas (hydrogen and carbon monoxide) from gaseous or liquid hydrocarbons for the ultimate production of high-purity hydrogen.

Synthetic crude oil (syncrude) A hydrocarbon product produced by the conversion of coal, oil shale, or tar sand bitumen that resembles conventional crude oil; can be refined in a crude oil refinery.

Tar The volatile, brown to black, oily, viscous product from the destructive distillation of many bituminous or other organic materials, especially coal; a name used for crude oil in ancient texts.

Tar sand (bituminous sand) Another name for oil sand. A sandstone deposit that is saturated with bitumen. The viscous hydrocarbonaceous mixtures found in Northern Alberta have historically been referred to (incorrectly) as tar, pitch, or asphalt; see bituminous sand.

Tcf Trillion cubic feet (of gas).

Technical reserves Accumulative production derived from a production forecast in which economic criteria are not applied.

Tertiary recovery (enhanced recovery) The third major phase of crude oil recovery that involves using more sophisticated techniques, such as steam flooding or injection of chemicals, to increase recovery.

Tertiary structure The three-dimensional structure of a molecule.

Tervahl H process A process in which the feedstock and hydrogen are heated and held in a soak drum as in the Tervahl T process.

Tervahl T process A process analogous to delayed coking in which the feedstock is heated to the desired temperature using a coil heater and held for a specified residence time in a soaking drum; see also Tervahl H process.

Tetraethyl lead (TEL) An organic compound of lead, $Pb(CH_3)_4$, which, when added in small amounts, increases the antiknock quality of gasoline.

Texaco gasification process See partial oxidation (Texaco gasification) process.

Thermal coke The carbonaceous residue formed as a result of a noncatalytic thermal process; the Conradson carbon residue; the Ramsbottom carbon residue.

Thermal cracking A process that decomposes, rearranges, or combines hydrocarbon molecules by the application of heat, without the aid of catalysts.

Thermal process Any refining process that utilizes heat, without the aid of a catalyst.

Thermal recovery See EOR process.

Thermal polymerization A thermal process to convert light hydrocarbon gases into liquid fuels.

Thermal reforming A process using heat (but no catalyst) to effect molecular rearrangement of low-octane naphtha into gasoline of higher antiknock quality.

Thermal stability (thermal instability) The ability (inability) of a liquid to withstand relatively high temperatures for short periods of time without the formation of carbonaceous deposits (sediment or coke).

Thermofor catalytic cracking A continuous, moving-bed catalytic cracking process.

Thermofor catalytic reforming A reforming process in which the synthetic, bead-type catalyst of coprecipitated chromia (Cr_2O_3) and alumina (Al_2O_3) flows down through the reactor concurrent with the feedstock.

Thermofor continuous percolation A continuous clay treating process to stabilize and decolorize lubricants or waxes.

Thief zone Any geologic stratum not intended to receive injected fluids in which significant amounts of injected fluids are lost; fluids may reach the thief zone due to an improper completion or a faulty cement job.

Tight gas Gas with very low flow rates. Found in sedimentary layers of rock that are cemented together so tightly that it greatly hinders the extraction. Getting tight gas out usually requires enhanced technology like "hydraulic fracturing" where fluid is pumped into the ground to make it more permeable.

Tool pusher Second-in-command of a drilling crew under the drilling superintendent. Responsible for the day-to-day running of the rig and for ensuring that all the necessary equipment is available.

Topped crude oil Crude oil that has had volatile constituents removed up to a certain temperature, e.g., 250°C+ (480°F+) topped crude; not always the same as a residuum.

Topping The distillation of crude oil to remove light fractions only.

Topside The superstructure of a platform.

Total thickness (h) Thickness from the top of the formation of interest down to a vertical boundary determined by a water level or by a change of formation.

Tower Equipment for increasing the degree of separation obtained during the distillation of oil in a still.

Trace element Those elements that occur at very low levels in a given system.

Tracer test A technique for determining fluid-flow paths in a reservoir by adding small quantities of easily detected material (often radioactive) to the flowing fluid and monitoring their appearance at production wells. Also used in cyclic injection to appraise oil saturation.

Transgression Geologic term used to define the immersion of one part of the continent under sea level, as a result of a descent of the continent or an elevation of the sea level.

Transmissibility (transmissivity) An index of producibility of a reservoir or zone, the product of permeability and layer thickness.

Transport liquefiables shrinkage factor (TLSF) The fraction obtained by considering the liquefiables obtained in transportation to the processing complexes.

Trap A stratigraphic or structural feature that ensures the juxtaposition of reservoir and seal such that hydrocarbons remain trapped in the subsurface, rather than escaping (due to their natural buoyancy) and being lost; the geologic geometry that permits the concentration of hydrocarbons; a sediment in which oil and gas accumulate from which further migration is prevented.

Treatment Any method, technique, or process that changes the physical and/or chemical character of petroleum.

Triaxial borehole seismic survey A technique for detecting the orientation of hydraulically induced fractures, wherein a tool holding three mutually seismic detectors is clamped in the borehole during fracturing; fracture orientation is deduced through the analysis of the detected microseismic perpendicular events that are generated by the fracturing process.

Trickle hydrodesulfurization A fixed-bed process for desulfurizing middle distillates.

Trillion 1×10^{12}.

True boiling point (true boiling range) The boiling point (boiling range) of a crude oil fraction or a crude oil product under standard conditions of temperature and pressure.

Tube-and-tank cracking An older liquid-phase thermal cracking process.

Ultimate analysis Elemental analysis.

Ultimate potential An estimate of recoverable reserves that will have been produced by the time all exploration and development activity is completed; includes production to date, remaining reserves, development of existing pools, and new discoveries.

Ultimate recovery The cumulative quantity of oil that will be recovered when revenues from further production no longer justify the costs of the additional production.

Ultrafining A fixed-bed catalytic hydrogenation process to desulfurize naphthas and upgrade distillates by essentially removing sulfur, nitrogen, and other materials.

Ultraforming A low-pressure naphtha reforming process employing onstream regeneration of a platinum-on-alumina catalyst and producing high yields of hydrogen and high-octane-number reformate.

Unassociated molecular weight The molecular weight of asphaltenes in a nonassociating (polar) solvent, such as dichlorobenzene, pyridine, or nitrobenzene.

Unconformity A surface of erosion that separates younger strata from older rocks.

Undeveloped proved area Plant projection of the extension drained by the future producing wells of a producing reservoir and located within the undeveloped proved reserve.

Undeveloped proved reserves Volume of hydrocarbons that is expected to be recovered through wells without current facilities for production or transportation and future wells. This category may include the estimated reserve of enhanced recovery projects, with pilot testing or with the recovery mechanism proposed in operation that has been predicted with a high degree of certainty in reservoirs that benefit from this kind of exploitation.

Undiscovered resource Volume of hydrocarbons with uncertainty but whose existence is inferred in geologic basins through favorable factors resulting from the geologic, geophysical, and geochemical interpretation. They are known as prospective resources when considered commercially recoverable.

Unicracking hydrodesulfurization (unicracking/HDS) process A fixed-bed, catalytic process for hydrotreating heavy feedstocks.

Unifining A fixed-bed catalytic process to desulfurize and hydrogenate refinery distillates.

Unisol process A chemical process for extracting mercaptan sulfur and certain nitrogen compounds from sour gasoline or distillates using regenerable aqueous solutions of sodium or potassium hydroxide containing methanol.

Universal viscosity See Saybolt universal viscosity.

Unstable Usually refers to a petroleum product that has more volatile constituents present or refers to the presence of olefin and other unsaturated constituents.

UOP alkylation A process using hydrofluoric acid (which can be regenerated) as a catalyst to unite olefins with *iso*butane.

UOP copper sweetening A fixed-bed process for sweetening gasoline by converting mercaptans to disulfides by contact with ammonium chloride and copper sulfate in a bed.

UOP fluid catalytic cracking A fluid process of using a reactor-over-regenerator design.

Upflow reactor A reactor in which the feedstock flows in an upward direction through the catalyst bed.

Upgrading The process of converting heavy oil, extra heavy oil, or tar sand bitumen into products.

Upper-phase microemulsion A microemulsion phase containing a high concentration of oil that, when viewed in a test tube, resides on top of a water phase.

Upstream The companies that explore, develop, and produce crude oil resources are known as the upstream sector of the crude oil industry.

Urea dewaxing A continuous dewaxing process for producing low-pour-point oils and using urea that forms a solid complex (adduct) with the straight-chain wax paraffins in the stock; the complex is readily separated by filtration.

Vacuum distillation Distillation under reduced pressure.

Vacuum residuum A residuum obtained by distillation of a crude oil under vacuum (reduced pressure); that portion of crude oil that boils above a selected temperature such as 510°C (950°F) or 565°C (1050°F).

Vapor-phase cracking A high-temperature, low-pressure conversion process.

Vapor-phase hydrodesulfurization A fixed-bed process for desulfurization and hydrogenation of naphtha.

Vapor pressure osmometry (VPO) A method for determining molecular weight.

Veba-Combi-Cracking process (VCC process) A thermal hydrocracking/hydrogenation process for converting heavy feedstocks.

Vertical sweep efficiency The fraction of the layers or vertically distributed zones of a reservoir that are effectively contacted by displacing fluids.

VGC (viscosity-gravity constant) An index of the chemical composition of crude oil defined by the general relation between specific gravity, sp gr, at 60°F and Saybolt universal viscosity, SUV, at 100°F $a = 10\,\text{sp gr} - 1.0752 \log(\text{SUV} - 38)/10\,\text{sp gr} - \log(\text{SUV} - 38)$ The constant, a, is low for the paraffinic crude oils and high for the naphthenic crude oils.

VI See viscosity index.

Visbreaking A process for reducing the viscosity of heavy feedstocks by controlled thermal decomposition.

Viscosity A measure of the ability of a liquid to flow or a measure of its resistance to flow; the force required to move a plane surface of area 1 m^2 over another parallel plane surface 1 m away at a rate of 1 m/s when both surfaces are immersed in the fluid.

Viscosity-gravity constant See VGC.

Viscosity index (VI) An arbitrary scale used to show the magnitude of viscosity changes in lubricating oils with changes in temperature.

Volatile organic compounds (VOCs) Gases and vapors, such as benzene, released by crude oil refineries, petrochemical plants, plastic manufacturing, and the distribution and use of gasoline; VOCs include carcinogens and chemicals that react with sunlight and nitrogen oxides to form ground-level ozone, a component of smog.

Volume flow The combined (fresh and unconverted) feedstock that is fed to a reactor $F_T = F_F + F_R$. where F_T is the total feedstock into the unit, F_F is the fresh feedstock, and F_R is the recycled feedstock.

Volumetric sweep The fraction of the total reservoir volume within a flood pattern that is effectively contacted by injected fluids.

VSP Vertical seismic profiling, a method of conducting seismic surveys in the borehole for detailed subsurface information.

Waterflood Injection of water to displace oil from a reservoir (usually a secondary recovery process).

Waterflood mobility ratio Mobility ratio of water displacing oil during waterflooding. (See *also* mobility ratio.)

Waterflood residual The waterflood residual oil saturation; the saturation of oil remaining after waterflooding in those regions of the reservoir that have been thoroughly contacted by water.

Watson characterization factor See characterization factor.

Wax See mineral wax and paraffin wax.

Wax distillate A neutral distillate containing a high percentage of crystallizable paraffin wax, obtained on the distillation of paraffin or mixed-base crude and on reducing neutral lubricating stocks.

Wax fractionation A continuous process for producing waxes of low oil content from wax concentrates; see also MEK deoiling.

Wax manufacturing A process for producing oil-free waxes.

Weathered crude oil Crude oil that, due to natural causes during storage and handling, has lost an appreciable quantity of its more volatile components; also indicates uptake of oxygen.

Well abandonment The final activity in the operation of a well when it is permanently closed under safety and environment preservation conditions.

Wellbore A hole drilled or bored into the earth, usually cased with metal pipe, for the production of gas or oil.

Well completion The complete outfitting of an oil well for either oil production or fluid injection; also the technique used to control fluid communication with the reservoir.

Wellhead That portion of an oil well above the surface of the ground.

Well logs The information concerning subsurface formations obtained by means of electric, acoustic, and radioactive tools inserted in the wells. The log also includes information about drilling and the analysis of mud and cuts, cores, and formation tests.

Wells to wheels The full product life cycle is considered—from production (wells) to the use of the fuel in a vehicle (wheels). Well-to-wheel analysis can be used to assess total life-cycle greenhouse gas emissions from production to combustion of different crude oils.

Wet gas Mixture of hydrocarbons obtained from processing natural gas from which nonhydrocarbon impurities or compounds have been eliminated and whose content of components that are heavier than methane is such that it can be commercially processed.

Wet scrubbers Devices in which a countercurrent spray liquid is used to remove impurities and particulate matter from a gas stream.

Wettability The relative degree to which a fluid will spread on (or coat) a solid surface in the presence of other immiscible fluids.

Wettability number A measure of the degree to which a reservoir rock is water-wet or oil-wet, based on capillary pressure curves.

Wettability reversal The reversal of the preferred fluid wettability of a rock, e.g., from water-wet to oil-wet or vice versa.

White oil A generic tame applied to highly refined, colorless hydrocarbon oils of low volatility and covering a wide range of viscosities.

Wildcat A well drilled in an area where no oil or gas production exists.

Workover Operations on a producing well to restore or increase production. A workover may be performed to stimulate the well, to remove sand or wax from the wellbore, to mechanically repair the well, or for other reasons.

WTI crude oil (West Texas Intermediate crude oil) A type of crude oil commonly used as a price benchmark; a 40° API light sweet crude based in Cushing, Oklahoma.

Wood alcohol See methyl alcohol.

Zeolite A crystalline aluminosilicate mineral used as a catalyst and having a particular chemical and physical structure.

Index

Note: Page numbers followed by *f* indicate figures and *t* indicate tables.

D

E

F

T

U

Kolibri 360 GmbH
Ferdinand-Jühlke-Straße 7
99095 Erfurt, Deutschland
produktsicherheit@kolibri360.de